Lecture Notes in Computer Science 9008

Commenced Publication in 1973
Founding and Former Series Editors:
Gerhard Goos, Juris Hartmanis, and Jan van Leeuwen

More information about this series at http://www.springer.com/series/7412

C.V. Jawahar · Shiguang Shan (Eds.)

Computer Vision – ACCV 2014 Workshops

Singapore, Singapore, November 1–2, 2014
Revised Selected Papers, Part I

Springer

Editors
C.V. Jawahar
Center for Visual Information Technology
International Institute of Information
   Technology
Hyderabad
India

Shiguang Shan
Institute of Computing Technology
Chinese Academy of Sciences
Beijing
China

ISSN 0302-9743   ISSN 1611-3349 (electronic)
Lecture Notes in Computer Science
ISBN 978-3-319-16627-8   ISBN 978-3-319-16628-5 (eBook)
DOI 10.1007/978-3-319-16628-5

Library of Congress Control Number: 2015934895

LNCS Sublibrary: SL6 – Image Processing, Computer Vision, Pattern Recognition, and Graphics

Printed on acid-free paper

Springer International Publishing AG Switzerland is part of Springer Science+Business Media
(www.springer.com)

Preface

The three-volume set of LNCS contains the carefully reviewed and selected papers presented at the 15 workshops that were held in conjunction with the 12th Asian Conference on Computer Vision, ACCV 2014, in Singapore, during November 1–2, 2014. These workshops were carefully selected from a large number of proposals received from almost all the continents.

This series contains 153 papers selected from 307 papers submitted to all the 15 workshops as listed below. A list of organizers for each of these workshops is provided separately.

1. Human Gait and Action Analysis in the Wild: Challenges and Applications
2. The Second International Workshop on Big Data in 3D Computer Vision
3. Deep Learning on Visual Data
4. Workshop on Scene Understanding for Autonomous Systems
5. RoLoD: Robust Local Descriptors for Computer Vision
6. Emerging Topics In Image Restoration and Enhancement
7. The First International Workshop on Robust Reading
8. The Second Workshop on User-Centred Computer Vision
9. International Workshop on Video Segmentation in Computer Vision
10. My Car Has Eyes - Intelligent Vehicles with Vision Technology
11. Feature and Similarity Learning for Computer Vision
12. The Third ACCV Workshop on e-Heritage
13. The Third International Workshop on Intelligent Mobile and Egocentric Vision
14. Computer Vision for Affective Computing
15. Workshop on Human Identification for Surveillance

Workshops in conjunction with ACCV have been emerging as a forum to present focused and current research in specific areas of interest within the broad scope of ACCV. This year, the workshops covered diverse research topics including both conventional ones (such as robust local descriptor) and newly emerging ones (such as deep feature learning). Besides direct submissions to the workshops, submissions rejected by the main conference were provided the opportunity to co-submit to the workshops, following the policy of previous ACCVs.

We would like to thank many people for their efforts in making this publication possible. General Chairs, Publication Chairs, and Local Organizing Chairs helped a lot in smoothly organizing the workshops and coming out with this proceedings. Reviewers of the individual workshops did an excellent job of selecting quality papers for the final presentation. They deserve credit for the excellent quality of the papers in this proceedings.

It is our pleasure to place these volumes in front of you.

November 2014

C.V. Jawahar
Shiguang Shan

Organization

ACCV 2014 Workshop Organizers

1. Human Gait and Action Analysis in the Wild: Challenges and Applications

Mark Nixon	University of Southampton, UK
Liang Wang	Chinese Academy of Sciences, China
Jian Zhang	University of Technology, Sydney, Australia
Qiang Wu	University of Technology, Sydney, Australia
Zhaoxiang Zhang	Beihang University, China
Yasushi Makihara	Osaka University, Japan

2. Second International Workshop on Big Data in 3D Computer Vision

Jian Zhang	University of Technology, Sydney, Australia; National ICT Australia, Australia
Mohammed Bennamoun	University of Western Australia, Australia
Fatih Porikli	NICTA, Australia
Ping Tan	National University of Singapore, Singapore
Hongdong Li	Australian National University, Australia
Lixin Fan	Nokia Research Centre, Finland
Qiang Wu	University of Technology, Sydney, Australia

3. Deep Learning on Visual Data

Wanli Ouyang	The Chinese University of Hong Kong, China
Xiaogang Wang	The Chinese University of Hong Kong, China
Kai Yu	Baidu, China
Quoc Le	Google, USA
Shuicheng Yan	National University of Singapore, Singapore

4. Workshop on Scene Understanding for Autonomous Systems (SUAS)

Sebastian Ramos	CVC, Universitat Autònoma de Barcelona, Spain
Raquel Urtasun	University of Toronto, Canada
Antonio Torralba	Massachusetts Institute of Technology, USA

Nick Barnes	NICTA and Australian National University, Australia
Markus Enzweiler	Daimler AG, Germany
David Vazquez	CVC, Universitat Autònoma de Barcelona, Spain
Antonio M. Lopez	CVC, Universitat Autònoma de Barcelona, Spain

5. RoLoD: Robust Local Descriptors for Computer Vision

Jie Chen	CMV, University of Oulu, Finland
Zhen Lei	NLPR, Chinese Academy of Sciences, China
Li Liu	VIP, University of Waterloo, Canada
Guoying Zhao	CMV, University of Oulu, Finland
Matti Pietikäinen	CMV, University of Oulu, Finland

6. Emerging Topics In Image Restoration and Enhancement

Zhe Hu	University of California Merced, USA
Oliver Cossairt	Northwestern University, USA
Yu-Wing Tai	KAIST, Korea
Sunghyun Cho	Samsung Electronics, Korea
Chih-Yuan Yang	University of California Merced, USA
Robby Tan	SIM University, Singapore

7. First International Workshop on Robust Reading

Masakazu Iwamura	Osaka Prefecture University, Japan
Dimosthenis Karatzas	CVC, Universitat Autònoma de Barcelona, Spain
Faisal Shafait	University of Western Australia, Australia
Pramod Sankar Kompalli	Xerox Research India, India

8. Second Workshop on User-Centred Computer Vision (UCCV 2014)

Gregor Miller	University of British Columbia, Canada
Darren Cosker	University of Bath, UK
Kenji Mase	Nagoya University, Japan

9. International Workshop on Video Segmentation in Computer Vision

| Michael Ying Yang | Leibniz University Hannover, Germany |
| Jason Corso | University of Michigan, Ann Arbor, USA |

10. My Car Has Eyes - Intelligent Vehicles with Vision Technology

Xue Mei	Toyota Research Institute North America, Ann Arbor, USA
Andreas Geiger	Max Planck Institute for Intelligent Systems, Germany
Michael James	Toyota Research Institute North America, Ann Arbor, USA
Yi-Ping Hung	National Taiwan University, Taiwan
Fatih Porikli	Australian National University, Australia
Danil Prokhorov	Toyota Research Institute North America, Ann Arbor, USA

11. Feature and Similarity Learning for Computer Vision

Jiwen Lu	Advanced Digital Sciences Center, Singapore
Shenghua Gao	ShanghaiTech University, China
Gang Wang	Nanyang Technological University, Singapore
Weihong Deng	Beijing University of Posts and Telecommunications, China

12. Third ACCV Workshop on e-Heritage

Takeshi Oishi	University of Tokyo, Japan
Ioannis Pitas	Aristotle University of Thessaloniki, Greece
Bo Zheng	University of Tokyo, Japan
Manjunath Joshi	DA-IICT, Gandhinagar, India
Anupama Mallik	Indian Institute of Technology, Delhi, India

13. Third International Workshop on Intelligent Mobile and Egocentric Vision (IMEV2014)

Chu-Song Chen	Academia Sinica, Taiwan, China
Mohan Kankanhalli	National University of Singapore, Singapore
Shang-Hong Lai	National Tsing Hua University, Taiwan, China
Joo Hwee Lim	Institute for Infocomm Research, Singapore
Vijay Chandrasekhar	Institute for Infocomm Research, Singapore
Liyuan Li	Institute for Infocomm Research, Singapore
Yu-Chiang Frank Wang	Academia Sinica, Taiwan, China
Shuicheng Yan	National University of Singapore, Singapore

14. Computer Vision for Affective Computing (CV4AC)

Abhinav Dhall	University of Canberra/Australian National University, Australia

| Roland Goecke | University of Canberra/Australian National University, Australia |
| Nicu Sebe | University of Trento, Italy |

15. Workshop on Human Identification for Surveillance (HIS)

Tao Xiang	Queen Mary University of London, UK
Nalini K. Ratha	IBM Research, USA
Venu Govindaraju	University at Buffalo, USA
Meina Kan	Chinese Academy of Sciences, China
Wei-Shi Zheng	Sun Yat-sen University, China
Marco Cristani	University of Verona, Italy

Contents – Part I

Second International Workshop on Big Data in 3D Computer Vision

Deep Learning on Visual Data

RoLoD: Robust Local Descriptors for Computer Vision

Contents – Part II

Second Workshop on User-Centred Computer Vision (UCCV 2014)

International Workshop on Video Segmentation in Computer Vision

My Car Has Eyes: Intelligent Vehicle with Vision Technology

Third ACCV Workshop on E-Heritage

Workshop on Computer Vision for Affective Computing (CV4AC)

Contents – Part III

Third International Workshop on Intelligent Mobile and Egocentric Vision (IMEV2014)

Human Gait and Action Analysis
in the Wild: Challenges and Applications

A New Gait-Based Identification Method Using Local Gauss Maps

Hazem El-Alfy[1,2]([⊠]), Ikuhisa Mitsugami[1], and Yasushi Yagi[1]

[1] The Institute of Scientific and Industrial Research, Osaka University, Osaka, Japan
{hazem,mitsugami,yagi}@am.sanken.osaka-u.ac.jp
[2] Alexandria University, Alexandria, Egypt

Abstract. We propose a new descriptor for human identification based on gait. The current and most prevailing trend in gait representation revolves around encoding body shapes as silhouettes averaged over gait cycles. Our method, however, captures geometric properties of the silhouettes boundaries. Namely, we evaluate contour curvatures locally using Gauss maps. This results in an improved shape representation, as contrasted to average silhouettes. In addition, our approach does not require prior training. We thoroughly demonstrate the superiority of our method in gait-based human identification compared to state-of-the-art approaches. We use the OU-ISIR Large Population dataset, with over 4000 subjects captured at different viewing angles, to provide statistically reliable results.

1 Introduction

Human identification using gait is gaining significant attention by computer vision researchers and crime prevention practitioners alike. The reason is that gait can serve as a biometric evidence for determining identities without the cooperation of the subjects. In addition, gait information can be collected at a distance. This is in contrast to traditional biometric methods, such as finger-printing, iris recognition or even face recognition which are both invasive and require subjects' cooperation. Unfortunately, collecting biometric information at a distance comes at the expense of performance degradation. Moreover, several parameters affect the accuracy of identification using gait such as variation of viewing angle [1–3], walking speeds [4,5], types of cloth [6] and walking surface [7], to mention a few.

Given the previously mentioned challenges, a considerable amount of research has been carried out in the area of gait analysis over the past two decades. The first decade focused on developing new techniques and devising appropriate descriptors. Among the first features used were contour signals [8], image sequence correlation [9], self similarity plots [10,11] and unwrapped contour signals [12,13]. The first decade concludes with the development of a simple, yet efficient and accurate descriptor, namely the average silhouette [14]. This latter

© Springer International Publishing Switzerland 2015
C.V. Jawahar and S. Shan (Eds.): ACCV 2014 Workshops, Part I, LNCS 9008, pp. 3–18, 2015.
DOI: 10.1007/978-3-319-16628-5_1

approach is currently widespread and represent the state-of-the-art in period based gait features. It has paved the way for the development of several variants such as Gait Energy Image [15], Frequency Domain Feature [2], Gait Entropy Image [16], Gait Flow Image [16] and Chrono-Gait [17].

Later approaches employed statistical tools, such as discriminant analysis, to enhance the identification rates of prior methods [18]. More recently, gait analysis methods have made use of new video capturing technology. With the development of affordable 3D capturing devices such as Kinect cameras, some approaches went about generalizing 2D methods to include depth information [19,20]. Lately, an approach that combines several existing gait features, using different fusion techniques, was presented in [21].

1.1 Motivation and Intuition

Our objective is to further investigate the geometry of the silhouettes. They contain more shape information than that derived from their average over a cycle. In particular, the local curvatures of a silhouette contour encode the body shape of a subject more robustly than the mere positions of the boundary pixels. Our motivation is the following: mild variations in the silhouette appearance resulting from, say, small gait fluctuations, will cause the average shape size or location to vary. On the other hand, the *curvature* of the body's outline will only change slightly.

We achieve our goal by introducing a novel gait representation, *histograms of boundary normal vectors*, in order to compute the curvature of body contours. Histograms of surface normal vectors have recently been used in object recognition in still depth images [22]. Unit vectors normal to a surface are linked to its curvature through *Gauss maps* [23,24]. We will further elaborate on Gauss maps in Sect. 3. For now, we talk about the intuition of using normal vectors. Generally speaking, two vectors are equal when their magnitude and direction are equal. If we fix the normal vectors magnitudes to unity, then "parallel" contours will have the same histograms of normal vectors. This is a desirable property that makes our descriptor robust to small changes in body dimensions, such as those resulting from minor gait fluctuations or slight weight gain or loss, for instance. However, different contours may have similar normal vectors histograms as we will highlight in Sect. 3. This leads us to use Gauss maps locally on small pieces of the contour.

1.2 Contributions

Our main contributions in this work are the following:

– We propose a novel gait descriptor that encodes body shape more robustly than existing methods and is less affected by subtle changes in appearance. Our feature is computed only on silhouettes contours and thus is more efficient in terms of storage requirements compared to prevalent approaches that use the entire silhouette area.

– We validate our approach achieves state-of-the-art performance in person recognition using the world's largest gait database. With over 4,000 individuals, a fair subject gender representation, a wide age range and a variable viewing angle, this dataset guarantees a statistically reliable performance evaluation.

The rest of this paper proceeds as follows. Section 2 reviews related work. Section 3 introduces the mathematical background necessary to explain how our feature works. Section 4 summarizes the process we follow to compute our descriptor. In Sect. 5, we describe the large population dataset used in our experiments then show our experimental results in Sect. 6. We conclude the paper with Sect. 7.

2 Related Work

Gait recognition has been studied extensively over the past two decades. Major approaches to gait representation broadly fit into two categories: model-based and image- or appearance-based (model-free). In model-based approaches, the observed human body parts are fit to a human body model. The work of Johansson using moving light displays [25] is considered to be one of the earliest model-based human gait recognition approaches. More recent approaches include the work by Bobick and Johnson [26] in which they use a three-linked model to fit the torso, leg lengths and strides. Yam *et al.* [27] extract joint angle sequences of legs and fit them to a pendulum-like model. Urtasun and Fua [28] employ a 3D model of links, and Yang *et al.* [29] exploit a 3D human model with cylindrical links. Model-based approaches have the advantage of being unaffected by variations in body shape. Unfortunately, they suffer from high costs of fitting images to models in addition to errors involved in the fitting process. Recently, as the technology of depth cameras evolved, Kumar and Babu [30] used Kinect cameras to capture 3D joint locations and hence build a more accurate skeleton. In that case, the covariance of joint location sequences is used as a feature. The same approach is used by Hussein *et al.* [31] in the wider area of motion recognition. This technology is still recent, and unavailable in most public surveillance systems. Moreover, large scale datasets have yet to be prepared. For all previous reasons, appearance-based approaches are widely spread and more commonly used.

Appearance-based approaches extract gait features directly from 2D images. The current trend is to use side-view silhouettes since those capture most variability in gait motion. No models need to be fit here. Again, this approach is further divided into two subgroups based on the gait features used: frame-based and period-based gait features. Frame-based gait features are matched frame by frame. For a successful evaluation, a synchronization step has to precede the matching step. Sequences have to be aligned (or be in phase) at a preprocessing stage. This approach was more frequently used a decade ago. Philips *et al.* [32] propose a direct silhouette sequence matching as a baseline method. Wang *et al.* [12] exhaustively search the phase shift with the minimal classification distance. Murase and Sakai [9] employ a parametric eigenspace to represent periodic gait silhouette sequences. Liu *et al.* [18] propose a gait dynamics normalization by using a population hidden Markov model to match two silhouettes at the same phase. Beyond the raw silhouette sequences, Cuntoor *et al.* [33] project the silhouettes into a width vector and Liu *et al.* [34] project it into a frieze pattern. The main disadvantage of frame-based image-based approaches is that frame-to-frame matching is very sensitive to noise and to slight phase shifts, specially when low frame rates are used. On top of that, phase synchronization has often been a time consuming process.

Finally, period-based gait features are evaluated by integrating individual frame features over a given period, commonly a computed gait cycle (two stances). This approach makes the extracted gait features more robust to noise and slight phase fluctuations when contrasted with frame-based features. Thus, it is no surprise to find such features commonly used in most current approaches. One of the earliest robust approaches still used frequently until now is the gait energy image (GEI) by Han and Bhanu [15] or average silhouette [14]. It is computed by averaging silhouettes values, pixel by pixel, over an entire gait cycle period. Given the periodic nature of gait motion, Makihara *et al.* [2] compute, pixel by pixel, amplitude spectra of lower frequency components elements in what they call the frequency domain feature (FDF). Other approaches have also been derived as variants of the GEI method [16,17,35]. Besides, other features have also been proposed, such as self-similarity plots [10,11], Gabor filter based feature [36], local auto-correlations of spatio-temporal gradients [37,38] and histograms of oriented gradients (HOG) [39,40]. Recently, motion and shape features were represented and learned separately, eventually combining both for identification [41,42].

3   Gauss Maps

The key tool in estimating the curvature of a silhouette contour is to evaluate its Gauss map. In this section, we introduce this notion, highlighting the necessary mathematical background. A detailed account of the differential geometry of curved surfaces is being referred to in [23].

Without loss of generality, a *Gauss map* is a mapping (function) g from a surface $M \in \mathbb{R}^3$ to the unit sphere $\mathbb{S}^2$ such that g associates to each point $p \in M$ the unit vector $\boldsymbol{n}_p$ normal to M at p.

$$g : M \to \mathbb{S}^2$$
$$p \mapsto \boldsymbol{n}_p \tag{1}$$

A Gauss map measures how curved a surface is. Suppose a surface M is a (flat) plane. All the normal vectors to M, i.e. the image of g, are parallel to each other, thus there is no variation between them. On the other hand, the normal vectors to an "overly" curved surface vary greatly from point to point. For that reason, it is reasonable to use the mapping g to investigate the curvature of the surface.

Given a subsurface $\Omega \subseteq M$, the *total curvature* of Ω is defined to be the area of the image of the Gauss map $g(\Omega)$. In modern literature, the curvature of M at a point is measured by what is called the *Gaussian curvature* k. It is defined as:

$$k = \lim_{\Omega \to 0} \frac{\text{area of } g(\Omega)}{\text{area of } \Omega} \tag{2}$$

Now, by looking at k as a function on M, we can define the total curvature of Ω by:

$$\text{total curvature of } \Omega = \int_\Omega k \, dA \tag{3}$$

where dA is a surface element on M.

The total curvature of a surface computed globally cannot always be used to discriminate different surfaces. The reason is that the Gaussian curvature of a surface $M \in \mathbb{R}^3$ is invariant under *local isometries*. A local isometry of a surface M is simply a deformation of the surface under which the lengths of *geodesics* (curves of shortest length that lie in M) are unchanged. A simple example of isometric surfaces is that of a cylinder and the plane that results from "unwrapping" its surface. Both will have the same total curvature.

For that reason, the Gauss maps are defined locally, i.e. on small pieces of the surface, then local curvatures are computed for each small surface and finally, all those local curvatures are stitched together in a long feature vector. Now, how is that related to gait recognition? We simply reduce one dimension. Instead of computing curvatures of surfaces, we compute them for the contours bounding the silhouettes. In that case, the Gauss map becomes a mapping from a curve to the unit circle as shown in Fig. 1.

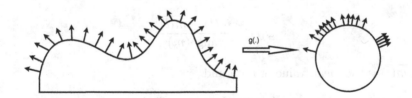

Fig. 1. Example of a Gauss map from a two-dimensional curve to the unit circle.

4   Approach Outline

Evaluating Gauss maps of digital images requires first to define a discretization approach. This is realized in the form of a *histogram of normal vectors*. The input to our feature extraction module are sequences of normalized *gait silhouettes volumes*, or GSV in short. Images of subjects undergo a sequence of preprocessing steps, namely segmentation, camera calibration and size normalization, as further presented in Sect. 5.1. We focus on developing the new descriptor and thus use the ready segmented silhouettes. In what follows, we describe our approach to compute the developed feature descriptor.

4.1   Gait Cycle Detection

Gait is mainly a periodic motion. Hence, capturing representative information requires to determine at least a full cycle, for example motion included between two double support phases that have the same leg on the front. This includes, half way through, a double support phase with the legs inverted. The viewing directions of the sequences used here (and in most literature as well) result in a variation that occurs chiefly in the horizontal direction. For that reason, we define a signal $m(t)$ that computes the second moments of the body masses around a vertical axis that passes through the center of the silhouettes:

$$m(t) = \sum_x b(x)|x - x_c|^2$$

$$\text{where } b(x) = \begin{cases} 1, & x \in \text{body} \\ 0, & \text{otherwise} \end{cases} \tag{4}$$

for all x-coordinates of pixels in the image; x_c is the vertical axis location. This signal has its peaks at double support phases (widest silhouettes) and its minima at single support phases. Thus, a full cycle starts at a local maximum, skips the following, then ends at the third one.

The computed moment signals are noisy, highly depending on the quality of the segmented silhouettes (Fig. 2 (a)). Those signals, hence, need to be smoothed first, before they can be used. We use the auto-correlation of $m(t)$ for that purpose [43]. Auto-correlation signals have the advantage of being smooth and maintain the same cycle length as the original signal. Figure 2 (b) shows the right half of that signal, since it is symmetric. To limit the range of the auto-correlation signal, the moment signals $m(t)$ are first normalized as follows:

$$w(t) = \frac{m(t) - \overline{m(t)}}{\text{range}(m)} \tag{5}$$

where $\overline{m(t)}$ is the mean value of $m(t)$ and

$$\text{range}(m) = \max_t\{m(t)\} - \min_t\{m(t)\} \tag{6}$$

Finally, the auto-correlation of $w(t)$ is computed:

$$R_w(k) = \sum_{t=1}^{n-k} w(t)w(t+k) \tag{7}$$

where n is the number of samples (frames) in the signal $w(t)$.

4.2   Histograms of Normal Vectors

This is where the main computation of our proposed descriptor takes place. Since the total curvature of the silhouette boundary is defined to be the area of the

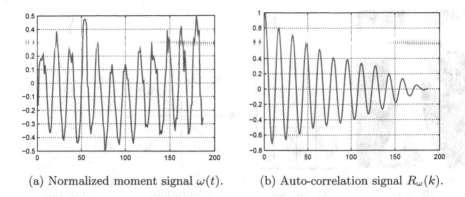

(a) Normalized moment signal $\omega(t)$.　　(b) Auto-correlation signal $R_\omega(k)$.

Fig. 2. Detecting the cycle length using second moments of silhouettes.

image of the Gauss map, we compute a discretized version of the Gauss map by accumulating the boundary normal vectors into a histogram. First, we extract the silhouette contour by a simple border following algorithm. Depending on the segmentation results, there might be discontinuities in the contour. Those are fixed by stitching contour segments which extremities are within a tolerable proximity. Images of silhouettes contours typically contain lots of noise, which affects the quality of the estimated Gauss map (Fig. 3 (a)). We overcome this artifact by smoothing them, using cubic spline interpolation, before computing the normal vectors.

It is very important to determine the orientation (sign) of the normal vectors to point outside the body of the silhouette. As shown in Fig. 3 (d), we first select a counterclockwise contour orientation by evaluating the cross product of a vector t tangent to a convex region (we choose the head) with a vector c pointing to the centroid of the silhouette. If the value is positive, the tangent vector points in the counterclockwise direction, otherwise, we flip the contour orientation. The direction of the normal vectors n can now be determined by computing their cross-product with the tangent vector t at the same contour point. A positive value indicates an outward pointing direction, otherwise, the normal vector is flipped.

The quantization of normal vectors orientations is done by assigning them to the two closest histogram bins, with linear weighting (Fig. 3 (c)). As a simple example, given an 8-bin histogram with 45°-wide bins and an orientation of 54°, and since $54/45 = 1.2$, that orientation will be assigned to bins 1 and 2 with respective weights 0.8 and 0.2. This reduces the quantization error that is involved with approaches that assign the orientation to a single bin [38].

As indicated earlier, local Gauss maps have the desirable property of uniquely identifying boundary segments. We approximate locality by dividing the silhouettes (actually their bounding boxes) into a regular grid as in Fig. 3 (b). Within each grid box, we compute the histograms of normal vectors separately. The feature descriptor is finally formed by concatenating the histograms of all grid boxes into a single vector. We call it the histogram of normal vectors, or HoNV, in short.

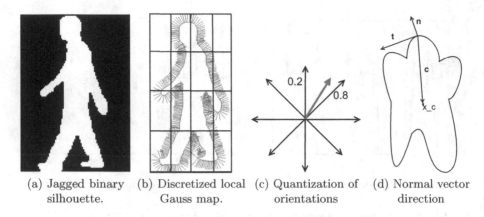

(a) Jagged binary   (b) Discretized local   (c) Quantization of   (d) Normal vector
silhouette.   Gauss map.   orientations   direction

Fig. 3. Some steps involved in computing the histograms of normal vectors.

4.3 Feature Matching

We propose two variants to our approach: a frame-based one and a period-based
one. In the frame-based variant, we choose a subset of representative key frames
to match their feature vectors, one-to-one, to feature vectors of key frames of
other sequences. First, key frames are selected by downsampling the number of
frames in a full gait cycle to some fixed value for all sequences. Then, a long
feature vector is formed by joining the vectors for individual key frames. As for
the period-based variant, we compute the histograms for each frame as previously
described and then average them over the period. Following the discussion in
Sect. 2, we expect the period-based approach to perform better, specially in cases
were the frame rate is low, such as young children sequences. We will further
investigate the difference between both approaches in the experimental results
section.

Feature matching is then performed by evaluating, for all pairs of sequences
(gallery and probe), a simple and efficient Euclidean distance metric. Given a
probe feature vector P_i and a gallery feature vector G_j, the distance between
them is computed as:

$$D_{i,j} = \| P_i - G_j \|_2 \tag{8}$$

In the dataset we are using, only one single sequence is captured per subject.
Therefore, learning based methods, such as statistical discriminant analysis tools
cannot be applied here. It is definitely expected, as demonstrated elsewhere, that
such approaches will improve performance in case datasets with more training
samples per class are used.

5 Large Population Dataset

We have evaluated our approach using the OU-ISIR Large Population dataset
[44]. The main significance of using this dataset lies in the large number of

subjects it contains. This provides statistically reliable results that cannot be verified with other available smaller datasets. In all, it has 4,007 subjects. However, in our experiments, we have used subsets of sizes ranging from 3,141 to 3,706 subjects. This is by far larger than any other available dataset we know of.

Other than being a large scale gait database, the used dataset balances the gender representation with 2,135 male subjects and 1,872 female subjects. It also has a wide range of ages spanning from 1 to 94 years old subjects. The presence of children, in particular, provides a unique and challenging testing situation, where the number of frames per gait cycle drops significantly. An additional advantage of using this dataset is that silhouette extraction is performed accurately. For the purpose of this paper, the tasks of background subtraction and silhouette segmentation are out of scope. Our aim is to validate our new feature descriptor, separately from errors that may arise from poorly extracted silhouettes.

Moreover, this dataset captures subjects from different viewing angles. Currently available datasets simply present their subjects from a "side view" angle. The subjects in the OU-ISIR dataset, however, are viewed from an angle that is carefully evaluated, gradually changing from about 50° to 90°. Based on that angle, each sequence is split into 4 subsequences, centered at 55, 65, 75 and 85 degrees respectively as illustrated in Fig. 4. This allows us to perform a more rigorous performance evaluation that verifies, first, the robustness of our approach on each view angle separately, then, on entire sequences as well.

(a) 55° subsequence. (b) 65° subsequence. (c) 75° subsequence. (d) 85° subsequence.

Fig. 4. Typical frames from a gait sequence that illustrate the four different viewing angles.

5.1 Preprocessing

In this section, we briefly note the preliminary stages required to reach the gait silhouettes volumes (GSV) input used by our feature evaluation code.

1. Silhouette extraction: Background subtraction is performed along with a graph-cut approach for segmentation [45]. The contour is visualized to fix any errors manually.
2. Camera calibration: Intrinsic and extrinsic camera parameters are estimated to correct the distortion and carry out the necessary camera rotation.
3. Image registration: A moving-average filter is applied to the extracted silhouette images then, their sizes are normalized to 88 × 128 pixels.

Some of the previous steps are illustrated in Fig. 5.

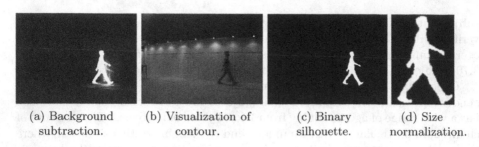

(a) Background subtraction. (b) Visualization of contour. (c) Binary silhouette. (d) Size normalization.

Fig. 5. The output of four preprocessing steps performed on the input sequences to extract gait silhouette volumes (GSV).

6 Experimental Results

We present here the results of evaluating our new feature descriptor on the OU-ISIR Large Population gait dataset [44]. This dataset is divided into five subsets, based on viewing angle, A-55, A-65, A-75, A-85 and A-All, where the last one contains full sequences with the viewing angle gradually varying from roughly 50 to 90 degrees. The number of subjects in each subset is shown in Table 1. We implemented the proposed method using Matlab without any code optimization. Under an Intel Core i7 processor running at 3.5 GHz, our code processes about 60 frames per second, the equivalent of two gait cycles per second given the used dataset.

Table 1. Number of subject in each subset of the OU-ISIR dataset.

Dataset	A-55	A-65	A-75	A-85	A-All
No. of subjects	3706	3770	3751	3249	3141

As mentioned earlier in Sect. 4.3, we have developed a frame-based variant and a period-based variant. In addition, several parameters, such as the number of histogram bins and the number of grid boxes can be tuned to enhance the performance. In this section, we compare the variants of our approach and study the effect of varying some parameters on the results. Finally, we compare it to other recent methods.

6.1 Evaluation of Descriptor Parameters

We consider first the effect of changing the number of grid boxes on the performance. A grid with one box corresponds to a global Gauss map, computed for the entire silhouettes. As the number of boxes increases, we have more and more local maps. We illustrate the verification performance using Receiver Operating Characteristic (ROC) curves. This is a common tool used in biometric applications. It denotes the trade-off between false rejection rate (FRR) and false

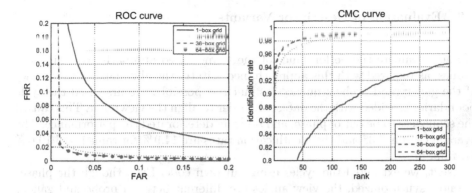

Fig. 6. Performance comparison for varying the number of grid boxes.

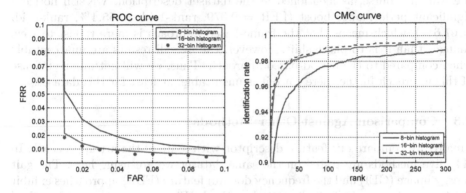

Fig. 7. Performance comparison for varying the number of histogram bins.

acceptance rate (FAR) as the acceptance threshold is being varied. We also use Cumulative Match Characteristic (CMC) curves to illustrate the identification performance. This tool estimates the probability that a correct match is observed within the top k "closest" matches.

We use the A-85 subset as a study case, employ the period-based approach and fix the number of histogram bins to 16. The effect of changing the number of grid boxes (1, 16, 36, 64) is illustrated in Fig. 6. As expected, the performance of the global Gauss map (1 box) is the lowest. Next, we vary the number of histogram bins (8, 16, 32) for a fixed grid of 36 boxes. The results are shown in Fig. 7. Further performance enhancements are achievable but cannot be shown at the used scale. Instead, we use the Equal Error Rate (EER) value from the ROC curve, that is the value at which the false acceptance and rejection rates (FAR and FRR) are equal. We also use the rank-1 and rank-5 identification rates as representative measures from the CMC curve. For the 88×128 pixel silhouettes that we use, best performance (EER = 0.013, rank-1 id. = 93.0 %, rank-5 id. = 96.7 %) is obtained using 32-bin histograms and a grid of 169 (13×13) boxes.

6.2   Evaluation of Descriptor Variants

The results shown earlier are for the period-based approach, as pointed out. We now evaluate the performance of the frame-based approach. Matching key frames, as required with this approach, necessitates the alignment of the phases of the gallery and probe sequences as much as possible. We first used the cycle boundaries provided with the dataset in order to align the viewpoint. This caused the sequences to be out of phase, and severely deteriorated the performance. The average EER is 0.38 and the identification rates are 14.9 % for rank-1 and 19.6 % for rank-5.

Next, we extracted the cycles using our own code. Even though the phases are now synchronized, the view angles are different between probe and gallery sequences. That is due to the fact that the view angle gradually changes over the walking course, as mentioned in the dataset description. We still notice a significant performance boost (EER = 0.079, rank-1 id. = 55.4 %, rank-5 id. = 67.6 %), which means that the frame-based approach is more robust to view changes than it is to phase shift. However, the overall results are unacceptable when compared to the period-based approach. Thus, we will not pursue the use of this approach further here since it is unsuitable, at least for this dataset.

6.3   Comparison Against Other Methods

Finally, we compare our feature descriptor with state-of-the-art approaches. In [44], five methods are tested on the same dataset we are using here. The gait energy image (GEI) and the frequency domain feature (FDF) approaches exhibit the best performance of all the tested methods. First, we compare our method against those two. The significant parts of the ROC and CMC curves (A-85 subset as above) are shown in Fig. 8. At these high identification rates, visualizing the performance improvements becomes significantly more difficult, and so we will use the EER metric, the rank-1 and the rank-5 identification rates as we did earlier. We also compare our method to the average of histograms of oriented

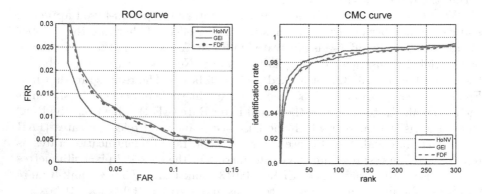

Fig. 8. Performance comparison versus other approaches.

gradients (HOG). That is a very recent approach [40] that employs HOG, a powerful human detection feature [16]. We implement the method according to the details in [40,40] and test it on the OU-ISIR dataset. We show all results in Table 2. Our method is abbreviated as HoNV for histogram of normal vectors. The best performance out of GEI and FDF is recorded in the GEI column. Which method performs best depends on the used dataset. Finally, the performance of the average HOG technique can be found in the HOG column. It is evident that our approach supersedes all other methods on almost all values of EER and on all values of ranking.

Table 2. Comparison of the performance of our approach (HoNV) versus GEI/FDF and HOG methods.

Dataset	EER [$\times 10^{-2}$]			Rank-1 [%]			Rank-5 [%]		
	HoNV	HOG	GEI	HoNV	HOG	GEI	HoNV	HOG	GEI
A-55	1.70	**1.59**	2.06	**91.6**	90.4	84.7	**95.1**	95.0	92.4
A-65	**1.43**	1.54	1.83	**92.1**	90.9	86.6	**95.4**	95.1	92.8
A-75	**1.31**	1.70	1.97	**93.3**	91.5	86.9	**96.4**	95.3	92.9
A-85	**1.32**	1.78	2.00	**93.0**	89.9	85.9	**96.7**	95.0	92.8
A-All	**0.58**	0.81	1.13	**97.5**	96.2	94.2	**98.7**	98.0	97.1

7   Conclusion

We presented a new feature descriptor for human identification using gait: histograms of normal vectors. We developed the method, explained the theory of Gauss maps on which it relies and demonstrated how it can exploit boundary curvature information. We verified that the new feature encodes enhanced shape information when compared with state-of-the-art approaches. Our superior results are validated through rigorous experiments using the world's largest gait database. In addition, our method does not require any training which makes it also valid when a small number of sequences per subject is available.

We are currently working on several improvements to the presented approach. When a large training sample is available, we are studying to which extent some statistical tools, such as discriminant analysis, can enhance the identification results. The main drawback of our approach is that it relies on the quality of the extracted silhouettes. In that regard, we will carry out further comparisons with other contour-based approaches. On a different note, recent work [21] has suggested that combining gait features can improve performance. Thus, we would like to study how does fusing our method with other approaches improve the discriminative power of the combined method.

Acknowledgement. This work was supported in part by the JST CREST "Behavior Understanding based on Intention-Gait Model" project.

References

1. Kusakunniran, W., Wu, Q., Zhang, J., Li, H.: Gait recognition under various viewing angles based on correlated motion regression. IEEE Trans. Circ. Syst. Video Technol. **22**, 966–980 (2012)
2. Makihara, Y., Sagawa, R., Mukaigawa, Y., Echigo, T., Yagi, Y.: Gait recognition using a view transformation model in the frequency domain. In: Leonardis, A., Bischof, H., Pinz, A. (eds.) ECCV 2006. LNCS, vol. 3953, pp. 151–163. Springer, Heidelberg (2006)
3. BenAbdelkader, C., Cutler, R., Davis, L.: View-invariant estimation of height and stride for gait recognition. In: Tistarelli, M., Bigun, J., Jain, A.K. (eds.) ECCV 2002. LNCS, vol. 2359, pp. 155–167. Springer, Heidelberg (2002)
4. Tsuji, A., Makihara, Y., Yagi, Y.: Silhouette transformation based on walking speed for gait identification. In: 2010 IEEE Conference on Computer Vision and Pattern Recognition (CVPR), pp. 717–722. IEEE (2010)
5. Kusakunniran, W., Wu, Q., Zhang, J., Li, H.: Speed-invariant gait recognition based on procrustes shape analysis using higher-order shape configuration. In: 2011 18th IEEE International Conference on Image Processing (ICIP), pp. 545–548. IEEE (2011)
6. Hossain, A., Makihara, Y., Wang, J., Yagi, Y., et al.: Clothing-invariant gait identification using part-based clothing categorization and adaptive weight control. Pattern Recogn. **43**, 2281–2291 (2010)
7. Sarkar, S., Phillips, P.J., Liu, Z., Vega, I.R., Grother, P., Bowyer, K.W.: The humanid gait challenge problem: data sets, performance, and analysis. IEEE Trans. Pattern Anal. Mach. Intell. **27**, 162–177 (2005)
8. Niyogi, S.A., Adelson, E.H.: Analyzing and recognizing walking figures in xyt. In: Proceedings of the 1994 IEEE Computer Society Conference on Computer Vision and Pattern Recognition, CVPR 1994, pp. 469–474. IEEE (1994)
9. Murase, H., Sakai, R.: Moving object recognition in eigenspace representation: gait analysis and lip reading. Pattern Recognit. Lett. **17**, 155–162 (1996)
10. BenAbdelkader, C., Cutler, R., Nanda, H., Davis, L.: EigenGait: motion-based recognition of people using image self-similarity. In: Bigun, J., Smeraldi, F. (eds.) AVBPA 2001. LNCS, vol. 2091, pp. 284–294. Springer, Heidelberg (2001)
11. BenAbdelkader, C., Cutler, R., Davis, L.: Motion-based recognition of people in eigengait space. In: Proceedings of the Fifth IEEE International Conference on Automatic Face and Gesture Recognition, pp. 267–272. IEEE (2002)
12. Wang, L., Hu, W., Tan, T.: A new attempt to gait-based human identification. In: Proceedings of the 16th International Conference on Pattern Recognition, vol. 1, pp. 115–118. IEEE (2002)
13. Wang, L., Tan, T., Ning, H., Hu, W.: Silhouette analysis-based gait recognition for human identification. IEEE Trans. Pattern Anal. Mach. Intell. **25**, 1505–1518 (2003)
14. Liu, Z., Sarkar, S.: Simplest representation yet for gait recognition: averaged silhouette. In: Proceedings of the 17th International Conference on Pattern Recognition, ICPR 2004, vol. 4, pp. 211–214. IEEE (2004)
15. Han, J., Bhanu, B.: Individual recognition using gait energy image. IEEE Trans. Pattern Anal. Mach. Intell. **28**, 316–322 (2006)
16. Bashir, K., Xiang, T., Gong, S.: Gait recognition using gait entropy image (2009)

17. Wang, C., Zhang, J., Pu, J., Yuan, X., Wang, L.: Chrono-gait image: a novel temporal template for gait recognition. In: Daniilidis, K., Maragos, P., Paragios, N. (eds.) ECCV 2010, Part I. LNCS, vol. 6311, pp. 257–270. Springer, Heidelberg (2010)
18. Liu, Z., Sarkar, S.: Improved gait recognition by gait dynamics normalization. IEEE Trans. Pattern Anal. Mach. Intell. **28**, 863–876 (2006)
19. Sivapalan, S., Chen, D., Denman, S., Sridharan, S., Fookes, C.: Gait energy volumes and frontal gait recognition using depth images. In: 2011 International Joint Conference on Biometrics (IJCB), pp. 1–6. IEEE (2011)
20. Hofmann, M., Bachmann, S., Rigoll, G.: 2.5d gait biometrics using the depth gradient histogram energy image. In: 2012 IEEE Fifth International Conference on Biometrics: Theory, Applications and Systems (BTAS), pp. 399–403. IEEE (2012)
21. Makihara, Y., Muramatsu, D., Iwama, H., Yagi, Y.: On combining gait features. In: 2013 10th IEEE International Conference and Workshops on Automatic Face and Gesture Recognition (FG), pp. 1–8. IEEE (2013)
22. Tang, S., Wang, X., Lv, X., Han, T.X., Keller, J., He, Z., Skubic, M., Lao, S.: Histogram of oriented normal vectors for object recognition with a depth sensor. In: Lee, K.M., Matsushita, Y., Rehg, J.M., Hu, Z. (eds.) ACCV 2012, Part II. LNCS, vol. 7725, pp. 525–538. Springer, Heidelberg (2013)
23. Gauss, K.F.: General investigations of curved surfaces of 1827 and 1825, translated with notes and a bibliography by J.C. Morehead and A.M. Hiltebeitel. The Princeton University Library (1902)
24. Hazewinkel, M.: Encyclopaedia of Mathematics, vol. 13. Springer, New York (2001)
25. Johansson, G.: Visual motion perception. Sci. Am. **232**, 76–88 (1975)
26. Bobick, A.F., Johnson, A.Y.: Gait recognition using static, activity-specific parameters. In: Proceedings of the 2001 IEEE Computer Society Conference on Computer Vision and Pattern Recognition, CVPR 2001, vol. 1, p. I-423. IEEE (2001)
27. Yam, C., Nixon, M.S., Carter, J.N.: Automated person recognition by walking and running via model-based approaches. Pattern Recogn. **37**, 1057–1072 (2004)
28. Urtasun, R., Fua, P.: 3d tracking for gait characterization and recognition. In: Proceedings of the Sixth IEEE International Conference on Automatic Face and Gesture Recognition, pp. 17–22. IEEE (2004)
29. Yang, H.-D., Lee, S.-W.: Reconstruction of 3D human body pose for gait recognition. In: Zhang, D., Jain, A.K. (eds.) ICB 2005. LNCS, vol. 3832, pp. 619–625. Springer, Heidelberg (2005)
30. Kumar, M., Babu, R.V.: Human gait recognition using depth camera: a covariance based approach. In: Proceedings of the Eighth Indian Conference on Computer Vision, Graphics and Image Processing, p. 20. ACM (2012)
31. Hussein, M.E., Torki, M., Gowayyed, M.A., El-Saban, M.: Human action recognition using a temporal hierarchy of covariance descriptors on 3d joint locations. In: Proceedings of the Twenty-Third International Joint Conference on Artificial Intelligence, pp. 2466–2472. AAAI Press (2013)
32. Phillips, P.J., Sarkar, S., Robledo, I., Grother, P., Bowyer, K.: The gait identification challenge problem: data sets and baseline algorithm. In: Proceedings of the 16th International Conference on Pattern Recognition, vol. 1, pp. 385–388. IEEE (2002)
33. Cuntoor, N., Kale, A., Chellappa, R.: Combining multiple evidences for gait recognition. In: Proceedings of the 2003 IEEE International Conference on Acoustics, Speech, and Signal Processing, (ICASSP 2003), vol. 3, p. III-33. IEEE (2003)

34. Liu, Y., Collins, R., Tsin, Y.H.: Gait sequence analysis using frieze patterns. In: Heyden, A., Sparr, G., Nielsen, M., Johansen, P. (eds.) ECCV 2002, Part II. LNCS, vol. 2351, pp. 657–671. Springer, Heidelberg (2002)

35. Lam, T.H., Cheung, K.H., Liu, J.N.: Gait flow image: a silhouette-based gait representation for human identification. Pattern Recogn. **44**, 973–987 (2011)

36. Tao, D., Li, X., Maybank, S.J., Wu, X.: Human carrying status in visual surveillance. In: 2006 IEEE Computer Society Conference on Computer Vision and Pattern Recognition, vol. 2, pp. 1670–1677. IEEE (2006)

37. Kobayashi, T., Otsu, N.: A three-way auto-correlation based approach to human identification by gait. In: IEEE Workshop on Visual Surveillance, vol. 1, p. 4. Citeseer (2006)

38. Kobayashi, T., Otsu, N.: Three-way auto-correlation approach to motion recognition. Pattern Recogn. Lett. **30**, 212–221 (2009)

39. Hofmann, M., Rigoll, G.: Improved gait recognition using gradient histogram energy image. In: 19th IEEE International Conference on Image Processing (ICIP), pp. 1389–1392. IEEE (2012)

40. Hofmann, M., Rigoll, G.: Exploiting gradient histograms for gait-based person identification. In: IEEE International Conference on Image Processing (ICIP), pp. 4171–4175 (2013)

41. Lombardi, S., Nishino, K., Makihara, Y., Yagi, Y.: Two-point gait: decoupling gait from body shape. In: 2013 IEEE International Conference on Computer Vision (ICCV), pp. 1041–1048. IEEE (2013)

42. Lee, C.S., Elgammal, A.: Style adaptive contour tracking of human gait using explicit manifold models. Mach. Vis. Appl. **23**, 461–478 (2012)

43. Boulgouris, N.V., Plataniotis, K.N., Hatzinakos, D.: Gait recognition using dynamic time warping. In: 2004 IEEE 6th Workshop on Multimedia Signal Processing, pp. 263–266. IEEE (2004)

44. Iwama, H., Okumura, M., Makihara, Y., Yagi, Y.: The OU-ISIR gait database comprising the large population dataset and performance evaluation of gait recognition. IEEE Trans. Inform. Forensics Secur. **7**, 1511–1521 (2012)

45. Makihara, Y., Yagi, Y.: Silhouette extraction based on iterative spatio-temporal local color transformation and graph-cut segmentation. In: 19th International Conference on Pattern Recognition, ICPR 2008, pp. 1–4. IEEE (2008)

46. Dalal, N., Triggs, B.: Histograms of oriented gradients for human detection. In: IEEE Computer Society Conference on Computer Vision and Pattern Recognition, CVPR, vol. 1, pp. 886–893. IEEE (2005)

Hand Detection and Tracking in Videos
for Fine-Grained Action Recognition

Nga H. Do[✉] and Keiji Yanai

Department of Informatics, The University of Electro-Communications,
Tokyo, 1-5-1 Chofugaoka, Chofu, Tokyo 182-8585, Japan
dohang@mm.cs.uec.ac.jp

Abstract. In this paper, we develop an effective method of detecting and tracking hands in uncontrolled videos based on multiple cues including hand shape, skin color, upper body position and flow information. We apply our hand detection results to perform fine-grained human action recognition. We demonstrate that motion features extracted from hand areas can help classify actions even when they look familiar and they are associated with visually similar objects. We validate our method of detecting and tracking hands on VideoPose2.0 dataset and apply our method of classifying actions to the playing-instrument group of UCF-101 dataset. Experimental results show the effectiveness of our approach.

1 Introduction

In recent years, using low-level features such as optical flows and spatio-temporal features to represent actions has become the most popular framework for action recognition. Several researches have employed high level features like human poses and human-object interactions. However, most of these high-level feature based approaches work only on still images and do not take advantage of motion characteristics of actions [1–4]. Among a couple of works which handle videos, Prest *et al.*'s work [5] learns human actions by using interactions between persons and objects. They proposed to localize in space and track over time both objects and persons, and represent actions as the trajectories of objects with respect to persons. Since their approach relies on object detection, they have to learn object detectors of all related objects. This process requires costly annotations. Moreover, they do not consider the case when the objects are visually similar. For instance, since violin and cello share the same visual characteristics, their detectors are supposed to fail to distinguish them. Consequently, it may be easy for Prest *et al.*'s method [5] to confuse "play violin" action and "play cello" action. In this paper, we propose to represent actions involved with objects solely based on how people perform them with the objects using their hands. According to our method, disparate actions associated with different but visually similar objects can be classified (see Fig. 1 for the illustration). We show that hand related motion features are discriminative and representative enough for human actions.

© Springer International Publishing Switzerland 2015
C.V. Jawahar and S. Shan (Eds.): ACCV 2014 Workshops, Part I, LNCS 9008, pp. 19–34, 2015.
DOI: 10.1007/978-3-319-16628-5_2

Fig. 1. An example which shows that actions with objects involved may not be recognized by object detection. The video shots are from UCF-101 dataset. We can see that cello and violin look very similar, since they are in the same class of musical instruments (string instruments). Therefore, it is not an easy task to distinguish the actions related with them ("play" in this case) using the two instrument detectors. However, while playing them, people put their arms/hands in different positions and move them in different directions. Consequently, exploiting motion features of arms/hands can be expected to be able to help classify "play cello" and "play violin".

In fact, in many cases of human actions, especially those that are involved with objects, people tend to move only their arms/hands to operate the actions. In this paper, we focus on human actions which require hand movements during the time the actions are operated. Thus, we propose to take all possible arm/hand motions into consideration to represent the actions. Since motions of hands also contain those of arms, and in some cases, not the entire arm but only hands move, in order to handle as many cases as possible, here we focus only on movements of hands.

The detection of hands has been known as a tremendously challenging task since hands are the most flexible human body parts compared to others. Their appearance can change unpredictably since they can be closed or open, and the fingers can have various articulations. Moreover, in videos, they are naturally the fastest moving body parts. This means that unlike in still images where their shape can be recognized quite clearly, in videos sometimes they are very hard to detect due to motion blurs caused by their movements. In this paper, we propose to exploit multiple cues including hand shape, skin color, upper body position and flow information to detect hands in videos. Our objective is to obtain 2D+t sequences of bounding boxes which tightly bound hands in the videos. We demonstrate that using motion features extracted only from hand regions can achieve comparable performance to using motion features extracted from the whole frame. That means hand motion features are the most informative representation of human actions involved with hand movements. Moreover, we further enhance action recognition precision by exploiting displacement features of hands which belong to the most reliable hand tracks. To the best of our knowledge, we are the first to classify human actions using only hand related motion features.

Our contributions can be summarized as follows: (i) we develop an efficient hand detector for hand detection in videos based on multiple cues; (ii) we propose to recognize actions by exploiting the information of how people operate the actions with their hands; (iii) we propose a discriminative hand displacement feature which improves action recognition; (iv) we confirmed the efficiency of our proposed method on uncontrolled videos. To validate the effectiveness of our hand detector, we use VideoPose2 dataset[1]. This dataset was originally developed for the challenging task of upper body estimation. To test the efficiency of our method on action recognition, we conduct experiments on playing-instrument group of UCF-101 [6] dataset. UCF-101 is one of the most challenging action dataset up to date with the large variations in human pose, object appearance, viewpoint, background and illumination conditions. Experimental results show the effectiveness of our approach.

The rest of this paper is organized as follows. Section 2 introduces some more related work. Section 3 describes our proposed method of hand detection and tracking. Section 4 explains how we apply the detecting and tracking results to action recognition. Experiments and discussions about their results are presented in Sect. 5. Finally Sect. 6 gives conclusions.

2   Related Work

Here we introduce some related work on detection and tracking of hands and some on recognition of actions involved with objects.

Hand detection and tracking: Hand detection is a topic which has a quite long history and a wide range of applications such as Human Computer Interaction, Sign Language translators, human pose recognition and surveillance. In the early stage of development, hand detection technique required markers or colored gloves to make the recognition easier. Second generation methods used low-level features such as color (skin based detection) [7,8] or shape [9]. Most recent works on hand detection in videos are performed in 3D [10–13]. They employ depth information provided by depth cameras. As one in a few recent 2D hand detectors for videos, the hand detector proposed by Sapp *et al.* [14] exploits flow field. They propose to extract motion discontinuities by computing the gradient magnitude of the flow field, and learn a linear filter via SVM using this motion discontinuity magnitude cue specific to hands. Hands are detected as regions with the max response from the detector at each frame location over a discrete set of hand orientations. In their work, the results of hand detection are only used as additional cues for limb localization since their final purpose is not hand detection but upper body pose estimation.

Most of hand tracking methods assume that hands are the most moving objects in an image frame. In [15], Yuan *et al.* proposed to use a temporal filter to select the most likely trajectory of hand locations among multiple candidates obtained by "block flow" matching. In [16], Baltzakis *et al.* proposed a skin color based tracker which allows the utilization of additional information cues such as image background model, expected spatial location, velocity and shape of the detected and tracked segments. The benefit of their trackers is that they can track hands in real time. However, their trackers only work under constrained environments where the background is unchanged, so that simply substracting

---

[1] http://vision.grasp.upenn.edu/cgi-bin/index.php?n=VideoLearning.VideoPose2.

background can bring them enough cues to infer the most moving objects which refer to hands. In this paper, we detect and track hand in realistic videos where maybe there are multiple moving objects and there also exists camera motion that can cause noise.

Recognition of actions related to objects: Approaches for recognizing object related actions have been developed for both static images and videos. Unlike works on static images [1,3,4], works on videos [17–19] generally take motion characteristics into account. Filipovych *et al.* [17,18] modeled human-object interactions based on the trajectories and appearance of spatio-temporal interest points. Their approach was applied only to controlled videos taken from viewpoint of the actor by a static camera against a uniform background.

One of the most related work to ours is Gupta *et al.*'s work [19]. They employed hand trajectories to model the objects and the human-object motions for classifying interactions between humans and objects. In their work, the motion can be simply extracted based on background subtraction since they worked only on videos with constrained environment (static and fixed background). On the contrary, our approach tackles the problem in uncontrolled videos. Moreover, while Gupta *et al.*'s work requires annotation efforts for building training data, including the annotation for the locations of the objects in all video frames, our method does not require such time consuming efforts.

3   Hand Detection and Tracking

3.1   Hand Detection

Here we aim to automatically estimate the hand locations using flow information and two trained detectors: a upper body detector and a static hand detector. As for flow estimation, we use DeepFlow proposed by [20]; for upper body detection, we employ Cavin's upper body detector[2]; for detection of static hands, we apply a state-of-the-art hand detector in still images proposed by Mittal *et al.* [21]. We improve their hand detector, originally developed only for hand detection in still images, to become a hand detector in videos by exploiting motion information and introducing upper body pose based spatial constraints.

Method of the Baseline. We, first, briefly describe the method of static hand detection in [21], which is used as the baseline of our hand detector. According to [21], hand hypotheses are first proposed by three dependent detectors: a sliding window hand shape detector, a context based detector, and a skin based detector. Then, the proposals are scored by all three detectors and a trained model for scores is used to verify them. The hand shape detector was trained using Felzenszwalb *et al.* [22]'s part based deformable model with HOG features. The contexts here refer to the cues captured around the hands, especially the wrists. In order to learn the contexts, another part based deformable model [22]

---

[2] http://groups.inf.ed.ac.uk/calvin/calvin_upperbody_detector/.

was trained from the hand bounding boxes which were extended to cover the wrists. The skin detector first builds a skin mask based on the skin color of face(s) detected by OpenCV face detector. It then detects skin regions by fitting lines using Hough transform and finding the medial axis of the blob-shaped regions. The hands are hypothesized at the ends of the lines.

The hand bounding boxes proposed by above three detectors are scored and combined as follows:

Hand detector score: the score obtained directly from hand detector.

$$\alpha_1 = \beta_{HD}(b) \tag{1}$$

where β_{HD} is the scoring function of the hand detector [22].

Context detector score: the score obtained by max-pooling over all bounding boxes which overlap with given hand boxes. The overlap threshold is set as 0.5.

$$\alpha_2 = \max_{b_h \in B_h} (\beta_{CD}(b_h)) \tag{2}$$

B_h refers to the set of context bounding boxes overlapping with the hand bounding box b_h. β_{CD} is the scoring function of the context detector [22].

Skin detector score: the score calculated by the fraction of pixels belong to skin regions in a given bounding box and denoted as α_3.

The three scores are combined into a single feature vector $(\alpha_1, \alpha_2, \alpha_3)$. This vector is then classified by a trained linear SVM classifier [23]. Finally, bounding boxes are suppressed depending on their overlap with other highly scored boxes using super pixel based non-maximum suppression. The superpixels are obtained by Arbelaez et al.'s method of image segmentation [24].

Mittal et al. trained their detector by using the data which was collected by themselves from various public image datasets including PASCAL VOC 2007[3], PASCAL VOC2010[4], Poselet [25], Buffy stickman[5], INRIA pedestrian [26] and Skin dataset [27], with 2861 hand instances for training and 660 hand instances for test in total. According to their experimental results, 48.2 % of the test instances were correctly detected.

Proposed Method. Even though Mittal et al.'s hand detector achieved good performance, it needs two conditions about the data to work well: first, image resolution should be high and second, face should be easy to detect. Hands in images with good resolution commonly have clear shape, so that shape detector can be effectively employed. Moreover, most of faces in their test data can be seen from front view, so that face skin based hand detection is possible. However, here we have to deal with more complex and totally unconstrained data. In our

---

[3] http://pascallin.ecs.soton.ac.uk/challenges/VOC/voc2007/.
[4] http://pascallin.ecs.soton.ac.uk/challenges/VOC/voc2010/.
[5] http://www.robots.ox.ac.uk/~vgg/data/stickmen/.

data, many videos have low resolution or are taken under bad light condition, and faces are sometimes hard to be recognized. In such cases, we cannot find any of shape and/or color cues to detect hands. Thus, instead of employing Mittal *et al.*'s detector as it is, we propose to make it possible to work in such videos by introducing upper body based spatial constraints and motion information. The pose and position of detected upper body are used for two purposes: to estimate face region and to refine final detection results. On the other hand, flow information is exploited in two directions: to select upper body and to rescore hand hypotheses.

Our proposed method of hand detection is a three-step method which can be summarized as follows: (1) Detecting upper body by employing upper body detector and motion information, (2) Finding hand hypotheses based on multiple static cues, (3) Inferring the best hand hypotheses by exploiting motion cue and upper body based spatial constraints. Refer to Fig. 2 for the illustration of our proposed method of hand detection in videos.

For a given frame, at the first step, we apply Calvin *et al.*'s upper body detector and flow information to detect the most dominant upper body. This upper body detector has been demonstrated as a powerful human pose estimator and applied by many approaches recently. One of this detector's benefits is that it can estimate rather precisely the head position even when the face is hard to be detected. This detector returns several results, each of them contains position of head, torso and two limbs, and scores for each result. However, the problem is that not all results returned by this detector are perfect, and the good ones are sometimes not highly scored. Moreover, we found that even when the faces and the torsos are quite precisely localized, it is not going that well for the limbs.

Assume that there exists at least one good prediction among the results, we infer it by introducing motion information and spatial constraints. We postulate the two following holistic hypotheses: (i) hypothesis about hands: hands are the most moving body parts in a upper body, and generally looked not big compared to the upper body from common views; (ii) hypothesis about the main actor: the main actor is generally in motion and captured in the easiest way to recognize. That means his or her upper body is likely in the middle of the image frame, and/or bigger than the others. Based on the first assumption, "good" upper body should cover moving regions, and these regions are supposed to include hands. For each detected upper body, we first segment it to regions with different movements by the gradient magnitude of the flow field. Regions that are smaller than upper body area multiplied by predefined area threshold $thresh_a$ then become motion based hand hypotheses of that upper body. Score of a upper body is redefined as the ratio between areas of hand hypotheses which lie inside and outside that upperbody and finally normalized by area of that upper body. In the case that there are no significant movements (no moving regions with average flow magnitude being larger than flow threshold $thresh_f$), "good" upper body is simply selected based on the second assumption: the more centered and the bigger, the higher probability to be selected. In our experiments, $thresh_a$ is fixed as 0.5 and $thresh_f$ is fixed as 1.

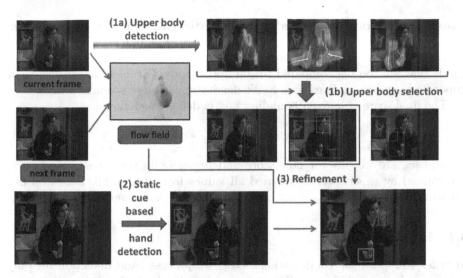

Fig. 2. Illustration for our proposed method of detecting hands. (1) For a given frame, we first apply Calvin *et al.*'s upper body pose estimator to obtain proposals of upper body pose. (1a) Each proposal consists of sticks with different colors: pink, red, green and yellow which respectively refer to position of head, torso, upper arm and lower arm. To infer the best detection, we employ the motion of flow field. We segment the frame by magnitude of flow field to obtain regions with disparate movements. For each detected upper body, its score is redefined as the ratio of regions which are supposed to be hands. These regions should, first, be in motion, and second, be not too large or too small compared to the upper body. We show these regions by the red bounding boxes. The detection at the right side contains no motion, thus it is not supposed to be the good one. (1b) The middle detection is considered to be better than the left one since it contains more motion based hand hypotheses. Upper bodies and faces are marked by yellow and light blue bounding boxes respectively. (2) The face of the selected upper body (the middle one), along with hand shape and context, are then used as static cues for hand detection following Mittal *et al.* The three best results obtained by hand detection using static cues are shown in the middle image of the last row. The best is represented by the red bounding box, the second best is green and the third best is blue. As we can see, the second best is a failed detection even though it has hand shape. (3) We refine the detection results by considering motion information and upper body position and obtain the final results as shown in the right image of the last row (Color figure online).

The second step is hand detection based on multiple static cues using Mittal *et al.*'s detector. The face of chosen upper body is used to detect skin regions. At the final step, detection results of the previous step are rescored by introducing following flow score and upper body score.

Flow score: calculated as the average of gradient magnitude of flow of pixels in detection result. This score is normalized to have value between 0 and 1. We denote it as α_4.

Upper body score: determined by using spatial constraints based on position and area of upper body. It is calculated as percentage of area within the detected hand which overlaps with the upperbody. We also give penalties for detected hands that are too big compared to the upper body. For such detections, their upper body scores are fixed as -1. We denote upper body score as α_5.

The final score of a given bounding box is defined as follows:

$$\text{Mittal's detector score} + w_f * \alpha_4 + w_u * \alpha_5 \tag{3}$$

w_f and w_u are weights for flow score and upper body score respectively and determined by experiments. We tried all values from 0.1 to 0.9. Based on our experiments, $w_f = 0.7$, $w_u = 0.2$ obtained the best performance.

3.2   Hand Tracking

In order to reduce the computational cost, we process hand detection for only one frame in every k frames. Thus, we need to track obtained detections and automatically link detecting and tracking results over time. We also want to compensate for missing detections as well as search for the most reliable hand tracks.

We track h highest scored bounding boxes of every detection through L frames forwardingly. Since we obtain one detection in every k frames, we need to consider $L * h/k$ bounding boxes. We capture the persistence of hands over time with simple flow based tracking. We take the average flow of a bounding box to propagate it from a frame to the next. A reliable bounding box should overlap with many others during its propagation. A track of a detected bounding box will be employed if the bounding box overlaps more than 50 % with any of h bounding boxes of at least n frames which have hand detection processed among L frames. In our experiments, $h = 2, L = 15, k = 3, n = 2$. Some example results of our method of hand detection and tracking are shown in Fig. 3. As shown in Fig. 3, we are able to not only compensate missing or undone detections but also remove false detections.

4   Application on Action Recognition

Here we describe our approach of classifying action videos of a given action dataset employing the results of hand detection and tracking obtained by our above method.

4.1   Overview of Our Approach

We first apply our proposed hand detector on each video in the dataset. To reduce computational cost, we do not perform hand detection for all frames but for only one in every k frames. In our experiments, k is set to 3. Next, based on the detection results, we track all highly scored bounding boxes through L frames

Fig. 3. Example results of our method of hand detection and tracking on the group of playing instruments in UCF-101 dataset. From the top, first the detection results, then the tracking results of "playing daf", "playing guitar" and "playing violin" are respectively shown. Among 15 consecutive frames, there are 5 frames which have hand detection processed. Only 2 top scored bounding boxes are shown. We track hands and keep hand tracks which overlap with at least 2 detection results. As the results, we can eliminate some failed detections and missing detections as well as obtain quite good hand bounding box sequences.

to obtain connected and more reliable hand regions. In our experiments, L is set to 15. We then apply Wang *et al.* [28]'s method to extract dense trajectory aligned motion features and our hand displacement features from the detecting and tracking results. We conduct a Fisher vector for each type of extracted features. To combine different features, we concatenate their Fisher vectors. We train a multiclass linear SVM to classify the videos. By focusing only on regions which are expected to be the most related to the actions in stead of considering the whole frame, we can improve action recognition precision.

In this paper, we apply Fisher encoding methodology as described in [28]. The descriptor dimensionality is reduced by half using Principal Component Analysis (PCA). A subset of 256,000 features are randomly sampled from the training set to estimate the GMM and the number of Gaussians is set to 256. Each video is represented by a 2DK dimensional Fisher vector for each feature type, where D is the feature dimension after performing PCA. The following subsection explains in detail about the features.

4.2 Feature Extraction

We extract features based on 2D+t sequences of hand bounding boxes obtained by our proposed method of detecting and tracking hands. We apply Wang *et al.*'s method [28] to extract dense trajectories and their aligned motion features: HOF (Histograms of Optical Flow) and MBH (Motion Boundary Histograms) from all detecting and tracking results. Their method recently became the state of the arts for action recognition. According to their method, dense trajectories are obtained by tracking sampled points using optical flow fields for multiple spatial scales. HOF and MBH descriptors are computed within space-time volumes around the trajectories. HOF directly quantizes the orientation of flow vectors. MBH splits the optical flow into horizontal and vertical components, and quantizes the derivatives of each component.

In this paper, we extract dense trajectories for only points which lie inside detected and tracked bounding boxes. If a frame has hand detection processed, its h highest scored detections will be used, otherwise, tracking results will be employed. We demonstrate that hand movements are discriminative and representative enough for actions operated by hands.

Beside dense trajectories and their aligned motion features, we extract hand track feature which describe the shape of hand trajectory by using average flow magnitude of hand regions in complete hand tracks. Given a trajectory of length L, its shape is described by a sequence $S = (\Delta P_t, ..., \Delta P_{t+L-1})$ of displacement vectors $\Delta P_t = (P_{t+1} - P_t)$. Here $P_t = (x_t, y_t)$ indicates the location of point P at frame t. The vector is normalized by the sum of the magnitudes of the displacement vectors:

$$S' = \frac{(\Delta P_t, ..., \Delta P_{t+L-1})}{\sum_{j=t}^{t+L-1} ||\Delta P_j||} \tag{4}$$

This vector is refered to descriptor for trajectory of point P. For our hand track features, only the center points of consecutive hand bounding boxes are taken into account. As the result, we obtain one descriptor for each hand track. While dense trajectories are extracted from all detection and tracking results, our descriptors are obtained from only reliable hand tracks. Thus even though they seem to be less informative than dense trajectories, they are expected to be a useful representation for actions as well.

5 Experiments

We conducted experiments to validate the efficiency of first, our method of detecting and tracking hands and second, our method of classifying actions based on our results of hand detection and tracking. Experiment results show the effectiveness of our approach.

5.1 Experiments on Hand Detection

Here we want to show how our proposed utilization of static cues and motion information can improve hand detection in videos. We compare detection performance between our detector, Mittal et $al.$'s detector [21] which uses only static cues and Sapp et $al.$'s detector [14] which employs only motion information.

We validated our proposed method of hand detection on VideoPose2.0 dataset. The dataset consists of 14 video shots collected from movie source. It was originally developed only for the task of upper body estimation. Therefore, the exact locations of hands are not provided. We had to annotate hands in every frame by ourselves. There are 2453 frames and 3814 hands in total.

In these experiments, we detected hands in every frame. The performance is evaluated using average precision following Mittal et $al.$ [21]. A detection is considered true if its overlap score is more than 0.5. The overlap score of a detected bounding box B_d is defined as $O = \frac{area(B_g \bigcap B_d)}{area(B_g \bigcup B_d)}$, where B_g is the annotated grouth-truth bounding box. The results are summerized in Table 1 and some detection examples are shown in Fig. 4.

First we validated the effectiveness of using faces of selected upper bodies instead of OpenCV face detector. As we can see in Table 1, the result was slightly improved. This is because VideoPose2.0 dataset has high resolution so that faces are usually big and clear enough for OpenCV detector to detect. The precision was significantly enhanced by introducing flow score. The first three rows of Fig. 4 show the effectiveness of our detector over our baseline and flow based detector. While Mittal et $al.$'s detector sometimes failed to detect moving hands, mostly due to their unclear shape, our detector, by considering motion information, could detect them. This demonstrates that motion cue is extremely important for detecting hands in videos. However, employing only motion information can not robustly detect hands as Sapp et $al.$'s flow based detector could achieve only 18.6 % precision. Their flow based detector only concentrates on detection regions moving similarly to trained hands. Our proposed method which utilizes static cues and motion information achieved the best results. By adding upper body based spatial constraints, the precision was further improved. Our method of hand detection improved the baseline approximately 5 %.

Fig. 4. Some examples of our detection results. We show two detections with best scores for each image frame. The best is shown in red, the second best is shown in green bounding box. The three upper rows of this figure show some detection examples in VideoPose2.0 dataset to compare the performance of the baseline, flow based detector and our detector (from the top, respectively). As we can see, our detector can detect more hands, especially hands blurred by their movements. Especially, in the case that there are more than one character (the second image from the right), our detector tends to detect moving hands since they are expected to belong to the main character. On the other hand, using only static cues gives higher scores for static hands which may belong to the character in supporting role (the second example from the right). Using only motion cues (flow) concentrates on detecting moving regions (which sometimes belong to other body parts or background objects). The last row of this figure shows some detection results for the group of playing instruments in UCF-101 dataset.

Table 1. Results of hand detection. We conducted experiments on VideoPose2.0 video dataset and compared our hand detector with our baseline (Mittal *et al.*'s hand detector) and Sapp *et al.*'s flow based hand detector. Our (+upper body) means using face of selected upper body for skin detector. Our (+flow) means adding flow score to refine detection results. Our (+flow+body) means using our full proposed method which employs both flow information and body position based constraints to improve the final results.

Method	Precision
Mittal *et al.* [21]	41.7 %
Sapp *et al.* [14][a]	18.6 %
Our (+upper body)	42.6 %
Our (+flow)	45.5 %
Our (+flow+body)	46.3 %

[a]Their flow based hand detector

5.2 Experiments on Action Recognition

Here we applied the results of hand detection and trackings to action classification. We aimed to classify actions based on how persons move hands to operate them. The actions should have hand movements involved throughout the time they are performed. However, there was too few public data which matches our purposes. We found only the group of playing instruments in UCF-101 dataset as suitable data for us to validate our method. UCF-101 is a very challenging action data set as its video shots are collected from Web source. The data set has 5 action groups, but only the group of playing instruments is suitable for the purpose of fine-grained action classification.

The group of playing instruments in UCF-101 dataset consists of 1428 video shots of actions of playing 10 types of musical instruments: cello, guitar, violin, daf, dhol, piano, tabla, sitar, flute and drum. The shots in each action category are grouped into 25 groups, where each group can consist of 4–7 shots of the action. The video shots from the same group may share some common features, such as similar background and similar viewpoint. We followed evaluation set up as suggested in the ICCV2013 workshop on large-scale action recognition[6]. We adopted their provided three standard train/test splits to conduct experiments. In each split, clips from 7 of the 25 groups are used as test samples, and the rest for training. The result of each experiment reported here is calculated as the mean of average accuracies over the three provided test splits. We train multiclass linear SVMs [29] to perform action recognition.

For data from UCF-101 dataset, to reduce computational cost, we performed hand detection for only one in every three frames. To compensate the detections through the video as well as to find reliable hand tracks, we tracked hands as described in the Sect. 3.2. Since UCF-101 is a large dataset without hand annotations, we could not validate the performance of our method of hand detection and tracking on this data in details. However, based on experimental results, we demonstrate that extracting features from regions specified to hands can achieve comparable performance to extracting from the whole frame. Our baseline in the experiments here refers to the method of extracting dense trajectories proposed by Wang *et al.* [28]. The results of our experiments are shown in Table 2.

As shown in Table 2, using only hand displacement features obtained 36 % accuracy and using dense trajectories with their aligned motion features which were extracted from detected hand regions achieve comparable recognition performance to using original dense trajectories which were extracted from more regions. Even though precision rate of hand detection is not significant, imprecisely detected regions do not affect the final results that much since they are also informative (they are detected and employed by the baseline). The baseline, improved dense trajectory based method, extracts features only from foreground regions which move robustly. Instead of using all of those regions, in our method, we concentrate only on hand regions. The point is, despite of the fact that we use less information, we achieved comparable results to the baseline. That means our detection results are representative enough for the actions. Moreover, by combining multiple motion features considering hand positions, we could improve the baseline. This result demonstrates that the proposed method can extract the features which have different characteristics from the conventional features.

---

[6] http://crcv.ucf.edu/ICCV13-Action-Workshop/.

Table 2. Results of classification of actions in videos. DT means dense trajectories originally proposed in [28]. HDT means dense trajectories restricted to detected hand regions. HOF_{dt} and HOF_{hdt} refer to HOF features aligned with DT and HDT respectively (similarly with MBH). HT means our proposed hand track based displacement features. + means concatenating descriptors to a single descriptor before training and testing (early fusion).

Method	Precision
DT	66.7%
HOF_{dt}	83.8%
MBH_{dt}	86.6%
$MBH_{dt} + HOF_{dt} + DT$	87.3%
HDT	66.1%
HOF_{hdt}	81.4%
MBH_{hdt}	85.7%
$MBH_{hdt} + HOF_{hdt} + HDT$	86.2%
HT	**36.0%**
$MBH_{dt} + HOF_{dt} + DT + HT$	87.6%
$\mathbf{MBH_{hdt} + HOF_{hdt} + HDT + HT}$	**88.5%**

We also could prove that hand related motion features are particularly useful to recognize human actions.

6   Conclusions

In conclusions, we developed an effective hand detector in uncontrolled videos and obtained promising results. Furthermore, we proposed to improve action recognition precision by additionally considering hand movements. Our experiment results showed that this consideration is effective. We try to deeply consider hand movements for the problem of improving action classification in uncontrolled videos. To the best of our knowledge, we are the first to do that. This is the largest contribution of this paper. Even if hand detection accuracy was only about 50%, employing the hand detection could help improve action recognition accuracy. This is a meaningful result even though the improvement is not remarkably significant. If hand detection accuracy is further enhanced, the benefit which action recognition gains from that enhancement can be expected to be larger.

Acknowledgement. This work was supported by JSPS KAKENHI Grant Number 26011435.

References

1. Yao, B., Fei-Fei, L.: Modeling mutual context of object and human pose in human-object interaction activities. In: Proceedings of IEEE Computer Vision and Pattern Recognition, pp. 17–24 (2010)
2. Yao, B., Fei-Fei, L.: Discovering object functionality. In: Proceedings of IEEE International Conference on Computer Vision, pp. 2512–2519 (2013)
3. Delaitre, V., Sivic, J., Laptev, I.: Learning person-object interactions for action recognition in still images. In: Advances in Neural Information Processing Systems (2011)
4. Prest, A., Schmid, C., Ferrari, V.: Weakly supervised learning of interactions between humans and objects. IEEE Trans. Pattern Anal. Mach. Intell. **34**, 601–614 (2012)
5. Prest, A., Ferrari, V., Schmid, C.: Explicit modeling of human-object interactions in realistic videos. IEEE Trans. Pattern Anal. Mach. Intell. **35**, 835–848 (2013)
6. Khurram, S., Amir, R., Mubarak, S.: UCF101: A dataset of 101 human actions classes from videos in the wild. CoRR abs/1212.0402 (2012)
7. Binh, N.D., Shuichi, E., Ejima, T.: Real-time hand tracking and gesture recognition system. In: Proceedings of International Conference on Graphics, Vision and Image Processing, pp. 19–21 (2005)
8. Manresa, C., Varona, J., Mas, R., Perales, F.: Hand tracking and gesture recognition for human-computer interaction. Electron. Lett. Comput. Vis. Image Anal. **5**, 96–104 (2005)
9. Angelopoulou, A., Rodríguez, J.G., Psarrou, A.: Learning 2d hand shapes using the topology preservation model GNG. In: Proceedings of European Conference on Computer Vision, pp. 313–324 (2006)
10. Ren, Z., Yuan, J., Zhang, Z.: Robust hand gesture recognition based on finger-earth mover's distance with a commodity depth camera. In: Proceedings of ACM International Conference on Multimedia, pp. 1093–1096 (2011)
11. Van den Bergh, M., Van Gool, L.: Combining RGB and ToF cameras for real-time 3d hand gesture interaction. In: IEEE Workshop on Applications of Computer Vision, pp. 66–72 (2011)
12. Cerlinca, T.I., Pentiuc, S.G.: Robust 3D hand detection for gestures recognition. In: Brazier, F.M.T., Nieuwenhuis, K., Pavlin, G., Warnier, M., Badica, C. (eds.) Intelligent Distributed Computing V. SCI, vol. 382, pp. 259–264. Springer, Heidelberg (2011)
13. Oikonomidis, I., Lourakis, M.I., Argyros, A.: Evolutionary quasi-random search for hand articulations tracking. In: Proceedings of IEEE Computer Vision and Pattern Recognition (2014)
14. Sapp, B., Weiss, D., Taskar, B.: Parsing human motion with stretchable models. In: Proceedings of IEEE Computer Vision and Pattern Recognition, pp. 1281–1288 (2011)
15. Yuan, Q., Sclaroff, S., Athitsos, V.: Automatic 2d hand tracking in video sequences. In: IEEE Workshops on Application of Computer Vision, vol. 1, pp. 250–256 (2005)
16. Baltzakis, H., Argyros, A.A., Lourakis, M.I.A., Trahanias, P.: Tracking of human hands and faces through probabilistic fusion of multiple visual cues. In: Gasteratos, A., Vincze, M., Tsotsos, J.K. (eds.) ICVS 2008. LNCS, vol. 5008, pp. 33–42. Springer, Heidelberg (2008)
17. Filipovych, R., Ribeiro, E.: Recognizing primitive interactions by exploring actor-object states. In: Proceedings of IEEE Computer Vision and Pattern Recognition, pp. 1–7 (2008)

18. Filipovych, R., Ribeiro, E.: Robust sequence alignment for actor-object interaction recognition: discovering actor-object states. Comput. Vis. Image Underst. **115**, 177–193 (2011)
19. Gupta, A., Kembhavi, A., Davis, L.: Observing human-object interactions: using spatial and functional compatibility for recognition. IEEE Trans. Pattern Anal. Mach. Intell. **31**, 1775–1789 (2009)
20. Weinzaepfel, P., Revaud, J., Harchaoui, Z., Schmid, C.: DeepFlow: large displacement optical flow with deep matching. In: Proceedings of IEEE International Conference on Computer Vision (2013)
21. Mittal, A., Zisserman, A., Torr, P.H.: Hand detection using multiple proposals. In: Proceedings of British Machine Vision Conference, pp. 1–11 (2011)
22. Felzenszwalb, P.F., Girshick, R.B., McAllester, D., Ramanan, D.: Object detection with discriminatively trained part based models. IEEE Trans. Pattern Anal. Mach. Intell. **32**, 1627–1645 (2010)
23. Burges, C.J.: A tutorial on support vector machines for pattern recognition. Data Min. Knowl. Disc. **2**, 121–167 (1998)
24. Arbelaez, P., Maire, M., Fowlkes, C., Malik, J.: Contour detection and hierarchical image segmentation. IEEE Trans. Pattern Anal. Mach. Intell. **33**, 898–916 (2011)
25. Bourdev, L., Malik, J.: Poselets: body part detectors trained using 3d human pose annotations. In: Proceedings of IEEE International Conference on Computer Vision (2009)
26. Dalal, N., Triggs, B.: Histograms of oriented gradients for human detection. In: Proceedings of IEEE Computer Vision and Pattern Recognition, vol. 1, pp. 886–893 (2005)
27. Karlinsky, L., Dinerstein, M., Harari, D., Ullman, S.: The chains model for detecting parts by their context. In: Proceedings of IEEE Computer Vision and Pattern Recognition, pp. 25–32 (2010)
28. Wang, H., Schmid, C.: Action recognition with improved trajectories. In: Proceedings of IEEE International Conference on Computer Vision, pp. 3551–3558 (2013)
29. Tsochantaridis, I., Joachims, T., Hofmann, T., Altun, Y.: Large margin methods for structured and interdependent output variables. J. Mach. Learn Res. **6**, 1453–1484 (2005)

Enhancing Person Re-identification by Integrating Gait Biometric

Zheng Liu[1], Zhaoxiang Zhang[1(✉)], Qiang Wu[2], and Yunhong Wang[1]

[1] Laboratory of Intelligence Recognition and Image Processing,
Beijing Key Laboratory of Digital Media, School of Computer Science
and Engineering, Beihang University, Beijing 100191, China
zhaoxiang.zhang@ieee.org
[2] School of Computing and Communications,
University of Technology, Sydney, Australia

Abstract. This paper proposes a method to enhance person re-identification by integrating gait biometric. The framework consists of the hierarchical feature extraction and matching methods. Considering the appearance feature is not discriminative in some cases, the feature in this work composes of the appearance feature and the gait feature for shape and temporal information. In order to solve the view-angle change problem and measuring similarity, metric learning to rank is adopted. In this way, data are mapped into a metric space so that distances between people can be measured accurately. Then two fusion strategies are proposed. The score-level fusion computes distances of the appearance feature and the gait feature respectively and combine them as the final distance between samples. Besides, the feature-level fusion firstly installs two types of features in series and then computes distances by the fused feature. Finally, our method is tested on CASIA gait dataset. Experiments show that gait biometric is an effective feature integrated with appearance features to enhance person re-identification.

1 Introduction

With the growing demand for security surveillance, how to accurately identify people has drawn increasing attentions. Person re-identification is an important part of surveillance. The problem can be defined as recognizing an individual who has already appeared in another camera in a non-overlapped multi-camera system.

The research on person re-identification often concentrates on two main aspects: the extraction of features and matching methods, in which the former one focuses on how to describe individuals and matching methods try to measure distances between samples.

The selected feature should be robust to the variation of illumination, posture and view. Some excellent appearance-based feature extraction methods have been proposed. Farenzena *et al.* [4] used color histograms, MSCR [5] as well as RHSP with the symmetry property to describe a local patch. Gray and Tao [6] let

© Springer International Publishing Switzerland 2015
C.V. Jawahar and S. Shan (Eds.): ACCV 2014 Workshops, Part I, LNCS 9008, pp. 35–45, 2015.
DOI: 10.1007/978-3-319-16628-5_3

a machine learning algorithm find the best representation. Bazzani *et al.* [1] condensed a set of frames of an individual into a highly informative signature, called the Histogram Plus Epitome (HPE). Zhao *et al.* [17] found salience of people based on SIFT [12] and color histogram and then match patches with constraint. However, if there are several people having similar appearance, or appearance of the same person is quite different in different cameras, these appearance-based methods would have a poor performance (Figs. 1 and 2).

Fig. 1. Example of people observed from different views with different appearance in CASIA dataset. Each row is the same individual that (a) is condition with bag, (b) is condition with coat and (c) is normal condition.

In those proposed approaches, biometric is seldom mentioned, which we think should also be taken into account. Biometric gradually developed into an important kind of features which has a strong individual discrimination. It would play a significant role in person re-identification problem if used properly. Even though many mature methods of traditional biometrics have been proposed, such as fingerprint and face, they have to be obtained by getting close to target people, or making contact with them. In recent years, gait features are proposed and gradually attracted the attention of many scholars. There are many methods have been proposed about Gait Recognition. Little and Boyd [10] developed the shape of motion which is a model-free description of instantaneous motion, and used it to recognize individuals by their gait. Sarkar *et al.* [16] measured the similarity between the probe sequence and the gallery sequence directly by computing the correlation of corresponding frame pairs.

The gait feature is an ideal way for application in security surveillance as a biometric, because getting gait features is just using a camera from a long

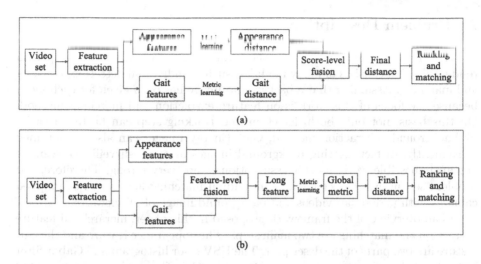

Fig. 2. Overview of proposed method. (a) is framework with score-level fusion and (b) is with feature-level fusion.

distance away and then doing image processing. In addition, carrying temporal information is the other good quality of gait features. Instead of using the single-shot image, image sequences grabbed from videos are used to generate temporal feature, considering inputs of cameras are sequences of images. Thus, a spatio-temporal analysis for person re-identification can be performed by integrating the gait feature with appearance features.

For these reasons, the gait feature can be used to enhance sequence based person re-identification. The gait feature is robust even if people change their appearance, where appearance features are helpless. The gait feature is not such strong discriminative as traditional biometric. However, it is much more convenient to fuse gait with appearance features extracted from surveillance videos. From this perspective, the gait biometric is more suitable for person re-identification than recognizing problems. Since a fusion step has to be processed, the selection of features and fusion strategies are important and hard work.

In this paper, we try to solve the person re-identification problem by integrating the gait feature with appearance features. For appearance features, HSV histogram is widely used to describe the color information, and texture information can be well represented by Gabor feature [11]. Gait Energy image (GEI) [7] is a popular feature that used to represent the gait biometric. So our descriptor composed of HSV histogram and Gabor feature as the appearance feature and GEI as the gait feature. Then these two type features are fused by two strategies, namely score-level fusion and feature-level fusion. After the data are modeled by descriptive features, a metric learning method is adopted for similarity measurement. The idea of metric learning is comparing descriptors by a learned metric instead of Euclidean distance. Finally, we test our method on CASIA dataset which is a gait dataset created by Chinese Academy of Sciences.

2   Problem Description

In the real application scene, the framework for video-based person re-identification consists of many modules, such as individual detection, tracking and matching. Assuming that a background image has been given for each view, because the focus of this work is on feature extraction and matching method. On this basis, not only both detection and tracking step can be taken easily by background subtraction method, Gait Energy Images can also be obtained conveniently. In fact, getting background images in many surveillance scene is not an impossible work, so this assumption is not very strong. Therefore, our problem setting can be set forth as individual matching across non-overlapped cameras with pedestrian videos and background image of all views.

As an overview of the framework proposed in this paper, hierarchical feature extraction and matching are two main steps. The appearance feature and the gait feature are two parts of the descriptor. The HSV color histogram and Gabor filter are used to describe the appearance of people. Gait Energy Image is generated to obtain the spatio-temporal information and Principal Component Analysis (PCA) [15] is used to obtain the low-dimension GEI feature. The gait feature is integrated with the appearance feature by two different fusion strategies. In the matching step, there are different procedures for different fusion strategies. For score-level fusion, two metrics are trained for the appearance feature and GEI feature, respectively. After that, two distances are computed and fused to obtain the final distance between samples. For feature-level fusion, a global metric matrix is trained by metric learning to matching the fused feature and then similarities are measured by this global metric.

3   Proposed Framework

3.1   Hierarchical Feature Extraction

Gait Feature. Gait Energy Image (GEI) is the gait feature used in proposed method. In general, GEI is generated by silhouette images extracted from the view of 90°. Nevertheless, GEI is considered still contains some global information of individuals such as body shape and spatio-temporal changes of the human body in other views in addition to 90°. So in this work, GEI generated of all 11 views is adopted from 0° to 180° for cross-view analysis.

Firstly, background subtraction is used to extract the connected area of foreground. Then a series of foreground images are grabbed from each frame in the video, which form a sequence of silhouette images for each person of each view. Meanwhile, each image is normalized to the size of 64 × 32 for the convenience to be processed. Given the binary gait silhouette images $S_{ijt}(x, y)$ at time t in a sequence of person i from view j, Gait Energy Images then can be obtained as follows:

$$G_{ij}(x,y) = \frac{1}{N_{ij}} \sum_{t=1}^{N_{ij}} S_{ijt}(x,y) \tag{1}$$

where N_{ij} is the total number of frames in view j of person i. It may be different for each person and view because of the different length of videos t in the frame index in the sequence and x and y are the coordinates of pixels in the image.

As the value of a pixel in the Gait Energy Image represents the possibility that the human body appears in the position, the feature can be described as a 1D vector that composed of values of all pixels. Because of the normalization step, the GEI feature is a 2048-dimension vector, so that dimensionality reduction has to be taken. Principal Component Analysis (PCA) is adopted since it is a classical approach that can reduce dimension of GEI feature significantly (Fig. 3).

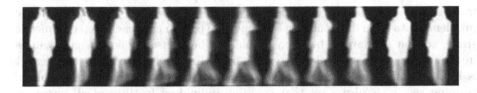

Fig. 3. Example of Gait Energy Images of all 11 views.

Appearance Features. Multiple frames help little for extraction of appearance features but increase calculation largely according to our experiments. So that a single frame for each person is randomly picked at this step to generate appearance features. The first part of the appearance feature is Hue-Saturation-Value (HSV) space color histogram. The reason why choosing HSV color space instead of RGB color space is that HSV color space is more adapting to the human perception of color and HSV model is not sensitive to illumination variation. Each person image is transformed from RGB color space to HSV color space in normalization step, then histogram equalization is carried out to reinforce the robustness to illumination further. Each dimension of HSV is divided into 128 bins to count the number of pixels whose value falls into the corresponding bin.

The other part is Gabor features [11]. The frequency and direction of Gabor filter are close to what in the human visual system so that the Gabor filter is often used to generate the texture feature in many fields, such as face and fingerprint recognizing. A 2-dimension Gabor filter is a Gaussian kernel function modulated by a complex sinusoidal plane wave in spatial domain. A Gabor filter can be defined as follows [3,9,13]:

$$\psi_{u,v}(z) = \frac{||k_{u,v}||^2}{\sigma^2} e^{-\frac{||k_{u,v}||^2 ||z||^2}{2\sigma^2}} [e^{ik_{u,v}z} - e^{-\frac{\sigma^2}{2}}] \tag{2}$$

where u is the orientation and v is the scale of the Gabor filters, $z = (x, y)$, and $k_{u,v}$ is defined as:

$$k_{u,v} = k_v e^{i\phi_\mu} \tag{3}$$

where $k_v = k_{max}/f^v$ and $\phi_\mu = \pi\mu/8$. k_{max} is the maximum frequency, and f is the spacing factor between kernels in the frequency domain [9].

Gabor filter of five different scales and eight orientations is used with the following parameters: $\sigma = \pi, k_{max} = \pi/2$ and $f = \sqrt{2}$.

3.2   Metric Learning to Rank

With the assumption that the view of observation is known, a view-independent method is used to measure similarities. A metric is trained by Metric Learning to Rank (MLR) [14] with data of all views. MLR is a general metric learning algorithm based on structural SVM [8] framework and view metric learning as problem of information retrieval. Its purpose is to learn a metric so that data can be well ranked by distances. In this paper, a metric is trained to map data to another space in which data from the same person locate closer than those from different people. By this metric, we try to solve the view-angle change problem. Data from different views of the same person can be clustered in metric space, so that similarities between people can be measured simply as distances in the metric space, regardless of view angles. In training step, the training set is built as a feature set $X = \{F_i\}_1^n$ where n is the total number of people in training set. F_i is also a set that contains individual features of all views, as $F_i = \{f_{ij}\}_{j=1}^v$ where v is the number of views which is 11 in CASIA dataset, $f_{ij} = (x_{1ij}, x_{2ij}, \ldots, x_{mij})^T$ is feature of person i view j and m is the length of the feature. Then a normalization step is taken. Data are normalized as follows:

$$Z(X) = \frac{X - \mu}{S} \tag{4}$$

where μ is the mean of features, S is the standard deviation and $Z(X)$ is normalized data matrix. After data are ready, a metric $W \in \mathbb{R}^{m \times m}$ is trained by MLR. Here as parameters of MLR, the area under the ROC curve(AUC) [2] is adopted to be the ranking measure and the slack trade-off parameter is 0.01 in MLR training.

When use metric W to solve person re-identification matching problem, the distance can be computed as follows:

$$D(f_1, f_2) = (f_1 - f_2)^T W (f_1 - f_2) \tag{5}$$

where f_1 and f_2 are features of probe person and gallery person. $D(f_1, f_2)$ is the distance between f_1 and f_2, which would be small if the probe image and the gallery image are similar. In test step, we have a gallery set A and a probe set B in general. Associating each person of set B to all people of A is the purpose of person re-identification. So for each probe person b_i, distances between b_i and all gallery people are computed, then ranking is processed.

3.3   Fusion Strategy

In order to effectively use the GEI feature and appearance features, two fusion methods in our framework will be introduced in the following. The first one is the score-level fusion that fuse distances which are calculated by GEI feature and the appearance feature, respectively. The second one is the feature-level fusion that installs two type features in series.

Score-Level Fusion. This fusion strategy is a view-independent global function, formulated as follows:

$$D_{FIN}(S_1, S_2) = D_{APP}(S_1, S_2) + D_{GEI}(S_1, S_2) \tag{6}$$

where $D_{FIN}(S_1, S_2)$ is the final relative distance of two samples get from two different views. $D_{APP}(S_1, S_2)$ is the distance that derived from the appearance feature and $D_{G}EI(S_1, S_2)$ is distance derived from GEI feature. The reason why just simply adding two distances without a weighting factor is that distances computed by metric are already optimized. Therefore, putting a factor into the equation is unreasonable which is also proved by experiment. If a weighting factor is added into equation as follows:

$$D_{FIN}(S_1, S_2) = D_{APP}(S_1, S_2) + \theta D_{GEI}(S_1, S_2) \tag{7}$$

where θ is the weighting factor. Repeat experiments with the value of θ range from 0.1 to 10, and the best result is obtained when the value of θ is 1.

Feature-Level Fusion. This strategy is to fusion two type features before calculating distance. Suppose we have an appearance feature vector $A \in \mathbb{R}^m$, and GEI feature vector $G \in \mathbb{R}^n$. Then these two features can be combined as follows:

$$F = [A, G] \tag{8}$$

where the operator [X,Y] is defined as installing X and Y in series. $F \in \mathbb{R}^{m+n}$ is the final feature vector.

4 Experiment

4.1 Dataset

The proposed method is tested on the CASIA Gait Database B created by The Institute of Automation, Chinese Academy of Sciences (CASIA). This dataset contains eleven views of 124 individuals with three conditions. These three conditions are 'bag', which means the pedestrian appears with a bag in the video, 'clothes', which means the pedestrian appears with coat in the video and 'normal' means pedestrians appear without coat or bag. The view contains the angle of $0°, 18°, 36°, 54°, 72°, 90°, 108°, 126°, 144°, 162°$ and $180°$. The CASIA dataset provides a background video for each view so that the foreground can be extracted conveniently.

4.2 Results

In the first step of the experiment, only three original conditions are involved, in which people do not change their appearance. When learning the metric matrix, data of all views are used. For testing, the view of each probe image in probe set is randomly chosen, respectively, and the same random process is conducted to gallery set. Data are operated like that to fit the real application scenario of person re-identification most. Under these experiment constraints, all two type fusion strategies of features are implemented comparing to situations of only one type feature is used. SDALF method is also put into comparison. It is a very classical and effective appearance-based method for person re-identification problem. Those methods are tested on three conditions in CASIA dataset include 'bag', 'clothes' and 'normal'. The data are separated to training data and testing data, each contains half of randomly picked samples, and the same separation is adopted to all experiments.

The results shown in Fig. 4 primarily verify the availability of gait in this initial experiment. It shows that the GEI feature can well describe people integrating with appearance features. On 'bag' condition, the performance of our method with feature-level fusion is the best. On 'clothes' condition, our method with two fusion strategies are better than single-feature situations and on 'normal' condition, method with score-level performs a little better.

The next step of the experiment is implemented on two challenging cross conditions that include 'bag-clothes' and 'clothes-normal'. We try to challenge this harder task because in person re-identification problem, cameras are

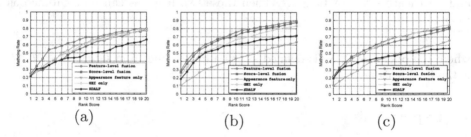

Fig. 4. CMC curves on the single condition in CASIA dataset. (a) is on 'bag' condition, (b) is on 'clothes' condition and (c) is on 'normal' condition.

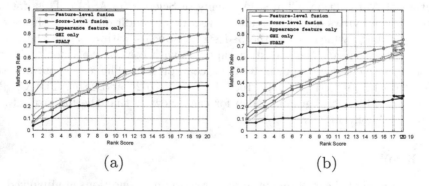

Fig. 5. CMC curves on cross conditions. (a) is on 'bag-clothes' condition and (b) is on 'clothes-normal' condition.

non-overlapped so that pedestrians have a certain probability to change their appearance. We would also like to indicate that gait can get rid of the limit of traditional appearance features and become a kind of robust features to enhance person re-identification. To this end, data that contain two appearance conditions are used to train a metric matrix. The probe set and the gallery set are built with different conditions. In other words, we try to match individuals from one view with one condition to another view in another condition. Testing views are also picked randomly and training set and testing set are separated randomly as well (Tables 1 and 2).

Table 1. Comparison of the average Matching Rate (%) on the single condition.

Rank	r = 1	r = 5	r = 10	r = 15	r = 20
Score-level	23.53	51.76	67.84	77.71	84.51
Feature-level	23.79	52.22	66.67	75.42	82.88
Appearance	21.41	48.01	64.67	74.15	82.25
GEI	15.27	33.68	50.50	60.25	68.58
SDALF	19.93	43.14	52.29	58.82	64.71

Table 2. Comparison of the average Matching Rate (%) on the cross condition.

Rank	r = 1	r = 5	r = 10	r = 15	r = 20
Score-level	8.53	27.94	44.51	57.25	68.43
Feature-level	25.00	48.43	60.88	70.69	77.84
Appearance	6.27	27.25	42.25	56.96	66.57
GEI	9.02	28.04	41.37	54.31	63.14
SDALF	5.39	15.20	22.55	28.92	33.33

As shown in Fig. 5, our method with feature-level fusion is far beyond others. Results show that both in traditional person re-identification and cross-condition person re-identification, the fused feature has an acceptable performance. Moreover, results show that the gait biometric can be an effective feature to enhance person re-identification.

The result in Table 1 shows a holistic perspective. In traditional application scene, fused features are as good as the appearance feature. On cross conditions, only the feature-level fusion method maintains the good performance. As shown from Table 2, neither the appearance feature nor the gait feature can handle the cross condition alone, which shows our method that integrating gait biometric with appearance features is effective.

Probe   **Rank in gallery set**

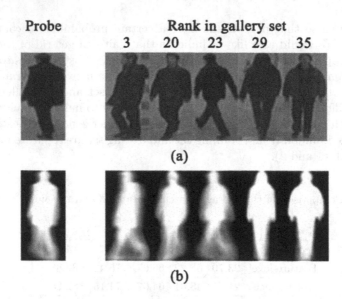

Fig. 6. Example of the effect of GEI feature. The appearance of Gallery images shown in (a) are similar to the probe image. They are all dressed in the black coat and cannot be distinguished by appearance features. Instead, Gait Energy Images shown in (b) are quite discriminative and help to re-identify the person in the gallery set.

5   Conclusion

The work of this paper is attempting to enhance person re-identification by integrating gait biometric. The proposed method contains hierarchical feature extraction and similarity measurement. Our experiments indicate that appearance features are not discriminative in some cases and other information should be integrated. Experiments show that gait biometric can be applied well in person re-identification. Even though GEI is not such discriminative, the attempt to enhance person re-identification by integrating gait biometric is successful in this paper. Our method is also verified as an effective way to combine gait biometric to appearance features, and can enhance person re-identification in many conditions.

References

1. Bazzani, L., Cristani, M., Perina, A., Murino, V.: Multiple-shot person re-identification by chromatic and epitomic analyses. Pattern Recogn. Lett. **33**(7), 898–903 (2012)
2. Bradley, A.P.: The use of the area under the roc curve in the evaluation of machine learning algorithms. Pattern Recogn. **30**(7), 1145–1159 (1997)
3. Daugman, J.G.: Two-dimensional spectral analysis of cortical receptive field profiles. Vis. Res. **20**(10), 847–856 (1980)

4. Farenzena, M., Bazzani, L., Perina, A., Murino, V., Cristani, M.: Person re-identification by symmetry driven accumulation of local features. In: Proceedings of the 2010 IEEE Computer Society Conference on Computer Vision and Pattern Recognition (CVPR 2010), IEEE Computer Society, San Francisco (2010)
5. Forssén, P.-E.: Maximally stable colour regions for recognition and matching. In: IEEE Conference on Computer Vision and Pattern Recognition, 2007. CVPR'07, pp. 1–8. IEEE (2007)
6. Gray, D., Tao, H.: Viewpoint invariant pedestrian recognition with an ensemble of localized features. In: Forsyth, D., Torr, P., Zisserman, A. (eds.) ECCV 2008, Part I. LNCS, vol. 5302, pp. 262–275. Springer, Heidelberg (2008)
7. Han, J., Bhanu, B.: Individual recognition using gait energy image. IEEE Trans. Pattern Anal. Mach. Intell. 28(2), 316–322 (2006)
8. Joachims, T.: A support vector method for multivariate performance measures. In: Proceedings of the 22nd international conference on Machine learning, pp. 377–384. ACM (2005)
9. Lades, M., Vorbruggen, J.C., Buhmann, J., Lange, J., von der Malsburg, C., Wurtz, R.P., Konen, W.: Distortion invariant object recognition in the dynamic link architecture. IEEE Trans. Comput. 42(3), 300–311 (1993)
10. Little, J., Boyd, J.: Recognizing people by their gait: the shape of motion. Videre J. Comput. Vis. Res. 1(2), 1–32 (1998)
11. Liu, C., Wechsler, H.: Gabor feature based classification using the enhanced fisher linear discriminant model for face recognition. IEEE Trans. Image Process. 11(4), 467–476 (2002)
12. Lowe, D.G.: Distinctive image features from scale-invariant keypoints. Int. J. Comput. Vis. 60(2), 91–110 (2004)
13. Marčelja, S.: Mathematical description of the responses of simple cortical cells*. JOSA 70(11), 1297–1300 (1980)
14. McFee, B., Lanckriet, G.R.: Metric learning to rank. In: Proceedings of the 27th International Conference on Machine Learning (ICML-10), pp. 775–782 (2010)
15. Preisendorfer, R.W., Mobley, C.D.: Principal Component Analysis in Meteorology and Oceanography. Elsevier Science Ltd, Amsterdam (1988)
16. Sarkar, S., Jonathon Phillips, P., Liu, Z., Robledo Vega, I., Grother, P., Bowyer, K.W.: The humanid gait challenge problem: data sets, performance, and analysis. IEEE Trans. Pattern Anal. Mach. Intell. 27(2), 162–177 (2005)
17. Zhao, R., Ouyang, W., Wang, X.: Unsupervised salience learning for person re-identification. In: 2013 IEEE Conference on Computer Vision and Pattern Recognition (CVPR), pp. 3586–3593. IEEE (2013)

Real Time Gait Recognition System Based on Kinect Skeleton Feature

Shuming Jiang[1], Yufei Wang[2], Yuanyuan Zhang[1], and Jiande Sun[2](✉)

[1] Information Research Institute, Shandong Academy of Sciences,
Jinan 250014, China
[2] School of Information Science and Engineering, Shandong University,
Jinan 250100, China
jd_sun@sdu.edu.cn

Abstract. Gait recognition is a kind of biometric feature recognition technique, which utilizes the pose of walking to recognize the identity. Generally people analyze the normal video data to extract the gait feature. These days, some researchers take advantage of Kinect to get the depth information or the position of joints for recognition. This paper mainly focus on the length of bones namely static feature and the angles of joints namely dynamic feature based on Kinect skeleton information. After preprocessing, we stored the two kinds of feature templates into database which we established for the system. For the static feature, we calculate the distance with Euclidean distance, and we calculated the distance in dynamic time warping algorithm (DTW) for the dynamic distance. We make a feature fusion for the distance between the static and dynamic. At last, we used the nearest neighbor (NN) classifier to finish the classification, and we got a real time recognition system and a good recognition result.

1 Introduction

Gait recognition which uses the walking posture of the people for people identification is a new biometric identification technology. This technology is non-invasive, long-distance identification and it is difficult to hide the biological characteristics, which other biometric identification technologies do not have. Based on these advantages, gait recognition has board application prospects in entrance guard system, security surveillance, human-computer interaction, medical diagnostics, and other fields. Because the research is still in the primary stage, the existing methods are generally utilize all kinds of moving object detection algorithm such as Inter-frame Difference, background subtraction to separate the moving people from the background, and then extract the feature which can best distinguish the person, and finally make the classification.

Lee et al. [1] characterize the human body as seven ellipses. Through analyzing the change of centroid and eccentricity, they extract the feature of the gait. Zhang et al. [2] propose a model called 3DHW (3-Dimension Human Walking), which abstract the whole body into a connected rigid body. The body is made

© Springer International Publishing Switzerland 2015
C.V. Jawahar and S. Shan (Eds.): ACCV 2014 Workshops, Part I, LNCS 9008, pp. 46–57, 2015.
DOI: 10.1007/978-3-319-16628-5_4

of 21 points, 14 line segments and 12 joints. Wang et al. [3] make use of the distance between centroid and the edge points of silhouette to get the feature of walking people. He et al. [4] model for the moving leg of human, and propose a method of gait recognition based on the angles of joints. Bobick et al. [5] take the length of body, torso, legs and step to recognize the gait. Kale et al. [6] make the extracted moving people into binary image, and use the width of side silhouette as features for classification. In these general methods, is the feature extraction good or not, and is classification successful or not always depends on the gait extraction. However, due to the dynamic changes of background, such as weather, illumination, shadow, etc. the gait extraction becomes a difficult mission. Kinect can be barely affected by these factors. Now some researchers take advantage of Kinect into the research of gait recognition. Sivapalan et al. [7] use the deep image information captured by Kinect, and analyze the gait energy image to identify the gait. Gabel et al. [8] make use of 20 joints' positions collected by Kinect for gait recognition.

This paper proposed a real time gait recognition system based on Kinect skeleton feature. Kinect can track and extract the joints of human, and provide the 3D coordinates of 20 joints. Because of this, we can complete the gait extraction easily, and can also get static gait feature and dynamic gait feature by calculating the coordinates of joints. After fitting the original data during the preprocessing, we established a system fitted data base. We calculated the distance between each templates and test samples in DTW (Dynamic Time warping), and made a feature fusion in matching layer. Finally we get the classification by using NN classifier.

2 Kinect

Kinect is a line of motion sensing input devices by Microsoft for Xbox 360 and Windows PCs. It has a normal webcam and a depth sensor which can provide RGB-D image. The depth sensor consists of an infrared laser projector combined with a monochrome CMOS sensor, which captures video data in 3D under any ambient light conditions. The detecting range of depth sensor is 0.8 m–4.0 m. It can utilize the machine learning technique to analyze and track the position of 20 skeleton joints in real time. Each joint includes X, Y, and Z, three coordinates information, whose accuracy can reach the millimeter level. The reasons of taking using Kinect to get through the gait detection and feature extraction are following:

1. Kinect is barely affected by the illumination and background, which often affect the recognition rate in general methods. As shown in Fig. 1
2. Kinect is barely affected by wearing. According to the result of experiment, Kinect can even recognize the joints of women's legs under the condition of wearing dress or skirt.
3. Depth information can separate the moving people from background conveniently. When the background includes some moving objects, depth information can also work.

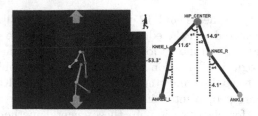

Fig. 1. Experiment in the dark condition.

3 Feature Extraction

3.1 Static Feature

The static feature is a kind of feature that barely changes during the process of walking, such as body height, the length of leg, the length of arm, etc. According to anthropometry [9], the position of each joint, the measurement of bone, and the scale of skeleton can distinguish each person theoretically. As the above suggest, Kinect is able to get the 3D coordinate of 20 joints, so each segment of skeleton can be used as static feature in theory. This paper selected the static feature according to the following principles:

1. The feature which can distinguish each person effectively.
2. On the premise of the first condition, the dimensions of feature vectors should be as small as possible, so that can reduce the computation complexity.

During the experiment, we perceived that some joint points which are hardly influenced by people's wearing and noise make a bigger contribution to the recognition, so we would like to use these joints. Based on the principles mentioned above, this paper selected following static feature. This paper defined the distance between HIP_RIGHT and KNEE_RIGHT as length of thigh, denoted by d_1, defined the distance between KNEE_RIGHT and ANKLE_RIGHT as length of calf, denoted by d_2, defined the distance between SHOULDER_RIGHT and ELBOW_RIGHT added the distance between ELBOW_RIGHT and WRIST_RIGHT as length of arm, denoted by d_3, defined the distance between HEAD and (FOOT_RIGHT+FOOT_LEFT)/2 as body height, denoted by d_4, defined the radio of thigh and height, namely d_1/d_4, denoted by d_5. As shown in Fig. 2. We can get the 3D coordinates of those mentioned joints provided by Kinect SDK, and according to the following formula,

$$d = \sqrt{(x - x_1)^2 + (y - y_1)^2 + (z - z_1)^2} \tag{1}$$

We can calculate the value of d_1, d_2, d_3, d_4, d_5, establishing the feature vector $D = (d_1, d_2, d_3, d_4, d_5)$. We stored this vector into the database as feature template.

Due to the existing gait databases are mostly based on video frame data, such as CASIA [10], are not suitable for our system, we had to build up our own database. There are 6 tables in the database, respectively are *peopleInfo*,

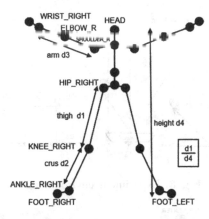

Fig. 2. Schematic of static feature.

staticSkeletonInfo, angleOne, angleTwo, angleThree, andangleFour. Here, we stored the feature vector $D = (d_1, d_2, d_3, d_4, d_5)$ into the *staticSkeletonInfo*, and built a connection with *peopleInfo* by key (*peopleID*).

3.2 Dynamic Feature

The dynamic features are changing at any time during the process of walking, and the changing always presents a periodicity. Tian et al. [11] analyzed the changing angle of leg during human walking on a treadmill through the video data captured by professional 3D camera VICON MX. The result of experiment provided a theoretical basis for the angle of leg can be used as a kind of dynamic feature. He et al. [12] separated the people from the video image, and extracted the leg angle for gait recognition.

In this paper, we mainly selected two kinds of dynamic features, including angles of swinging legs and angles of swinging arms. In the experiment, the Kinect placed on the platform of 1.3 m. Due to the viewing angle of Kinect is $\pm 28.5°$, to make sure that Kinect can record the whole process of walking, test people are suggested to stand 3.3 m far away from Kinect, and start walking behind optical axis at least 1.8 m. As shown in Fig. 3.

People walk along the negative X-axis of Kinect. The leg far away from Kinect would be covered by another leg, causing inaccuracy in data collecting, so we decided to get the joint angle data from the near leg. As shown in Fig. 4, this paper defined the angle between the right thigh and vertical direction as a_1, defined the angle between the right calf and vertical direction as a_2, defined the angle between the right arm and vertical direction as a_3, and defined the angle between the right forearm and vertical direction as a_4. Assuming the coordinate of HIP_CENTER is (x, y), and the coordinate of KNEE_RIGHT is (x_1, y_1), so we can get a_1 according to the following formula:

$$\tan \angle a_1 = (\frac{x - x_1}{y - y_1}) \quad \Rightarrow \quad a_1 = \arctan(\frac{x - x_1}{y - y_1}), \qquad (2)$$

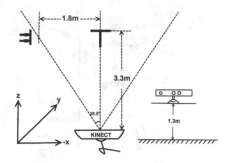

Fig. 3. Experiment environment.

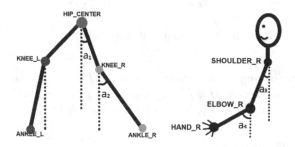

Fig. 4. Schematic of dynamic feature.

We can also get a_2, a_3, a_4 in the same way. These four angles can made up dynamic feature vector $a = (a_1, a_2, a_3, a_4)$. We built up four arrays in order to add corresponding angle in each frame, so we got four sequences of angles, $a_1(t), a_2(t), a_3(t), a_4(t)$. The dynamic feature vector is finally expressed by $A[a(t)]$.

4   Data Processing

4.1   Gait Cycle Analysis

Human's gait information shows some certain periodicity. A gait cycle contains a large number of human gait features [13]. Figure 5 shows one person's gait cycle which is defined as the gait motion set that the same foot touches the ground twice adjacently. Since the gait is periodic, the gait cycle feature could be expressed through extracting the feature vector of one gait cycle. So we should extract one gait cycle from the data we got.

Through analyzing the data of the video and angle collected by the Kinect, the paper regard the state when two feet get the farthest distance with each other and meanwhile the right foot is forward as the starting point of the gait cycle. Also, the paper regard the next similar state as the ending point of the gait cycle. We separately record the frame number of the two points

Fig. 5. One gait cycle, we can conspicuously notice the periodism from the leg and the arm separately marked by the green and the red (Color figure online).

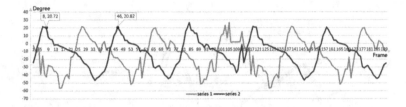

Fig. 6. Gait cycle analysis, two marked points are separately the starting point and ending point.

and get the data between the starting point and ending point from the original data as a gait cycle. The 4 angle sequences $a_1(t), a_2(t), a_3(t), a_4(t)$ turn into $a_1(T), a_2(T), a_3(T), a_4(T)$. As shown in Fig. 6.

4.2 Curve Fitting

In this paper, the original discrete data will be using least squares polynomial fitting method. Fitting function is $f(x) = a_0 + a_1x + a_2x^2 + \cdots + a_nx^n$. According to the experiments, when $n = 7$, the fitting result would be the best. Fitting result is shown in Fig. 7. The 4 angle sequences $a_1(T), a_2(T), a_3(T), a_4(T)$, after the curve fitting, they would turn into $a_1(T_{fitted}), a_2(T_{fitted}), a_3(T_{fitted}), a_4(T_{fitted})$. We respectively store these four sequences as template in the database.

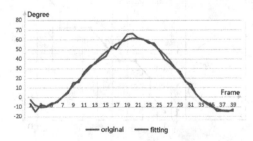

Fig. 7. Curve fitting.

4.3   Dynamic Time Warping (DTW)

Since we had already gotten the template, we want to calculate the distance between the tester and templates. However, the walking cycle is not entirely consistent even for the same person. Since the length of the sequence is inconsistent, we could not directly calculate the Euclidean distance. As shown in the Fig. 8. We can see that the curves are not completely the same even through the same person. But we can also notice that the tendency of three curves are the same.

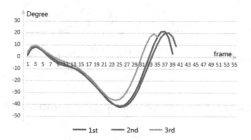

Fig. 8. Three times walking result of the same person.

The paper utilize the DTW which is widely applied in the speech recognition. DTW is able to compare the relationship between two templates which have the different time scale. Finally, the shortest cumulative distance obtained is the DTW distance between two templates, shown in Fig. 9.

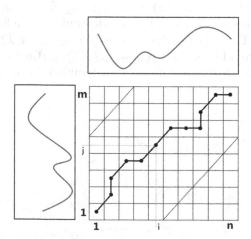

Fig. 9. Dynamic Time Warping schematics, the black segment is the path with the shortest cumulative distance.

5 Classification and Recognition

5.1 Recognition of Static Features

As mentioned above, Kinect can be used to extract static features based on bone length and store static feature vectors as templates in the user database. When tester is passing in front of Kinect, the same method can be used to extract static feature vectors. In this paper, we used NN classifier to perform the final classification. Here we selected Euclidean distance as the criteria for calculating the distance between sample and template. Templates for each row in the database and the test samples calculated Euclidean distance and returned the minimum distance corresponding to *peopleID*, then utilized *peopleID* to retrieve the corresponding user name.

5.2 Recognition of Dynamic Features

For the recognition of dynamic features, we still used NN classifier for classification. However, unlike the static feature classification, when calculating the distance we used the aforementioned DTW algorithm to calculate the sum of cumulative minimum distance between sample and template, namely the DTW distance, and this distance is set as the distance criteria between sample and template.

We used the four angle sequences generated by test samples and the statistics in corresponding angle table in database to calculate DTW distance, separately denoted as $d_{a1}, d_{a2}, d_{a3}, d_{a4}$, then linearly normalized these four distances, a.k.a. formula:

$$\hat{d} = \frac{d}{max(D)}, \tag{3}$$

where $\hat{d}$ is normalized value, d is the original distance to be normalized, D is the set of all the calculated distance for each row of corresponding angle table in test samples and database. $\hat{d}_{a1}, \hat{d}_{a2}, \hat{d}_{a3}, \hat{d}_{a4}$ denote the normalized distance, D_{total} denotes the sum, i.e. $D_{total} = \hat{d}_{a1} + \hat{d}_{a2} + \hat{d}_{a3} + \hat{d}_{a4}$. And it was characterized as final distance and then sorted ascendingly. Then categorized the test sample into the training sample set of minimum distance.

5.3 Recognition of Feature Fusion

In this paper, we extracted static feature based on bone length and dynamic feature based on angle change of lower limb and arm. As indicated by the experiments, using these two features separately to perform gait recognition is not satisfactory, thus we use feature fusion technique to classify and recognize these two features after feature fusion.

Biometric recognition can be broadly divided into three procedures: feature extraction, template matching and classification. According to these three procedures, multi-biometric features fusion can be correspondingly divided into three levels [14]: feature level fusion, matching level fusion, decision-making level fusion.

Distance calculated by different methods using static and dynamic features is used as matching score for each part respectively, attempting to perform feature fusion of static and dynamic features in matching level.

Normalization of matching scores of each component is cardinal to feature fusion in matching level, here we use linear normalization method.

$$\hat{s} = \frac{s - min(S)}{max(S) - min(S)},\tag{4}$$

Where S is the matching score matrix before normalization, and its elements are s. Scoring matrix after normalization is $\hat{S}$, and its element is $\hat{s}$. Matching scores after normalization are mapped to the range $[0, 1]$, enabling the later fusion of the static and dynamic skeletal features. There are many rules in data fusion, whose purpose is to perform calculation using matching score of each feature according to certain fusion rules, to obtain a matching score with higher separability. In this paper, the addition rule is used to add up normalized scores to get the final score, and perform classification. The addition rules:

$$F = \frac{1}{R} \sum_{i=1}^{R} s_i^n,\tag{5}$$

Where F represents the score after fusion, s_i^n denotes the ith normalized matching score, $i = 1, 2, \ldots, R$. We then reused NN classifier to classify and recognize after getting the matching score F.

6   Experiment and Result

The experiment was proceeded in the environment mentioned above, as shown in Fig. 3. The paper established a Kinect fitted database which included 10 persons' gait information. Each person walks for 5 times as training data, in other words, there are 50 series of gait templates in the database. When the test begin, each person also walks in front of Kinect for 5 times as testing data. In order to estimate the performance of system, the paper made a statistics about the CCR (Correct Classification Rate) of using static feature (STA) alone, using dynamic feature (DYN) alone, and after feature fusion (DYN+STA). The results are shown in Fig. 10.

In the figure we can obviously find that the CCR of feature fusion (DYN+STA) is higher than STA and DYN. In order to compare the verification performance of three features, paper also made the ROC of three features under the same classification. As shown in Fig. 11.

In the ROC, the line which is closer to the upper left corner has the better classification method. Due to using the same classification, we can indirectly proved that the line is closer, the feature is better. In the figure we can see that the line of feature fusion (DYN+STA) is much closer than STA and DYN.

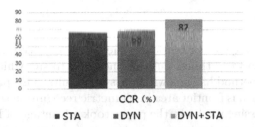

Fig. 10. CCR of 3 different features.

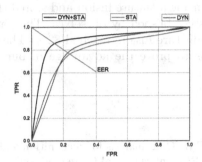

Fig. 11. ROC of three different features under the same classification.

In the research of gait recognition, different gait database would get the different recognition rate. Because of using our own database, it's hard to compare with other methods. In order to compare with others, this paper learned the holistic thinking and algorithm flow of paper [4,15], and transplanted their methods into our database for comparison experiment. These two papers are all based on the general video methods. Paper [4] proposed that make use of leg angles, transform the angle sequence into Fourier series, search the coefficient of each harmonic by genetic algorithm, and generate the feature vector. Paper [15] extracted the key frame in the period of gait, locate the position of leg joints, and calculate the angle of leg joints. Table 1 shows the comparison of CCR.

Table 1. The comparison of three papers in CCR.

Algorithm	CCR
Weihua H et al. [4]	78.5%
AI-HUA WANG et al. [15]	80%
This paper	82%

The experiment result shows that the CCR of this paper is better because of adding static feature and making a feature fusion.

7   Conclusion

People has been paying more attention to the biometric recognition. It has many kinds of methods to work things out, and no need to say which method is the best because each method has its own merits and faults. This paper focus on the gait recognition which is frontier area in biometric recognition sector. Other than general video processing methods, the paper took advantage of Kinect to extract the gait feature. The paper extracted the length of skeleton as the static feature, and extracted the angle of swing legs and arms as the dynamic feature by Kinect. On the basis of this, we made a feature fusion and stored the feature vector into the database which was established by ourselves. The paper proved the CCR is much higher after the feature fusion, reached 82 %. The paper also compared with other two methods to prove the advantage of our feature extracted by Kinect.

References

1. Lee, L., Grimson, W.E.L.: Gait analysis for recognition and classification. In: Proceedings of the Fifth IEEE International Conference on Automatic Face and Gesture Recognition 2002, pp. 148–155. IEEE (2002)
2. Bofeng, Z., Jingru, Z., Ke, Y., et al.: Research on gait feature extracting methods based on human walking model. Comput. Appl. Softw. **26**(5), 198–201 (2009)
3. Wang, L., Tan, T., Ning, H., et al.: Silhouette analysis-based gait recognition for human identification. IEEE Trans. Pattern Anal. Mach. Intell. **25**(12), 1505–1518 (2003)
4. Weihua, H., Ping, L., Haijun, Y.: Gait recognition using the information about Crura's joint angle. In: 13th National Conference on Image and Graphics (NCIG 2006), pp. 411–414 (2006)
5. Bobick, A.F., Davis, J.W.: The recognition of human movement using temporal templates. IEEE Trans. Pattern Anal. Mach. Intell. **23**(3), 257–267 (2001)
6. Kale, A., Cuntoor, N., Chellappa, R.: A framework for activity-specific human identification. In: 2002 IEEE International Conference on Acoustics, Speech, and Signal Processing (ICASSP), vol. 4, pp. IV-3660–IV-3663. IEEE (2002)
7. Sivapalan, S., Chen, D., Denman, S., et al.: Gait energy volumes and frontal gait recognition using depth images. In: 2011 International Joint Conference on Biometrics (IJCB), pp. 1–6. IEEE (2011)
8. Gabel, M., Gilad-Bachrach, R., Renshaw, E., et al.: Full body gait analysis with Kinect. In: 2012 Annual International Conference of the IEEE Engineering in Medicine and Biology Society (EMBC), pp. 1964–1967. IEEE (2012)
9. Wang, G.: Anthropometric and application. The Anthropology Department of Biology Department of Fudan University
10. Powered by Institute of Automation, Chinese Academy of Sciences. http://www.cbsr.ia.ac.cn/china/Gait%20Databases%20CH.asp
11. Tian, W., Cong, Q., Yan, Z., et al.: Spatio-temporal characteristics of human gaits based on joint angle analysis. In: 2010 3rd IEEE International Conference on Computer Science and Information Technology (ICCSIT), vol. 6, pp. 439–442. IEEE (2010)

12. He, W., Li, P.: Gait recognition using the temporal information of leg angles. In: 2010 3rd IEEE International Conference on Computer Science and Information Technology (ICCSIT), vol. 5, pp. 78–83. IEEE (2010)
13. Makihara, Y., Mannami, H., Yagi, Y.: Gait analysis of gender and age using a large-scale multi-view gait database. In: Kimmel, R., Klette, R., Sugimoto, A. (eds.) ACCV 2010, Part II. LNCS, vol. 6493, pp. 440–451. Springer, Heidelberg (2011)
14. Fang, Z., Ye, W.: A survey on multi-biometrics. Comput. Eng. 29(9), 140–142 (2003)
15. Wang, A.-H., Liu, J.-W.: Gait recognition method based on position humanbody joints. In: Proceedings of the 2007 International Conference on Wavelet Analysis and Pattern Recognition, Beijing, China, 2–4 November 2007

2-D Structure-Based Gait Recognition in Video Using Incremental GMM-HMM

Rui Pu[✉] and Yunhong Wang

Laboratory of Intelligence Recognition and Image Processing,
Beijing Key Laboratory of Digital Media,
School of Computer Science and Engineering,
Beihang University, Beijing 100191, China
purui520@126.com

Abstract. Gait analysis is a feasible approach for human identification in intelligent video surveillance. However, the effectiveness of the dominant silhouette-based approaches are severely affected by dressing, bag, hair style and the like. In this paper, we propose a useful 2-D structural feature, named skeleton-based feature, effective improvements for human pose estimation in human walking environment and a recognition framework based on GMM-HMM using incremental learning, which can greatly improve the availability of gait traits in intelligent video surveillance. Our skeleton-based feature uses a 15-DOFs, which is effective in eliminating the interference of dressing, bag, hair style and the like, to represent the torso. In addition, to imitate the natural way of human walking, a Hidden Markov Model (HMM) representing the gait dynamics of human walking incrementally evolves from an average human walking model that represents the average motion process of human walking. Our work makes the gait recognition more robust to noise. Experiments on widely adopted databases prove that our proposed method achieves excellent performance.

1 Introduction

Gait, as a promising biometric characteristic, has attracted many researchers in recent years. In intelligent surveillance, the advantage of accessibility at a distance makes gait a promising biometric characteristic for human recognition. The silhouette has been regarded as the starting line of gait analysis because some databases provide silhouette directly and many gait researchers [1–3] managed to identify human by individual walking styles using silhouette-based methods. However, all the related methods are severely affected by dressing, bag, hair style and the like. Consequently, if someone changes his/her dressing or hair style, these methods perform badly. In this paper, we propose a new robust 2-D structural feature, effective improvements for human pose estimation in human walking environment and a recognition framework based on GMM-HMM using incremental learning. Furthermore, we assume that there is only one person walking in videos. Or if there are several persons superimposing each other, we cannot get skeleton-based feature, as a result, we cannot perform the identification.

© Springer International Publishing Switzerland 2015
C.V. Jawahar and S. Shan (Eds.): ACCV 2014 Workshops, Part I, LNCS 9008, pp. 58–70, 2015.
DOI: 10.1007/978-3-319-16628-5_5

In terms of feature, there have been some other efforts at gait analysis on 2-D structural features. Guoshuang Huang [4] employed different blocks, which represent the solid silhouettes, and fitted the blocks with ellipse. Then, they performed recognition after merging the ellipse parameters of different view angles. Baofeng Guo [5] utilized the maximum mutual information (MMI) algorithm to select gait features, aiming at abandoning the redundant information in high-dimensional features and extracting the most important parts for identification. They applied the MMI to gait features, such as the size and position of each part of body and motion parameters like speed, then they performed recognition using Support Vector Machine (SVM). Their method achieved better performance than correlation analysis and variance analysis. But, in summary, these 2-D structural features are almost represented by shapes such as triangle, ellipse, polygon among others. Obviously, these shapes will be different when someone walks wearing thick clothes or carrying bags. Furthermore, these 2-D structural features are almost attained by background subtraction which is clumsy and rigorous to the video surveillance environment. On the other hand, skeleton-based feature is just human skeleton represented by 15-DOFs, which can reflect the eigen gait characteristics more thoroughly.

With regard to identification framework, there are several time series modeling methods such as Dynamic Time Wrap (DTW), Hidden Markov Model (HMM) and Conditional Random Field (CRF). First of all, DTW has a deadly limitation that it demands the same frequency between gallery set and probe set. Secondly, normal CRF is so complicated and unsuitable for gait analysis. Although the linear CRF is suitable, it is more sophisticated than HMM but not better than HMM on effectiveness for gait analysis. Consequently, we choose HMM. But, the normal HMM demands a mass of gait sequences as training samples. However, there are not enough samples in many cases. To conquer this problem, we get the individual HMMs evolving from an average HMM. In addition, the same person may walk at different time or different places under real circumstance so that the gait samples cannot be available in one shot, so the offline learning method is not suitable. On the contrary, incremental learning is rather useful to this problem. As a result, we get our incremental GMM-HMM evolving from an average GMM-HMM.

Incremental learning has been widely applied to many video-based applications, especially face tracking. For example, David A. Ross [6] proposed an online method based on incremental algorithm for Principal Component Analysis (PCA). They updated the eigen dynamics using incremental learning. Most work mainly consider the variances of statistical features, since the motion dynamics seems less useful for recognition in their applications, such as face tracking. However, dynamics modeling is the core of gait analysis. In this paper, we attempt to incrementally learn the periodic gait dynamics, and exploit spatiotemporal relationships for recognition. Similar to [7], gait dynamics is regarded as the outward manifestation of stance transitions. Unlike some existing tracking methods such as particle filters [8] that depend on the similarities of appearances between frames, this work aims to recover and compare the periodic dynamics based on stance transitions. Furthermore, Maodi Hu [9] proposed an approach based on incremental learning, which achieved a good performance. An incremental learning method for HMM with Gaussian Mixture Model (GMM) representation (denoted as iGMM-HMM afterwards) is proposed, which shows promising performance in recognition experiments. The overall framework of the incremental learning process is shown in Fig. 1.

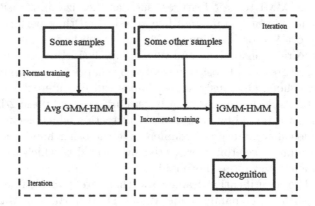

Fig. 1. Overall framework for incremental learning process

The remainder of this paper is organized as follows. Section 1.1 simply presents skeleton-based feature, human pose estimation method and its results. Section 2 is the technical details about human pose estimation method and iGMM-HMM. At last, Sect. 3 is the experiment results in CASIA-B gait database.

1.1 Skeleton-Based Feature

As we know, human motion can be represented using Degree of Freedom nodes (DOFs) model. In this paper, we adopt the 15-DOFs, which is relatively easy to detect and track and enough to represent skeleton-based feature (see Fig. 2).

To get the 15-DOFs for each walking stance, there are two main approaches, which are human pose estimation and human pose tracking. Marcus A. Brubaker [10,11] proposed a useful approach about human pose tracking for human in walking. But his method just tracks the lower body, which is insufficient to gait recognition. Furthermore, because the current human pose tracking methods are not good enough to achieve our goal and so complicated, we choose the first.

Vittorio Ferrari [12] proposed an approach for human pose estimation which achieved a good performance. But its method needs to label human upper body artificially. And then, he utilised another method [13] proposed by Navneet Dalal to detect upper body. But the upper-body detection method performed badly in low resolution images. Finally, Vittorio Ferrari [14] proposed a fully automated method for human pose estimation in uncontrolled environment. So we choose

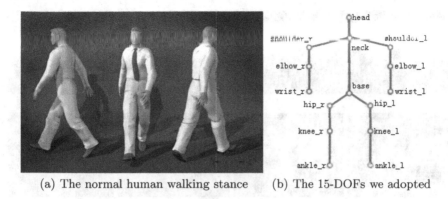

(a) The normal human walking stance   (b) The 15-DOFs we adopted

Fig. 2. The walking stance and the DOFs we adopted

this method to be the base of our first part algorithm to get skeleton-based feature. In addition, their method is performed in still images, so we have to convert videos into images at first. Based on Ferrari's method, we made some improvements which can improve the upper-body detection accuracy and the image parsing speed. The difference between our's and Ferrari's is shown in (see Fig. 3).

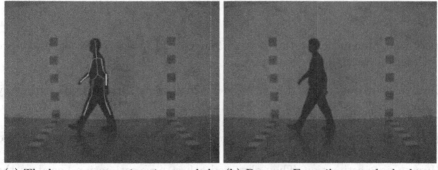

(a) The human pose estimation result by (b) Because Ferrari's upper-body detec-
our method   tion method cannot detect the upper
                                        body, nothing is attained

Fig. 3. The difference between our's and Ferrari's

The human pose estimation results by our method are shown in (see Fig. 4).

2   Technical Details

2.1   Skeleton-Based Feature

The main idea of the human pose estimation method proposed by Vittorio Ferrari is to progressively reduce the search space for body parts, greatly improving the

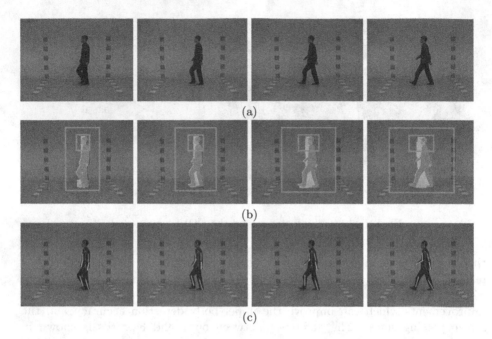

Fig. 4. The human pose estimation results step by step. (a) Is the original images in gait database CASIA-B. (b) Is the human detection results. (c) Is the human pose estimation results.

chances that human pose estimation will succeed. In their approach, there are three stages in total.

1. Human detection. They started by detecting human upper bodies in every frame, using a sliding window detection based on Histograms of Oriented Gradients [13], and associate detections over time.
2. Foreground highlighting. At this stage, the search for body parts is limited in the detected regions.
3. Human pose estimation. They obtained a first pose estimation based on the image parsing technique of Ramanan [15]. The area to be parsed is restricted to the regions attained by foreground highlighting.

In their approach, the first two stages use a weak human model. This weak model only determines the approximate location and scale of the person, and roughly where the torso and head should lie. The last stage switches to a stronger model, a pictorial structure composed of body parts tied together in a tree-structured conditional random field. Parts, l_i are oriented patches of fixed size, and their positions are parameterized by location and orientation. The posterior of a configuration of parts $L = l_i$ given an image I can be written as a log-linear model

$$P(L|I) \propto exp(\sum_{(i,j)\in} E)\psi(l_i, l_j) + \sum_i \phi(l_i)) \tag{1}$$

The binary potential $\psi(l_i, l_j)$ corresponds to a spatial prior on the relative positions of parts and embeds the kinematic constraints (e.g. the upper arms must be attached to the torso). The unary potential $\phi(l_i)$ corresponds to the local image evidence for a part in a particular position (likelihood). Since the model structure E is a tree, inference is performed exactly and efficiently by sum-product Belief Propagation.

Furthermore, there are some rules to utilise in the walking environment.

1. The upper arm must be above the lower arm, and the thigh must be above the shank.
2. The left arm and right leg, right arm and left leg must be in the same direction.
3. The left arm and right arm, left leg and right leg must be symmetric.
4. If the camera is fixed, the stance in the next frame must be near the pervious stance.

Based on their approach, by taking the advantage of the four rules, we make some effective improvements in the last stage.

First of all, according to the first three rules, we uses prior physiological characteristics of human gait and effectively limits the search space, which ameliorates the efficiency.

Secondly, in terms of the fixed camera, we initialize the position in current frame using the result of the previous one, and it only focuses on the person to be studied and largely reduces the number of false candidate, thus improving the performance.

Then, according to the human pose estimation results, we extracted skeleton-based feature, including the lengthes and angles of every two connective joints (see Fig. 2). The lengthes should be divided by the body height in the images. Since this feature is totally structural, it is hardly influenced by dressing, bag and hair style.

2.2   Incremental GMM-HMM

In gait analysis, stances are usually used to indicate the periodical latent states over gait cycles. After years of researches, human gait is widely accepted to be an identifiable periodic pattern with several stance phases. Consequently, a HMM that models the representation within and between states is very suitable for this application.

We will simply review the development of incremental learning for HMM below. Besides the offline Expectation Maximization (EM) algorithm and the batch learning Baum-Welch (BW) algorithm, the parameters of HMM can also be estimated incrementally with improved convergence and reduced memory requirements [16]. Krishnamurthy [17] derived online EM algorithm by using random approximations to maximize the Kullback-Leibler information. Stenger [18] proposed the Incremental Baum-Welch (IBW) algorithm, in which each latent state of their HMM includes a single Gaussian model. It is further derived to a discrete model with a new backward procedure based on a one-step lookahead by Florez-Larrahondo [16], which is known as the improved Incremental

Baum-Welch (IBW+) algorithm, which achieved a better performance. In the purpose of learning gait dynamics for recognition, the models mentioned above should be enhanced. First, the model including the IBW+ is discrete. On the other hand, the model including the IBW involves only one Gaussian model for each latent state. But the model we need is continuous and may includes several Gaussian models for each state. Consequently, we apply the idea of IBW+ to our iGMM-HMM and learn the updating approach for GMM from the IBW.

About the symbol notation, we use iGMM-HMM to represent the incremental GMM-HMM we proposed and oGMM-HMM to represent the normal GMM-HMM gained by the offline BW algorithm. The feature vector extracted from t^{th} frame is indicated as O_t. There are some parameters in incremental learning. The Θ is used for the model representation, which is composed of the transition probability matrix A between latent states and the observable representations B. Each single stance within a gait cycle is represented by a latent state in HMM, and the probability density function (pdf) of each latent state is modeled by a GMM. Considering a HMM consisting of Q latent states with M Gaussian mixture components, $A = \{\alpha_{ij}\}_{1 \leq i \leq Q, 1 \leq j \leq Q}$ denotes the transition probability from latent state i to latent state j, and $B = \{\phi_{ik}, \mu_{ik}, \sigma_{ik}\}_{1 \leq i \leq Q, 1 \leq k \leq M}$ denotes the mixing coefficient, mean vector, and covariance matrix of component k in latent state i.

$\alpha_T(i) = P(O_1, \ldots, O_T, q_T = i|\Theta)$ is the forward cumulative probability of being in state i,

$$\alpha_T(i) = \begin{cases} (\sum_{j=1}^{Q} \alpha_{T-1}(j)a_{ji})b_T(i) & T > 1, \\ b_T(i) & T = 1, \end{cases} \tag{2}$$

and $\beta_T(i) = P(O_T, O_{T+1}, q_T = i|\Theta)$ is the backward one proposed in IBW+ [16],

$$\beta_T(i) = \sum_{j=1}^{M} a_{ij}b_{T+1}(j). \tag{3}$$

Since the real $\beta_T(i)$ is based on an exponential decay function computed via the backward procedure, for large T this approximation seems to be appropriate. In any case, it provides a better approximation than $\forall_T \forall_i \beta_T(i) = 1.0$.

This backward procedure of IBW+ algorithm reduces the training complexity of β in backward procedure of BW algorithm in discrete model from $O(n^2T)$ to $O(n^2)$. Although it does not improve the global time complexity, the experimental results in [16] show that IBW+ converges faster than BW and IBW. Note that it requires a one-step look ahead in the sequence of observations.

$b_T(i) = P(q_T = i|O_T, \Theta)$ is the pdf of O_T at state i, which indicates the fitness of a single frame for an averaged walking stance. Because of the usage of IBW+, both $b_T(i)$ and $b_{T+1}(i)$ are updated in the Tth iteration.

$$b_T(i) = \sum_{k=1}^{M} \phi_{ik}\mathcal{N}(O_T; \mu_{ik}, \sigma_{ik}), \tag{4}$$

$$b_{t+1}(i) = \sum_{k=1}^{M} \psi_{ik} \mathcal{N}(O_{T+1}, \mu_{ik}, \sigma_{ik}).$$ (5)

$c_T(i, k)$ is the probability of O_T being in component k at state i,

$$c_T(i, k) = \frac{\phi_{ik}\mathcal{N}(O_T; \mu_{ik}, \sigma_{ik})}{b_T(i)},$$ (6)

$\gamma_T(i) = P(q_T = i | O_1, \ldots, O_{T+1}, \Theta)$ is the probability of being in state i,

$$\gamma_T(i) = \frac{\alpha_T(i)\beta_T(i)}{\sum_{i=1}^{Q} \alpha_T(i)\beta_T(i)},$$ (7)

$\xi_{T-1}(i, j) = P(q_{T-1} = i, q_T = j | O_1, \ldots, O_{T+1}, \Theta)$ is the probability of $T-1^{th}$ frame being in state i and T^{th} frame being in state j,

$$\xi_{T-1}(i, j) = \begin{cases} \frac{\alpha_{T-1}(i)a_{ij}b_T(j)\beta_T(j)}{\sum_{i=1}^{Q}\sum_{j=1}^{M}\alpha_{T-1}(i)a_{ij}b_T(j)\beta_T(j)} & T > 1, \\ 0 & T = 1. \end{cases}$$ (8)

The estimation of $\xi_{T-1}(i, j)$ is improved by the approximation of β_T [16]. We use the parameters of an average GMM-HMM (denoted as avgGMM-HMM afterwards) to serve as the model representation Θ of the iGMM-HMM in the 0th iteration. Given the Tth and $T+1^{th}$ frames, $b_T(i)$, $b_{T+1}(i)$, $c_T(i, k)$, $\alpha_T(i)$, $\beta_T(i)$, $\gamma_T(i)$, and $\xi_{T-1}(i, j)$ can be calculated in order, based on Θ in the $T-1^{th}$ iteration.

Below we will introduce our incremental updating algorithm. At first, supposing there are N frames in the training group of avgGMM-HMM, we number them as $x_{-N+1}, \ldots, x_0$ to differentiate them from the frames in incremental learning. Given the values of model parameters estimated in the previous frames, the equations suitable for T^{th} updating are shown in Eqs. (9) to (12).

$$\bar{a}_{ij}^T = \frac{\bar{a}_{ij}^{T-1}(\sum_{t=-N+1}^{T-2} \gamma_t(i)) + \xi_{T-1}(i, j)}{\sum_{t=-N+1}^{T-1} \gamma_t(i)},$$ (9)

$$\bar{\phi}_{ik}^T = \frac{\sum_{t=-N+1}^{T} \gamma_t(i)c_t(i, k)}{\sum_{t=-N+1}^{T} \gamma_t(i)},$$ (10)

$$\bar{\mu}_{ik}^T = \frac{\bar{\mu}_{ik}^{T-1}(\sum_{t=-N+1}^{T-1} \gamma_t(i)c_t(i, k))}{\sum_{t=-N+1}^{T-1} \gamma_t(i)c_t(i, k) + \gamma_T(i)c_T(i, k)}$$
$$+ \frac{\gamma_T(i)c_T(i, k)O_t}{\sum_{t=-N+1}^{T-1} \gamma_t(i)c_t(i, k) + \gamma_T(i)c_T(i, k)},$$ (11)

$$\bar{\sigma}_{ik}^T = (\bar{\sigma}_{ik}^{T-1} + (\bar{\mu}_{ik}^{T-1} - \bar{\mu}_{ik}^T)(\bar{\mu}_{ik}^{T-1} - \bar{\mu}_{ik}^T)^H)$$
$$\cdot \frac{\sum_{t=-N+1}^{T-1} \gamma_t(i)c_t(i, k)}{\sum_{t=-N+1}^{T-1} \gamma_t(i)c_t(i, k) + \gamma_T(i)c_T(i, k)}$$
$$+ \frac{\gamma_T(i)c_T(i, k)(O_T - \bar{\mu}_{ik}^T)(Ox_T - \bar{\mu}_{ik}^T)^H}{\sum_{t=-N+1}^{T-1} \gamma_t(i)c_t(i, k) + \gamma_T(i)c_T(i, k)},$$ (12)

Compared to previous studies on incremental HMM [16,18], such as IBW and IBW+, the proposed updating rules make it possible to model the state representations of the HMM by several Gaussian models.

3   Our Experiment

Recognition approaches based on HMM is straight-forward. Let Θ^{id} denote the HMM trained by the gallery set of subject id. Given the data-case $probe$, the recognition process can be simply solved by Maximal A Posterior (MAP) rule,

$$\operatorname{argmax}_{id} P(probe|\Theta^{id}). \tag{13}$$

where $P(probe|\Theta^{id})$ is the probability of the observation sequence $probe$ given Θ^{id}.

3.1   The Database Introduction

The database we used is CASIA-B gait database. There are 124 persons in total, 11 view angles for each person, three types for each view angles. The view angles are 0, 18, 36, 54, 72, 90, 108, 126, 144, 162 and 180 degree respectively. The types are nm, bg and cl, respectively standing for dressing normally, wearing thick clothes and carrying bag. There are only two gait sequences for bg and cl and six for nm.

Before we choose to use the iGMM-HMM, we attempted to build oGMM-HMM for each person, each angle and each type. In a small probe set including 622 gait sequences, we experimented and concluded that the oGMM-HMM for each type performs better than that for each angle and each type. The experimental results are shown in Table 1. The person oGMM-HMM, angle oGMM-HMM and type oGMM-HMM stand for the oGMM-HMM for each person, angle and type respectively.

Table 1. Rank 1 recognition performance with view angle unknown (%).

Approaches	Person oGMM-HMM	Angle oGMM-HMM	Type oGMM-HMM
Accuracy	74.65 %	76.85 %	80.06 %

Consequently, building GMM-HMM for each type is the best. However, there are only two gait sequences for bg and cl and six for nm in CISIA-B gait database. And also we have to extract at least one gait sequence for each type as probe set, so that the training samples are too small to make the GMM-HMM convergent. As a result, the initial parameter settings are far away from the true values, then the errors will slow down the convergence process [18]. Therefore, we trained an average GMM-HMM using some gait samples, whose parameters are estimated using the offline EM algorithm and the BW algorithm. Then, with incremental adjustments of the iGMM-HMM parameters, the fitness and validity of specific individuals increase simultaneously.

3.2 The Contrastive Methods We Used

In our experiment, except for our proposed method, we also take three other methods as contrastive methods.

Skeleton-Based Feature Plus Sub-sequence DTW. The traditional dynamic time warping (DTW) algorithm is to compute the distance from the probe sequence to the gallery sequence. But in many cases, we need the minimum distance from the sub-sequences of the probe sequence to the sub-sequences of the gallery sequence, so that we shouldn't compute the distance from the probe sequence to the gallery sequence. In our experiment, for simplicity, we make sure that the probe sequence is shorter than the gallery sequence. Consequently, we just need to compute the minimum distance from the probe sequence to the sub-sequences of the gallery sequence. So, we call this method sub-sequence DTW (denoted as subDTW afterwards). The recognition results are based on the distance between the *gallery* set and the *probe* one.

$$\mathrm{argmin}_{id} d(probe, gallery_{id}) \tag{14}$$

$d(S_1, S_2)$ represents the Euclidean distance between two sequences denoted by S_1 (in the gallery set) and S_2 (in the probe set). Let T_1 and T_2 denote the lengths of S_1 and S_2 respectively.

$$d(S_1, S_2) = \mathrm{min}_{s=1}^{T_1-T_2+1} \| \sum_{t_1=s}^{s+T_2} S_1(t_1) - \sum_{t_2=1}^{T_2} S_2(t_2) \|. \tag{15}$$

Skeleton-Based Feature Plus Offline GMM-HMM. The traditional oGMM-HMM for each type using Skeleton-based feature may not converge because of the small amount of samples. In this method, we initialize the prior probability and the initial transition probability from latent states to observable states with uniform distribution. In addition, the initial transition possibility from latent states to latent states is stochastic.

Gait Energy Image (GEI) Plus PCA Plus Nearest-Neighbor Classifier. This method uses the classical feature GEI. Then, after PCA process, the nearest-neighbour classifier can achieve a very good performance. This method is denoted as GEI-PCA-NN afterwards.

3.3 Experimental Results

In our experiment, we choose 40 persons randomly, including 20 women and 20 men respectively, to train the avgGMM-HMM (only the iGMM-HMM uses this). Then we split the left gait sequences into probe set and gallery set. The probe set includes one gait sequence for each type and the left is the gallery set. The results are shown in Table 2.

Table 2. Rank 1 recognition performance with view angle unknown in type to type (%)

Approaches	subDTW	oGMM-HMM	iGMM-HMM	GEI-PCA-NN
nm-nm	95.89	93.71	98.53	**98.83**
nm-bg	47.80	49.27	**70.97**	53.67
nm-cl	30.79	28.74	**43.99**	26.10
bg-nm	39.31	38.71	**53.96**	36.66
bg-bg	88.27	84.59	91.79	**95.60**
bg-cl	16.13	17.72	17.89	**20.23**
cl-nm	22.58	20.31	**23.17**	17.60
cl-bg	16.67	16.42	**20.23**	12.90
cl-cl	83.87	81.33	86.80	**97.07**

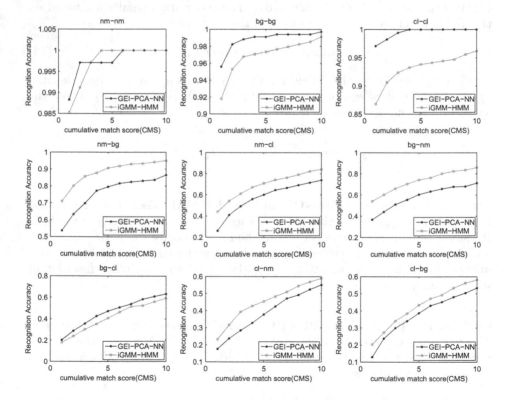

Fig. 5. Comparison between iGMM-HMM and GEI-PCA-NN in CMS

In terms of the parameters setting, there are two controllable parameters which are the number Q of latent states and the number M of Gaussian mixture components for each latent state. The experiments prove that iGMM-HMM performs better when Q is between 6 to 8 and M is between 3 to 4.

Obviously, the iGMM-HMM is better than the oGMM-HMM and subDTW with the same skeleton-based feature we proposed. Furthermore, although iGMM-HMM plus skeleton-based feature performs a little worse than GEI-PCA-NN in nm-nm, bg-bg and cl-cl, the results with cumulative match score (CMS) (see Fig. 5) prove that iGMM-HMM catches up with GEI-PCA-NN quickly. Anyway, the recognition accuracy of iGMM-HMM is still high in nm-nm, bg-bg, cl-cl. In addition, iGMM-HMM performs a little bit worse in bg-cl, which is mainly because the human pose estimation results are relatively worse in bg and cl than nm. But in the majority of cross-type recognitions such as nm-bg, nm-cl, bg-nm and so on, iGMM-HMM is obviously better than GEI-PCA-NN, moreover, the superiority keeps the same with CMS. In a word, our skeleton-based feature is a better feature than GEI in cross-type recognition. However, the cross-type recognition accuracy is still a little low, especially in bg-cl, cl-nm and cl-bg, whose reason is that there are still some errors in the human pose estimation method.

4 Conclusions and Future Work

In this paper, a novel 2-D structural feature, effective improvements for human pose estimation in human walking environment and an incremental identification framework for gait dynamics are proposed. The experiments prove that our skeleton-based feature can eliminate the interference of dressing, bag, hair style and the like effectively. However, only structural feature is not enough to human identification problem in gait analysis. As a result, our future work should be fusing the skeleton-based feature with some other features to cover this shortage. In addition, in spite of the improvement we do in human pose estimation, the human pose estimation results are still not that much good because of some detection errors. But, as the human pose estimation or human pose tracking improves, our approach must achieve better performance. Whatsoever, the iGMM-HMM is really a good framework for spatiotemporal problems.

References

1. Liu, Z., Malave, L., Sarkar, S.: Studies on silhouette quality and gait recognition. In: CVPR (2004)
2. Liu, Z., Sarkar, S.: Simplest representation yet for gait recognition: averaged silhouette. In: ICPR (2004)
3. Yu, S., Tan, T., Huang, K., Jia, K., Wu, X.: A study on gait-based gender classification. IEEE J. IP 18, 1905–1910 (2009)
4. Huang, G., Wang, Y.: Gender classification based on fusion of multi-view gait sequences. In: Yagi, Y., Kang, S.B., Kweon, I.S., Zha, H. (eds.) ACCV 2007, Part I. LNCS, vol. 4843, pp. 462–471. Springer, Heidelberg (2007)
5. Guo, B., Nixon, M.S.: Gait feature subset selection by mutual information. IEEE Trans. Syst. Man Cybern. 39, 36–46 (2009)
6. Ross, D.A., Lim, J., Lin, R.S.: Incremental learning for robust visual tracking. IJCV 77, 125–141 (2008)

7. Liu, Z., Sarkar, S.: Improved gait recognition by gait dynamics normalization. IEEE J. PAMI **28**, 863–876 (2006)
8. Ristic, B., Arulampalam, S., Gordon, N.: Beyond the Kalman Filter: Particle Filters for Tracking Applications. Artech House, Norwood (2004)
9. Hu, M., Wang, Y., Zhang, Z., Zhang, D., Little, J.J.: Incremental learning for video-based gait recognition with LBP flow. IEEE Trans. Cybern. **43**, 77–89 (2013)
10. Brubaker, M.A., Fleet, D.J., Hertzmann, A.: Physics-based person tracking using simplified lower-body dynamics. In: CVPR (2007)
11. Brubaker, M.A., Fleet, D.J.: The kneed walker for human pose tracking. In: CVPR (2008)
12. Ferrari, V., Zisserman, A.: Progressive search space reduction for human pose estimation. In: CVPR (2008)
13. Dalal, N., Triggs, B.: Histograms of oriented gradients for human detection. In: CVPR (2005)
14. Ferrari, V., Marín-Jiménez, M., Zisserman, A.: 2D human pose estimation in TV shows. In: Cremers, D., Rosenhahn, B., Yuille, A.L., Schmidt, F.R. (eds.) Visual Motion Analysis. LNCS, vol. 5604, pp. 128–147. Springer, Heidelberg (2009)
15. Ramanan, D.: Learning to parse images of articulated bodies. In: NIPS (2006)
16. Florez-Larrahondo, G., Bridges, S., Hansen, E.A.: Incremental estimation of discrete hidden Markov models based on a new backward procedure. In: AAAI, vol. 1, pp. 758–763 (2005)
17. Krishnamurthy, V., Moore, J.B.: On-line estimation of hidden markov model parameters based on the kullback-leibler information measure. IEEE J. SP **41**, 2557–2573 (1993)
18. Stenger, B., Ramesh, V., Paragios, N., Coetzee, F., Buhmann, J.: Topology free hidden markov models: application to background modeling. In: ICCV, vol. 1, pp. 294–301 (2001)

Unsupervised Temporal Ensemble Alignment for Rapid Annotation

Ashton Fagg[1,2]([✉]), Sridha Sridharan[2], and Simon Lucey[2,3]

[1] CSIRO, Brisbane, QLD, Australia
[2] Queensland University of Technology, Brisbane, QLD, Australia
ashton@fagg.id.au, s.sridharan@qut.edu.au, slucey@cs.cmu.edu
[3] Carnegie Mellon University, Pittsburgh, PA, USA

Abstract. This paper presents a novel framework for the unsupervised alignment of an ensemble of temporal sequences. This approach draws inspiration from the axiom that an ensemble of temporal signals stemming from the same source/class should have lower rank when "aligned" rather than "misaligned". Our approach shares similarities with recent state of the art methods for unsupervised images ensemble alignment (e.g. RASL) which breaks the problem into a set of image alignment problems (which have well known solutions i.e. the Lucas-Kanade algorithm). Similarly, we propose a strategy for decomposing the problem of temporal ensemble alignment into a similar set of independent sequence problems which we claim can be solved reliably through Dynamic Time Warping (DTW). We demonstrate the utility of our method using the Cohn-Kanade+ dataset, to align expression onset across multiple sequences, which allows us to automate the rapid discovery of event annotations.

1 Introduction

Time series alignment is an important problem for many areas of research - including speech processing, activity recognition, sensor networks and computer vision. Of particular interest is the alignment of time series which describe human motion. This problem is particularly challenging as the motions themselves may have disparate appearance. Such disparity may include differences in event speed and duration, physical differences between subjects and different presentation of the events themselves. These problems are amplified when considering the alignment of a set of sequences. If we were to attempt the alignment multiple sequences naively, a simple method would be to select a template from the sequences available and align all remaining sequences to that sequence. However this approach inherits several problems. For example, which sequence should be picked as a template? Can it be assured that this template produces reliable alignment across all of the sequences? This problem has been explored thoroughly in the image alignment domain, and has led to the proposal of methods known as ensemble alignment. For a set of semantically similar images, ensemble alignments aims to solve the alignment globally by finding a set of alignments which best align every image within the ensemble relative to all other images.

© Springer International Publishing Switzerland 2015
C.V. Jawahar and S. Shan (Eds.): ACCV 2014 Workshops, Part I, LNCS 9008, pp. 71–84, 2015.
DOI: 10.1007/978-3-319-16628-5_6

Taking the insights presented by image ensemble alignment [1–4], and recent work in temporal alignment [5,6], this paper will consider the application of ensemble alignment methodologies to multiple temporal sequences of the same modality. We propose that treating the set of sequences as an ensemble will enable an optimal alignment to be discovered, following the methodology proposed by [3,4]. We make the assertion that semantically similar sequences, when aligned, should exist within a common, low rank subspace, which can be discovered using Robust PCA [7].

The alignment of a set of sequences lends itself to the automation of what is usually a tedious and time consuming task - event annotation. By solving for the alignment of the ensemble, we are able to rapidly discover event annotations for all sequences in the set. In this paper, we shall present an example which shows our method recovering approximate annotation for a set of sequences depicting facial expression onset.

1.1   Contributions

In this paper we shall present the following contributions:

- We present a novel framework for unsupervised alignment of an ensemble of temporal sequences.
- We demonstrate the use of RPCA [7] and DTW [8] to uncover a common low rank subspace for semantically similar temporal sequences.
- We demonstrate promising initial results on Cohn-Kanade+ for alignment of broad expression sequences for annotation of expression onset.

1.2   Notation Used in this Paper

Sets are notated as follows: $\mathbb{B}, \mathbb{R}$. Lower case bold letters denote column vectors. For example, an M dimensional column vector is denoted as $\mathbf{x}$, such that $\mathbf{x} \in \mathbb{R}^{M \times 1}$. Scalars are denoted by upper case non-bold letters. Upper case bold letters denote a matrix, e.g. $\mathbf{A} \in \mathbb{R}^{M \times M}$. Operations are denoted by a special font, e.g. the Lagrangian operator is denoted as $\mathcal{L}$, and the soft thresholding operator is denoted as $\mathcal{S}$.

2   Prior Art

This section will review current literature in the areas of ensemble alignment and time series alignment techniques.

2.1   Ensemble Alignment

Ensemble alignment, at its core, attempts to minimize misalignment over a set of samples. In the spatial domain, there has been significant interest in the area of multi-image alignment [1,2,4]. Of particular interest, is the RASL objective [4] which decomposes the problem to be a set of simple problems, which are solvable using Augmented Lagrangian Methods. The motivation behind ensemble alignment is to exploit redundancies within the set of samples to recover a common, low rank subspace [7] in which all examples reside. The RASL objective [3] posits that an aligned ensemble of linearly correlated images can be formulated as:

$$\arg \min_{\mathbf{L},\mathbf{E},\mathbf{P}} \text{rank}(\mathbf{L}) + \lambda \|\mathbf{E}\|_0$$
$$\text{s.t. } \mathbf{D}(\mathbf{P}) = \mathbf{L} + \mathbf{E} \tag{1}$$

where $\mathbf{L}$ describes a low rank subspace, $\mathbf{E}$ models sparse errors, and $\mathbf{D}(\mathbf{P})$ being the aligned ensemble, given a set of transformations represented by $\mathbf{P}$ and the original images $\mathbf{D}$. In effect, $\mathbf{L}$ describes a base image, where appearance variations are modelled by $\mathbf{E}$.

When solved using Augmented Lagrangian Methods [4], at each iteration the ensemble alignment problem is decomposed to set of discrete image alignment problems. Each image within the ensemble is warped with respect to the current estimate of the base image, solved using the Lucas-Kanade algorithm [9]. If one were to make the same assumptions about a set of similar time series, it is possible to posit the ensemble alignment problem in the same manner, where the temporal warping is discretely solved using proven time warping theory.

2.2   Time Series Alignment

In the area of time series alignment, there has been significant work based upon Dynamic Time Warping (DTW) [8]. DTW allows for the computation of a temporal warping which minimises the misalignment of two sequences. DTW is a powerful framework for time series alignment as it can be considered optimal when considering the distance between two sequences. The alignment path produced by DTW aims to reduce the distance between the sequences as much as possible.

If we define two 1D time series of different lengths, $\mathbf{x} \in \mathbb{R}^{N \times 1}$ and $\mathbf{y} \in \mathbb{R}^{M \times 1}$, the DTW objective which minimises the misalignment of $\mathbf{x}$ with respect to $\mathbf{y}$ can be formulated as:

$$\text{DTW}(\mathbf{x}, \mathbf{y}) = \min_{\mathbf{P} \in \mathbb{B}} S(\mathbf{P}\mathbf{x}, \mathbf{y}) \tag{2}$$

Where, $\mathbf{P} \in \mathbb{B}$ encodes the alignment path between $\mathbf{x}$ and $\mathbf{y}$. The set $\mathbb{B}$ represents the set of all valid alignments, such that $\mathbb{B} \in \{0, 1\}^{M \times N}$. A valid alignment is defined as continuous and increasing in unitary increments. An optimal alignment can be efficiently drawn from $\mathbb{B}$ through the use of Dynamic Programming. S is a measure of cost, typically the least squares distance: $S(\mathbf{x}, \mathbf{y}) = \|\mathbf{x} - \mathbf{y}\|_2^2$.

To understand the set $\mathbb{B}$, we visualise two examples of valid alignment paths computed using DTW. Figures 1a and b show alignment paths computed for random signals. In this instance, $\mathbf{P} \in \mathbb{B}^{320 \times 240}$. However, one of the difficulties encountered in solving for valid alignment, is that the set $\mathbb{B}$ is non-convex. For a convex set, it would be expected that any linear combination of valid elements of the set, would also lie within the set. For $\mathbb{B}$, this is assumption does not hold. In Fig. 1c, the average of the paths shown in Fig. 1a and b is illustrated. It is apparent the result does not lie with the set of valid alignments as the path is not causal and does not lie within the set of $\{0, 1\}^{M \times N}$.

When combined with Augmented Lagrangian Methods, we assert that the use of DTW will ensure that optimal solutions can be drawn from $\mathbb{B}$ as DTW provides an efficient means of traversing the non-convex set and enforcing the alignment constraints.

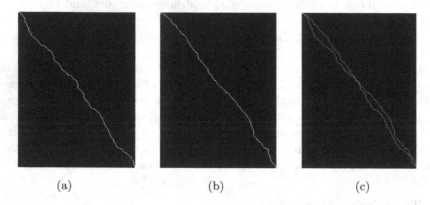

    (a)   (b)   (c)

Fig. 1. (a) and (b) illustrate random examples from the set of valid alignments, $\mathbb{B}$. Note that each of the paths is continuous, causal and increasing in unitary increments. In (c), we visualise the average of the two paths in (a) and (b) in order to demonstrate that $\mathbb{B}$ is non-convex. The path in (c) violates the constraints of $\mathbb{B}$ indicating non-convexity.

DTW features extensively in many existing time series alignment frameworks. Recent work in Time Series Kernels [10–12] highlights an important insight into sequences of common modality - that is, sequence similarity can be measured using DTW. This insight is applied in the formulation of the Time Series Kernel proposed by [11], which is in essence a measure of relative alignment. When applied to a temporal detection problem [12] it was shown that the Time Series Kernel provides excellent detection performance for broad expression detection on the Cohn-Kanade+ dataset [13]. Whilst the Time Series Kernel is in essence a measure of relative alignment between sequences, the representation is able to avoid a fundamental problem of temporal detection - variable event length.

Whilst DTW offers great utility to the temporal alignment problem, it has several drawbacks. Firstly, the alignment computed by DTW whilst considered optimal with respect to the pair of sequences, does not guarantee that the alignment will be meaningful. DTW makes the assumption that the sequence similarity is indicated by Euclidean distance. For many computer vision problems, the Euclidean distance has been shown to be an unreliable measure of similarity due to the effects of corruption - for instance, a small amount of error in spatial registration of the features, or a cross-subject variability.

To compensate for these drawbacks, recent work has extended the DTW framework to incorporate tolerance towards sequence variability [5, 6, 14]. Canonical Time Warping (CTW) [14] attempts to account for cross-subject variability and mild amounts of sequence corruption by incorporating Canonical Correlation Analysis (CCA) into the objective. CTW parameterizes the temporal warping to a set of basis functions which maximise the correlation between the sequences. The work presented in [14] demonstrates superior performance when compared with regular DTW for sequences which are semantically similar, but contain variations in appearance.

Furthermore, CTW was extended to allow for the alignment of multiple sequences [5] of different modalities. The Generalized Time Warping (GTW) algorithm of [5] places emphasis on aligning sequences of different modalities. For example, given a sequence consisting of camera, motion capture and accelerometer data, the GTW algorithm is able to discern meaningful alignment of all modalities.

A further extension is the recent work proposed by [6] which intends to recover a common, low rank subspace for a pair of sequences which are corrupted by noise. For two sequences, a low rank projection is used to recover clean, aligned sequences from a pair of corrupted, but semantically similar sequences.

In this work we shall draw upon the insights from [4,6], but apply them in a different manner. Rather than focusing on alignment of corrupted sequences and subsequent noise removal, we shall focus on using the power of DTW to uncover commonality on a larger scale - across many sequences where the definition of a reliable template may be difficult or impossible. In a similar manner to [4], we seek to minimize misalignment across a set of sequence by decomposing the global alignment problem to a set of independent alignment problems which are easily solved.

Using the alignment computed for the ensemble, we propose that the alignment can be used to rapidly generate sequence annotations, specifically for the onset of an expression. By aligning the whole set of sequences in time, we are required only to perform a minimal amount of annotation manually. Once the ensemble is aligned in time, in the best case all sequences will adhere to the same temporal profile, and expression onset can be annotated based on a single point in time.

3   Unsupervised Temporal Ensemble Alignment

Our method poses the ensemble alignment objective as an RPCA [7] problem:

$$\arg\min_{\mathbf{L},\mathbf{E},\mathbf{P}} \operatorname{rank}(\mathbf{L}) + \lambda\|\mathbf{E}\|_0$$

$$\text{s.t. } \mathbf{L} + \mathbf{E} = \mathbf{D}(\mathbf{P})$$

$$\mathbf{P}_i \in \mathbb{B} \,\forall i = 1, \ldots, N \tag{3}$$

Where $\mathbf{L}$ describes a low rank subspace, $\mathbf{E}$ is the sparse error estimate and $\mathbf{D}(\mathbf{P})$ represents a set of DTW warps ($\mathbf{P}$) applied to the raw sequence ensemble ($\mathbf{D}$), such that:

$$\mathbf{D}(\mathbf{P}) = [\text{vec}(\mathbf{P}_1\mathbf{D}_1), \ldots, \text{vec}(\mathbf{P}_n\mathbf{D}_n)] \tag{4}$$

Each sequence, $\mathbf{D}_i \in \mathbb{R}^{F_i \times D}$ is warped to a predefined sequence length, F_0, by application of a temporal warping $\mathbf{P}_i \in \mathbb{B}^{F_0 \times F_i}$.

Similarly, $\mathbf{L}$ is defined such that:

$$\mathbf{L} = [\text{vec}(\mathbf{L}_1), \ldots, \text{vec}(\mathbf{L}_n)] \tag{5}$$

where, $\mathbf{L}_i \in \mathbb{R}^{F_0 \times D}$. Hence, $\mathbf{D}(\mathbf{P}), \mathbf{L}, \mathbf{E} \in \mathbb{R}^{DF_0 \times N}$.

Equation 3 is considered difficult to solve efficiently due the non-convexity of the rank operation and L0 norm. Fortunately, a convex surrogate can be used in place of these operations to allow a solution to be found efficiently.

$$\arg\min_{\mathbf{L},\mathbf{E},\mathbf{P}} \|\mathbf{L}\|_* + \lambda\|\mathbf{E}\|_1$$
$$\text{s.t. } \mathbf{L} + \mathbf{E} = \mathbf{D}(\mathbf{P})$$
$$\mathbf{P}_i \in \mathbb{B} \ \forall i = 1, \ldots, N \tag{6}$$

The substitution of the rank term for the nuclear (trace) norm enforces a convex lower bound on rank. We adopt the L1 norm to promote error sparsity.

This objective can be solved efficiently through the use of Augmented Lagrangian Methods (ALM). For purposes of simplicity, we express the ALM in scaled form [15]. The final objective is thus:

$$\arg\min_{\mathbf{L},\mathbf{E},\mathbf{P},\mathbf{X}} \|\mathbf{L}\|_* + \lambda\|\mathbf{E}\|_1$$
$$\text{s.t. } \mathbf{L} + \mathbf{E} = \mathbf{X}$$
$$\mathbf{X} = \mathbf{D}(\mathbf{P})$$
$$\mathbf{P}_i \in \mathbb{B} \ \forall i = 1, \ldots, N \tag{7}$$

For purposes of simplicity, we expression the Lagrangian in scaled form [15]:

$$\mathcal{L}(\mathbf{L}, \mathbf{E}, \mathbf{X}, \mathbf{U}) = \|\mathbf{L}\|_* + \lambda\|\mathbf{E}\|_1 + \frac{\rho}{2}\|\mathbf{X} - \mathbf{L} - \mathbf{E} + \mathbf{U}\|_2^2 \tag{8}$$

where $\mathbf{U}$ are the scaled Lagrange multipliers, such that:

$$\mathbf{U} = \frac{1}{\rho} \times \mathbf{Y} \tag{9}$$

The algorithm can be summarized according to Algorithm 1.

3.1 Valid Solutions

For traversing the set $\mathbb{B}$, we assert that the use of DTW allows for an optimal solution to be gleaned for the alignment parameters. At each iteration, the problem of updating the ensemble alignment parameters is decomposed to individual alignment problems. Hence, we assert that utilizing DTW in a similar fashion to LK in [3,4] allows for an acceptable solution to be found.

A large number of iterations typically allows a reasonable solution to be found. The following heuristics were used empirically to determine the feasibility of a solution:

$$\|\mathbf{X}^k - \mathbf{L}^k - \mathbf{E}^k\|_F < \alpha \tag{10}$$

where α is a small tolerance.

$$\|\mathbf{X}^k - \mathbf{X}^{k-1}\|_F == 0 \tag{11}$$

Data: $\mathbf{D}$ (raw ensemble), $\mathbf{P}$ (arbitrary time warps), $\mathbf{X} = \mathbf{D}(\mathbf{P})$
Result: $\mathbf{P}$ (optimal alignment paths), $\mathbf{X}, \mathbf{L}, \mathbf{E}$

Initialize $\mathbf{L}, \mathbf{E}, \mathbf{U}$ to zero matrices of appropriate dimensionality, $k = 0$, λ and ρ as appropriate.
while *not* ***converged*** **do**
    Update $\mathbf{L}$ using singular value soft thresholding:
    $(\Gamma, \Sigma, \theta) = \text{svd}(\mathbf{X}^k - \mathbf{E}^k + \mathbf{U}^k)$
    $\mathbf{L}^{k+1} = \Gamma \times \mathcal{S}_{\frac{2}{\rho}}[\Sigma] \times \theta^T$

    Update $\mathbf{E}$ using soft thresholding:
    $\mathbf{E}^{k+1} = \mathcal{S}_{\frac{2\lambda}{\rho}}[\mathbf{X}^k - \mathbf{L}^{k+1} + \mathbf{U}^k]$

    Update $\mathbf{P}$ and $\mathbf{X}$ using DTW:
    $\mathbf{P}_i^{k+1} = \text{DTW}(\mathbf{D}_i, \mathbf{L}_i^{k+1}))\forall i = 1, \ldots, N$
    $\mathbf{X}^{k+1} = \mathbf{D}(\mathbf{P}^{k+1})$

    Lagrangian update:
    $\mathbf{U}^{k+1} = \mathbf{U}^k + \rho(\mathbf{X}^{k+1} - \mathbf{L}^{k+1} - \mathbf{E}^{k+1})$
    $k = k + 1$
end

Algorithm 1. Algorithm for Unsupervised Temporal Ensemble Alignment for Rapid Annotation.

4 Experimental Evaluation

4.1 Performance Metrics

There are two key areas of performance which were considered for the evaluation of this work. First, we considered the number of sequences within the ensemble which are aligned to ground truth at a given point in time. The point at which the most sequences correspond to their ground truth frame is considered to be the "consensus" point. To evaluate the performance over the set of sequences we define a measure of global alignment. We define x to be an alignment tolerance threshold, and evaluate the error for a given threshold as:

$$\mathcal{E}(x) = \frac{\text{Number of sequences which are within } x \text{ frames of ground truth}}{\text{Total Sequences}} \quad (12)$$

The overall performance measure is the area under the curve produced when x is varied from 0 (aligned) to the maximum possible misalignment. Misalignment is measured with respect to the target ensemble - that is, we evaluate the misalignment given the consensus point and the first appearance of the ground truth frame within the aligned sequence.

Using $\mathcal{E}$, we define qualitative measures of performance for each of the methods. These qualitative measures are to allow for a small degree of tolerance for misalignment across the set. These measures are:

– Perfect - Sequence matches ground truth exactly (highlighted below in green).
– Acceptable - No more than 10 % of F_0 error (highlighted below in blue).
– Critical - More than 10 % of F_0 error (highlighted below in orange).

An indicator of good performance would be a large number of "Perfect" alignments, with no "Critical" alignments. Bad performance would be indicated by the presence of "Critical" errors, no matter how many "Perfect" alignments are presented.

4.2   Cohn-Kanade+

For evaluation, we used the Cohn-Kanade+ dataset [13] for aligning ensembles of sequences which are labelled as the same expression category. We utilised spatially normalised 2D landmark data which describe the appearance of the face. All 68 landmarks were used. As the sequences are of different lengths, we compute an initialisation for each sequence which consists of a random path computed using DTW to initialise the sequences to the chosen ensemble length (100 frames). The random initialization was used so as not to bias the initial alignment in favour of any particular sequence and to demonstrate worst-case performance where no sequence annotation is provided. For evaluation, the emotion categories of "Anger", "Surprise" and "Disgust" were selected.

The proposed ensemble method was evaluated against two sequence to template techniques, DTW and CTW. As a template, we randomly selected an initialised sequence from each class. Subsequently, we aligned all sequences in the class to this selected template using each method.

For the ensemble method, we initialised using the strategy above and aligned all sequences within each category.

For all three methods, we randomly selected a subset of sequences for evaluation and manually annotated the onset of the expression. We used this ground truth to evaluate the "unsupervised" alignment. To ensure integrity of ground truth selection, the subject selection and annotation was undertaken separately to the evaluation of alignment results.

The entire "Surprise" category consisting of 83 sequences was reduced to a rank 3 basis by our ensemble method. The ground truth for the six sequences evaluated is shown in Fig. 2a. In Fig. 2b and c, it can be observed that there is little correspondence across the sequences. Most sequences are behind the ground truth, with DTW and CTW presenting the most error - a maximum of 82 frames and 74 frames respectively. Our method (shown in Fig. 2d) recovers optimal synchronisation for three of the sequences. Whilst alignment is not consistent across the selected samples, the maximum error present is 6 frames, within the defined tolerance. The error curve shown in Fig. 3 shows that the ensemble method (shown in green) offers superior performance than DTW (red) and CTW (blue), not only in accurately synchronising the most sequences but also in minimising the misalignment for the entire ensemble.

The "Anger" category consisting of 45 sequences was reduced to a 3 basis by our method. The ground truth for the six selected sequences evaluated is shown in Fig. 4a. In Fig. 4b and c, it can be observed that there is little consistency across

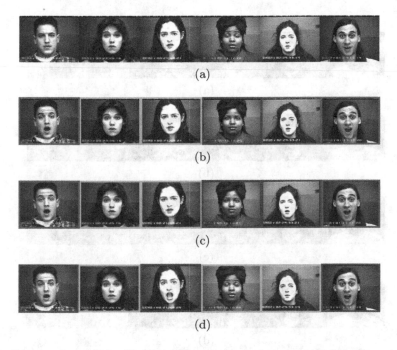

(a)

(b)

(c)

(d)

Fig. 2. Results from Cohn-Kanade+ Surprise category. Error categories are indicated for each sequence by green (Perfect), blue (Acceptable) and orange (Critical). (a) Ground truth. (b) Sequence to template alignment computed with DTW. (c) Sequence to template alignment computed with CTW. (d) Alignment computed using the ensemble method. Note that the ensemble method perfectly aligns 3 sequences (as opposed to 2) and produces acceptable alignment for the remaining sequences. Meanwhile, both DTW and CTW contain alignment errors which can be deemed critical (Color figure online).

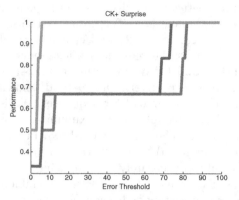

Fig. 3. Error curve for Cohn-Kanade+ Surprise category. DTW in red (AUC = 0.691), CTW in blue (AUC = 0.733), ensemble method in green (AUC = 0.969) (Color figure online).

Fig. 4. Results from Cohn-Kanade+ Anger category. Error categories are indicated for each sequence by green (Perfect), blue (Acceptable) and orange (Critical). (a) Ground truth. (b) Sequence to template alignment computed with DTW. (c) Sequence to template alignment computed with CTW. (d) Alignment computed using the ensemble method. The ensemble method successfully aligned three of the sequences with those remaining being in acceptable alignment. DTW and CTW do not reach a consensus point, and all sequences are critically misaligned (Color figure online).

the sequences in terms of expression progression with respect to ground truth. Both methods do not successfully align any of the ground truth frames, and return a maximum error of 86 (DTW) and 82 (CTW) frames. The results from the ensemble method are shown in Fig. 4d, which show three ground truth frames in correspondence. Whilst the other sequences are not in correspondence, the maximum error returned by the ensemble method is 5 frames. The error curves shown in Fig. 5 shows the ensemble method (shown in green) offers superior performance to both sequence to template methods.

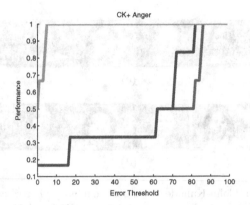

Fig. 5. Error curve for Cohn-Kanade+ Anger category. DTW in red (AUC = 0.441), CTW in blue (AUC = 0.488), ensemble method in green (AUC = 0.976) (Color figure online).

The "Disgust" category, consisting of 59 sequences was reduced to a rank 4 basis by our method. The ground truth for the six selected sequences is shown in Fig. 6a. In Fig. 6b and c, both DTW and CTW have successfully aligned two of the sequences in accordance with ground truth. However, both sequence to template methods have significant error across the remaining sequences - 90 frames for DTW and 76 frames for CTW. The ensemble method also successfully aligned two of the sequences, with a maximum observed error of 4 frames.

5   Discussion

5.1   Alignment Consistency

The experiments performed highlight the effectiveness of our method over template based methods. The ensemble alignment outperforms sequence to template alignment for all three selected CK+ categories. Whilst the number of sequences in alignment after processing may not necessarily be greater than the alignment recovered by template based methods, the overall misalignment across the sequences is greatly reduced (often by an order of magnitude). This results in a set of sequences which are vastly more synchronised.

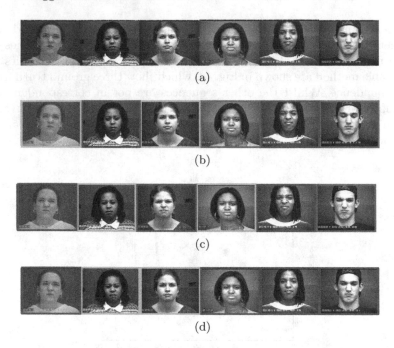

Fig. 6. Results from Cohn-Kanade+ Disgust category. Error categories for each sequence are indicated for each sequence by green (Perfect), blue (Acceptable) and orange (Critical). (a) Ground truth. (b) Sequence to template alignment computed with DTW. (c) Sequence to template alignment computed with CTW. (d) Alignment computed using the ensemble method. Note that the ensemble method does not return any critical levels of error (Color figure online).

Across all three emotion categories, the ensemble method performs adequately, with no errors in the "Critical" category. Raw CTW and DTW, whilst able to recover adequate alignment in some instances, encounter some "Critical" errors. In the case of Surprise and Anger, the ensemble method returns more "Perfect" alignments.

Disgust, however, yields interesting results across all three methods. For all three methods, a maximum of 2 "Perfect" alignments are returned (Fig. 6). However, it is of note that the ensemble method does not encounter any "Critical" levels of error, whilst both DTW and CTW yield some "Critical" errors. Disgust is considered to be a more difficult category, as the presentation of the expression is more varied than other expression categories.

Whilst not perfect, the ensemble method is shown to outperform DTW and CTW for approximating expression onset annotation (Fig. 7). It is possible that a different initialisation strategy for our method may result in better performance. For example, rather than initialising every sequence against a random alignment path, if a subset of the ensemble was correctly aligned a priori, this may be sufficient to boost alignment performance over the entire ensemble (Fig. 7).

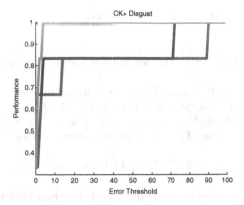

Fig. 7. Error curve for Cohn-Kanade+ Disgust category. DTW in red (AUC = 0.812), CTW in blue (AUC = 0.863), ensemble method in green (AUC = 0.98) (Color figure online).

5.2 Convergence and Scalability

Convergence of large ensembles (such as those representing an entire category of CK+) typically occurs within 10,000–15,000 iterations (a few hours on a single CPU using MATLAB). However, convergence of smaller ensembles can occur within a few hundred iterations. It is worth noting that the scalability of the algorithm may be affected as the number and length of sequences grows. It is possible that modification of the objective as demonstrated in [16] may improve performance.

6 Conclusion

In this paper, we have proposed an ensemble-based approach for the alignment of semantically similar time series and its application to the discovery of approximate event annotation. Through the use of Dynamic Time Warping, we have demonstrated the application of insights from image ensemble alignment to be reasonably effective in the time domain. The proposed method delivers promising results for alignment and annotation generation of sequences consisting of facial expression onset.

Acknowledgements. This research was supported by an Australian Research Council (ARC) Discovery Research Grant DP140100793.

References

1. Learned-Miller, E.G.: Data driven image models through continuous joint alignment. PAMI **28**, 236–250 (2006)
2. Cox, M., Sridharan, S., Lucey, S., Cohn, J.: Least squares congealing for unsupervised alignment of images. In: CVPR, pp. 1–8. IEEE (2008)

3. Peng, Y., Ganesh, A., Wright, J., Ma, Y.: RASL: Robust alignment via sparse and lowrank decomposition. In: CVPR. IEEE (2010)
4. Peng, Y., Ganesh, A., Wright, J., Xu, W., Ma, Y.: RASL: Robust alignment by sparse and low-rank decomposition for linearly correlated images. PAMI **34**, 2233–2246 (2012)
5. Zhou, F., De la Torre, F.: Generalized time warping for multi-modal alignment of human motion. In: CVPR, pp. 1282–1289. IEEE (2012)
6. Panagakis, Y., Nicolaou, M.A., Zafeiriou, S., Pantic, M.: Robust canonical time warping for the alignment of grossly corrupted sequences. In: CVPR, pp. 540–547. IEEE (2013)
7. Candès, E.J., Li, X., Ma, Y., Wright, J.: Robust principal component analysis? JACM **58**, 11 (2011)
8. Müller, M.: Dynamic time warping. In: Müller, M. (ed.) Information Retrieval for Music and Motion, pp. 69–84. Springer, Heidelberg (2007)
9. Lucas, B.D., Kanade, T., et al.: An iterative image registration technique with an application to stereo vision. In: IJCAI, vol. 81, pp. 674–679 (1981)
10. Cuturi, M., Vert, J.P., Birkenes, Ø., Matsui, T.: A kernel for time series based on global alignments. In: ICASSP, vol. 2, pp. 413–416. IEEE (2007)
11. Cuturi, M.: Fast global alignment kernels. In: ICML, pp. 929–936 (2011)
12. Lorincz, A., Jeni, L.A., Szabó, Z., Cohn, J.F., Kanade, T.: Emotional expression classification using time-series kernels. In: CVPRW, pp. 889–895. IEEE (2013)
13. Lucey, P., Cohn, J.F., Kanade, T., Saragih, J., Ambadar, Z., Matthews, I.: The Extended Cohn-Kanade Dataset (CK+): A complete dataset for action unit and emotion-specified expression. In: CVPRW, pp. 94–101. IEEE (2010)
14. Zhou, F., De la Torre, F.: Canonical time warping for alignment of human behavior. In: NIPS, pp. 2286–2294 (2009)
15. Boyd, S., Parikh, N., Chu, E., Peleato, B., Eckstein, J.: Distributed optimization and statistical learning via the alternating direction method of multipliers. Found. Trends Mach. Learn. **3**, 1–122 (2011)
16. Dai, Y., Li, H., He, M.: A simple prior-free method for non-rigid structure-from-motion factorization. In: CVPR, pp. 2018–2025. IEEE (2012)

Motion Boundary Trajectory for Human Action Recognition

Sio-Long Lo[✉] and Ah-Chung Tsoi

Faculty of Information Technology,
Macau University of Science and Technology, Macau, China
sllo@must.edu.mo

Abstract. In this paper, we propose a novel approach to extract local descriptors of a video, based on two ideas, one using motion boundary between objects, and, second, the resulting motion boundary trajectories extracted from videos, together with other local descriptors in the neighbourhood of the extracted motion boundary trajectories, histogram of oriented gradients, histogram of optical flow, motion boundary histogram, can be used as local descriptors for video representations. The motion boundary approach captures more information between moving objects which might be caused by camera movements. We compare the performance of the proposed motion boundary trajectory approach with other state-of-the-art approaches, e.g., trajectory based approach, on a number of human action benchmark datasets (YouTube, UCF sports, Olympic Sports, HMDB51, Hollywood2 and UCF50), and found that the proposed approach gives improved recognition results.

1 Introduction

Recognizing human action in a video is a commonly studied topic in computer vision and machine learning [1–4]. Broadly speaking, a popular approach is to first extract a set of local descriptors, and then use a bag-of-features model for matching those local descriptors obtained in the set of labeled training video clips, to those as yet unlabelled in the testing dataset [5–7].

Laptev [8] introduced space-time interest points (STIPs) using an extension of the Harris corner detection method [9] from image to video. Other detectors are also used to detect interest points in videos, e.g., Willems et al. [10] proposed using the determinant of the spatiotemporal Hessian matrix for interest point detection, Dollar et al. [11] proposed a 1D Gabor filter in the time dimension with a 2D Gaussian in the spatial dimensions to detect the underlying periodic frequency components for interest point detection.

Based on the detected interest points in a video, a descriptor is proposed to describe the information of sub-regions of the video as local features. Several descriptors have been proposed for describing these spatiotemporal local features, e.g., higher order derivatives (local jets) [8], histogram of oriented gradient (HOG) [12] for capturing object shape, these are called the appearance descriptors; histogram of optical flow (HOF) [12] for capturing object motion

© Springer International Publishing Switzerland 2015
C.V. Jawahar and S. Shan (Eds.): ACCV 2014 Workshops, Part I, LNCS 9008, pp. 85–98, 2015.
DOI: 10.1007/978-3-319-16628-5_7

information, a spatiotemporal version of HOG, called HOG3D [13] which extends the idea of HOG to the 3D case, histogram of oriented flows (HOF), a way of representing movements across time [12], and motion boundary histograms (MBH) [14] to cope with the camera motion. This detector/descriptor approach can be considered as a kind of bag-of-features video representation.

In contrast from detecting interest points in a 3D volume data, another approach to obtaining local features from a video is the trajectory approach, so called dense trajectory approach, as the patch is represented by a large number of interest points, [4,15]. In this approach, a set of local interest points is first detected using the 2D Harris condition [9] from video frames and an optical flow field is then used to track these interest points temporally to form the patch trajectories in the video [4]. The trajectory descriptor, together with the local descriptors, can be used to represent the video under a bag-of-features framework.

However, it is difficult to detect the actual moving objects in a complex background scene with severe camera motion using the 2D Harris corner condition [9] as the local patch detector. In this paper, we wish to show that the motion patterns of objects are important and will help detect informative patch trajectories for action recognition. In [16], the authors also introduced a motion boundary based sampling for action recognition, though it is different from the one which we proposed in this paper. The fact that motion provides important cue for grouping objects is well known [17]. On the other hand, to cope with camera motion, Dalal et al. introduced the motion boundary histogram (MBH) [14] as an effective local descriptor. MBH encodes the gradients of optical flow, which are helpful for canceling constant camera motion. Despite the importance of MBH as clearly shown in Dalal et al. [14], it appears that no one has yet explored the idea of a motion boundary in the dense trajectory approach [4]. It is expected that if we can embed the motion boundary concept in the dense trajectory approach [4], then it can handle issues related to camera motion, and thus would result in improved recognition rate, for datasets which may have taken while the camera might be moving. In this paper, we propose to use the motion boundary between objects for detecting local patches within the dense trajectory approach [4]. The motion boundary can capture more informative information between moving objects which might be caused because the camera was moving. With the motion boundary defined, the motion boundary trajectory can be extracted and can be used for the video representation. We compare the performances of various approaches on a number of standard benchmark datasets [18–22] and achieve better results using the proposed approach.

The rest of this paper is organized as follows. Section 2 discusses related work; Sect. 3.3 briefly introduces the concept of local descriptor extractions from videos, which include motion boundary trajectories (in Sect. 3.2), appearance based descriptors and motion descriptors; Sect. 4 provides approaches to classification; experimental results are shown in Sect. 5. Finally, some conclusions are drawn in Sect. 6.

Contribution: This paper establishes the deployment of motion boundary determination in the dense-trajectory approach for action recognition. The motion boundary between objects is determined and then those points in this motion boundary are tracked to form the motion boundary trajectories for video representation. Experimental results show that this idea can improve the performance of recognition significantly.

2 Relative Works

The most popular approach for action recognition is the well known bag-of-feature model [19,23,24]. In this model, the selection of local features of a video is important for the video representation. There are two broad approaches within this tradition: the detector/descriptor approach [18] and the trajectory approach [4]. In the detector/descriptor approach [18], the detector is used to detect interesting sub-regions of a video, contained within such sub-regions are typically the intensity values that have significant local variations in both space and time. For these sub-regions, the descriptors are applied to describe the spatial-temporal local features of the video [18]. The dense trajectory approach [4] tracks the detected local patches in the video frames through time. Then patch trajectories can be extracted from these sub-regions of the video. In the dense trajectory approach, the extracted spatial-temporal local features are significant [4]. It can be explained that the detected/extracted features are specifically based on object appearances and, to some extent, on motions (as the motion boundary histogram is used to represent the motion).

Some related work can be found in motion segmentation and video co-segmentation [25]. Motion segmentation is the problem of decomposing a video and to detect moving objects and background based on the idea of coherent regions with respect to motion and appearance properties [25]. Motion information provides an important cue for identifying the surfaces in a scene and for differentiating image texture from physical structures. In [17], long term point trajectories based on dense optical flow are used to spatial–temporal cluster the feature points into temporally consistent segmentations of moving objects. The quality of motion segmentation depends significantly on the pair of frames with a clear motion difference between the objects [26]. The advantage of motion segmentation derives from the fact that it combines motion estimation with segmentation. For segmenting multiple objects in the scene, the layered model for motion segmentation is proposed [27]. Typically, the scene consists of a number of moving objects and representing each moving object by a layer that allows the motion of each layer to be described [27]. Such a representation can model the occlusion relationships among layers making the detection of occlusion boundaries possible [28,29]. Typically, the background/foreground segmentation is a special case of binary object segmentation in this layered model [30].

In [25], multiple objects and multi-class video co-segmentation task is proposed to segment objects in videos. Object co-segmentation [25] is to segment a prominent object based on an image pair in which it appears in both images. With this idea, video co-segmentation segments the objects that are shared between videos, therefore co-segmentation can be encouraged. With this approach, object boundaries can be detected [28,29].

Based on the idea of motion segmentation, objects may be segmented from the background in the action recognition. Inspired by the idea of motion boundary histogram descriptor in the bag-of-feature framework, in this paper, we propose to use the boundary between objects as a descriptor in the dense-trajectory approach. The motion boundary can then be tracked frame by frame and then deployed as a descriptor, very much in the same manner as the patch trajectories in the dense trajectory approach [4] and then used for action recognition. This has the advantage of not requiring to perform the segmentation or co-segmentation task which are very time consuming tasks, where there is no significant occlusion of the objects involved.

3   Motion Boundary Trajectories

In this section, we will describe the proposed motion boundary dense trajectory approach. We will first describe the dense trajectory approach [4] briefly, and then we will show how motion boundary trajectories can be extracted from the video.

3.1   Dense Trajectories

The idea of a trajectory is based on interest points tracking [4]; the interest points are tracked frame by frame and then the corresponding trajectory can be extracted based on the tracked points [4]. For the motion boundary trajectories, we first detect the motion boundary on video frames and then track the detected motion boundary through time to form the motion boundary trajectories of a video.

Consider a video which consists of $I^{(t)}, t = 1, 2, \ldots, T$ and $I^{(t)}$ is a 2D pixel intensity array with dimensions $W \times H$. The optical flow field is computed over a two-frame sequence $I^{(t)}$ and $I^{(t+1)}$, $\omega^{(t)} = (u^{(t)}, v^{(t)})$, where, $u^{(t)}$, $v^{(t)}$ are respectively the optical flow in the horizontal and vertical directions. We apply a median filtering on the optical flow field $\omega^{(t)} = (u^{(t)}, v^{(t)})$ within a 3×3 patch. The resulting optical flow field is denoted by $\bar{\omega}^{(t)} = (\bar{u}^{(t)}, \bar{v}^{(t)}) = \omega^{(t)} \star M_{3 \times 3}$, where $M_{3 \times 3}$ is the median filter kernel and $\bar{\omega}^{(t)}$ is the filtered result of the optical flow field and $\star$ is the convolution operator.

In the dense trajectory approach [4], the Harris corner condition [9]. With this selection, a set of interest points, determined using a 2D Harris corner condition [9] on the object appearance, is then tracked frame by frame to form the dense trajectories.

In other to cope with the camera motion, a matching of feature points using SURF descriptors and dense optical flow is applied to estimate a homography between two subsequent frames by RANSAC algorithm as in [31]. Based on the reason of human action is in general different from camera motion. A human detector is employed to remove matches from human regions to improve the camera motion estimation. Finally, the trajectories consistent with the camera motion are then removed which are no longer useful for the tracking process [31].

3.2   Motion Boundary Trajectories

Different from using object appearances, motion boundary trajectory approach is based on the motion boundary between objects. To detect the motion boundary, we extract its location using the optical flow. Assume each object will have different flow directions and velocities, we detect their boundaries using the derivative of the optical flow field which captures the discontinuity, e.g., edges, of the optical flow field. For the point $P_i^{(t)} \in I^{(t)}$, the measurement of its boundary is given by

$$H_{P_i^{(t)}} = ||\nabla \bar{u}_{P_i^{(t)}}||^2 + ||\nabla \bar{v}_{P_i^{(t)}}||^2$$

where, $(\bar{u}_{P_i^{(t)}}, \bar{v}_{P_i^{(t)}})$ is the flow vector of point $P_i^{(t)}$.

The determination of the motion boundary trajectories is very similar to that proposed in [4] in the dense trajectory approach. Given a dense grid of frame $I^{(t)}$, we can densely sample points on a grid spaced by w pixels. In our case, the dense grid is set to 5×5. Sampling is carried out on each spatial scale separately. Different scales can be obtained by simply re-sizing the video to different resolutions, with a scaling factor of $\frac{1}{\sqrt{2}}$. In our setting, there are at most 8 spatial scales in total [4]. To obtain the motion boundary trajectories, we first select the points based on Harris corner condition

$$T_{corner}^{(t)} = C_1 \times \max_{P_i^{(t)} \in I^{(t)}} \min(\lambda_{P_i^{(t)}}^1, \lambda_{P_i^{(t)}}^2)$$

where, $(\lambda_{P_i^{(t)}}^1, \lambda_{P_i^{(t)}}^2)$ are the eigenvalues of the auto-correlation matrix of point $P_i^{(t)}$ in frame $I^{(t)}$. We then threshold the motion boundary based on the threshold $T_{corner}^{(t)}$ as

$$\tilde{H}_{P_i^{(t)}} = \begin{cases} H_{P_i^{(t)}} & \min(\lambda_{P_i^{(t)}}^1, \lambda_{P_i^{(t)}}^2) \geq T_{corner}^{(t)} \\ 0 & otherwise \end{cases}$$

We then use another threshold condition for which a point is of interest (i.e., significant enough for further consideration):

$$T_{motion}^{(t)} = C_2 \times \max_{P_i^{(t)} \in I^{(t)}} \tilde{H}_{P_i^{(t)}} + C_3$$

The point $P_i^{(t)}$ will be selected, if its magnitude is greater than the threshold, i.e., $\tilde{H}_{P_i^{(t)}} > T_{motion}^{(t)}$, while those points which do not satisfy this condition will not be considered further. In our setting, we set $C_1 = 0.0001$, $C_2 = 0.01$ and $C_3 = 0.002$. From the above process, we will know which sub-sampled point $P_i^{(t)}$ will need to be considered for the trajectory tracking. We then track the selected points using optical flow field $\bar{w}^{(t)} = (\bar{u}^{(t)}, \bar{v}^{(t)})$. Consider a point $P_i^{(t)} = (x_i^{(t)}, y_i^{(t)})$ in frame $I^{(t)}$, the tracked point $P_i^{(t+1)} = (x_i^{(t+1)}, y_i^{(t+1)})$ of $P_i^{(t)}$ in the next frame $I^{(t+1)}$ is computed by:

$$P_i^{(t+1)} = P_i^{(t)} + \bar{\omega}_{t,P_i^{(t)}}$$
$$= (x_i^{(t)}, y_i^{(t)}) + (\bar{u}_t, \bar{v}_t) \mid_{(x_i^{(t)}, y_i^{(t)})}$$

The tracked points of subsequent frames are then concatenated temporally to form a trajectory, $\text{Traj}_i = (P_i^{(t)}, P_i^{(t+1)}, P_i^{(t+2)}, \ldots)$. For each frame, if no tracked point is found in the neighborhood, a new point $P_{i*}^{(t)}$ is sampled and added to the tracking process. If the length of a trajectory has reached a maximum length $L = 15$, a post-processing stage is then performed to remove the static trajectories [4].

In order to obtain a better motion boundary, we follow [31] and estimate the homography of two subsequent frames, and then warp the second frame with the estimated homography. Based on the warped frame, the Harris cornerness is computed by the warped second frame and the optical flow is computed between, the first and the warped second frame. To obtain more interest points surrounding the moving objects, we apply a Gaussian filter and then a median filter on the motion boundary map, i.e., $\check{H}$. We then select and track the points for extracting the motion boundary trajectories. For the optical flow, we use the Farneback optical flow algorithm [32], which employs a polynomial expansion to approximate the pixel intensities in the neighborhood to obtain a good quality flow field as well as capturing some fine details [4]. Figure 1 shows the results of the motion boundary as well as the motion boundary trajectory obtained from some selected videos.

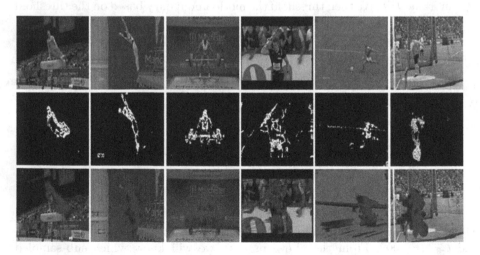

Fig. 1. The first row shows the original images; the second row shows the detected motion boundaries; and the third row shows the corresponding motion boundary trajectories

It is observed that the motion boundary trajectories can capture the motion quite well.

3.3 Motion Boundary Descriptors

Local descriptors are features which describe the spatial temporal behaviours of humans in the video. There are a number of such descriptors proposed by various researchers: [4]. The essential idea is to find good descriptors which will describe the spatial temporal behaviours of pixel values in a small neighborhood of a volume consisting of two dimensional space and time [4]. Some of these methods were extended from image processing techniques, while others were constructed explicitly for spatial temporal behaviours [4].

Several descriptors can be obtained to encode either the shape of a trajectory or the local motion [4] and appearance within a space-time volume [14] around the trajectory. The trajectory shape descriptor encodes local motion patterns by using the displacement vectors of a trajectory [4]. HOG (Histogram of oriented gradient) along a trajectory focuses on the static part of the appearance of a local patch of the video. For encoding the motion information, HOF (Histograms of optical flow) captures the local motion information based on the optical flow field; MBH (Motion boundary histogram) uses the gradient of the optical flow to cancel out most of the effects of camera motion [14]. These descriptors give a state-of-the-art performance for representing local information.

In this paper, we will add the motion boundary trajectories as the descriptors for the motion in the time axis. The motion trajectory descriptor can be formed by considering the shape of the trajectories, in a manner very similar to that proposed in [4]. Given a trajectory of length L, a sequence $(\Delta P_i^{(t)}, \ldots, \Delta P_i^{(t+L-1)})$ of the displacement vectors $\Delta P_i^{(t)} = P_i^{(t+1)} - P_i^{(t)} = (x_i^{(t+1)} - x_i^{(t)}, y_i^{(t+1)} - y_i^{(t)})$ is used for describing the trajectory shape. The normalized concatenation of the displacement vectors will become the feature vector of the trajectory shape:

$$\text{Shape}_i = \frac{(\Delta P_i^{(t)}, \ldots, \Delta P_i^{(t+L-1)})}{\sum_{k=t}^{t+L-1} \| \Delta P_i^{(k)} \|}$$

With the motion boundary trajectory, $\text{Traj}_i = (P_i^{(t)}, P_i^{(t+1)}, P_i^{(t+2)}, \ldots)$, the corresponding HOG, HOF and MBH descriptors can also be extracted based on the motion boundary trajectory as the trajectory based HOG, HOF and MBH descriptors (please see Fig. 2 for an illustration of these concepts). We follow [31], motion descriptors (HOF and MBH) are computed on the warped optical flow. The trajectory shape descriptor and HOG descriptor remain unchanged.

4 Classification

We apply the standard bag-of-features approach to convert the local descriptors from a video into a fixed-dimensional vector. We first construct a codebook for the trajectory descriptor (Sect. 3.3) using the k-mean clustering algorithm, and

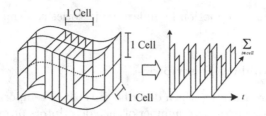

Fig. 2. Illustration of motion boundary trajectory descriptor. The motion boundary trajectory is represented by relative point coordinates, $\text{Traj}_i = (P_i^{(t)}, P_i^{(t+1)}, P_i^{(t+2)}, \ldots)$; based on the motion boundary trajectories, the HOG, HOF and MBH descriptors are computed along the trajectories.

then the clusters will serve as visual words. We fix the number of visual words to $V = 4,000$. To limit the complexity of the problem, we cluster a subset of 100,000 randomly selected from the training features in the k-mean clustering algorithm. Descriptors are then assigned to their closest vocabulary word using an Euclidean norm. The resulting histograms of visual word occurrences are used as video representations.

We apply the linear and non-linear SVM for action recognition. For the linear SVM [33], we first scale the value of each visual word feature to $[0, 1]$, and then the feature vector of a video is normailzied by a norm-2 normalization. For the nonlinear SVM [12], we normalize the histogram using the RootSIFT approach [34], i.e., square root each dimension after L1 normalization, and then apply the standard RBF (radial basis function)-χ^2 kernel [4] as the baseline algorithm in our experiments.

$$K_{\chi^2}(H_i, H_j) = \exp\left(-\frac{1}{2A}\sum_{k=1}^{V}\frac{(h_{ik} - h_{jk})^2}{h_{ik} + h_{jk}}\right)$$

where $H_i = \{h_{ik}\}_{k=1}^{V}$ and $H_j = \{h_{jk}\}_{k=1}^{V}$ are the frequency histograms of word occurrences and V is the vocabulary size. A is the mean value of distances between all training samples [18]. In the case of multi-class classification, the one-against-all approach is applied, we select the class with the highest score. Typically, the approach for integrating the contribution of different descriptors is the multiple channel SVM [7,12], which is a special case of multiple kernel learning [35]. We simply average the kernels computed from different representations to combine different channels using the idea of multiple channel SVM.

We also apply the Fisher vector [36] encoding for video representation. Fisher vector encodes both first and second order statistics between the video descriptors and a Gaussian Mixture Model (GMM). We follow [31], first reduce the descriptor dimensionality by Principal Component Analysis (PCA), as in [31]. We set the number of Gaussians to $K = 256$ and randomly sample a subset of 256,000 features from the training set to estimate the GMM [31]. As a result, for each type of descriptor, each video is represented by a $2DK$ dimensional Fisher vector, where D is the dimension of the descriptor after performing PCA. Finally, we apply power and the RootSIFT approach normalization to the Fisher vector. For integrating different descriptor types, we concatenate their normalized Fisher vectors, and a linear SVM is used for classification.

5 Experiments

This section evaluates the proposed motion boundary trajectories as a descriptor. We run the experiments at least 3 times for descriptor-classifier pairs. We will report the average accuracy of those experiments.

5.1 Datasets

We evaluate our proposed motion boundary descriptor on six standard benchmark datasets, viz., UCF-Sports [20], YouTube dataset [19], Olympic Sports dataset [21], the HMDB51 dataset [22], the Hollywood2 datasets, and the UCF50 datasets.

The UCF-Sports dataset contains 150 videos from ten action classes, diving, golf swinging, kicking, lifting, horse riding, walking, running, skating, swinging (on the pommel horse and on the floor), and swinging (at the high bar). These videos are taken from real sports broadcasts and the bounding boxes around the subjects are provided for each frame. We follow the protocol proposed in [37,38] using the same training/testing samples for our experiments; by taking one third of the videos from each action category to form the test set, and the rest of the videos are used for training. Average accuracy over all classes is reported as the performance measure.

The YouTube dataset contains 11 action categories: basketball shooting, biking/cycling, diving, golf swinging, horse back riding, soccer juggling, swinging, tennis swinging, trampoline jumping, volleyball spiking, and walking with a dog. For each category, the videos are grouped into 25 groups with more than 4 action clips in it. The dataset contains a total of 1,168 sequences. We follow the original setup [19], using leave-one-out cross-validation for a pre-defined set of 25 groups. Average accuracy over all classes is reported as the performance measure.

The Olympic Sports dataset [21] consists of athletes practising different sports, which are collected from YouTube and annotated using the Amazon Mechanical Turk technique. There are 16 sports actions: high jump, long jump, triple jump, pole vault, discuss throw, hammer throw, javelin throw, shot put, basketball layup, bowling, tennis serve, platform (diving), springboard (diving), snatch (weight lifting), clean and jerk (weight lifting) and vault (gymnastics), represented by a total of 783 video sequences. We adopt the train/test split from [21]. The mean average precision (mAP) over all classes [12,39] is reported as the performance measure.

The HMDB51 contains 51 distinct action categories, each containing at least 101 clips for a total of 6,766 video clips extracted from a wide range of sources. We follow the original evaluation protocol using three train-test splits [22]. For every class and split, there are 70 videos for training and 30 videos for testing. We report the average accuracy over three-splits as performance measure.

The Hollywood2 dataset [40] has been collected from 69 different Hollywood movies and includes 12 action classes. It contains 1,707 videos split into a training set (823 videos) and a test set (884 videos). Training and test videos come from different movies. The performance is measured by mean average precision (mAP) over all classes, as in [40].

The UCF50 dataset [41] has 50 action categories, consisting of real-world videos taken from YouTube. There are 50 categories in UCF50 dataset, the videos are split into 25 groups. For each group, there are at least 4 action clips. In total, there are 6,618 video clips. We apply the leave-one-group-out cross-validation as recommended by the authors and report average accuracy over all classes.

5.2 Experimental Results

The experimental results using bag-of-feature histogram are shown in Table 1. We also list the results of improved dense trajectory approach [4] in our experiments, under the name Dense Trajectory in Table 1. For the dense trajectory approach, the 2D interest points are detected based on corner condition [4], and then track the detected points frame by frame to form the dense trajectories. From the results listed in Table 1, we note that the best performance is achieved using our motion boundary trajectory descriptor.

Table 1. Experimental results of motion boundary trajectory on different datasets.

	UCF Sport				YouTube			
	Dense Trajectory		Motion Boundary		Dense Trajectory		Motion Boundary	
	Linear	χ^2 SVM	Linear	χ^2 SVM	Linear	χ^2 SVM	Linear	χ^2 SVM
Traj. Shape	73.1	79.4	70.6	**83.8**	66.0	76.4	71.2	**78.6**
HOG	71.6	74.5	72.8	**80.0**	69.0	74.4	69.5	**74.6**
HOF	75.9	82.3	85.1	**91.5**	76.8	80.9	78.0	**82.2**
MBH	78.0	80.9	80.4	**84.2**	77.4	**85.1**	78.1	84.2
Combined	82.3	85.1	90.2	**90.6**	86.6	87.1	**87.9**	87.4
	Olympic Sports				HMDB51			
	Dense Trajectory		Motion Boundary		Dense Trajectory		Motion Boundary	
	Linear	χ^2 SVM	Linear	χ^2 SVM	Linear	χ^2 SVM	Linear	χ^2 SVM
Traj. Shape	65.8	73.3	67.7	**76.7**	19.2	34.8	23.4	**39.1**
HOG	66.0	70.8	68.1	**73.4**	22.8	**33.5**	20.9	32.9
HOF	73.9	78.2	78.9	**80.6**	26.8	42.2	30.3	**45.3**
MBH	80.1	81.6	**83.9**	83.2	28.9	46.6	30.8	**50.0**
Combined	85.4	84.0	**86.5**	84.7	49.7	53.6	52.5	**56.7**

We found that on the UCF Sports dataset, the motion boundary trajectory descriptor together with HOF as well as MBH obtain very good results. The UCF Sports dataset contains videos which are typically featured on broadcast television channels, e.g., BBC and ESPN; these videos are recorded by professional cameramen and camera movement is relatively smooth. As a result, the detected motion boundary is much more meaningful, which is shown in Fig. 3. This observation is also true with the Olympic Sports dataset, in which the motion boundary trajectory with MBH descriptor obtain good results.

The videos of YouTube dataset are collected from YouTube and are personal videos. This dataset is very challenging due to large variations in camera motion. In this case, the motion boundary trajectories are not very accurate. As a result, the performance of motion boundary trajectory only improve slightly that compare with dense trajectory.

Fig. 3. Comparison between the dense trajectories and motion boundary trajectories (the first row shows dense trajectory; the second shows motion boundary trajectory)

We also evaluated the performance of combining representations named Combined as listed in Table 1. We evaluated two different classifiers, viz., the linear SVM and the χ^2 SVM. We simply average the kernel matrices computed from different representations to obtain the aggregated results. The motion boundary trajectory also improves the performance at least 1 % on the UCF Sports and HMDB51 datasets and slightly improves on YouTube and Olympic Sports datasets.

Figure 3 show the motion boundary trajectories and the dense trajectories. In Fig. 3, we note that the motion boundary detected in some videos is significant, the motion boundary can capture the trajectories around the moving objects when compare with those obtained from the dense trajectory approach.

Comparison to the state of the art. In [31], Wang introduced improved dense trajectory feature for action recognition. Together with the Fisher vector encoding for video representation, Wang obtained state-of-the-art results. We use the same setting as in [31] but instead of extracting dense trajectory, we extract the motion boundary trajectory. We also use the human boundary boxes provided by authors [31] for better eastimation of homography between two subsequent frames. The experimental result in Table 2, we also listed the result from [31], named as IDT (improved dense trajectory). In Table 2, we noted that the Olympic Sports dataset, the motion boundary trajectory (MBT) approach obtains at least 2 % improvement. We obtain 93.5 % mAP. For the HMDB51 dataset, we obtain at least 5 % improvement and obtain 63.8 accuracy. For the Hollywood2 dataset, the improvement is not too much, only 0.1 % improvement. For the UCF50 dataset, we get 1 % improvement and obtain 92.2 % accuracy. Those results show that the motion boundary is useful for describing the motion information and significantly improve the recognition accuracy in action recognition.

Table 2. Experimental results of motion boundary trajectory on different datasets using Fisher vector video representation; IDT means Improved Dense Trajectory, and MBT means Motion Boundary Trajectory; The results listed in IDT here are from [31].

	Olympic Sports		HMDB51		Hollywood2		UCF50	
	IDT [31]	MBT	IDT [31]	MBT	IDT [31]	MBT	IDT [31]	MBT
Traj. Shape	77.2	**81.5**	32.4	**35.9**	**48.5**	45.8	**75.2**	74.5
HOG	78.8	**82.1**	40.2	**43.2**	**47.1**	44.3	82.6	**83.9**
HOF	**87.6**	87.5	48.9	**53.2**	**58.8**	58.1	85.1	**87.1**
MBH	89.1	**92.2**	52.1	**58.2**	60.5	**60.7**	88.9	**90.5**
Combined	91.1	**93.5**	57.2	**63.8**	64.3	**64.4**	91.2	**92.2**

6 Conclusion

In this paper, we propose a novel approach based on two ideas, one using motion boundary between objects, and, second, the resulting motion boundary trajectories extracted from videos as the local descriptors. These resulted in a new descriptor, the motion boundary descriptor. We compare the performance of the proposed approach with other state-of-the-art approaches, e.g., trajectory based approach, on six human action recognition benchmark datasets, and found that the proposed approach gives better recognition results.

Acknowledgment. This work was financially supported by Fundo para o Desenvolvimento das Ciencia das e da Tecnologia, Macau SAR Grant Number 034/2011/A2. The authors would like to thank Associate Prof. Markus Hagenbuchner, University of Wollongong and Prof. Franco Scarselli, University of Siena, for many helpful comments on the proposed approach.

References

1. Brendel, W., Todorovic, S.: Learning spatiotemporal graphs of human activities. In: ICCV, pp. 778–785 (2011)
2. Niebles, J.C., Wang, H., Fei-Fei, L.: Unsupervised learning of human action categories using spatial-temporal words. IJCV **79**, 299–318 (2008)
3. Guo, K., Ishwar, P., Konrad, J.: Action recognition in video by covariance matching of silhouette tunnels. In: Brazilian Symposium on Computer Graphics and Image Processing, pp. 299–306 (2009)

4. Wang, H., Kläser, A., Schmid, C., Liu, C.L.: Dense trajectories and motion boundary descriptors for action recognition. IJCV **103**, 60–79 (2013)
5. Wallraven, C., Caputo, B., Graf, A.: Recognition with local features: the kernel recipe. In: ICCV, pp. 257–264 (2003)
6. Willamowski, J., Arregui, D., Csurka, G., Dance, C.R., Fan, L.: Categorizing nine visual classes using local appearance descriptors. In: ICPR Workshop on Learning for Adaptable Visual Systems (2004)
7. Zhang, J., Lazebnik, S., Schmid, C.: Local features and kernels for classification of texture and object categories: a comprehensive study. IJCV **73**, 213–238 (2007)
8. Laptev, I.: On space-time interest points. IJCV **64**, 107–123 (2005)
9. Harris, C., Stephens, M.: A combined corner and edge detector. In: Proceedings of the Alvey Vision Conference, pp. 147–151 (1988)
10. Willems, G., Tuytelaars, T., Van Gool, L.: An efficient dense and scale-invariant spatio-temporal interest point detector. In: Forsyth, D., Torr, P., Zisserman, A. (eds.) ECCV 2008, Part II. LNCS, vol. 5303, pp. 650–663. Springer, Heidelberg (2008)
11. Dalal, N., Triggs, B.: Histograms of oriented gradients for human detection. In: CVPR, pp. 886–893 (2005)
12. Laptev, I., Marszałek, M., Schmid, C., Rozenfeld, B.: Learning realistic human actions from movies. In: CVPR, pp. 1–8 (2008)
13. Kläser, A., Marszałek, M., Schmid, C.: A spatio-temporal descriptor based on 3d-gradients. In: BMVC, pp. 995–1004 (2008)
14. Dalal, N., Triggs, B., Schmid, C.: Human detection using oriented histograms of flow and appearance. In: Leonardis, A., Bischof, H., Pinz, A. (eds.) ECCV 2006. LNCS, vol. 3952, pp. 428–441. Springer, Heidelberg (2006)
15. Matikainen, P., Hebert, M., Sukthankar, R.: Trajectons: action recognition through the motion analysis of tracked features. In: ICCV Workshop on Video-oriented Object and Event Classification (2009)
16. Peng, X., Qiao, Y., Peng, Q., Qi, X.: Exploring motion boundary based sampling and spatial-temporal context descriptors for action recognition. In: BMVC (2013)
17. Brox, T., Malik, J.: Object segmentation by long term analysis of point trajectories. In: Daniilidis, K., Maragos, P., Paragios, N. (eds.) ECCV 2010, Part V. LNCS, vol. 6315, pp. 282–295. Springer, Heidelberg (2010)
18. Schuldt, C., Laptev, I., Caputo, B.: Recognizing human actions: a local SVM approach. In: ICPR, vol. 3, pp. 32–36 (2004)
19. Liu, J., Luo, J., Shah, M.: Recognizing realistic actions from videos in the wild. In: CVPR (2009)
20. Rodriguez, M., Ahmed, J., Shah, M.: Action mach a spatio-temporal maximum average correlation height filter for action recognition. In: CVPR, pp. 1–8 (2008)
21. Niebles, J.C., Chen, C.-W., Fei-Fei, L.: Modeling temporal structure of decomposable motion segments for activity classification. In: Daniilidis, K., Maragos, P., Paragios, N. (eds.) ECCV 2010, Part II. LNCS, vol. 6312, pp. 392–405. Springer, Heidelberg (2010)
22. Kuehne, H., Jhuang, H., Garrote, E., Poggio, T., Serre, T.: HMDB: a large video database for human motion recognition. In: ICCV (2011)
23. Sivic, J., Zisserman, A.: Video Google: a text retrieval approach to object matching in videos. In: ICCV, vol. 2, pp. 1470–1477 (2003)
24. Lazebnik, S., Schmid, C., Ponce, J.: Beyond bags of features: Spatial pyramid matching for recognizing natural scene categories. In: Proceedings of CVPR 2006, pp. 2169–2178 (2006)

25. Chiu, W.C., Fritz, M.: Multi-class video co-segmentation with a generative multi-video model. In: CVPR (2013)
26. Wang, J.Y., Adelson, E.H.: Representing moving images with layers (1994)
27. Sun, D., Sudderth, E.B., Black, M.J.: Layered segmentation and optical flow estimation over time. In: CVPR, pp. 1768–1775 (2012)
28. Black, M.J., Fleet, D.J.: Probabilistic detection and tracking of motion boundaries. IJCV **38**, 231–245 (2000)
29. Feghali, R., Mitiche, A.: Spatiotemporal motion boundary detection and motion boundary velocity estimation for tracking moving objects with a moving camera: a level sets pdes approach with concurrent camera motion compensation. IEEE Trans. Image Process. **13**, 1473–1490 (2004)
30. Sun, D., Wulff, J., Sudderth, E., Pfister, H., Black, M.: A fully-connected layered model of foreground and background flow. In: CVPR (2013)
31. Wang, H., Schmid, C.: Action recognition with improved trajectories. In: ICCV (2013)
32. Farnebäck, G.: Two-frame motion estimation based on polynomial expansion. In: Bigun, J., Gustavsson, T. (eds.) SCIA 2003. LNCS, vol. 2749, pp. 363–370. Springer, Heidelberg (2003)
33. Chang, C.C., Lin, C.J.: Libsvm: a library for support vector machines. ACM Trans. Intell. Syst. Technol. **2**, 1–27 (2011)
34. Arandjelović, R., Zisserman, A.: Three things everyone should know to improve object retrieval. In: CVPR (2012)
35. Gönen, M., Alpaydin, E.: Multiple kernel learning algorithms. JMLR **12**, 2211–2268 (2011)
36. Perronnin, F., Sánchez, J., Mensink, T.: Improving the fisher kernel for large-scale image classification. In: Daniilidis, K., Maragos, P., Paragios, N. (eds.) ECCV 2010, Part IV. LNCS, vol. 6314, pp. 143–156. Springer, Heidelberg (2010)
37. Shapovalova, N., Vahdat, A., Cannons, K., Lan, T., Mori, G.: Similarity constrained latent support vector machine: an application to weakly supervised action classification. In: Fitzgibbon, A., Lazebnik, S., Perona, P., Sato, Y., Schmid, C. (eds.) ECCV 2012, Part VII. LNCS, vol. 7578, pp. 55–68. Springer, Heidelberg (2012)
38. Lan, T., Wang, Y., Yang, W., Robinovitch, S., Mori, G.: Discriminative latent models for recognizing contextual group activities. PAMI **34**(8), 1549–1562 (2012)
39. Everingham, M., Van Gool, L., Williams, C.K.I., Winn, J., Zisserman, A.: The PASCAL Visual Object Classes Challenge 2007 (VOC 2007) Results
40. Marszałek, M., Laptev, I., Schmid, C.: Actions in context. In: CVPR (2009)
41. Reddy, K.K., Shah, M.: Recognizing 50 human action categories of web videos. Mach. Vis. Appl. **24**, 971–981 (2013)

Action Recognition Using Hybrid Feature Descriptor and VLAD Video Encoding

Dong Xing[✉], Xianzhong Wang, and Hongtao Lu

Key Laboratory of Shanghai Education Commission for Intelligent Interaction
and Cognitive Engineering, Department of Computer Science and Engineering,
Shanghai Jiao Tong University, Shanghai, China
xingdong0625@gmail.com

Abstract. Human action recognition in video has found widespread
applications in many fields. However, this task is still facing many challenges due to the existence of intra-class diversity and inter-class overlaps
among different action categories. The key trick of action recognition lies
in the extraction of more comprehensive features to cover the action,
as well as a compact and discriminative video encoding representation.
Based on this observation, in this paper we propose a hybrid feature
descriptor, which combines both static descriptor and motional descriptor to cover more action information inside video clips. We also adopt
the usage of VLAD encoding method to encapsulate more structural
information within the distribution of feature vectors. The recognition
effects of our framework are evaluated on three benchmark datasets:
KTH, Weizmann, and YouTube. The experimental results demonstrate
that the hybrid descriptor, facilitated with VLAD encoding method, outperforms traditional descriptors by a large margin.

1 Introduction

The task of action recognition in video can be divided into five procedures:
extracting Space-Time Interest Points (STIPs), describing STIPs, building visual
words, encoding video clips and finally classifying action categories. Recent
advances in action recognition show that the enhancement of feature description [6,16–18] as well as video encoding methods [1,2,24–26] can significantly
improve the correct recognition rate. This paper takes the advantage of both two
approaches. A hybrid feature descriptor is built, which combines both the static
and motional information inside each STIPs to represent local feature points.
Then Vector of Locally Aggregated Descriptor (VLAD) [1,2] encoding method is
adopted, which is proved to be a compact and discriminative encoding method
in image representation, to encode the distribution of high-dimensional feature
vectors. Figure 1 illustrates the work flow of our action recognition framework.

While in general, motional feature descriptors perform better than static ones
in action recognition [3], we strongly believe that motional and static features
should be complementary to each other in realistic settings. For example, many
two-player ball games, such as badminton and tennis, share similar motional

© Springer International Publishing Switzerland 2015
C.V. Jawahar and S. Shan (Eds.): ACCV 2014 Workshops, Part I, LNCS 9008, pp. 99–112, 2015.
DOI: 10.1007/978-3-319-16628-5_8

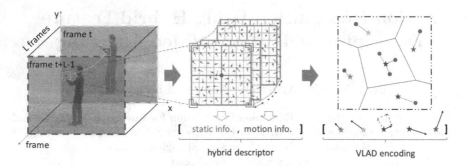

Fig. 1. An illustration of *STIPs extraction* (left), *hybrid feature description* (middle) and *VLAD video encoding* (right). Here we adopt dense trajectory sampling in [6] to extract STIPs. The hybrid descriptor, which is composed of two different descriptors, HOG (static information) and MBH (motional information), is used to cover the features around STIPs. Then, VLAD encoding is adopted to encapsulate the distribution of all hybrid descriptors.

features like waving rackets and jumping. It can be confusing to distinguish these actions simply by their motional features. However, as we all know, our human vision system can easily recognize these sports even with a single static frame according to their appearance. Yet, for some actions like running and jogging which share similar static appearance, it can be hard to distinguish one from another simply by the static information. In these cases, the motional information is required to represent the features.

Based on the complementary idea mentioned above, two different types of feature descriptors are carefully chosen and combined to form our hybrid descriptor. One is histogram of oriented gradient (HOG) [4], which accumulates the static oriented gradient information inside each frame around the feature point; the other is motion boundary histogram (MBH) [5], which focuses on the dynamic motion boundary information within two or more neighbouring frames. Both of these two descriptors were originally used on the pedestrian detection [4,5]. However, they also find their place in many other fields respectively. We compare the ability of our hybrid descriptor with the separated individual descriptors on several datasets, and the experimental result shows that our hybrid feature descriptor can achieve a state of the art recognition result.

The impact of different video encoding methods is also considered. Instead of traditional Bag of Words (BoW) encoding, VLAD [1,2] is chosen to encode the distribution of feature vectors. The idea of VLAD is to aggregate all the differences between feature vectors and their corresponding visual words to form signatures, then concatenate all the signatures to construct the video representation. Although VLAD requires more dimensions than BoW to form the encoding vector of each video clip, the experimental result shows that VLAD encodes more details inside each video, and yields a better result than BoW even with a smaller codebook.

This paper is organized as follows. Section 2 talks about the related work on action recognition. In Sect. 3, the details about hybrid descriptor, VLAD encoding as well as other implementation details in action recognition framework are explained. In Sect. 4, we discuss our experimental result over several public datasets, including YouTube [7], Weizmann [8] and KTH [9]. Finally in Sect. 5, we make a brief summary on action recognition.

2 Related Work

Successful extraction of more comprehensive features from video clips is the precondition for action recognition. Poppe [10] divides feature representations into two categories: global features and local features. Global feature extraction obtains an top-down fashion, which captures the information, such as silhouettes [11], edges [12], shapes [8], optic flows [13] of the human body as a whole, to form an overall representation. Although global features extract much of the action information, this kind of methods generally rely on accurate body localization, foreground extraction or tracking as preprocessing, and the result is sensitive to many environmental disturbance such as various viewpoints, lighting conditions and occlusions. On the other extreme, local feature extraction proceeds in a bottom-up fashion, which describes the observation as a combination of independent local portions. Comparing to global feature extraction, local feature extraction does not strictly rely on the effect of background subtraction or tracking, and is less sensitive to noise and partial occlusion. Due to these advantages, local feature extraction attracts more and more focus in the field of action recognition in recent years.

A wide range of local feature descriptors have been evaluated for action recognition. Based on different extraction methods, local feature descriptors can be divided into two groups: (i) motional feature descriptors, such as motion boundary histogram (MBH) [5] and histogram of optic flow (HOF) [14], which extract information from neighbouring frames through tracking the optic flows or other motional information around feature points; (ii) static feature descriptors, mostly originated from image processing, such as histogram of oriented gradient (HOG) [4] and 2D-SIFT [15], which regard the video clip as a sequence of frames and extract action information inside each frame respectively. Some methods extend the 2D image descriptors into 3D version, such as 3D-HOG [16], 3D-SIFT [17] and eSURF [18], by taking the temporal dimension as the third spatial axis to form a space-time video volume.

All these descriptors, static and motional, have various feature describing emphases, which offers a chance for us to evaluate combinations of different descriptors in order to cover more information about action characteristics. Several previous works [7,19–21] have shown the effect of combining multiple features or visual cues. However, random combination of different descriptors does not always work. A descriptor of poor quality may drag down the effect of a descriptor of high quality, as our experiment result shows. How to align different descriptors to evoke the potentiality remains a problem, and our approach of combining static and motional descriptors offers a clue to solve this question.

Video encoding encapsulates the distribution of local feature descriptors. Bag of words (BoW) [22,23] is one of the most popular encoding method, which assigns each feature vector to its nearest neighbouring visual word. The vector frequency of each visual word is accumulated, which is further concatenated directly as the video representation. Although BoW is proved to be a simple but valid encoding method, it omits lots of structural information inside the distribution of high-dimensional feature vectors, which is expected to have the ability of indicating the difference among each action classes to a large extent. Several novel encodings have been proposed to improve the BoW, including locality-constrained linear coding (LLC) [24], improved Fisher encoding [25], super vector encoding [26], VLAD [1,2] and so on. Among all these encodings, VLAD maintains a simplicity of computational complexity as well as a quality of discrimination.

3   The Proposed Recognition Framework

This section explains in detail the formation of hybrid feature descriptor, the mechanism of VLAD encoding scheme as well as other implementation details in our action recognition framework.

The crux of action recognition lies in the procedure of feature extracting as well as video encoding. Feature extraction should extract features which are relevant to their corresponding action classes from video clips, and video encoding should encapsulate more of the action information inside each video clips into a compact and discriminative representation.

3.1   Hybrid Feature Descriptor

Two different descriptors, HOG [4] (static) and MBH [5] (motional), are combined to form our hybrid feature descriptor directly. Although the idea is simple, we found that this direct combination, facilitated with VLAD encoding, is capable of achieving an advanced recognition result without adding too much complexity.

The essential thought of HOG is to describe action appearance as the distribution of intensity gradients inside each localized portions of the video clip. Gradient values of each pixels inside the local portion are computed firstly frame by frame to describe the local patch appearance, then all the pixels inside each portion cast a weighted vote for an orientation-based histogram according to their amplitudes and orientations.

Unlike HOG or other static descriptors, MBH focuses on the motional information along the boundary of different depth of fields. Optic flows of neighbouring frames are computed first to indicate the motional information. Then a pair of x- and y-derivative differential flow images are obtained, on which large value indicates drastic motion changing. These two differential flow images cast the corresponding orientation-based histograms.

Several spatio-temporal grid combinations of size $n_\sigma \times n_\sigma \times n_\tau$ are evaluated to subdivide the local patch in order to embed the structural information of local portion descriptors. However, denser grid leads to descriptors with more dimensionality and extra computational burden, which should be taken into account when applying action recognition to more realistic situations. Here we set $n_\sigma = 2$ and $n_\tau = 3$ as in [6], which has shown to be the optimal choice on most cases in our experiments, meanwhile maintaining a moderate complexity. The orientations inside each grid are quantized into 8 bins, producing the final 96 dimension HOG descriptor and 192 dimension MBH descriptor.

Some papers [7,27] also discuss the seamlessly combination over different features. Here we consider the dimensionality balance issue, caused by obvious dimension difference between HOG and MBH. We evaluate the usage of PCA to balance the dimension of different descriptors in order to even the impact of each descriptors. However, the size of million feature points makes PCA not feasible. Picking some dimensions randomly to equalize two descriptors is also tested, which works well in some cases, but the result is not stable and controllable. Therefore, we make a trade-off, and directly combine HOG and MBH to form the hybrid descriptors.

3.2   Video Encoding

VLAD [1] was firstly proposed in 2010 on the application of massive image searching. Unlike traditional BoW encoding, which requires a large size of codebook to achieve a good encoding effect, VLAD can achieve a better result even with a smaller codebook. Besides, VLAD representations are more discriminative than other encoding methods such as local linear constraint (LLC) [24], sparse coding based methods [28], etc.

The idea of VLAD is very simple. A codebook $D = \{\mu_1, \mu_2, ..., \mu_K\}$ of size K is learned using clustering methods (here we adopt k-means clustering). Then for each video clips, the differences between feature vectors and their belonging visual words are aggregated to form the signatures $\{v_1, v_2, ..., v_K\}$ of all visual words. The signature v_i is initialized with zero, and then being accumulated as Eq. 1 does:

$$v_i = \sum_{x_t : \text{NN}(x_t) = i} x_t - \mu_i \qquad (1)$$

where, $\text{NN}(x_t)$ is a function indicating the index of visual words in the codebook D, which should be the nearest neighbour to x_t. The VLAD representation is then further normalized with power-low normalization [25] followed by L2-normalization.

3.3   Other Implementation Details

We adopt regular dense trajectory sampling of space-time features used in [6] to detect the STIPs inside each video clips. Wang et al. [3] has proved that dense trajectory sampling outperforms other commonly used feature detectors such as Harris3D [29], Cuboid [19] and Hessian [18] detectors in realistic settings. Meanwhile, dense sampling also maintains a simplicity to scale up the sampling density with a pre-computed dense optic flow fields.

Algorithm 1. Our Algorithm: Hybrid Feature Descriptor with VLAD Encoding

Input:

$TrainVideo : \{a_1, a_2, ..., a_M\}$ is the set of training videos with size M;

$TrainVideoLabel : \{l_1, l_2, ..., l_M\}$ is the set of action label of each training videos;

$TestVideo : \{a_{M+1}, a_{M+2}, ..., a_{M+N}\}$ is the set of testing videos with size N;

Output:

$TestVideoLabel : \{l_{M+1}, l_{M+2}, ..., l_{M+N}\}$ is the set of action label of each testing videos.

1: $X := \{X_1, X_2, ..., X_{M+N}\}$
2: **for** $d := 1$ to $M + N$ **do**
3:     $X_d := \{\}$
4:     $P := \{p_1, p_2, ..., p_S\}$ is the set of STIPs of video a_d using dense sampling detector
5:     **for** $s := 1$ to $size(P)$ **do**
6:         $hogVector := HOG(ps)$
7:         $mbhVector := MBH(ps)$
8:         $hybridVector := [hogVector, mbhVector]$
9:         $X_d := \{X_d, hybridVector\}$
10:    **end for**
11: **end for**
12: $D : \{\mu_1, \mu_2, ..., \mu_K\} := kMeans(X, K)$
13: $V := \{V_1, V_2, ..., V_{M+N}\}$
14: **for** $d := 1$ to $M + N$ **do**
15:    $Y := X_d$
16:    **for** $k := 1$ to K **do**
17:        $v_k := 0_d$
18:    **end for**
19:    **for** $t := 1$ to $size(Y)$ **do**
20:        $i := \arg\min_j Dist(Y_t, \mu_j)$
21:        $v_i := v_i + Y_t - \mu_i$
22:    **end for**
23:    $V_d := [v_1^T, v_2^T, ..., v_K^T]$
24:    **for** $k := 1$ to K **do**
25:        $v_k := sign(v_k) |v_k|^\alpha$
26:    **end for**
27:    $V_d := V_d / \|V_d\|_2$
28: **end for**
29: $SVMClassifier := InitializeSVM(\{< V_1, l_1 >, < V_2, l_2 >, ..., < V_M, l_M >\})$
30: **for** $d := M + 1$ to $M + N$ **do**
31:    $l_d = SVMClassifier(V_d)$
32: **end for**
33: **return** $\{l_{M+1}, l_{M+2}, ..., l_{M+N}\}$;

For each sampling point, a list of predefined discrete spatial scale parameters have been covered to maintain the scale-invariant virtue. The trajectory
neighbourhood is divided into a spatial-temporal grid of size $n_\sigma \times n_\sigma \times n_\tau$, then
being described into a vector using our hybrid feature descriptor as well as other
describing methods for comparison.

Once we derive all the feature vectors of each video clips in the dataset, we
use k-means clustering over the whole feature vectors to quantize the standard of
video representation. The centroids produced by k-means clustering is regarded
as the visual words, which is further used in VLAD to form the encoding vector.
Based on the encoding vector representing each video clips, action classification
is finally performed with a one-vs-rest linear SVM classifier [30].

A whole pipeline of our algorithm is illustrated in Algorithm 1.

4 Experiments

In this section, we present a detailed analysis of our action recognition result
based on the hybrid feature descriptor as well as VLAD encoding method on
several datasets. We evaluate the performance among different descriptors to
justify our choices. We also make a comparison between our results and the
previously published works.

We choose three publicly available standard action datasets to report our
recognition result, which are: YouTube [7], KTH [9] and Weizmann [8] datasets.
Figure 2 shows some sample frames from these datasets. For each action classes,
mean of average precision is calculated as performance measure. The experimental results show that our action recognition framework is competitive and can
achieve a state of the art result.

4.1 Datasets

YouTube: The YouTube dataset [7] contains 1168 video clips from 11 action
types, which are: basketball shooting (*B_Shooting*), biking, diving, golf swinging
(*G_Swinging*), horse back riding (*H_Riding*), soccer juggling (*S_Juggling*), swinging, tennis swinging (*T_Swinging*), trampoline jumping (*T_Jumping*), volleyball
spiking (*V_Spiking*), and walking with a dog (*Walking*). This is one of the challenging datasets due to its wild camera vibrations, cluttered background, viewpoint transitions and complicated recording conditions. Videos for each action
type are wrapped into 25 groups, and each group contains four or more video
clips sharing common features like same actor, similar background and viewpoint. We evaluate the classification accuracy by leave one out cross validation
over the predefined 25 groups.

KTH: The KTH dataset [9] contains 600 video clips from six action types,
which are: walking, jogging, running, boxing, hand waving and hand clapping.
Each action type is performed by 25 persons in four different scenarios: outdoors,

Fig. 2. Some samples from video sequences on *YouTube* (the first row), *KTH* (the second row) and *Weizmann* (the third row) datasets. Among all, YouTube dataset contains large variation, KTH dataset has homogeneous indoor and outdoor backgrounds, and Weizmann dataset records all videos on a static camera.

outdoors with scale variation, outdoors with different clothes and indoors. All sequences are taken over homogeneous backgrounds shot by a static camera, and have an average length of four seconds. We follow the split setup in [9], and choose 16 persons for training, and remaining 9 persons for testing.

Weizmann: The Weizmann dataset [8] contains 93 video clips from 10 action types, which are: bending, jumping jack (*J_Jump*), jumping forward (*F_Jump*), jumping in place (*P_Jump*), jumping sideways (*S_Jump*), skipping, running, walking, waving with two hands (*Wave2*), and waving with one hand (*Wave1*). Each action type is performed by 9 different persons. All the video clips are recorded in homogeneous outdoor background with a static camera, and have an average length of two seconds. We evaluate the classification accuracy by leave one out cross validation over 9 different persons repeatedly, and take the mean of average precision as the final correction rate.

4.2   Experiments on YouTube

We firstly decide the size of visual vocabulary generated by k-means clustering. A list of exponential increasing vocabulary sizes are evaluated over several types of descriptors, including optic Trajectories [6] (motional), HOG (static),

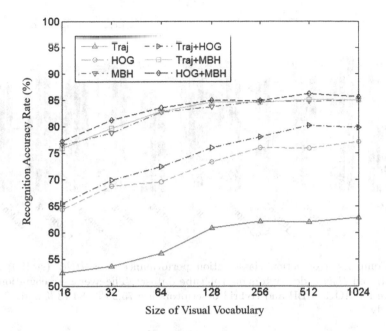

Fig. 3. Comparison among different sizes of visual vocabulary over several descriptors on YouTube dataset. Smaller vocabulary size leads to lower recognition accuracy rate, while larger vocabulary size brings more computational burden. Here we choose 512 visual words, which keeps a balance between recognition accuracy and computational efficiency.

MBH (motional) and their pairwise combinations. The result, as Fig. 3 illustrates, indicates that smaller vocabulary size generally leads to lower accuracy rate. However, picking a vocabulary size too large brings more computational burden, and since we choose dense sampling to find STIPs, the situation is even aggravated because the nearest visual word of all feature points should be required in VLAD. From Fig. 3 we observe that accuracy changing between 512 and 1024 is very limited, which gives us a chance to choose a size of 512 visual words over all descriptors so as to make a balance between recognition accuracy and computational efficiency.

A comparison of action classification effect among HOG, MBH and hybrid descriptors is performed to evaluate the improvement of hybrid descriptor over separated individual descriptors, and the result is shown in Fig. 4. Among these descriptors, the hybrid descriptor achieves a 86.23 % recognition accuracy rate, which is about 1.50 % improvement over MBH, and 10.28 % over HOG. Of all eleven actions in YouTube dataset, seven action classes gain improvement of recognition accuracy using hybrid descriptor. We also compare our recognition result to several previously published works [6,7,31–35] in Table 1, and the comparison shows our method obtains a state of the art result.

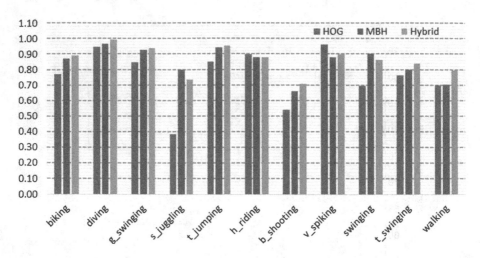

Fig. 4. Comparison of action classification performance over *HOG* (static), *MBH* (motional) and *Hybrid* descriptors on YouTube dataset. The average recognition precision rate for HOG, MBH and Hybrid descriptors are 75.95 %, 84.73 % and 86.23 %, respectively.

Table 1. Performance comparison between our method and some previously published works on YouTube dataset.

	Proposed	Liu *et al.* [7]	Zhang *et al.* [31]	Reddy *et al.* [32]
mAP	**86.2 %**	71.2 %	72.9 %	73.2 %
	Ikizler *et al.* [33]	Le *et al.* [34]	Brendel *et al.* [35]	Wang *et al.* [6]
mAP	75.2 %	75.8 %	77.8 %	84.2 %

4.3   Experiments on KTH and Weizmann

The strategy of combining static and motional descriptors is further evaluated on KTH dataset. Figure 5 shows the comparison between hybrid feature descriptors and the separated individual descriptors. We choose four different types of descriptors in [6] to make the combinations, which are HOG (static), TRAJ (motional), HOF (motional) and MBH (motional). Among all the combinations, the ones generated from static and motional descriptors (see row 1 in Fig. 5) gain considerable improvement, while we can hardly find this improvement within the ones generated from two motional descriptors (see row 2 in Fig. 5).

We perform action recognition on KTH and Weizmann respectively using hybrid feature descriptor and VLAD encoding method, and achieves a recognition accuracy rate of 95.4 % on KTH and 97.8 % on Weizmann. Figure 6 shows the confusion matrix of our recognition result in these two datasets. From Fig. 6 we can see that errors in KTH are mostly caused by mislabelling running to jogging, and errors in Weizmann are caused by mislabelling jumping forward and

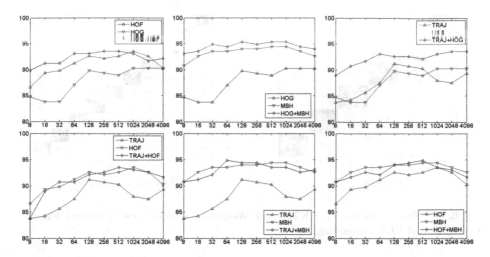

Fig. 5. Comparison between *hybrid feature descriptors* (blue line) and *the separated individual descriptors* (red lines). The *x*- and *y*-axis are the size of visual vocabulary and the recognition accuracy rate, respectively. Row 1 contains 3 different combinations of static and motional descriptors, and row 2 contains 3 combinations of two motional descriptors. We can observe that in row 1, hybrid descriptors gain considerable improvement, while in row 2, this improvement can hardly be found (Color figure online).

jumping in place to skipping. If the "skip" action class is expelled from Weizmann dataset, the recognition accuracy rate can be further raised to 100.0 %.

Table 2 shows a comparison between our proposal and several previous published works [21,36–41]. From Table 2 we can see that our proposed method achieves a state of the art result on KTH dataset, and maintains a competitive recognition result on Weizmann dataset.

Table 2. Performance comparison between our method and some previously published works on KTH and Weizmann dataset.

	KTH	Weizmann		KTH	Weizmann
Proposed	**95.4 %**	**97.8 %**	Cao et al. [36]	93.5 %	94.6 %
Grundmann et al. [37]	93.5 %	94.6 %	Fathi et al. [38]	90.5 %	100.0 %
Lin et al. [21]	93.4 %	100.0 %	Schindler et al. [39]	92.7 %	100.0 %
Cai et al. [40]	94.2 %	98.2 %	Liu et al. [41]	94.8 %	100.0 %

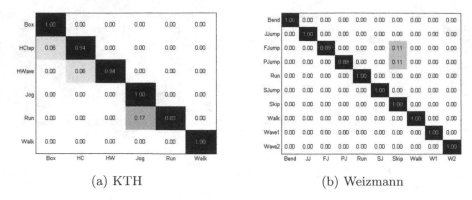

(a) KTH   (b) Weizmann

Fig. 6. Confusion matrices for KTH and Weizmann datasets using hybrid feature descriptor and VLAD encoding.

5   Conclusion

This paper presents a hybrid descriptor to describe local features extracted from video clips, which takes the advantage of both static and motional information to cover more details inside neighbourhood of spatial-temporal interest points. Besides, VLAD encoding is adopted for each video clips to encapsulate more structural information on the distribution of feature vectors. We evaluate the effect of our action recognition framework over several datasets, and the experimental result shows the usage of hybrid feature descriptor as well as VLAD encoding can significantly improve the average recognition accuracy.

Acknowledgement. This work is supported by NSFC (No.61272247 and 60873133), the Science and Technology Commission of Shanghai Municipality (Grant No.13511500200), 863 (No.2008AA02Z310) in China and the European Union Seventh Frame work Programme (Grant No.247619).

References

1. Jégou, H., Douze, M., Schmid, C., Pérez, P.: Aggregating local descriptors into a compact image representation. In: 2010 IEEE Conference on Computer Vision and Pattern Recognition (CVPR), pp. 3304–3311. IEEE (2010)
2. Arandjelovic, R., Zisserman, A.: All about vlad. In: 2013 IEEE Conference on Computer Vision and Pattern Recognition (CVPR), pp. 1578–1585. IEEE (2013)
3. Wang, H., Ullah, M.M., Klaser, A., Laptev, I., Schmid, C., et al.: Evaluation of local spatio-temporal features for action recognition. In: BMVC 2009-British Machine Vision Conference (2009)
4. Dalal, N., Triggs, B.: Histograms of oriented gradients for human detection. In: IEEE Computer Society Conference on Computer Vision and Pattern Recognition, CVPR 2005, vol. 1, pp. 886–893. IEEE (2005)
5. Dalal, N., Triggs, B., Schmid, C.: Human detection using oriented histograms of flow and appearance. In: Leonardis, A., Bischof, H., Pinz, A. (eds.) ECCV 2006. LNCS, vol. 3952, pp. 428–441. Springer, Heidelberg (2006)

6. Wang, H., Klaser, A., Schmid, C., Liu, C.L.: Action recognition by dense trajectories. In: 2011 IEEE Conference on Computer Vision and Pattern Recognition (CVPR), pp. 3169–3176. IEEE (2011)
7. Liu, J., Luo, J., Shah, M.: Recognizing realistic actions from videos in the wild. In: IEEE Conference on Computer Vision and Pattern Recognition, CVPR 2009, pp. 1996–2003. IEEE (2009)
8. Gorelick, L., Blank, M., Shechtman, E., Irani, M., Basri, R.: Actions as space-time shapes. IEEE Trans. Pattern Anal. Mach. Intell. **29**, 2247–2253 (2007)
9. Schuldt, C., Laptev, I., Caputo, B.: Recognizing human actions: a local svm approach. In: Proceedings of the 17th International Conference on Pattern Recognition, ICPR 2004, vol. 3, pp. 32–36. IEEE (2004)
10. Poppe, R.: A survey on vision-based human action recognition. Image Vis. Comput. **28**, 976–990 (2010)
11. Bobick, A.F., Davis, J.W.: The recognition of human movement using temporal templates. IEEE Trans. Pattern Anal. Mach. Intell. **23**, 257–267 (2001)
12. Carlsson, S., Sullivan, J.: Action recognition by shape matching to key frames. In: Workshop on Models versus Exemplars in Computer Vision, vol. 1, p. 18 (2001)
13. Efros, A.A., Berg, A.C., Mori, G., Malik, J.: Recognizing action at a distance. In: Ninth IEEE International Conference on Computer Vision, Proceedings, pp. 726–733. IEEE (2003)
14. Laptev, I., Marszalek, M., Schmid, C., Rozenfeld, B.: Learning realistic human actions from movies. In: IEEE Conference on Computer Vision and Pattern Recognition, CVPR 2008, pp. 1–8. IEEE (2008)
15. Lowe, D.G.: Object recognition from local scale-invariant features. In: The Proceedings of the Seventh IEEE International Conference on Computer vision, vol. 2, pp. 1150–1157. IEEE (1999)
16. Klaser, A., Marszalek, M.: A spatio-temporal descriptor based on 3d-gradients (2008)
17. Scovanner, P., Ali, S., Shah, M.: A 3-dimensional sift descriptor and its application to action recognition. In: Proceedings of the 15th International Conference on Multimedia, pp. 357–360. ACM (2007)
18. Willems, G., Tuytelaars, T., Van Gool, L.: An efficient dense and scale-invariant spatio-temporal interest point detector. In: Forsyth, D., Torr, P., Zisserman, A. (eds.) ECCV 2008, Part II. LNCS, vol. 5303, pp. 650–663. Springer, Heidelberg (2008)
19. Dollár, P., Rabaud, V., Cottrell, G., Belongie, S.: Behavior recognition via sparse spatio-temporal features. In: 2nd Joint IEEE International Workshop on Visual Surveillance and Performance Evaluation of Tracking and Surveillance, pp. 65–72. IEEE (2005)
20. Laptev, I., Pérez, P.: Retrieving actions in movies. In: IEEE 11th International Conference on Computer Vision, ICCV 2007, pp. 1–8. IEEE (2007)
21. Lin, Z., Jiang, Z., Davis, L.S.: Recognizing actions by shape-motion prototype trees. In: 2009 IEEE 12th International Conference on Computer Vision, pp. 444–451. IEEE (2009)
22. Liu, J., Ali, S., Shah, M.: Recognizing human actions using multiple features. In: IEEE Conference on Computer Vision and Pattern Recognition, CVPR 2008, pp. 1–8. IEEE (2008)
23. Liu, J., Shah, M.: Learning human actions via information maximization. In: IEEE Conference on Computer Vision and Pattern Recognition, CVPR 2008, pp. 1–8. IEEE (2008)

24. Wang, J., Yang, J., Yu, K., Lv, F., Huang, T., Gong, Y.: Locality-constrained linear coding for image classification. In: 2010 IEEE Conference on Computer Vision and Pattern Recognition (CVPR), pp. 3360–3367. IEEE (2010)
25. Perronnin, F., Sánchez, J., Mensink, T.: Improving the fisher kernel for large-scale image classification. In: Daniilidis, K., Maragos, P., Paragios, N. (eds.) ECCV 2010, Part IV. LNCS, vol. 6314, pp. 143–156. Springer, Heidelberg (2010)
26. Zhou, X., Yu, K., Zhang, T., Huang, T.S.: Image classification using super-vector coding of local image descriptors. In: Daniilidis, K., Maragos, P., Paragios, N. (eds.) ECCV 2010, Part V. LNCS, vol. 6315, pp. 141–154. Springer, Heidelberg (2010)
27. Kovashka, A., Grauman, K.: Learning a hierarchy of discriminative space-time neighborhood features for human action recognition. In: 2010 IEEE Conference on Computer Vision and Pattern Recognition (CVPR), pp. 2046–2053. IEEE (2010)
28. Yang, J., Yu, K., Gong, Y., Huang, T.: Linear spatial pyramid matching using sparse coding for image classification. In: IEEE Conference on Computer Vision and Pattern Recognition, CVPR 2009, pp. 1794–1801. IEEE (2009)
29. Laptev, I.: On space-time interest points. Int. J. Comput. Vis. **64**, 107–123 (2005)
30. Chang, C.C., Lin, C.J.: Libsvm: a library for support vector machines. ACM Trans. Intell. Syst. Technol. (TIST) **2**, 27 (2011)
31. Zhang, Y., Liu, X., Chang, M.-C., Ge, W., Chen, T.: Spatio-temporal phrases for activity recognition. In: Fitzgibbon, A., Lazebnik, S., Perona, P., Sato, Y., Schmid, C. (eds.) ECCV 2012, Part III. LNCS, vol. 7574, pp. 707–721. Springer, Heidelberg (2012)
32. Reddy, K.K., Shah, M.: Recognizing 50 human action categories of web videos. Mach. Vis. Appl. **24**, 971–981 (2013)
33. Ikizler-Cinbis, N., Sclaroff, S.: Object, scene and actions: combining multiple features for human action recognition. In: Daniilidis, K., Maragos, P., Paragios, N. (eds.) ECCV 2010, Part I. LNCS, vol. 6311, pp. 494–507. Springer, Heidelberg (2010)
34. Le, Q.V., Zou, W.Y., Yeung, S.Y., Ng, A.Y.: Learning hierarchical invariant spatio-temporal features for action recognition with independent subspace analysis. In: 2011 IEEE Conference on Computer Vision and Pattern Recognition (CVPR), pp. 3361–3368. IEEE (2011)
35. Brendel, W., Todorovic, S.: Activities as time series of human postures. In: Daniilidis, K., Maragos, P., Paragios, N. (eds.) ECCV 2010, Part II. LNCS, vol. 6312, pp. 721–734. Springer, Heidelberg (2010)
36. Cao, X., Zhang, H., Deng, C., Liu, Q., Liu, H.: Action recognition using 3d daisy descriptor. Mach. Vis. Appl. **25**, 159–171 (2014)
37. Grundmann, M., Meier, F., Essa, I.: 3d shape context and distance transform for action recognition. In: 19th International Conference on Pattern Recognition, ICPR 2008, pp. 1–4. IEEE (2008)
38. Fathi, A., Mori, G.: Action recognition by learning mid-level motion features. In: IEEE Conference on Computer Vision and Pattern Recognition, CVPR 2008, pp. 1–8. IEEE (2008)
39. Schindler, K., Van Gool, L.: Action snippets: How many frames does human action recognition require? In: IEEE Conference on Computer Vision and Pattern Recognition, CVPR 2008, pp. 1–8. IEEE (2008)
40. Cai, Q., Yin, Y., Man, H.: Learning spatio-temporal dependencies for action recognition, ICIP (2013)
41. Liu, L., Shao, L., Zhen, X., Li, X.: Learning discriminative key poses for action recognition (2013)

Human Action Recognition Based on Oriented Motion Salient Regions

Baoxin Wu[1], Shuang Yang[1], Chunfeng Yuan[1], Weiming Hu[1（✉）],
and Fangshi Wang[2]

[1] NLPR, Institute of Automation, Chinese Academy of Sciences, Beijing, China
{bxwu,syang,cfyuan,wmhu}@nlpr.ia.ac.cn
[2] School of Software, Beijing Jiaotong University, Beijing, China
fshwang@bjtu.edu.cn

Abstract. Motion is the most informative cue for human action recognition. Regions with high motion saliency indicate where actions occur and contain visual information that is most relevant to actions. In this paper, we propose a novel approach for human action recognition based on oriented motion salient regions (OMSRs). Firstly, we apply a bank of 3D Gabor filters and an opponent inhibition operator to detect OMSRs of videos, each of which corresponds to a specific motion direction. Then, a new low-level feature, named as oriented motion salient descriptor (OMSD), is proposed to describe the obtained OMSRs through the statistics of the texture in the regions. Next, we utilize the obtained OMSDs to explore the oriented characteristics of action classes and generate a set of class-specific oriented attributes (CSOAs) for each class. These CSOAs provide a compact and discriminative middle-level representation for human actions. Finally, an SVM classifier is utilized for human action classification and a new compatibility function is devised for measuring how well a given action matches to the CSOAs of a certain class. We test the proposed approach on four public datasets and the experimental results validate the effectiveness of our approach.

1 Introduction

Traditional approaches for human action recognition are based on either local or global features. The former [2,3,5] is usually extracted from a sparse set of local salient regions and tends to lose useful global information about the action. Contrastively, the latter [6,7] treats the video as a whole. It contains all the information of the sequence but is sensitive to occlusion and background variation. Regions with high motion saliency are of great significance because they indicate where actions occur in videos and contain the most relevant information about the actions. In this paper, we propose a novel approach for human action recognition based on the oriented motion salient regions (OMSRs).

Much effort has been devoted to estimate motion using successive frames, but detecting motion in specific directions is still a challenge. In this paper, we apply a bank of 3D Gabor filters with multiple directions and an opponent inhibition

© Springer International Publishing Switzerland 2015
C.V. Jawahar and S. Shan (Eds.): ACCV 2014 Workshops, Part I, LNCS 9008, pp. 113–128, 2015.
DOI: 10.1007/978-3-319-16628-5_9

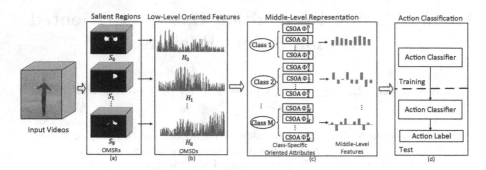

Fig. 1. The flowchart of our proposed approach. (a) The detected OMSRs. (b) The OMSDs extracted from OMSRs. (c) Learned CSOAs and the middle-level attribute vectors. (d) Action classification based on CSOAs.

operator to detect the motion salient region in a video sequence. Specifically, the detected region is decomposed into a set of OMSRs, each of which corresponds to a specific motion direction.

We extract a new low-level descriptor from each OMSR, which is named as oriented motion salient descriptor (OMSD). The OMSDs are obtained by the statistics of the texture information in OMSRs and when computing them, a two-layer polar coordinate system is used to encode the spatial distribution information of the OMSRs. Each OMSD is actually a semi-holistic feature for human action representation because it contains all the information about an action in a specific direction.

Taking advantage of OMSDs, a series of class-specific oriented attributes (CSOAs) are learnt for each action class. The CSOAs reflect the characteristics of action classes in different motion directions. Mapping an action represented by a set of low-level OMSDs into the space of CSOAs related to an class generates a middle-level attribute feature, which has high discriminative power for human action recognition. Finally, an SVM formulation is utilized for the action classification problem, where action classes are characterized by a set of class-specific attributes.

The flowchart of our proposed approach is shown in Fig. 1. The main contributions of this paper are as follows:

- 3D Gabor filters, incorporating an opponent inhibition operator, are used to detect the oriented motion salient regions which contain the most relevant information about actions with respect to different motion directions.
- A new low-level descriptor OMSD, encoding both appearance and motion information of OMSRs, is proposed to represent the extracted regions.
- A set of CSOAs are learned for recognizing human actions. The CSOAs characterize action classes in different directions and provide an middle-level representation for human actions.

2   Related Work

Features used for human action representation can be roughly categorized into two groups: local features and holistic features. The local feature based approaches represent an action as a sparse set of extracted local spatiotemporal features. Dollár et al. [2] experiment with both pixel gradients and optical flows to describe the extracted spacetime cuboids. Laptev et al. [3] apply histograms of gradients (HoG) and histograms of optical flows (HoF) as local descriptors, capturing both motion and structure information of local regions of interest points. These features are extracted from the local regions around the interest points and the information in other regions is usually ignored. Compared with these local features, the proposed OMSDs focus on the whole motion salient regions with respect to a specific direction, capturing more visual information which is useful for human action recognition.

The global feature based approaches represent an action by treating the video sequence as a whole. Bobick and Davids [6] introduce motion energy image (MEI) and motion history image (MHI) to describe the spatial distribution of motion energy in a given sequence. Wang et al. [7] construct average motion energy (AME) and mean motion shape (MMS) based on the human body silhouettes and shapes respectively, to characterize an action. Ikizler et al. [8] describe a pose as a histogram of oriented rectangles and then sum up the histograms from all frames to form a compact representation for the whole video sequence. These holistic features capture sufficient visual information. However, they are highly sensitive to shift and background variations. Our OMSDs, based on the oriented motion salient regions, are more robust to these variations.

In addition, inspired by the formulation on object recognition [11,13], a number of researchers show great interest in the attribute based representation for human actions. Yao et al. [14] use attributes and parts for human action recognition in still images. Liu et al. [12] combine manually predefined attributes with data-driven ones obtained by clustering local features, and use a latent SVM to learn the importance of each part. There are two main differences between our CSOAs and these sematic attributes. First, our CSOAs are all learnt automatically without any manual annotation. Second, our CSOAs are class-specific attributes and they are more discriminative for classifying actions from different action classes.

Over the years, another group of methods perform human action recognition by aggregating the responses of 3D directional filters and these methods is quite related to our approach. Derpanis et al. [17,22] propose to use 3D Gaussian third derivative filters for human action recognition. In their work a marginalization process is used to discount spatial orientation component. In our work, a more simple 3D Gabor filter and a opponent inhibition operator are designed for OMSR detection. On the other hand, we do not aggregate the motion energies of different directions simply, but give a mid-level representation of actions through exploiting the obtained CSOAs.

3   Oriented Motion Salient Regions

In this section, we introduce how to detect OMSRs in videos. Adelson and Bergen [10] have demonstrated that motion can be perceived as orientation in space-time and spatiotemporally oriented filters can be used to detect it. Inspired by this, we apply a bank of 3D Gabor filters [9] with multiple directions and an opponent inhibition operator for motion analysis in videos.

A 3D Gabor filter is formed as a product of a Gaussian window and complex sinusoid. It consists of two parts: the real part and the imaginary part. These two parts are defined as

$$g_r^{3d}(x,y,t) = \hat{g}(x,y,t)\cos[\frac{2\pi}{\lambda}(\eta_x x + \eta_y y + \eta_t t)], \tag{1}$$

and

$$g_i^{3d}(x,y,t) = \hat{g}(x,y,t)\sin[\frac{2\pi}{\lambda}(\eta_x x + \eta_y y + \eta_t t)], \tag{2}$$

respectively, where

$$\hat{g}(x,y,t) = \exp[-(\frac{x^2 + y^2 + t^2}{2\sigma^2})], \tag{3}$$

σ controls the scale of the Gaussian, λ is the wavelength of the sinusoidal factor, and (η_x, η_y, η_t), which satisfies $\eta_x^2 + \eta_y^2 + \eta_t^2 = 1$, determines the direction of the filter. The response of a 3D Gabor filter on a video sequence I is expressed as

$$R = (I * g_r^{3d})^2 + (I * g_i^{3d})^2, \tag{4}$$

where $*$ denotes the convolution operator. Through squaring and summing the outputs of two part filters which are 90 degrees out of phase, the 3D Gabor filter gives a phase-independent measurement of local motion strength.

To capture motions towards multiple directions, we design a filter bank which contains nine 3D Gabor filters with different directions. Let $\{g_k^{3d}\}_{k=0}^8$ denote the filters in the bank. These filters are sensitive to motions with any of directions: flicker, up, down, left, right and four diagonals. Table 1 shows the filters and their corresponding motions. By convoluting a video sequence with these 3D Gabor filters, we obtain the responses $\{R_k\}_{k=0}^8$, which are actually a series of oriented motion energy measurements on this sequence.

Each 3D Gabor filter responds to motion with a specific direction independently. However, the motion detection should be inherently opponent [10]. That is motions with two opposite directions cannot occur at the same place and time within the same frequency band. Accordingly, we apply an opponent inhibition operator on the original responses $\{R_k\}_{k=0}^8$, which will decrease the influence of opposite motion. The opponent inhibition operator is defined as the half-wave-rectified difference between the oriented motion energies corresponding to opposite motion directions

$$\begin{aligned} \bar{R}_0(\boldsymbol{x}) &= R_0(\boldsymbol{x}), \\ \bar{R}_i(\boldsymbol{x}) &= |R_i(\boldsymbol{x}) - aR_{i+4}(\boldsymbol{x})|^+, \quad 1 \le i \le 4, \\ \bar{R}_j(\boldsymbol{x}) &= |R_j(\boldsymbol{x}) - aR_{j-4}(\boldsymbol{x})|^+, \quad 5 \le j \le 8, \end{aligned} \tag{5}$$

Table 1. The nine 3D Gabor filters with different directions and their corresponding motions. For example, the 3D Gabor filter g_1^{3d} with (η_w, η_y, η_z) $(\frac{\sqrt{2}}{2}, 0, \frac{\sqrt{2}}{2})$ is sensitive to motion towards left.

Motion	Flicker	Left	Left-Up
Filter	g_0^{3d}:(0,0,1)	g_1^{3d}:($\frac{\sqrt{2}}{2}, 0, \frac{\sqrt{2}}{2}$)	g_2^{3d}:($\frac{1}{2}, -\frac{1}{2}, \frac{\sqrt{2}}{2}$)
Motion	Up	Right-Up	Right
Filter	g_3^{3d}:($0, -\frac{\sqrt{2}}{2}, \frac{\sqrt{2}}{2}$)	g_4^{3d}:($-\frac{1}{2}, -\frac{1}{2}, \frac{\sqrt{2}}{2}$)	g_5^{3d}:($-\frac{\sqrt{2}}{2}, 0, \frac{\sqrt{2}}{2}$)
Motion	Right-Down	Down	Left-Down
Filter	g_6^{3d}:($-\frac{1}{2}, \frac{1}{2}, \frac{\sqrt{2}}{2}$)	g_7^{3d}:($0, \frac{\sqrt{2}}{2}, \frac{\sqrt{2}}{2}$)	g_8^{3d}:($\frac{1}{2}, \frac{1}{2}, \frac{\sqrt{2}}{2}$)

where $\boldsymbol{x} = (x, y, t)$ is a 3D position in the space-time, a controls the weight of opposite motion and is set to 1 here, and $|\cdot|^+$ is defined as $|z|^+ = \max(0, z)$.

The motion salience E is measured by the summation of all the oriented motion energies

$$E(\boldsymbol{x}) = \sum_{k=0}^{8} \bar{R}_k(\boldsymbol{x}). \tag{6}$$

A threshold ϵ_s is used to detect region with high motion saliency and generate a binary motion salient region (MSR)

$$S(\boldsymbol{x}) = \begin{cases} 1 & \text{if } E(\boldsymbol{x}) > \epsilon_s, \\ 0 & \text{otherwise.} \end{cases} \tag{7}$$

The definition of MSR involves all the oriented motion energies. In order to emphasize motion along a specific direction, we define the oriented motion salient region (OMSR) as

$$S_k(\boldsymbol{x}) = \begin{cases} 1 & \text{if } S(\boldsymbol{x}) = 1 \text{ and } \bar{R}_k(\boldsymbol{x}) > \epsilon_k, \\ 0 & \text{otherwise.} \end{cases} \tag{8}$$

where ϵ_k is a threshold and $0 \le k \le 8$. The OMSRs $\{S_k\}_{k=0}^{8}$ decompose S, but there may be overlaps between different OMSRs.

4   Oriented Motion Salient Descriptors

Having detected a set of OMSRs for a video sequence, we construct a low-level OMSD to describe each OMSR. To compute an OMSD, we first extract the texture information of each pixel in the OMSR. We then create a texture polar histogram to describe the salient region in each frame and combine all the polar histograms from the video sequence together to form the final OMSD.

The texture information of the pixels in the OMSR is captured by a bank of 2D Gabor filters. A 2D Gabor filter consists of the real part and imaginary part, which are defined as

$$g_r^{2d}(x,y) = \exp(-\frac{x'^2 + \gamma y'2}{2\sigma^2})\cos(\frac{2\pi}{\lambda}x') \tag{9}$$

and

$$g_j^{2d}(x,y) = \exp(-\frac{x'^2 + \gamma y'2}{2\sigma^2})\sin(\frac{2\pi}{\lambda}x') \tag{10}$$

respectively, where $x' = x\cos\theta + y\sin\theta$, and $y' = -x\sin\theta + y\cos\theta$, θ controls the direction of the 2D Gabor filter, and γ is spatial respect ratio. An overall response of a 2D Gabor filter is obtained by squaring and summing the outputs of the two part filters. There are 5 scales: $\sigma \in \{5, 7, 9, 11, 13\}$ and 6 directions: $\theta \in \{0°, 30°, 60°, 90°, 120°, 150°\}$ in the bank. There are in total 30 filters in this bank. Through convoluting the video sequence with all the filters, a 30 dimensional response vector is obtained for each pixel.

For the tth frame, a polar coordinate system is applied to model the spatial distribution of the salient regions in this frame. The origin of the polar coordinate system is set as the geometric center l^t of the salient region of MSR S in the tth frame. We divide the polar coordinate system into N_1 cells. For each cell, we build a histogram of the 2D Gabor filter responses at different orientations and scales. The histogram is computed as a summation of response vectors of pixels in the region belonging to this cell. The final polar histogram for the whole salience region in this frame is computed as a concatenation of the histograms from all the cells. There are total $30*N_1$ bins in the polar histogram. So, for this frame, we obtain (h_k^t, l^t), where h_k^t denotes obtained polar histogram, and $l^t = (x_t, y_t)$ is the geometric center. Figure 2 shows the process of the construction of the texture polar histogram in a frame.

The global representation for OMSR S_k, is computed as a summation of all the h_k^t from the sequence, taking into acount the spatial distribution of the l^t

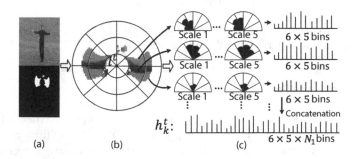

Fig. 2. (a) The tth frame of a video sequence and its one OMSD S_k in this frame. (b) A polar coordinate system. (c) The construction of the polar histogram in this frame.

in each frame. Similarly, we apply another polar coordinate system to describe the relative positions of all the geometric centers $\{l^1, l^2, ...\}$ in all the frames. The origin of the polar coordinate is set as the mean of $\{l^1, l^2, ...\}$. The polar coordinate is used to divided the plane into N_2 cells. We sum up the h_k^t through the sequence whose center point l^t is in the ith cell to generate a global vector

$$H_{k\_i} = \sum_t h_k^t \, \delta(i, cell(l^t)) \tag{11}$$

where $\delta(\cdot, \cdot)$ is a Dirac kernel and $cell(l^t)$ returns the index of the cell where l^t is located. The OMSD is expressed as the concatenation of H_{k_i} from all the cells

$$H_k = [H_{k\_1}^T, H_{k\_2}^T, ..., H_{k\_N_2}^T]^T,$$

with a dimension of $30*N_1*N_2$. Figure 3 gives a illustration of how to combine all the polar histograms from the video sequence into the OMSD.

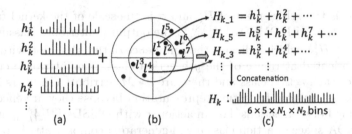

Fig. 3. The process of combining all the polar histograms $\{h_k^1, h_k^2, ...\}$ into OMSD H_k.

The obtained OMSD H_k can be viewed as a two-layered texture polar histogram. The texture information of pixels and spatial distribution information of the OMSR are both included in this descriptor. In fact OMSDs fuse both appearance and motion information in a way quite different from other descriptors, i.e., though encoding only texture information, each OMSD itself corresponds to a specific motion direction. The OMSDs $\{H_k\}_{k=0}^8$ give an effective and informative representation for an action, based on the main motion directions in the action.

5   Class-Specific Oriented Attributes

For an action, we have extracted a set of low-level OMSDs, which describe the action with respect to different motion directions. A traditional way to deal with these OMSDs is to concatenate them together to form a single complex descriptor for action representation. Then the human action recognition problem is transformed into a problem of learning a classification function to assign an action class label to the complex action descriptor. However, actions are so complex that the low-level action descriptors and the action labels cannot capture all the intrinsic characteristics of actions. We do not use these obtained OMSDs directly for action classification, but instead use them to obtain the class-specific oriented attributes (CSOAs) of each action class. A CSOA describe a specific action class in a specific motion direction, and it is only shared by the actions from this class. So the CSOAs can be used to infer action class labels.

A one-versus-all attribute classifier is trained for each CSOA. Given N training action examples from M action classes, we denote them as $\{(\{H_k^i\}_{k=0}^8, y_i)\}_{i=1}^N$, where $\{H_k^i\}_{k=0}^8$ are the OMSDs for the ith action and $y_i \in \{1, 2, ..., M\}$ is the action class label. Let Φ_y^k denote the attribute classifier for the yth action class on the kth motion direction. The positive examples for this classifier are a set of OMSD $\{H_k^i \mid y_i = y\}$ and the negative ones are $\{H_k^i \mid y_i \neq y\}$. The applied attribute classifier is a simple SVM classifier with a χ^2-kernel to measure the similarity between two OMSDs H_k^i and H_k^j

$$k(H_k^i, H_k^j) = \exp\{-\frac{1}{2\sigma^2} \sum_r \frac{[(H_k^i)_r - (H_k^j)_r]^2}{[(H_k^i)_r + (H_k^j)_r]}\}, \tag{12}$$

where $(H_k^i)_r$ is the rth element of H_k^i and σ is the scale of the kernel function. For the ith action with OMSDs $\{H_k^i\}_{k=0}^8$, the output of attribute classifier Φ_y^k is denoted as $\Phi_y^k(H_k^i)$, which indicates the confidence of the ith action with respect to the kth oriented attribute of the yth action class. Nine attribute classifiers are learnt for each action class and there are $9 * M$ attribute classifiers in total.

The CSOAs are used for action representation because they are effective for distinguishing between actions. For an action x with OMSDs $\{H_k\}_{k=0}^8$, mapping it to the CSOA space of action class y will generate a compact attribute feature vector with respect to class y. The elements of this vector are the outputs of oriented attribute classifiers related to this action class

$$\Psi(x, y) = [\Phi_y^0(H_0), \Phi_y^1(H_1), ..., \Phi_y^8(H_8)]. \tag{13}$$

The attribute feature vector $\Psi(x, y)$ is a middle-level representation for the action. It connects the low-level feature OMSDs $\{H_k\}_{k=0}^8$ and the action class label y together and indicates the confidence of action x belonging to class y with respects to multiple motion directions.

6   Action Classification Based on CSOAs

We apply the learned CSOAs for human action classification. CSOAs characterize each action class with respect to the different directions of the actions. If an action belongs to a certain action class, it should match the CSOAs related to this class well. In this way, the action classification problem is solved by finding the action class whose SCOAs match the given action best.

Given an action, we can obtain a class-specific attribute vector with the form of Eq. (13) by mapping its low-level OMSDs to the CSOA space of a specific action class. Intuitively, the class-specific attribute vector itself suggests whether an given action belongs to the specific action class, i.e., if all the entries of the vector have great values, it belongs to this class and vice versa. However, in practice only one or several oriented attributes are of great significance for actions. For example, in videos of "running left", the detected OMSRs are mainly focusing on the directions of left, left-up, left-down and flicker. It means that only the oriented attributes of the above mentioned directions have great discriminative power for recognizing the action "running left" while attributes of other directions have little relevance.

In such case, we define a compatibility function f to measure how well the given action x matches the CSOAs of action class y. We assume that f is a linear combination of the attribute feature vector

$$f(x, y) = \omega_y^T \Psi(x, y), \tag{14}$$

where ω_y is the parameter vector of f associated with class y. It emphasizes the importance of the oriented attributes of a specific class. The function f plays a role as a action classifier and the predicted label y^* for the action x is derived by maximizing f over all $y \in Y$

$$y^* = \arg\max_{y \in Y} f(x, y). \tag{15}$$

The compatibility function f can be learned in an SVM formulation. Given N training examples $\{(x_i, y_i)\}_{i=1}^N$ where x_i is an input video sequence and $y_i \in Y = \{1, 2, ..., M\}$ is the corresponding action label, the parameter vector ω_y is learned by solving a convex quadratic optimization problem which is expressed as

$$\min_{\omega_y} \frac{\gamma}{2} \parallel \omega_y \parallel^2 + \sum_{y_i=y} \xi_{1,i} + \sum_{y_j \neq y} \xi_{2,j} + \sum_{y_i=y, y_j \neq y} \xi_{3,ij} \tag{16}$$

$$\text{s.t.} \forall y_i = y, \quad \omega_y^T \Psi(x_i, y_i) \geq 1 - \xi_{1,i}, \ \xi_{1,i} \geq 0, \tag{17}$$

$$\forall y_j \neq y, \quad \omega_y^T \Psi(x_j, y_i) \leq -1 + \xi_{2,j}, \ \xi_{2,j} \geq 0, \tag{18}$$

$$\forall y_i = y, y_j \neq y, \quad \omega_y^T \Psi(x_i, y_j) \leq -1 + \xi_{3,ij}, \ \xi_{3,ij} \geq 0, \tag{19}$$

where $\xi_{1,i}$, $\xi_{2,j}$, and $\xi_{3,ij}$ are the slack variables and γ is a constant that controls the trade-off between training error minimization and margin maximization.

We analyze the above constrains for a better understanding of the learning formulation. For class y, constraint (17) requires that in the training stage, only the attribute vectors generated by mapping the input videos of class y to the CSOAs of class y are used as positive examples while constrains (18) and (19) indicates that when mapping the videos of class y into the CSOAs of other classes or mapping the videos of other classes into CSOAs of class y, the generated attribute vectors are regarded as negative examples. The whole process of learning the function f is quite similar to a multiclass SVM formulation. However, the input attribute vectors for each action class are different because they are determined not only by the low-level features, but also by the action class labels.

7   Experimental Results

We perform a set of experiments to evaluate the performance of our proposed approach on four publicly available datasets: the KTH [15], UCF sports [16], UCF films [16] and Hollywood2 [1]. Some frames extracted from the four datasets are shown in Fig. 4.

We conduct two groups of experiments on these datasets. In the first group, we evaluate the performance of our proposed low-level features OMSDs. The extracted OMSDs are directly used for human action representation and an SVM classifier is trained for action classification. In the second group, we evaluate the performance of our CSOAs based approach. After extracting the low-level features OMSDs, we obtain the CSOAs related to each action class and a compatibility classifier is applied for action classification. In addition, we also compare the performance of our CSOAs based approach with some state-of-the-art approaches.

Fig. 4. Some frames are extracted from the four datasets. The rows from top to down are the frames from the KTH dataset, UCF sports dataset, UCF films dataset, and Hollywood2 dataset respectively.

7.1   Datsets and Evaluation Protocol

The KTH dataset contains six types of human actions (walking, jogging, running, boxing, hand waving and hand clapping) performed several times by 25 subjects in four different scenarios. There are in total 599 sequences on this dataset. We perform leave-one-person-out cross validation to make the performance evaluation. In each run, 24 actors' video sequences are used for training and the remaining actor's videos for test.

The UCF sports dataset contains ten sports actions: diving, golf swinging, kicking, lifting, horse riding, running, skateboarding, swinging bench, swinging from side angle and walking. It consists of 150 video sequences taken from actual sporting activities from a variety of sources with a wide range of viewpoints and scene backgrounds. Leave-one-out cross validation is used to evaluate our approach. One video sequence is used for test and the remaining sequences are used for training.

The UCF feature films dataset provides a representative pool of nature samples of two action classes including kissing and hitting/slapping. It contains 92 samples of kissing and 112 samples of hitting/slapping which are extracted from a range of classic movies. The actions are captured in a wide range of scenes under different viewpoints with different camera movement patterns. The test for this dataset uses leave-one-out cross validation.

The Hollywood2 dataset contains 12 action classes collected from 69 different Hollywood movies. There are in total 1707 video sequences on this dataset, which are divided into a train set of 823 sequences and a test set of 884 sequences. We follow the standard evaluation protocol on this benchmark, i.e., computing the average precision (AP) for each class and using the mean of APs (mAP) for performance evaluation.

7.2 Evaluation of Low-Level OMSDs

We test our low-level feature OMSDs on both the KTH dataset and UCF sports dataset. Given a video sequence containing an action x, we detect the OMSRs and compute the corresponding OMSDs $\{H_k\}_{k=0}^8$. In our experiments, we first evaluate the performance of each OMSD. The whole video sequence is represented by only a single OMSD H_k. A simple SVM classifier with χ^2-kernel is applied for human action classification. Then we utilize a feature-level fusion approach in which all the OMSDs $\{H_k\}_{k=0}^8$ are concatenated together to form a large feature vector

$$H = (H_0^T, H_1^T, ..., H_8^T)^T \tag{20}$$

for action representation. Similarly, this large feature vector is supplied as input to a SVM with χ^2-kernel for action classification. The performance of each OMSD and the concatenation of OMSDs on both datasets are shown in Table 2. In Table 2, 'OMSD-k' means that video sequences are represented by the kth OMSD H_k, and 'Average of OMSDs' is computed as the mean of accuracies of the 9 OMSDs.

Three points can be drawn from Table 2. First using only a single low-level feature OMSD obtains relative good results, varying from 91.0 % to 95.6 % on the KTH dataset and from 80.7 % to 85.7 % on the UCF sports dataset. This demonstrates the effectiveness of the proposed OMSDs. Since the construction of a OMSD is based on a specific OMSR detected in a video sequence, it captures only a small part of the visual information of an action. However, the average accuracy of the OMSD still reaches 92.8 % on the KTH dataset and 82.4 % on the UCF sports dataset.

Second, the large feature H obtained by concatenating all the OMSDs achieves the best results on both datasets, reaching 96.5 % for the KTH dataset and 88.0 % for the UCF sports dataset. This is about 3.7 % and 5.6 % higher than the average accuracy of single OMSD on the KTH and UCF sports datasets respectively. We can see that the simple concatenation of all the OMSDs can improve the performance of human action recognition by a large amount.

Table 2. Comparison the performance of each low-level feature OMSD and the concatenation of OMSDs on the KTH and UCF sports datasets.

Low-Level Features	KTH	UCF sports
OMSD-0	**95.6 %**	**85.3 %**
OMSD-1	91.0 %	82.6 %
OMSD-2	91.2 %	83.3 %
OMSD-3	94.0 %	81.3 %
OMSD-4	93.5 %	80.7 %
OMSD-5	93.3 %	80.7 %
OMSD-6	92.1 %	84.0 %
OMSD-7	91.6 %	82.6 %
OMSD-8	92.6 %	81.3 %
Average of OMSDs	92.8 %	82.4 %
Concatenation of OMSDs	**96.5 %**	**88.0 %**

Third, when using a single OMSD for action representation, the 'OMSD-0' outperforms other 'OMSD-k's, reaching accuracies of 95.6 % and 85.3 % respectively on the two datasets. In our experiments, the 'OMSD-0' corresponded to the 3D Gabor filter with $(\eta_x, \eta_y, \eta_t) = (0, 0, 1)$. Except for the Gaussian scales, this 3D Gabor filter is equivalent to Dollár's linear separable filters designed for spatiotemporal interest points detection. This 3D Gabor filter generates a large response where motion occurs, regardless of the motion direction.

7.3   Evaluation of Middle-Level CSOAs

In this subsection, we evaluate the performance of our CSOAs based approach for human action recognition on the four datasets mentioned above. We extract the low-level OMSDs for each video sequence and utilize these OMSDs to train the CSOA classifiers for each action class. Then mapping actions into the CSOA space of each action class, a set of middle-level attribute features are constructed, which combine the low-level OMSDs and action class labels together. A compatibility function is learned to measure how well the low-level features match the CSOAs of action classes.

Table 3 presents a comparison of our proposed CSOA based approach with other approaches on the KTH and UCF sports datasets. Our CSOA based approach outperforms the other methods, achieving 97.2 % and 91.3 % on these two datasets respectively, which demonstrates the effectiveness of our CSOAs based approach. It is notable that the performance of our CSOA based approach is 0.7 % higher than that of the low-level concatenation of OMSDs on the KTH dataset and 3.3 % on the UCF sport dataset. This shows that firstly the CSOAs learnt from low-level OMSDs carry great discriminative power and improve the performance of human action recognition and secondly only the concatenation

Table 3. Comparison of our CSOA based approach with state-of-the-art approaches on the KTH and UCF sports datasets.

Algorithm	KTH	UCF sports
Derpanis *et al.* [17]	93.2%	81.5%
Wang *et al.* [4]	92.1%	85.6%
Kovashka *et al.* [18]	94.5%	87.3%
Le *et al.* [19]	93.9%	86.5%
Wang *et al.* [20]	94.2%	88.2%
Liu *et al.* [25]	94.8%	-
Wang *et al.* [23]	93.3%	-
Shi *et al.* [24]	93.0%	-
Concatenation of OMSDs	96.5%	88.0%
Our CSOAs	**97.2%**	**91.3%**

Table 4. Confusion table of our CSOA based approach on the KTH dataset.

	Box	Handclap	Handwave	Jog	Run	Walk
Box	**1.00**					
Handclap	0.01	**0.98**	0.01			
Handwave		0.01	**0.99**			
Jog				**0.95**	0.03	0.02
Run				0.08	**0.92**	
Walk				0.01		**0.99**

Table 5. Confusion table of our CSOA based approach on the UCF sprots dataset.

	Dive	Golf	Kick	Lift	Ride	Run	Skate	Swing1	Swing2	Walk
Dive	**1.00**									
Golf		**0.95**				0.05				
Kick			**1**							
Lift				**1**						
Ride		0.08			**0.83**	0.08				
Run		0.08			0.15	**0.69**		0.08		
Skate		0.16			0.08		**0.75**			
Swing1					0.05			**0.95**		
Swing2									**1.00**	
Walk						0.09				**0.91**

Table 6. The results of our approach on the UCF film dataset.

Algorithms	Kiss	Slap	Average
Rodriguez et al. [16]	66.4%	67.2%	66.8%
Yeffet et al. [21]	77.3%	84.2%	80.75%
Concatenation of OMSDs	92.2%	93.2%	92.7%
Our CSOAs	**95.6%**	**96.5%**	**96.1%**

Table 7. The results of our approach on the Hollywood2 dataset.

Algorithms	mAP
Wang et al. [4]	47.7%
Le et al. [19]	53.3%
Wang et al. [20]	58.3%
Concatenation of OMSDs	52.6%
Our CSOAs	**58.6%**

of low-level OMSDs achieves a good performance on the KTH dataset, because the actions on the KTH dataset are simple actions performed against static and un-cluttered backgrounds. The confusion matrixes of the proposed CSOA based approach on the KTH and UCF sports datasets are shown in Tables 4 and 5 respectively.

Tables 6 and 7 show the performance of our approach on both UCF films and Hollywood2 datasets. Our CSOAs based approach achieves 96.1% and 58.6% respectively on both datasets which is comparable to the listed approaches. It demonstrates the effectiveness of our CSOAs based approach on the realistic datasets. Meanwhile, the CSOAs based approach is 3.4% and 6.0% respectively higher than the simple concatenation of all low-level OMSDs. It indicates the CSOA based approach outperforms the low-level OMSDs based approach.

8   Conclusion

In this paper, we have proposed a novel approach for human action recognition based on the oriented motion salient regions. First, a 3D Gabor filter bank, incorporated with an opponent inhibition operator, has been applied to detect the OMSRs and a set of OMSDs have been extracted from these detected regions. Then, the obtained OMSDs have been used to explore the oriented characteristics of each action class, obtaining a series of CSOAs for each class. Taking advantage of these CSOAs, we have obtained a compact and discriminative middle-level feature to represent human actions. Finally, a compatibility function has been devised for action classification. We have tested our proposed approach on several public datasets. The experimental results have demonstrated that the proposed approach are effective in human action recognition.

Acknowledgement. This work is partly supported by the 973 basic research program of China (Grant No. 2014CB349303), the National 863 High-Tech R&D Program of China (Grant No. 2012AA012504), the Natural Science Foundation of Beijing (Grant No. 4121003), the Project Supported by Guangdong Natural Science Foundation (Grant No. S2012020011081) and NSFC (Grant No. 61100099, 61303086).

References

1. Marszalek, M., Laptev, I., Schmid, C.: Actions in context. In: CVPR, pp. 2929–2936 (2009)
2. Dollár, P., Rabaud, V., Cottrell, G., Belongie, S.: Behavior recognition via sparse spatio-temporal features. In: VSPTES, pp. 65–72 (2005)
3. Laptev, I., Marszalek, M., Schmid, C., Rozenfeld, B.: Learning realistic human actions from movies. In: CVPR, pp. 1–8 (2008)
4. Wang, H., Ullah, M., Klaser, A., Laptev, I., Schmid, C.: Evaluation of local spatio-temporal features for action recognition. In: BMVC (2009)
5. Scovanner, P., Ali, S., Shah, M.: A 3-dimensional SIFT descriptor and its application to action recognition. In: ICMM, pp. 357–360 (2007)
6. Bobick, A.F., Davis, J.W.: The recognition of human movement using temporal templates. PAMI **23**, 257–267 (2001)
7. Wang, L., Suter, D.: Informative shape representations for human action recognition. In: ICPR, vol. 2, pp. 1266–1269 (2006)
8. Ikizler, N., Duygulu, P.: Histogram of oriented rectangles: A new pose descriptor for human action recognition. IVC **27**, 1515–1526 (2009)
9. Reed, T.R.: Motion analysis using the 3-d gabor transform. SSC **1**, 506–509 (1996)
10. Adelson, E.H., Bergen, J.R.: Spatiotemporal energy models for the perception of motion. J. Opt. Soc. Am. A **2**, 284–299 (1985)
11. Farhadi, A., Endres, I., Hoiem, D., Forsyth, D.: Describing objects by their attributes. In: CVPR, pp. 1778–1785 (2009)
12. Liu, J., Kuipers, B., Savarese, S.: Recognizing human actions by attributes. In: CVPR, pp. 3337–3344 (2011)
13. Wang, Y., Mori, G.: A discriminative latent model of object classes and attributes. In: Daniilidis, K., Maragos, P., Paragios, N. (eds.) ECCV 2010, Part V. LNCS, vol. 6315, pp. 155–168. Springer, Heidelberg (2010)
14. Yao, B., Jiang, X., Khosla, A., Lin, A.L., Guibas, L., Fei-Fei, L.: Human action recognition by learning bases of action attributes and parts. In: ICCV, pp. 1331–1338 (2011)
15. Schuldt, C., Laptev, I., Caputo, B.: Recognizing human actions: a local SVM approach. In: ICPR, vol. 3, pp. 32–36 (2004)
16. Rodriguez, M., Ahmed, J., Shah, M.: Action mach a spatio-temporal maximum average correlation height filter for action recognition. In: CVPR, pp. 1–8 (2008)
17. Derpanis, K., Sizintsev, M., Cannons, K., Wildes, R.: Action spotting and recognition based on a spatiotemporal orientation analysis. PAMI **35**, 527–540 (2012)
18. Kovashka, A., Grauman, K.: Learning a hierarchy of discriminative space-time neighborhood features for human action recognition. In: CVPR, pp. 2046–205 (2010)
19. Le, Q.V., Zou, W.Y., Yeung, S.Y., Ng, A.Y.: Learning hierarchical invariant spatio-temporal features for action recognition with independent subspace analysis. In: CVPR, pp. 3361–3368 (2011)

20. Wang, H., Klaser, A., Schmid, C., Liu, C.: Action recognition by dense trajectories. In: CVPR, pp. 3169–3176 (2011)
21. Yeffet, L., Wolf, L.: Local trinary patterns for human action recognition. In: ICCV, pp. 492–497 (2009)
22. Derpanis, K., Lecce, M., Daniilidis, K., Wildes, R.P.: Dynamic scene understanding: The role of orientation features in space and time in scene classification. In: CVPR, pp. 1306–1313 (2012)
23. Wang, L., Qiao, Y., Tang, X.: Motionlets: mid-level 3D parts for human motion recognition. In: CVPR (2013)
24. Shi, F., Petriu, E., Laganiere, R.: Sampling strategies for real-time action recognition. In: CVPR, pp. 2595–2602 (2013)
25. Liu, L., Shao, L., Zhen, X., Li, X.: Learning discriminative key poses for action recognition. Cybernetics 43, 1314–1317 (2013)

3D Activity Recognition Using Motion History and Binary Shape Templates

Saumya Jetley$^{(\boxtimes)}$ and Fabio Cuzzolin

Oxford Brookes University, Oxford, UK
saumya.jetley@gmail.com

Abstract. This paper presents our work on activity recognition in 3D depth images. We propose a global descriptor that is accurate, compact and easy to compute as compared to the state-of-the-art for characterizing depth sequences. Activity enactment video is divided into temporally overlapping blocks. Each block (set of image frames) is used to generate Motion History Templates (MHTs) and Binary Shape Templates (BSTs) over three different views - front, side and top. The three views are obtained by projecting each video frame onto three mutually orthogonal Cartesian planes. MHTs are assembled by stacking the difference of consecutive frame projections in a weighted manner separately for each view. Histograms of oriented gradients are computed and concatenated to represent the motion content. Shape information is obtained through a similar gradient analysis over BSTs. These templates are built by overlaying all the body silhouettes in a block, separately for each view. To effectively trace shape-growth, BSTs are built additively along the blocks.

Consequently, the complete ensemble of gradient features carries both 3D shape and motion information to effectively model the dynamics of an articulated body movement. Experimental results on 4 standard depth databases (MSR 3D Hand Gesture, MSR Action, Action-Pairs, and UT-Kinect) prove the efficacy as well as the generality of our compact descriptor. Further, we successfully demonstrate the robustness of our approach to (impulsive) noise and occlusion errors that commonly affect depth data.

1 Introduction

Action recognition from video sequences is a widely explored challenge, which continues to be the subject of active research. With important applications in the areas of video surveillance, video indexing/retrieval, human-computer interaction (in particular for gaming consoles or surgical assistance), robotic navigation and many others, the task of human activity recognition has immense practical value. Research in this field began on RGB video sequences captured by single traditional cameras – proven successful approaches to static image analysis were extrapolated to cater to the additional time axis. Such techniques can be typically categorized into local descriptor [1–3] and global descriptor-based [4–6] approaches.

© Springer International Publishing Switzerland 2015
C.V. Jawahar and S. Shan (Eds.): ACCV 2014 Workshops, Part I, LNCS 9008, pp. 129–144, 2015.
DOI: 10.1007/978-3-319-16628-5_10

More recently, the advent of affordable depth cameras has shifted the focus onto depth images, which can now be easily recorded in real-time with good accuracies. Besides providing clear advantages in terms of illumination invariance and robust foreground extraction, range images also encode shape and motion information which (if represented efficiently) can facilitate much improved activity recognition results.

Depth images, however, need to be treated differently from their colored counterparts. Any attempt to use local differential operators tends to fail, due to false firing on discontinuous black regions of undefined depth values. As confirmed in [7], for MSR Daily Activity Dataset [8] 60 % of the identified Dollar interest points [2] were empirically observed at locations irrelevant to the action. Thus, interest-point based 2D video analysis approaches like STIP [1] have been shown inadequate to deal with 3D depth data.

The development of robust approaches to the quick and accurate estimation of 3D joint positions of human body from a single depth image [9] have spurred substantial research work, aimed at achieving action recognition through skeleton tracking [8,10–12]. Reference [8], for instance, proposes the use of Fourier pyramids to represent the temporal dynamics of 3D joint-positions, learned using discriminative actionlet pools and a multiple kernel based approach. Li Xia et al. [13] also employ 3D skeletal joint positions for action recognition. Although these approaches do yield good practical accuracies, 3D joint positions are prone to noise and even partial occlusion can drastically affect recognition. Also, until recently, due to the difficulty of efficient skeleton tracking for hands these approaches were not employable in hand gesture recognition. Although the approach in [14] has overcome such a limitation, the additional tracking step is still susceptible to occlusion problems.

Approaches which do not make use of skeleton tracking have been proposed. In [15] an 'action graph' based on bag of 3D contour points is used to identify action classes. Methods based on dense sampling have also been attempted, which perform a grid-based analysis of the 4D spatio-temporal volume [7] associated with a depth sequence for recognizing actions [16]. Although such methods can be improved by introducing random sampling of sub-volumes [17] to reduce computation cost and make the features more discriminative, they are still computationally demanding, both during training and testing.

Bobick & Davis [18] introduced the concepts of Motion Energy Templates (METs) and Motion History Templates (MHTs) in traditional videos. They recognized body movements in input RGB sequences using Hu-moments based statistical models built from above templates. Subsequently, Davis [19] improved on the approach by using an MHT pyramid to account for varying action-enactment speeds, and characterized the resulting motion field using a polar histogram. Tian et al. [20] also employed MHTs for action recognition in crowded scene videos, owing to its effectiveness in capturing continuous and recent foreground motion even in a cluttered and inconsistently moving background. Using Harris corner points to eliminate noisy motion from MHT, they computed gradient features from intensity image and MHT. The features were learnt by a Gaussian Mixture Model (GMM) based classifier. However, it was Yang et al. [21]

who first extended the idea to 3D depth images. They adopted METs and stacked motion regions for entire video sequences over three distinct views – front, side and top. Gradient features for the 3 consolidated binary templates were then used to represent depth sequences.

1.1 Contribution

Based on the conducted critical review of past approaches and their limitations we propose a highly discriminative ensemble of both shape and motion features, extracted over a novel temporal construct formed by overlapping video blocks, for 3D action recognition from depth sequences.

First of all, contrarily to [21] where a single template is built over the complete video sequence, we build and analyze motion and shape templates separately for overlapping temporal blocks. The intuition behind this novel temporal construct and its discriminative advantages are elaborated in Sect. 3.1. By dividing the video sequence in smaller temporal blocks, we prevent the loss of motion information which happens when a more recent action overwrites an old action at the same point. Temporal overlap across blocks maintains continuity in the templates.

Secondly, instead of using Depth Motion Maps (a form of METs) as in [21], we adopt Motion History Templates so as to effectively capture information on the direction of motion, in the form of gray level variations. This allows our approach to cope successfully with sequences like those in the 3D Action Pairs database [7], in which the direction of motion is the essential discriminating factor.

In an additional original feature of our proposal, METs are assembled over temporal blocks in an accumulative manner to capture shape information. The shape of the resulting black and white boundaries describes how the human body envelope grows in the course of a certain action.

Finally, histograms of oriented gradients extracted from these templates are concatenated to yield the final feature vector. The latter constitutes a novel, low-dimensional, computationally inexpensive and highly discriminative global descriptor for depth sequences. This descriptor is used to train a non-linear large-margin classifier (RBF kernel based SVM) to identify action classes. The complete process flow is shown in Fig. 1.

The generality of the approach is empirically demonstrated by evaluating it on 4 standard depth benchmarks – the 3D Hand Gesture, 3D Action, Action Pair, and UT-Kinect datasets. A vanilla implementation of our method outperforms the existing-best on 2 out of three 3 standard benchmarks, and has performance equal to the state of the art on the skeleton-biased UT-Kinect dataset. Note that our approach uses a block-size and an overlap-factor as parameters, that are in these experiments still manually selected by trial & error. By setting these parameters via cross validation our work has much scope for further performance gains.

Last but not least, in tests commonly neglected in the literature, we experimentally demonstrate the robustness of our approach to noise and occlusions that typically affect depth images by artificially corrupting sequences from the MSR 3D Action dataset. Results show that performance is remarkably robust to both nuisances, making our approach suitable for real-world deployment.

1.2 Paper Outline

The paper is organized as follows. Section 2 compares our approach in detail with the closest related work. Section 3 illustrates the various steps of the proposed approach. Section 4 recalls the benchmark datasets considered in our tests and details the experimental settings. The performance of our algorithm on the four considered benchmarks is illustrated in Sect. 5. Section 6 studies the robustness of the approach against common occlusion and noise errors. Section 7 concludes the paper with a word about the future directions of research.

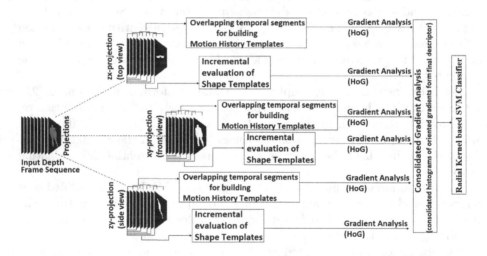

Fig. 1. Process flow for the proposed 3D activity recognition approach.

2   Relevant Related Work

Vennila et al. [22] follow a similar idea to that proposed in our approach, but in a rather more simplistic way. Motion and 3D shape information are not considered in conjunction. Their analysis of only frontal MHIs leaves out depth (3D shape) information, while their Average Depth and Difference Depth Images do not consider motion and thus capture background-clutter depth details also.

Weinland et al. [23] suggest the use of motion history volumes for analyzing voxelset sequences. Their normalization along the r & z axes may cause alignment failure, like confusion between a person spreading his arms and a person dropping them. The Fourier analysis that they apply in concentric circles at each height does not capture limb movement as well as our gradient analysis of MHTs. For composite actions such as the walking gait, only one temporal segment is used. All these cause confusion between well distinct actions such as sit down & pick-up, turn-around & majority-of-other-actions, walk & pick-up or kick. Also, the used 1D-FT overlooks motion symmetry around z axis and may fail to distinguish between single & two arm waves, forward & side punch, and forward & side kick.

Finally, Oreifej and Liu [7] represent the current state-of-the-art for the task at hand. Their method computes and concatenates the histograms of oriented surface normals for each grid cell and uses the final feature descriptor to train an SVM classifier. We compare ourselves to [7] empirically in Sect. 5.

3   Proposed Approach

In a depth image, each pixel contains a value which is a function of the depth/distance at which the camera encounters an object at the corresponding point in the real scene. For a depth camera recording from the front, the information of overlapped object points in the rear is not captured. Therefore, a depth image at any time does not contain an exact all-view model of the 3D objects present in the scene. Nonetheless, the data is sufficient to generate three approximate front, side and top views of the scene which, for a single 3D object, contain important, non-redundant and discriminative information both in terms of shape (over a single frame) and motion (over a sequence of frames). It is to obtain and leverage this enhanced 3D shape and motion information that we first project each frame of the incoming video sequence onto the three orthogonal Cartesian planes. The front, side and top views so obtained for a given frame are as illustrated in Fig. 2. Construction of motion and shape information from these projections is illustrated in Fig. 3, and described in the remainder of the Section.

3.1   Motion Information Evaluation

We split the input video sequence into overlapping temporal blocks. The split allows motion to be captured over a smaller number of frames thus enabling a more detailed analysis of motion. As seen in Fig. 3 (Set 1), if one motion history template is built for the complete action of 'draw tick', any old motion information gets overwritten when a more recent action occurs at the same point. Not only is part of the motion information lost, but the exact direction of motion (represented by the direction of increasing gray intensity) also gets distorted.

Using overlapping temporal blocks helps to ensure a smooth connectivity of the articulated motion across consecutive blocks. Note that the movement is more continuous in Fig. 3 (Set 3) as compared to Fig. 3 (Set 2), in which such a temporal overlap is not maintained.

As mentioned before, the three orthogonal views are handled separately. Firstly, for the complete video sequence we gather all the frame projections for each view. Then, separately for each view, we divide the frames into overlapping blocks and proceed to build their motion field. For each block, consecutive

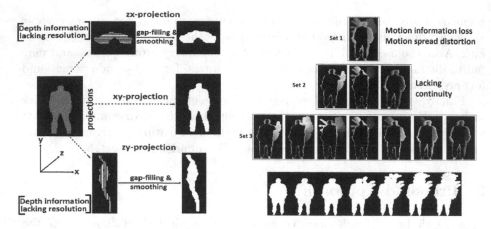

Fig. 2. Depth images projected onto three Cartesian planes to obtain front, side and top views. (Discontinuities resolved using gap-filling and smoothing.)

Fig. 3. (Top) Motion history templates for (Set 1) complete sequence, (Set 2) non-overlapping blocks, (Set 3) overlapping blocks; (Bottom) Binary Shape Templates for a 'draw tick' action focused in the top-right region.

frames are subtracted to identify the parts in motion. These parts are stacked together in a weighted manner, with the weight being proportional to the recency of motion, to obtain the desired Motion History Templates.

Namely, the following operations are performed on each block.

A motion template MT_f, the difference between two consecutive body silhouettes B_f, is computed as: $MT_f = B_{f+1} - B_f$ for each frame $f \in 1, \ldots, n-1$ in the block. The motion history template MHT_b for block b is computed iteratively, by first initializing it to null, and then updating it by applying for each $f \in 1, \ldots, n-1$ a pixel-wise maximum ($pmax$) of the current motion history template and the weighted motion template $f * MT_f$, namely: $MHT_b' = pmax(MHT_b, f * MT_f)$.

3.2   Shape Information Evaluation

Shape information is constructed in the form of gradient features of binary shape templates, compiled over temporal blocks of the input video sequence. As previously mentioned the three orthogonal views are processed separately. Frame projections for each view are collected and divided into overlapping blocks for which shape templates are constructed. Shape templates become clearer over longer intervals of time and are thus incrementally built, as shown in Fig. 3 (bottom). The shape template for block 2 is added over the shape template for block 1, the template for block 3 is built over the shape templates for blocks 1 and 2, and so-on.

Thus, the shape template for a given block is built by additive overlaying of body silhouettes, via a frame on frame pixel-wise OR operation, for all the frames in the current as well as the previous blocks.

First, the shape template is initialized to $ST_b = ST_{b-1}$, for $b > 1$ (a matrix of zeros otherwise). Then, ST_b is updated iteratively via a pixel-wise OR of the current shape template and the current frame B_f for all frames $f \in 1, \ldots, n$ as: $ST_b' = ST_b || B_f$.

3.3 Gradient Analysis

Motion history is contained in gray-level variations (MHTs) while shape is represented by white-object boundaries (BSTs). Extraction of shape and motion information therefore necessitates a form of gradient analysis. Herein lies our motivation behind the use of histograms of oriented gradients [24] to obtain the final feature descriptor for the input depth sequence.

For the 3D Hand Gesture Dataset, a single histogram of gradients is computed for a complete template, i.e., the standard implementation, where histograms are built over sets of grid cells and later concatenated, is not followed. We observe that different hand gesture enactments have different proportions of gradient values. Thus, without the need for greater discrimination, positions of gradient values could be bypassed.

For whole-body datasets, however, the standard HoG implementation is followed. Different regions or body parts can make the same gradient contribution. For examples, actions pairs like - "side boxing and side kick", "forward punch and forward kick", and "high arm wave and two hand wave"; show similar gradient proportions over the complete image but the contributing gradient regions are different. Thus, for a better discrimination between different action enactments a region-wise gradient analysis is essential.

4   Experimental Settings

4.1   Datasets

We have evaluated the performance of our proposed approach on 4 standard depth datasets - MSR 3D Gesture Dataset [25], MSR 3D Action Dataset [15], 3D Action Pair Dataset [7], and UT-Kinect Database [13]. Details of each database, their preprocessing operations, and the parameter settings for feature extraction are as elaborated below.

MSR 3D Gesture Dataset: The Gesture3D dataset [25] is a hand gesture dataset of depth sequences captured using a depth camera (more particularly a Kinect device). It contains a set of 12 dynamic American Sign Language (ASL) gestures, namely - "bathroom", "blue", "finish", "green", "hungry", "milk", "past", "pig", "store", "where", "j", "z". 10 subjects have performed each hand

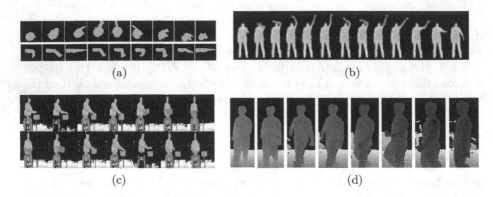

Fig. 4. Sample frames for (a) Hand gesture enactments of 'where', 'pig' (b) Action category of 'high arm wave' (c) Action pairs of 'pick up' & 'put down a box' (d) UT Kinect action category of 'walk'.

gesture 2–3 times yielding a total of 336 sample sequences. Sample frames for two gesture categories are shown in Fig. 4a. The dataset contains images of hand portions segmented above the wrist and is highly challenging due to self-occlusions in gesture enactments.

MSR 3D Action Dataset: MSR Action3D dataset [15] is an action dataset of depth sequences captured using a depth camera similar to the Kinect device. This dataset contains twenty action categories: "high arm wave", "horizontal arm wave", "hammer", "hand catch", "forward punch", "high throw", "draw x", "draw tick", "draw circle", "hand clap", "two hand wave", "side-boxing", "bend", "forward kick", "side kick", "jogging", "tennis swing", "tennis serve", "golf swing", "pickup & throw". Each action is performed around 2–3 times by 10 subjects, yielding a total of 567 action samples. Subjects face the camera in the depth maps which are captured at a rate of 15 fps and a resolution of 320 × 240. Sample frames for a 'high arm wave' action are shown in Fig. 4b.

The 20 actions were chosen in the context of using the actions to interact with gaming consoles. They reasonably cover various movements of arms, legs, torso and their combinations. In addition, for actions performed by a single arm or leg, the subjects were advised to use their right arm or leg. In this dataset, the background has been cleaned to remove the discontinuities due to undefined depth points. However, recognition remains highly challenging due to close similarities between different actions.

3D Action Pair Database: This database was introduced by Oreifej and Liu [7] to emphasize the point that two different actions may have similar shape and motion cues, but the correlation between these cues may vary. Quoting the example from [7], "Pick up" and "Put down" actions have similar motion and shape; however, the co-occurrence of object-shape and hand-motion is in different spatio-temporal order and that holds the key to distinguishing the two. Based on this idea, the database puts together 6 action pairs for classification, namely - "Pick up a box/Put down a box", "Lift a box/Place a box", "Push a chair/Pull a chair", "Wear a hat/Take off a hat", "Put on a backpack/Take off a backpack", "Stick a poster/Remove a poster". Each action is performed 3 times using 10 different actors yielding a total of 360 depth sequences. Sample frames presenting an action-pair are as shown in Fig. 4c.

UT Kinect Action Database: This database was compiled by Li Xia et al. using a single stationary Kinect camera. It includes 10 action types: "walk", "sit-down", "stand-up", "pick-up", "carry", "throw", "push", "pull", "wave-hands", "clap-hands". Each of the 10 subjects perform each action twice. The UT-Kinect database is highly challenging, with significant variations in single action enactment ('Throw' with left/right arm, 'Carry' with one/both hands, 'walk' - first front-ward and then side-ward as in Fig. 4d), action clips duration, frequent occlusions due to changing views, and an annoying shifting background clutter possibly due to camera trembling.

With the frames being recorded only when a skeleton is tracked, the database is biased towards skeleton based processing. The above frame selection also reduces the practical frame rate from 30fps to 15fps. Furthermore, around 2 % of the frames have multiple skeletons recorded with slightly different joint locations. Again, using skeleton information the main object can be easily segmented, while in general depth-information processing (as adopted by us) this introduces additional challenges.

4.2 Preprocessing

Images in the hand gesture dataset have varying sizes. For experimental purposes, the images are used as is without any cropping and/or resizing. The aim is to preserve the original shape and gradients in the image. Values in the histograms of gradients are normalized and carry information of the relative gradient proportion. Thus, image size does not affect this normalized distribution.

For the 3 full-body datasets (MSR Action, Action-Pair and UT-Kinect) images need to be uniformly sized to facilitate a more effective grid-based HoG analysis. Also, we can observe that depth information is discontinuous. As can be seen in Fig. 2, depth values appear in steps resulting in the black slit-like gaps. The gaps introduce irrelevant gradients. To remove the gaps, we divide each depth value by the average step-size (for gap-filling) and perform small-sized structuring element-based erosion and dilation (for smoothing). This is followed by image packing and resizing (if required). Finally, the front-view, side-view and top-view templates for the 3 databases are sized: $(240 \times 160, 240 \times 180, 180 \times 160)$; $(240 \times 160, 240 \times 480, 480 \times 160)$; and $(320 \times 240, 320 \times 240, 320 \times 320)$ respectively.

4.3 Parameter Settings

To compute our motion analysis descriptors, video sequences (say, of f frames) are divided into t units along the time axis. We move temporally in size of a unit (f/t frames) and in steps decided by the overlap factor (o), to assemble the blocks along the video sequence. Essentially, each block contains f/t frames, and shares $1/o$ amount of overlap with the previous block. Alternatively, this can be seen as dividing the video into $t*o$ frame sets and grouping o such sets starting at each, to yield a total of: $b = (t*o) - (o - 1)$ blocks.

For each block, motion as well as binary shape templates are constructed for each of the 3 orthogonal views. Thus, for any given video sequence, a maximum of $(3*2*b)$ templates are analyzed. For motion history templates, the direction of motion is stored in grey level variations: hence, all gradients in range $0 - 360°$ are considered. For our HoG-based analysis, this range is divided into b_1 gradient bins. For shape analysis, a $90°$ spread was found to be adequate and the $0 - 90°$ range is divided into b_2 gradient bins. Finally, all the histograms are normalized using the classical L2-norm [24] and concatenated to form the overall feature descriptor.

Dataset specific values for all these parameters are provided below. Presently these are empirically decided, however subsequently we aim to set these parameters via cross validation to boost our performance.

MSR 3D Gesture Dataset: With $t - 3$ and $o = 3$, the number of blocks $b = 7$. We build and analyze a total of 42 (3*2*7) templates to extract motion and shape information. As mentioned previously, for hand gesture templates gradient analysis is done once for the complete image without any division into smaller grids. b_1 and b_2 are set to 72 and 18 respectively.

MSR 3D Action Dataset: With $t = 3$ and $o = 3$, the number of blocks $b = 7$. We build and analyze a total of 42 (3*2*7) templates. The 7 motion history templates and 7 shape templates for front-view of 'draw tick' action are as shown in Fig. 3 (Set 3) & (bottom). Further, for HoG analysis, the front, side and top views of motion history templates are divided into 6×4, 6×4 and 4×4 non-overlapping grid blocks while for shape templates they are divided into 1×1, 1×1 and 1×1 sized grid blocks. b_1 and b_2 are set to 36 and 9 respectively.

3D Action Pair Database: With $t = 4$ and $o = 2$, the number of blocks $b = 7$. We build and analyze a total of 42 (3*2*7) templates. For HoG analysis, front, side and top views of motion history templates are divided into 6×4, 6×8 and 8×4 non-overlapping grid blocks while for shape templates they are divided into 1×1, 1×1 and 1×1 sized grid blocks. b_1 and b_2 are set to 36 and 9 respectively.

UT Kinect Action Database: With $t = 4$ and $o = 2$, the number of blocks $b = 7$. We build and analyze a total of 21 (3*1*7) templates. Due to significant variations in view and action enactment, as well frequent occlusions and camera motion, we observed shape templates to be counter-productive and hence shape templates were not considered. For HoG analysis, front, side and top views of motion history templates are divided into 3×4, 3×4 and 4×4 non-overlapping grid blocks. In place of shape templates, motion history templates are additionally analyzed over 1×1, 1×1 and 1×1 sized grid blocks. b1 is set to 36.

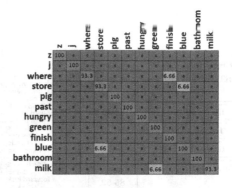

Approach	Accuracy (%)
Proposed (MHI + BST based Gradient Analysis)	96.6
Oreifej & Liu[7]	92.45
Jiang et al.[17]	88.5
Kurakin et al.[25]	87.77

Fig. 5. Accuracy comparison for 3D Hand Gesture Recognition.

Fig. 6. Confusion matrix for 3D hand gesture recognition using the proposed approach.

5 Recognition Results

We use SVM [26] discriminative classifier in one-versus-one configuration for the task of multi-class classification. In order to handle non-linear class boundaries SVM uses radial basis function kernel.

MSR 3D Gesture Dataset: As per the standard protocol, for every gesture category samples of first 5 subjects form the training set while those of next 5 subjects form the test set. With this cross-subject accuracy of 96.6 % our proposed method significantly outperforms the existing best [7]. Figure 5 lists the comparison with the previously attempted approaches for the given dataset. The related confusion matrix is presented in Fig. 6.

MSR 3D Action Dataset: Our approach yields a cross-subject accuracy of 83.8 %. Figure 7 presents the performance comparison with previously attempted approaches for the given dataset. Related confusion matrix is presented in Fig. 8. Jiang et al. [8] obtain an accuracy figure of 88.20 % with skeleton tracking approach. However, it has the drawback of being dependent on the precision of 3D joints positions making it susceptible to occlusion errors. Yang et al. [21], Vieira et al. [16], Li et al. [15] perform experiments on three subsets of the 3D Action dataset – AS1, AS2 and AS3, containing 8 action categories each. Confusion among a set of categories is always greater than or equal to the confusion in its subsets. Hence, the correct overall-accuracy for the approaches is not the average of the 3 subset accuracies but lesser than the minimum amongst the 3 values i.e. <84.1 %, <81.30 % and <71.90 % respectively. Considering the HoN4D approach on equal grounds i.e. without data-specific quantization refining [7], the accuracy of our approach is 2 % lower.

Given the performance of our approach on the other 3 datasets, we can argue that the relatively low accuracy is mainly due to inadequate data. The number of classes is the highest (20), but the number of sample videos per class is the same as in other datasets. We believe that with a higher number of videos our approach would generalize better to test set.

Approach	Accuracy (%)
Proposed (MHI + BST based Gradient Analysis)	83.8
Oreifej & Liu [7]	88.89
Jiang et al. [8]	88.2
Jiang et al. [17]	86.5
Yang et al. [21]/ [7]	<84.1/ 85.52
Vieira et al.[16]	<81.30
Li et al. [15]	<71.90

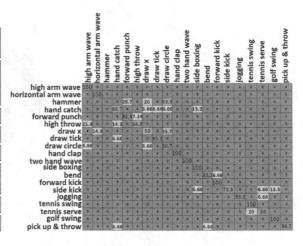

Fig. 7. Accuracy comparison for 3D Action Data Recognition.

Fig. 8. Confusion matrix for 3D action data recognition using the proposed approach.

3D Action Pair Database: A cross-subject accuracy of 97.22 % indicates that if sufficient amount of data is available per action category, our proposed approach has the potential to give top recognition results. Figure 9 presents the performance comparison with other approaches. The related confusion matrix is shown in Fig. 10.

UT Kinect Action Database: To the best of our knowledge, our approach is the only non-skeleton based technique attempted for UT Kinect Database. Despite the data being biased for skeletal analysis, as explained in Sect. 4.1, our method gives a competitive accuracy of 90 % (under the same testing conditions as in [13]) compared to the authors' 90.92 %.

Approach	Accuracy (%)
Proposed (MHI + BST based Gradient Analysis)	97.22
Oreifej & Liu [7]	96.67
Jiang et al. [8]	82.22
Yang et al. [21]	66.11

Fig. 9. Accuracy comparison for 3D action-pair data recognition.

Fig. 10. Confusion matrix for 3D action-pair data recognition using the proposed approach.

6   Performance Under Occlusion and Noise

The performance of the proposed approach has also been evaluated in the presence of two noise effects as elaborated in the subsections below. Results have been compiled on MSR 3D Action Database.

6.1   Occlusion

To analyze and compare the robustness of our method in the presence of occlusion, we carried out an experiment similar to that presented in [15]. As shown in Fig. 11, for a given occlusion setting, a single quadrant or a combination of 2 quadrants may be occluded in the depth image. Occlusion is incorporated in the test set and Fig. 12 shows the relative accuracy for each case. (Relative accuracy is defined as the recognition accuracy with occlusion, as a percentage of the accuracy without any occlusion).

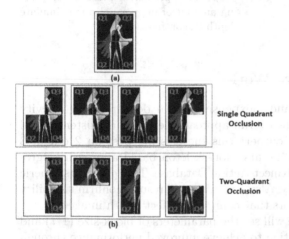

Occluded Quadrant(s)	Relative Accuracy (%)
none	100
Q1	91.57
Q2	95.99
Q3	64.66
Q4	96.39
Q1+Q3	43.376
Q3+Q4	62.25
Q4+Q2	85.948
Q2+Q1	65.87

Fig. 11. Occlusion of quadrant(s) for robustness testing.

Fig. 12. Recognition performance of the proposed approach in presence of occlusion.

With most of the actions focused on the top-right of the image, occlusion of Q3 gives the highest error rate. For all other single quadrant occlusions, accuracy reduction is no greater than 9%. Our method has the potential to handle occlusions primarily because it performs an independent region-wise gradient analysis over motion history and shape templates. Even if a particular part of the image is occluded, visible body regions are unaffected and can be independently analyzed.

6.2   Pepper Noise

The most common error that affects depth images, and consequently renders interest point approaches ineffective, is the presence of discontinuous black regions in the images. In a completely new performance measure, we deliberately introduce pepper noise in different percentages (of the total number of image pixels) in the depth images, as in Fig. 13. From Fig. 14, the recognition accuracy does not fall by more than 6 % even in up to 10 % pepper noise. This indicates the high robustness of our method to depth discontinuities.

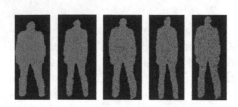

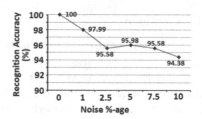

Fig. 13. Images affected by pepper noise in increasing percentages of total no. of pixels - 1 %, 2.5 %, 5 %, 7.5 % and 10 %.

Fig. 14. Recognition performance of the proposed approach in presence of varying amount of pepper noise (to simulate depth discontinuities).

7 Conclusion and Future Work

This paper presents a compact and inexpensive global descriptor for activity recognition from depth videos. The overall approach performs consistently well on 4 very different databases. It outperforms the existing best on 2 out of 3 standard depth databases and is the first non-skeleton based approach to give highly competitive results on UT Kinect Action Database. This proves its generic performance and wide applicability. Also, experimental results confirm its ability to handle occlusion and noise errors that commonly affect depth images.

As part of the future work, we will set the parameters of block-size $(1/t)$ and overlap-factor (o) via cross-validation to achieve improved performance through a database-specific setting. Also, we aim to assess the applicability of our approach to real-time action recognition from online depth videos. Temporal blocks can be treated as independent observation states. A model can be made to learn the observation-state-transition-probability to action-category mapping. Thus, temporal blocks and their sequence can help identify the action enacted in a live video. With every incoming block, the decision needs to be reconsidered in the light of the most recent input information.

References

1. Laptev, I., Lindeberg, T.: Space-time interest points. In: ICCV, pp. 432–439. IEEE Computer Society (2003)
2. Dollar, P., Rabaud, V., Cottrell, G., Belongie, S.: Behavior recognition via sparse spatio-temporal features. In: Proceedings of the 14th International Conference on Computer Communications and Networks, ICCCN 2005, pp. 65–72. IEEE Computer Society, Washington (2005)

3. Kläer, A., Marszalek, M., Schmid, C.: A spatio-temporal descriptor based on 3d-gradients. In: Everingham, M., Needham, C.J., Fraile, R. (eds.) BMVC. British Machine Vision Association (2008)

4. Yilmaz, A., Shah, M.: Actions sketch: A novel action representation. In: CVPR (1), pp. 984–989. IEEE Computer Society (2005)

5. Gorelick, L., Blank, M., Shechtman, E., Irani, M., Basri, R.: Actions as space-time shapes. IEEE Trans. Pattern Anal. Mach. Intell. **29**, 2247–2253 (2007)

6. Li, W., Zhang, Z., Liu, Z.: Expandable data-driven graphical modeling of human actions based on salient postures. IEEE Trans. Circuits Syst. Video Techn. **18**, 1499–1510 (2008)

7. Oreifej, O., Liu, Z.: Hon4d: Histogram of oriented 4d normals for activity recognition from depth sequences. In: 2013 IEEE Conference on Computer Vision and Pattern Recognition (CVPR), pp. 716–723 (2013)

8. Wang, J., Liu, Z., Wu, Y., Yuan, J.: Mining actionlet ensemble for action recognition with depth cameras. In: 2012 IEEE Conference on Computer Vision and Pattern Recognition (CVPR), pp. 1290–1297 (2012)

9. Shotton, J., Fitzgibbon, A., Cook, M., Sharp, T., Finocchio, M., Moore, R., Kipman, A., Blake, A.: Real-time human pose recognition in parts from single depth images. In: 2011 IEEE Conference on Computer Vision and Pattern Recognition (CVPR), pp. 1297–1304 (2011)

10. Yun, H., Sheng-Luen, C., Jeng-Sheng, Y., Qi-Jun, C.: Real-time skeleton-based indoor activity recognition. In: 2013 32nd Chinese Control Conference (CCC), pp. 3965–3970 (2013)

11. Shuzi, H., Jing, Y., Huan, C.: Human actions segmentation and matching based on 3d skeleton model. In: 2013 32nd Chinese Control Conference (CCC), pp. 5877–5882 (2013)

12. Yu, X., Wu, L., Liu, Q., Zhou, H.: Children tantrum behaviour analysis based on kinect sensor. In: 2011 Third Chinese Conference on Intelligent Visual Surveillance (IVS), pp. 49–52 (2011)

13. Xia, L., Chen, C.C., Aggarwal, J.: View invariant human action recognition using histograms of 3d joints. In: 2012 IEEE Computer Society Conference on Computer Vision and Pattern Recognition Workshops (CVPRW), pp. 20–27 (2012)

14. : (Efficient model-based 3d tracking of hand articulations using kinect)

15. Li, W., Zhang, Z., Liu, Z.: Action recognition based on a bag of 3d points. In: 2010 IEEE Computer Society Conference on Computer Vision and Pattern Recognition Workshops (CVPRW), pp. 9–14 (2010)

16. Vieira, A.W., Nascimento, E.R., Oliveira, G.L., Liu, Z., Campos, M.F.M.: STOP: space-time occupancy patterns for 3D action recognition from depth map sequences. In: Alvarez, L., Mejail, M., Gomez, L., Jacobo, J. (eds.) CIARP 2012. LNCS, vol. 7441, pp. 252–259. Springer, Heidelberg (2012)

17. Wang, J., Liu, Z., Chorowski, J., Chen, Z., Wu, Y.: Robust 3D action recognition with random occupancy patterns. In: Fitzgibbon, A., Lazebnik, S., Perona, P., Sato, Y., Schmid, C. (eds.) ECCV 2012, Part II. LNCS, vol. 7573, pp. 872–885. Springer, Heidelberg (2012)

18. Bobick, A.F., Davis, J.W.: The recognition of human movement using temporal templates. IEEE Trans. Pattern Anal. Mach. Intell. **23**, 257–267 (2001)

19. Davis, J.W.: Hierarchical motion history images for recognizing human motion. In: IEEE Workshop on Detection and Recognition of Events in Video, pp. 39–46 (2001)

20. Tian, Y., Cao, L., Liu, Z., Zhang, Z.: Hierarchical filtered motion for action recognition in crowded videos. IEEE Trans. Syst. Man Cybern. Part C: Appl. Rev. **42**, 313–323 (2012)
21. Yang, X., Zhang, C., Tian, Y.: Recognizing actions using depth motion maps-based histograms of oriented gradients. In: Proceedings of the 20th ACM International Conference on Multimedia, MM 2012, pp. 1057–1060. ACM, New York (2012)
22. Megavannan, V., Agarwal, B., Venkatesh Babu, R.: Human action recognition using depth maps. In: 2012 International Conference on Signal Processing and Communications (SPCOM), pp. 1–5 (2012)
23. Weinland, D., Ronfard, R., Boyer, E.: Free viewpoint action recognition using motion history volumes. Comput. Vis. Image Underst. **104**, 249–257 (2006)
24. Dalal, N., Triggs, B.: Histograms of oriented gradients for human detection. In: IEEE Computer Society Conference on Computer Vision and Pattern Recognition, CVPR 2005, vol. 1, pp. 886–893 (2005)
25. Kurakin, A., Zhang, Z., Liu, Z.: A real time system for dynamic hand gesture recognition with a depth sensor. In: 2012 Proceedings of the 20th European Signal Processing Conference (EUSIPCO), pp. 1975–1979 (2012)
26. Chang, C.C., Lin, C.J.: Libsvm: A library for support vector machines. ACM Trans. Intell. Syst. Technol. **2**, 27:1–27:27 (2011)

Gait Recognition Based Online Person Identification in a Camera Network

Ayesha Choudhary[1]([⊠]) and Santanu Chaudhury[2]

[1] School of Computer and System Sciences,
Jawaharlal Nehru University, New Delhi, India
ayeshac@mail.jnu.ac.in
[2] Department of Electrical Engineering,
Indian Institute of Technology Delhi, New Delhi, India
santanuc@ee.iitd.ac.in

Abstract. In this paper, we propose a novel online multi-camera framework for person identification based on gait recognition using Grassmann Discriminant Analysis. We propose an online method wherein the gait space of individuals are created as they are tracked. The gait space is view invariant and the recognition process is carried out in a distributed manner. We assume that only a fixed known set of people are allowed to enter the area under observation. During the training phase, multi-view data of each individual is collected from each camera in the network and their global gait space is created and stored. During the test phase, as an unknown individual is observed by the network of cameras, simultaneously or sequentially, his/her gait space is created. Grassmann manifold theory is applied for classifying the individual. The gait space of an individual is a point on a Grassmann manifold and distance between two gait spaces is the same as distance between two points on a Grassmann manifold. Person identification is, therefore, carried out on-the-fly based on the uniqueness of gait, using Grassmann discriminant analysis.

1 Introduction

In this paper, we propose a novel online distributed multi-camera person identification framework based on gait recognition. Gait recognition is a proven unique biometric for person identification. One of its main advantages over other biometrics such as iris recognition and fingerprint recognition, is that it is unobtrusive and requires no attention or cooperation from the person to be identified. It is also a preferable choice for surveillance applications over other biometrics because gait data can be captured from a far distance inconspicuously unlike face recognition data. Gait as a biometric also has typical challenges. One of the main challenges is to correctly identify a person using their gait signature as they are viewed from various angles in a camera network. Gait recognition works best in an environment where there are a fixed set of people that are allowed to enter the area under observation, a fixed set of entry/exit points. Moreover, all cameras together observe the complete area under observation at all times. In

© Springer International Publishing Switzerland 2015
C.V. Jawahar and S. Shan (Eds.): ACCV 2014 Workshops, Part I, LNCS 9008, pp. 145–156, 2015.
DOI: 10.1007/978-3-319-16628-5_11

our framework, during the training phase, a global gait space of each individual is constructed incrementally, as these people move in the area under observation using Incremental Principal Component Analysis (IPCA) [1]. This gait space is view invariant and is a point on the Grassmann manifold. During the identification phase, when a person enters the area under observation, his/her gait space is constructed incrementally. A gait space is considered as an element of a Grassmann manifold and Grassmann discriminant analysis [2] is used to identify the person. Grassmannian framework gives us the benefit of working within the non-linear structure of the data with the simplicity of the vector based computation. Moreover, because our framework is a distributed framework, each camera builds the gait space of the individual based on its view and the gait spaces are merged as the object moves from one camera to another in the network based on message passing in the network. One of the advantages of our framework is that it does not require the person to remain in the view of any one camera, requires no cooperation or time of the individual, the person is identified as they move around in the area. In case, the person is not identified within a fixed time from the first moment of entry into the area, he or she is labeled as *unknown* and the security official is notified about the person's presence in the scene and his/her whereabouts in the area. Our model is scalable in terms of both the number of cameras in the network as well as the number of people that are allowed to enter the area under observation.

2   Related Work

The process of gait recognition requires recording videos of people walking, extracting the silhouettes, then extracting gait features and finally classifying the individuals based on these features. In general, for automatic gait recognition, detection and extraction of silhouettes are performed using background subtraction [3]. After the silhouettes are extracted, based on the method to be employed for gait recognition, features are extracted and selected. There are two main approaches to gait recognition, namely, model-based approach and model-free approach.

In the model-based approach [4–7], static and dynamic features are extracted from the silhouettes. These features, in general, depict the position and pose of various body parts with respect to each other as a person moves in the scene. These features are extracted by tracking and modeling the body parts such as legs, head, arms, etc. The main advantage of forming a gait signature in this manner is that it is view as well as scale invariant. These invariants are necessary in practical situations since the training sequence and test sequence need not be taken from the same camera view. The drawback in this approach is that they are highly sensitive to the quality of gait sequences and the silhouette extraction. Moreover, these methods are in general offline methods since they require large computations.

Model-free approach does not model the whole body or body-parts, but focuses on the shape of the silhouettes or the whole body motions. The advantages of model-free approach are that it is less effected by the quality of silhouette extracted compared to model based approach, and have low cost of computation. However, in most cases, they are view and scale dependent. Methods such as [8] use silhouettes directly as features that are aligned and scaled. Authors in [9] propose and define motion-energy image and motion history image while [10] propose gait energy images for gait recognition. Hidden Markov Models (HMMs) [11] are also used for gait recognition as these models are able to represent the different phases of the gait cycle. Methods such as [12] based on K-Nearest neighbor classification do not take into consideration the temporal information in gait sequences. They work with a single key frame extracted from the gait sequence.

Multi-view gait recognition is gaining popularity mainly because single view gait recognition methods are many-a-time view dependent. The viewing angle at which the gait database is formed need not be the same as that used for obtaining test data. Multi-view gait recognition methods are either based on view invariant features [14,19] or based on multi-camera calibration that extract 3D structural information. However, calibration based systems require a fully calibrated multi-camera set-up which may not always be available. Approaches to multi-camera gait recognition such as [20,22] are based on view transformation and do not require camera calibration. Although these methods allow large changes in viewing angles by transforming the gallery and test data to the same direction, they suffer from lack of information present in views separated by large angles. Authors in [23] propose a novel gait recognition approach based on correlating gait sequences from different views using Canonical Correlation Analysis (CCA). The CCA model implicitly resolves the mapping relations between gait features from different views and projects gait sequences from different views into maximally correlated subspaces. The method in [17], forms the Eigen-gait space of training samples and for a test sample it uses k-nearest neighbor classification for classifying a test object.

In our method, we form a gait space of each individual based on the multiple view data that is recorded. We assume that cameras are mounted at various viewing angles and may observe the subjects either simultaneously or sequentially. During the training phase, the gait space is created online as the subject walks around in the camera network. The gait space is updated every time the subject walks in the view of a new camera. This gait space is view invariant and represents data from all angles in the input viewing space.

During the test phase, as a person walks through the area under observation, his/her gait space is created. Then, Grassmann Discriminant Analysis is applied to identify the person. If the person does not get identified from the first camera, his/her gait space is augmented as it moves in the views of other cameras and after a certain time interval GDA is again applied. However, we perform a cascaded classification, where in the next classification step, we use only those training classes with which the distance is less than a predefined threshold in the previous step. The probe subspace gets updated as the person walks through the network and the number of training subspaces considered for classification reduces making the system fast and online.

Authors in [15], have proposed a gait recognition method called Sparse Grassmannian Locality Preserving Discriminant Analysis (SGLPDA), where they form a gait energy image of each person. A set of gait energy images are then modeled as a collection of linear subspaces. They formulate the gait recognition problem through the graph embedding framework in [16]. They apply sparse representation along with locality preserving Grassmann Discriminant Analysis to find the inter-and intra-class variations and perform gait recognition. Our framework is an online system where a single global gait space is formed for each individual while theirs is an offline system where more than one subspace exits for the same individual. Moreover, our framework is completely distributed where each camera forms its own gait space based on its view and the gait spaces for the same person are merged to form the global gait space. GDA is applied for finding the distance of the probe gait space from the training gait spaces to be able to identify people who enter the area under observation. This identification is also carried out in real-time, as the individual is moving around in the area under observation and in a distributed manner.

3   Background

3.1   Grassmann Manifold

A Grassmann manifold denoted by, $\mathcal{G}(k, n)$ is a set of k-dimensional linear subspaces in $\mathbb{R}^n$. Each unique subspace is a point on the Grassmann manifold. Therefore, in our framework the gait space of each individual is a point on a Grassmann manifold. The distance between two gait spaces is well-defined and is computed as distance between two points on the Grassmann manifold. The basic premise is therefore, that if the test subject is one of the people allowed to enter the building, the probe gait space should be close to one of the training gait spaces.

In general, the distance between two subspaces is computed using the principle angles between the two subspaces. The distance between the two subspaces on the Grassmann manifold is calculated as the distance between two points on the manifold. Mathematically, a point on the Grassmann manifold, $\mathcal{G}(k, n)$ is represented by an orthonormal matrix $S \in \mathbb{R}^{n \times k}$, where the columns of S span the corresponding k-dimensional subspace in $\mathbb{R}^n$, denoted by $span(S)$.

Two subspaces, $span(S_1)$ and $span(S_2)$, are two points on the Grassmann manifold, S_1 and S_2 respectively. The distance between them is given by Eq. 1.

$$d_{proj}^2(Y_1, Y_2) = \frac{1}{2}\|Y_1 Y_1' - Y_2 Y_2'\|_F^2 \qquad (1)$$

where, Y_1 and Y_2 are the matrix representations of the subspaces $span(S_1)$ and $span(S_2)$. Y_1' and Y_2' are the transpose of matrices Y_1 and Y_2. This also shows that the matrix representation of S_1 and S_2 is directly used for computing the projection distances between the two gait spaces. The advantage of using the projection distance is that it is an unbiased measure as it uses all the principle angles. This is specially beneficial since we have no prior knowledge of the data and all principle angles may be important.

3.2   Grassmann Discriminant Analysis

The Grassmann discriminant analysis [2] framework is specially focused on the problems where the data consist of linear subspaces instead of vectors. As mentioned in Sect. 3.1, a Grassmann manifold is a collection of all linear subspaces of a Euclidean space such that the dimension of all the subspaces is the same. More formally, $\mathcal{G}(k, n)$ is the collection of all linear k-dimensional subspaces of $\mathcal{R}^n$. An element of $\mathcal{G}(k, n)$ is an orthogonal $n \times k$ matrix X. Therefore, X is a point on the Grassmann manifold. Formally, the distance between two points on the Grassmann manifold is the length of the shortest geodesic connecting the two points. However, principal angles between two subspaces provide a more computationally efficient method of defining distances between two points on this manifold.

Let S_1 and S_2 be two points on the Grassmann manifold, then the distance between the two points is computed as the projection distance given by Eq. 1. Then, the projection kernel given by Eq. 2

$$k_P(S_1, S_2) = \|S_1'S_2\|_F \tag{2}$$

is a Grassmann kernel [2]. The Grassmann discriminant analysis algorithm uses the projection kernel in Eq. 2 to perform Kernel LDA using Grassmann kernel.

The GDA algorithm assumes that the subspace bases S_i are already computed. The authors in [2] assume that the subspace bases are computed from the sets in the data using SVD. However, we use the bases computed during the gait space construction phase discussed in Sect. 5. During the training phase, the algorithm finds distances between subspaces S_i and S_j using the projection kernel given by Eq. 2 for all subspaces S_i and S_j in the training set. These distances are stored as a matrix K_{train}. The next step is to solve for the Rayleigh quotient α using Eigen-decomposition and calculate the (C-1)-dimensional coefficients, $F_{train} = \alpha K_{train}$, where C are the class labels.

During the testing phase, first the distance between the test subspaces and all the training subspaces are calculated. Then, the (C-1)-dimensional coefficients, F_{test} are calculated. Finally, 1-NN classification from the Euclidean distance between F_{train} and F_{test} is carried out for classifying the test cases. In Sect. 6, we describe how we adapt GDA for gait recognition in a camera network.

4   Overview of the Framework

We assume that the area under observation is observed by multiple cameras, such that at a time one or more cameras may be observing the person under consideration. We also assume that only a certain set of people are allowed to enter and exit the area under observation.

During the training phase, as these people walk in the area under observation, their gait space is constructed incrementally in each camera of the camera network as discussed in Sect. 5. We apply background subtraction [3] and Incremental Principal Component Analysis (IPCA [1]) to create the gait space on-the-fly for each object. IPCA also gives the advantage of adding new information to the already existing gait space in case new cameras are added to the system. We assume that the gallery data is present in each of the cameras and that the network has a known topology. We define neighbors of a camera C_i as those cameras that can simultaneously view the individual under consideration or view the person as it moves out of the view of C_i. When a person is about to get out of a camera's view, it passes the person's gait space and other relevant information to its neighbors. Then, the new camera augments the person's gait space incrementally using its own view. In case the new camera gets information from more than one camera about the same person, it merges the gait bases to form a single gait space of the person using the method discussed in Sect. 5.1 and then augments it. In such a manner, a global gait space of an individual is formed while the person is tracked across all the cameras in the network. This global gait space is used for classification using Grassmann discriminant analysis.

During the test phase, as the object enters the area under observation, it is tracked and its gait space is formed on-the-fly. Then, Grassmann Discriminant Analysis is applied to identify the person based on the gait signatures created during the training phase. The details are given in Sect. 3.2. We define a confidence measure for identification of the person. However, in case the person does not get recognized from one view, as his/her gait space is augmented by various views, the identification process is carried out periodically. However, for each next identification step, only those training classes are used with which the distance is less than a pre-defined threshold in the previous step. This cascaded recognition makes the identification process fast. Any object that is not recognized with a certain confidence level even after the person has been in the network for a certain time period is flagged as an *unknown* person.

Using IPCA and GDA makes the system robust to addition and deletion of cameras and therefore, makes the system scalable. Deletion of camera does not affect the gallery since extra information does not mislead the system. Moreover, on addition of a camera, IPCA is used to update the gallery. Since IPCA is also used to form the probe's gait space, data of the new camera can be easily incorporated if the probe subject enters its view. Another important feature of our system is that recognition occurs online as the subject is tracked in the views of the various cameras in the network.

5   Forming the Gait Space

We form the gait space using incremental PCA [1] for creating the gait spaces as the person is tracked in each camera. In [17], background subtraction [3] is done to extract the moving object and track it in each frame. The extracted silhouette is then aligned and scaled to obtain a uniform height. This is done for taking into account the errors in background subtraction and height changes when the object moves away from the camera or towards the camera. Then, using a sequence of silhouettes the self-similarity plot (SP) of the person are detected. Then, the *Units of Self-Similarity* (USS), that is, a set of normalized feature vectors that are extracted using these self-similarity plots.

We modify this method to create the gait space incrementally. As the person is tracked, and the foreground silhouette extracted, the corresponding silhouette is scaled to a uniform height. Then, the self-similarity plot is obtained by Eq. 3 between consecutive frames.

$$S(t_1, t_2) = \min_{|dx, dy| < r} \sum_{(x,y) \in B_{t_1}} |O_{t_1}(x + dx, y + dy) - O_{t_2}(x, y)| \qquad (3)$$

where, $1 \leq t_1, t_2 \leq N$, B_{t_1} is the bounding box of the silhouette in frame t_1, r is a small search radius and $O_{t_1}, O_{t_2}, \ldots, O_{t_N}$ are the scaled silhouettes. Since a person's gait is periodic and continuous, the similarity plot is tiled into rectangular blocks, known as Units of Self-Similarity (USS). These USS consists of self-similarity over two periods of gait for each person. Each USS is the gait feature vector corresponding to N frames. We construct the USS's and apply IPCA on these feature vectors to incrementally find the d most significant eigenvectors that contain maximum information about the person's gait from one view. This creates the gait space of the person from one view. Each camera that observes the person creates its gait space in a similar manner. When the object is about to get out of a camera's view, the camera sends the gait space along with the identity of the person to all its neighboring cameras.

If the neighboring camera receives more than one gait space for a particular person, it merges the gait spaces as discussed in Sect. 5.1. Otherwise, the new camera creates its own gait space for the person and merges with the gait space(s) it received for creating a global gait space. In this manner, a global gait space is created for each individual in the training set.

5.1   Merging Two Gait Basis

Our method for merging two gait spaces is based on the method for merging two subspaces as proposed in [18]. Let the two sets of observations be $\mathbf{A}_{n \times N}$ and $\mathbf{B}_{n \times M}$. Then, their corresponding Eigenspace models are denoted by $\Omega = (a, S_{np}, \Lambda_{pp}, N)$ and $\Psi = (b, T_{nq}, \Delta_{qq}, M)$, respectively. The goal is to merge the two spaces and to compute the combined Eigenspace $\Phi = (c, U_{nr}, \Pi_{rr}, P)$ for the combined observation $C_{n(N+M)} = [A_{nN} B_{nM}]$ using only Ω and Ψ. Then, using the Gram-Schmidt orthonormalization [21], we first construct the orthonormal basis set γ_{ns} that spans both Ω and Ψ and $x - y$. The basis Γ_{ns} differs from the required basis U_{ns} by a rotation R_{ss} as given in Eq. 4

$$U_{ns} = \Gamma_{ns} R_{ss} \qquad (4)$$

We then derive another Eigenproblem using the basis Γ_{ns} whose solution gives the eigenvalues Π_{ss} that are required for the merged model. The corresponding eigenvectors R_{ss} form the rotation matrix that is required in Eq. 4. Using R_{ss}, we compute the eigenvectors U_{ns} as given by Eq. 4. The r non-negligible eigenvalues and their corresponding eigenvectors form U_{nr}. Thus, the merged Eigenspace is computed in this manner.

6   Person Identification Through Gait Recognition

Each person walks with a gait that cannot be replicated by another person, making gait a unique biometric. In our framework, during the training phase, the gait space of all the people allowed in the area under observation is formed and stored in each of the cameras.

Person identification is performed online as the person moves in the area under observation. When a person comes into the view of a camera, it starts getting tracked and its gait space is formed incrementally as described in Sect. 5. As a person walks in the area observed by multiple cameras, we assume that the subject is viewed by multiple cameras simultaneously or sequentially. For each of the individuals that are allowed to be present in the area under observation, their gait space is created in the training phase using IPCA. As mentioned before, IPCA gives us the flexibility of adding new cameras without having to re-compute the complete gait space of an individual. For Grassmann discriminant analysis, we form the matrix $K_{train} = k_P(X_i, X_j)$ for all training subspaces X_i and X_j. We then compute the Rayleigh quotient and F_{train} as described in Sect. 3.2.

During the test phase, when an object is detected to have entered the area under observation, its gait space constructed using the method outlined in Sect. 5. After a certain time interval, the distance between this test subspace and the training subspaces of all the people allowed in the area is computed, and the (C-1)-dimensional coefficients, F_{test} are computed. The Euclidean distance between each F_{train} and F_{test} is computed as $d(F_{test}, F_{train}(i))$. We define a confidence measure, given by Eq. 5, as a distance decay function based on the distance between F_{train} and F_{test}.

$$CM(i) = e^{-(d(F_{test}, F_{train}(i)))} \tag{5}$$

The dimension of the gait space of the probe individual changes as the person moves in the area under observation and gets recorded by various cameras in the system. We assume a match with the ith individual, if the confidence measure is above a threshold. However, we perform recognition in a cascaded manner wherein, as the person moves in the area under observation, recognition is performed at fixed time intervals. All the training classes are used in the first attempt, after that in every interval only those training classes are used with which the confidence measure of match is above a per-defined threshold, t_{elim}. In this manner, as the test data increases the dimension of the gait space formed on-the-fly also increases and the chances of a correct match increases. An individual is said to be correctly recognized if the confidence measure of a match is above another pre-defined threshold t_{match}.

An important point to be taken into consideration is that the distance on the Grassmann manifold is measured between two subspaces of the same dimension only. However, the dimension of the training gait spaces is much larger than the test gait space. Therefore, we consider only the first n basis vectors of all the training gait spaces where, n is the dimension of the test gait space in that time interval.

7   Experimental Results

We used the CASIA gait dataset B [13] for our experiments. The dataset consists of 124 subjects. This dataset is a multi-view gait dataset as it was captured from 11 viewing angles. The viewing angles are 0°, 18°, 36°, 54°, 72°, 90°, 108°, 126°, 144°, 162° and 180°. Moreover, 6 gait sequences are captured for each individual under each viewing angle. Therefore, there are a total of 11 × 124 × 6 or 8184 gait sequences. For each person, we form the gait space using two gait sequence for each angle. We started with 0° to form the gait space and then, incrementally create the gait space using two gait sequences for all the views.

For the identification phase, we use a gait sequence of each person that was not used for training. We start with the 0° view and create the gait space of the test subject. We check for identification using the GDA algorithm and then, add data from each of the viewing angles and re-checking the identification using the algorithm in Sect. 6. We find that the identification rate improves as the number of views are increased as shown in Fig. 1. We calculate the Ambiguity resolution measure as given by Eq. 6

$$\text{Ambiguity resolution} = \frac{\text{no. of objects correctly classified}}{\text{total no. of objects}} \tag{6}$$

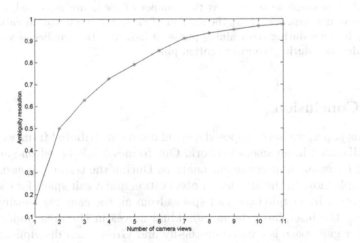

Fig. 1. The graph shows that as the number of views are increased, the recognition rate increases. It can be seen that initially there is a steep rise in the number of subjects correctly classified as the number of views increase, however, after a certain number of views have been considered the graph saturates. The x-axis shows the number of views considered and the y-axis represents the ambiguity resolution calculated by Eq. 6.

We also use the Cumulative Match Characteristic (CMC) to present our results for the GDA based gait recognition. We find that for each person the identification rate increases as the number of views taken in the creating the test gait space is increased. This can be seen by the different curves in Fig. 2. In general, we see that for all the 124 subjects, not all 11 views are required for the identification. In most cases, 5 views were the maximum that was required for recognition while in a few cases 7 views were required for the person to be correctly classified.

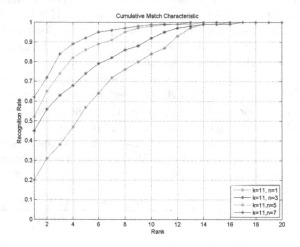

Fig. 2. The graph shows that as the number of views are increased, the recognition rate also increases. $k = 11$ are the number of views taken into consideration for forming the gait space during the training phase. n indicates the number of views taken into consideration during the identification phase.

8   Conclusion

In this paper, we have proposed a novel online, distributed framework for person identification in a camera network. Our framework is based on gait recognition using Grassmann discriminant analysis. During the training phase, a known set of people move in the area under observation and a gait space for each individual is created, by merging the gait spaces from all the cameras viewing the person. During the test phase, as an individual moves in the area under observation, his/her gait space is created on-the-fly and Grassmann discriminant analysis is applied for classifying the individual. Therefore, as a person moves in the area under observation, our system is capable of identifying him/her. In case, the person does not get identified after a certain time interval, we label the person as an *unknown* person.

References

1. Weng, J., Zhang, Y., Hwang, W.S.: Candid covariance-free incremental principal component analysis. IEEE Trans. Pattern Anal. Mach. Intell. **25**(8), 1034–1040 (2003)

2. Hamm, J., Lee, D.D: Grossmann discriminant analysis: a unifying view on subspace based learning. In: International Conference on Machine Learning, pp. 376–383 (2008)
3. Stauffer, C., Grimson, W.E.L.: Adaptive background mixture models for real-time tracking. In: IEEE Conference on Computer Vision and Pattern Recognition (CVPR), vol. 2, pp. 2246–2252 (1999)
4. Bobick, A.F., Johnson, A.Y.: Gait recognition using static, activity-specific parameters. In: IEEE International Conference on Computer Vision and Pattern Recognition, pp. 423–430 (2001)
5. Jang-Hee, Y., Doosung, H., Ki-Young, M., Nixon, M.S: Automated human recognition by gait using neural network. In: First Workshop on Image Processing Theory, Tools and Applications, pp. 1–6 (2008)
6. Boulgouris, N.V., Chi, Z.X.: Human gait recognition based on matching of body components. Pattern Recogn. **40**, 1763–1770 (2007)
7. Xuelong, L., Maybank, S.J., Shuicheng, Y., Dacheng, T., Dong, X.: Gait components and their application to gender recognition. IEEE Trans. Syst. Man Cybern. Part C: Appl. Rev. **38**, 145–155 (2008)
8. Sarkar, S., Phillips, P.J., Liu, Z., Vega, I.R., Grother, P., Bowyer, K.W.: The humanID gait challenge problem: data sets, performance, and analysis. IEEE Trans. Pattern Anal. Mach. Intell. **27**, 162–177 (2005)
9. Bobick, A.F., Davis, J.W.: The recognition of human movement using temporal templates. IEEE Trans. Pattern Anal. Mach. Intell. **23**, 257–267 (2001)
10. Han, J., Bhanu, B.: Individual recognition using gait energy image. IEEE Trans. Pattern Anal. Mach. Intell. **28**, 316–322 (2006)
11. Rabiner, L.R.: A tutorial on hidden Markov models and selected applications in speech recognition. Proc. IEEE **77**(2), 257–286 (1989)
12. Collins, R.T., Gross, R., Jianbo, S.: Silhouette-based human identification from body shape and gait. In: Fifth IEEE International Conference on Automatic Face and Gesture Recognition, pp. 366–371 (2002)
13. CASIA Gait Database. http://www.cbsr.ia.ac.cn/english/Gait%20Databases.asp
14. Han, J., Bhanu, B., Chowdhury, A.K.R.: A study on view-insensitive gait recognition. In: International Conference on Image Processing, pp. 297–300 (2005)
15. Tee, C., Goh, M.K.O., Teoh, A.B.J.: Gait recognition using sparse grassmannian locality preserving discriminant analysis. In: IEEE International Conference on Acoustics, Speech and Signal Processing, pp. 2989–2993 (2013)
16. Yan, S., Xu, D., Zhang, B., Zhang, H.-J., Yang, Q., Lin, S.: Graph embedding and extensions: a general framework for dimensionality reduction. IEEE Trans. Pattern Anal. Mach. Intell. **29**(1), 40–51 (2007)
17. BenAbdelkader, C., Cutler, R., Davis, L.: Motion-based recognition of people in EigenGait space. In: Fifth IEEE International Conference on Automatic Face and Gesture Recognition, pp. 267–272 (2002)
18. Hall, P., Marshall, D., Martin, R.: Merging and splitting eigenspace models. IEEE Trans. Pattern Anal. Mach. Intell. **22**(9), 1042–1049 (2000)
19. Jean, F., Bergevin, R., Branzan, A.: Trajectories normalization for viewpoint invariant gait recognition. In: International Conference on Pattern Recognition, pp. 1–4 (2008)
20. Kusakunniran, W., Wu, Q., Li, H., Zhang, J.: Multiple views gait recognition using view transformation model based on optimized gait energy image. In: International Conference on Computer Vision (ICCV) Workshop, pp. 1058–1064 (2009)
21. Golub, G.H., Van Loan, C.F.: Matrix Computations Johns Hopkins (1983)

22. Makihara, Y., Sagawa, R., Mukaigawa, Y., Echigo, T., Yagi, Y.: Gait recognition using a view transformation model in the frequency domain. In: Leonardis, A., Bischof, H., Pinz, A. (eds.) ECCV 2006. LNCS, vol. 3953, pp. 151–163. Springer, Heidelberg (2006)
23. Bashir, K., Xiang, T., Gong, S.: Cross-view gait recognition using correlation strength. In: British Machine Vision Conference, pp. 1–11 (2010)

Gesture Recognition Performance Score:
A New Metric to Evaluate Gesture
Recognition Systems

Pramod Kumar Pisharady[✉] and Martin Saerbeck

Institute of High Performance Computing (IHPC),
A*STAR, Level 16-16 Connexis, 1 Fusionopolis Way,
Singapore 138632, Singapore
pramodkp@mit.edu, Saerbeckm@ihpc.a-star.edu.sg

Abstract. In spite of many choices available for gesture recognition algorithms, the selection of a proper algorithm for a specific application remains a difficult task. The available algorithms have different strengths and weaknesses making the matching between algorithms and applications complex. Accurate evaluation of the performance of a gesture recognition algorithm is a cumbersome task. Performance evaluation by recognition accuracy alone is not sufficient to predict its successful real-world implementation. We developed a novel Gesture Recognition Performance Score ($GRPS$) for ranking gesture recognition algorithms, and to predict the success of these algorithms in real-world scenarios. The $GRPS$ is calculated by considering different attributes of the algorithm, the evaluation methodology adopted, and the quality of dataset used for testing. The $GRPS$ calculation is illustrated and applied on a set of vision based hand/ arm gesture recognition algorithms reported in the last 15 years. Based on $GRPS$ a ranking of hand gesture recognition algorithms is provided. The paper also presents an evaluation metric namely Gesture Dataset Score (GDS) to quantify the quality of gesture databases. The $GRPS$ calculator and results are made publicly available (http://software.ihpc.a-star.edu.sg/grps/).

1   Introduction

Successful research efforts in gesture recognition within the last two decades paved the path for natural human-computer interaction systems. Challenges like identification of gesturing phase, sensitivity to size, shape, and speed variations, and issues due to occlusion and complex backgrounds keep gesture recognition research still active. One ongoing goal in human-machine interface design is to enable effective and engaging interaction. For example, vision based gesture recognition systems can enable contactless interaction in sterile environments such as hospital surgery rooms, or simply provide engaging controls for entertainment and gaming applications. Other applications of gesture recognition systems include human-robot interaction, augmented reality, surveillance systems,

C.V. Jawahar and S. Shan (Eds.): ACCV 2014 Workshops, Part I, LNCS 9008, pp. 157–173, 2015.
DOI: 10.1007/978-3-319-16628-5_12

behavior analysis systems, and smart phone applications. However current gesture recognition systems are not as robust as standard keyboard and mouse interaction.

Hand gestures are one of the most common category of body language used for communication and interaction. Hand gestures are distinguished based on temporal relationships, into two types; *static* and *dynamic* gestures. Static hand gestures (*aka* hand postures/hand poses) are those in which the hand position does not change during the gesturing period. Static gestures mainly rely on the shape and flexure angles of the fingers. In dynamic hand gestures, the hand position changes continuously with respect to time. Dynamic gestures rely on the hand trajectories and orientations, in addition to the shape and fingers flex angles. Dynamic gestures, which are actions composed of a sequence of static gestures, can be expressed as a temporal combination of static gestures [1].

1.1   Taxonomy of Gesture Recognition Systems

The initial attempts in hand gesture recognition utilized contact sensors that directly measure hand and/or arm joint angles and spatial position, using glove-based devices [2]. Later vision based non-contact methods developed. Based on feature extraction, vision-based gesture recognition systems are broadly divided into two categories, appearance-based methods and three dimensional (3D) hand model-based methods. Appearance-based methods utilize features of training image/video to model the visual appearance, and compare these parameters with the features of test image/video. Three-dimensional model-based methods rely on a 3D kinematic model, by estimating the angular and linear parameters of the model. Appearance based methods are the more widely used approach in gesture recognition with RGB cameras, whereas model based methods are more suitable for the use with new generation RGB-D cameras having skeletal tracking capability.

Mitra *et al.* [3] provided a survey of different gesture recognition methods, covering hand and arm gestures, head and face gestures, and body gestures. The hand gesture recognition methods investigated in the survey include Hidden Markov Models (HMM), particle filtering and condensation algorithms, Finite State Machines (FSM), and Artificial Neural Networks (ANN). Hand modeling and 3D motion based pose estimation methods are reviewed in [4]. An analysis of sign languages, grammatical processes in sign gestures, and issues relevant to the automatic recognition of sign languages are discussed in [5]. The review concluded that the methods studied are experimental and their use is limited to laboratory environments.

1.2   Performance Characterization in Gesture Recognition

A major cause which limits the utility of gesture recognition systems (hardware and software) in real-world applications is the lack of user's expertise to make the right choice of the algorithm, for a specific application in mind. Proper guidance on the type of gestures to be used and algorithms to recognize them is limited. In spite of the vast number of gesture recognition algorithms proposed in recent years, the availability of off-the-shelf gesture recognition softwares and standard APIs remains limited. There exist no standards for hardware or software for gesture recognition systems.

The difficulty in predicting how a given algorithm perform on a new problem makes the performance characterization in computer vision challenging. Thacker et al. [6] provided a review of performance characterization approaches in computer vision. Performance characterization is referred as 'obtaining a sufficiently quantitative understanding of performance that the output data from an algorithm can be interpreted correctly'. The paper reviewed good practices in assessing the performance of essential stages such as sensing, feature detection, localization and recognition in computer vision systems. Some specific topics, face recognition, structural analysis in medical imaging, coding, optical flow, and stereo vision, are explored in depth. The evaluation methods explored for recognition performance characterization include true-false detection metric, receiver-operating characteristics, confusion matrix, and recognition rate. The paper concluded that accurate quantitative performance characterizations should be application specific and it is impossible to define one single measure applicable in all domains.

Ward et al. [7] proposed a set of specific performance metrics for action recognition systems, highlighting the failure of standard evaluation methods borrowed from other related pattern recognition problems. The metrics attempted to capture common artifacts such as event fragmentation, event merging and timing offsets. They extended the standard confusion matrix notion to include eight new error categories. The new metrics are evaluated on a limited set of three algorithms (string matching, Hidden Markov Models, and decision trees).

In this paper we focus on performance characterization in gesture recognition. A new metric called Gesture Recognition Performance Score (GRPS) is proposed which considers a wide range of factors for performance evaluation of gesture recognition algorithms (Sect. 2). Based on GRPS the gesture recognition algorithms are ranked. GRPS is calculated by considering three groups of factors, (i) the algorithm performance, (ii) evaluation methodology followed, and (iii) the quality of the dataset utilized (how challenging the dataset is?) to test the algorithm. GRPS predicts the possibility of an algorithm to be successful in its real-world implementation. It helps the algorithm designer to follow the best practices to make the algorithm effective in real applications. Based on the proposed scoring strategy, we ranked hand gesture and posture recognition algorithms published in the last 15 years and provided a list of 10 top-performing algorithms in each category. Both GRPS calculator and algorithm rankings are publicaly available (http://software.ihpc.a-star.edu.sg/grps/). The dataset evaluation components of the GRPS are utilized to rank a list of publicly available hand gesture and posture datasets (Sect. 3). The paper also discusses possible improvements of the proposed metric and its customization for other related pattern recognition tasks (Sect. 4).

2   The Gesture Recognition Performance Score

Evaluation of a gesture recognition algorithm with recognition accuracy alone is not sufficient to predict its success in real-world applications. Factors such as the number of classes the algorithm can recognize, its person independence, and its robustness to noise and complex environments are also to be considered while evaluating gesture recognition algorithms. Multi-problem benchmarks exist for the performance comparison of hardware and software components like CPUs and compilers. Such application based benchmarks provide a better measure of the real-world performance of a given system. We derived the factors (Table 1) affecting the effectiveness of a gesture recognition system from a survey[1] of algorithms reported in the past 15 years. The proposed *GRPS* is based on the factors listed.

2.1   Components of *GRPS*

We considered 14 component factors in the calculation of *GRPS* (Table 1). The components are divided into three groups based on the factor they depend on (algorithm, methodology, and dataset). Fourteen index scores are calculated from the 14 components and the *GRPS* is calculated as the weighted average of these index scores. The different levels of weight assignment are shown in Fig. 1. The description of each of these components and calculation of index scores are provided in the following subsections.

Accuracy Index. The *GRPS* is proposed due to the limited expressiveness of recognition accuracy of the algorithm about its effectiveness in real world applications. However we considered the recognition accuracy as one component in the *GRPS*, together with other factors affecting the recognition accuracy. The accuracy index X_1 of *GRPS* is calculated from the reported recognition accuracy of the algorithm (1).

$$X_1 = \frac{Number\ of\ correctly\ classified\ samples}{Total\ number\ of\ samples} \times 100 \qquad (1)$$

Spotting Index. The spotting index X_2 of the *GRPS* provides credit to algorithms which can spot (detect) gestures. X_2 is a binary variable representing whether the algorithm has the capability to spot gestures ($X_2 = 1$) or not ($X_2 = 0$).

Class Index. The number of classes a recognition algorithm can discriminate between is a major factor in multi-class pattern recognition. Algorithm which can recognize more number of classes are to be given higher performance scores as

---

[1] We are in the process of publishing a detailed survey on the topic of gesture recognition.

Table 1. Different components of the $GRPS$

No.	Component	Depends on[a]	Deciding factor
1	Accuracy index (X_1)	Algorithm	Recognition accuracy of the algorithm
2	Spotting index (X_2)	Algorithm	Ability of the algorithm to spot gestures
3	Class index (X_3)	Method./Data.	Number of classes considered
4	Subjects index (X_4)	Method./Data.	Number of subjects in the test set
5	Samples index (X_5)	Method./Data.	Number of test samples per class per subject
6	Complexity index (X_6)	Algorithm	Computational complexity of the algorithm
7	Cross validation index (X_7)	Methodology	Cross validation or not
8	Dataset index (X_8)	Dataset	Public or private dataset
9	Availability index (X_9)	Methodology	System availability
10	Background index (X_{10})	Dataset	Complex or simple background
11	Noise index (X_{11})	Dataset	Presence of other human in the background
12	Scale index (X_{12})	Dataset	Variation in scale/size considered or not
13	Lighting index (X_{13})	Dataset	Variation in lighting considered or not
14	Extensibility index (X_{14})	Algorithm	Online or offline learning

[a]Components depend on factors such as algorithm performance, evaluation methodology, and quality of dataset. The factors shown as Method./Data. depend on methodology for algorithm evaluation (Sect. 2.2) and on dataset itself for dataset evaluation (Sect. 3).

those algorithms have better versatility. The class index X_3 of $GRPS$ represents the number of classes the algorithm can handle; the number of classes considered while testing the algorithm. X_3 varies in a non-linear and saturating manner with respect to the number of classes. The value of X_3 saturates at large values of the number of classes. A scaled sigmoidal logistic function (2) is used for the calculation of X_3.

$$X_3 = 2 \times \left(\frac{1}{1 + e^{-4lc}} - 0.5 \right) \tag{2}$$

where c is the number of classes and l represents the slope of the logistic function at the origin. The parameter l is calculated using (3).

$$l = \frac{1}{N_c^{max}} \tag{3}$$

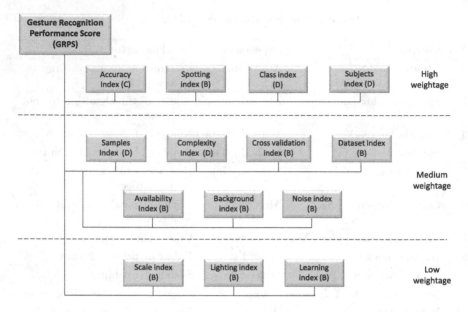

Fig. 1. Components of the *GRPS*. The components are divided into three levels and are given weightage in the ratio 4 : 2 : 1 (top to bottom) for the calculation of *GRPS* (Sect. 2.2). **B** - Binary, **C** - Continuous, **D** - Discrete.

where N_c^{max} is the maximum of the number of classes reported in the literature surveyed (Table 2).

Table 2. Parameters[a] in the calculation of *GRPS*

Parameter	Value	
	Hand gesture	Hand posture
Maximum of the number of classes (N_c^{max})	**120** [8]	**30** [9]
Maximum of the number of subjects (N_s^{max})	**75** [8]	**40** [1]
Maximum of the number of test samples (N_t^{max})	**80** [10]	**100** [11]

[a]The parameters for hand gestures and postures are identified separately (by reviewing the literature), to make the comparison using *GRPS* precise.

The number of classes considered for calculation of *GRPS* is discrete and finite. Identifying continuous blends of discrete gestures is out of scope for this study. Typical applications of gesture recognition systems only require a finite number of classes (for example 26 gestures are needed to represent alphabets in English language). Figure 2 shows the logistic function utilized (corresponding to $N_c^{max} = 120$, the maximum number of dynamic gesture classes reported in the literature [8]). The selection of parameter l as the inverse of maximum number of classes N_c^{max} is intuitive as the component X_3 achieves a value of 0.964 when

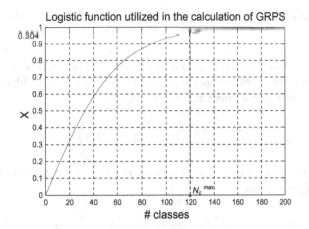

Fig. 2. Logistic function (2) utilized in the calculation of class index X_3 of $GRPS$. The slope of the logistic function at origin is tuned according to the maximum number of classes N_c^{max} (= 120 here). The component X_3 attains a value 0.964 when the number of classes is 120, providing space to consider more number of classes. The logistic functions for calculation of subjects index (X_4) and samples index (X_5) are tuned in a similar manner.

number of classes is 120. This assignment provides future researchers the space to consider more number of classes[2].

Subjects Index. Gesture recognition algorithms are trained using gestures performed by one or more subjects. In order to ensure the person independence and generality of the algorithm, the testing is to be done with the data from multiple persons. The subjects index X_4 of $GRPS$ represents the number of persons from the test data of the algorithm is acquired. A logistic function (4) similar to that for class index is utilized for the calculation of subjects index.

$$X_4 = 2 \times \left(\frac{1}{1 + e^{-4ms}} - 0.5 \right) \tag{4}$$

where s is the number of subjects in the test data. The desired slope m of the logistic function at the origin is calculated by (5).

$$m = \frac{1}{N_s^{max}} \tag{5}$$

where N_s^{max} is the maximum of the number of subjects reported in the literature surveyed (Table 2).

---

[2] Considering larger number of classes will not increase the score much. This is reasonable as the number of gestures used in interaction applications is limited.

Samples Index. The number of samples in the test data of the algorithm is another important factor which decides the reliability of reported recognition performance. Successful recognition of more number of samples with variations shows the algorithm's generality and robustness. The logistic function (6) is utilized to extract the samples index X_5 of the $GRPS$. X_5 considers the number of samples per class per subject in the test data.

$$X_5 = 2 \times \left(\frac{1}{1 + e^{-4nt}} - 0.5 \right) \tag{6}$$

where t is the number of test samples/class/subject. The slope n of the logistic function at origin is given by (7).

$$n = \frac{1}{N_t^{max}} \tag{7}$$

where N_t^{max} is the maximum of the number of test samples/class/subject reported in the literature surveyed (Table 2).

Complexity Index. The computational complexity of the gesture recognition algorithm with respect to the number of classes is a major factor which decides its success in real world implementation. For simplicity we only consider seven categories of worst case complexities (Table 3). The complexity index X_6 is calculated as the inverse of the complexity class number C_N (8).

$$X_6 = \frac{1}{C_N} \tag{8}$$

Cross Validation Index. The cross validation index X_7 of the $GRPS$ provides credit for algorithms tested through cross validation. X_7 is a binary variable representing whether the reported results are average accuracies on cross validation ($X_7 = 1$) or not ($X_7 = 0$).

Table 3. Complexity classes considered in the calculation of $GRPS$

Complexity class number (C_N)	Complexity type
1	Constant
2	Logarithmic
3	Liner
4	Quadratic polynomial
5	Cubic polynomial
6	Higher order (>3) polynomial
7	Exponential or higher

Dataset Index. Sharing data and code is important for replication of systems and the community needs to build on the work of others to make advancements, as in the case of any other scientific discipline [6]. The availability of public (downloadable) gesture datasets was limited till the year 2007 and has been increased recently. Publishing the dataset used to test the algorithm helps other researchers to verify the results and to utilize the database for their own research. Reporting the performance of the algorithm by testing it using a publicly available dataset increases the authenticity of reported results. Dataset index X_8 of the $GRPS$ provides credit to algorithms tested using publicly available datasets. X_8 is a binary variable representing whether the algorithm is tested using publicly available dataset ($X_8 = 1$, credit is also given if the authors of the paper published the dataset used), or whether the algorithm is tested using a dataset private to the authors ($X_8 = 0$).

Availability Index. The availability index X_9 of $GRPS$ provides credit to publicly available algorithms. Making the algorithm available helps other researchers to recreate the study and evaluate the results objectively. X_9 is a binary variable representing whether the source code (or binaries) of the algorithm is available for download ($X_9 = 1$) or not ($X_9 = 0$). This component is included to motivate researchers and developers to make their algorithm available to the community, in spite of the current limited availability of testable gesture recognition algorithms.

Background Index. Backgrounds in real visual scenes are complex. To ensure success in real world application, developers of gesture recognition algorithms should consider complex and cluttered backgrounds with the gesture patterns to be recognized. To provide better ranking to algorithm which can handle complex backgrounds[3], the background index X_{10} is included in the $GRPS$ ($X_{10} = 1$ if the algorithm is tested with complex background data, $X_{10} = 0$ otherwise).

Noise Index. Practical use of gesture recognition systems may need its implementation in crowded places or in places where humans other than the gesturer are present. The noise index X_{11} of the $GRPS$ provides credit ($X_{11} = 1$) to algorithms which are tested using samples with noises such as full or partial human body, and hands or faces of other human in the background.

Scale Index. The size and scale of the gesture varies with relative position of the sensor with respect to gesturer. Algorithms having robustness against size and scale variations of the gestures are given higher credit in the scale index (binary) X_{12} of the $GRPS$. Algorithms which are tested using size/scale variations of the gestures have $X_{12} = 1$ whereas $X_{12} = 0$ for other algorithms.

---

[3] The complexity due to the presence of other objects is considered in background index. The complexity due to the presence of other human (which is more challenging due to skin colored backgrounds) is considered in noise index.

Lighting Index. The practical use of gesture recognition systems requires its operation in indoor and outdoor environments with various lighting conditions. The robustness of the algorithm against lighting variations is another important factor which decides its success in real-world implementation. The lighting index (binary) X_{13} of the $GRPS$ provides credit ($X_{13} = 1$) to algorithm which can handle lighting variations in the scene.

Extensibility Index. Gesture recognition algorithm which can be trained online for new gesturers have flexibility and better utility compared to algorithm which is to be trained offline. The extensibility index (binary) X_{14} of $GRPS$ consider this factor. It provides higher score ($X_{14} = 1$) to algorithm which can be trained online, than algorithm which is to be trained offline ($X_{14} = 0$).

2.2   Calculation of $GRPS$

There are 14 indices in the $GRPS$ as detailed in Sect. 2.1, which collectively decides the overall effectiveness of a gesture recognition algorithm. The influence of different indices of the $GRPS$ in the effectiveness of the algorithm are different. To consider the different levels of influence of the components, three levels of weightage are given to the 14 $GRPS$ indices (Fig. 1). The $GRPS$ is calculated as the weighted mean of the 14 indices in three levels (9–12).

$$GRPS_{c1} = w_1 \times \sum_{i=1}^{i=4} X_i \tag{9}$$

$$GRPS_{c2} = w_2 \times \sum_{i=5}^{i=11} X_i \tag{10}$$

$$GRPS_{c3} = w_3 \times \sum_{i=12}^{i=14} X_i \tag{11}$$

$$GRPS = \frac{GRPS_{c1} + GRPS_{c2} + GRPS_{c3}}{n_1 \times w_1 + n_2 \times w_2 + n_3 \times w_3} \times 100 \tag{12}$$

where,

X_i i^{th} index of $GRPS$,
w_1, w_2, w_3 the three level weights of the components, $= 4, 2,$ and 1 respectively,
n_1, n_2, n_3 the number of indices in $GRPS_{c1}$, $GRPS_{c2}$, and $GRPS_{c3}$, $= 4, 7,$
         and 3 respectively.
     The ideal (maximum possible) value of $GRPS$ is 100. The weights w_1, w_2 and w_3 are selected as per the division of GRPS components into three levels with high, medium, and low weightages (Fig. 1, refer Sect. 4 for a discussion on weight selection).

2.3 Online Web-Portal for *GRPS* Calculation and Algorithm Ranking

A web-portal (http://software.ihpc.a-star.edu.sg/grps/) (Fig. 3) is created to provide the gesture recognition researchers and algorithm developers the facility to calculate the *GRPS* of their algorithm online. The users are prompted to input the values of 14 components of the *GRPS* to calculate the corresponding index scores and the *GRPS*. In addition to the GRPS calculator tool, the portal provides a list[4] of ten top-ranked hand gesture and posture recognition algorithms. Table 4 provides the list of ten top-ranked algorithms at the time this paper is submitted for publication.

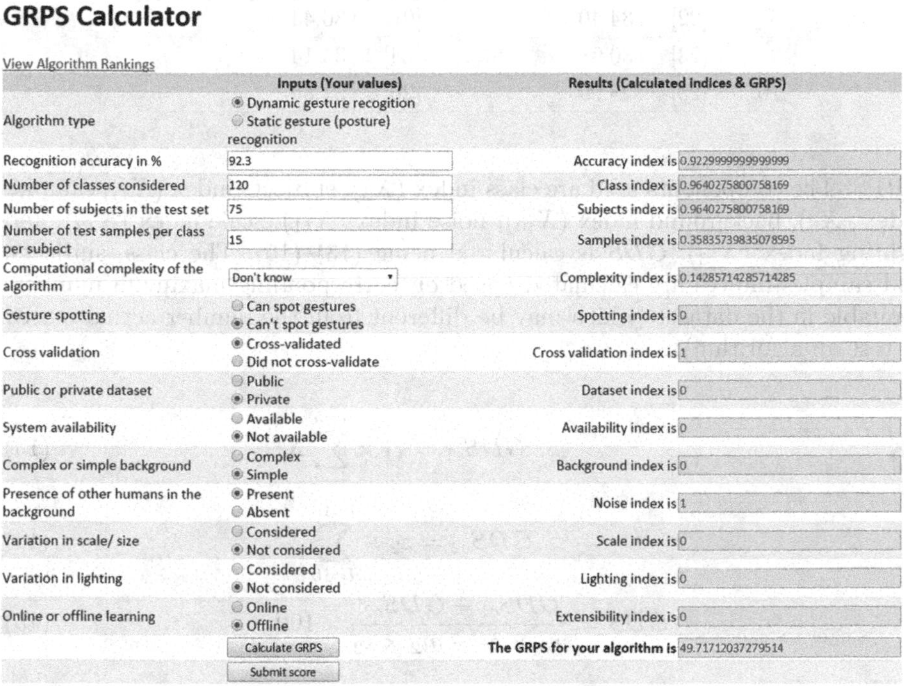

GRPS Calculator

View Algorithm Rankings

	Inputs (Your values)	Results (Calculated Indices & GRPS)
Algorithm type	● Dynamic gesture recognition ○ Static gesture (posture) recognition	
Recognition accuracy in %	92.3	Accuracy index is 0.922999999999999
Number of classes considered	120	Class index is 0.9640275800758169
Number of subjects in the test set	75	Subjects index is 0.9640275800758169
Number of test samples per class per subject	15	Samples index is 0.35835739835078595
Computational complexity of the algorithm	Don't know ▾	Complexity index is 0.14285714285714285
Gesture spotting	○ Can spot gestures ● Can't spot gestures	Spotting index is 0
Cross validation	● Cross-validated ○ Did not cross-validate	Cross validation index is 1
Public or private dataset	○ Public ● Private	Dataset index is 0
System availability	○ Available ● Not available	Availability index is 0
Complex or simple background	○ Complex ● Simple	Background index is 0
Presence of other humans in the background	● Present ○ Absent	Noise index is 1
Variation in scale/ size	○ Considered ● Not considered	Scale index is 0
Variation in lighting	○ Considered ● Not considered	Lighting index is 0
Online or offline learning	○ Online ● Offline	Extensibility index is 0
	Calculate GRPS	The GRPS for your algorithm is 49.71712037279514
	Submit score	

Fig. 3. A screen shot of the GRPS web-portal showing the score calculator.

3 The Gesture Dataset Score

We propose a score namely Gesture Dataset Score (*GDS*) to evaluate quality of publicly available gesture datasets (Table 5). The score quantifies how challenging a dataset is. *GDS* is calculated using the dataset depended components of

---

[4] The list will be maintained and updated regularly. The portal provides authors of research papers a provision to submit their GRPS score and paper details to be included in the ranking list.

Table 4. Top-ranked algorithms and their *GRPS*

Rank	Hand gesture recognition		Hand posture recognition	
	Work	*GRPS*	Work	*GRPS*
1	[8]	56.57	[1]	67.06
2	[12]	41.20	[13]	53.08
3	[10]	40.31	[11]	51.94
4	[14]	37.49	[15]	43.85
5	[16]	35.92	[17]	42.28
6	[18]	34.79	[19]	39.95
7	[20]	34.50	[21]	38.06
8	[22]	34.40	[9]	36.44
9	[23]	30.98	[24]	34.14
10	[25]	28.99	[26]	31.66

GRPS. The components used are class index (X_3), subjects index (X_4), samples index (X_5), background index (X_{10}), noise index (X_{11}), scale index (X_{12}), and lighting index (X_{13}). *GDS* is calculated using (13)–(15). The class, subjects, and samples indices are calculated based on corresponding maximum numbers available in the dataset (which may be different from the number actually used to test an algorithm).

$$GDS_{c1} = z_1 \times \sum_{i=3}^{i=5} X_i \tag{13}$$

$$GDS_{c2} = z_2 \times \sum_{i=10}^{i=13} X_i \tag{14}$$

$$GDS = \frac{GDS_{c1} + GDS_{c2}}{m_1 \times z_1 + m_2 \times z_2} \times 100 \tag{15}$$

where,

X_i i^{th} index of *GRPS*,

z_1, z_2 the two level weights of the components = 2 and 1 respectively,

m_1, m_2 the number of indices in GDS_{c1} and GDS_{c2}, = 3 and 4 respectively.

   The ideal (maximum possible) value of *GDS* is 100. Table 5 provides the *GDS* based rank list of gesture datasets. The score varied from 53.87 to 78.32. On comparison with the competence of algorithms the datasets are more competitive and challenging considering the high values of *GDS*.

4   Discussion

The proposed Gesture Recognition Performance Score evaluates gesture recognition algorithms more effectively than the evaluation using recognition accuracy

Table 5. List of publicly available hand gesture databases and their GDS

Rank	Name, Year	Works	GDS
1	ChaLearn gesture gata, 2011	[10, 11, 27–30]	78.32
2	MSRC-12 Kinect gesture dataset, 2012	[31]	72.71
3	ChaLearn multi-modal gesture data, 2013	[32]	70.93
4	NUS hand posture dataset-II, 2012	[1]	69.37
5	6D motion gesture database, 2011	[33]	66.91
6	Sebastien Marcel interact play database, 2004	[22, 34]	65.10
7	NATOPS aircraft handling signals database, 2011	[35]	62.44
8	Sebastien Marcel hand posture and gesture datasets, 2001	[13, 36–38]	61.73
9	Gesture dataset by Shen *et al.*, 2012	[39]	59.94
10	Gesture dataset by Yoon *et al.*, 2001	[14]	57.82
11	ChAirGest multi-modal dataset, 2013	[40]	56.73
12	Sheffield Kinect Gesture (SKIG) Dataset, 2013	[41]	55.21
13	Keck gesture dataset, 2009	[42]	54.98
14	NUS hand posture dataset-I, 2010	[43]	54.19
15	Cambridge hand gesture data set, 2007	[44]	53.87

alone. For example Patwardhan and Roy [45] reported a recognition accuracy of 100 % for their algorithm. However the experiments are conducted on an 8 class private dataset, collected from only one subject, without considering complex backgrounds/ noises, and without a cross validation. The $GRPS$ rated the algorithm with a score of 20.77. The algorithm by Ramamoorthy *et al.* [18] received a score of 34.79, even though it provided only 81.71 % recognition accuracy. The higher $GRPS$ of the algorithm is due to the experiments with more number of subjects (5) and test samples (14/class/subject), in extreme testing conditions (complex backgrounds, lighting variations), and by considering external noises (face of the posturer, other human in the background).

The maximum values of reported $GRPS$ are 56.57 and 67.06 for gesture and posture recognition algorithms respectively. This points out the scope for improving current gesture recognition systems and the testing methodology followed. For example the person independence of algorithms is to be improved and the algorithm testing is to be conducted in environments outside laboratory, to enhance its performance in complex scenarios. The different factors to be considered while evaluating the performance of a gesture recognition system are listed in the paper, motivating researchers to develop and test new algorithms using competitive methodology, in challenging environments.

4.1   Selection of Weights

The weights w_1, w_2 and w_3 in the $GRPS$ calculation (12) are selected based on the preferences given to the different $GRPS$ indices. The three weights are selected such that $w_1 > w_2 > w_3$ which gives high, medium, and low weightages to the three classes of indices (Fig. 1). The reported results are achieved with $w_1 = 4, w_2 = 2$, and $w_3 = 1$. Our experiments have shown that there is no major changes in the algorithm comparison and ranking with variations in weights, provided the rule $w_1 > w_2 > w_3$ is followed.

4.2   Possible Improvements

The effectiveness of proposed $GRPS$ in comparing gesture recognition algorithms can be improved by including objective measures of problem size (or problem difficulty) in the $GRPS$ calculation. For example including measures of interclass similarity and speed of gestures will help to give credit to algorithms which are discriminative (which can discriminate classes with higher interclass similarity), and which can recognize gestures in spite of its high speed. Another possible improvement of the $GRPS$ is its modification by considering different levels of noises, scale and lighting variations to refine on its components X_{11}, X_{12} and X_{13} respectively.

4.3   Customization for Specific Applications and Other Recognition Tasks

The $GRPS$ measure could be customized for the evaluation of gesture recognition algorithms for specific applications, by adjusting the weights of its constituent indices. For example the accuracy index (X_1) could be given higher weightage to the class index (X_3) in the case of vision system for doctor-computer interaction in a surgery room, whereas X_3 could be given higher weightage to X_1 in the case of vision system for a social robot operating in a supermarket.

The proposed performance score could be extended to other recognition tasks like face, object, and action recognition with necessary modifications in the constituent components. For example the presence of complex backgrounds is not relevant for the evaluation of a face recognition algorithm (as robust face detection algorithms are available), and the number of subjects is not applicable for object recognition.

5   Conclusion

The quantitative performance characterization of pattern recognition systems is a challenging task. We took an initial step in this direction and proposed novel evaluation methods for gesture recognition algorithms and gesture datasets. The proposed scores provided ranking for both algorithms and datasets. We are currently preparing a detailed survey of gesture recognition algorithms with qualitative comparison of the ranked algorithms. The quantitative comparison using

GRPS will be supported by testing the top ranked algorithms under same conditions to extract reliable scientific conclusions on gesture recognition systems.

Acknowledgement. The authors would like to thank Mr. Joshua Tan Tang Sheng for helping in the implementation of online web-portal for the calculation of *GRPS*.

References

1. Pisharady, P.K., Vadakkepat, P., Loh, A.P.: Attention based detection and recognition of hand postures against complex backgrounds. Int. J. Comput. Vision **101**, 403–419 (2013)
2. Dipietro, L., Sabatini, A.M., Dario, P.: A survey of glove-based systems and their applications. IEEE Trans. Syst. Man Cybern. Part C: Appl. Rev. **38**, 461–482 (2008)
3. Mitra, S., Acharya, T.: Gesture recognition: A survey. IEEE Trans. Syst. Man Cybern. Part C: Appl. Rev. **37**, 311–324 (2007)
4. Erol, A., Bebis, G., Nicolescu, M., Boyle, R.D., Twombly, X.: Vision-based hand pose estimation: A review. Comput. Vis. Image Underst. **108**, 52–73 (2007)
5. Ong, S.C.W., Ranganath, S.: Automatic sign language analysis: A survey and the future beyond lexical meaning. IEEE Trans. Pattern Anal. Mach. Intell. **27**, 873–891 (2005)
6. Thacker, N.A., Clark, A.F., Barron, J.L., Beveridge, J.R., Courtney, P., Crum, W.R., Ramesh, V., Clark, C.: Performance characterization in computer vision: A guide to best practices. Comput. Vis. Image Underst. **109**, 305–334 (2008)
7. Ward, J.A., Lukowicz, P., Gellersen, H.W.: Performance metrics for activity recognition. ACM Trans. Intell. Syst. Technol. 02, 6:01–6:23 (2011)
8. Lichtenauer, J.F., Hendriks, E.A., Reinders, M.J.T.: Sign language recognition by combining statistical dtw and independent classification. IEEE Trans. Pattern Anal. Mach. Intell. 30 (2008)
9. Teng, X., Wu, B., Yu, W., Liu, C.: A hand gesture recognition system based on local linear embedding. J. Vis. Lang. Comput. **16**, 442–454 (2005)
10. Lui, Y.M.: A least squares regression framework on manifolds and its application to gesture recognition. In: IEEE Conference on Computer Vision and Pattern Recognition Workshops (CVPRW), pp. 13–18 (2012)
11. Keskin, C., Kirac, F., Kara, Y., Akarun, L.: Randomized decision forests for static and dynamic hand shape classification. In: IEEE Conference on Computer Vision and Pattern Recognition Workshops (CVPRW), pp. 31–46 (2012)
12. Yang, M.H., Ahuja, N., Tabb, M.: Extraction of 2d motion trajectories and its application to hand gesture recognition. IEEE Trans. Pattern Anal. Mach. Intell. **24**, 1061–1074 (2002)
13. Triesch, J., Malsburg, C.: A system for person-independent hand posture recognition against complex backgrounds. IEEE Trans. Pattern Anal. Mach. Intell. **23**, 1449–1453 (2001)
14. Yoon, H.S., Soh, J., Bae, Y.J., Yang, H.S.: Hand gesture recognition using combined features of location, angle, and velocity. Pattern Recogn. **34**, 1491–1501 (2001)
15. Huang, D.Y., Hu, W.C., Chang, S.H.: Gabor filter-based hand-pose angle estimation for hand gesture recognition under varying illumination. Expert Syst. Appl. **38**, 6031–6042 (2011)

16. Chen, F.S., Fu, C.M., Huang, C.L.: Hand gesture recognition using a real-time tracking method and hidden markov models. Image Vis. Comput. **21**, 745–758 (2003)
17. Zhou, R., Junsong, Y., Zhengyou, Z.: Robust hand gesture recognition based on finger-earth movers distance with a commodity depth camera. In: Proceedings of ACM Multimeida (2011)
18. Ramamoorthy, A., Vaswani, N., Chaudhury, S., Banerjee, S.: Recognition of dynamic hand gestures. Pattern Recogn. **36**, 2069–2081 (2003)
19. Licsar, A., Sziranyi, T.: User-adaptive hand gesture recognition system with interactive training. Image Vis. Comput. **23**, 1102–1114 (2005)
20. Lai, K., Konrad, J., Ishwar, P.: A gesture-driven computer interface using kinect. In: IEEE Southwest Symposium on Image Analysis and Interpretation (SSIAI), pp. 185–188 (2012)
21. Van den Bergh, M., Carton, D., De Nijs, R., Mitsou, N., Landsiedel, C., Kuehnlenz, K., Wollherr, D., Van Gool, L., Buss, M.: Real-time 3d hand gesture interaction with a robot for understanding directions from humans. In: IEEE International Symposium on Robot and Human Interactive Communication (IEEE RO-MAN) (2011)
22. Just, A., Marcel, S.: A comparative study of two state-of-the-art sequence processing techniques for hand gesture recognition. Comput. Vis. Image Underst. **113**, 532–543 (2009)
23. Frolova, D., Stern, H., Berman, S.: Most probable longest common subsequence for recognition of gesture character input. IEEE Trans. Cybern. **43**, 871–880 (2013)
24. Ge, S.S., Yang, Y., Lee, T.H.: Hand gesture recognition and tracking based on distributed locally linear embedding. Image Vis. Comput. **26**, 1607–1620 (2008)
25. Shin, M.C., Tsap, L.V., Goldof, D.B.: Gesture recognition using bezier curves for visualization navigation from registered 3-d data. Pattern Recogn. **37**, 1011–1024 (2004)
26. Zhao, M., Quek, F.K.H., Wu, X.: Rievl: Recursive induction learning in hand gesture recognition. IEEE Trans. Pattern Anal. Mach. Intell. **20**, 1174–1185 (1998)
27. Guyon, I., Athitsos, V., Jangyodsuk, P., Hamner, B., Escalante, H.: Chalearn gesture challenge: Design and first results. In: IEEE Conference on Computer Vision and Pattern Recognition Workshops (CVPRW), pp. 1–6 (2012)
28. Malgireddy, M.R., Inwogu, I., Govindaraju, V.: A temporal bayesian model for classifying, detecting and localizing activities in video sequences. In: IEEE Conference on Computer Vision and Pattern Recognition Workshops (CVPRW), pp. 43–48 (2012)
29. Di, W., Fan, Z., Ling, S.: One shot learning gesture recognition from rgbd images. In: IEEE Conference on Computer Vision and Pattern Recognition Workshops (CVPRW) (2012)
30. Mahbub, U., Imtiaz, H., Roy, T., Rahman, M., Ahad, M.: A template matching approach of one-shot-learning gesture recognition. Pattern Recogn. Lett. (2012). http://dx.doi.org/10.1016/j.bbr.2011.03.031
31. Simon, F., Helena, M.M., Pushmeet, K., Sebastian, N.: Instructing people for training gestural interactive systems. In: International Conference on Human Factors in Computing Systems, CHI, pp. 1737–1746. ACM (2012)
32. Escalera, S., Gonzlez, J., Bar, X., Reyes, M., Lopes, O., Guyon, I., Athitsos, V., Escalante, H.: Multi-modal gesture recognition challenge 2013: Dataset and results. In: Proceedings of the 15th ACM International Conference on Multimodal Interaction (ICMI), Sydney, Australia (2013)

33. Chen, M., AlRegib, G., Juang, B.H.: 6dmg: A new 6d motion gesture database. In: IEEE Conference on Computer Vision and Pattern Recognition Workshops (CVPRW) (2011)
34. Just, A., Bernier, O., Marcel, S.: Hmm and iohmm for the recognition of mono- and bi-manual 3d hand gestures. In: Proceedings of the British Machine Vision Conference (BMVC) (2004)
35. Song, Y., Demirdjian, D., Davis, R.: Tracking body and hands for gesture recognition: Natops aircraft handling signals database. In: Proceedings of the 9th IEEE Conference on Automatic Face and Gesture Recognition (FG 2011), Santa Barbara, CA, pp. 500–506 (2011)
36. Triesch, J., Malsburg, C.: Robust classification of hand postures against complex backgrounds. In: Proceedings of the Second International Conference on Automatic Face and Gesture Recognition, Killington, VT, USA, pp. 170–175 (1996)
37. Triesch, J., Malsburg, C.: A gesture interface for human-robot-interaction. In: Proceedings of the Third IEEE International Conference on Automatic Face and Gesture Recognition, Nara, Japan, pp. 546–551 (1998)
38. Marcel, S.: Hand posture recognition in a body-face centered space. In: Proceedings of the Conference on Human Factors in Computer Systems (CHI) (1999)
39. Shen, X.H., Hua, G., Williams, L., Wu, Y.: Dynamic hand gesture recognition: An exemplar-based approach from motion divergence fields. Image Vis. Comput. **30**, 227–235 (2012)
40. Ruffieux, S., Lalanne, D., Mugellini, E.: Chairgest: A challenge for multimodal mid-air gesture recognition for close hci. In: Proceedings of the 15th ACM on International Conference on Multimodal Interaction (ICMI) (2013)
41. Liu, L., Shao, L.: Learning discriminative representations from rgb-d video data. In: Proceedings of International Joint Conference on Artificial Intelligence (IJCAI) (2013)
42. Zhuolin, J., Davis, L.S.: Recognizing actions by shape-motion prototype trees. In: IEEE International Conference on Computer Vision (ICCV), pp. 444–451 (2009)
43. Pisharady, P.K., Vadakkepat, P., Loh, A.P.: Hand posture and face recognition using a fuzzy-rough approach. Int. J. Humanoid Rob. **07**, 331–356 (2010)
44. Kim, T.K., Wong, S.F., Cipolla, R.: Tensor canonical correlation analysis for action classification. In: IEEE Conference on Computer Vision and Pattern Recognition (CVPR), pp. 1–8 (2007)
45. Patwardhan, K.S., Roy, S.D.: Hand gesture modelling and recognition involving changing shapes and trajectories, using a predictive eigentracker. Pattern Recogn. Lett. **28**, 329–334 (2007)

Second International Workshop on Big Data in 3D Computer Vision

Object Recognition in 3D Point Cloud of Urban Street Scene

Pouria Babahajiani[1]([✉]), Lixin Fan[1], and Moncef Gabbouj[2]

[1] Nokia Research Center, Tampere, Finland
ext-pouria.babahajiani@nokia.com, fanlixin@ieee.org
[2] Tampere University of Technology, Tampere, Finland
moncef.gabbouj@tut.fi

Abstract. In this paper we present a novel street scene semantic recognition framework, which takes advantage of 3D point clouds captured by a high-definition LiDAR laser scanner. An important problem in object recognition is the need for sufficient labeled training data to learn robust classifiers. In this paper we show how to significantly reduce the need for manually labeled training data by reduction of scene complexity using non-supervised ground and building segmentation. Our system first automatically segments grounds point cloud, this is because the ground connects almost all other objects and we will use a connect component based algorithm to oversegment the point clouds. Then, using binary range image processing building facades will be detected. Remained point cloud will grouped into voxels which are then transformed to super voxels. Local 3D features extracted from super voxels are classified by trained boosted decision trees and labeled with semantic classes e.g. tree, pedestrian, car, etc. The proposed method is evaluated both quantitatively and qualitatively on a challenging fixed-position *Terrestrial Laser Scanning* (TLS) Velodyne data set and two *Mobile Laser Scanning* (MLS), Paris-rue-Madam and NAVTEQ True databases. Robust scene parsing results are reported.

1 Introduction

Automatic urban scene objects recognition refers to the process of segmentation and classifying of objects of interest into predefined semantic labels such as building, tree or car etc. This task is often done with a fixed number of object categories, each of which requires a training model for classifying scene components. While many techniques for 2D object recognition have been proposed, the accuracy of these systems is to some extent unsatisfactory because 2D image cues are sensitive to varying imaging conditions such as lighting, shadow etc. In this work, we propose a novel automatic scene parsing approach which takes advantage of 3D geometrical features extracted from Light Detection And Ranging (LiDAR) point clouds. Since such 3D information is invariant to lighting and shadow, as a result, significantly more accurate parsing results are achieved.

© Springer International Publishing Switzerland 2015
C.V. Jawahar and S. Shan (Eds.): ACCV 2014 Workshops, Part I, LNCS 9008, pp. 177–190, 2015.
DOI: 10.1007/978-3-319-16628-5_13

While a laser scanning or LiDAR system provides a readily available solution for capturing spatial data in a fast, efficient and highly accurate way, the enormous volume of captured data often come with no semantic meanings. We, therefore, develop techniques that significantly reduce the need for manual labelling of training data and apply the technique to the all data sets. Laser scanning can be divided into three categories, namely, *Airborne Laser Scanning* (ALS), *Terrestrial Laser Scanning* (TLS) and *Mobile Laser Scanning* (MLS). The proposed method is evaluated both quantitatively and qualitatively on a challenging TLS Velodyne data set and two MLS, Paris-rue-Madam and NAVTEQ True databases.

1.1 Literature Review

Automatic scene parsing is a traditional computer vision problem. Many successful techniques have used single 2D image appearance information such as color, texture and shape [1,2]. By using just spatial cues such as surface orientation and vanishing points extracted from single images considerably more robust results are achieved [3]. In order to alleviate sensitiveness to different image capturing conditions, many efforts have been made to employ 3D scene features derived from single 2D images and thus achieving more accurate object recognition [4]. For instance, when the input data is a video sequence, 3D cues can be extracted using Structure From Motion (SFM) techniques [5]. With the advancement of LiDAR sensors and Global Positioning Systems (GPS), large-scale, accurate and dense point cloud are created and used for 3D scene parsing purpose. In the past, research related to 3D urban scene analysis had been often performed using 3D point cloud collected by airborne LiDAR for extracting vegetation and building structures [6]. Hernndez and Marcotegui use range images from 3D point clouds in order to extract k-flat zones on the ground and use them as markers for a constrained watershed [7]. Recently, classification of urban street objects using data obtained from mobile terrestrial systems has gained much interest because of the increasing demand of realistic 3D models for different objects common in urban era. A crucial processing step is the conversion of the laser scanner point cloud to a voxel data structure, which dramatically reduces the amount of data to process. Yu Zhou and Yao Yu (2012) present a voxel-based approach for object classification from TLS data [8]. Classification using local features and descriptors such as Spin Image [9], Spherical Harmonic Descriptors [10], Heat Kernel Signatures [11], Shape Distributions [12], and 3D SURF feature [13] have also demonstrated successful results to various extent.

1.2 Overview of the Proposed Framework

In this work, the ground is first segmented and building facades are subsequently detected based on range image morphological operations. We use voxel segmentation that relies on local features and descriptors, to successfully classify different segmented objects in the urban scene.

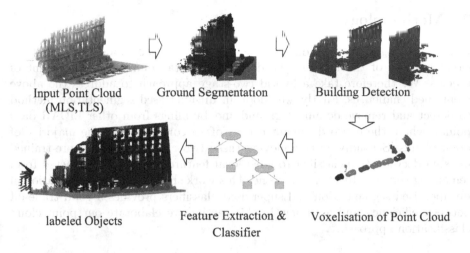

Fig. 1. Overview of the proposed framework

Figure 1 shows the overview of the proposed street scene object recognition pipeline, in which LiDAR Point Cloud (PC) is the input of the processing pipeline and result is PC segments assigned with different class labels. At the outset, the proposed parsing pipeline finds ground points by fitting a ground plane to the given 3D point cloud of urban street scene. Then, non-ground point cloud are projected to range images because they are convenient structure for visualization. Remaining data are processed subsequently to segment building facades. When this process is completed, range images are projected to the 3D point cloud in order to make segmentation on other remained vertical objects. We use a connect component based algorithm to voxilisation of data. The voxel based classification method consists of three steps, namely, (a) voxilisation of point cloud, (b) merging of voxels into super-voxels and (c) the supervised scene classification based on discriminative features extracted from super-voxels.

Using a trained boosted decision tree classifier, each 3D feature vector is then designated with a semantic label such as tree, car, pedestrian etc. The offline training of the classifier is based on a set of 3D features, which are associated with manually labeled super-voxels in training point cloud. Main contributions of this work are the following:

- Develop a novel street object recognition method which is robust to different types of LiDAR point clouds acquisition methods.
- Proposed two-stage (supervised and non-supervised) classification pipeline which requires only small amount of time for training.
- Propose to use novel geometric features leads to more robust classification results (see Sect. 3).

2   Methodology

It is a challenging task to directly extract objects from mobile LiDAR point cloud because of the noise in the data, huge data volume and movement of objects. We therefore take a hybrid two-stage approach to address the above mentioned challenges. Firstly, we adopt an unsupervised segmentation method to detect and remove dominant ground and buildings from other LiDAR data points, where these two dominant classes often correspond to the majority of point clouds. Secondly, after removing these two classes, we use a pre-trained boosted decision tree classifier to label local feature descriptors extracted from remaining vertical objects in the scene. This work shows that the combination of unsupervised segmentation and supervised classifiers provides a good trade-off between efficiency and accuracy. In this section we elaborate our point cloud classification approach.

2.1   Ground Segmentation

The aim of the first step is to remove points belonging to the scene ground including road and sidewalks, and as a result, the original point cloud are divided into ground and vertical object point clouds (Fig. 2). The scene point cloud is first divided into sets of 10 m × 10 m regular, non-overlapping tiles along the horizontal x–y plane. Then the following ground plane fitting method is repeatedly applied to each tile. We assume that ground points are of relatively small z values as compared to points belonging to other objects such as buildings or trees (see Fig. 2). The ground is not necessarily horizontal, yet we assume that there is a constant slope of the ground within each tile. Therefore, we first find the *minimal-z-value* (MZV) points within a multitude of 25 cm × 25 cm grid cells at different locations. For each cell, neighboring points that are within a z-distance threshold from the MZV point are retained as *candidate ground points*. Subsequently, a RANSAC method is adopted to fit a plane to candidate ground points that are collected from all cells. Finally, 3D points that are within certain distance (d2 in Fig. 2) from the fitted plane are considered as ground points of each tile. The constant slope assumption made in this approach is valid for our data sets as demonstrated by experimental results in Sect. 3. The approach is fully automatic and the change of two thresholds parameters do not lead to dramatic change in the results. On the other hand, the setting of grid cell size as 25 cm × 25 cm maintains a good balance between accuracy and computational complexity.

2.2   Building Segmentation

After segmenting out the ground points from the scene, we present an approach for automatic building surface detection. High volume of 3D data impose serious challenge to the extraction of building facades. Our method automatically extract building point cloud (e.g. facades) based on two assumptions: (a) building facades are the highest vertical structures in the street; and (b) other

Fig. 2. Ground Segmentation. Left image: Segmented ground and remained vertical objects point cloud are illustrated by red and black color respectively. Right figure: sketch map of fitting plane to one tile (Color online figure).

non-building objects are located on the ground between two sides of street. As can be seen in Fig. 3, our method projects 3D point clouds to range images because they are convenient structures to process data. Range images are generated by projecting 3D points to horizontal x–y plane. In this way, several points are projected on the same range image pixel. We count the number of points that falls into each pixel and assign this number as a pixel *intensity* value. In addition, we select and store the maximal height among all projected points on the same pixel as *height* value. We define range images by making threshold and binarization of I, where I pixel value is defined as Eq. (1)

$$I_i = \frac{P_{intensity}}{Max\_P_{intensity}} + \frac{P_{height}}{Max\_P_{height}} \tag{1}$$

where I_i is grayscale range image pixel value, $P_{intensity}$ and P_{height} are intensity and height pixel value and $Max\_P_{intensity}$ and Max_P_{height} represent the maximum intensity and height value over the grayscale image.

In the next step we use morphological operation (e.g. close and erode) to merge neighboring point and filling holes in the binary range images (see middle image in Fig. 3). Then we extract contours to find boundaries of objects. In order to trace contours, Pavlidis contour-tracing algorithm [14] is proposed to identify each contour as a sequence of edge points. The resulting segments are checked on aspects such as size and diameters (height and width) to distinguish building from other objects. More specifically, Eq. (2) defines the geodesic elongation E(X), introduced by Lantuejoul and Maisonneuve (1984), of an object X, where S(X) is the area and L(X) is the geodesic diameter.

$$E(\pi) = \frac{\pi L^2(X)}{4S(X)} \tag{2}$$

The compactness of the polygon shape based on equation (2) can be applied to distinguish buildings from other objects such as trees. Considering the sizes and shape of buildings, the extracted boundary will be eliminated if its size is less than a threshold. The proposed method takes advantage of priori knowledge about urban scene environment and assumes that there are not any important objects laid on the building facades. While this assumption appears to be oversimplified, the method actually performs quite well with urban scenes as demonstrated in the experimental results (see Sect. 3).

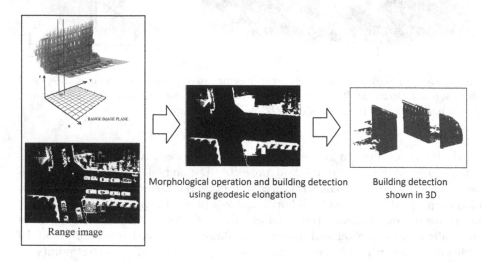

Fig. 3. Building Segmentation

The resolution of range image is the only projection parameter during this point cloud alignment that should be chosen carefully. If each pixel in the range image cover large area in 3D space too many points would be projected as one pixel and fine details would not be preserved. On the other hand, selecting large pixel size compared to real world resolution leads to connectivity problems which would no longer justify the use of range images. In our experiment, a pixel corresponds to a square of size .05 m^2.

2.3 Voxel Based Segmentation

After quick segmenting out the ground and building points from the scene, we use an inner street view based algorithm to cluster point clouds. Although top view range image analysis generates a very fast segmentation result, there are a number of limitation to utilize it for the small vertical object such as pedestrian and cars. These limitations are overcome by using inner view (lateral) or ground based system in which, unlike top view the 3D data processing is done more precisely and the point view processing is closer to objects which provides a more detailed sampling of the objects.

However, this leads to both advantages and disadvantages when processing the data. The disadvantage of this method s includes the demand for more processing power required to handle the increased volume of 3D data. The 3D point clouds by themselves contain a limited amount of positional information and they do not illustrate color and texture properties of object. According to voxel based segmentation, points which are merely a consequence of a discrete sampling of 3D objects are merged into clusters voxels to represent enough discriminative features to label objects. 3D features such as intensity, area and normal angle are extracted based on these clustersvoxels. The voxel based classification method consists of three steps, voxilisation of point cloud, merging of voxels into super-voxels and the supervised classification based on discriminative features extracted from super-voxels.

2.3.1   Voxelisation of Point Cloude

In the voxelisation step, an unorganized point cloud p is partitioned into small parts, called voxel v. The middle image in Fig. 4 illustrates an example of voxelisation results, in which small vertical objects point cloud such as cars are broken into smaller partition. Different voxels are labelled with different colors. The aim of using voxelisation is to reduce computation complexity by and to form a higher level representation of point cloud scene. Following [8], a number of points is grouped together to form a variable size voxels. The criteria of including a new point p_{in} into an existing voxel i is essentially determined by the crucial minimal distance threshold d_{th} which is defined as Eq. (3):

$$\min(\|p_{im} - p_{in}\|_2) \le d_{th}, 0 \le m, n \le N, m \ne n \tag{3}$$

where p_{im} is an existing 3D point in voxel, p_{in} is a candidate point to merge to the voxel, i is the cluster index, d_{th} is the maximum distance between two point, and N is the maximum point number of a cluster. If the condition is met, the new point is added and the process repeats until no more point that satisfies the condition is found (see Algorithm 1). Equation (3) ensures that the distance between one point and its nearest neighbors belonging to the same cluster is less than d_{th}. Although the maximum voxel size is predefined, the actual voxel sizes depend on the maximum number of points in the voxel (N) and minimum distance between the neighboring points.

repeat
   Select a 3D point for Voxelisation;
   Find all neighboring points to be included in the voxel, with this condition that:
   a point p_{in} directly merge to voxel if its distance to any point p_{in} the voxel will not be farther away than a given distance (d_{th});
until *all 3D points are used in a voxel or the size of cluster is less than (N)*;

Algorithm 1. Voxelisation

2.3.2 Super Voxelisation

For transformation of a voxel to super voxel we propose an algorithm to merge voxels via region growing with respect to the following properties of clusters:

Input Point CLoud   voxel   S-voxel

Fig. 4. Voxelisation of Point Cloud. from left to right: top view row point cloud, voxelisation result of objects point cloud after removing ground and building, s-voxelisation approach of point cloud

- **If the minimal geometrical distance, D_{ij}, between two voxels is smaller than a given threshold**, where D_{ij} is defined as Eq. (4):

$$D_{ij} = \min(\|p_{ik} - p_{jl}\|_2), k \in (1, m), l \in (1, n) \qquad (4)$$

where voxels v_i and v_j have m and n points respectively, and p_{ik} and p_{jl} are the 3D point belong to voxel v_i and v_j.

- **If the angle between Normal vectors of two voxels is smaller than a threshold:** In this work, normal vector is calculated using PCA (Principal Component Analysis) [15]. The angle between two s-voxels is defined as angle between their normal vectors Eq. 5:

$$\Theta_{ij} = \arccos(< n_i, n_j >) \qquad (5)$$

where n_i and n_j are normal vectors at v_i and v_j respectively.

The proposed grouping algorithm merges the voxels by considering the geometrical distance ($M < d_{th}$) and normal features of clusters ($\Theta_{ij} < \Theta_{th1}$). All these Voxelisation steps then would be used in grouping these super-voxels (from now onwards referred to as s-voxels) into labeled objects. The advantage of this approach is that we can now use the reduced number of super voxels instead of using thousands of points in the data set, to obtain similar results for classification. The right image in Fig. 4 illustrates an example of s-voxelisation results, in which different s-voxels are labelled with different colors.

2.3.3 Feature Extraction

For each s-voxel, seven main features are extracted to train the classifier. The seven features are *geometrical shape, height above ground, horizontal distance to center line of street, density, intensity, normal angle* and *planarity*. In order to classify these s-voxels, we assume that the ground points have been segmented well. The object types are so distinctly different however these features as mentioned are sufficient to make a classification. Along with the above mentioned features, geometrical shape descriptors plays an important role in classifying objects. These shape-related features are computed based on the projected bounding box to x - y plane (ground).

Geometrical shape: Projected bounding box has effective features due to the invariant dimension of objects. We extract four feature based on the projected bonding box to represent the geometry shape of objects.

- **Area:** the area of the bounding box is used for distinguishing large-scale objects and small ones.
- **Edge ratio:** the ratio of the long edge and short edge.
- **Maximum edge:** the maximum edge of bounding box.
- **Covariance:** is used to find relationships between point spreading along two largest edges.

- **Height above ground:** Given a collection of 3D points with known geographic coordinates, the median height of all points is considered as the height feature of the s-voxel. The height information is independent of camera pose and is calculated by measuring the distance between points and the road ground.
- **Horizontal distance to center line of street:** Following [16], we compute the horizontal distance of the each s-voxel to the center line of street as second geographical feature. The street line is estimated by fitting a quadratic curve to the segmented ground.
- **Density:** Some objects with porous structure such as fence and car with windows, have lower density of point cloud as compared to others such as trees and vegetation. Therefore, the number of 3D points in a s-voxel is used as a strong cue to distinguish different classes.
- **Intensity:** Following [17], LiDAR systems provide not only positioning information but also reflectance property, referred to as intensity, of laser scanned objects. This intensity feature is used in our system, in combination with other features, to classify 3D points. More specifically, the median intensity of points in each s-voxel is used to train the classifier.
- **Normal angle:** Following [18], we adopt a more accurate method to compute the surface normal by fitting a plane to the 3D points in each s-voxel.
- **Planarity:** Patch planarity is defined as the average square distance of all 3D points from the best fitted plane computed by RANSAC algorithm. This feature is useful for distinguishing planar objects with smooth surface like cars form non planar ones such as trees.

2.3.4   Classifier

The Boosted decision tree [19] has demonstrated superior classification accuracy and robustness in many multi-class classification tasks. Acting as weaker learners, decision trees automatically select features that are relevant to the given classification problem. Given different weights of training samples, multiple trees are trained to minimize average classification errors. Subsequently, boosting is done by logistic regression version of Adaboost to achieve higher accuracy with multiple trees combined together. In our experiments, we boost 20 decision trees each of which has 6 leaf nodes. This parameter setting is similar to those in [3], but with slightly more leaf nodes since we have more classes to label. The number of training samples depends on different experimental settings, which are elaborated in Sect. 3.

3   Experimental Result

The LiDAR technology has been used in the remote sensing urban scene understanding by two main technology: Terrestrial Laser Scanning (TLS), useful for large scale buildings survey, roads and vegetation, more detailed but slow in urban surveys in outdoor environments; Mobile Laser Scanning (MLS), less precise than TLS but much more productive since the sensors are mounted on a vehicle; In order to test our algorithm both type of data sets were used:

1. 3D Velodyne LiDAR as TLS data set [20]
2. Paris-rue-Madame [21] and NAVTAQ True as MLS datasets [17]

We train boosted decision tree classifiers with sample 3D features extracted from training s-voxels. Subsequently we test the performance of the trained classifier using separated test samples. The accuracy of each test is evaluated by comparing the ground truth with the scene parsing results. We report *global accuracy* as the percentage of s-voxel correctly classified, *per-class accuracy* as the normalized diagonal of the confusion matrix and class average which represents the average value of per class accuracies.

3.1   Evaluation Using the Velodyne LiDAR Database

The database includes ten high accurate 3D point cloud scenes collected by a Velodyne LiDAR mounted on a vehicle navigating through the Boston area. Each scene is a single rotation of the LIDAR, yielding a point cloud of nearly 70,000 points. Scenes may contain objects including cars, bicycles, buildings, pedestrians and street signs. Finding ground and building points is discussed in Sects. 2.1 and 2.2, and the recognition accuracy is approximately 98, 4 % and 95, 7 % respectively. We train our classifier using seven scene datasets, selected randomly, and test on the remaining three scenes. Table 1 presents the confusion matrices between the six classes over all 10 scenes. Our algorithm performs well on most per class accuracies with the heights accuracy 98 % for Ground and the lowest 72 % for sign-symbol. The global accuracy and per-class accuracy are about 94 % and 87 % respectively.

We also compare our approach to the method described by Lai in [20]. Table 1 shows its quantitative testing result. In terms of per class accuracy, we achieve 87 % in comparison to 76 %. Figure 5 shows some of the qualitative results of the test scene, achieved by our approach (Table 2).

Table 1. Confusion matrix Velodyne LiDAR database

	Tree	Car	Sign	Person	Fence	Ground	Building
Tree	0.89	0.00	0.07	0.00	0.04	0.00	0.00
Car	0.03	0.95	0.00	0.00	0.02	0.00	0.00
Sign	0.17	0.00	0.72	0.11	0.00	0.00	0.00
Person	0.03	0.00	0.27	0.78	0.00	0.00	0.00
Fence	0.03	0.00	0.00	0.00	0.85	0.00	0.12
Ground	0.00	0.00	0.00	0.00	0.00	0.98	0.02
Building	0.00	0.00	0.00	0.00	0.04	0.00	0.96

Test Result PC   Misclassified Map

Fig. 5. Left image shows scene object recognition qualitative results, right image represent misclassified points.

Table 2. Comparison of the class accuracy of our approach and Lais approach

Results	Tree	Car	Sign	Person	Fence	Ground	Building
Lais	0.83	0.91	0.80	0.41	0.61	0.94	0.86
Our	0.89	0.95	0.72	0.88	0.85	0.98	0.95

3.2 Evaluation Using Paris-rue-Madame and NAVTAQ True Datasets

Paris-rue-Madame and NAVTAQ True datasets contains 3D MLS data. The Paris-rue-Madame point cloud is collected from rue Madame Street with 160 m long. The dataset contains 20 million points, 642 objects categorized in 26 classes. Its noteworthy that several objects such as wall sign and wall light are considered as building facades. The second MLS dataset is collected by NAVTAQ True system consisting of point cloud form New York streets. This LiDAR data was collected using terrestrial scanners and contains approximately 710 million points covering 1.2 km. These point clouds hold additional information such as RGB color, time step and etc. which is ignored here as our focus remained on using the pure geometry and intensity for the classification of objects. Same as TLS evaluating test we use 11 dominant categories: Building, Tree, Bike, Car, Sign-Symbol, Ground, Building. The Paris-rue-Madame and NAVTAQ True data sets are divided into two portions: the training set, and the testing set. The 70 % long of each data set are randomly selected and mixed for training of classifier and

Table 3. Confusion matrix of Paris-rue-Madame and NAVTAQ true database

	Tree	Car	Sign	Person	Bike	Ground	Building
Tree	0.75	0.07	0.10	0.00	0.00	0.00	0.08
Car	0.11	0.73	0.00	0.00	0.05	0.00	0.11
Sign	0.09	0.00	0.78	0.13	0.00	0.00	0.00
Person	0.07	0.00	0.21	0.58	0.14	0.00	0.00
Bike	0.03	0.00	0.00	0.04	0.81	0.00	0.12
Ground	0.00	0.00	0.00	0.00	0.00	0.97	0.03
Building	0.05	0.00	0.00	0.00	0.04	0.00	0.95

Building   Ground   Car   Sign   person   Bike   Tree

Fig. 6. Scene object recognition qualitative results in different view

30 % remained long of point cloud is used for testing. Table 3 shows the quantities results achieved by our approach.

Comparing to Terrestrial Laser Scanning, our results are not as good as in shown in Table 1. Since mixing two data sets captured from different cities poses serious challenges to the parsing pipeline. Furthermore, 3D street object detection is a much harder task than reconstructing walls or road surface. Because street objects can have virtually any shape and due to small resolution and the fact that the LiDAR only scans one side of the object, the detection is sometimes impossible. Moving objects are even harder to reconstruct based solely on LiDAR data. As these objects (typically vehicles, people) are moving through the scene, which make them appear like a long-drawn shadow in the registered point cloud. The long shadow artifact is not appear in TLS system because in which we face to one point as exposure point to scan the street objects. Figure 6 shows some of the qualitative results of the test scene.

4 Conclusion

We have proposed a novel and comprehensive framework for semantic parsing of street view 3D MLS and TLS point cloud based on geometrical features. First, ground are segmented using a heuristic approach based on the assumption of constant slope group plane. Second, building points are then extracted by tracing contours of projections of 3D points onto the x - y plane. Using this segmentation huge amount of data (more than 75 % of points) are labeled, and only small amount of point cloud which have complex shape remained to be segmented. During the offline training phase 3D features are extracted at s-voxel level and are used to train boosted decision trees classifier. For new scene, the same unsupervised ground and building detection are applied and geometrical features are extracted and semantic labels are assigned to corresponding point cloud area. The proposed two-stage method requires only small amount of time for training while the classification accuracy is robust to different types of LiDAR point clouds acquisition methods. To our best knowledge, no existing methods have demonstrated the robustness with respect to variety in LiDAR point data.

References

1. Liu, C., Yuen, J., Torralba, A.: Nonparametric scene parsing: label transfer via dense scene alignment. In: IEEE Conference on Computer Vision and Pattern Recognition, 2009, CVPR 2009, pp. 1972–1979. IEEE (2009)
2. Csurka, G., Perronnin, F.: A simple high performance approach to semantic segmentation. In: BMVC, pp. 1–10 (2008)
3. Hoiem, D., Efros, A.A., Hebert, M.: Recovering surface layout from an image. Int. J. Comput. Vision **75**, 151–172 (2007)
4. Floros, G., Leibe, B.: Joint 2d–3d temporally consistent semantic segmentation of street scenes. In: 2012 IEEE Conference on Computer Vision and Pattern Recognition (CVPR), pp. 2823–2830. IEEE (2012)
5. Zhang, G., Jia, J., Wong, T.T., Bao, H.: Consistent depth maps recovery from a video sequence. IEEE Trans. Pattern Anal. Mach. Intell. **31**, 974–988 (2009)
6. Lu, W.L., Murphy, K.P., Little, J.J., Sheffer, A., Fu, H.: A hybrid conditional random field for estimating the underlying ground surface from airborne lidar data. IEEE Trans. Geosci. Remote Sens. **47**, 2913–2922 (2009)
7. Hernández, J., Marcotegui, B., et al.: Filtering of artifacts and pavement segmentation from mobile lidar data. In: ISPRS Workshop Laserscanning 2009 (2009)
8. Zhou, Y., Yu, Y., Lu, G., Du, S.: Super-segments based classification of 3d urban street scenes. Int. J. Adv. Rob. Syst. **9**, 1–8 (2012)
9. Johnson, A.: Spin-Images: A Representation for 3-D Surface Matching. Ph.D. thesis, Robotics Institute, Carnegie Mellon University, Pittsburgh, PA (1997)
10. Kazhdan, M., Funkhouser, T., Rusinkiewicz, S.: Rotation invariant spherical harmonic representation of 3 d shape descriptors. In: Symposium on Geometry Processing, vol. 6 (2003)

11. Sun, J., Ovsjanikov, M., Guibas, L.: A concise and provably informative multi-scale signature based on heat diffusion. In: Computer Graphics Forum, vol. 28, pp. 1383–1392. Wiley Online Library (2009)
12. Osada, R., Funkhouser, T., Chazelle, B., Dobkin, D.: Shape distributions. ACM Trans. Graph. (TOG) **21**, 807–832 (2002)
13. Knopp, J., Prasad, M., Van Gool, L.: Orientation invariant 3d object classification using hough transform based methods. In: Proceedings of the ACM Workshop on 3D Object Retrieval, pp. 15–20. ACM (2010)
14. Pavlidis, T.: Algorithms for Graphics and Image Processing. Computer Science Press, Rockville (1982)
15. Klasing, K., Althoff, D., Wollherr, D., Buss, M.: Comparison of surface normal estimation methods for range sensing applications. In: IEEE International Conference on Robotics and Automation, 2009, ICRA 2009, pp. 3206–3211. IEEE (2009)
16. Zhang, C., Wang, L., Yang, R.: Semantic segmentation of urban scenes using dense depth maps. In: Daniilidis, K., Maragos, P., Paragios, N. (eds.) ECCV 2010, Part IV. LNCS, vol. 6314, pp. 708–721. Springer, Heidelberg (2010)
17. Babahajiani, P., Fan, L., Gabbouj, M.: Semantic parsing of street scene images using 3d lidar point cloud. In: Proceedings of the 2013 IEEE International Conference on Computer Vision Workshops, vol. 13, pp. 714–721 (2013)
18. Xiao, J., Quan, L.: Multiple view semantic segmentation for street view images. In: 2009 IEEE 12th International Conference on Computer Vision, pp. 686–693. IEEE (2009)
19. Collins, M., Schapire, R.E., Singer, Y.: Logistic regression, adaboost and bregman distances. Mach. Learn. **48**, 253–285 (2002)
20. Lai, K., Fox, D.: Object recognition in 3d point clouds using web data and domain adaptation. Int. J. Rob. Res. **29**, 1019–1037 (2010)
21. Serna, A., Marcotegui, B.: Attribute controlled reconstruction and adaptive mathematical morphology. In: Hendriks, C.L.L., Borgefors, G., Strand, R. (eds.) ISMM 2013. LNCS, vol. 7883, pp. 207–218. Springer, Heidelberg (2013)

Completed Dense Scene Flow in RGB-D Space

Yucheng Wang[1,3]([envelope]), Jian Zhang[1], Zicheng Liu[2], Qiang Wu[1], Philip Chou[2], Zhengyou Zhang[2], and Yunde Jia[3]

[1] Advanced Analytics Institute, University of Technology, Sydney, Australia
yucheng.wang@student.uts.edu.au
[2] Microsoft Research, Redmond, WA, USA
[3] Beijing Lab of Intelligent Information Technology,
Beijing Institute of Technology, Beijing, China

Abstract. Conventional scene flow containing only translational vectors is not able to model 3D motion with rotation properly. Moreover, the accuracy of 3D motion estimation is restricted by several challenges such as large displacement, noise, and missing data (caused by sensing techniques or occlusion). In terms of solution, there are two kinds of approaches: local approaches and global approaches. However, local approaches can not generate smooth motion field, and global approaches is difficult to handle large displacement motion. In this paper, a completed dense scene flow framework is proposed, which models both rotation and translation for general motion estimation. It combines both a local method and a global method considering their complementary characteristics to handle large displacement motion and enforce smoothness respectively. The proposed framework is applied on the RGB-D image space where the computation efficiency is further improved. According to the quantitative evaluation based on Middlebury dataset, our method outperforms other published methods. The improved performance is further confirmed on the real data acquired by Kinect sensor.

1 Introduction

Dense scene flow (3D motion) estimation is a challenging research task in computer vision. Consumer RGB-D cameras like Kinect, which provide relatively reliable depth information, promote a trend to estimate scene flow from RGB-D data. Unlike most conventional scene flow methods [1–5] generating only translation vectors, completed scene flow methods can acquire both rotation and translation information, which is more favorable for two main reasons. The first reason is that it can model the general 3D rotational motion in the physical world. The second reason is that it provides abundant temporal information for high-accuracy vision tasks (e.g. 3D reconstruction).

However, it is very challenging to estimate completed scene flow from RGB-D data. The first challenging problem is that large displacement motion. Large displacement motion often indicates the searching dimension and range for scene flow are both large. Without good initial or candidate values, it is difficult to obtain accurate and robust estimates. The second problem is that there usually

© Springer International Publishing Switzerland 2015
C.V. Jawahar and S. Shan (Eds.): ACCV 2014 Workshops, Part I, LNCS 9008, pp. 191–205, 2015.
DOI: 10.1007/978-3-319-16628-5_14

exists noises and missing data in the captured RGB-D data. The RGB-D data may be affected or even disappeared from the reference image to the target image.

Currently, the solutions for motion estimation can be divided into two types: local approaches and global approaches. Local approaches only focus on feature consistency between the corresponding points (or their *local* supporting areas) on two neighboring frames on the time domain. Some local approaches [6–8] can address the displacement issue, since they can employ a random search strategy. However, they can not generate very accurate and smooth motion field. Global approaches are able to further consider the spatial relation of all points in the image, such as occlusion and smoothness. Since global approaches model the complex spatial relation, a limitation is that they often trap into local minima and require good initial values to achieve accurate performance [9].

In this paper, we propose a new scene flow estimation framework to address these challenging issues. Different from previous methods, our framework fully combines the complementary advantages of a local method and a global method, and avoid their corresponding drawbacks. The local method is utilized to provide good candidate values for the global method to overcome large displacement motion. The global method combines these candidate values by explicitly modeling occlusion and enforcing smoothness for good-quality results. In addition, we further handle the missing data issue caused by sensing techniques and occlusion. Our contributions can be summarized as follows: (1) We present a framework to combine the advantages of local and global approaches, i.e. handling large displacement and enforcing smoothness, respectively. (2) We give a new formulation of scene flow estimation that is able to further handle missing data caused by various reasons. (3) We propose compute the matching cost for each point in a 3D local supporting area with adaptive weights, which is more robust to noise. (4) We convert 2D motion as initial values and reduce the searching dimension in the optimization, which improves the accuracy and efficiency.

2   Related Work

Scene flow is 3D motion in the physical world. Compared with optical flow, scene flow has view-independent characteristics, which is preferred in many vision applications like action recognition [11]. We refer the readers to optical flow [12] and scene flow literatures [13] for more details about the similarity and difference.

Most scene flow methods [14–16] employ only RGB stereo images. Until recent years, some RGB-D image-based methods have emerged thanks to the development of consumer RGB-D cameras. The classification for existing scene flow methods are shown in Table 1. Zhang *et al.* [5] proposed a two-step framework consisting a global optimization and a bilateral filtering to compute scene flow. Hadfield and Bowden [1] estimated the scene flow using a particle filtering technique. Gottfried *et al.* [2] presented an extended optical flow framework for the estimation of range flow fields from RGB-D video sequences captured by Kinect. Herbst *et al.* [3] presented a variational method for dense 3D motion

Table 1. Classification for scene flow according to conventional or completed scene flow modeling, and local or global approach employed in the method.

RGB-D scene flow	Conventional scene flow	Completed scene flow
Local method	Hadfield and Bowden [1]	Hornáček et al. [10]
Global method	Gottfried et al. [2]	Our work
	Herbst et al. [3]	
	Quiroga et al. [4]	
	Zhang et al. [5]	

estimation for rigid motion segmentation. Quiroga et al. [4] solved the scene flow problem in a variational framework combining local and global constraints.

These conventional scene flow methods only employ translation in the motion modeling and are not able to handle large displacement motion, since large displacement motion usually contains complex components including rotation and translation. Completed scene flow of rotation and translation information can model 3D motion better and generate more precise results than using translation only. Recently, Hornáček et al. [10] proposed a completed scene flow method. However, this method estimated the scene flow relying heavily on a local method, which may introduce error in occlusion detection and can not generate very accurate motion field. Our framework only derive good initial and candidate values from the local method, and estimate the scene flow with explicitly modeling the occlusion and smoothness in the global method.

3 Our Framework

Figure 1 shows the overview of the framework. Our framework starts with a local optical flow named NRDC [17] to generate completed optical flow from RGB image pair. We transform the optical flow into scene flow as good initial values for our local scene method. Based on the initial values, our local method combine cross-modal RGB color and depth information to refine the scene flow. Next, we derive a set of candidate motion patterns from the local scene flow results. Finally, the set of candidate motion patterns are fused by further modeling occlusion and enforcing 3D smoothness in a global approach. Details will be given in following sections.

Given two RGB-D images $\{I, D\}$ and $\{I', D'\}$, we aim to compute motion from the reference image $\{I, D\}$ to the target image $\{I', D'\}$. Each pixel $\mathbf{p}$ in the

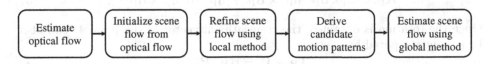

Fig. 1. Framework overview.

reference image has RGB color $I(\mathbf{p})$ and depth $D(\mathbf{p})$. A pixel $\mathbf{p}$ is considered to be valid if its depth value is provided in the depth map. Thus, each valid pixel $\mathbf{p}$ can be deemed as a 3D point $\mathbf{P}$ with color information in the scene.

The 3D coordinate of $\mathbf{P} = \{X_\mathbf{P}, Y_\mathbf{P}, Z_\mathbf{P}\}$ is compute by back-projecting $\mathbf{p}$ using its depth value $D(\mathbf{p})$ and intrinsic camera parameters $\mathbf{K}$ using $\mathbf{P} = \Pi^{-1}(\mathbf{p}) = D(\mathbf{p}) \cdot \mathbf{K}^{-1}\tilde{\mathbf{p}}$, and vice versa $\mathbf{p} = \Pi(\mathbf{P})$. Here, Π is the projection operation, while Π^{-1} means back-projection operation. Let $\mathbf{V} = (\mathbf{R}, \mathbf{T}) \in SE(3)$ denotes a 6-DoF (Degree of Freedom) motion in 3D, where $\mathbf{R} \in SO(3)$ and $\mathbf{T} \in \mathbb{R}^3$. This completed scene flow is employed in our framework. Our goal is to assign such a 6-DoF scene flow $\mathbf{V}_\mathbf{P}$ to each point $\mathbf{P}$ in the reference RGB-D image. The predicted 3D position of point $\mathbf{P}$ with motion $\mathbf{V}_\mathbf{P} = (\mathbf{R}_\mathbf{P}, \mathbf{T}_\mathbf{P})$ denotes $\mathbf{P}' = \mathbf{V}_\mathbf{P}(\mathbf{P}) = \mathbf{R}_\mathbf{P}\mathbf{P} + \mathbf{T}_\mathbf{P}$.

3.1   Initialization from 2D Optical Flow

Some 2D optical flow methods deal with large displacement on the image plane, since the search dimension is smaller than the scene flow situations. Thus, we choose an efficient method named NRDC [17] to generate initial values. NRDC can generate 2D motion field includes 2-DoF translational vectors $\mathbf{t}_\mathbf{p}$ (see Fig. 2). However, the required motion parameters for our method is $\mathbf{V}_\mathbf{P} = \{\mathbf{R}_\mathbf{P}, \mathbf{T}_\mathbf{P}\}$. We give a simple approach to enable the conversion from 2D motion field into 3D completed scene flow.

In order to compute rotation matrix, we intuitively define each point $\mathbf{P}$ having corresponding 2D and 3D principal directions. 3D principal direction $\mathbf{d}_\mathbf{P}$ on the 3D object surface which is orthogonal to its normal $\mathbf{n}_\mathbf{P}$, and 2D principal direction is the projection of 3D principal direction on the image plane. Inspired by [18], we adopt the prominent orientation in SIFT feature detection [19] as the 2D principal directions, i.e. $[\sin(\theta_\mathbf{P}), \cos(\theta_\mathbf{P})]$ for the point $\mathbf{P}$ and $[\sin(\theta_{\mathbf{P}'}), \cos(\theta_{\mathbf{P}'})]$ for the point $\mathbf{P}'$. According to our definition, 3D principal direction vectors can be then computed by

$$\mathbf{d}_\mathbf{P} = \text{orthonorm}([\sin(\theta_\mathbf{P}), \cos(\theta_\mathbf{P}), 0]^\mathrm{T}, \mathbf{n}_\mathbf{P}), \tag{1}$$

$$\mathbf{d}_{\mathbf{P}'} = \text{orthonorm}([\sin(\theta_{\mathbf{P}'}), \cos(\theta_{\mathbf{P}'}), 0]^\mathrm{T}, \mathbf{n}_{\mathbf{P}'}), \tag{2}$$

where $\text{orthonorm}(\cdot, \cdot)$ is the Gram-Schmidt orthonormalization procedure. The rotation variation of a point is reflected by the variations of its normal and principal directions: $\mathbf{n}_{\mathbf{P}'} = \mathbf{R}_\mathbf{P}\mathbf{n}_\mathbf{P}$ and $\mathbf{d}_{\mathbf{P}'} = \mathbf{R}_\mathbf{P}\mathbf{d}_{\mathbf{P}'}$. Thus, we can calculate the 3D rotation matrix $\mathbf{R}_\mathbf{P}$ of the point $\mathbf{P}$ by

$$\mathbf{R}_\mathbf{P} = [\mathbf{n}_{\mathbf{P}'}, \mathbf{d}_{\mathbf{P}'}, \mathbf{n}_{\mathbf{P}'} \times \mathbf{d}_{\mathbf{P}'}] \cdot [\mathbf{n}_\mathbf{P}, \mathbf{d}_\mathbf{P}, \mathbf{n}_\mathbf{P} \times \mathbf{d}_\mathbf{P}]^{-1}. \tag{3}$$

Once the rotation is obtained, the translational vector of the point $\mathbf{P}$ can also be simply computed by

$$\mathbf{T}_\mathbf{P} = \mathbf{P}' - \mathbf{R}_\mathbf{P} \cdot \mathbf{P}. \tag{4}$$

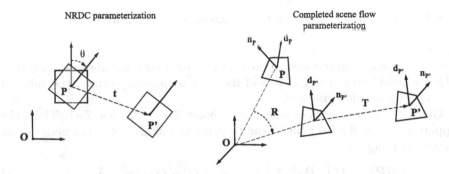

Fig. 2. NRDC and complete scene flow parameterization. To clearly see the rotation change for a point, we use a square patch with a principal direction vector to represent the point.

3.2   Refinement Using Local Method

The motion initial value from the optical flow only concern RGB color information. Thus, the major principle for optimizing the local scene flow estimates is that multi-modal RGB-D features (descriptors) consistency for a point in the reference image and its corresponding position in the target image. To address the noise and missing data issue, for a point $\mathbf{P}$ with a motion $\mathbf{V_P}$, we aggregate cost values of points using adaptive weights and reliability in a corresponding 3D supporting area. Our goal in the local is to reduce overall matching cost of all the points in the reference image:

$$E_{\text{local}}(\mathbf{V}) = \sum_{\mathbf{P}} C_{\text{local}}(\mathbf{P}, \mathbf{V_P}). \tag{5}$$

where $C_{\text{local}}(\mathbf{P}, \mathbf{V_P})$ is the 3D supporting patch-based matching cost for the point $\mathbf{P}$ with the motion $\mathbf{V_P}$, and it is defined by

$$C_{\text{local}}(\mathbf{P}, \mathbf{V_P}) = \frac{\sum\limits_{\mathbf{Q} \in S(\mathbf{P})} \omega(\mathbf{P}, \mathbf{Q}) \cdot R(\mathbf{Q}) \cdot R'(\mathbf{Q}') \cdot C(\mathbf{Q}, \mathbf{V_P})}{\sum\limits_{\mathbf{Q} \in S(\mathbf{P})} \omega(\mathbf{P}, \mathbf{Q}) \cdot R(\mathbf{Q}) \cdot R'(\mathbf{Q}')}, \tag{6}$$

where $S(\mathbf{P})$ is the 3D supporting area for the point $\mathbf{P}$, $\omega(\mathbf{P}, \mathbf{Q})$ is the weighting function which gives the probability of points $\mathbf{P}$ and $\mathbf{Q}$ on a same surface, R and R' are the reliability maps for the reference and target RGB-D image respectively, and $C(\mathbf{Q}, \mathbf{V_P})$ is the point-based matching cost for the point $\mathbf{P}$ with the motion $\mathbf{V_P}$.

3D Supporting Area. Due to noises on RGB-D data, the features (descriptors) of a single point is usually unstable. To deal with noises, we assume local rigidity for each point, and aggregate cost values of local neighboring points on the same surface for a robust matching cost. Unlike [10], 3D geodesic distance is a better choice to judge whether 3D neighboring points are on a same surface than Euclidean distance. However, it is expensive to compute geodesic distance between all the points. We propose a new 3D patch representation as an approximation by using the normal information n_P, which is capable of selecting such neighboring points on the same surface for a point. Our basic observation is that if Q is a neighboring point of P and they are on a same surface, the value of $(P - Q) \cdot n_P$ should close to 0.

Given the 3D world coordinates of a point $P = \{X_P, Y_P, Z_P\}$. Thus, the supporting patch of a 3D point P can be expressed as the set of the neighboring points satisfying

$$S(P) = \{Q \mid ||P - Q||_2 < \epsilon \cdot Z_P \wedge (P - Q) \cdot n_P < \delta \cdot Z_P\} \qquad (7)$$

where ϵ is a threshold ratio using in the previous normal estimation, and δ is usually a small threshold ratio decided by the sensor noise.

Weighting Function. In the supporting area of a point P, its neighboring points $Q \in S(P)$ should have higher probability if they are closer in the 3D space. Thus, we utilize an adaptive weight $\omega(P, Q)$ based on Euclidean distance to aggregate the cost values of neighboring points Q in the support area $S(P)$. The weighting function is

$$\omega(P, Q) = \exp(-||P - Q||_2/\gamma). \qquad (8)$$

where γ is a parameter to control the weight function.

Reliability Map. Considering the fact that the depth channel of RGB-D data often contains missing data and noises, we introduce reliability of each pixel (point) in the RGB-D data. We observe that depth noise often occur predominantly near depth discontinuities. Therefore we apply an edge detector on the depth map, and use the 2D spatial distance ρ_p to the closest depth edge as a reliability measure for p. The reliability of p is

$$R(p) = \begin{cases} \exp(-\frac{2 \cdot \min(\rho_{max} - \rho_p, 0)}{\rho_{max}}) & \text{if } D(p) \text{ is valid} \\ 0 & \text{otherwise} \end{cases} \qquad (9)$$

where ρ_{max} are constant scaling parameters.

Point-Based Matching Cost. Given a point in P in the reference image and a 6-DoF motion V_P, the motion is of high probability for this point if we can find a position with similar appearance and geometrical information in the target image. We assume brightness constancy and use color difference $||I(p) - I'(p')||_2$ to measure appearance similarity. For geometrical similarity, we use difference of depth values $||Z_{P'} - D'(p')||_2$ as an approximation of 3D Euclidean distance. The matching cost of one single point P with motion $V_P = \{R_P, T_P\}$ is defined as

$$C(P, V_P) = ||I(p) - I'(p')||_2 + \alpha \cdot ||Z_{P'} - D'(p')||_2. \qquad (10)$$

where α is the parameter to control the ratio of the two components.

3D Searching. We modify the 2D PatchMatch method for our 3D scene flow ຫຼຸດ ､ due to its good characteristic to handle large displacement. Firstly, each point is assigned with the initial value provided by the local optical flow method. Next, we iteratively carry out two steps to refine the motion estimates for each point, i.e. spatial propagation and random search. In the spatial propagation, we use 6-DoF completed scene flow instead of 2-DoF translational optical flow. In the random search, the searching dimension is too large to efficiently obtain good results. We introduce a reduced-DoF random search by only generating a random 2-DoF translation t_p. We compute the 2D principal direction vectors of p in the reference image and $p+t_p$ in the target image by adopting the prominent orientations in SIFT feature detection [19]. Then, the following computation is similar with the situation when we convert 2D motion field to 3D completed scene flow in the section *Initialization from Optical flow*. We can finally acquire a 6-DoF motion from a 2-DoF random guess using this reduced-DoF random search. Thus, the dimension of random searching for 3D scene flow case is then significantly reduced from six to two.

3.3   Estimation Using Global Method

In spite of feature consistency assumption in the local method, we can further explicitly model the occlusion and enforce 3D smoothness in the global approach. The energy function of the global scene flow is

$$E_{\text{global}}(\mathbf{V}) = \sum_{\mathbf{P}} C_{\text{global}}(\mathbf{P}, \mathbf{V_P}) + \lambda \sum_{\mathbf{P,Q}} S_{\text{global}}(\mathbf{P}, \mathbf{Q}, \mathbf{V_P}, \mathbf{V_Q}) \quad (11)$$

where $\mathbf{Q} \in S(\mathbf{P}) \cap N(\mathbf{P})$, $S(\mathbf{P})$ is the set of points in the supporting patch of $\mathbf{P}$, and $N(\mathbf{P})$ is the set of 4(8) connected neighboring points on the image plane, $C_{\text{global}}(\mathbf{P}, \mathbf{V_P})$ is used for feature consistency, and $S_{\text{global}}(\mathbf{P}, \mathbf{Q}, \mathbf{V_P}, \mathbf{V_Q})$ promotes the 3D smoothness of the motion field. Note that C_{global} is different from C_{local} by further modeling occlusion.

Robust Matching Cost with Occlusion Modeling. To address the occlusion issue, we incorporate the occlusion in our matching cost computation. We deem the occluded points as outliers when finding correspondence, and use a constant cost value for matching outliers. The robust matching cost in the global method is

$$C_{\text{global}}(\mathbf{P}, \mathbf{V_P}) = (1 - O(\mathbf{P})) \cdot C_{\text{local}}(\mathbf{P}, \mathbf{V_P}) + O(\mathbf{P}) \cdot \xi \quad (12)$$

where ξ is set to be a minimal value for matching outliers, and $O(\mathbf{P})$ is the occlusion status of the point $\mathbf{P}$.

Points in the target image are occluded if there exist other points in front of them from the camera view. Previous methods [5, 10] usually estimate the motion without considering occlusion first, then use consistency check to detect occlusion, and refine the motion in occluded region as a postprocessing. This may introduce error if the estimated motion is incorrect. Our method can explicitly model the occlusion using the depth order, since occlusion relationship has been directly reflected in depth values. For robust performance, we also consider depth noises in the occlusion modeling. We assume the depth noises following a Gaussian distribution with mean zero and standard derivation σ. The occlusion status of a point is defined as

$$O(\mathbf{P}) = [Z_{\mathbf{P}'} > D(\mathbf{P}') + 3 \cdot \sigma(\mathbf{p})], \tag{13}$$

where $[\cdot]$ is the Iverson bracket which denotes a number that is 1 if the condition in square brackets is satisfied, and 0 otherwise, the computation of $\sigma(\mathbf{p})$ is given in the experiment section.

3D Smoothness. Instead of enforcing smoothness only on the translation vectors, we apply a 3D smoothness considering both translation and rotation. The basic idea is to promote one point to have similar 3D positions after applying the motion of itself or its neighbors. Thus, the smoothness term is the energy function can be expressed as

$$S_{\text{global}}(\mathbf{P}, \mathbf{Q}, \mathbf{V_P}, \mathbf{V_Q}) = \omega(\mathbf{P}, \mathbf{Q}) \cdot \left(\|\mathbf{V_P}(\mathbf{P}) - \mathbf{V_Q}(\mathbf{P})\|_2^2 + \|\mathbf{V_P}(\mathbf{Q}) - \mathbf{V_Q}(\mathbf{Q})\|_2^2 \right) \tag{14}$$

where $\mathbf{Q} \in S(\mathbf{P}) \cap N(\mathbf{P})$, $S(\mathbf{P})$ is the set of points in the supporting patch of $\mathbf{P}$, and $N(\mathbf{P})$ is the set of 4 connected neighboring points on the image plane, $\|\mathbf{V_P}(\mathbf{P}) - \mathbf{V_Q}(\mathbf{P})\|_2$ is the Euclidean distance of the point $\mathbf{P}$ with motion patterns $\mathbf{V_P}$ and $\mathbf{V_Q}$, and $\|\mathbf{V_P}(\mathbf{Q}) - \mathbf{V_Q}(\mathbf{Q})\|_2$ is the Euclidean distance of the point $\mathbf{Q}$ with motion patterns $\mathbf{V_P}$ and $\mathbf{V_Q}$.

Optimization. Given Eqs. 12 and 14, we minimize our energy function in Eq. 11 via the FusionMoves [20] method using QPBO [21]. The FusionMoves can efficiently combine two proposal labelings (candidates) in a theoretically sound way, which is in practice often globally optima. The key of achieving good results is to generate high-quality motion proposals for FusionMoves. One direct way is to use existing motion pattern directly from the result of our local method. However, the number of different motion patterns in the local result of is usually very limited. Thus, we not only include the motion patterns from the local result as proposals, but also add some random slight perturbation on them as new proposals. The perturbation can be combinations of changing the translation by jumping to its neighboring points (3 DoF), altering the rotation axis (2 DoF), or modifying the rotation angle (1 DoF). The algorithm stops when energy change in a period is less than a threshold, and outputs the final result.

4   Experimental Results

To analyze the performance of the proposed method, we apply our algorithm on the Middlebury dataset and some challenging RGB-D images captured by Kinect cameras as a complement. We use millimeter as the unit of distance measure, and $[0, 255]$ for color range. For the local optical flow NRDC [17], we use its default parameters. The threshold ratio ϵ for normal estimation is set to 0.05, the threshold ratio δ for 3D supporting area is 0.02, the ratio α in cost computation for a singe point is set to 1.0, the constant cost ξ for outliers is set to 30.0, the standard derivation of sensor noise $\sigma(\mathbf{p}) = k \cdot D(\mathbf{p})^2$, and the constant used in computing the weight of two point $\gamma = 10.0$. For Middlebury dataset, the ratio in the global optimization $\lambda = 100$, the parameters to model data reliability $\rho_{max} = 2$, and the parameters to model depth noise $k = 1.5 \times 10^{-4}$. For Kinect RGB-D data, $\lambda = 1$, $\rho_{max} = 4$ and $k = 1.5 \times 10^{-5}$.

4.1   Middlebury Dataset

We accordingly test the method on Middlebury dataset following Huguet and Devernay [16] in order to perform quantitative evaluation. The RGB-D images are captured by a set of cameras which are parallel and equally spaced along the X axis at the same time. The motion along Y and Z axis is always zero, and the ground truth of motion along X axis can be obtained from corresponding disparity, which is also available in the Middlebury dataset. We take the color images and ground truth disparity maps of frames 2 and 6 of the Middlebury Cones, Teddy, and Venus as the reference and target RGB-D images.

Our approach is compared with three optical flow methods [9,17,22], two stereo-based scene flow methods [15,16] and four RGB-D scene flow methods [1,4,5,10]. Following [1,15,16], we use end point error (RMS$_O$), disparity change error (RMS$_Z$) and average angular error (AAE) as the error measurement criteria. Results were computed over all valid pixels. For stereo-based methods [15,16], they jointly estimate the scene flow and disparity using frames 2, 4, 6 and 8 of the Middlebury Cones, Teddy, and Venus. For the two optical flow techniques [9,22], RMS$_Z$ was computed by estimating 3D translational flow by interpolating depth encoded at the start and end points given its 2D flow vector. The error values are given as reported in their papers or computed using provided codes with default parameters. From Table 2, our method is the top performer under most evaluation criteria among all the optical flow and scene flow algorithms.

An interesting observation is that the local optical flow (NRDC) and local scene flow employed in our framework perform worse than most competing methods while our global scene flow still can generate good-quality motion results. This is consistent with the qualitative results shown in Fig. 3. The estimated local optical flow by NRDC is quite false and noisy on the all the three RGB-D images pairs. The local scene flow improves the motion quality in some region, but the result is still incorrect especially on occlusion, textureless regions and repeated patterns. Our global scene flow can capture the correct motion patterns from the noisy input result of local scene flow, and overcome these issues to generate accurate results.

Table 2. The evaluated errors of compared methods.

Methods	Venus			Cones			Teddy		
	RMS_O	RMS_Z	AAE	RMS_O	RMS_Z	AAE	RMS_O	RMS_Z	AAE
Brox2011 [9]	0.72	0.14	1.28	2.83	1.75	0.39	3.20	0.47	0.39
Xu2012 [22]	0.30	0.22	1.43	1.66	1.15	**0.21**	1.70	0.50	0.28
Huguet2007 [16]	0.31	N/A	0.98	1.10	N/A	0.69	1.25	N/A	0.51
Basa2013 [15]	0.16	N/A	1.58	0.58	N/A	0.39	0.57	N/A	1.01
Zhang2013 [5]	**0.15**	N/A	1.15	1.04	N/A	0.69	0.73	N/A	0.66
Quiroga2013 [4]	0.31	**0.00**	1.26	0.57	0.05	0.42	0.69	0.04	0.71
Hadfield2014 [1]	0.36	0.02	1.03	1.24	0.06	1.01	0.83	0.03	0.83
Hornáček2014 [10]	0.26	0.02	**0.53**	0.54	0.02	0.52	**0.35**	0.01	**0.15**
NRDC [17]	5.65	N/A	16.2	15.5	N/A	18.3	17.7	N/A	14.3
Our local SF	3.35	0.27	14.5	7.91	1.29	7.10	11.4	0.30	10.9
Our global SF	**0.15**	**0.00**	1.17	**0.33**	**0.00**	0.39	0.40	**0.00**	0.50

4.2   Kinect RGB-D Data

We also apply our algorithm on two frames of the RGB-D video sequence *Tshirt4* recorded by a Kinect camera from [23] and RGB-D data captured by us as a complement. We compare the performance of our algorithm to a scene flow method called *RGB-D flow* method [3], a 3D surface tracking method [23], a large displacement optical flow method [9] based on RGB color images. For the optical flow, the scene flow can be computed by back-projecting the 2D optical flow to 3D space domain using camera intrinsic parameters and depth values.

Qualitative Evaluation. We visualize the motion results from different methods in two strategies for qualitative evaluation. The first strategy is that we create XY motion (Optical flow) map and Z motion map to show the motion projection on 2D image plane and the motion along depth Z direction respectively. These maps are illustrated based on middlebury color coding [12]. To visualize Z motion map, the values along x-axis of middlebury color coding map are employed. The second strategy is using the motion field to register the point cloud of reference image to the point cloud of the target image. A good motion estimation result should be able to register two point clouds to each other closely and smoothly.

Figure 4 shows the results of estimating the 3D motion field between the frame 58 and the frame 61 on the Tshirt4 sequence. As shown in the row 2 and 3, the RGB-flow method and the optical flow method fail in capturing the distinct motion of the right side of T-shirt marked by the green circle due to occlusion, the surface tracking method over-smooth the region, while our method robustly estimates the 3D motion field. This difference reflects the advantages of the completed scene flow parametrization and occlusion modeling employed by our framework. Row 4–6 depicts the registration results from three orthographic views.

Local optical flow NRDC   Our local scene flow   Our global scene flow   Ground Truth

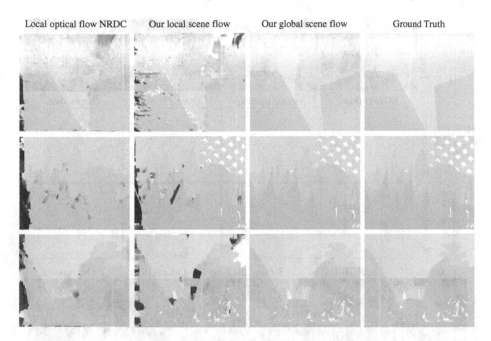

Fig. 3. 2D XY motion (optical flow) by projection of 3D displacements on image plane using middlebury color coding. From left to right, these images are the results of local optical flow, local scene flow and global scene flow in our motion estimation framework along with the ground truth. From up to down, these images are the results of Middlebury Cones, Teddy, and Venus. The optical flow maps are rendered using middlebury coloring method. For scene flow, 3D displacements are projected to image space to obtain 2D optical flow.

We can see that the other three methods fail in registering the reference point clouds to target point clouds smoothly on the regions marked by green circles, and our method works robustly on the deformable surfaces of T-shirt. The registration results are consistent with the XY- and Z-motion maps.

Figure 5 gives the motion results of two RGB-D images of a person waving his hands captured by us. The data is challenging since there is almost no texture on the clothes worn on the person. As shown in the row 2 and 3, other methods fail in estimating the motion field on the region marked by the green circle. In contrast, our method still works robustly against these competing methods.

Quantitative Evaluation. It is prohibitively expensive to label correspondences for every point in the two RGB-D sequences. Instead, we use a sparse set of hand-tracked points, approximately uniformly spaced in the first frame of each sequence. The position displacement of these points are served as ground truth to measure accuracy and robustness of the estimated motion results.

We evaluate the four methods on the two sequence under different two time intervals configuration of neighboring frames: $\Delta t = 1$ and $\Delta t = 3$. The motion

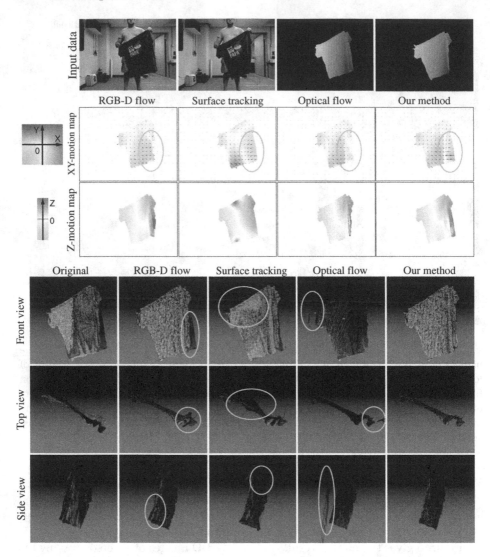

Fig. 4. Scene flow on two frames on Tshirt4 sequence. Row 1: Input reference and target RGB-D images. For clarity, we only show the depth values of the foreground. Row 2–3: XY-motion (optical flow) maps and Z-motion maps of the RGB-D flow, surface tracking, optical flow and our methods. Left images are extended Middlebury color coding maps for 3D motion visualization. Row 4–6: Three basic orthographic views of the two point clouds from reference and target data before and after registration using 3D motion field generated by different methods.

displacement in the time interval configuration $\Delta t = 3$ is approximately 3 times larger than $\Delta t = 1$. Thus, we can discriminate between two time interval configuration by considering them as small displacement ($\Delta t = 1$) and large displacement ($\Delta t = 3$) scenarios, respectively. Table 3 depicts mean and standard

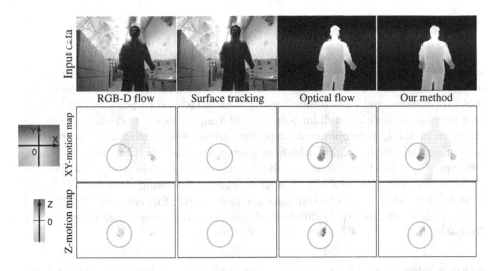

Fig. 5. 3D motion estimation on two RGB-D images captured by us. Row 1: Input reference and target RGB-D images. For clarity, we only show the depth values of the foreground. Row 2–3: XY-motion maps and Z-motion maps of the RGB-D flow, surface tracking, optical flow and our methods. Left images are extended Middlebury color coding maps for 3D motion visualization.

deviation of error (3D Euclidean distance from the ground truth) with small and large displacement scenarios in *Tshirt4* and *Human Hand Waving* sequences. From the table, we can observe that our method achieve comparative results compared with other three state-of-art methods in small displacement scenario. When it turns to the situation that there exists large displacement 3D motion in the scene, our method performs much better and reaches the lowest mean and standard deviation of error. This proves the ability of the proposed method in dealing with large displacement 3D motion estimation.

Table 3. The mean and standard deviation of error (mm) with small and large displacement in *Tshirt4* and *Human Hand Waving* sequences.

Methods	Tshirt4				Human Hand Waving			
	$\Delta t = 1$		$\Delta t = 3$		$\Delta t = 1$		$\Delta t = 3$	
	mean	std	mean	std	mean	std	mean	std
RGB-D flow [3]	3.5	2.0	18.8	56.4	6.6	**3.9**	11.3	14.9
Surface tracking [23]	7.4	4.2	71.9	128.8	11.0	15.1	39.0	129.0
Optical flow [9]	7.1	4.7	23.8	71.1	6.7	5.1	35.9	129.7
Our method	**2.8**	**1.8**	**8.9**	**4.2**	**5.9**	4.2	**7.7**	**10.4**

5   Conclusions

In this paper, we present a framework to address the challenging problems of scene flow estimation based on RGB-D data. In the framework, we efficiently initialize scene flow from a 2D motion method to address the large displacement motion problem, and then refine it using a local method to provide candidates, and fuse these motion candidates by considering occlusion and smoothness. In the local method, we propose calculate the matching cost using a 3D supporting area using adaptive weights which is robust to noise. In the global method, we explicitly model occlusion to jointly estimate occlusion and scene flow to address the occlusion problem. For the noise and missing data issues, RGB-D data reliability is also taken into account in the formulation. We showed compelling results on the Middlebury datasets as well as on challenging Kinect RGB-D data.

Acknowledgement. This work was supported by Microsoft Research, Redmond. We also acknowledge Minqi Li for recording the Kinect RGB-D data sequence in our experiment.

References

1. Hadfield, S., Bowden, R.: Scene particles: unregularized particle based scene flow estimation. IEEE Trans. Pattern Anal. Mach. Intell. **36**, 564–576 (2014)
2. Gottfried, J.-M., Fehr, J., Garbe, C.S.: Computing range flow from multi-modal Kinect data. In: Bebis, G., et al. (eds.) ISVC 2011, Part I. LNCS, vol. 6938, pp. 758–767. Springer, Heidelberg (2011)
3. Herbst, E., Ren, X., Fox, D.: Rgb-d flow: Dense 3-d motion estimation using color and depth. In: IEEE International Conference on Robotics and Automation (ICRA). IEEE (2013)
4. Quiroga, J., Devernay, F., Crowley, J.L., et al.: Local/global scene flow estimation. In: ICIP-IEEE International Conference on Image Processing (2013)
5. Zhang, X., Chen, D., Yuan, Z., Zheng, N.: Dense scene flow based on depth and multi-channel bilateral filter. In: Lee, K.M., Matsushita, Y., Rehg, J.M., Hu, Z. (eds.) ACCV 2012, Part III. LNCS, vol. 7726, pp. 140–151. Springer, Heidelberg (2013)
6. Barnes, C., Shechtman, E., Finkelstein, A., Goldman, D.: Patchmatch: A randomized correspondence algorithm for structural image editing. ACM Transactions on Graphics-TOG **28**, 24 (2009)
7. Barnes, C., Shechtman, E., Goldman, D.B., Finkelstein, A.: The generalized patchmatch correspondence algorithm. In: Daniilidis, K., Maragos, P., Paragios, N. (eds.) ECCV 2010, Part III. LNCS, vol. 6313, pp. 29–43. Springer, Heidelberg (2010)
8. Korman, S., Avidan, S.: Coherency sensitive hashing. In: 2011 IEEE International Conference on Computer Vision (ICCV), pp. 1607–1614. IEEE (2011)
9. Brox, T., Malik, J.: Large displacement optical flow: descriptor matching in variational motion estimation. IEEE Trans. Pattern Anal. Mach. Intell. **33**, 500–513 (2011)

10. Hornacek, M., Fitzgibbon, A., Carsten, R.: Sphereflow: 6 dof scene flow from rgb-d pairs. In: 2014 IEEE Conference on Computer Vision and Pattern Recognition (CVPR). IEEE (2014)
11. Junejo, I.N., Dexter, E., Laptev, I., Perez, P.: View-independent action recognition from temporal self-similarities. IEEE Trans. Pattern Anal. Mach. Intell. **33**, 172–185 (2011)
12. Baker, S., Scharstein, D., Lewis, J., Roth, S., Black, M.J., Szeliski, R.: A database and evaluation methodology for optical flow. Int. J. Comput. Vis. **92**, 1–31 (2011)
13. Vedula, S., Rander, P., Collins, R., Kanade, T.: Three-dimensional scene flow. IEEE Trans. Pattern Anal. Mach. Intell. **27**, 475–480 (2005)
14. Vogel, C., Schindler, K., Roth, S.: Piecewise rigid scene flow. In: 2013 IEEE International Conference on Computer Vision (ICCV). IEEE (2013)
15. Basha, T., Moses, Y., Kiryati, N.: Multi-view scene flow estimation: a view centered variational approach. Int. J. Comput. Vis. **101**, 6–21 (2013)
16. Huguet, F., Devernay, F.: A variational method for scene flow estimation from stereo sequences. In: IEEE 11th International Conference on Computer Vision, ICCV 2007, pp. 1–7. IEEE (2007)
17. HaCohen, Y., Shechtman, E., Goldman, D.B., Lischinski, D.: Non-rigid dense correspondence with applications for image enhancement. ACM Trans. Graph. (TOG) **30**, 70:1–70:9 (2011)
18. Eshet, Y., Korman, S., Ofek, E., Avidan, S.: Dcsh-matching patches in rgbd images. In: 2013 IEEE International Conference on Computer Vision (ICCV), pp. 89–96. IEEE (2013)
19. Lowe, D.G.: Distinctive image features from scale-invariant keypoints. Int. J. Comput. Vis. **60**, 91–110 (2004)
20. Lempitsky, V., Rother, C., Roth, S., Blake, A.: Fusion moves for markov random field optimization. IEEE Trans. Pattern Anal. Mach. Intell. **32**, 1392–1405 (2010)
21. Boros, E., Hammer, P.L.: Pseudo-boolean optimization. Discrete Appl. Math. **123**, 155–225 (2002)
22. Xu, L., Jia, J., Matsushita, Y.: Motion detail preserving optical flow estimation. IEEE Trans. Pattern Anal. Mach. Intell. **34**, 1744–1757 (2012)
23. Willimon, B., Hickson, S., Walker, I., Birchfield, S.: An energy minimization approach to 3d non-rigid deformable surface estimation using rgbd data. In: 2012 IEEE/RSJ International Conference on Intelligent Robots and Systems (IROS), pp. 2711–2717. IEEE (2012)

Online Learning of Binary Feature Indexing for Real-Time SLAM Relocalization

Youji Feng[1]([✉]), Yihong Wu[1], and Lixin Fan[2]

[1] Institute of Automation, Chinese Academy of Sciences, Beijing, China
yjfeng@nlpr.ia.ac.cn
[2] Nokia Research Center, Tampere, Finland

Abstract. In this paper, we propose an indexing method for approximate nearest neighbor search of binary features. Being different from the popular Locality Sensitive Hashing (LSH), the proposed method construct the hash keys by an online learning process instead of pure randomness. In the learning process, the hash keys are constructed with the aim of obtaining uniform hash buckets and high collision rates, which makes the method more efficient on approximate nearest neighbor search than LSH. By distributing the online learning into the simultaneous localization and mapping (SLAM) process, we successfully apply the method to SLAM relocalization. Experiments show that camera poses can be successfully recovered in real time even there are tens of thousands of landmarks in the map.

1 Introduction

Simultaneous localization and mapping (SLAM) has been extensively studied in both robotics and computer vision [1,2]. One important module in SLAM is the relocalization, *i.e.* the recovery of the camera pose after tracking failure. Since the camera pose can be estimated using correspondences between 3D points in the map and 2D features in the image, the key problem in the relocalization is to obtain the 3D-2D correspondences.

The problem of obtaining 3D-2D correspondences is typically treated as the problem of feature matching: during the mapping process, each 3D point in the map is associated with some image features; in the relocalization process, the same type of features are extracted from the image and matched against those of the 3D points. Some existing works [3–5] on image-based localization employ robust image features such as SIFT [6] and Daisy [7] in their systems. They are able to localize an image accurately even the scene contains millions of 3D points. However, due to the expensive computation of these features, they are not well suited for SLAM relocalization in which real time performance is of critical importance. Williams *et al.* [8] propose a relocalization approach using the adapted random ferns [9] to collect 3D-2D correspondences. While being performed in real time, their approach can hardly be extended to large scenes due to the large memory footprint. Recently, Straub *et al.* [10] have designed a relocalization module based on binary features [11–13]. Their system is demonstrated

C.V. Jawahar and S. Shan (Eds.): ACCV 2014 Workshops, Part I, LNCS 9008, pp. 206–217, 2015.
DOI: 10.1007/978-3-319-16628-5_15

on a scene containing tens of thousands 3D points and achieves near real time performance due to two attractive characteristics of binary features, (a) they are extracted faster than SIFT like features by two orders of magnitude; (b) the distance of two binary features can be computed by using fast SSE instructions. To further speed up feature matching, Straub *et al.* also use Locality Sensitive Hashing [14] to perform approximate nearest neighbor (ANN) search.

In this paper, we demonstrate a relocalization module which employs binary features. Being different from [10], we propose for ANN search an indexing method that is more efficient than LSH. The higher efficiency is achieved through an online learning process. Initially, the hash keys are randomly generated as in LSH, and then they are adapted incrementally during the SLAM process with the objective to attain more uniform hash buckets and higher collision rates. Experiment results show that using the proposed indexing method takes less time than using the original LSH to reach the same search accuracy, or reaches higher accuracy by taking the same time.

The rest of this paper is organized as follows: Sect. 2 gives an overview of the entire SLAM system; Sect. 3 introduces the relocalization module and the proposed indexing method; Experiment results are shown in Sect. 4 and Conclusions are drawn in Sect. 5.

2 System Overview

Our SLAM implementation is adapted from PTAM [2]. It consists of two threads, *i.e.* the background mapping thread and the foreground tracking thread. The mapping thread collects keyframes from the video sequence and performs structure from motion to build a map of the environment. The process is done incrementally. When a new keyframe is inserted, FAST corners [15] are first extracted, and then some of the corners are identified as the observations of the old 3D points, while the remaining are matched against the corners in the nearest keyframes to triangulate new 3D points. The tracking thread estimates the camera pose of each frame by tracking the 3D points in the map. Under the assumption of continuous camera motion, the observations of the 3D points can be searched around their predicted locations. Once the observations are found, *i.e.* the 3D-2D correspondences are established, the camera pose is trivially estimated by using a non-linear optimization routine. During the tracking process, tracking failure may happen due to sudden changes of the illumination, full occlusions, or extreme motion blur. Then a relocalization module is implemented to re-estimate the camera pose and restart the tracking process.

3 Relocalization

The aim of the relocalization is to estimate the camera pose of an image after the tracking failure. Being the same as in the tracking process, the core problem in the relocalization is also to find the observations of the 3D points, except that the observations can no longer be predicted from the camera poses of previous frames. Once the observations are found, the camera pose can be estimated by RANSAC [16] and Perspective-n-Points algorithms.

Obtaining the observations of the 3D points is treated as the problem of feature matching. In the mapping process, whenever the observation of a 3D point is found in a keyframe, a binary descriptor at this observation is extracted. Consequently, each 3D point corresponds to a set of binary descriptors which are referred to as database features. In an image being relocalized, Fast corners [15] along with their binary descriptors, referred to as query features, are first extracted. Then, the top two nearest neighbors of each query are searched among the database features. If the ratio test [6] is past, the nearest neighbor is deemed as the match. To be efficient, we use an indexing method to perform fast ANN search instead of brute-force search. The indexing method is more efficient than LSH due to an online learning process, which is presented below after a brief description on LSH.

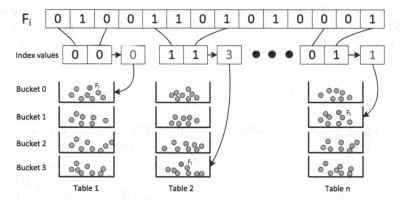

Fig. 1. An example of indexing a database feature. Each hash key consists of two randomly selected bits.

3.1   LSH for ANN Search of Binary Features

LSH uses hash keys to index a feature. It assigns an index value to the feature under each key. When performing ANN search, database features are stored in multiple hash tables. Each hash table corresponds to a hash key. The designated table entry, *i.e.* hash bucket, in which a database feature should be stored is determined by the index value. Given a query feature, the index values are first assigned and the corresponding buckets are found in the same way as database features. Then a linear search is performed among the database features in these buckets to retrieve the approximate nearest neighbors. For binary features, the hash key is simply a set of bits randomly selected from the descriptor bits. Consequently, a binary code can be assigned to a feature by concatenating the values at these bits, and the index value of the feature is the integer converted by the binary code. Figure 1 gives a simple example on how a database feature is indexed and stored in the hash tables.

3.2   The Proposed Indexing Method

There are two factors closely related to the efficiency of the indexing scheme for ANN search: the distribution of the bucket sizes and the collision rate of matched feature pairs. The size of a bucket indicates the number of the database features stored in the bucket. Usually, the more uniform the bucket sizes are, the faster ANN search will be, which have been noticed by Rublee *et al.* [12]. In this paper (see analysis below), we illustrate that maintaining uniform bucket sizes is equivalent to minimizing the total ANN search time. The collision rate of matched feature pairs is the probability of that two matched features collide in the same bucket. Higher collision rate leads to higher search accuracy. With both factors being considered, our indexing method aims to attain uniform buckets and high collision rates, so that ANN search can be performed fast and accurately. We reach this aim through a learning process, in which the bits composing the hash keys are selected to minimize a corresponding cost function. In the remainder of the section, we present the learning process together with the derivation of the cost function.

To alleviate the latency in the system and handle the ever-changing set of database features, the bit selection is distributed into the SLAM process. Initially, the bits composing each hash key are selected randomly, as in LSH. Whenever a new keyframe is inserted and the set of the database features is updated, one bit of each hash key is re-selected to minimize the cost function. Then the database features are re-indexed using the new hash keys.

Recalling the two factors that affect the efficiency, the cost function should contain terms about the uniformity of the bucket sizes and the collision rate:

The uniformity of the bucket sizes. Let the normalized bucket sizes of a hash table be $\{s_n, n = 1, 2, \ldots, N\}$ subject to $\sum s_n = 1$, where N is the total number of buckets, then the extent to which the bucket sizes are uniform can be expressed by

$$u = \sum_{n=1}^{N} \left(s_n - \frac{1}{N}\right)^2 = \sum_{n=1}^{N} s_n^2 - \frac{1}{N}. \tag{1}$$

The smaller u is, the more uniform the bucket sizes are. This expression is well compatible with the relation between the uniformity of the bucket sizes and the speed of ANN search. Since the time cost of the linear search in the buckets dominates the ANN search process, and under the assumption that the query features have similar distribution to the database features, the time cost is proportional to $\sum_{n=1}^{N} s_n^2$. As only one bit at a time is re-selected, the uniformity can also be expressed by the form

$$u' = \frac{\sum_{n=1}^{N} s_n^2}{\sum_{m=1}^{N/2} \tilde{s}_m^2}, \tag{2}$$

where $\{\tilde{s}_m, m = 1, 2, \ldots, \frac{N}{2}\}$ are constant, representing the normalized bucket sizes of the hash table which is obtained by indexing the database features with a key consisting of the unchanged bits. It can be verified that the value of u'

belongs to [0.5 1]. $u' = 0.5$ indicates that with the newly selected bit in the hash key, the time cost of the linear search can be reduced by a half compared to with only the unchanged bits in the hash key. While $u' = 1$ indicates that the newly selected bit does not bring any time saving. In the cost function, the term about the uniformity is set as $\frac{1}{1-u'}$ to encourage a small u' and to avoid the situation that $u' = 1$.

The collision rate. The collision rate of matched feature pairs is actually the probability that two matched features coincide with each other at all bits of the hash key. Refer to the probability that two matched features coincide at a certain bit as the stability p_c of the bit, then according to the principle of greedy algorithms, the most stable one is preferred when selecting a bit for the hash key. Thus the cost term about the collision rate is simply set as $1 - p_c$. Since the features of a 3D point are actually the matches of each other, matched feature pairs can be easily obtained in the database and p_c can be estimated by making statistics on these pairs.

Given the above two terms about the collision rate and the uniformity, the cost function C reads as:

$$C = \lambda(1 - p_c) + \frac{1}{1 - u'}, \tag{3}$$

where λ is a preset weight.

Algorithm 1. Bit Selection

Input: $\mathbb{H}_K = \{b_{s_1}, \ldots, b_{s_k}, \ldots, b_{s_K}\}$, with b_{s_k} being re-selected; $\mathbb{F}$, all the database features.
  begin
    Remove b_{s_k} from $\mathbb{H}_K$ to get $\tilde{\mathbb{H}}_K = \{b_{s_1}, \ldots, b_{s_K}\}$;
    Index $\mathbb{F}$ using $\tilde{\mathbb{H}}_K$;
    Count the bucket sizes $\{\tilde{s}_m, m = 1, 2, \ldots, \frac{N}{2}\}$;
    Generate 40 random numbers$\{r_1, r_2, \ldots, r_{40}\}$,
    $r_i \in [1\ D]$;
    Compute the stability of each b_{r_i};

    for i=1; i≤40; i++; **do**
      Replace b_{s_k} in $\mathbb{H}_K$ with b_{r_i} to get $\mathbb{H}'_K$;
      Index $\mathbb{F}$ using $\mathbb{H}'_K$;
      Compute the cost function C;
    end
    Select $r*$ from $\{r_i\}$ so that C is minimized;
    Replace b_{s_k} with b_{r*} in $\mathbb{H}_K$;
  end
  Output: $\mathbb{H}_K = \{b_{s_1}, \ldots, b_{r*}, \ldots, b_{s_K}\}$.

For a better understanding of the proposed method, the process of selecting one bit for a hash key is summarized in Algorithm 1. The hash key is denoted as

$\mathbb{H}_K = \{b_{s_1}, b_{s_2}, \ldots, b_{s_K}\}$, where b_{s_k} is the $s_k th$ bit of the descriptor, $s_k \in [1\ D]$ and D the length of the descriptor.

Notes on the implementation. As can be seen from Algorithm 1, the time cost of the bit selection process is linear with the number of the database features, namely the number of the elements in $\mathbb{F}$. To make Algorithm 1 be scalable, we actually use only a subset, which is randomly composed and has up to 80,000 elements, of $\mathbb{F}$ in Algorithm 1. We find this strategy brings about a constant time cost for the bit selection process while retaining the ability to discover 'good bits', *i.e.* the bits producing uniform bucket sizes and high collision rates.

To further reduce the training time, we perform the bit selection process on not all but a half of the hash keys when a new keyframe is inserted, and the two halves are selected in turn. This strategy would slow down the discovering of good bits but effectively reduces the system latency.

4 Experiments

In this section, we first evaluate the efficiency of the proposed indexing method on ANN search, and then show the comprehensive performance of the relocalization module. The entire system is implemented in C++ and runs on a laptop with an Intel Core i3-2310 2.1 GHz CPU. The binary feature employed in the system is BRISK [13].

4.1 The ANN Search Efficiency of the Indexing Method

Dataset. We construct a dataset to evaluate the ANN search efficiency of the proposed indexing method. We first take a long video containing 12,821 frames around a building, and then run the SLAM system described in Sect. 2 on this video, from which 293 keyframes are extracted and 37,641 3D points along with 175,207 BRISK descriptors are obtained. These descriptors will serve as the database features in the experiment. To obtain query features with known ground truth matches, we select 500 well tracked frames from the video and reproject 400 visible 3D points on each of them. The binary descriptors at these reprojections are extracted as query features. Consequently, in each of the 500 frames we have 400 query features of which the corresponding 3D points are known.

Evaluation Criterion. The ANN search efficiency is demonstrated by the search accuracy and the search time. For a query feature, if the obtained approximate nearest neighbor corresponds to the same 3D point as itself, the ANN search is deemed as successful. The search accuracy is then defined as the ratio of successful ANN searches. The search time is simply the time cost of the whole ANN search process.

Setups. To demonstrate the advantage of the proposed indexing method, namely the learned LSH, we compare the ANN search performance of the method to that of the original LSH and the setups are listed below.

- **Original LSH.** Three configurations of the number of hash tables, *i.e.* 2, 6 and 10 respectively, are used in the experiment. In each configuration the length of the hash key ranges from 17 to 13, resulting in different search accuracies and timings.
- **Learned LSH.** The configurations of the number of hash tables and the length of the hash key are identical to the original LSH. Each hash key in the learned LSH is the same as that in the Original LSH initially, and is changed during the SLAM process. The final hash keys used for the ANN search are the ones learned from all the 293 keyframes.

Results. Figures 2(a) to (c) demonstrate the ANN search results in terms of search accuracies and search timings obtained by using 2, 6, 10 hash tables respectively. It can be seen that under these configurations, the ANN search performances of the learned LSH are consistently improved over the original LSH. Using the learned LSH takes less time than using the original LSH to reach the same search accuracy. For example, when 2 hash table is used, the time cost can be reduced by about a half. These results indicate that the bit selection process is effective. Rather than being randomly generated, better hash keys can be learned by exploiting the features collected in the SLAM process. To be noted that, since only real time performance is of interest, the comparisons are performed in relatively low accuracy regions.

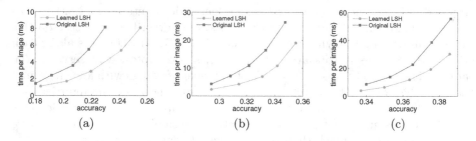

(a)   (b)   (c)

Fig. 2. The ANN search results of the original LSH and the learned LSH. (a) The results obtained by using 2 hash tables. (b) The results obtained by using 6 hash tables. (c) The results obtained by using 10 hash tables.

The evolution of the search efficiency. Since hash keys of the learned LSH are changed incrementally during the SLAM process, we may want to know how the search efficiency of the learned LSH evolves. To figure out this, some experiments are carried out as follows. We stop the bit selection process after k keyframes are inserted and then use the resulted hash keys to perform ANN search for query features in the above dataset. The results are shown in Fig. 3(a),

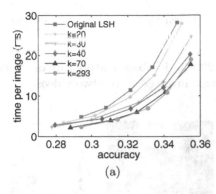
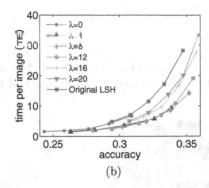

(a)   (b)

Fig. 3. (a) The ANN search results of the learned LSH with hash keys learned from the top k keyframes. 6 hash tables are used. (b) the ANN search results of the learned LSH with different values of λ. 6 hash tables are used.

from which we can see that the search efficiency roughly increases as k increases. This is reasonable since as k increases, more bit selection processes are performed to increase the chance of find good bits, and wider range of training data, *i.e.* larger number of database features, is also available. It can also be seen from Fig. 3(a) that as k increase to some extent, *e.g.* 70, the performance nolonger improves. This result suggests that it is not necessary for endless learning and we may stop the learning after enough bit selection processes are performed.

The choice of λ. There is a parameter λ in the cost function (Eq. 3) needed to be determined. Figure 3(b) shows its impact on the efficiency of the indexing method. It can be found that either a too small value, *e.g.* 0, or a too big value *e.g.* 20, degrades the search efficiency, and a moderate value is preferred. So in this paper, we empirically set λ to be 12.

The timing of learning. The learning time cost is dominated by the bit selection process. Benefited from the strategy described in Sect. 3.2, the time cost of the bit selection process is constant. We use 10 hash tables with key length of 14 and find that selecting one bit for a hash key takes about 53 ms even the number of database features increases to 175207. Since we perform bit selection on a half of the hash keys at a time, the learning time cost at a time is about 260 ms. This time cost is favorable since the learning process runs on the background thread and 260 ms is much less than the time interval, usually several seconds, between two keyframes.

4.2 The Relocalization Performance

To evaluate the performance of the proposed relocalization module, we run this module on two video segments and try to relocalize each frame of them. The first video segment (V1) contains 8000 frames and is clipped from the video which has been used in Sect. 4.1 to evaluate the efficiency of the indexing method. The scene map involved in this segment contains 37,641 3D points and 175,207 database features. The second video segment (V2) is taken by moving the camera above an office desk and contains 1063 frames. The map involved in the second video segment is built by running SLAM on another video, from which 31 keyframes are extracted and 4173 3D points along with 20,846 database features are obtained. Figure 4 gives a shot on the scenes and maps involved in the two videos.

Fig. 4. Left, the scene in which the second video segment is taken. Right, the scene in which the first video segment is taken.

We follow the pipeline described in Sect. 3 to relocalize each frame of V1 and V2. Since the camera intrinsic parameters are fixed and have been calibrated in advance, we use RANSAC and EPnP [17] followed by a non-linear optimization to estimate the camera poses from 3D-2D correspondences. The relocalization is deemed as successful if more than 12 inlier correspondences are found in the RANSAC process. The learned LSH is used for ANN search in the process of establishing 3D-2D correspondences. For comparison, the original LSH is also used. In V1, both the original LSH and the learned LSH has 10 hash tables with key length of 14, and the hash keys in the learned LSH are learned from all the keyframes. In V2, the original LSH has 10 hash tables with key length of 14 and the learned LSH has 10 hash tables with key length of 13. Again the hash keys in the learned LSH are learned from all the keyframes.

Table 1. The relocalization results.

Video segments	V1		V2	
# query frames	8000		1063	
# 3D points	37,641		4173	
# database features	175,207		20,846	
Indexing method	Original LSH	Learned LSH	Original LSH	Learned LSH
# relocalized frames	*7832.0*	*7859.0*	*927.8*	*953.2*
Timings (ms)				
Corner detection	8.3	8.5	7.8	7.8
Descriptor extraction	8.3	8.3	10.03	10.03
ANN search	*30.4*	*17.5*	*3.7*	*3.5*
RANSAC	0.5	0.5	1.2	1.2
Total	48.3	35.9	24.8	24.6

Comprehensive Results. Table 1 summarizes the relocalizaiton results which are averaged over 5 repetitions to alleviate the effect of the randomness in RANSAC. It is shown that in V1, 7859 frames out of 8000 frames, *i.e.* 98.2 % are successfully relocalized at a speed of the frame rate, which demonstrates the good scalability of the relocalization module for large maps. Besides, due to the higher efficiency of the learned LSH compared to the original LSH, the time cost of ANN search can be reduced by 42 %, and a slightly more, *i.e.* 0.3 %, frames can be successfully relocalized. In V2, the view point differences between the query frames and the keyframes are larger than that in V1, thus a lower registration rate, *i.e.* 89.6 %, is obtained. Since the map in V2 is relatively small, the time cost saving by using the learned LSH is not as prominent as that in V1. However the improvement of the registration rate, 2.4 %, is now more noticeable. This improvement should also be ascribed to the higher efficiency of the learned LSH: by taking the same time, using the learned LSH produces higher search accuracy than using the original LSH.

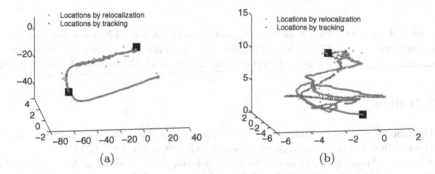

(a)   (b)

Fig. 5. The camera locations estimated by the relocalization module (the blue dots) and the tracking module (the red dots). (a) Camera locations of the first video segment. (b) Camera locations of the second video segment (color figure online).

Relocalization accuracy. The aim of the relocalization is to restart the tracking. If the camera pose of a frame obtained by the relocalization module is close enough to the pose that should have been obtained by the tracking module, the tracking usually restarts successfully. So the accuracy of the relocalization should be measured by the differences between poses from the relocalization and poses from the tracking. In real situations, the relocalization is performed on frames suffering from tracking failure, and thus poses from the tracking can not be obtained. To evaluate the relocalization accuracy, we instead run the relocalization module on frames that are tracked without failure. Again, we use V1 and V2 in which all frames can actually be well tracked. Figure 5 depicts the camera locations estimated by both modules. The black squares represent two special camera locations from the tracking module, one of them is the initial location and the other is the location farthest from the initial one. After dividing the average distance between the tracked locations and the relocalized locations by the distance of the two special locations, we get a relative location error of 0.05 % in v1 and 0.7 % in v2. Defining the rotation error between two rotation matrices R_1 and R_2 as the angle of the rotation $R_1 R_2^T$, as in [4], we also get a average rotation error of 0.25 degrees in v1 and 0.4 degrees in v2. These results indicate that the relocalization is relatively accurate.

5   Conclusion

In this paper, we have presented a binary feature indexing method for real-time SLAM relocalization. The core of the indexing method lies in that hash keys are learned online with the aim of attaining uniform hash buckets and high collision rates. By implementing the learning process on the background mapping thread and activating it only when the map is updated, little latency is brought about to the system. Experiments show that the proposed indexing method is more efficient than LSH and the relocalization module is able to handle large maps with tens of thousands of landmarks.

Acknowledgement. This work was supported by the National Natural Science Foundation of China under Grant No. 61421004, the National Basic Research Program of China under grant No. 2012CB316302 and Nokia Research Grant No. LF14011659182.

References

1. Davison, A., Reid, I., Molton, N., Stasse, O.: Monoslam: Real-time single camera slam. IEEE Trans. Pattern Anal. Mach. Intell. **29**, 1052–1067 (2007)
2. Klein, G., Murray, D.: Parallel tracking and mapping for small ar workspaces. In: Proceedings of the 6th IEEE and ACM International Symposium on Mixed and Augmented Reality, pp. 225–234 (2007)
3. Sattler, T., Leibe, B., Kobbelt, L.: Fast image-based localization using direct 2d-to-3d matching. In: Proceedings of the IEEE International Conference Computer Vision, pp. 667–674 (2011)
4. Lim, H., Sinha, S., Cohen, M., Uyttendaele, M.: Real-time image-based 6-dof localization in large-scale environments. In: Proceedings of the IEEE Conference on Computer Vision and Pattern Recognition, pp. 1043–1050 (2012)
5. Li, Y., Snavely, N., Huttenlocher, D., Fua, P.: Worldwide pose estimation using 3D point clouds. In: Fitzgibbon, A., Lazebnik, S., Perona, P., Sato, Y., Schmid, C. (eds.) ECCV 2012, Part I. LNCS, vol. 7572, pp. 15–29. Springer, Heidelberg (2012)
6. Lowe, D.: Distinctive image features from scale-invariant keypoints. Int. J. Comput. Vis. **60**, 91–110 (2004)
7. Tola, E., Lepetit, V., Fua, P.: Daisy: An efficient dense descriptor applied to wide-baseline stereo. IEEE Trans. Pattern Anal. Mach. Intell. **32**, 815–830 (2010)
8. Williams, B., Klein, G., Reid, I.: Automatic relocalization and loop closing for real-time monocular slam. IEEE Trans. Pattern Anal. Mach. Intell. **33**, 1699–1712 (2011)
9. Ozuysal, M., Calonder, M., Lepetit, V., Fua, P.: Fast keypoint recognition using random ferns. IEEE Trans. Pattern Anal. Mach. Intell. **32**, 448–461 (2010)

10. Straub, J., Hilsenbeck, S., Schroth, G., Huitl, R., Moller, A., Steinbach, E.: Fast relocalization for visual odometry using binary features. In: Proceedings of the IEEE International Conference on Image Processing, pp. 2548–2552 (2013)
11. Calonder, M., Lepetit, V., Strecha, C., Fua, P.: BRIEF: binary robust independent elementary features. In: Daniilidis, K., Maragos, P., Paragios, N. (eds.) ECCV 2010, Part IV. LNCS, vol. 6314, pp. 778–792. Springer, Heidelberg (2010)
12. Rublee, E., Rabaud, V., Konolige, K., Bradski, G.: Orb: An efficient alternative to sift or surf. In: Proceedings of the IEEE International Conference on Computer Vision, pp. 2564–2571 (2011)
13. Leutenegger, S., Chli, M., Siegwart, R.: Brisk: Binary robust invariant scalable keypoints. In: Proceedings of the IEEE International Conference on Computer Vision, pp. 2548–2555 (2011)
14. Gionis, A., Indyk, P., Motwani, R.: Similarity search in high dimensions via hashing. In: Proceedings of the 25th International Conference on Very Large Data Bases, pp. 518–529 (1999)
15. Rosten, E., Drummond, T.W.: Machine learning for high-speed corner detection. In: Leonardis, A., Bischof, H., Pinz, A. (eds.) ECCV 2006, Part I. LNCS, vol. 3951, pp. 430–443. Springer, Heidelberg (2006)
16. Fischler, M.A., Bolles, R.C.: Random sample consensus: a paradigm for model fitting with applications to image analysis and auto cartography. Commun. ACM 24, 381–395 (1981)
17. Lepetit, V., Moreno-Noguer, F., Fua, P.: Epnp: An accurate o(n) solution to the pnp problem. Int. J. Comput. Vis. 81, 155–166 (2009)

Depth-Based Real-Time Hand Tracking with Occlusion Handling Using Kalman Filter and DAM-Shift

Kisang Kim and Hyung-Il Choi[✉]

School of Media, Soongsil University, Seoul, Korea
illusion1004@gmail.com, hic@ssu.ac.kr

Abstract. In this paper, we propose real-time hand tracking with a depth camera by using a Kalman Filter and an improved DAM-Shift (Depth-based adaptive mean shift) algorithm for occlusion handling. DAM-Shift is a useful algorithm for hand tracking, but difficult to track when occlusion occurs. To detect the hand region, we use a classifier that combines a boosting and a cascade structure. To verify occlusion, we predict in real time the center position of the hand region using Kalman Filter and calculate the major axis using the central moment of the preceding depth image. Using these factors, we measure real-time hand thickness through a projection and the threshold value of the thickness using a 2nd linear model. If the hand region is partially occluded, we cut the useless region. Experimental results show that the proposed approach outperforms the existing method.

1 Introduction

In the last few decades, various studies have been conducted on automatic analysis of human behavior. The most sophisticated research on the subject is being carried out in HCI (human-computer interaction). Human gesture recognition is an important area in this field. A gesture is a simple and effective nonverbal communication tool that assists complex human interactions. In many fields such as sign language, hand gesture recognition is a primary method for those with hearing impairment to smart devices for effective interactions. Several methods have been proposed for gesture recognition, which includes hand region detection and hand feature extraction. Existing research on the subject includes analyzing hand images using a data glove [1–3], color data [4,5], a combination of color and depth data [6–8], and depth data alone [9–13]. It is difficult to devise an interface for a data glove because it requires a line to connect to the entire system. Various studies that integrate depth and color data seek to mitigate the sensitivity of the color data method to environmental changes. Under the assumption that the hand lies before the body. Park et al. [6] generated a histogram from a depth image of the Kinect motion sensing input device to detect candidate hand regions, and located a final hand region by using Bayes rule and skin color to find the precise hand region. This method executes significantly better than the

© Springer International Publishing Switzerland 2015
C.V. Jawahar and S. Shan (Eds.): ACCV 2014 Workshops, Part I, LNCS 9008, pp. 218–226, 2015.
DOI: 10.1007/978-3-319-16628-5_16

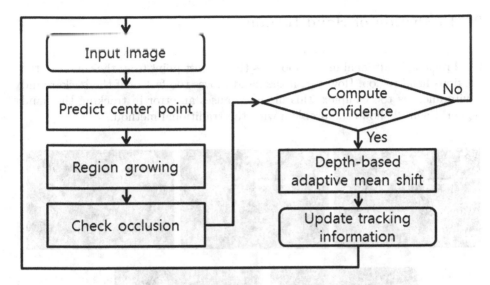

Fig. 1. Structure of occlusion handling tracking system.

sole use of the color and depth data, however performance decreases in darkness due to its basic assumption; that is, the hand always lies before the body and the use of color. Van den Bergh and Van Gool [7] suggested a combined method that locates a face in a color image, removes the background by using the threshold value along with the distance of the detected face, and searches the hand region in the remaining region. Furthermore, this method is more accurate than ones that use either a depth or a color image, but requires more processing time and is more difficult to use in dim lighting. Trindade et al. [8] proposed skin color filtering by using an RGB-D sensor prior of detecting the face, body, and hand regions, distributing the histogram with the depth axis, and filtering out the hand region based on a threshold value. The outliers are thus removed by k-means clustering to find the center of the hand region, which becomes the base point for detecting the hand region and for pose recognition. This method, with a mixed use of color and depth data, can improve detection accuracy by deleting outliers during filtering and applying a segmentation technique. However, this method is difficult to use in varying lighting conditions and is sensitive to errors because it undergoes several processes prior to hand region detection.

To overcome these problems, we only use depth data from the Kinect camera sensor. To track the hand region, we use a boosting and cascading algorithm to detect region. Figure 1 shows the structure of our proposed system which consists of main two steps: prediction and verification of occlusion. The remainder of this paper is structured as follows: in Sect. 2, we explain hand region prediction using Kalman Filter. Section 3 explains improved DAM-Shift [14], comparing with traditional DAM-Shift. Section 4 describes our testing environment along with experimental results that confirm the effectiveness of the proposed algorithm. We present our conclusions in Sect. 5.

2   Prediction of Hand Region

Prior to the region growing process, we need to find the point, which is the one of hand region. Traditional method locates the nearest point from the center of the previous hand region. However, if the hand moves too fast or if the background environment is too complex, this method causes an error in tracking the hand region. Figure 2 shows the problem with the traditional method.

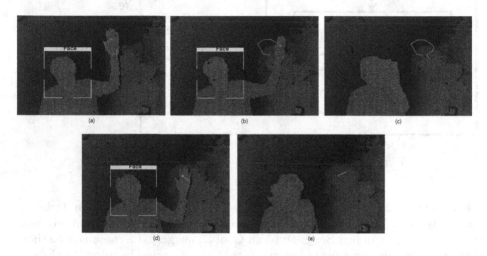

Fig. 2. Problem with the traditional tracking method. (a), (b), (c) is a set that fail to track. (d), (e) is also a set.

To solve this problem, we use a prediction method that combines Kalman Filter, previous hand moving velocity and 2nd polynomial model. Using only velocity, it is too risky because in except situation, this method occur the error due to the prediction point moves too much. Thus, in order to reduce the error rate, we use 3D Kalman Filter. Furthermore, when the distance between the hand and the camera is small, hand tends to move considerably in a corresponding image; otherwise, it only moves slightly. We use this observation in a 2nd polynomial model in order to predict hand region size and use 2nd polynomial model to predict hand region movement [15].

$$r = \begin{bmatrix} 1 & x & x^2 \end{bmatrix} \cdot \begin{bmatrix} \alpha_1 \\ \alpha_2 \\ \alpha_3 \end{bmatrix} \tag{1}$$

$$\alpha = (P^T \cdot P)^{-1} \cdot P^T \cdot y \tag{2}$$

$$P = \begin{bmatrix} 1 & x_1 & x_1^2 \\ 1 & x_2 & x_2^2 \\ \vdots & \vdots & \vdots \\ 1 & x_n & x_n^2 \end{bmatrix}, y = \begin{bmatrix} r_1 \\ r_2 \\ \vdots \\ r_n \end{bmatrix} \tag{3}$$

Fig. 3. Prediction of center point. Rectangle: Predicted center, circle: Previous center.

Equation (1) is to presume the radius of the including circle of a hand region. x represents the depth of a candidate hand region in the current image. To make the value r, we need the coefficients $\alpha = [\alpha_1\ \alpha_2\ \alpha_3]$. These coefficients can be calculated by (2) and (3). The value of x_i and r_i are manually collected during the learning phase. We check the size of a hand region by varying the depth values of the hand region. We assume that the size of the hand could be represented as the 2nd polynomial function of depth. Therefore, if α is determined by the learning data, the hand region size can be predicted by using (1).

$$P_i(x, y, z) = KF\{C_{i-1} + \frac{\beta \cdot r}{\sqrt{w^2 + h^2}}(2 \times C_{i-1} - 3 \times C_{i-2} + C_{i-3})\} \quad (4)$$

Equation (4) estimates of the hand region with the predicted hand region size. $KF(\cdot)$ represents the Kalman Filter. $P_i(x, y, z)$ is the predicted hand center. C_i is the hand center of i-th image. w and h represent the image width and height, respectively and β is the weight value, mostly it use from 0.3 to 0.7. Figure 3 shows the predicted point and the previous center point of the hand region.

3   DAM-Shift with Occlusion Handling

3.1   Detecting and Handling Occlusion

The major assumptions underlying hand occlusion prevention are that the tracking arm region always lies below the hand region, and that the thickness of the arm is always less than the thickness of hand. Given these assumptions, the occluded image can be found and the hand region can be revised. After the occlusion frames, this method is easy to re-track the hand region. To calculate

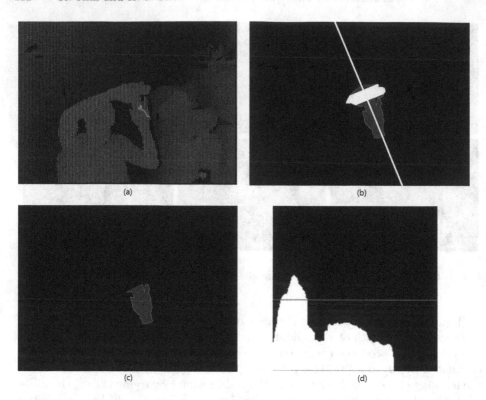

Fig. 4. Elimination of occluded region. (a) Input image, (b) Found occluded region, (c) Hand region, (d) Thickness projection.

thickness, the major axis of the hand region is required. We use central moment to measure this axis.

$$\theta = \frac{1}{2}tan^{-1}(\frac{2\mu'_{11}}{\mu'_{20} - \mu'_{02}}) \tag{5}$$

$$C_x = \frac{M_{10}}{M_{00}}, C_y = \frac{M_{01}}{M_{00}} \tag{6}$$

In Eq. (5), θ is the angle of the major axis. μ'_{11} is the value of μ_{11}/M_{00}, μ'_{20} is that of μ_{20}/M_{00} and μ'_{02} is that of μ_{02}/M_{00}. In Eq. (6), C_x and C_y are the center point of each coordinate of this region. Following this measurement, we make a projection histogram using this axis. Figure 4(b) shows a projected image with the major axis.

In Fig. 4(d), the gray line in the projection histogram is the value of r, the radius of the hand region in Eq. (1). If the thickness value is greater than the radius, it means the region is occluded by other objects. Therefore, if we detect a region as occluded, we remove it, as it is useless. Figure 4(c) shows an example of the removal of the occluded region.

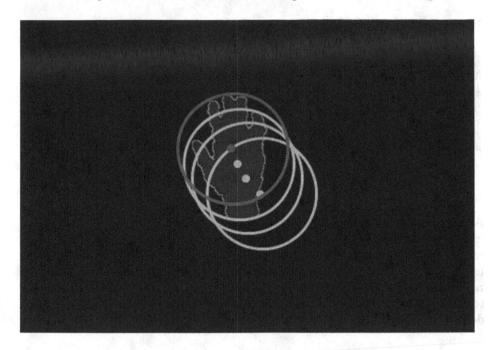

Fig. 5. Example of the movement of DAM-Shift.

3.2 DAM-Shift

DAM-Shift is defined similarly to Mean Shift [16], but its kernel size changes according to the depth values and the iteration time. Equation (7) represents the DAM-Shift algorithm.

$$TP_{i+1}^{mean} = \frac{\sum_{p \in \Omega} p \cdot K(p, TP_i^{mean}, DTP_{i-1}^{mean}, i)}{\sum_{p \in \Omega} K(p, TP_i^{mean}, DTP_{t-1}^{mean}, i)} \tag{7}$$

$$K(p, s, d, i) = \begin{cases} 1 & \|p - s\| < R(d, i) \\ 0 & otherwise, \end{cases} \tag{8}$$

$$R(d, i) = \begin{cases} SPM(d) & 2SPM(d) - i \cdot T_{rc} < SPM(d) \\ 2SPM(d) - i \cdot T_{rc} & otherwise, \end{cases} \tag{9}$$

$$SPM(d) = \alpha_1 + \alpha_2 d + \alpha_3 d^2 \tag{10}$$

TP_{i+1}^{mean} in the above represents the tracking point coordinates at the $i + 1$ iteration. These coordinates are updated as the iteration continues. $K(\cdot)$ presents a kernel function whose size changes depending on the depth of the tracking point. DTP_t^{mean} depicts the depth value of the tracking point in the $t - 1$th frame. p represents a point that belongs to set Ω and denotes the set of valid borderlines VB obtained during the region growing. The coordinates of the nearest point TP_t^{seed} is acquired as in region growing, and it is substituted for TP_0^{mean}.

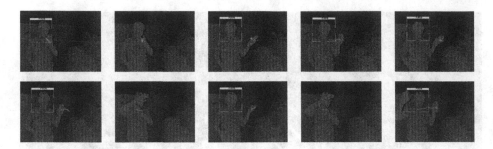

Fig. 6. Results of hand tracking.

As the iteration continues, TP_{t+1}^{mean} is alternated with TP_t^{mean} for the next iteration. The process is repeated until the point of convergence. In Eq. (8), The radius $R(d, i)$ of the kernel function changes according to the depth value and the number of repetition. Equation (9) extracts the radius according to depth and repetition. For this purpose, the function $SPM(d)$ is used, which corresponds to the 2nd polynomial model defined in Sect. 2. Figure 6 shows the process of determining the tracking point using the DAM-Shift algorithm.

4   Experimental Results

For experimental evaluation, we used a computer with an Intel(R) Core(TM) i5-3470 CPU and an 8 GBbyte memory. We have used Microsoft Kinect camera, (320 × 240 pixels) at 30 fps to acquire depth images.

In Fig. 6 shows the results of hand tracking in input frames at one second interval. The point shows the center point of the hand region and the tail represents the tracked line.

Figure 7 shows the features of the hand region. The white line shows the axis of the hand region and the rectangle is its predicted center point. This shows that it solved the occlusion problem to track. Further, occluded region can checked at Fig. 8, shows the results of hand projection.

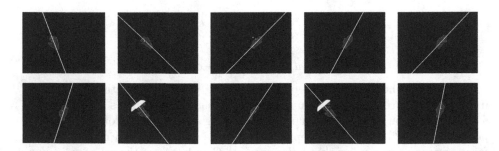

Fig. 7. Results of hand features.

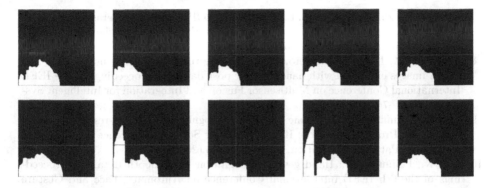

Fig. 8. Results of hand projection with checking occlusion.

5   Conclusions

We proposed a hand tracking method that works well in the real world environment. For tracking a hand, we have developed improved DAM-Shift to handle the occlusion. Our 2nd polynomial model and Kalman Filter work well to predict the center of the hand, which plays an important role in confining a search area. To handle occlusion, we developed an improved DAM-Shift, which consists of axis extraction and make hand projection to verify and remove occlusion region. As the experimental results, our method eliminates the occlusion region well and reduce problem encountered in hand tracking.

Acknowledgement. This study was supported by the Basic Science Research Program through the National Research Foundation of Korea (NRF) funded by the Ministry of Science, ICT and Future Planning (2013R1A1A2012012).

References

1. Quam, D.L.: Gesture recognition with a DataGlove. In: Proceedings of the IEEE National Aerospace and Electronics Conference, pp. 755–760 (1990)
2. Wang, R.Y., Popovic, J.: Real-time hand-tracking with a color glove. ACM Trans. Graph. **28**(3), 63:1–63:8 (2009)
3. Lamberti, L., Camastra, F.: Handy: a real-time three color glove-based gesture recognizer with learning vector quantization. Expert Syst. Appl. **39**(12), 10489–10494 (2012)
4. Suk, H.I., Sin, B.H.: Dynamic Bayesian network based two-hand gesture recognition. J. KIISE: Softw. Appl. **35**(4), 265–279 (2008)
5. Bhuyan, M.K., Neog, D.R., Kar, M.K.: Fingertip detection for hand pose recognition. Int. J. Comput. Sci. Eng. **4**(3), 501–511 (2012)
6. Park, M.S., Hasan, M., Kim, J.M., Chae, O.S.: Hand detection and tracking using depth and color information. In: Proceedings of the International Conference on Image Processing, Computer Vision, and Pattern Recognition, pp. 779–785 (2012)

7. Van den Bergh, M., Van Gool, L.: Combining RGB and ToF cameras for real-time 3D hand gesture interaction. In: Proceedings of the IEEE Workshop on Applications of Computer Vision, pp. 66–72 (2011)
8. Trindade, P., Lobo, J., Barreto, J.P.: Hand gesture recognition using color and depth images enhanced with hand angular pose data. In: Proceedings of the IEEE International Conference on Multisensor Fusion and Integration for Intelligent Systems, pp. 71–76 (2012)
9. Mo, Z., Neumann, U.: Real-time hand pose recognition using low-resolution depth images. In: Proceedings of the IEEE Computer Society Conference on Computer Vision and Pattern Recognition, pp. 1499–1505 (2006)
10. Liu, X., Fujimura, K.: Hand gesture recognition using depth data. In: Proceedings of the 6th IEEE International Conference on Automatic Face and Gesture Recognition, pp. 529–534 (2004)
11. Malassiotis, S., Strintzis, M.G.: Real-time hand posture recognition using range data. Image Vis. Comput. $26(7)$, 1027–1037 (2008)
12. Suryanarayan, P., Subramanian, A., Mandalapu, D.: Dynamic hand pose recognition using depth data. In: Proceedings of the 20th International Conference on Pattern Recognition, pp. 3105–3108 (2010)
13. Oikonomidis, I., Kyriazis, N., Argyros, A.A.: Efficient model-based 3D tracking of hand articulations using Kinect. In: Proceedings of the British Machine Vision Conference (2011)
14. Joo, S.-I., Weon, S.-H., Choi, H.-I.: Real-time depth-based hand detection and tracking. Sci. World J. **2014**, 17 p. (2014)
15. Park, S., Yu, S., Kim, J., Kim, S., Lee, S.: 3D hand tracking using Kalman filter in depth space. EURASIP J. Adv. Sig. Process. **2012**, 18 (2012)
16. Bradski, G.R.: Computer vision face tracking for use in a perceptual user interface. In: IEEE Workshop on Applications of Computer Vision, Princeton, NJ, pp. 214–219 (1998)

Evaluation of Depth-Based Super Resolution on Compressed Mixed Resolution 3D Video

Michal Joachimiak[1]([✉]), Payman Aflaki[1], Miska M. Hannuksela[2], and Moncef Gabbouj[1]

[1] Tampere University of Technology, Tampere, Finland
michal.joachimiak@tut.fi
[2] Nokia Research Center, Tampere, Finland

Abstract. The MVC+D standard specifies coding of Multiview Video plus Depth (MVD) data for enabling advanced 3D video applications. MVC+D defines that all views are coded with H.264/MVC encoder at equal spatial resolution. To improve compression efficiency it is possible to use mixed resolution coding in which part of texture views are coded at reduced spatial resolution. In this paper we evaluate the performance of Depth-Based Super Resolution (DBSR) on compressed mixed resolution MVD data. Experimental results show that for sequences with accurate depth data the objective coding performance metric increases. Even though some sequences, with poor depth quality, show slight decrease in coding performance with respect to objective metric, subjective evaluation shows that perceived quality of DBSR method is equal to symmetric resolution case. We also show that depth re-projection consistency check step of the DBSR can be changed to simpler consistency check method. In this way the DBSR computational complexity is reduced by 26 % with 0.2 % dBR average bitrate reduction for coded views and 0.1 % average bitrate increase for synthesized views. We show that proposed scheme outperforms the anchor MVC+D coding scheme by 7.2 % of dBR on average for total coded bitrate and by 10.9 % of dBR on average for synthesized views.

1 Introduction

3D video consumer devices, including video cameras and displays, start to emerge on the market. To store 3D video data efficiently new compression methods are required. As a response to the growing need for 3D video compression the Moving Picture Experts Group (MPEG) initiated 3D video standardization process [5], that has been continued by the Joint Collaborative Team on 3D Video Coding (JCT-3V) since July 2012. 3D video consists of a set of 2D video sequences, registered by cameras synchronized in time. Video acquisition using many cameras is challenging and some displays like autostereoscopic displays (ASD) require many views on input. However, encoding and transmission of many views would require a great amount of processing power and bandwidth. With the help of DIBR [16] techniques it is possible to register and encode a lower amount of views and synthesize missing views from decoded texture and corresponding depth data.

© Springer International Publishing Switzerland 2015
C.V. Jawahar and S. Shan (Eds.): ACCV 2014 Workshops, Part I, LNCS 9008, pp. 227–237, 2015.
DOI: 10.1007/978-3-319-16628-5_17

In addition, depth-enhanced 3D video allows for baseline adjustment on the receiver side. Variable baseline might be required to adjust to properties of some displays, viewing conditions or user preferences [20]. The Advanced Video Coding (H.264/AVC) standard [3] was earlier amended by a Multiview Video Coding (MVC) extension [3,13], and has been recently amended by a multiview-and-depth coding extension (MVC+D) [3,12]. MVC+D uses the MVD format and specifies that texture and depth data are coded independently but put in the same bitstream. In addition, it specifies that texture views have an equal resolution, while the depth views also share the same resolution, which may differ from the texture resolution.

In mixed resolution, stereoscopic video coding, first introduced in [21], one of the two views is coded at lower spatial resolution than the other view. According to the binocular suppression theory [9] the Human Visual System (HVS) is able to fuse stereoscopic images in way that the perceived quality is close to that of the higher quality view. The theory has been verified by the systematic subjective viewing experiments, such as [7,24]. Mixed resolution stereoscopic video coding allows a reduction of computational and storage resources due to one of the views having a lower spatial resolution. Thus, mixed resolution stereoscopic video makes it possible to keep the video quality close to that of full-resolution, symmetric, stereoscopic video, while the computational complexity is reduced.

In the case when the MVD format is used with texture views having mixed resolution, it is possible to improve the perceived quality by upsampling low-resolution views using a Depth-Based Super Resolution (DBSR). Since all views represent the same scene observed from different viewpoints, the video content, present in each view, is highly correlated. Thus, in case of mixed resolution 3D video, a low-resolution view can be enhanced using a neighboring high-resolution view. The first approach to improve low-resolution images in mixed resolution, stereoscopic image sequences was proposed in [22]. The method was based on a weighted averaging of the pixel intensity values coming from corresponding pixel positions in a high-resolution view. Another approach [17,19] assumes that low-frequency components can be restored with high fidelity during the upsampling process of the low-resolution view. After that, missing high-frequency components can be extracted from the high-resolution neighboring view. Both methods project a high-resolution view to the position of the low-resolution view and use the projected view data to improve the upsampling of the low-resolution view. The resulting stereoscopic pair has one view in full resolution without any modifications and one low-resolution view upsampled with the use of projected pixel data from the full-resolution view.

Even though the first DBSR approach was introduced in [17], its performance was not evaluated on compressed mixed resolution 3D video. In this paper we evaluate the performance of the DBSR upsamling [17] on mixed resolution 3D video, encoded with a modified MVC+D codec [8], based on the recent 3D video coding standard [12]. We show that this coding and upsampling concept provides objective gains by means of the Bjontegaard delta bitrate reduction (dBR) [10] when compared to both the MVC+D coding with symmetric resolution between

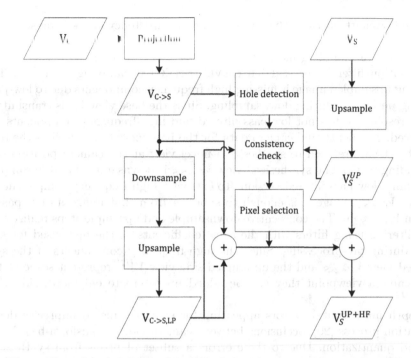

Fig. 1. Flow chart of the modified depth based super resolution algorithm.

views and mixed resolution coding with conventional upsampling. Moreover, we propose a simplified consistency check and show that depth re-projection consistency check is not only slower but, in some cases, can impact the quality of upsampled view. Finally, we show that even if the objective quality in terms of Peak Signal-to-Noise Ratio (PSNR) does not show improvement for sequences with low quality depth maps, the subjective evaluation proves that DBSR produces mixed resolution 3D video with quality equal to the symmetric resolution 3D video.

The rest of the paper is organized as follows. Section 2 presents the DBSR upsampling with the proposed consistency check. Section 3 describes coding simulation and subjective evaluation conditions. Section 4 summarizes simulation and evaluation results. Section 5 concludes the paper.

2 Depth-Based Super Resolution

The rendering of 3D video on a 3D display requires views with the same resolution. In order to render mixed resolution 3D video at high resolution, the decoded low-resolution, dependent views have to be upsampled to the base-view resolution. Since depth data and camera parameters corresponding to each view are encoded in a 3D video bitstream it is possible to use the DBSR to improve the quality of upsampling.

The flowchart for the DBSR with modified consistency check is presented in Fig. 1. The dependent, low-resolution, side view V_S is first upsampled using a conventional upsampling method, either with usage of the H.264/AVC motion interpolation filter [4] or scalable HEVC (SHVC) upsampling filter [11]. The resulting upsampled image is missing high frequency components due to low-pass filtering performed during downsampling. Since the base view V_C is transmitted at full resolution it is not low-pass filtered and high-frequency components are preserved. The depth and camera data for the base view V_C, as well as the base view V_C itself, are used to synthesize a novel view at the camera position corresponding to V_S. The synthesized view $V_{C->S}$ is downsampled and upsampled to obtain a low-pass filtered version. To extract a high-frequency component for pixel in $V_{C->S}$ image it is enough to subtract from that image its low-passed version $V_{C->S,LP}$. The consecutive downsample and upsample steps realize low-pass filtering using filters with the same coefficients as the ones used for side views during pre-processing. Since the high-frequency components of the synthesized view $V_{C->S}$ and the upsampled side view V_S^{UP} represent scene at the same camera viewpoint they can be added up to create enhanced, side view V_S^{UP+HF}.

Depth maps contain errors appearing for example due to imprecise depth estimation process [20], occlusions between camera views, transform-based coding and quantization. Due to these errors a subset of pixels from synthesized view might be projected to wrong locations. To mitigate the impact of these errors on the view synthesis quality and, consequently, on the super resolution enhancement quality, the consistency check based on minimizing the depth map reprojection error was proposed in [17].

However, the consistency check based on depth has two limitations. First - the depth re-projection is computationally costly and second - instead of measuring the inconsistencies in the image domain the depth domain is taken into account. We propose an alternative, computationally simple, solution for a consistency check realized in two consecutive steps. Firstly, hole areas detected during view synthesis are excluded since they do not contain projected pixels that can be compared to their correspondents in the dependent view. After that, a pixel-wise similarity, derived from luminance component, between the upsampled version of side view V_S^{UP} and the low-pass filtered version of center view, projected to the side view position $V_{C->S,LP}$ is calculated. The motivation behind this procedure is that these images contain corresponding pixels at the same low frequency spectrum. The similarity check is calculated within the matching window according to the following formula:

$$\sum_{x=x_0-w}^{x_0+w} \sum_{y=y_0-h}^{y_0+h} |V_{C->S,LP}(x,y) - V_S^{UP}(x,y)| < T, \qquad (1)$$

where $2w$ and $2h$ are width and height of the matching window. The pixels for which the absolute difference is smaller than a threshold T are considered similar and the enhancement, using high-frequency component extracted from the synthesized view, is executed. The pixels for which similarity condition does not hold are kept intact and copied from V_S^{UP}. Our empirical study shows that threshold T is not a function of a quantization parameter and thus it is kept constant over all quantization parameters. The consistency check based on pixel-wise similarity check does not rely solely on depth re-projection and it is expected to have much lower computational complexity. Simulation results confirm these expectations.

3   Subjective Test Setup and Simulation Environment

The modified 3DV-ATM reference software [8] was used to encode mixed resolution 3D video. The base view is coded at original resolution and dependent views are coded at half resolution in vertical and horizontal direction. Hence, the base views of the created bitstreams conform to H.264/AVC, while the dependent views are compatible with the MVC extension [3] except for the re-sampling of the decoded base view pictures for inter-view prediction, as described in [8]. Texture and depth views were coded into the same bitstream using the MVC+D bitstream syntax. The coding and view synthesis simulations were executed under the Common Test Conditions (CTC) [1] on the test sequences specified by the same document. The three-view (C3) coding scenario was selected in which three texture views accompanied by corresponding depth data are encoded with center view coded as the base one. A set of three additional views between each pair of adjacent coded views was synthesized using the view synthesis reference software (VSRS) described in [23], intended to serve for multiview autostereoscopic displays. The synthesized views PSNR is calculated against the views at same camera positions, rendered from uncompressed texture and depth data.

The anchor bitstreams were created with the 3DV-ATM software [6,18], using the MVC+D configuration with full resolution for the base and dependent views, in C3 scenario. The major settings for 3DV-ATM are summarized in Table 1. To obtain low resolution versions of side views they were downsampled with use of the 12-tap low-pass filter [15] with a cut-off frequency of 0.9π and the following coefficients:

$$[2 \quad -3 \quad -9 \quad 6 \quad 39 \quad 58 \quad 39 \quad 6 \quad -9 \quad -3 \quad 2 \quad 0]/128. \tag{2}$$

After that, the coding process of mixed resolution 3D video was conducted using the modified 3DV-ATM software [8] with the same settings as specified in Table 1. The only difference was to set width and height of the dependent, texture views to half comparing to the anchor bitstream settings specified in CTC [1].

Two types of upsampling algorithms were executed on the decoded dependent views of the mixed resolution 3D video. First type was 8-tap upsampling filter used in the scalable HEVC [11] with the following filter coefficients:

$$[-1 \quad 4 \quad -11 \quad 40 \quad 40 \quad -11 \quad 4 \quad -1]/64. \tag{3}$$

In the second upsampling scheme, the modified DBSR algorithm presented in Sect. 2 was used. In the first upsampling step that creates V_S^{UP}, the same 8-tap scalable HEVC filter [11] was used. This approach guarantees fair comparison between the two tested upsampling schemes. A threshold $T = 15$ for the proposed consistency check serving for the DBSR scheme was empirically found the best from the set {5, 10, 15, 20}.

The execution times of the tested algorithms were measured in seconds on the same computer with 1 process per CPU unit. The simulation framework was set up on the Linux operating system with minimal set of services and graphical user interface shut down.

The subjective evaluation experiment was carried out using four sequences: Poznan Hall2, Poznan Street, Kendo and Balloons [1]. Four quantization parameters (QP) were used according to CTC. For the purpose of subjective test the sequences encoded with QPs 26 and 36, namely the lowest and second highest, were selected. The naive subjects compared 3D stereoscopic videos generated from views specified according to [1]. The polarized Sony Bravia 55" 3D display was used for subjective viewing. The viewing distance was equal to 4 times the displayed image height (2.72 m). Subjective quality assessment was conducted according to the Double Stimulus Impairment Scale (DSIS) method [2] with a discrete scale from 0 to 10 for quality assessment. Prior to each test, subjects were familiarized with the test task, the test sequences, and the expected variation in quality. The subjects were instructed that 0 stands for the lowest quality and 10 for the highest. Moreover, duration of each session was limited to half an hour to prevent the subjects from experiencing fatigue or eye strain. The test sequences were played in a random order and each video clip was played twice to increase the accuracy of the evaluation. Subjective viewing was conducted with 20 subjects, (14 males and 6 females), aged between 23 and 32 years with mean 27 years. All subjects passed the stereovision test prior to the 3D viewing.

4   Results and Discussion

To compare the objective quality of the test sequences the Bjontegaard delta bitrate (dBR) [10] and Peak Signal-To-Noise (PSNR) ratio were used. The dBR results, for each scheme tested, are shown in two columns and were calculated as

Table 1. Major 3DV-ATM configuration settings.

Coding parameter	Setting
Texture : Depth width ratio	1 : 0.5
Texture : Depth height ratio	1 : 0.5
Inter-view prediction structure	PIP
Inter prediction structure	HierarchicalB, GOP8
QP settings for texture and depth	26, 31, 36, 41
Encoder optimization settings	RDO ON, VSO ON
View Synthesis in post-processing	Fast 1D VSRS [19]

Table 2. Performance of mixed resolution 3D video coding with use of the SHEVC upsampling method

Sequence	Index	Coded views		Synthesized views	
		dBR [%]	dPSNR [dB]	dBR [%]	dPSNR [dB]
Poznan Hall2	S01	−18.36	0.65	−20.33	0.79
Poznan Street	S02	0.61	−0.10	−5.77	0.16
Undo Dancer	S03	21.26	−0.91	3.49	−0.24
Ghost Town Fly	S04	1.75	−0.40	−5.46	0.05
Kendo	S05	−12.61	0.60	−15.14	0.70
Balloons	S06	−13.54	0.69	−16.08	0.80
Newspaper	S08	−2.74	0.07	−6.58	0.22
Shark	S10	−3.13	0.07	−9.37	0.37
Average		**−3.35**	**0.08**	**−9.41**	**0.35**

following. For the 'coded views' column an aggregated bitrate of coded texture and depth data versus average PSNR of the coded texture was used. For the 'synthesized views' column an aggregated bitrate of coded texture and depth data versus average PSNR of three synthesized views between each adjacent pair of coded views for a given sequence was used. The objective results of mixed resolution coding are presented in Tables 2 and 3. The anchor was generated using the 3DV-ATM software [6,18] that uses symmetric resolution for texture coding. The results in Table 2 correspond to the scalable HEVC upsampling. In Table 3 results pertain to the DBSR method [17] and its variation presented in Sect. 2. It can be seen that both DBSR upsampling methods improve coding efficiency significantly, for coded and synthesized views. Our simulations showed that the proposed DBSR with simplified consistency check executes 26 % faster, independently on the sequence used in the test. Furthermore, gain for sequences with high accuracy depth, such as Undo Dancer and Ghost Town Fly, is higher in case of the proposed consistency check, which proves better performance of the proposed consistency check.

The results tabularized in Tables 2 and 3 show also that the DBSR upsampling performance, measured by the objective dBR metric, does not show improvement for sequences with less accurate depth quality such as Poznan Hall2, Kendo, Balloons and Newspaper. In order to verify the subjective performance of the DBSR upsampling for these sequences the subjective quality evaluation test was executed with results presented in Fig. 2. The values shown, correspond to mean opinion scores (MOS) with 95 % confidence interval. The reference sequences, abbreviated as R, were generated using original, not compressed, texture and depth data. The symmetric resolution sequences, abbreviated as S were generated using unmodified 3DV-ATM software [6,18] with same resolution parameters for all texture views. The mixed resolution sequences,

Table 3. Performance of mixed resolution 3D video coding with use of the two versions of DBSR.

Seq.	DBSR [17]				Proposed DBSR			
	Coded views		Synthesized views		Coded views		Synthesized views	
	dBR [%]	dPSNR [dB]	dBR [%]	dPSNR [dB]	dBR [%]	dPSNR [dB]	dBR [%]	dPSNR [dB]
S01	−17.21	0.59	−19.19	0.74	−17.14	0.59	−19.02	0.73
S02	−0.46	−0.05	−4.46	0.11	−0.72	−0.03	−4.40	0.11
S03	0.75	−0.09	−7.67	0.23	−0.53	−0.04	−8.04	0.25
S04	−7.26	0.19	−11.71	0.42	−9.22	0.30	−12.54	0.46
S05	−11.64	0.54	−13.86	0.63	−11.02	0.50	−13.32	0.60
S06	−12.83	0.64	−14.96	0.74	−12.17	0.60	−14.34	0.70
S08	−1.68	0.02	−5.41	0.17	−0.99	−0.01	−4.75	0.15
S10	−5.86	0.26	−10.74	0.46	−5.76	0.25	−10.52	0.45
Average	**−7.02**	**0.26**	**−11.00**	**0.44**	**−7.19**	**0.27**	**−10.87**	**0.43**

Table 4. Wilcoxon's signed rank test. Pairwise comparison between symmetric and mixed resolution MOS scores with statistical difference level $p < 0.05$.

Sequence	p - value	
	Low bitrate	High bitrate
Poznan Hall2	0.00	0.46
Poznan Street	0.02	0.20
Kendo	0.65	0.22
Balloons	0.15	0.00

abbreviated as M, were generated using the modified 3DV-ATM software [8] with mixed resolution texture views. The results are presented for low bitrate (LB) with QP parameter equal to 36, and high bitrate with QP equal to 26. Results in Fig. 2 show that subjective quality of the DBSR upsampling is perceived equal or, in some cases slightly better then symmetric resolution coding. To test the significant difference between the scores the Wilcoxon's signed rank test [25] was used. The significance level was set to $p = 0.05$. The results of the Wilcoxon's test, tabularized in Table 4 show that statistical difference is detected for 3 cases. In these 3 cases, according to Fig. 2, mixed resolution was perceived better than symmetric resolution. In all other cases there was no statistically significant difference in perceived quality between tested schemes. Based on these results it can be concluded that perceived quality between two schemes: symmetric resolution and mixed resolution with DBSR upsampling, cannot be subjectively differentiated.

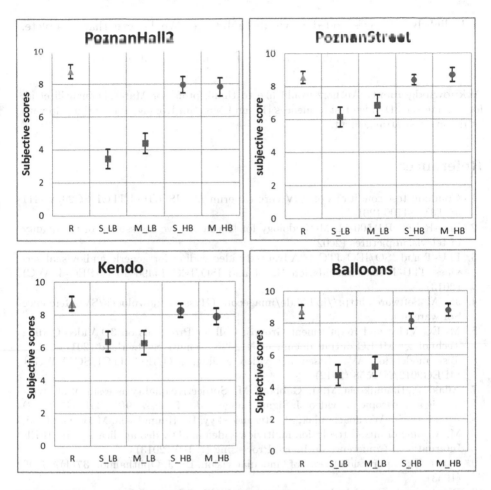

Fig. 2. Viewing experience MOS with 95 % confidence interval. Abbreviations correspond to R: Reference, S: Symmetric resolution, M: Mixed resolution, LB: Low bitrate, HB: High bitrate.

5   Conclusions

Since introduction of the DBSR algorithm [17] its performance was not analyzed on compressed mixed resolution 3D video. This paper presents the objective and subjective evaluation of the DBSR upsampling scheme executed on compressed, mixed resolution 3D video. The experiments show that even though, some sequences with low accuracy depth maps, show slight decrease in coding efficiency, the perceived quality of DBSR-upsampled views is equal to the symmetric case. In addition, an improvement for the consistency check step is introduced that speeds up the total execution time of the DBSR algorithm by 26 %, and slightly improves coded views delta bitrate. We believe that complete, objective and subjective, performance analysis of the DBSR upsampling approach shows that it is a viable solution for the mixed resolution 3D video coding.

Acknowledgement. Authors would like to thank professor Marek Domanski et al. for sharing the 3D video test sequences [14] and Maryam Homayouni for the help with subjective test arrangement.

References

1. Common test conditions of 3DV core experiments. ISO/IEC JTC1/SC29/WG11 JCT3V-E1100 (2013)
2. ITU-R Rec. BT.500-11, Methodology for the subjective assessment of the quality of television pictures (2002)
3. ITU-T and ISO/IEC JTC 1: Advanced video coding for generic audiovisual services. ITU-T Recommendation H.264 and ISO/IEC 14496–10 (MPEG-4 AVC) (2013)
4. JSVM Software. http://ip.hhi.de/imagecom_G1/savce/downloads/SVCReference Software.htm
5. MPEG Video and Requirement Groups: Call for Proposals on 3D Video Coding Technology. MPEG output document N12036, Geneva, Switzerland (2011)
6. Test model for AVC based 3D video coding, ISO/IEC JTC1/SC29/WG11 MPEG2012/N12558 (2012)
7. Aflaki, P., Hannuksela, M.M., Gabbouj, M.: Subjective quality assessment of asymmetric stereoscopic 3D video. J. Signal Image Video Process. **9**(2), 331–345 (2013)
8. Aflaki, P., Su, W., Joachimiak, M., Rusanovskyy, D., Hannuksela, M.M., Gabbouj, M.: Coding of mixed-resolution multiview video in 3D video application. In: IEEE International Conference on Image Processing (ICIP) (2013)
9. Asher, H.: Suppression theory of binocular vision. Br. J. Ophthalmol. **37**(1), 37–49 (1953)
10. Bjontegaard, G.: Calculation of average PSNR differences between RD-Curves. ITU-T SG16 Q.6 document VCEG-M33 (2001)
11. Chen, J., Boyce, J., Ye, Y., Hannuksela, M.: Scalable HEVC (SHVC) Test Model 4 (SHM 4) ISO/IEC JTC1/SC29/WG11 MPEG2013/N13939 (2013)
12. Chen, Y., Hannuksela, M.M., Suzuki, T., Hattori, S.: Overview of the MVC+D 3D video coding standard. J. Vis. Commun. Image Represent. **25**(4), 679–688 (2013)
13. Chen, Y., Wang, Y.K., Ugur, K., Hannuksela, M.M., Lainema, J., Gabbouj, M.: The emerging MVC standard for 3D video services. EURASIP J. Appl. Signal Process. **8** (2009)
14. Domanski, M., Grajek, T., Klimaszewski, K., Kurc, M., Stankiewicz, O., Stankowski, J., Wegner, K.: Poznan Multiview Video Test Sequences and Camera Parameters. ISO/IEC JTC1/SC29/WG11 MPEG 2009/M17050 (2009)
15. Dong, J., He, Y., Ye, Y.: Downsampling filters for anchor generation for scalable extensions of HEVC. ISO/IEC JTC1/SC29/WG11 MPEG2012/M23485 (2012)
16. Fehn, C.: Depth-image-based rendering (DIBR), compression and transmission for a new approach on 3D-TV. In: Proceedings of the SPIE Conference on Stereoscopic Displays and Virtual Reality Systems XI, CA, vol. 5291, pp. 93–104 (2004)

17. Garcia, D.C., Dorea, C., Queiroz, R.L.: Super resolution for multiview images using depth information. IEEE Trans. Circuits Syst. Video Technol. 22(9), 1249–1256 (2012)
18. Hannuksela, M.M., Rusanovskyy, D., Su, W., Chen, L., Li, R., Aflaki, P., Lan, D., Joachimiak, M., Li, H., Gabbouj, M.: Multiview-video-plus-depth coding based on the advanced video coding standard. IEEE Trans. Image Process. 22(9), 3449–3458 (2013)
19. Joachimiak, M., Hannuksela, M.M., Gabbouj, M.: View synthesis quality mapping for depth-based super resolution on mixed resolution 3D video. In: 3DTV-Conference: The True Vision-Capture, Transmission and Display of 3D Video (3DTV-CON) (2014)
20. Kauff, P., Atzpadin, N., Fehn, C., Muller, M., Schreer, O., Smolic, A., Tanger, R.: Depth map creation and image-based rendering for advanced 3DTV services providing interoperability and scalability. EURASIP Int. J. Signal Process. 22(2), 217–234 (2007)
21. Perkins, M.G.: Data compression of stereopairs. IEEE Trans. Commun. 40(4), 684–696 (1992)
22. Sawhney, H.S., Guo, Y., Hanna, K., Kumar, R., Adkins, S., Zhou, S.: Hybrid stereo camera: an IBR approach for synthesis of very high resolution stereoscopic image sequences. In: Proceedings of the 28th Conference on Computer Graphics and Interactive Techniques, New York, pp. 451–460 (2001)
23. Schwarz, H., et al.: Description of 3D Video Technology Proposal by Fraunhofer HHI (MVC compatible). ISO/IEC JTC1/SC29/WG11 MPEG2011/M22569 (2011)
24. Tam, W.J.: Image and depth quality of asymmetrically coded stereoscopic video for 3D-TV. Joint Video Team document JVT-W094 (2007)
25. Wilcoxon, F.: Individual comparisons by ranking methods. Biometrics 1, 80–83 (1945)

Global Volumetric Image Registration Using Local Linear Property of Image Manifold

Hayato Itoh[1]([✉]), Atsushi Imiya[2], and Tomoya Sakai[3]

[1] Graduate School of Advanced Integration Science,
Chiba University, Chiba, Japan
hayato-itoh@graduate.chiba-u.jp
[2] Institute of Management and Information Technologies,
Chiba University, Yayoicho 1-33, Inage-ku, Chiba 263-8522, Japan
[3] Graduate School of Engineering, Nagasaki University,
Bunkyo-cho 1-14, Nagasaki 852-8521, Japan

Abstract. We propose a three-dimensional global image registration method for a sparse dictionary. To achieve robust and accurate registration, which based on template matching, a large number of transformed images are prepared and stored in the dictionary. To reduce the spatial complexity of this image dictionary, we introduce a method of generating a new template image from a collection of images stored in the image dictionary. This generated template image allows us to achieve accurate image registration even if the population of the image dictionary is relatively small and the template has a small pattern perturbation. To further reduce the complexity, we compute a matching process in a low-dimensional Euclidean space projected by a random projection.

1 Introduction

We propose a three-dimensional global image registration method for a compressed sparse dictionary. Our method allows us to achieve accurate volumetric image registration even if the population of pregenerated images is relatively small. Furthermore, our method also achieves robust registration for template images with a small pattern perturbation.

Three-dimensional image registration mainly focuses on registration of point clouds and volumetric data. The former problem is applied to a point cloud representing depth map. Since this depth map expresses a terrain surface, registration is achieved as surface registration in three-dimensional space [1,2]. For this surface registration, iterative closest algorithm is a well established method [3,4]. The later problem mainly deals with volumetric medical data obtained by computed tomography, magnetic resonance imaging and positron emission tomography [5,6].

Medical image registration categorised into linear and nonlinear methods. For nonlinear image registration, global image registration is used as preprocess because nonlinear image registration is mainly valid for local deformation between images. Nonlinear registration is used to detect optimally local transform

© Springer International Publishing Switzerland 2015
C.V. Jawahar and S. Shan (Eds.): ACCV 2014 Workshops, Part I, LNCS 9008, pp. 238–253, 2015.
DOI: 10.1007/978-3-319-16628-5_18

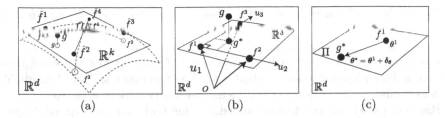

Fig. 1. (a) Nearest neighbours of g searched for by the k-nearest-neighbour search on a manifold. Our method of projecting the manifold to a low-dimensional subspace. (b) Generation of a new entry in a dictionary. The input image g is projected onto the subspace spanned by three nearest neighbours. (c) Interpolation of parameter. For the new entry g^*, we interpolate the parameter θ^* of the image g^*. Here, Π represents the parameter space of the transform.

between a template and the reference images, if the difference between these two images are small and local [6]. Linear registration, however, detects global geometric relations between a template and the Ref. [7].

In global image registration, the best geometrical transform between reference and template images is accomplished by computing the best geometrical matching between these two images. Therefore, for accurate registration, we are required to prepare as many template images as possible in the dictionary [8]. This implies that the spatial complexity of global image registration depends on the population of templates in the dictionary. The generation of new template images from existing templates reduces the spatial complexity of global image registration. However, in image registration with a sparse dictionary, we are required to generate a new entry and to compute the transform from this new entry simultaneously [9,10].

Since the image pattern space is a curved manifold in higher-dimensional space, on the tangent space of this curved manifold, an image pattern is expanded to a finite Fourier series using local bases. This expansion means that local bases span a local part of the manifold. Figure 1(a) illustrates a manifold on a low-dimensional subspace [11,12]. In the neighbourhood of an image pattern, image patterns can be expressed as a linear combination of this image and derivatives of this image. We call this property of an image manifold the local linear property. Figure 1(b) shows the generation of a template image. Combining these two local expressions for an image pattern implies two advantages for pattern generation. One is that we can compute the parameters of the transform. The other is that we can compute a new pattern that is sufficiently close to a reference image. Figure 1(c) shows the relation between a generated image g^* and the nearest neighbour in a local subspace. In Fig. 1(c), the perturbation δ_θ is small because the generated image g^* is close to the nearest neighbour. If the rotation angle, scaling factor and shear ratio are all small, these transforms can be expressed as a linear sum of the identity transform and linear transforms with parameters that define the type of transform. Therefore, we can decompose the affine matrix into a rotation, scaling and shear to estimate each transform.

To use the local linear property for image registration, a curved manifold of image patterns is generated from reference images. For the template image, we generate this curved manifold using the k-nearest-neighbour search (k-NNS) [13]. To reduce the temporal and spatial complexities of the search of the k-neighbourhood, a dimension-reduction method is used to generate the curved manifold. We adopt random projection as the dimension-reduction method [8–10].

Random projection is a metric-embedding method that approximately preserves distances between points in the original space [14]. Therefore, the random projection is used in the nearest-neighbour search (NNS) to speed up numerical computation [15]. Furthermore, the random projection preserves manifolds [16]. The validity of random projection for manifold learning of noiseless images, noisy images and text data is shown in [17]. We use the random projection to generate a manifold in a low-dimensional subspace.

In Sect. 2, we summarise the linear subspace method for pattern recognition, the random projection for dimension reduction and the local linear property of an image pattern space. Then, in Sect. 3, we derive relations between the global transform of an image and the local linear properties of an image pattern space. The relations in Sect. 3 are used to derive a method of performing the global image registration using a dictionary with a small population of entries.

2   Mathematical Preliminaries

2.1   Global Image Registration

Setting Π to be an appropriate parameter space for image generation, we assume that images are expressed as $f(\boldsymbol{x}, \boldsymbol{\theta}_i)$ for $\exists \boldsymbol{\theta}_i \in \Pi$, $\boldsymbol{x} \in \mathbb{R}^3$. The parameter $\boldsymbol{\theta}_i$ generates a transform for $f(\boldsymbol{x})$. We call the set of generated images $f(\boldsymbol{x}, \boldsymbol{\theta}_i)$ and parameters $\{\boldsymbol{\theta}_i\}_{i=1}^N$ a dictionary.

For the global alignment of images with respect to the region of interest Ω, we find the linear transformation $\boldsymbol{x}' = \boldsymbol{A}\boldsymbol{x} + \boldsymbol{t}$ that minimises the criterion

$$R(f, g) = \sqrt{\int_\Omega |f(\boldsymbol{x}') - g(\boldsymbol{x})|^2 d\boldsymbol{x}} \tag{1}$$

for functions $f(\boldsymbol{x})$ and $g(\boldsymbol{x})$ defined on $\mathbb{R}^3$ such that

$$\int_{\mathbb{R}^3} |f(\boldsymbol{x})|^2 d\boldsymbol{x} < \infty, \quad \int_{\mathbb{R}^3} |g(\boldsymbol{x})|^2 d\boldsymbol{x} < \infty. \tag{2}$$

In image registration, we assume that the parameter $\boldsymbol{\theta}_i$ in Π generates the affine coefficients $\boldsymbol{A}$ and $\boldsymbol{t}$. Solving the NNS problem using the dictionary, we can estimate the transform $\boldsymbol{A}$ and $\boldsymbol{t}$ as $\boldsymbol{\theta}_i$. The computational cost of a naive approach for the NNS is $\mathcal{O}(Nd)$, where N and d are the cardinality of the set of points in the metric space and the dimension of the metric space, respectively. The factor d in the nearest-neighbour search [14] is reduced by using the random projection. Furthermore, using the local linear property in Sect. 2.2, we can also reduce N in the nearest NNS.

2.2   Local Eigenspace

Setting the Hilbert space H to be the space of patterns, we assume that the inner product (f, g) is defined in H. Let $f \in H$ and P be a pattern and an operator for a class, respectively. We then define the class $C = \{f \mid Pf = f, P^*P = I\}$. For recognition, we construct P for $f \in C$ while minimising $E[\|f - Pf\|_2]$ with respect to $P^*P = I$, where $f \in C$ is the pattern for a class, I is the identity operator and E is the expectation in H. This methodology is known as the subspace method [11, 12]. For the practical calculation of P, we adopt the Karhunen-Loeve expansion for the construction of the eigenspace.

We deal with images $f(\boldsymbol{x})$ defined in the three-dimensional Euclidean space $\boldsymbol{x} = (x, y, z)^\top \in \mathbb{R}^3$. We assume that a small perturbation of the parameter causes a small geometrical transform of the image pattern, that is, we assume the relation $f(\boldsymbol{x}+\boldsymbol{\delta}, \boldsymbol{\theta}) = f(\boldsymbol{x}, \boldsymbol{\theta}+\boldsymbol{\psi})$. Therefore, a small perturbation of an image caused by parameter perturbation is replaced with a geometrical perturbation of the image, that is,

$$f(\boldsymbol{x}, \boldsymbol{\theta} + \boldsymbol{\psi}) = f(\boldsymbol{x} + \boldsymbol{\delta}) = f(\boldsymbol{x}) + \boldsymbol{\delta}^\top \nabla f(x, y), \tag{3}$$

where $\boldsymbol{\delta} \in \mathbb{R}^3$ is a perturbation vector. Setting f_x, f_y and f_z to $\partial_x f(\boldsymbol{x}), \partial_y f(\boldsymbol{x})$ and $\partial_z f(\boldsymbol{x})$, respectively, since

$$\int_{\mathbb{R}^3} f f_x d\boldsymbol{x} = 0, \int_{\mathbb{R}^3} f f_y d\boldsymbol{x} = 0, \int_{\mathbb{R}^3} f f_z d\boldsymbol{x} = 0, \tag{4}$$

$$\int_{\mathbb{R}^3} f_x f_y d\boldsymbol{x} = 0, \int_{\mathbb{R}^3} f_y f_z d\boldsymbol{x} = 0, \int_{\mathbb{R}^3} f_z f_x d\boldsymbol{x} = 0, \tag{5}$$

for images $g(\boldsymbol{x}) = f(\boldsymbol{A}\boldsymbol{x} + \boldsymbol{t})$ with a small perturbation affine transform $\boldsymbol{A}$ and a small translation vector $\boldsymbol{t}$, we can assume the relation

$$g(\boldsymbol{x}) = a_0 f + a_1 f_x + a_2 f_y + a_3 f_z, \quad \boldsymbol{x} \in \mathbb{R}^3. \tag{6}$$

Equation (6) implies that the number of independent images among the collections of images,

$$L(f) = \{f_{ij} \mid f_{ij}(\boldsymbol{x}) = \lambda f(\boldsymbol{A}_i \boldsymbol{x} + \boldsymbol{t}_j)\}_{i,j=1}^{p,q} \tag{7}$$

is four, if the domain of the image is $\mathbb{R}^3$. We can use the first four principal vectors of $L(f)$ as the local basis for image expression for a three-dimensional image. We call this property the local linear property and the space spanned by $\{f, f_x, f_y, f_z\}$ the local eigenspace. Figure 1(b) shows the projection of the input image g to the three-dimensional local subspace.

2.3   Three-Dimensional Affine Transform

To avoid the estimation of a translation, we set the origin of the coordinates to be the centre of an image. We assume small rotations around the x, y and z axes of angles ϕ_1, ϕ_2 and ϕ_3, given by the transform matrices such that

$$\boldsymbol{R}_x = \boldsymbol{I} + \begin{pmatrix} 0 & 0 & 0 \\ 0 & 0 & -\phi_1 \\ 0 & \phi_1 & 0 \end{pmatrix} = \boldsymbol{I} + \boldsymbol{R}'_x, \tag{8}$$

$$\boldsymbol{R}_y = \boldsymbol{I} + \begin{pmatrix} 0 & 0 & \phi_2 \\ 0 & 0 & 0 \\ -\phi_2 & 0 & 0 \end{pmatrix} = \boldsymbol{I} + \boldsymbol{R}'_y, \tag{9}$$

$$\boldsymbol{R}_z = \boldsymbol{I} + \begin{pmatrix} 0 & -\phi_3 & 0 \\ \phi_3 & 0 & 0 \\ 0 & 0 & 0 \end{pmatrix} = \boldsymbol{I} + \boldsymbol{R}'_z. \tag{10}$$

Multiplying $\boldsymbol{R}_x, \boldsymbol{R}_y$ and $\boldsymbol{R}_z$, and ignoring terms of order larger than one, we have an arbitrary rotation expressed as

$$\boldsymbol{R}(\boldsymbol{\phi}) = \begin{pmatrix} 1 & -\phi_3 & \phi_2 \\ \phi_3 & 1 & -\phi_1 \\ -\phi_2 & \phi_1 & 1 \end{pmatrix} = \boldsymbol{I} + \boldsymbol{R}'_x + \boldsymbol{R}'_y + \boldsymbol{R}'_z = \boldsymbol{I} + [\boldsymbol{R}]_\times, \tag{11}$$

where $[\boldsymbol{R}]_\times$ is the outer-product operator of vector $\boldsymbol{\phi} = (\phi_1, \phi_2, \phi_3)^\top$.

For small scaling factors ϕ_4, ϕ_5 and ϕ_6, and small shearing ratios ϕ_7, ϕ_8, ϕ_9, ϕ_{10}, ϕ_{11} and ϕ_{12}, we have the scaling matrix and shearing matrix

$$\begin{pmatrix} 1+\phi_4 & 0 & 0 \\ 0 & 1+\phi_5 & 0 \\ 0 & 0 & 1+\phi_6 \end{pmatrix} = \boldsymbol{I} + \boldsymbol{\Lambda}, \tag{12}$$

$$\begin{pmatrix} 1 & \phi_7 & \phi_8 \\ \phi_9 & 1 & \phi_{10} \\ \phi_{11} & \phi_{12} & 1 \end{pmatrix} = \boldsymbol{I} + \boldsymbol{S}, \tag{13}$$

respectively.

Combining these rotation, scaling and shearing matrices in Eqs. (11)–(13), we can define all affine transforms except translation. Multiplying these three matrices and ignoring terms of order larger than one, we have the affine transform matrix

$$\boldsymbol{A} = \boldsymbol{I} + [\boldsymbol{R}]_\times + \boldsymbol{\Lambda} + \boldsymbol{S} = \boldsymbol{I} + \boldsymbol{A}_\delta, \tag{14}$$

where the rotation, scaling and shear transforms are commutative since their transform matrices consist of only small-value elements. Here, $\boldsymbol{A}_\delta$ represents a small affine transform.

2.4   Neighbours of Template Image

For the reference image f and template image g in Hilbert space H, applying affine transforms $\{\boldsymbol{A}_i\}_{i=1}^N$ except for translation to f, we have the finite collection $\{f_i | \boldsymbol{A}_i \boldsymbol{x}\}_{i=1}^N$. For $0 < k \ll N$, let $\pi(i)$ be one-to-one injection from $1 \le i \le N$

to $1 \leq \pi(i) \leq k$ such that $\pi(i) \neq \pi(j)$ for $i \neq j$. Using $\pi(i)$, we define the k-neighbourhood $KN(g) \in L(J)$ of g. For a finite collection of images $\{J_i\}_{i=1}^{u}$, $KN(g)$ is a collection $\{f_{\pi(i)}\}_{i=1}^{N}$ that satisfies the inequalities

$$\|g - f_{\pi(1)}\|_2 \leq \|g - f_{\pi(2)}\|_2 \leq \cdots \leq \|g - f_{\pi(N)}\|_2 \tag{15}$$

where $\|\cdot\|_2$ is the L_2 metric on H.

2.5   Manifold Generation by Random Projection

We construct the image manifold of entries in the dictionary using the nearest-neighbour method. To reduce the time complexity of the nearest neighbourhood mesh on the image manifold, we adopt the random projection.

The random projection reduces the dimension of the discrete vector space while preserving both local and global topologies and geometries. The random projection satisfies the following theorem. For a set $X = \{x_i\}_{i=1}^{N}$ of N points in d-dimensional Euclidean space, consider a mapping onto the set $\hat{X} = \{\hat{x}_i\}_{i=1}^{N}$ in k-dimensional Euclidean space. For the vector $x = (x_1, \ldots, x_d)^{\top}$, we define the Euclidean norm as $\|x\|_2 = \left(\sum_{i=1}^{d} x_i\right)^{1/2}$. The Johnson-Lindenstrauss lemma indicates that there is a mapping approximately preserving the Euclid distance between two arbitrary points [18]. Setting $|x - y|_2$ to be the Euclidean distance between two points x and y in appropriate dimensional Euclidean space, the next therem is satisfied [19].

Theorem 1. *(Johnson-Lindenstrauss lemma). For a subspace with dimension $\hat{d} \geq \hat{d}_0 = \frac{9 \log N}{\epsilon^2 - \frac{2}{3}\epsilon^3} + 1 = \mathcal{O}(\epsilon^{-2} \log N)$, where ϵ is a real number such that $0 < \epsilon < \frac{1}{2}$, a set X of N d-dimensional points $\{x_i\}_{i=1}^{N}$ and an integer $\hat{d}$ with $\hat{d} \ll d$, there exists a mapping f from $\mathbb{R}^d$ to $\mathbb{R}^{\hat{d}}$ such that*

$$(1 - \epsilon)|x_j - x_i|_2 \leq |\hat{x}_j - \hat{x}_i|_2 \leq (1 + \epsilon)|x_j - x_i|_2, \tag{16}$$

for all $i, j = 1, 2, \ldots, N$.

Therefore, setting R to be the random projection from $\mathbb{R}^d$ to $\mathbb{R}^{\hat{d}}$, Theorem 1 implies the relation

$$P(|\,|x - y|_2 - |Rx - Ry|_2\,| < \varepsilon) > 1 - \delta \tag{17}$$

for small positive constants ε and δ, where P is a probability distribution. To use these topological and geometrical properties for fast computation in the nearest neighbour method, we sampled images to construct the image manifold of data in the dictionary.

Let $\mathbb{Z}^d$ be the integer grid in $\mathbb{R}^d$, Setting $\mathbf{D}$ and Δ to be a finite subset of $\mathbb{Z}^d$ and a positive number that defines the resolution of sampling, respectively, the distance

$$D(f, g) = \sqrt{\int_{\mathbb{R}^d} |f(x) - g(x)|^2 dx} \tag{18}$$

is approximately computed as

$$D(f,g) = \sqrt{\sum_{\boldsymbol{z} \in \mathrm{D} \subset \mathbb{Z}^d} |f(\Delta \boldsymbol{x}) - g(\Delta \boldsymbol{z}))|^2 \Delta}, \tag{19}$$

for functions $f(\boldsymbol{x})$ and $g(\boldsymbol{x})$ defined on $\mathbb{R}^d$.

By expressing $\{f(\Delta \boldsymbol{z})\}_{\boldsymbol{z} \in \mathrm{D}}$ and $\{g(\Delta \boldsymbol{z})\}_{\boldsymbol{z} \in \mathrm{D}}$ as finite vectors $\boldsymbol{f}$ and $\boldsymbol{g}$, respectively, Eq. (19) is expressed as

$$D(f,g) = |\boldsymbol{f} - \boldsymbol{g}|_2 \tag{20}$$

if we set $\Delta = 1$. Using the random projection, the distance between $\boldsymbol{f}$ and $\boldsymbol{g}$ is computed as

$$D(f,g) \approx |\boldsymbol{R}\boldsymbol{f} - \boldsymbol{R}\boldsymbol{g}|, \tag{21}$$

for functions $f(\boldsymbol{x})$ and $g(\boldsymbol{x})$ defined on $\mathbb{R}^n$ such that

$$\int_{\mathbb{R}^n} |f(\boldsymbol{x})|^2 d\boldsymbol{x} < \infty, \quad \int_{\mathbb{R}^n} |g(\boldsymbol{x})|^2 d\boldsymbol{x} < \infty. \tag{22}$$

Therefore, by searching $KN(g)$ in $\hat{d}$-dimensional Euclidean space with the random projection, we obtain the discrete version of $KN(g)$. For practical computation, we adopt an efficient random projection [20].

3   Local Linear Method

Using the local linear property of images in the image space, we first generate an image in a sparse dictionary. To register a template g, using the generated image g^*, we next estimate the small affine transform between the generated image g^* and the nearest neighbour f^1 of g in the dictionary. From the generated image and the estimated transform, the local linear method can generate new entries in the dictionary. Figure 1 shows a flow of this local linear method.

For image generation, we use the k nearest neighbours of g in the dictionary. Let $\{f^i\}_{i=1}^k \in \mathcal{L}(g)$, be the ith neighbour of g. The random projection preserves the pairwise distances between vectorised images. Therefore, f^i is searched for in a random projected space. For a template $g(\boldsymbol{x})$, we assume $g(\boldsymbol{x}) = f^1(\boldsymbol{A}\boldsymbol{x}, \theta) + \epsilon$, where $\boldsymbol{A}$ gives the best matching between g and f^1, and ϵ is a small difference between the reference pattern and the registered template pattern. For three-dimensional images, using the local linear property, we can approximate the space spanned by $\{u_i\}_{i=1}^4$ using the space spanned by $\{g\} \cup \{f^i\}_{i=1}^4$ if the data space $\mathcal{L}(g)$ is not extremely sparse. Using Gram-Schmidt orthonormalisation for $\{f^i\}_{i=1}^4$, we obtain the basis $\{u_i\}_{i=1}^4$. Projecting the template to the space spanned by $\{u_i\}_{i=1}^4$, we obtain a new image,

$$g^* = \sum_{i=1}^4 b_i u_i, \tag{23}$$

from a triplet of prepared entries in the dictionary. Here, $\{b_i\}_{i=1}^4$ represents the coefficients of the linear combination.

For the projected template image and its nearest neighbour $f^1(x, \theta)$, using the Taylor expansion, we have the relation

$$g^* = f^1(x + \delta, \theta) = f^1(Ax, \theta) = f^1((I + A_\delta)x, \theta)$$
$$= f^1(x, \theta) + (A_\delta x)^\top \nabla f^1(x, \theta) \qquad (24)$$

if the higher order terms with respect *delta* is sufficiently small. For the transform matrix A_δ, we have the relation

$$(A_\delta x)^\top \nabla f^1(x, \theta) = g^* - f^1(x, \theta). \qquad (25)$$

Representing the left side of Eq. (25) in terms of the variables that generate each transform, we can decompose the small affine transform between the reference and template. Using matrices $[R]_\times, \Lambda$ and S, and coefficients $\gamma_i \in \{0, 1\}$, $i = 1, 2, \ldots, 9$, we can represent the left side of Eq. (25) as

$$x^\top \left(\begin{pmatrix} 0 & \gamma_3 & \gamma_2 \\ \gamma_3 & 0 & \gamma_1 \\ \gamma_2 & \gamma_1 & 0 \end{pmatrix} \circ [R]_\times^\top + \mathrm{diag}(\gamma_4, \gamma_5, \gamma_6)\, \Lambda + \mathrm{diag}(\gamma_7, \gamma_8, \gamma_9)\, S^\top \right) \nabla f^1,$$
$$(26)$$

where $A \circ B$ is the Hadamard product of matrices A and B. Furthermore, setting

$$\alpha_1 = yf_z^1 - zf_y^1, \alpha_2 = zf_x^1 - xf_z^1, \alpha_3 = yf_x^1 - xf_y^1, \qquad (27)$$
$$\alpha_4 = xf_x^1, \alpha_5 = yf_y^1, \alpha_6 = zf_z^1, \qquad (28)$$
$$\alpha_7 = yf_y^1, \alpha_8 = zf_z^1, \alpha_9 = xf_x^1, \alpha_{10} = zf_y^1, \alpha_{11} = xf_z^1, \alpha_{12} = yf_z^1, \qquad (29)$$

we rewrite Eq. (25) as

$$\sum_{i=1}^6 \gamma_i \alpha_i \phi_i + \sum_j^3 (\alpha_{2(j-1)+7}\phi_{2(j-1)+7} + \alpha_{2(j-1)+8}\phi_{2(j-1)+8}) = g^* - f^1(x, \theta). \qquad (30)$$

Equation (30) contains 12 unknowns in a single equation. The sum of coefficients $\sum_{i=1}^6 \gamma_i + \sum_{i=7}^9 2\gamma_i$ is greater or equal to one even though we have only one template. We adopt the surface integration for Eq. (30) for this template image. Selecting different surfaces of a surface integration, we obtain more than one independent equation. For the centre $\mu = (\mu, \mu, \mu)^\top$ of a template image and radius $\{r_i\}_{i=1}^n$, $r_i \neq r_j$, we define surface of a sphere as

$$S_3(r) = \{x \mid \|x - \mu\|_2 = r\}. \qquad (31)$$

For the centre $\mu = (\mu, \mu, \mu)^\top$ of a template image, radius r, rotation angles $\phi_i = (\phi_{i1}, \phi_{i2}, \phi_{i3})^\top$ and vectors $p_1 = (x-\mu, y-\mu, -\mu)^\top$, $p_2 = (-\mu, y-\mu, z-\mu)^\top$ and $p_3 = (x - \mu, -\mu, z - \mu)^\top$, we define the surface comprising three planes as

$$\mathcal{P}_3(r, \phi) = \{\mu + R(\phi)p_1, \mu + R(\phi)p_2, \mu + R(\phi)p_3 \mid \mu - r \leq x, y, z, \leq \mu + r\}. \qquad (32)$$

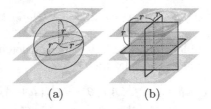

(a)   (b)

Fig. 2. Surfaces used for surface integration to obtain independent equations. (a) and (b) show surface $\mathcal{S}_3(r)$ of the sphere and surface $\mathcal{P}_3(r, 0)$ comprising three planes. Integration of the volume gives a equation for an image. The integration of different surfaces, such as a different spheres and orthogonal square planes, gives several independent equations for an image.

For a set of radius $\{r_i\}_{i=1}^n$, we obtain sets of $\{\mathcal{S}_3(r_i)\}_{i=1}^n$ and $\{\mathcal{P}_3(r_i)\}_{i=1}^n$. We adopt $\{\mathcal{S}_3(r_i)\}_{i=1}^n$ and $\{\mathcal{P}_3(r_i)\}_{i=1}^n$ as surfaces $\{\Omega_i\}_{i=1}^n$ for the surface integration. Figure 2 shows the surfaces used for surface integration. For $\{\Omega_i\}_{i=1}^n$ and $\beta_{ij} = \int_{\Omega_i} \alpha_j(\boldsymbol{x})d\boldsymbol{x}, j = 1, 2, \ldots, 12$, we set the coefficient vector

$$\chi_i = (\beta_{i\,1}, \beta_{i\,2}, \ldots, \beta_{i\,12}). \tag{33}$$

Here, we have the relations $\chi_i \neq \chi_j$ and $h_i \neq h_j$ for $i \neq j$. Setting $n \geq \sum_{i=1}^6 \gamma_i + \sum_{i=7}^9 2\gamma_i$,

$$\begin{pmatrix} \beta_{11} & \beta_{1\,2} & \ldots & \beta_{1\,12} \\ \beta_{21} & \beta_{2\,2} & \ldots & \beta_{2\,12} \\ & & \vdots & \\ \beta_{n\,1} & \beta_{n\,2} & \ldots & \beta_{n\,12} \end{pmatrix} \boldsymbol{\zeta} = \begin{pmatrix} h_1 \\ h_2 \\ \vdots \\ h_n \end{pmatrix}. \tag{34}$$

where

$$\boldsymbol{\zeta} = (\gamma_1\phi_1, \gamma_2\phi_2, \ldots, \gamma_6\phi_6, \gamma_7\phi_7, \gamma_7\phi_8, \gamma_8\phi_9, \gamma_8\phi_{10}, \gamma_9\phi_{11}, \gamma_9\phi_{12})^\top, \tag{35}$$

and

$$h_i = \int_{\mathcal{S}_i} (g^* - f^1)d\boldsymbol{x}, \tag{36}$$

then, we can estimate the transforms as a solution to the linear system of equations.

4   Numerical Examples

Three experiments evaluate the performance of our local linear method. The first and second experiments show the accuracy of estimation for a single transform and multiple transforms, respectively. The third experiment shows the robust estimation of templates with small pattern perturbations.

The first and second experiments use volumetric data obtained by MRI simulation of human brain [21]. Figure 3 shows slice images of the volumetric data.

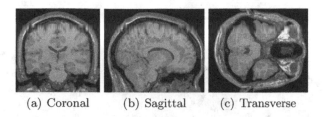

(a) Coronal   (b) Sagittal   (c) Transverse

Fig. 3. Slice images extracted from volumetric data. (a)–(c) Slice images extracted from a voxel image obtained by MRI simulation of a human brain [21]. The size of the voxel image is $181 \times 217 \times 181$ voxels. The slice images (a), (b) and (c) are extracted from the $z = 45$, $x = 90$ and $y = 100$ planes, respectively. In experiments, we embed the voxel image in a background image of $308 \times 308 \times 308$ voxels. The intensities of the background images are 0.

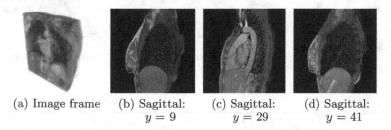

(a) Image frame   (b) Sagittal: (c) Sagittal: (d) Sagittal:
                  $y = 9$ $y = 29$ $y = 41$

Fig. 4. Volumetric spatiotemporal MRI lung data [22]. (a) Voxel image of a frame of a sequence. (b)–(d) Sagittal slices of the frame. The spatial and time resolutions of the data are $50 \times 224 \times 224$ and 200, respectively. The time between frames is 331 ms. In the experiments, we embed a volumetric image of a frame on a background image of $316 \times 316 \times 316$ voxels. Each voxel value in the background image is 0.

Furthermore, for the first and second experiments, we generate smooth images from these slice images by linear filtering of the convolution with Gaussian kernel of standard deviation τ. For third experiment, we use volumetric spatiotemporal MRI lung data. [22]. Figure 4 shows a few frame of the volumetric spatiotemporal MRI lung data. In a sequence, the volumetric data gradually changes with the breathing of the patient. Table 1 summarises the parameters for the first and second experiments. Table 2 summarises the data for the third experiment.

Figure 5 shows the results of the first experiment for the estimation of rotation angle. In Fig. 5(a), (b) and (c), for displacements of less than 4 voxels, the estimation errors are smaller than $2.5°$ if we use the surfaces $\{\mathcal{S}_3(r_i)\}_{i=1}^{10}$ for surface integration. Figure 5(d), (e) and (f) shows that for displacements of less than 4 voxels, the estimation errors are smaller than $3.5°$ if we use the surfaces $\{\mathcal{P}(r_i, \mathbf{0})\}_{i=1}^{10}$ for surface integration. In Fig. 5(g), (h), (i), (j), (k) and (l), for displacements of greater than 4 voxels in smooth images, our method estimates rotation angles with errors smaller than $1°$.

The second experiment evaluates estimation errors for multiple transforms. Figure 6 shows the results of the second evaluation. In Fig. 6, the results show

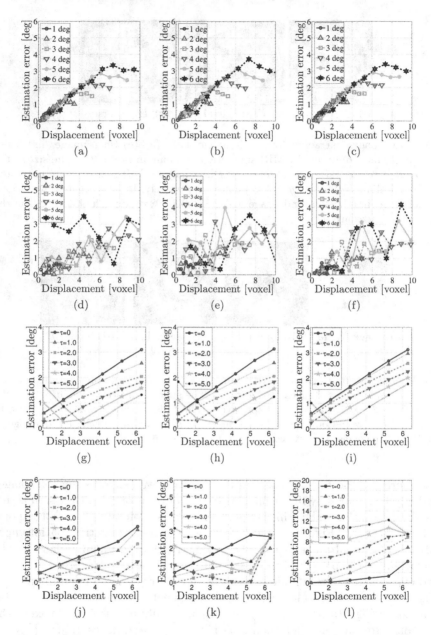

Fig. 5. Accuracy of estimation for a spatial rotation. We estimate the rotation angles ϕ_1, ϕ_2 and ϕ_3 independently. The first, second and third columns represent the accuracy of estimation for rotation around the x, y and z axes, respectively. (a) and (d), (b) and (e), and (c) and (f) show the accuracy of estimation without Gaussian filtering. (g) and (j), (h) and (k), and (i) and (l) show the accuracy of estimation for smooth images, for the rotation around x, y and z axes, respectively. In the first and third rows and the second and fourth rows, we adopt $\mathcal{S}_3(r)$ and $\mathcal{P}_3(r, \phi)$ as the surfaces for the surface integration, respectively. Displacements are given by $r\phi_1, r\phi_2$ and $r\phi_3$.

Table 1. Parameters for the first and second experiments using voxel image of human brain data.

A_δ	Pregeneration	Template	Filtering	Dimension	Lines
R_1	$-60 < \phi_1 < 60$	$1 < \phi_1 < 6$	$0 < \tau < 5$	$\hat{d} = 1024$	$S_3(r)$
	step of 12	step of 1	step of 0.1		$10 < r < 110$ step of 10
R_2	$-60 < \phi_2 < 60$	$1 < \phi_2 < 6$	$0 < \tau < 5$	$\hat{d} = 1024$	$S_3(r)$
	step of 12	step of 1	step of 0.1		$10 < r < 110$ step of 10
R_3	$-60 < \phi_3 < 60$	$1 < \phi_3 < 6$	$0 < \tau < 5$	$\hat{d} = 1024$	$S_3(r)$
	step of 12	step of 1	step of 0.1		$10 < r < 110$ step of 10
R	$-7 < \phi_1, \phi_2, \phi_3 < 7$	$1 < \phi_1, \phi_2, \phi_3 < 3$	$0 < \tau < 5$	$\hat{d} = 1024$	$S_3(r)$
	step of 7	step of 1	step of 0.1		$10 < r < 50$ step of 10

Table 2. Data for the third experiment using the volumetric spatiotemporal data.

A_δ	Pregeneration with 22nd frame	Template with 22nd, 23rd, 24th and 34th frame	Filtering	Dimension	Lines
R_3	$-60 < \phi_3 < 60$	$1 < \phi_1 < 6$	not used	$\hat{d} = 1024$	$S_3(r)$
	step by 12	step 1			$10 < r < 20$ step by 5

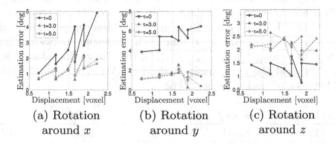

(a) Rotation around x (b) Rotation around y (c) Rotation around z

Fig. 6. Accuracy of estimation for multiple transforms. For the estimation, we adopt combinations of three rotations around the x, y and z axes. The left, middle and right graphs show the results of estimation for rotation around the x, y and z axes with Gaussian filtering with standard deviation τ, respectively. For the surface integration, we adopt surfaces $\{\mathcal{P}_i^1\}_{i=1}^n$. For the rotations around the x, y and z axes, the displacements are given by $r\sqrt{\phi_3^2 + \phi_2^2}$, $r\sqrt{\phi_3^2 + \phi_1^2}$ and $r\sqrt{\phi_1^2 + \phi_2^2}$ with radius r in the surface integral, respectively.

that the estimation of multiple transforms is unstable. Furthermore, the estimation errors are larger than 1 degree even for small displacements of one voxel. However, for smoothed images, the mean estimation error of the multiple transforms is about 1.5° for the three rotation axes.

The third experiment evaluates the accuracy and robustness of estimation of rotation for a template with a small pattern perturbation. Figure 7(a) and (b) shows the differences between the 22nd frame and the 23rd-200th frames of the four-dimensional data. Figure 7(c) shows the results of the estimation. In Fig. 7(c), curves represent absolute values of the estimation error plotted against the displacement caused by the rotation. For differences from $-\infty$ to -6.94 dB,

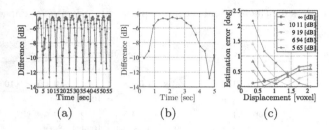

(a)   (b)   (c)

Fig. 7. Estimation for a template with small pattern perturbation. (a) Difference between 22nd frame and 23rd-200th frames of four-dimensional MRI lung data. (b) Scaled-up graph of (a) showing difference between 22nd frame and 23rd-38th frames. (c) Accuracy of estimation for rotation angle ϕ_3 around the z axis. The differences between the 22nd frame and the 22nd, 23rd, 24th, 25th and 34th frames are $-\infty, -10.11, -9.19, -6.94$ and -5.65 [dB], respectively. For surface integration, we adopt the surface $\mathcal{S}_3(r)$. The displacement is given by $r\phi_3$.

Table 3. Evaluation of approximation for generated new entries. We generate new entries for rotated images with small pattern perturbation. For a generation of a new entry, we use 4-neighbours of a template. As templates, we use rotated images of the 22nd, 23rd, 24th and 25th frame of data with angle ϕ_3. For a template g, we first compute the difference between g and its nearest neighbour in pregenerated images as $10\log_{10}\left(\|f^1 - g\|_2/\|g\|_2\right)$. Second, we compute the difference between g and a generated new entry g^* as $10\log_{10}\left(\|g^* - g\|_2/\|g\|_2\right)$. In this Table, the columns for the nearest neighbour (NN) and the local linear method (LLM) show the difference between f^1 and g and between g^* and g, respectively.

Angle [degree]	Difference between f^1 and g[dB]									
	22nd		23rd		24th		25th		34th	
ϕ_3	NN	LLM	NN	LLM	NN	LLM	NN	LLM	NN	LLM
2	-4.77	-5.15	-4.85	-5.15	-4.78	-5.06	-4.39	-4.70	-4.60	-4.90
4	-3.19	-3.95	-3.31	-3.96	-3.26	-3.91	-3.21	-3.81	-3.25	-3.87
6	-2.57	-3.71	-2.70	-3.72	-2.66	-3.68	-2.69	-3.62	-2.68	-3.65

the estimation errors are smaller than $1.5°$. Table 3 shows the difference between a template and generated g^*. In Table 3, the distance between the generated g^* and template is smaller than one between template and its nearest neighbour.

Table 4 summarises the accuracy, the number of pregenerated images and the dimension of the search space in Figs. 5 and 6. The results in Fig. 5 imply that integration with the surfaces $\{\mathcal{S}_3(r_i)\}_{i=1}^{10}$ leads to more accurate and stable estimation than integration with the surfaces $\{\mathcal{P}(r_i,\mathbf{0})\}_{i=1}^{10}$ for the case of a rotation. For the estimation of a single transform, our method requires 16.7% of the number of pregenerated images of naive NNS. Furthermore, for the estimation of multiple transforms, our method requires 2.1% of the number of pregenerated images of the naive NNS. Moreover, our method reduces size of search space to $4.0 \times 10^{-3}\%$. For both the estimations, the dimension of the search space is $3.5 \times 10^{-3}\%$ of the original dimension of the images. Moreover, the results

Table 4. Accuracy and compression ratio for volumetric data obtain by MRI simulation of human brain. First column shows given accuracy in the estimation. Second column shows necessary step sizes in pregeneration, which give the accuracy in first column, for the nearest neighbour search (NNS) and the local linear method. Third column shows dimensions of search space for NNS and LLM. Fourth column illustrates compression ration of the LLM compared with the NNS.

A_δ	Accuracy	Necessary step size		Dimension		Compression ratio
		NNS	LLM	Original	Search space	in pregeneration
R_1	1 [degree]	2 [degree]	12 [degree]	29218112	1024	16.7 [%]
R_2	1 [degree]	2 [degree]	12 [degree]	29218112	1024	16.7 [%]
R_3	1 [degree]	2 [degree]	12 [degree]	29218112	1024	16.7 [%]
R	1.5 [degree]	3 [degree]	12 [degree]			
	2.0 [degree]	4 [degree]	12 [degree]	29218112	1024	2.1 [%]
	1.5 [degree]	3 [degree]	12 [degree]			

of third experiment show that our method estimate transform for the template image with small pattern perturbation.

5   Conclusions

For volumetric images, we first defined the local linear property of the image manifold for a small geometrical perturbation. We then introduced an algorithm based on the local linear property for three-dimensional affine image registration to reduce the time and spatial complexity of computation. The algorithm first generates a new image for a template using a small number of images preproduced from the reference image. Second, using the new image, the proposed method finds a small affine transform between the new image and the best matching image in the dictionary. Finally, our method estimates transforms using the new image and its neighbours. This algorithm reduces the computational cost of preprocessing and the size of the images used in the nearest-neighbour search. In the numerical examples, using volumetric data of the human brain obtained by simulated MRI, we show that our method can accurately estimate single and multiple transforms using a small number of pregenerated images. Furthermore, using four-dimensional MRI lung data, we show that our method can robustly estimate single transform for a template with small pattern perturbation.

    This research was supported by the "Computational Anatomy for Computer-Aided Diagnosis and Therapy: Frontiers of Medical Image Sciences" and "Multidisciplinary Computational Anatomy and Its Application to Highly Intelligent Diagnosis and Therapy" projects funded by a Grant-in-Aid for Scientific Research on Innovative Areas from MEXT, Japan, and by Grants-in-Aid for Scientific Research funded by the Japan Society for the Promotion of Science.

References

1. Nishino, K., Ikeuchi, K.: Robust Simultaneous Registration of Multiple Range Images. In: Digitally Archiving Cultural Objects, pp. 71–88. Springer, New York (2008)
2. Salvi, J., Matabosch, C., Fofi, D., Forest, J.: A review of recent range image registration methods with accuracy evaluation. Image Vis. Comput. **25**, 578–596 (2007)
3. Besl, P., McKay, N.D.: A method for registration of 3-D shapes. IEEE Trans. Pattern Analy. Mach. Intell. **14**, 239–256 (1992)
4. Daniel, F.H., Hebert, M.: Fully automatic registration of multiple 3D data sets. Image Vis. Comput. **21**, 637–650 (2003)
5. Markelj, P., Tomaževič, D., Likar, B., Pernus, F.: A review of 3D/2D registration methods for image-guided interventions. Med. Image Anal. **16**, 642–661 (2012)
6. Klein, A., Andersson, J., Ardekani, B.A., Ashburner, J., Avants, B.B., Chiang, M.C., Christensen, G.E., Collins, D.L., Gee, J.C., Hellier, P., Song, J.H., Jenkinson, M., Lepage, C., Rueckert, D., Thompson, P.M., Vercauteren, T., Woods, R.P., Mann, J.J., Parsey, R.V.: Evaluation of 14 nonlinear deformation algorithms applied to human brain MRI registration. NeuroImage **46**, 786–802 (2009)
7. Capekm, M.: Optimisation strategies applied to global similarity based image registration methods. In: Proceedings of the 7th International Congerence in Central Europoe on Computer Graphic, pp. 369–374 (1999)
8. Itoh, H., Lu, S., Sakai, T., Imiya, A.: Global image registration by fast random projection. In: Bebis, G., Boyle, R., Parvin, B., Koracin, D., Wang, S., Kyungnam, K., Benes, B., Moreland, K., Borst, C., Di Verdi, S., Yi-Jen, C., Ming, J. (eds.) ISVC 2011, Part I. LNCS, vol. 6938, pp. 23–32. Springer, Heidelberg (2011)
9. Itoh, H., Lu, S., Sakai, T., Imiya, A.: Interpolation of reference images in sparse dictionary for global image registration. In Proceedings of the 8th International Symposium on Visual Computing, pp. 657–667 (2012)
10. Itoh, H., Sakai, T., Kawamoto, K., Imiya, A.: Global image registration using random projection and local linear method. In: Wilson, R., Hancock, E., Bors, A., Smith, W. (eds.) CAIP 2013, Part I. LNCS, vol. 8047, pp. 564–571. Springer, Heidelberg (2013)
11. Cock, K.D., Moor, B.D.: Subspace angles between ARMA models. Syst. Control Lett. **46**, 265–270 (2002)
12. Hamm, J., Lee, D.D.: Grassmann discriminant analysis: a unifying view on subspace-based learning. In: Proceedings of the International Conference on Machine Learning, pp. 376–383 (2008)
13. Altman, N.S.: An introduction to kernel and nearest-neighbor nonparametric regression. Am. Stat. **46**, 175–185 (1992)
14. Vempala, S.S.: The Random Projection Method, vol. 65. American Mathematical Society, Providence (2004)
15. Arya, S., Mount, D.M., Netanyahu, N.S., Silverman, R., Wu, A.Y.: An optimal algorithm for approximate nearest neighbor searching in fixed dimensions. In Proceedings of ACM-SIAM Symposium on Discrete Algorithms, pp. 573–582 (1994)
16. Baraniuk, R.G., Wakin, M.B.: Random projections of smooth manifolds. Found. Comput. Math. **9**, 51–77 (2009)
17. Bingham, E., Mannila, H.: Random projection in dimensionality reduction: applications to image and text data. In: Proceedings of the International Conference on Knowledge Discovery and Data Mining, pp. 245–250 (2001)

18. Johnson, W., Lindenstrauss, J.: Extensions of Lipschitz maps into a Hilbert space. Contomp. Math. 26, 180 206 (1984)
19. Frankl, P., Maehara, H.: The Johnson-Lindenstrauss lemma and the sphericity of some graphs. Comb. Theory Ser. B **44**, 355–362 (1988)
20. Sakai, T., Imiya, A.: Practical algorithms of spectral clustering: toward large-scale vision-based motion analysis. In: Wang, L., Zhao, G., Cheng, L., Pietikäinen, M. (eds.) Machine Learning for Vision-Based Motion Analysis, pp. 3–26. Springer, London (2011)
21. Cocosco, C., Kollokian, V., Kwan, R.S., Evans, A.: Brainweb. Online interface to a 3D MRI simulated brain database. NeuroImage **5**, 425 (1997)
22. Boye, D., Samei, G., Schmidt, J., Székely, G., Tanner, C.: Population based modeling of respiratory lung motion and prediction from partial information. In: Proceedings of SPIE, vol. 8669, Medical Imaging 2013: Image Processing 8669 (2013)

A Comparative Study of GPU-Accelerated Multi-view Sequential Reconstruction Triangulation Methods for Large-Scale Scenes

Jason Mak[✉], Mauricio Hess-Flores, Shawn Recker, John D. Owens, and Kenneth I. Joy

University of California, Davis, USA
jwmak@ucdavis.edu

Abstract. The angular error-based triangulation method and the parallax path method are both high-performance methods for large-scale multi-view sequential reconstruction that can be parallelized on the GPU. We map parallax paths to the GPU and test its performance and accuracy as a triangulation method for the first time. To this end, we compare it with the angular method on the GPU for both performance and accuracy. Furthermore, we improve the recovery of path scales and perform more extensive analysis and testing compared with the original parallax paths method. Although parallax paths requires sequential and piecewise-planar camera positions, in such scenarios, we can achieve a speedup of up to 14x over angular triangulation, while maintaining comparable accuracy.

1 Introduction

Recently, there has been a great deal of work dealing with multi-view reconstruction of scenes, for example in applications such as robotics, surveillance and virtual reality. One specific scenario for reconstruction is aerial video. Accurate models derived from aerial video can form a base for large-scale multi-sensor networks that support activities in detection, surveillance, tracking, registration, terrain modelling and ultimately semantic scene analysis. Time-effective, accurate and in some cases dense scene models are needed for such purposes. In addition, unmanned aerial vehicles may become common tools for government and commercial use in the future, and allowing them to detect the underlying environment will enable increased autonomy and the ability to perform the type of useful analysis mentioned previously.

For aerial reconstruction, and reconstruction in general, performance scalability is a growing concern. Improved technology has allowed for the collection of numerous images at very high resolutions. To address this problem, researchers have developed new algorithms that can perform faster while yielding accurate reconstructions. Another approach is to leverage modern hardware, specifically parallel architectures, that include Graphics Processing Units (GPUs), which are now widely used to speed up a variety of computational problems.

© Springer International Publishing Switzerland 2015
C.V. Jawahar and S. Shan (Eds.): ACCV 2014 Workshops, Part I, LNCS 9008, pp. 254–269, 2015.
DOI: 10.1007/978-3-319-16628-5_19

GPU hardware is designed to compute large amounts of work simultaneously, which makes GPUs ideal for high-performance image processing.

To this end, the parallax paths method is a promising framework developed by Hess-Flores et al. [1] for aerial and turntable reconstruction. It uses the path of a moving camera as a strong constraint that can be applied to various stages in reconstruction including camera calibration, feature track correction, and final scene reconstruction. For each feature track of the reconstruction, a scale value is computed within the framework, which is a direct function of perceived parallax for the corresponding scene point. The method, however, requires that the camera path used in the reconstruction to be piecewise planar, and that it does not intersect the set of viewed scene points.

The main advantage of this method is that it allows for the correction of inaccurate feature tracks given constraints arising from the path of the moving camera and the projected path of a feature track as a replica of the camera path up to a scale. However, it is not clear from the original method if there is a direct way to compute accurate scale values in general, nor what the effect of scale actually is on the final computed 3D position. Also, the performance of triangulation based on the corrected tracks is not directly analyzed, and no attempts at parallelization are made. Given the way feature tracks are corrected in this method, it provides the advantages that the final triangulation can be performed efficiently, but this was not exploited by Hess-Flores et al. [1].

The main contribution of this paper is to evaluate and compare the performance of triangulation based on the parallax paths framework with another algorithm used for reconstruction, Recker's angular error-based triangulation algorithm [2], which we refer to from now on as *fast triangulation*. Recker's triangulation method is both accurate and one of the fastest known in the literature, and has been successfully parallellized on the GPU. This is the first comparison analysis between these two promising tools for solving the structure-from-motion problem. To perform the study on the most state-of-the-art high-performance hardware, we develop the first GPU implementation of the parallax paths method to compare it with the GPU implementation of the fast triangulation method. Parallax paths is more parallelizable and performs faster than fast triangulation, and with good starting feature tracks and the camera path as a reliable constraint, we can obtain comparable accuracy. Furthermore, the effect of different path scales is further defined and evaluated with respect to the original method. In the first section, we discuss related work, then in the following sections, we provide an overview of the fast triangulation method and the parallax paths method, including new insights for the latter. Next, we discuss GPU implementations, show the results of our comparison experiments, and finally end with our conclusions.

2   Related Work

The input to scene reconstruction is typically a set of images and in some cases camera calibration information, while the output is typically a 3D point cloud along with color and/or normal information, representing scene structure. For general reconstruction algorithms in the literature, there are comprehensive overviews and comparisons given in Seitz et al. [3] and Strecha et al. [4]. As for sequential reconstruction algorithms, Pollefeys et al. [5] provides a method for reconstruction from hand-held cameras, Nistér [6] deals with reconstruction from trifocal tensor hierarchies, while Fitzgibbon et al. [7] provides an approach for turntable sequences. State-of-the-art software packages such as *VisualSfM* [8] and *Bundler* [9] provide very accurate feature tracking, camera poses and scene structure, based mainly on sparse feature detection and matching, such as with the SIFT algorithm [10] and others inspired by its concept.

This paper focuses on one of the final stages of reconstruction, known as *triangulation* [11], where 3D positions for scene points are computed. The accuracy of triangulation is a function of previously-computed feature tracking, camera intrinsic calibration, and pose estimation [11]. Typically, 3×4 projection matrices are used to encapsulate all camera intrinsic and pose information. The most widely-used method in the literature is *linear triangulation* [11], where a system of the form $AX = 0$ is solved by eigen-analysis or Singular Value Decomposition (SVD). The data matrix A is a function of feature track and camera projection matrix values. The obtained solution is a direct, best-fit, and non-optimal solve, where numerical stability issues arise with near-parallel cameras. Another simple method is the midpoint method [11], but it is very inaccurate in general. A second class of algorithms is based on optimizing a cost function based on an initial direct solution. In general, these methods lack solid experimental results as far as error and processing time against different noise and camera configurations. Agarwal et al. [12] use fractional programming and a *branch and bound* algorithm to determine the global optimum. Hartley and Kahl [13] as well as Min [14] perform convex optimization on an L_∞ cost function making use of second-order cone programming (SOCP). Dai et al. [15] use a L_∞ optimization method based on gradually contracting a region of convexity towards computing the optimum.

Additional triangulators have been developed with not only accuracy but also performance scalability in mind. Recker et al.'s *fast triangulation* method [2] obtains an initial position through the midpoint method and applies adaptive gradient descent [16] on an angular error-based L_1 cost function. This function is shown to have a large basin in the vicinity of the global optimum, which avoids converging to unwanted local minima. Also, the L_1 cost function is more robust to outliers than the L_2 norm of reprojection error. Furthermore, it introduces a statistical sampling component to increase efficiency without sacrificing accuracy. This results in a significant speed increase and better reprojection errors than with other triangulators, such as linear triangulation. Hess-Flores [1] developed the parallax paths method for reconstruction in scenarios where the camera path can be modeled by piecewise-planar segments. The procedure results in an updated set of feature tracks, such that the speed and accuracy of reconstruction and bundle adjustment is improved. However, it was not tested as a standalone triangulator nor compared against any other known triangulation algorithms for speed or accuracy. Furthermore, no attempts at parallelization were made. Sánchez et al. [17,18] developed a triangulator on the GPU using an algorithm based on Monte Carlo simulations. They achieve good speedup over Levenberg-Marquardt [19] and have comparable accuracy. However, their implementation was not tested on large-scale data. Mak et al. [20] developed a GPU implementation of Recker's fast triangulation method and achieve up to 40x speedup over a CPU implementation on large-scale data. To the best of our knowledge, this is one of the fastest GPU triangulators in existence that still maintains good accuracy. In this work, we implement parallax paths on the GPU for the first time, and compare it with the GPU fast triangulation method for speed and accuracy.

3   A Summary of Fast Triangulation

Recker et al. [2] propose a L_1 triangulation cost function based on an angular error measure for a candidate 3D position, p, with respect to its feature track t. Its inputs are feature tracks across N images and their respective 3×4 camera projection matrices P_i. For each camera in a track, the cost function measures the dot product between the ray from the camera center C_i to p, known as unit vector v_i, and the ray from C_i through its 2D feature t, known as w_{ti}. This concept is displayed in Fig. 1(a). The total cost function value is the sum of these dot products across all cameras, subtracted from the total number of cameras to ensure that the absolute global minimum is zero, and then averaged. Dot products can vary from $[-1, 1]$, but only points that lie in front of the cameras need to be taken into account, corresponding to the range $[0, 1]$. Given C_i cameras, the set of all feature tracks T, and a 3D evaluation position $p = (X, Y, Z)$, the cost function for p with respect to a track $t \in T$ is given by Eq. 1 [2].

$$f_{t \in T}(p) = \frac{\sum_{i \in I}(1 - \hat{v}_i \cdot \hat{w}_{ti})}{||I||}. \tag{1}$$

In their nomenclature, $I = \{C_i | t$ "appears in" $C_i\}$, $v_i = (p - C_i)$, and $w_{ti} = P_i^+ t_i$. The right pseudo-inverse of P_i is given by P_i^+, while t_i is the homogeneous coordinate of track t in camera i. The normalized vectors are defined as $\hat{v}_i = \frac{v_i}{||v_i||}$ and $\hat{w}_{ti} = \frac{w_{ti}}{||w_{ti}||}$. Gradient values are defined in Eqs. 5–7 of Recker et al. [2].

To visualize the smooth variation of this L_1 cost function, as analyzed and discussed in Recker et al. [2], Fig. 1(b) shows a scalar field, where each dense grid position encodes the cost function value at that specific 3D location. Values are color-coded such that redder values represent higher cost function values, while blue represents values closer to zero. Notice that in this particular case, the global minimum, not explicitly displayed, lies in the center of the displayed bounding box. The variation within this box, however, is very smooth, indicating that the function is a sink with convergence likely even from distant locations. This allows for simple methods such as adaptive gradient descent [16] to be used for optimization, starting from an initial midpoint estimate [2]. For more details on the overall method, the reader is referred to Recker et al. [2].

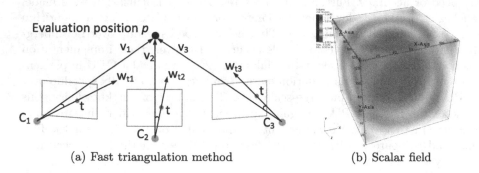

(a) Fast triangulation method   (b) Scalar field

Fig. 1. (a) In fast triangulation, rays are shot through a candidate point and through feature locations. (b) A volume view of a scalar field representing an L_1 cost function [2] evaluated at a dense grid inside a bounding box encasing a position in the reconstruction, with blue closer to zero cost.

3.1   Degeneracies in Fast Triangulation

There are specific degeneracies that can affect fast triangulation. The first is an initial midpoint estimate which is very inaccurate. Despite the sink behavior of the cost function, a very inaccurate starting position can lead adaptive gradient descent in the wrong direction. Though Recker et al. [2] proved that this seldom occurs, very erroneous feature tracks may need to be evaluated via RANSAC [21] or other robust methods before triangulating. The second degeneracy occurs with small baselines. For near-parallel cameras and/or small baselines, the obtained midpoint estimate can also be very inaccurate, and similar convergence issues can result. Generally, triangulating with very short baselines should be avoided, and algorithms such as frame decimation [22] can be used for this purpose.

4   Parallax Paths—A Further Analysis

The parallax paths method was developed by Hess-Flores et al. [1] to help yield more accurate 3D reconstructions in aerial and turntable sequences. However, the concept of scale was not generalized or further analyzed, and this paper helps enhance and improve on this concept. First, the method for reconstructing a single point will be summarized. It is assumed that a set of coplanar cameras and a set of feature tracks beginning at the first camera are the input. For a given feature track, a ray is shot from each camera center position through the point's pixel feature location in that camera's image plane. The intersection of this ray with a pre-selected *reconstruction plane* 'beneath' the scene, which is parallel to the camera plane, yields a parallax path position. The set of all ray-plane intersections for a given feature track results in its *parallax path*. There are two insights to this method: first, if a feature track is accurate, all rays should intersect at a common scene point; and second, the ray-plane intersections should

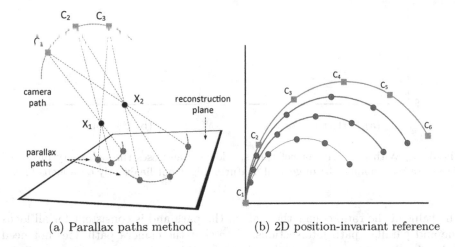

(a) Parallax paths method (b) 2D position-invariant reference

Fig. 2. (a) Rays from cameras $C_i \ldots C_n$ through a scene point X_i intersect a plane, creating a parallax path, which is a scaled version of the camera path. Points closer to the cameras create bigger paths. (b) The camera path and parallax paths are translated to a position-invariant reference, with a track's path origin coinciding with the anchor camera for the track.

be an exact yet scaled projection of the camera's path projected onto the plane, as shown in Fig. 2(a). The concept of scale is easily visualized when translating all parallax paths to a 2D position-invariant reference, as shown in Fig. 2(b).

Once the camera path and parallax paths are translated to this position-invariant reference, with all paths beginning at a common origin as shown in Fig. 2(b), *locus lines* (shown in light green) can be traced from the origin through all the parallax path points. In this case, it is assumed that the first camera is the *anchor camera* and is used as reference to provide this origin. However, any camera can be chosen as the anchor. For perfect feature tracks, a locus line should perfectly intersect every path point corresponding to a feature seen by that camera, as shown in Fig. 3(a). In this perfect setting, the *scale* of a parallax path is defined as the intersection between a locus line and the parallax path. Notice that scale values grow when moving from the reconstruction plane towards the camera plane. The original work by Hess-Flores et al. [1] did not mathematically define a direct way to obtain this scale, and we provide an efficient way to compute its value. For the first locus line in Fig. 3(a), the scale of the parallax path is the ratio of the lengths of two line segments: the segment $\overrightarrow{P_1 P_2}$ from the parallax path origin point P_1 and a second path point P_2; and the segment $\overrightarrow{C_1 C_2}$ between the anchor camera C_1 and the next camera C_2 corresponding to the second path point. This is applicable to all the locus lines, as shown in Eq. 2.

$$scale = \frac{|\overrightarrow{P_1 P_2}|}{|\overrightarrow{C_1 C_2}|} = \frac{|\overrightarrow{P_1 P_3}|}{|\overrightarrow{C_1 C_3}|} = \frac{|\overrightarrow{P_1 P_4}|}{|\overrightarrow{C_1 C_4}|} = \frac{|\overrightarrow{P_1 P_5}|}{|\overrightarrow{C_1 C_5}|} = \cdots = \frac{|\overrightarrow{P_1 P_N}|}{|\overrightarrow{C_1 C_N}|} . \qquad (2)$$

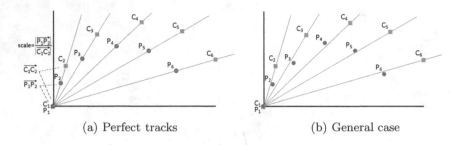

(a) Perfect tracks   (b) General case

Fig. 3. (a) With perfectly correct tracks, a locus line passes through every projected feature (a path point). (b) In general, features might not lie exactly on a locus line.

The value of the ratio equals the scale of the path and is consistent for all locus line and parallax path intersections. Note that the camera path does not need to be circular or any determinable shape for this to be true, as long as all the cameras can be fitted by a common plane (are coplanar) by segments. For long camera trajectories that are non-planar, parallax paths must be computed and concatenated across segments to obtain the final reconstruction.

4.1   Obtaining the Correct Scale

The scale value is significant because it tells us how each feature track—and therefore each point in the reconstructed scene—relates to the camera path. In practice, there are errors in the feature tracks, and so the projected camera path or ray-plane intersections are incorrect, as shown in Fig. 3(b). If the correct scale value for a parallax path is known, this fixes the locations of its parallax path positions along respective locus lines. For example, if we know a parallax path has a 0.5 scale of the camera path, each parallax path position on the position-invariant plot should lie halfway along the locus line segment traced between the origin and the respective camera projection. Once the correct scale value and position along locus lines have been determined, we can easily triangulate the correct 3D scene point as follows. First, the parallax path is translated back to its original position on the reconstruction plane. We then pick any two points on the path, shoot a ray from each point back to its associated camera, and compute the intersection point. This intersection is guaranteed to be unique, since all point-to-camera segments must intersect at a common 3D point given correct parallax paths, as shown in Fig. 2(a).

In practice, there is no easy way of obtaining the absolute correct scale. However, we now propose two simple methods to approximately obtain the correct scale. The first involves averaging all the scales derived from a potentially incorrect track. In this case, the ratios in Eq. 2 would likely not be equal across the track, but the average of all ratios approximates the scale value. We then use this consensus scale value to correct this track. If the cameras used in the reconstruction are too numerous, this approach could hinder performance, but we can employ statistical sampling the way fast triangulation does and use only a random subset of the features in each track to obtain an average scale. Note that there are robust methods such as RANSAC [21] that can be applied to detect highly inaccurate feature tracks. However, this adds undesired overhead to the method, and our main focus is on runtime performance.

In the second method, rather than averaging the scales derived from all cameras, or a randomly sampled subset, we only average scales for the first M cameras of the track. The reasoning behind this approach is that long feature tracks are known to sometimes experience degradation [1]. Therefore, if we assume that the first M feature track positions are more likely to be accurate, using a sequence of early cameras would yield a more accurate scale. In addition, parallax paths does not suffer from degeneracy problems when the baseline between cameras is small because an intersection is enforced given the constraints no matter what the camera baseline is, so using adjacent cameras is less of a problem.

The parallax paths method is very powerful because it provides additional constraints to yield an accurate reconstruction, which are not present in bundle adjustment [19] or traditional multi-view reconstruction. However, its application space is more limited than that of fast triangulation, since parallax paths is constrained to certain types of scenes. First, the cameras used in the reconstruction must all lie on a common plane, a case that can often be found in aerial image and turntable datasets, but that is only a subset of all possible reconstruction scenarios. Second, a proper reconstruction plane parallel to the camera trajectory must be chosen, and the scene cannot intersect either plane. The method also needs accurate camera calibration, both extrinsics and intrinsics, since the method relies exclusively on camera information to create parallax paths and correct them. It is potentially sensitive to very inaccurate feature tracks as well. However, state-of-the-art packages such as *VisualSfM* [8] and *Bundler* [9] can provide accurate feature tracking and camera projection matrices, so this has become less of a concern. Also, accurate camera positions can be obtained from external tools like GPS and for aircraft, IMU. In addition to triangulation, parallax paths can potentially be used for other purposes such as pose estimation and compression of scene information, but these are outside the scope of this work.

5   Methods on the GPU

In this section, we discuss an existing GPU fast triangulation implementation, followed by the introduction of a novel GPU implementation of the parallax paths framework, where we discuss high-level implementation details. We use the CUDA programming model to implement code and analyze performance on the GPU. As an overview of the model, CUDA *threads* are divided into *blocks* that run in parallel, each of which contains up to 1024 threads. Each thread runs the same CUDA program called a *kernel*. Blocks of threads are assigned to *streaming multiprocessors* (SMs). An SM runs a group of 32 threads called *warps* in a SIMD manner. Like a CPU, a GPU has a large, slow main memory, as well as caches. A faster, more local memory called shared memory is also available that allows threads in the same block to share data. Certain types of applications are more suitable for the GPU. They must be parallel enough to require an enormous number of threads, and since threads within a warp run in SIMD, the program control flow must not cause threads to frequently diverge (perform different operations due to conditional statements). A more detailed description of the CUDA programming model is given by Nickolls et al. [23].

5.1   Fast Triangulation GPU Implementation

Mak et al. [20] provide a GPU implementation of Recker's fast triangulation algorithm using two different approaches. The first approach, *one-thread-per-track*, parallelizes across tracks and assigns one thread to each track to perform gradient descent for that track. This approach can potentially lead to high thread divergence. In one scenario, different tracks can vary widely in length, so the gradient term may be more expensive to recompute for some tracks than for others. The authors propose that this problem can be mitigated to an extent by a prior sorting of the tracks, which increases the likelihood that threads within the same warp will be assigned tracks of similar length. Another case is when some tracks converge in fewer iterations of gradient descent than others. Both of these load-balancing problems cause threads in a warp that have finished processing their work to have to wait for other unfinished threads in the same warp. The authors also propose another approach to parallelizing the triangulator: *one-block-per-track*. This approach assigns a block of threads to process each track, which makes it more appropriate for datasets with long feature tracks. Each thread in a block computes one per-feature term in the gradient computation, and a parallel reduction sums these terms to GPU shared memory to obtain the final gradient value. In terms of the amount of parallelism during execution, this approach is an improvement over the previous.

Although the fast triangulation method obtains large speedups when run on the GPU, it still has issues fully utilizing the highly-parallel GPU programming model. The method relies on gradient descent, an iterative algorithm, making it hard to predict the amount of work needed per feature track until convergence. The step size for gradient descent must also be carefully considered due to its impact on the convergence rate and the stability of the algorithm. Furthermore, the one-block-per-track implementation can leave threads idling uselessly in a block if the track lengths are not long enough to fill a block [20], which must be a size that is a multiple of the warp size (32).

5.2   Parallax Paths GPU Implementation

The parallax paths method is a highly parallelizable method because the bulk of the computation involves two main stages: (1) computing ray-plane intersections for determining an initial set of parallax paths; and (2) computing all the scale values to be used in the per-track average scale. If N is the number of tracks and C is the number of cameras, there would be max $N \times C$ ray-plane intersections and $N \times C$ scale values. For computing ray-plane intersections and individual

Fig. 4. Parallax paths stages on the GPU, including parallelism P per stage.

scale values, we can compute each work-item completely independently and have a maximum $N \times C$-way parallelism running on a highly-parallel GPU. In the third stage, to compute the average scales, we need to sum all the scales within each track. Although it is possible to parallelize a sum reduction, we opt to have each thread compute the sum in serial, since we only need to perform the reduction once, and it is an insignificant portion of the runtime. Next, we correct the parallax path for each track. In practice, we only need to correct two points on the path because in the next and last stage, we recover the 3D position by intersecting two corrected rays from two corrected path points. Figure 4 shows a high-level overview of the parallax paths streaming workflow on the GPU. Although the last three stages are shown as separate, they can be combined into one GPU kernel to preserve data locality, since they all operate per track and therefore all exhibit N-way parallelism. Unlike gradient descent in fast triangulation, parallax paths on the GPU does not require multiple iterations and multiple sum reductions, instead providing a faster, more direct solution.

6   Results

We compare the processing times and general behavior of fast triangulation and parallax paths on both synthetic and real data. Our test computer has 2 Intel Xeon E5-2637 v2 CPUs, each with 4 cores clocked at 3.5 GHz, for a total of 8 cores that we use for multicore tests. Our GPU is an NVIDIA Tesla K40c, which features 15 SMs, for a total of 2880 ALUs. For running the serial tests on real data, we use a different CPU, the Intel Core i7-3630QM at 2.4 GHz, since we found it had the best single-core performance. We use the OpenMP programming model to implement a multi-core parallelization of parallax paths by partitioning the set of feature tracks among the CPU threads. The following abbreviations are used throughout the section: *PP* stands for parallax paths with scale determined from an average across an entire track; *PP2* indicates parallax paths with scaling determined from only the first two features of a track; and *FT* refers to fast triangulation. *MC* denotes multi-core, while *NU* indicates that the tracks are of non-uniform length. We do not perform statistical sampling for any tests, except for some *FT* error tests, where sampling can help avoid degeneracies of close adjacent cameras.

6.1   Synthetic Tests

The goal of synthetic testing is to compare fast triangulation versus parallax paths runtime performance on large-scale data and their accuracy in a ground-truth sense, with ground-truth not typically being available in real datasets.

Figure 5(a) shows runtime performance scaling with an increasing number of tracks. In this test, we use the one-thread-per-track GPU implementation of *FT*. With increasing tracks, Fig. 5(a) shows that *PP* on the GPU scales better than its multi-core version and also scales better than *FT* on the GPU. We do not display *FT* on multi-core because its runtime is much higher than other tests. *PP2* unsurprisingly has an insignificant runtime since it only triangulates with the first 2 cameras. For the *FT NU* test, we sort the tracks to aid in load balancing. Even so, compared to *FT* on the GPU, *PP* on the GPU has a much higher improvement in runtime (a max 55 % vs 22 % drop) when processing non-uniform (*NU*) tracks instead of uniform tracks. The reason is that *PP* has more parallelism, with more independent work across tracks. Unlike *FT*, it does not have load-balancing issues that nullify some of the runtime reduction expected due to an overall decrease in the number of features to process.

Figure 5(b) shows runtime performance scaling with an increasing number of cameras. In this test, we use the one-block-per-track GPU implementation of *FT*, since it is more suitable for longer track lengths. *PP2* is left out here because it only uses 2 cameras regardless of track length. Figure 5(b) shows that *PP* also scales better than *FT* with increasing cameras.

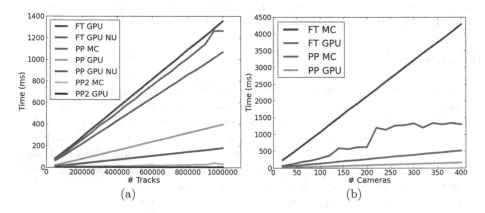

(a)   (b)

Fig. 5. (a) Runtime performance with an increasing number of tracks. The number is increased up to 1,000,000, in increments of 50,000. Track length is fixed at 100 cameras, except for the *NU* cases, where it is varied from 2–100. (b) Runtime performance with an increasing number of cameras. Cameras are varied up to 400, in increments of 50, and track length is fixed at 100,000.

In the error tests shown in Fig. 6, we use three types of camera configurations: *circle*, where cameras were placed on a circular configuration above the scene, *line* for a linear camera configuration, and *random*, where cameras are randomly placed in 2 dimensions above the scene, while all still lying on a common flat plane. Track length is fixed at 100, and the number of tracks is fixed at 10,000.

Figure 6(a) shows ground-truth error versus feature track error, where all features in a track are subject to error. Error is introduced to the perfect synthetic

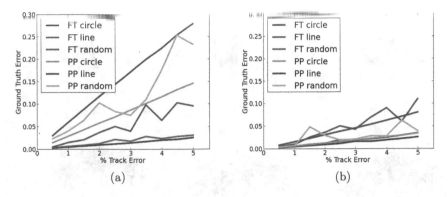

Fig. 6. Ground truth error vs. feature track error for synthetic data. (a) All features in each track subject to error. (b) No error in first feature of each track.

Table 1. Times in milliseconds for serial, multi-core, and GPU with number of tracks N and total number of cameras C. Speedup is the speedup of the GPU over the serial implementation and ϵ is the average reprojection error in pixels. Some runtimes for *Horse* are left out because they were too small to measure.

(a) Fast triangulation

Data set	N	C	serial	multicore	GPU	Speedup	ϵ
Dinosaur	4983	36	8	2	3	3x	0.467
Canyon	103,153	90	272	70	14	19x	0.226
Canyon Dense	997,115	2	1258	273	23	55x	1.838
Horse	9509	73	27	7	7	3.8x	0.770

(b) Parallax paths

Data set	N	C	serial	multicore	GPU	Speedup	ϵ	Speedup vs. FT
Dinosaur	4983	36	2	0.7	0.13	15x	0.668	23x
Canyon	103,153	90	75	16	1	75x	0.354	14x
Canyon Dense	997,115	2	351	64	3	117x	1.847	7x
Horse	9509	73	7	1.8	–	–	8.6	–

(c) Parallax paths first 2 cameras

Data set	N	C	serial	multicore	GPU	Speedup	ϵ	Speedup vs. FT
Dinosaur [24]	4983	36	2	0.5	0.07	28x	1.246	42x
Canyon [2]	103,153	90	37	7	0.36	102x	0.863	39x
Canyon Dense [2]	997,115	2	351	64	3	117x	1.847	7x
Horse [25]	9509	73	3.4	0.8	–	–	1.232	–

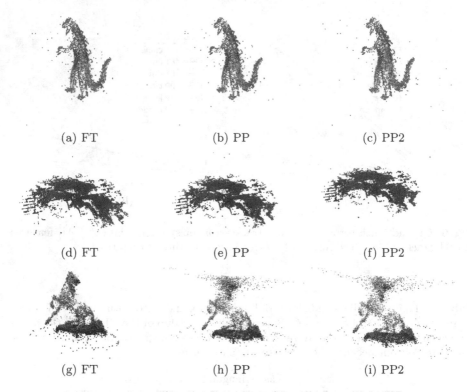

(a) FT   (b) PP   (c) PP2

(d) FT   (e) PP   (f) PP2

(g) FT   (h) PP   (i) PP2

Fig. 7. Reconstructions of three scenes: (a)–(c) *Dinosaur* [24]. (d)–(f) *Canyon* [2]. (g)–(i) *Horse* [25]. Parallax paths performs poorly on *Horse* due to the camera plane intersecting part of the scene. To obtain good parallax path reconstructions, the camera plane should be separate from the scene, as is the case in *Dinosaur* and *Canyon*.

tracks by adding noise of 0.5–5 % of a unit on the uncalibrated image plane diagonal in random directions. For all camera configurations, *PP* is less accurate than *FT*. Figure 6(b) shows the results for the same analysis, but in this case the first feature in each track (feature first seen in the anchor camera) is kept noiseless, which is a more realistic scenario. Now, we observe an improved accuracy in *PP* comparable to that of *FT*. This test demonstrates that having accurate features in anchor frames is critical for good parallax path reconstructions.

6.2   Tests on Real Datasets

For real datasets, we measure performance and reprojection error, including speedup across implementations. It is important to note that the concept of reprojection error may not be applicable for parallax paths. The reason is that the camera path constraint in parallax paths enables it to be used as a means to correct feature tracks [1]. Once the scales are obtained, the tracks can be corrected and reprojected back into images, changing the features themselves and leading to a zero reprojection error. Although in the table we show reprojection error versus original feature tracks, it is not a good indicator of parallax path accuracy given that it can be forced to 0, but it's the best that can be done in the absence of ground-truth information. For the tests, the real datasets were rotated to align with a vertical axis to make it easier to select a reconstruction plane for parallax paths. Table 1(a) displays results for fast triangulation (FT), Table 1(b) for parallax paths (PP), and Table 1(c) for PP2. For all three triangulators, larger datasets lead to larger speedups of the GPU over a serial implementation. *PP* and *PP2* are both faster than *FT*, with up to 14x and 39x speedup respectively for a meaningfully sized dataset (Canyon). However, they have higher reprojection error, though this may not be a meaningful comparison.

Finally, Fig. 7(a)–(c) shows the reconstruction of the *Dinosaur* dataset [24] using respectively *FT*, *PP*, and *PP2* from left to right. Figure 7(d)–(f) displays the same but for the *Canyon* dataset [2]. For the smaller *Dinosaur* dataset, there are no obvious major differences for different methods. One limitation of parallax paths versus fast triangulation is that the scene is not allowed to intersect the plane of the cameras. To display the problems that occur, Fig. 7(g) shows a good reconstruction obtained from *FT* for the *Horse* [25] dataset, whereas Fig. 7(h)–(i) show the bad result obtained from parallax paths. For this scene of a horse, the camera plane intersects the top of the scene, causing some rays to be nearly parallel to the reconstruction plane, which leads to ill-conditioned problems and inaccurate reconstructed points. To obtain good parallax path reconstructions, the camera plane should be separate from the scene.

7 Conclusion

In this paper, we present a comparison of a novel GPU implementation of a triangulator based on the parallax paths method versus the state-of-the-art multiview triangulation method, angular error-based ('fast') triangulation. The main contributions of the paper are the following. We map the parallax paths method to the GPU and analyze its performance as an efficient triangulation method for the first time. To this end, we compare it with the existing fast triangulation GPU implementation for both performance and accuracy. We develop the parallax paths further than in the original method, with more analysis on the effect of scaling. We also demonstrate the importance of having an accurate first feature in a feature track to yield an accurate parallax path reconstruction. Overall, the parallax paths method is highly parallelizable and efficient, but requires that the cameras used in reconstruction be piecewise-planar and not intersect the scene itself. Though limited to applications with sequential camera motion, such as aerial video or turntable sequences, it yields a substantial speedup over fast triangulation, as demonstrated on real and synthetic testing, while maintaining comparable accuracy. If accuracy is absolutely critical, fast triangulation may still be a more preferable method. Future work involves mainly attempting to obtain more accurate scales in parallax paths, by taking into account further constraints such as intensity consensus at candidate scales.

References

1. Hess-Flores, M., Duchaineau, M.A., Joy, K.I.: Sequential reconstruction segment-wise feature track and structure updating based on parallax paths. In: Lee, K.M., Matsushita, Y., Rehg, J.M., Hu, Z. (eds.) ACCV 2012, Part III. LNCS, vol. 7726, pp. 636–649. Springer, Heidelberg (2013)
2. Recker, S., Hess-Flores, M., Joy, K.I.: Statistical angular error-based triangulation for efficient and accurate multi-view scene reconstruction. In: Workshop on the Applications of Computer Vision (WACV), pp. 68–75 (2013)
3. Seitz, S.M., Curless, B., Diebel, J., Scharstein, D., Szeliski, R.: A comparison and evaluation of multi-view stereo reconstruction algorithms. In: Proceedings of the 2006 IEEE Conference on Computer Vision and Pattern Recognition, pp. 519–528 (2006)
4. Strecha, C., von Hansen, W., Gool, L.J.V., Fua, P., Thoennessen, U.: On benchmarking camera calibration and multi-view stereo for high resolution imagery. In: Proceedings of the 2008 IEEE Conference on Computer Vision and Pattern Recognition (2008)
5. Pollefeys, M., Van Gool, L., Vergauwen, M., Verbiest, F., Cornelis, K., Tops, J., Koch, R.: Visual modeling with a hand-held camera. Int. J. Comput. Vis. **59**, 207–232 (2004)
6. Nistér, D.: Reconstruction from uncalibrated sequences with a hierarchy of trifocal tensors. In: Vernon, D. (ed.) ECCV 2000. LNCS, vol. 1842, pp. 649–663. Springer, Heidelberg (2000)
7. Fitzgibbon, A.W., Cross, G., Zisserman, A.: Automatic 3D model construction for turn-table sequences. In: Koch, R., Van Gool, L. (eds.) SMILE 1998. LNCS, vol. 1506, p. 155. Springer, Heidelberg (1998)
8. Wu, C.: VisualSfM: A visual structure from motion system (2011). http://ccwu.me/vsfm/
9. Snavely, N., Seitz, S.M., Szeliski, R.: Photo tourism: exploring photo collections in 3D. ACM Trans. Graph. **25**, 835–846 (2006)
10. Lowe, D.: Distinctive image features from scale-invariant keypoints. Int. J. Comput. Vis. **60**, 91–110 (2004)
11. Hartley, R.I., Zisserman, A.: Multiple View Geometry in Computer Vision, 2nd edn. Cambridge University Press, Cambridge (2004)
12. Agarwal, S., Chandraker, M.K., Kahl, F., Kriegman, D.J., Belongie, S.: Practical global optimization for multiview geometry. In: Leonardis, A., Bischof, H., Pinz, A. (eds.) ECCV 2006, Part I. LNCS, vol. 3951, pp. 592–605. Springer, Heidelberg (2006)
13. Hartley, R.I., Kahl, F.: Optimal algorithms in multiview geometry. In: Yagi, Y., Kang, S.B., Kweon, I.S., Zha, H. (eds.) ACCV 2007, Part I. LNCS, vol. 4843, pp. 13–34. Springer, Heidelberg (2007)
14. Min, Y.: L-Infinity norm minimization in the multiview triangulation. In: Wang, F.L., Deng, H., Gao, Y., Lei, J. (eds.) AICI 2010. LNCS (LNAI), vol. 6319, pp. 488–494. Springer, Heidelberg (2010)
15. Dai, Z., Wu, Y., Zhang, F., Wang, H.: A novel fast method for L_∞ problems in multiview geometry. In: Fitzgibbon, A., Lazebnik, S., Perona, P., Sato, Y., Schmid, C. (eds.) ECCV 2012, Part V. LNCS, vol. 7576, pp. 116–129. Springer, Heidelberg (2012)
16. Snyman, J.A.: Practical Mathematical Optimization: An Introduction to Basic Optimization Theory and Classical and New Gradient-Based Algorithms. Applied Optimization, vol. 97, 2nd edn. Springer-Verlag New York, Inc., Secaucus (2005)

17. Sánchez, J.R., Álvarez, H., Borro, D.: GFT: GPU fast triangulation of 3D points. In: Bolc, L., Tadeusiewicz, R., Chmielewski, L.J., Wojciechowski, K. (eds.) ICCVG 2010, Part II. LNCS, vol. 6375, pp. 235–242. Springer, Heidelberg (2010)

18. Sánchez, J.R., Álvarez, H., Borro, D.: GPU optimizer: a 3D reconstruction on the GPU using Monte Carlo simulations - how to get real time without sacrificing precision. In: Proceedings of the 2010 International Conference on Computer Vision Theory and Applications, pp. 443–446 (2010)

19. Lourakis, M.I.A., Argyros, A.A.: The design and implementation of a generic sparse bundle adjustment software package based on the Levenberg-Marquardt algorithm. Technical Report FORTH-ICS TR-340-2004, Institute of Computer Science - FORTH (2004)

20. Mak, J., Hess-Flores, M., Recker, S., Owens, J.D., Joy, K.I.: GPU-accelerated and efficient multi-view triangulation for scene reconstruction. In: Proceedings of the IEEE Winter Conference on Applications of Computer Vision, WACV 2014, pp. 61–68 (2014)

21. Fischler, M.A., Bolles, R.C.: Random sample consensus: a paradigm for model fitting with applications to image analysis and automated cartography. In: Fischler, M.A., Firschein, O. (eds.) Readings in Computer Vision: Issues, Problems, Principles, and Paradigms, pp. 726–740. Morgan Kaufmann Publishers Inc., San Francisco (1987)

22. Nistér, D.: Frame decimation for structure and motion. In: Pollefeys, M., Van Gool, L., Zisserman, A., Fitzgibbon, A.W. (eds.) SMILE 2000. LNCS, vol. 2018, p. 17. Springer, Heidelberg (2001)

23. Nickolls, J., Buck, I., Garland, M., Skadron, K.: Scalable parallel programming with CUDA. ACM Queue 6, 40–53 (2008)

24. Oxford Visual Geometry Group: Multi-view and Oxford Colleges building reconstruction (2009). http://www.robots.ox.ac.uk/~vgg/

25. Moreels, P., Perona, P.: Evaluation of features detectors and descriptors based on 3D objects. In: Proceedings of the Tenth IEEE International Conference on Computer Vision (ICCV 2005), vol. 1, pp. 800–807 (2005)

Indoor Objects and Outdoor Urban Scenes Recognition by 3D Visual Primitives

Junsheng Fu[1,3]($\boxtimes$), Joni-Kristian Kämäräinen[1],
Anders Glent Buch[2], and Norbert Krüger[2]

[1] Vision Group, Tampere University of Technology, Tampere, Finland
junsheng.fu@tut.fi
http://vision.cs.tut.fi
[2] CARO Group, University of Southern Denmark, Odense, Denmark
http://caro.sdu.dk
[3] Nokia Research Center, Tampere, Finland

Abstract. Object detection, recognition and pose estimation in 3D images have gained momentum due to availability of 3D sensors (RGB-D) and increase of large scale 3D data, such as city maps. The most popular approach is to extract and match 3D shape descriptors that encode local scene structure, but omits visual appearance. Visual appearance can be problematic due to imaging distortions, but the assumption that local shape structures are sufficient to recognise objects and scenes is largely invalid in practise since objects may have similar shape, but different texture (e.g., grocery packages). In this work, we propose an alternative appearance-driven approach which first extracts 2D primitives justified by Marr's primal sketch, which are "accumulated" over multiple views and the most stable ones are "promoted" to 3D visual primitives. The 3D promoted primitives represent both structure and appearance. For recognition, we propose a fast and effective correspondence matching using random sampling. For quantitative evaluation we construct a semi-synthetic benchmark dataset using a public 3D model dataset of 119 kitchen objects and another benchmark of challenging street-view images from 4 different cities. In the experiments, our method utilises only a stereo view for training. As the result, with the kitchen objects dataset our method achieved almost perfect recognition rate for $\pm 10°$ camera view point change and nearly 80 % for $\pm 20°$, and for the street-view benchmarks it achieved 75 % accuracy for 160 street-view images pairs, 80 % for 96 street-view images pairs, and 92 % for 48 street-view image pairs.

1 Introduction

Over the past few decades, object and scene recognition have achieved great success using 2D image processing methods. Recently, with the increasing popularity of Kinect sensors and the emergence of dual-camera mobile phone, researchers are motivated to approach the traditional image recognition problem with 3D computer vision methods. Compared with the successful 2D methods, 3D approaches

© Springer International Publishing Switzerland 2015
C.V. Jawahar and S. Shan (Eds.): ACCV 2014 Workshops, Part I, LNCS 9008, pp. 270–285, 2015.
DOI: 10.1007/978-3-319-16628-5_20

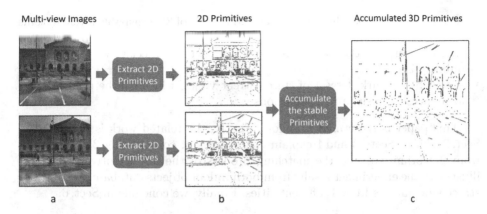

Fig. 1. Construct the 3D primitives from Multi-view images.

are not limited to image 2D appearance as the cue for detection and recognition [1,2]. A number of 3D methods for object and scene recognition have been proposed [3–5] to extract global or local shape descriptors that encode scene structure, however, they do not take the advantage of 2D visual appearance, e.g. colour and texture.

In accord with the recent trend of 3D object detection and recognition research, we propose in this paper an approach that utilizes both the 2D appearance and 3D structure from the multi-view images. The most important and novel processing of the proposed method, in our view, is the construction of the 3D primitive, i.e. 3D classified features derived from multi-view images. Figure 1 shows the work-flow of the 3D primitive construction: Firstly, for each multi-view input, the pipeline computes the 2D visual primitives [6] using the intrinsic dimension by Kalkan et al. [7]. Secondly, the stable 2D primitives are matched across multi-view images and triangulated to 3D primitives, as shown in Fig. 1c (see Sect. 3 for details). Then the 3D primitives are used for matching 3D objects primitives stored in a database.

To evaluate the proposed method, we tested our pipeline with both indoor objects and outdoor urban scenes. With the indoor objects dataset, our method achieved almost perfect recognition rate for ±10° camera view point change and nearly 80 % for ±20°, and for the real world street-view dataset from 4 different cities, our method achieved 75 % accuracy for 160 street-view images pairs, 80 % for 96 street-view images pairs, and 92 % for 48 street-view image pairs.

Our main contributions are as follows:

- A novel 3D primitive extraction method for object recognition: 2D appearance primitives are extracted and promoted to 3D based on matching results across multi-view images.
- A simple random sampling based recognition to match observed 3D primitives to database objects. The training is based on a single recorded view.
- Novel results on the effect of primitive accumulation vs. no accumulation and 3D matching vs. 2D matching for object recognition in 3D.

- A semi-synthetic benchmark dataset and toolkit of 3D graspable kitchen items captured in the KIT[1]. This can be used for further analysis in a controlled environment, and the code for rendering novel KIT object views will be made publicly available.
- A real benchmark dataset of stereo street views, which can be used for performances analysis in real conditions.

This paper is structured as follow. Firstly, the related work is presented in Sect. 2. Then, Sects. 3 and 4 explain the process of constructing 3D primitives from 2D primitives and the matching process of the 3D primitives. Section 5 illustrates the experiment results from both indoor objects database and outdoor street-view images from 4 different cities. Finally, we conclude in Sect. 6.

2   Related Work

The object detection and recognition approaches can be roughly divided into 2D-to-2D (genuine 2D), 3D-to-2D (or 2D-to-3D) and 3D-to-3D (genuine 3D) methods, where the first term defines whether a model (and training data) are 2D or 3D and the latter whether objects are detected from 2D or 3D images. The most successful approach is part-based: local features are extracted and the object described as the parts and their location. Successful results have been reported for detection of visual classes and specific objects in 2D-to-2D [1,2] and 3D-to-2D [8–10], and many of the methods provide state-of-the-art classification accuracy on common benchmarks.

Our main interest, however, are genuine 3D methods which have not yet reached a mature stage as the aforementioned methods. Next, we give a brief survey on the most recent works, but omit methods based on global description (e.g., [11]), those using temporal information [12,13] and those tailored for a specific application, such as 3D face recognition [14,15].

Two notable works related to our method are the ones by Papazov and Burschka [16] and Drost et al. [17]. Papazov and Burschka utilise a random sample principle while Drost et al. use Hough-like voting, but the main commonality is in the fact that they both directly use 3D point clouds, which ties their methods to the selected 3D capturing method. We use local primitives extracted from 2D RGB images. Similar vision primitives were used in Detry et al. [18,19], but their method do not retain 3D structure, and recognition is performed by Markov process message passing utilising pairs of the primitives similar to [17].

The popular 2D interest point detectors and descriptors have also been extended to 3D, for example 3D SURF by Knopp et al. [20], local surface histograms [21] in Pham et al. [22], HOG and DoG by Zaharescu et al. [23] and kernel descriptors [24]. Special 3D shape detectors and descriptors have also been proposed [25,26] along with neighbourhood processing to improve the robustness of shape descriptors [3,5]. There are many local 3D shape descriptors (see [27,28]), but their main limitation is that they select the points based on

---

[1] http://i61p109.ira.uka.de/ObjectModelsWebUI/.

local shape information and discard appearance which, after all, is the low-level source of information in the human visual system and used in the Marr's primal sketch [6]. The shape descriptors have been recently evaluated in [4]. One exception is Lee et al. [29] who utilise lines, but that is particularly suitable for their objects of interest (boxes). Hybrids of 3D shape and 2D texture descriptors were proposed by Hu and Zhu [30] and Kang et al. [31].

3 Constructing 3D Primitives from 2D Primitives

The visual primitives used in this work derive from the primitives found in various layers of the "deep vision hierarchy" [32]. Starting from the pixels (retinal image) we extract low level primitives which are re-sampled (added), deleted, combined (grouped) and promoted through bottom-up processing in the hierarchy. We refer to the operations with a single term, "accumulation". Various computational models of the hierarchy have been proposed [33–35]. out of which we adopt the "cognitive vision model" hierarchy by Pugeault et al. [35]. The main goal of their hierarchy is a symbolic 3D description of a scene, but we form primitives that construct a part-based 3D object model.

On the lowest hierarchy level, 2D primitives are extracted from the left and right images of a stereo pair (see Fig. 1). The primitives are extracted on a regular spatial grid where circular patches are extracted and assigned to one of four low-level classes: a constant colour region, edge/line, junction or texture. The classification is based on computational intrinsic dimensionality [7]. The computational intrinsic dimension, ifD, defined by a real number f measures the effective texture patch dimension similar to the fractal dimension [36], but can be computed fast with linear quadrature filters [37]. The ifD space forms a triangular region where basic perceptual classes map to distinct locations (Fig. 2):

- Constant colour: $ifD \approx i0D$
- Edge/line: $ifD \approx i1D$
- Junction: $i1D << ifD < i2D$
- Textured region $ifD \approx i2D$

The extracted 2D primitives are encoded as

$$\pi = (x, \theta, \phi, c) \tag{1}$$

where x is the 2D image position, θ is the local orientation angle of an edge or line, ϕ is the local phase of an edge/line, and c is the RGB colour vector of the left, middle and right edge colours.

The accumulation of 2D primitives to 3D primitives Π is based on multiple views with known calibration: $accumulation : (\pi, \pi') \rightarrow \Pi$. In order to be promoted, the 2D primitive descriptors—colour, orientation and phase—must match, the primitives must lie on their corresponding epipolar lines, and finally the spatial constraints must hold. For putative matches for a primitive π at x in the left image, the epipolar line $x' \in l' = e' \times H_\pi x$, where $e' \times H_\pi = F$

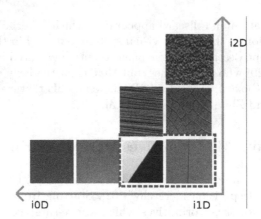

Fig. 2. Texture characterisation in the intrinsic dimension space [7], the 2D line and edge primitives used in our work marked with the dashed line.

is the fundamental matrix [38], in the right image is searched for π'. Since the 2D primitives are computed sparsely on a grid, the matches within the distance of 1.5 times the patch size are accepted. The accumulated 3D primitives are encoded as

$$\Pi = (X, n, \Theta, \Phi, C) \tag{2}$$

where X is the 3D location in space, n is the surface normal, Θ the line/edge orientation, Φ the line/edge phase and C the colour vector constructed by the weighted average of the corresponding 2D colours.

In this work, we use the line/edge primitives (see Fig. 2). The 2D primitive extraction can be adjusted by three quadrature filter parameters [37]. The first parameter is the highest filter frequency (or image resolution). The second parameter is the minimum required energy within the circular patches (normalised to $[0, 1]$) and the third parameter is the maximum variance (normalised to $[0, 1]$), i.e. whether primitives must come from clearly isolated points (low variance). The descriptor match is a weighted sum of colour (weight 0.5), orientation (0.3) and phase (0.06) differences, all normalised to $[0, 1]$, and the match threshold set to 0.3. Moreover, a spatial constraint, "external confidence", similar to stereo algorithms was added to ensure that the accepted 3D primitives are supported by their neighbourhood. By changing the values of the parameters we can affect the number of extracted 2D and 3D primitives and their robustness. Several settings are demonstrated in Table 1 for the first 12 KIT objects.

For the setting 1 approximately 50 % of the 2D primitives are promoted. For other settings, the number of 2D primitives is much larger, but due to the accumulation there is not much difference between the number of 3D primitives for the settings 1–3. This is further illustrated in Fig. 3 where the 3D primitives (bottom) look alike for all settings. Note, however, that for Setting 2 and Setting 3 the new primitives are less reliable and therefore more noise appears. By using higher frequencies (a larger image), the number of primitives increases

Table 1. Various 3D primitive extraction Parameter settings and the corresponding numbers of produced 3D primitives.

Parameter	Setting 1	Setting 2	Setting 3	Setting 4
Image size	300x300	300x300	300x300	400x400
Min. energy	0.4	0.4	0.4	0.4
Max. variance	0.2	0.6	0.2	0.2
Ext. conf.	0.1	0.1	-1.0	0.1

Object	Setting 1 (2D)	Setting 1	Setting 2	Setting 3	Setting 4
OrangeMarmelade	324	120	243	219	244
BlueSaltCube	410	251	326	315	433
YellowSaltCube	380	201	289	293	338
FruitTea	282	168	258	227	265
GreenSaltCylinder	246	72	158	166	140
MashedPotatoes	424	223	374	329	387
YellowSaltCylinder	355	168	236	247	329
Rusk	503	234	393	303	381
Knaeckebrot	372	186	269	242	300
Amicelli	414	276	384	384	509
HotPot	376	131	200	216	193
YellowSaltCube2	380	210	278	303	396
Avg.	372	187	284	270	326

Fig. 3. Top: extracted 2D primitives (stereo left) with Settings 1–4 from the left to right. Bottom: the corresponding 3D primitives after the accumulation.

Fig. 4. The 3D primitives at the bottom of Fig. 3 re-drawn using the detected scales. See the last paragraph of Sect. 3 for details.

"naturally", i.e., more details are added to places where also the depth information is reliable. That is illustrated Fig. 4, where 3D primitives are plotted in 3D space with their detected scale.

4   Matching 3D Primitives

The 3D primitive based object description in Sect. 3 represents object appearance in the primitive descriptors Θ, Φ and C and object location in the 3-vectors X. The two popular approaches to match descriptors in space are voting and random sampling. A variant of the random sampling appears in Papazov and Burschka [16] and voting (Hough transform) in Drost et al. [17].

The random sampling and voting have certain distinct properties as compared to each other. In the voting approach every primitive is processed once and they cast votes for multiple objects and for multiple poses. The best hypothesis is the one with the highest number of votes. A disadvantage is the size of the vote (accumulator) space, which can become huge without coarse discretisation. In the sampling approach, no accumulation is needed since every random sample generates one hypothesis of an object and its pose. The obvious disadvantage is that the required number of random samples may be large. In other words, the voting is more storage intensive and the sampling more computationally intensive. There exists studies to improve storage requirements and to reduce the number of samples (e.g., [39]), but in this work we select the sampling approach due to its simplicity.

Algorithm 1. Random sample consensus matching.

1: Compute the match matrix between each observed primitive $\Pi_{i=1...N}$ and each model primitive $\Pi_{i=1...M}$: $D_{N \times M}$.

2: Sort and select the K best matches for each observation primitive $\rightarrow \hat{D}_{N \times K}$.

3: **for** R iterations **do**

4:    Randomly select 3 observation primitives from $1 \ldots N$ and their correspondences in $1 \ldots K$ in $\hat{D}_{N \times K}$.

5:    Estimate the linear 3D transformation (isometry/similarity) T using the Umeyama method [40].

6:    Transform the all N observation primitives to the model space with T.

7:    Select the geometrically closest matches (within the K best) and compute the match score s.

8:    Update the best match (s_{best}, T_{best}) if necessary.

9: **end for**

10: Return s_{best} and T_{best}.

We randomly sample from the primitives of an object model i (object databreakbase), select corresponding primitives from an observed scene, and then compute the transformation T which brings the observed scene and database model primitives in correspondence. The method is similar to Papazov and Burschka [16], except that they directly use dense point cloud points which

are sensitive to a selected 3D acquisition process. Additionally, to avoid computational explosion (every observation point is a candidate match to every model
point), they utilise heuristics. Our method selects the best match using the 3D
primitive descriptors. To estimate the 3D transformation (isometry) we use the
linear method by Umeyama [40]. A high level algorithm for our matching method
is given in Algorithm 1.

There are two important considerations for Algorithm 1: the number of iterations R and a method to compute the match score s. Since the colour plays the
most important role in the accumulation, we omit Θ and Φ and use the colour
vector C to compute the match matrix D. C is a 9-vector of the RGB values
for the edge/line left, middle and right which are uniquely defined. The match
is the Euclidean distance between the vectors which is fast to compute. Also the
colour covariances are available, but using them is computationally inefficient.
L^2-normalisation makes the colour descriptors semi illumination invariant.

The number of iterations R is an important parameter since a sufficient number of samples is needed to guarantee that the correct combination is found with
high confidence. To derive a formula for R we can consider the ideal case that
each N observation point has a correct match in the model. The total number of
points is not important, but the number of possible candidates. In Algorithm 1
this is K and we further assume that a correct correspondence is within the
K best matches. Now, the probability of randomly selecting a correct combination of three point correspondences (the minimum for 3D isometry/similarity
estimation) is

$$P(K) = \frac{1}{K} \cdot \frac{1}{K} \cdot \frac{1}{K}. \tag{3}$$

Note that this would be $1/K(K-1)(K-2)$ if the points are shared. The probability that after R iterations no correct triplets have been drawn is $(1-P(K))^R$, and
thus, the probability that at least one correct has been drawn is $1-(1-P(K))^R$.
The analytical formula for the number of samples in order to pick at least one
correct match with the probability P_S is

$$R = \frac{\log(1 - P_S)}{\log(1 - P(K))}. \tag{4}$$

For example, with $P_S = 0.9$ (90 % confidence level), we get $R = 287$ for $K = 5$
and $R = 2302$ for $K = 10$. In practise, some primitives have no matches at all,
but on the other hand, representation is typically dense in the most informative
areas and any primitive near the correct one may succeed. In any case, K should
not be more than 10 to limit computational burden ($R \leq 2000$).

To select the best strategy to compute the match score s, we run preliminary
tests with the first 12 objects in the KIT dataset (see Table 2 for the results).
More details are in Sect. 5, but here we focus only on the recognition accuracy.
The rank order statistics rules, such as *median matching*, are superior due to their
robustness to outliers and still computationally affordable. There is no major
differences between the median (best 50 %) and best 25 %, with the number of
samples doubled (2×iterations) and isometry vs. similarity, and therefore we

selected the median rule. Note that the reverse matching (from models to the scene), is clearly inferior.

Table 2. Recognition accuracies for the first 12 KIT objects using variants of the match score s in Algorithm 1. $K = 10$ best matches and $R = 1000$ random samples (Setting 1, pure chance 8 %).

s Method	El-Az 5°	El-Az 10°	El-Az 20°	El-Az 30°	El-Az 40°
Mean match	84 %	74 %	50 %	34 %	18 %
Med match	100 %	100 %	98 %	77 %	46 %
Med match (reverse)	100 %	97 %	65 %	49 %	28 %
Best 25 % match	100 %	100 %	93 %	70 %	44 %
Med match (2× iters)	100 %	100 %	98 %	78 %	46 %
Med match (simil.)	100 %	100 %	96 %	77 %	45 %

5   Experiments

In this Section, we evaluate our pipeline with both the **indoor objects dataset** and the **outdoor urban street-view images**.

A dataset was collected in Karlsruhe Institute of Technology (KIT): KIT Object Models Web Database[2]. The KIT dataset provides full high-quality 3D models, so we use the KIT dataset as the indoor objects database for testing the pipeline. For evaluation, we implemented a synthetic view generator that can be used to evaluate methods in controlled view points and illumination. To further evaluate the robustness of our pipeline, we gathered 160 street-view images pairs with the known camera poses from 4 different cities. The datasets and experiment results are discussed in the following two Subsections.

5.1   Indoor Object Dataset

Toolkit for semi-synthetic KIT Objects – The KIT object dataset contains 119 3D captured kitchen items (marmalade packages, mugs, tea packages etc.) suitable for robot grasping and manipulation [41] and stored as high-quality textured 3D polygon models. Using the KIT models (Fig. 5) we provide a public toolkit to generate arbitrary views points, ground truth, and benchmark recognition algorithms.

The toolkit was used to render the training images in roughly frontal pose (Fig. 5), automatically adjust the camera distance to fit objects' bounding boxes to the visible image area, generate stereo pairs (Fig. 6) and output the stereo camera matrices and bounding box world coordinates.

For our experiments, the object database (training set) was made by storing primitives from only one view per object: the frontal views shown in Fig. 5.

---

[2] http://i61p109.ira.uka.de/ObjectModelsWebUI/.

Fig. 5. Examples of the 119 KIT object models in frontal (training) pose. Note that some objects differ only by details in their appearance (colour or texture).

Fig. 6. The stereo pair frontal views of "Amicelli" (left) and "MashedPotatoes" (right). The camera baseline is fixed to 50 world units (1wu ≈ 1 mm).

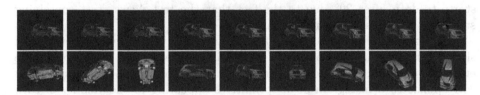

Fig. 7. Variation in the "ToyCarYellow" test images (stereo left): El-Az 5° (top row, the simplest set), and El-Az 40° (bottom, the most difficult set) (Color figure online).

The test set images were generated by geometrically transforming the same objects by adjusting the camera azimuth and elevation angles. A total of five different test sets were generated using gradually increasing angles: $\{-40°, -30°, \ldots + 40°\}$ This results to 9 test images per object and $119 \times 9 = 1071$ images in total for each test set. The test sets are referred to as Ez-Al-5° ... Ez-Al-40°. The two extremal test set images for an object are illustrated in Fig. 7 and the stereo pairs of each were used to extract the primitives and match them to the all database (training set) objects with Algorithm 1.

Results – The recognition accuracies for all experimental scenarios are presented in Table 3 for the primitive extraction settings Setting 1 and Setting 2 (see Sect. 3). To compare 2D and 3D matching we utilised directly the 2D primitives with and without the accumulation.

Table 3. Recognition accuracies for the KIT object models (tot. of 1071 test image per set) using median matching (pure chance 0.08 %).

Method	El-Az 5°	El-Az 10°	El-Az 20°	El-Az 30°	El-Az 40°
Med match - Sett. 1	98 %	93 %	78 %	55 %	33 %
Med match - Sett. 1 (2D)	98 %	94 %	78 %	51 %	28 %
Med match - Sett. 1 (2D, no acc.)	79 %	72 %	52 %	34 %	23 %
Med match - Sett. 4	99 %	97 %	87 %	63 %	38 %
Med match - Shape descr. [42]	88 %	75 %	47 %	33 %	19 %

Using more primitives achieved by, for example, higher resolution images, is beneficial as the Setting 4 provides the best results. However, the Setting 1 is not significantly worse being much faster (ten seconds vs. minutes in our Matlab implementation). Moreover, the importance of the accumulation process is verified as the 2D matching with accumulated 2D primitives is almost the same to the accumulated 3D matching. 3D primitives are more beneficial with large view point changes where 2D transformation cannot represent the view anymore.

Overall, for small view angle variation (azimuth and elevation $\leq 10°$) our recognition rate is almost perfect and for 20° still almost 80 %. The accuracy starts to drop after 20° due to the fact that the test views start containing structures not present in the training view.

To compare our method with other descriptors, we implemented the local shape context, originally proposed for 2D in [43], extended to 3D by Frome et al. [42] and similar to the heuristic approach in [16]. The local shape context corresponds to a histogram of 3D primitives appearing in the vicinity of each primitive. The local shape context is simple and efficient to compute. The bin size was optimised by cross-validation and the results are shown in the last row of Table 3. For KIT objects, the local shape context descriptors are clearly inferior to the colour matching, but still perform well with the smaller angles and are thus promising for applications and imaging conditions where the colour is not informative.

Fig. 8. Extracted 3D primitives (yellow dots) and database object bounding box and 3D primitives (green) projected by the estimated T (Color figure online).

It is noteworthy that since our approach is genuine 3D it also produces the object pose T as a side product. The detected poses are coarse (Fig. 8), but provide good initial guesses for more accurate pose optimisation.

5.2   Outdoor Street View Scenes

In this part of experiment, 160 street-view image pairs at various locations from 4 different cities were used as benchmark database. These database consists of 40 different urban scenes, where each urban scene has 4 street-view pairs, see Fig. 9(a) as an example.

The ground truth camera pose recorded in the metadata of the street-view images were used to estimate approximate camera extrinsics. For each urban scene, we selected one pair of images for training and the rest 3 pairs for testing. Otherwise, all method settings were the same as in the previous experiment. Without any parameter tuning, we achieved satisfactory results as shown in Table 4.

- For 12 classes (or urban scenes) with 48 street-view pairs, the pipeline achieved 92 % accuracy, and 97 % of the results ranked the correct class within the 5 best candidates produced by the algorithm.
- For 24 classes with 96 street-view pairs, the pipeline achieved 80 % accuracy, and 94 % of the results ranked the correct class within the 5 best candidates produced by the algorithm.
- For 40 classes with 160 street-view pairs, the pipeline achieved 75 % accuracy, and 85 % of the results ranked the correct class within the 5 best candidates produced by the algorithm.

The result shows that our 3D promoted primitives and the simple matching algorithm also work with realistic data of moderate occlusion and viewpoint changes.

Table 4. Recognition accuracies for outdoor urban scenes using median matching.

Three Sets	Set1	Set2	Set3
Number of classes	12	24	40
Number of street-view pairs	48	96	160
By pure chance to find the correct class	8 %	4 %	2 %
Accuracy	92 %	80 %	75 %
The correct class within the best 5 candidates	97 %	94 %	85 %

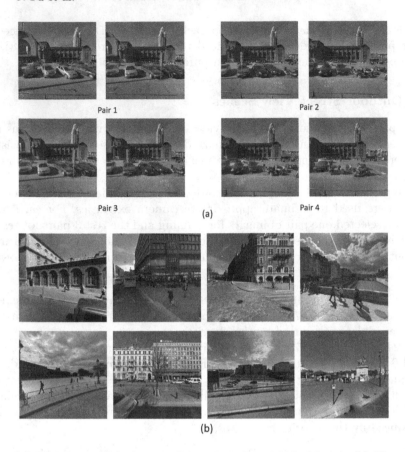

Fig. 9. (a) Here are 4 pairs of street-view images for one urban scene. (b) These are 8 examples of urban scenes from our street-view database.

6   Conclusions

This paper proposes an approach that utilizes both the 2D appearance and 3D structure from the multi-view images for 3D object detection and recognition. We introduced novel 3D primitives for indoor objects and urban scenes recognition in 3D. The 3D primitive extraction is based on low level visual 2D primitives selected by computational intrinsic dimension that classifies them according to Marr's primal sketch. The 2D primitives are matched across multi-view images and triangulated to 3D primitives. For matching the primitives, we introduced a simple but effective random sampling procedure that achieved 80 % accuracy for the view angle variation up to ±20° with indoor objects dataset and satisfactory accuracy for the street-view dataset. Our future work will include investigation of other primitive types, such as local texture and higher level primitives, such as constant colour regions.

Acknowledgement. The authors would like to give thanks to Dr. Lixin Fan for the valuable discussions.

References

1. Philbin, J., Chum, O., Isard, M., Sivic, J., Zisserman, A.: Object retrieval with large vocabularies and fast spatial matching. In: CVPR (2007)
2. Chum, O., Matas, J.: Unsupervised discovery of co-occurrence in sparse high dimensional data. In: CVPR (2010)
3. Rodola, E., Albarelli, A., Bergamasco, F., Torsello, A.: A scale independent selection process for 3d object recognition in cluttered scenes. Int. J. Comput. Vis. **102**, 129–145 (2013)
4. As'ari, M., Supriyanto, U.S.E.: 3d shape descriptor for object recognition based on kinect-like depth image. Image Vis. Comput. **32**, 260–269 (2014)
5. Buch, A., Yang, Y., Krüger, N., Petersen, H.: In search of inliers: 3d correspondence by local and global voting. In: CVPR (2014)
6. Marr, D.: Vision. A Computational Investigation into the Human Representation and Processing of Visual Information. W.H. Freeman and Company, New York (1982)
7. Kalkan, S., Wörgötter, F., Krüger, N.: Statistical analysis of local 3d structure in 2d images. In: CVPR (2006)
8. Glasner, D., Galun, M., Alpert, S., Basri, R., Shakhnarovich, G.: Viewpoint-aware object detection and pose estimation. In: ICCV (2011)
9. Sattler, T., Leibe, B., Kobbelt, L.: Fast image-based localization using direct 2d-to-3d matching. In: ICCV (2011)
10. Zia, M., Stark, M., Schiele, B., Schindler, K.: Detailed 3d representations for object recognition and modeling. IEEE PAMI **35**, 2608–2623 (2013)
11. Dorai, C., Jain, A.: Shape spectrum based view grouping and matching of 3D free-form objects. T-PAMI **19**, 1139–1145 (1997)
12. Fayad, J., Russell, C., Agapito, L.: Automated articulated structure and 3D shape recovery from point correspondences. In: ICCV (2011)
13. Sharma, A., Horaud, R., Cech, J., Boyer, E.: Topologically-robust 3D shape matching based on diffusion geometry and seed growing. In: CVPR (2011)
14. Bronstein, A., Bronstein, M., Kimmel, R.: Three-dimensional face recognition. Int. J. Comput. Vis. **64**, 5–30 (2005)
15. Gökberg, B., Irfanoglu, M., Akarun, L.: 3D shape-based face representation and feature extraction for face recognition. Image Vis. Comput. **24**, 857–869 (2006)
16. Papazov, C., Burschka, D.: An efficient RANSAC for 3D object recognition in noisy and occluded scenes. In: Kimmel, R., Klette, R., Sugimoto, A. (eds.) ACCV 2010, Part I. LNCS, vol. 6492, pp. 135–148. Springer, Heidelberg (2011)
17. Drost, B., Ulrich, M., Navab, N., Ilic, S.: Model globally, match locally: Efficient and robust 3D object recognition. In: CVPR (2010)
18. Detry, R., Pugeault, N., Piater, J.: A probabilistic framework for 3D visual object representation. T-PAMI **31**, 1790–1803 (2009)
19. Baseski, E., Pugeault, N., Kalkan, S., Kraft, D., Wörgötter, F., Krüger, N.: A scene representation based on multi-modal 2d and 3d features. In: ICCV Workshop on 3D Representation for Recognition (2007)

20. Knopp, J., Prasad, M., Willems, G., Timofte, R., Van Gool, L.: Hough transform and 3D SURF for robust three dimensional classification. In: Daniilidis, K., Maragos, P., Paragios, N. (eds.) ECCV 2010, Part VI. LNCS, vol. 6316, pp. 589–602. Springer, Heidelberg (2010)
21. Tombari, F., Salti, S., Di Stefano, L.: Unique signatures of histograms for local surface description. In: Daniilidis, K., Maragos, P., Paragios, N. (eds.) ECCV 2010, Part III. LNCS, vol. 6313, pp. 356–369. Springer, Heidelberg (2010)
22. Pham, M.T., Woodford, O., Perbert, F., Maki, A., Stenger, B., Cipolla, R.: A new distance for scale-invariant 3D shape recognition and registration. In: ICCV (2011)
23. Zaharescu, A., Boyer, E., Horaud, R.: Keypoints and local descriptors of scalar functions on 2d manifolds. Int. J. Comput. Vis. **100**, 78–98 (2012)
24. Bo, L., Lai, K., Ren, X., Fox, D.: Object recognition with hierarchical kernel descriptors. In: CVPR (2011)
25. Sun, J., Ovsjanikov, M., Guibas, L.: A concise and provably informative multi-scale signature based on heat diffusion. In: Eurographics Symposium on Geometry Processing (2009)
26. Bronstein, A., Bronstein, M., Guibas, L., Ovsjanikov, M.: Shape google: geometric words and expressions for invariant shape retrieval. ACM Trans. Graph. **30**, 1–20 (2011)
27. Ahmed, N., Theobalt, C., Rössl, C., Thrun, S., Seidel, H.P.: Dense correspondence finding for parameterization-free animation reconstruction from video. In: CVPR (2008)
28. Mian, A., Bennamoun, M., Owens, R.: On the repeatability and quality of key-points for local feature-based 3D object retrieval from cluttered scenes. Int. J. Comput. Vis. **89**, 348–361 (2010)
29. Lee, S., Lu, Z., Kim, H.: Probabilistic 3D object recognition with both positive and negative evidences. In: ICCV (2011)
30. Hu, W., Zhu, S.C.: Learning a probabilistic model mixing 3d and 2d primitives for view invariant object recognition. In: CVPR (2010)
31. Kang, H., Hebert, M., Kanade, T.: Discovering object instances from scenes of daily living. In: ICCV (2011)
32. Krüger, N., Janssen, P., Kalkan, S., Lappe, M., Leonardis, A., Piater, J., Rodriguez-Sanchez, A., Wiskott, L.: Deep hierarchies in the primate visual cortex: what can we learn for computer vision? IEEE PAMI **35**, 1847–1871 (2013)
33. Fidler, S., Boben, M., Leonardis, A.: Similarity-based cross-layered hierarchical representation for object categorization. In: CVPR (2008)
34. Mutch, J., Lowe, D.: Object class recognition and localization using sparse features with limited receptive fields. Int. J. Comput. Vis. **80**, 45–57 (2008)
35. Pugeault, N., Wörgötter, F., Krüger, N.: Accumulated visual representation for cognitive vision. In: BMVC (2008)
36. Chaudhuri, B., Sarkar, N.: Texture segmentation using fractal dimension. T-PAMI **17**, 72–76 (1995)
37. Felsberg, M., Sommer, G.: Image features based on a new approach to 2D rotation invariant quadrature filters. In: Heyden, A., Sparr, G., Nielsen, M., Johansen, P. (eds.) ECCV 2002, Part I. LNCS, vol. 2350, pp. 369–383. Springer, Heidelberg (2002)
38. Hartley, R., Zisserman, A.: Multiple View Geometry in Computer Vision. Cambridge University Press, Cambridge (2003)
39. Chum, O., Matas, J.: Optimal randomized RANSAC. T-PAMI **30**, 1472–1482 (2008)

40. Umeyama, S.: Least-squares estimation of transformation parameters between two point patterns. T-PAMI **13**, 376–380 (1991)
41. Xue, Z., Kasper, A., Zoellner, J., Dillmann, R.: An automatic grasp planning system for service robots. In: ICAR (2009)
42. Frome, A., Huber, D., Kolluri, R., Bülow, T., Malik, J.: Recognizing objects in range data using regional point descriptors. In: Pajdla, T., Matas, J.G. (eds.) ECCV 2004. LNCS, vol. 3023, pp. 224–237. Springer, Heidelberg (2004)
43. Belongie, S., Malik, J., Puzicha, J.: Shape matching and object recognition using shape context. T-PAMI **24**, 509–522 (2002)

3D Reconstruction of Planar Surface Patches: A Direct Solution

József Molnár, Rui Huang, and Zoltan Kato[✉]

Institute of Informatics, University of Szeged,
Arpad ter 2, Szeged 6720, Hungary
kato@inf.u-szeged.hu

Abstract. We propose a novel solution for reconstructing planar surface patches. The theoretical foundation relies on variational calculus, which yields a closed form solution for the normal and distance of a 3D planar surface patch, when an affine transformation is known between the corresponding image region pairs. Although we apply the proposed method to projective cameras, the theoretical derivation itself is not restricted to perspective projection. The method is quantitatively evaluated on a large set of synthetic data as well as on real images of urban scenes, where planar surface reconstruction is often needed. Experimental results confirm that the method provides good reconstructions in real-time.

1 Introduction

Wide baseline multi view stereo has important role in image-based urban scene reconstruction [1]. Classical approaches are either based on sparse point correspondences or dense stereo matching [2]. Then a 3D point cloud is obtained, which is the basis for scene objects' mesh modeling. Recently Poisson surface reconstruction [3] became widely used for this purpose. This method uses point coordinates as well as normal vectors to construct a smooth and detailed polygon mesh. Recently, region-based methods has been gaining more attention, in particular affine invariant detectors [4]. This affine invariance is closely related to the normal of the observed surface patch as we will see in this paper in a general context.

The most frequently used volumetric 3D object representation obtained by space carving [5] or variational level set methods [6] requires bounded objects. The accuracy of the reconstruction is determined by the resolution of the spatial grid used to define the smallest distinguishable elements. These methods would not fit for large open scenes. Multiple depth map [7] is a possible alternative 3D object representation, but it requires complicated registration steps in a later stage assuring the consistency and accuracy. Patch-based scene representation is proved to be efficient [8] and consistent with region-based correspondence-search methods.

The importance of piecewise planar object representation in 3D stereo has been recognized by many researchers. Habbecke and Kobbelt used a small plane, called 'disk', for surface reconstruction [9,10]. They proved that the normal is a linear function of the camera matrix and homography. By minimizing the difference of the warped images, the surface is reconstructed. In this paper, we give a

© Springer International Publishing Switzerland 2015
C.V. Jawahar and S. Shan (Eds.): ACCV 2014 Workshops, Part I, LNCS 9008, pp. 286–300, 2015.
DOI: 10.1007/978-3-319-16628-5_21

closed form solution to surface normal and distance. Kannala and Brandt also started from a seed region which is obtained by point detector or blob detector [11]. An affine transformation is then applied to the seed region for further propagation. In our method, we determine planar perspective transformation which provides the surface normal and distance in a closed form. Furukawa proposed using a small patch for better correspondence [12]. The surface is then grown with the expansion of the patches. The piecewise planar stereo method of Sinha et al. [13] uses shape from motion to generate an initial point cloud, then a best fitting plane is estimated, and finally an energy optimization problem is solved by graph cut for plane reconstruction. Combining the work by Furukawa and Sinha [12,13], Kowdle et al. introduced learning and active user interaction for large plane objects [14]. Hoang et al. also started from a point cloud [15] which was subsequently used for creating a visibility consistent mesh. In our approach, planes are directly reconstructed from image region(s) rather than a point cloud. Fraundorfer et al. [16] used MSER regions to establish corresponding regions pairs. Then a homography is calculated using SIFT detector inside the regions. Planar regions are then grown until the reprojection error is small. Zhou et al. assumed the whole image is a planar object, and proposed a short sequence SFM framework called TRASAC [17]. The homography is calculated using optical flow. Although the role of planar regions in 3D reconstruction has been noticed by many researchers, the final reconstruction is still obtained via triangulation for most state-of-the-art methods. Planar objects are only used for better correspondences or camera calibration.

In this paper we will develop a direct method to reconstruct whole planar patches using only the camera matrices and an affine or homography map between the image region pairs corresponding to the 3D scene patch. Since we use the correspondence-less approach of Domokos et al. [18] to estimate planar homography directly between image regions, our method doesn't require any point correspondences between stereo image pairs. Another important advantage of the proposed method is its real-time performance due to the closed form solution while also maintaining robustness. This opens the way to use our reconstruction algorithm on mobile or embedded devices.

The theoretical derivation of the general formula for 3D plane reconstruction is presented in Sect. 2, where we also discuss numerical stability of the formulas based on geometric consideration, and a simple recipe is also proposed to avoid unstable situations. Section 3 contains comprehensive numerical test results both for normal and distance calculation using synthetic and real data.

2 Normal and Distance Computation

We now derive a simple, closed form solution to reconstruct the normal and distance of a 3D planar surface patch from a pair of corresponding image regions and the camera matrices. Although differential geometric approaches were used to solve various problems in projective 3D reconstruction, the approach proposed here is unique to the best of our knowledge. For example, [19,20] are about generic surface normal reconstruction using point-wise orientation- or spatial frequency disparity maps, while our method avoids point correspondences and reconstructs both normal and distance of a planar surface from the induced planar homography between image regions. Unlike [19,20], which consider only projective camera and uses a parameterization dependent, non-invariant representation; we use a very general camera model and invariant representation.

The notation used in this section follows [21] and is widely used in continuum mechanics and classical differential geometry. For vectors and tensors we use bold letters. We use the symbol "·" for dot product, between tensors (in their matrix representation this is the usual matrix-matrix product). A simple sequence of vectors represents their dyadic product. The transpose of a dyad is the reversed sequence of the constituent vectors. A short "dictionary" is provided here for quick reference: $\mathbf{a}^T\mathbf{b} \to \mathbf{a} \cdot \mathbf{b}$, $\mathbf{ab}^T \to \mathbf{ab}$, $\mathbf{Ab} \to \mathbf{A} \cdot \mathbf{b}$, $\mathbf{b}^T\mathbf{A} \to \mathbf{b} \cdot \mathbf{A}$, $\mathbf{AB} \to \mathbf{A} \cdot \mathbf{B}$, where $\mathbf{a}, \mathbf{b}$ are column vectors and $\mathbf{A}, \mathbf{B}$ are second order tensors represented by two-dimensional matrices. Note that in this notation $(\mathbf{ab})^T = \mathbf{ba}$.

2.1   Basic Equations for Normal Computation

Herein, after briefly summarizing the theoretical backround based on [21], we will show how these results can be applied to compute the normal of a 3D scene plane from corresponding observed image regions. Let us consider the visible part of the scene objects as reasonably smooth surfaces embedded into the ambient 3D space. An image of the scene is a 3D-2D mapping given by two smooth projection functions: $x = x(X, Y, Z)$, $y = y(X, Y, Z)$, with x, y being the image coordinates. Hereafter we don't assume any special form of these coordinate-functions, except their differentiability w.r.t. spatial coordinates X, Y, Z of a world coordinate system given in standard basis $\mathbf{i}$, $\mathbf{j}$, $\mathbf{k}$. For the surface representation we use general Gauss-coordinates:

$$\mathbf{S}(u, v) = X(u, v)\mathbf{i} + Y(u, v)\mathbf{j} + Z(u, v)\mathbf{k} \tag{1}$$

If the projected spatial points are on the surface too, the image coordinates depend on the general parameters as well:

$$x(u, v) = x(X(u, v), Y(u, v), Z(u, v))$$
$$y(u, v) = y(X(u, v), Y(u, v), Z(u, v)) \tag{2}$$

We suppose that the surface point $\mathbf{u_0} = \begin{bmatrix} u_0 & v_0 \end{bmatrix}^T$ with a neighborhood constituting a small open patch are visible, therefore its mapping to the camera image is a bijection. The differential $d\mathbf{u} = \begin{bmatrix} du & dv \end{bmatrix}^T$ represents a point shift on the surface with its effect on the image being $d\mathbf{x} \approx \mathbf{J} \cdot d\mathbf{u}$ where $d\mathbf{x} = \begin{bmatrix} dx & dy \end{bmatrix}^T$ and the Jacobian $\mathbf{J}$ of the mapping is invertible. Now consider a stereo camera pair (distinguishing them with indices i, j). Since $\mathbf{J}$ is invertible, we can establish correspondences between the images having the same point-shift $d\mathbf{u} = \mathbf{J}_i^{-1} \cdot d\mathbf{x}_i$:

$$d\mathbf{x}_j = \mathbf{J}_j.\mathbf{J}_i^{-1} \cdot d\mathbf{x}_i = \mathbf{J}_{ij} \cdot d\mathbf{x}_i \tag{3}$$

where $\mathbf{J}_{ij}$ is the Jacobian of the $\mathbf{x}_i \rightarrow \mathbf{x}_j$ mapping. Considering the derivative of a composite function f

$$\frac{\partial f}{\partial u} = \frac{\partial X}{\partial u}\frac{\partial f}{\partial X} + \frac{\partial Y}{\partial u}\frac{\partial f}{\partial Y} + \frac{\partial Z}{\partial u}\frac{\partial f}{\partial Z} = \mathbf{S}_u \cdot \nabla f \tag{4}$$

where ∇f is the gradient of f w.r.t. the spatial coordinates, and $\mathbf{S}_u$ is the local basis vector alongside parameter line u. Applying this result to the projection functions, the Jacobians take the following form:

$$\mathbf{J}_k = \begin{bmatrix} \mathbf{S}_u \cdot \nabla x_k & \mathbf{S}_v \cdot \nabla x_k \\ \mathbf{S}_u \cdot \nabla y_k & \mathbf{S}_v \cdot \nabla y_k \end{bmatrix}, \quad k = i, j \tag{5}$$

After substitution, the products of the above quantities appear in $\mathbf{J}_{ij}$. For example, the determinant

$$\det(\mathbf{J}_i) = (\mathbf{S}_u \cdot \nabla x_i)(\mathbf{S}_v \cdot \nabla y_k) - (\mathbf{S}_v \cdot \nabla x_i)(\mathbf{S}_u \cdot \nabla y_i) \tag{6}$$

which can be expressed by dyadic products equivalent to the surface normal's cross-tensor as

$$\det(\mathbf{J}_i) = \nabla x_i \cdot (\mathbf{S}_u\mathbf{S}_v - \mathbf{S}_v\mathbf{S}_u) \cdot \nabla y_i$$
$$= -\nabla x_i \cdot [\mathbf{N}]_\times \cdot \nabla y_i = -|\mathbf{N}||\nabla x_i \mathbf{n} \nabla y_i|, \tag{7}$$

where $\mathbf{N}$ is the surface normal, $\mathbf{n}$ is the unit normal, and $|\nabla x_i \mathbf{n} \nabla y_i|$ is the triple scalar product of the gradients and the normal. Finally, we get [21]

$$\mathbf{J}_{ij} = \frac{1}{|\nabla x_i \mathbf{n} \nabla y_i|} \begin{bmatrix} |\nabla x_j \mathbf{n} \nabla y_i| & |\nabla x_i \mathbf{n} \nabla x_j| \\ |\nabla y_j \mathbf{n} \nabla y_i| & |\nabla x_i \mathbf{n} \nabla y_j| \end{bmatrix} \tag{8}$$

The above quantities are all invariant first-order differentials: the gradients of the projections and the surface unit vector. Note that (8) is a general formula: neither a special form of projections, nor a specific surface is assumed here, hence it can be applied for any camera type and for any reasonably smooth surface.

The formula derived above can be used for different purposes:

1. an affine transformation can be established between the images of a known surface using known projection functions;
2. if the projections are known and the parameters of the affine mapping acting between corresponding regions of a stereo image pair are estimated, then the normal of the corresponding 3D surface patch can be computed;
3. if the 3D surface normal is known and the affine mapping parameters are estimated, then the gradients of one of the projection functions can be computed.

Case (1) is addressed in [21]. Herein, we will show how to use this formula in case (2) for normal vector computing. Let us write the matrix components - estimated either directly with affine estimator or taking the derivatives of an estimated homography - with:

$$\mathbf{J}_{ij\,est} = \begin{bmatrix} a_{11} & a_{12} \\ a_{21} & a_{22} \end{bmatrix} \tag{9}$$

To eliminate the common denominator we may use ratios, which can be constructed using either row, column, or cross ratios. Without loss of generality, we deduce the equation for the 3D surface normal using cross ratios:

$$\frac{\mathbf{n} \cdot (\nabla y_i \times \nabla x_j)}{\mathbf{n} \cdot (\nabla y_j \times \nabla x_i)} = \frac{a_{11}}{a_{22}}, \quad \frac{\mathbf{n} \cdot (\nabla x_j \times \nabla x_i)}{\mathbf{n} \cdot (\nabla y_i \times \nabla y_j)} = \frac{a_{12}}{a_{21}} \tag{10}$$

After rearranging:

$$\mathbf{n} \cdot [a_{22} (\nabla y_i \times \nabla x_j) - a_{11} (\nabla y_j \times \nabla x_i)] = 0$$
$$\mathbf{n} \cdot [a_{21} (\nabla x_j \times \nabla x_i) - a_{12} (\nabla y_i \times \nabla y_j)] = 0 \tag{11}$$

Here we have two (known) vectors, both perpendicular to the normal:

$$\mathbf{p} = [a_{22} (\nabla y_i \times \nabla x_j) - a_{11} (\nabla y_j \times \nabla x_i)]$$
$$\mathbf{q} = [a_{21} (\nabla x_j \times \nabla x_i) - a_{12} (\nabla y_i \times \nabla y_j)] \tag{12}$$

Thus the 3D surface normal can readily be computed as

$$\mathbf{n} = \frac{\mathbf{p} \times \mathbf{q}}{|\mathbf{p} \times \mathbf{q}|}$$

2.2   Specialization to Perspective Camera

Let us now apply our general results to the case of perspective cameras. The camera matrix of the i-th camera $\mathbf{P}^{(i)}$ is a 3×4 rank 3 matrix with row vectors $\pi_k^{(i)T} = \begin{bmatrix} p_{k1}^{(i)} & p_{k2}^{(i)} & p_{k3}^{(i)} & p_{k4}^{(i)} \end{bmatrix}$, $k = 1, 2, 3$. Furthermore, spatial coordinates are represented as homogeneous four-vectors $\hat{\mathbf{X}} = \begin{bmatrix} X & Y & Z & 1 \end{bmatrix}^T$ and the projection functions become rational functions due to projective division:

$$x_i = \frac{\pi_1^{(i)} \cdot \hat{\mathbf{X}}}{s_i}, \quad y_i = \frac{\pi_2^{(i)} \cdot \hat{\mathbf{X}}}{s_i} \tag{13}$$

with $s_i = \pi_3^{(i)} \cdot \hat{\mathbf{X}}$. Using these notations, the gradients become

$$\nabla x_i = \frac{1}{s_i} \begin{bmatrix} p_{11}^{(i)} - p_{31}^{(i)} x_i & p_{12}^{(i)} - p_{32}^{(i)} x_i & p_{13}^{(i)} - p_{33}^{(i)} x_i \end{bmatrix}^T$$
$$\nabla y_i = \frac{1}{s_i} \begin{bmatrix} p_{21}^{(i)} - p_{31}^{(i)} y_i & p_{22}^{(i)} - p_{32}^{(i)} y_i & p_{23}^{(i)} - p_{33}^{(i)} y_i \end{bmatrix}^T \tag{14}$$

Observing that each coefficient composed by cross product has exactly one gradient of projection i and one gradient of projection j, the scaled vectors $\mathbf{P} = s_i s_j \mathbf{p}$ and $\mathbf{Q} = s_i s_j \mathbf{q}$ yields the same result

$$\mathbf{n} = \frac{\mathbf{P} \times \mathbf{Q}}{|\mathbf{P} \times \mathbf{Q}|}$$

with denominators s_i, s_j eliminated.

2.3 Using Homography

It is well known from projective geometry that images of a planar surface patch are related by planar homography, which is given by a 3×3 matrix realizing the mapping between homogeneous coordinates. It follows that if this matrix is known, one can accurately determine the affine parameters calculating the homography's partial derivatives. Denoting the components of the homography matrix $\mathbf{H}_{ij}$ with h_{kl} $(k, l = 1, 2, 3)$ acting between images i and j, the elements of $\mathbf{J}_{ij}$ become

$$a_{11} = \frac{1}{r}\left(h_{11} - h_{31}x_j\right), \; a_{12} = \frac{1}{r}\left(h_{12} - h_{32}x_j\right)$$

$$a_{21} = \frac{1}{r}\left(h_{21} - h_{31}y_j\right), \; a_{22} = \frac{1}{r}\left(h_{22} - h_{32}y_j\right) \tag{15}$$

with scale factor $r = h_{31}x_i + h_{32}y_i + h_{33}$.

2.4 Discussion

For the sake of simplicity we suppose that our cameras have zero skew (*e.g.* camera with CCD sensor), hence the calibration matrix is

$$\mathbf{K} = \begin{bmatrix} \alpha & 0 & x_0 \\ 0 & \beta & y_0 \\ 0 & 0 & 1 \end{bmatrix} \tag{16}$$

The projection functions expressed in the camera coordinate system are then $x = \alpha \frac{X}{Z} + x_0$, $y = \beta \frac{Y}{Z} + y_0$ and the gradients (scaled by the common multipliers) become

$$\frac{1}{\alpha}Z^2 \nabla x = \begin{bmatrix} Z & 0 & -X \end{bmatrix}^T$$

$$\frac{1}{\beta}Z^2 \nabla y = \begin{bmatrix} 0 & Z & -Y \end{bmatrix}^T \tag{17}$$

This result shows that ∇x is on the $Y = 0$ plane relative to camera (*i.e.* perpendicular to its momentary "up-down" direction), and perpendicular to the direction of the object's projection onto that plane. Similarly, ∇y is on the $X = 0$ plane (*i.e.* perpendicular to its momentary "left-right" direction), and perpendicular to the direction of the object's projection onto that plane.

Since we use these gradients in cross products, the "perpendicularities on the planes clause" can be lifted and finally we have a very important condition for the cross products involved in (8): they admit parallelism *if and only if the two camera centers and the observed point of interest are on the same line*. Note that in the case of video sequences, this parallax-less condition renders the motion toward the observed object as critical motion. Furthermore, the basic equations (8) include triple scalar products with surface normal involved. The parallelism of any gradient and the normal (the case to be excluded) means that the observed point is imaged as contour point.

Nevertheless, in practice the following algebraic consideration is usually sufficient. As discussed in Sect. 2.1, the normal can be expressed by three different ratios. Of course, in theory these ratios yield exactly the same normal vector: Clearly, taking row ratios is equivalent to column ratios of the inverse transformation and vice versa; while cross ratios are equivalent for both. In practice, however, affine parameters are subject to noise, inherent to any image processing algorithm, causing slight numerical differences in the normals provided by the 3 ratios. To choose the numerically most stable one, we recommend to follow this three-step procedure:

1. Determine the estimated transformation's Jacobian (see (9)) and choose the two components having the smallest absolute values
2. If these values are both significantly less than the next in order (element having the 3-rd smallest absolute value), then the equations expressed with that particular ratios (i.e. where either $\mathbf{P}$ or $\mathbf{Q}$ are close to zero vector) should be excluded from step (3)
3. Choose the expression serving the biggest weighted value for $|\mathbf{P} \times \mathbf{Q}|$.

The weight we recommend is based on our numerical experience and not yet theoretically investigated. According to that, expression based on cross ratio seems to be the most reliable and accurate in practice. Therefore if it is not dropped prior to step (3), we recommend its weighting with a greater number than 1 (say 4) as its "effective" magnitude for comparison. If more than two cameras are involved then we can repeat the above procedure to choose the most favorable camera pair.

2.5   Distance Calculation

While for the surface normal, only an affine mapping is needed between the image pairs, knowing the normal and the plane-induced homography allows us to determine the distance from an observed planar patch too. It is well known that a plane-induced homography encapsulates the plane's unit normal and perpendicular distance from the origin [2]. Furthermore, our homography matrix is a homogeneous entity, therefore the ratio of its any two components gives one equation for distance - leading to a highly overdetermined system which can be solved in the least square sense. Within the cameras' relative coordinate system, the world coordinate system can be canonically attached to one of the cameras, and the transformation between normalized points $\mathbf{X}_i = \mathbf{K}_i^{-1}\mathbf{x}_i$ in camera i and $\mathbf{X}_j = \mathbf{K}_j^{-1}\mathbf{x}_j$ in camera j can be described as

$$\mathbf{X}_j = (\mathbf{R} + \frac{1}{d}\mathbf{tn}) \cdot \mathbf{X}_i \tag{18}$$

where d is the perpendicular distance of the plane to the camera center i, $\mathbf{R}$ and $\mathbf{t}$ are relative rotation and translation of the two camera coordinate frames, and $\mathbf{n}$ is the normal of the 3D plane. Using homogeneous coordinates, the above equation is satisfied up to an arbitrary non-zero scale factor, hence the homography

H can be expressed as

$$\mathbf{H} = d\mathbf{It} + t_n \tag{19}$$

Note that the only unknown of the above equation is d. In order to set the scale of the above relation, the last element of the homography matrix can be fixed to 1 by dividing **H** with its last element, assuming it is non-zero. If it would be 0 – which is theoretically possible – then **H** would map points to infinity, which is usually excluded by physical constraints in real applications.

When the camera poses are given in an arbitrary world coordinate frame, then relative rotation and translation can be computed as

$$\mathbf{R} = \mathbf{R}_j \cdot \mathbf{R}_i^{-1}$$
$$\mathbf{t} = \mathbf{R}_j \cdot (\mathbf{C}_i - \mathbf{C}_j) \tag{20}$$

where $\mathbf{R}_i$ and $\mathbf{R}_j$ are the orientations while $\mathbf{C}_i$ and $\mathbf{C}_j$ are the positions of camera i and j in the world coordinate system. Furthermore, the surface normal in relative coordinates can be expressed in terms of its world coordinates $\mathbf{n}_w$ as $\mathbf{n} = \mathbf{R}_i \cdot \mathbf{n}_w$, and the distance $d = d_w - \mathbf{n}_w \cdot \mathbf{C}_i$, where d_w is the distance expressed in the world coordinate frame. Finally, when the estimated homography $\mathbf{H}_{i,j}$, mapping the corresponding regions from camera i to j, is given in unnormalized image coordinates, then $\mathbf{H}_{i,j} = \mathbf{K}_j \mathbf{H} \mathbf{K}_i^{-1}$, where **H** is from (19). We thus get the following general relation between the homography $\mathbf{H}_{i,j}$, camera and plane relative poses:

$$\mathbf{H}_{i,j} \cong (d_w - \mathbf{n}_w \cdot \mathbf{C}_i) \cdot \mathbf{R}_j \cdot \mathbf{R}_i^{-1} + \mathbf{R}_j \cdot (\mathbf{C}_i - \mathbf{C}_j)(\mathbf{R}_i \cdot \mathbf{n}_w) \tag{21}$$

The only unknown in the above equation is d_w, which can be obtained by minimizing the geometric error of the transferred points over the image regions:

$$\arg\min_{d_w} = \sum_p \|\mathbf{H}_{i,j}\mathbf{p} - \mathbf{A}\mathbf{p}\|^2 \tag{22}$$

where **A** is the right hand side of (21). The minimizer of the above expression is easily obtained in a closed form as the position of the zero first order derivative w.r.t. d_w.

3   Experimental Results

The proposed method was tested on an Intel i7 3.4 GHz CPU with 8 GB memory. A total of 300 synthetic examples were generated by selecting 15 templates introduced by [18]. The camera intrinsic matrix was derived from a real world camera. The extrinsic parameters were randomly set with the orientation between $-\pi/6$ to $\pi/6$ for each axis, the translation chosen from -20 to 20 in x and y directions, and from -10 to -20 in z directions. The z component was set to be negative so the scene was in front of the camera. The normal of the plane was a random selection with the only assumption that it points out of the image plane.

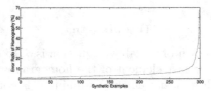

Fig. 1. Homography error for our synthetic dataset (the test cases are sorted on the x-axis).

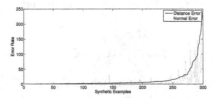

Fig. 2. Distance error and normal error plot for our synthetic dataset (test cases are sorted on the x-axis based on distance error)

The first step of our algorithm is homography estimation between the corresponding region pairs. For that purpose, we use the correspondence-less method of [18] using the publicly available implementation. For a detailed evaluation of the method, see [18]. For reference, we show the homography error on our synthetic dataset in terms of the percentage of non overlapping area sorted in increasing order in Fig. 1. The registration method has less than 5 % error for more than 250 examples. Obviously, this error directly affects the reconstruction error of our method - as we will see later.

Once the planar homography between the corresponding region pair is estimated, we can compute the 3D surface normal and distance using the closed form formulas derived in Sect. 2. A sample 3D reconstruction for synthetic data is shown in Fig. 3. The red surface is the ground truth surface and the blue one is the recovered surface. We also show the error map of the reconstruction. The color bar gives the index of the distance error in percentage - the error rate was less than 0.3%. The different colours also indicate that the two normals of the two surfaces are not perfectly parallel. Figure 2 shows the error plots for the whole synthetic dataset. It is clear that distance error plot runs together with the normal error, hence our method provides reliable reconstructions for most test cases, giving low error rates for both surface parameters.

It is important to note that the proposed method can reach real time speed due to the closed form solution of the surface parameters.

3.1 Comparison with Classical Methods

Herein, we perform an experimental comparison of well known classical plane reconstruction methods and quantitatively demonstrate the performance of our

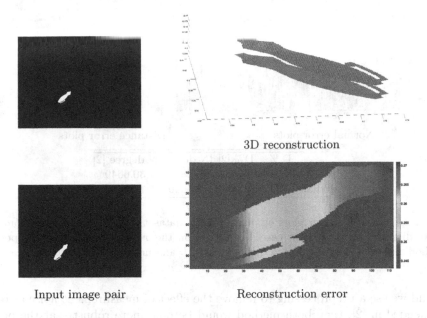

3D reconstruction

Input image pair   Reconstruction error

Fig. 3. 3D reconstruction of a synthetic image pair

method with respect to these algorithms on our synthetic dataset. The comparison is done first with the plane from homography method described by Hartley and Zisserman [2] for accuracy of the plane parameters, and then with the triangulation method of Fraundorfer *et al.* [16] for reconstruction accuracy. We remark, that Fraundorfer *et al.* originally use Harris corners for point correspondences to estimate homography in [16]. While the accuracy of their homography estimation (around 6%) is comparable to our method, we used the same homography for all methods to guarantee a fair comparison.

In Fig. 4, distance error and normal error plots are shown for the proposed method and the plane from homography direct method [2] (the Matlab code is available from http://www.robots.ox.ac.uk/~vgg/hzbook/code/codevgg_plane_from_2P_H.m). The purpose of this experiment is to compare our direct method derived via differential geometric considerations with a classical direct methods derived via projective geometric considerations, as a basis. Of course, in our experiments, we work with the estimated scene plane induced homography, which is theoretically correct but subject to numerical errors (see Fig. 1 for the homography errors). More than 200 examples gave less than 2.5 % in distance error and 5 degree in normal error. Indeed, the proposed method performs an order of magnitude better than the classical method. Let us stress again that both methods used exactly the same input, so the results show the (very) different behaviour of these direct formulas in case of realistic image measurements! These experiments show that our direct method can tolerate slight errors in the homography, while the formula obtained via projective geometry is extremely sensitive to the smallest amount of numerical error, as it is also noted in [2].

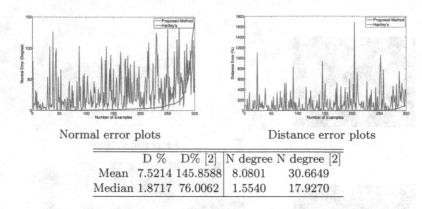

	D %	D% [2]	N degree	N degree [2]
Mean	7.5214	145.8588	8.0801	30.6649
Median	1.8717	76.0062	1.5540	17.9270

Fig. 4. Comparative error plots on our synthetic dataset with the plane from homography direct method [2] (test cases are sorted on the x-axis based on the proposed method's error). The table shows distance error D and normal error N statistics.

Would we use a preprocessing to remove the effect of measurement noise as recommended in [2], then both method would become more robust – at the price of an increased computational compexity, of course.

Table 1. Reconstruction accuracy in terms of 3D point distance

	Distance	Distance [16]
Mean	0.0358	0.2095
Median	0.0350	0.2008

Table 1 shows the mean and median 3D distance error of reconstructed points for the proposed method and the triangulation method of Fraundorfer *et al.* [16]. The proposed method performs again and order of magnitude better. In addition, our method recovers the whole surface patch in one step, while [16] gives only a point cloud for matched point pairs.

3.2 Robustness

Table 2. Normal error w.r.t. rotation error in different axes

Noise %	0	1	2	5	10	15	20
x	3.325	5.222	7.861	17.73	33.54	45.23	53.06
y	3.325	5.414	8.81	20.03	36.79	51.79	67.03
z	3.325	4.638	6.997	15.17	29.73	43.72	53.51

Table 3. Distance error w.r.t. rotation error in different axes

Noise %	0	1	2	5	10	15	20
x	1.706	2.642	3.891	8.753	15.91	19.6	20.69
y	1.706	2.624	4.072	9.495	16.48	20.98	23.92
z	1.706	2.005	3.501	6.703	14.14	19.13	20.55

Table 4. Distance error w.r.t. translation error

Noise %	0	1	2	5	10	15	20
	1.706	2.264	3.362	7.109	14.53	22.27	29.75

The accuracy of the proposed method depends not only on the quality of homography estimation, but also on the camera pose parameters which are used to compute relative rotation and translation as described in Sect. 2. Obviously, normal estimation is only affected by the rotation matrix, while distance calculation depends on both rotation and translation. To characterize the robustness of our method against errors in these parameters, we added various percent of noise to the original values and quantitatively evaluated the reconstruction error on our synthetic dataset. Tables 2 and 3 show that normal is slightly more sensitive to this type of error, but its error is still below 10 % up to 2 % noise. Distance estimation can tolerate up to 5 % noise in both rotation and translation.

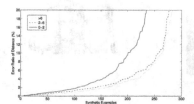 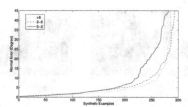

Distance error w.r.t. different baselines Normal error w.r.t. different baselines

Fig. 5. Error plots w.r.t. different baselines (test cases are sorted on the x-axis based on the error).

Baseline is also an important parameter for 3D reconstruction. Short baseline is often seen in short sequence images such as video. With the distance to the plane set to be around 15 m, 3 different baselines were tested. The shortest baseline is within 0–2 m, the medium one is between 2–6 m, and large baseline is considered larger than 6 m. Figure 5 shows the error with respect to each baseline range. Of course, shorter baseline has higher error rate, which is a well known fact for stereo reconstruction. In addition, our method is also affected by larger homography error in case of decreasing baseline (see Fig. 6). Nevertheless, the proposed method still have robust performance within a large range of baselines.

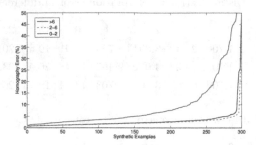

Fig. 6. Homography error w.r.t. different baselines

Fig. 7. 3D reconstruction result using MSER regions

Fig. 8. 3D reconstruction results using regions extracted by color-based clustering

Finally, we test our method on real world objects. There are various ways to extract corresponding regions from real image pairs. For example, we can use standard MSER regions - such a reconstruction result is presented in Fig. 7. Another possibility is to extract larger regions corresponding to building facades using *e.g.* color-based clustering such as in Fig. 8. Note that all real examples contain various patch orientations, the color labels denote corresponding image regions (Table 4).

4   Conclusion

We proposed an efficient 3D reconstruction method, which allows the reconstruction of complete planar surface patches from a homography map between corresponding image regions and calibrated cameras. The theoretical foundation relies on variational calculus, which leads to a closed form solution for the surface normal and distance parameters. Being a direct solution, it runs in real-time which can be particularly useful for mobile and embedded vision systems. Another advantage is that it works without point correspondences by making use of segmented regions. Quantitative experiments on a large synthetic dataset confirm the superior performance w.r.t. classical plane reconstruction algorithms, while reconstruction of whole building facades from real images confirm the applicability of our approach for real-life problems. In our future work, the focus will be on reliable planar segmentation methods for urban environments.

Acknowledgement. This research was supported by the European Union and the European Social Fund through project FuturICT.hu (grant no.: TAMOP-4.2.2.C-11/1/KONV-2012-0013).

References

1. Musialski, P., Wonka, P., Aliaga, D.G., Wimmer, M., van Gool, L., Purgathofer, W.: A survey of urban reconstruction. In: EUROGRAPHICS 2012 State of the Art Reports, PP. 1–28. Eurographics Association (2012)
2. Hartley, R.I., Zisserman, A.: Multiple View Geometry in Computer Vision. Cambridge University Press, Cambridge (2004). ISBN: 0521540518
3. Kazhdan, M., Bolitho, M., Hoppe, H.: Poisson surface reconstruction. In: Proceedings of the fourth Eurographics symposium on Geometry processing, pp. 61–70. Eurographics Association (2006)
4. Mikolajczyk, K., Tuytelaars, T., Schmid, C., Zisserman, A., Matas, J., Schaffalitzky, F., Kadir, T., Gool, L.V.: A comparison of affine region detectors. Int. J. Comput. Vis. **65**, 43–72 (2005)
5. Laurentini, A.: The visual hull concept for silhouette-based image understanding. IEEE Trans. Pattern Anal. Mach. Intell. **16**, 150–162 (1994)
6. Faugeras, O., Keriven, R.: Variational principles, surface evolution, PDE's, level set methods and the stereo problem. IEEE Trans. Image Process. **7**, 336–344 (1999)
7. Fuhrmann, S., Goesele, M.: Fusion of depth maps with multiple scales. ACM Trans. Graph. **30**, 148:1–148:8 (2011)

8. Furukawa, Y., Ponce, J.: Accurate, dense, and robust multiview stereopsis. IEEE Trans. Pattern Anal. Mach. Intell. **32**, 1362–1376 (2010)

9. Habbecke, M., Kobbelt, L.: Iterative multi-view plane fitting. In: VMV06, pp. 73–80 (2006)

10. Habbecke, M., Kobbelt, L.: A surface-growing approach to multi-view stereo reconstruction. In: 2007 IEEE Conference on Computer Vision and Pattern Recognition, CVPR 2007, pp. 1–8 (2007)

11. Kannala, J., Brandt, S.: Quasi-dense wide baseline matching using match propagation. In: 2007 IEEE Conference on Computer Vision and Pattern Recognition, CVPR 2007, pp. 1–8 (2007)

12. Furukawa, Y., Ponce, J.: Accurate, dense, and robust multi-view stereopsis. In: CVPR, pp. 1362–1376 (2007)

13. Sinha, S., Steedly, D., Szeliski, R.: Piecewise planar stereo for image-based rendering. In: 2009 IEEE 12th International Conference on Computer Vision, pp. 1881–1888 (2009)

14. Kowdle, A., Chang, Y.J., Gallagher, A., Chen, T.: Active learning for piecewise planar 3d reconstruction. In: Proceedings of the 2011 IEEE Conference on Computer Vision and Pattern Recognition. CVPR 2011, Washington, DC, USA, pp. 929–936. IEEE Computer Society (2011)

15. Hiep, V.H., Keriven, R., Labatut, P., Pons, J.P.: Towards high-resolution large-scale multi-view stereo. In: 2009 IEEE Conference on Computer Vision and Pattern Recognition, CVPR 2009, pp. 1430–1437 (2009)

16. Fraundorfer, F., Schindler, K., Bischof, H.: Piecewise planar scene reconstruction from sparse correspondences. Image Vis. Comput. **24**, 395–406 (2006)

17. Zhou, Z., Jin, H., Ma, Y.: Robust plane-based structure from motion. In: Proceedings of the 2012 IEEE Conference on Computer Vision and Pattern Recognition (CVPR), CVPR 2012, Washington, DC, USA, pp. 1482–1489. IEEE Computer Society (2012)

18. Domokos, C., Nemeth, J., Kato, Z.: Nonlinear shape registration without correspondences. IEEE Trans. Pattern Anal. Mach. Intell. **34**, 943–958 (2012)

19. Devernay, F., Faugeras, O.: Computing differential properties of 3-D shapes from stereoscopic images without 3-D models. In: Proceedings of International Conference on Computer Vision and Pattern Recognition, pp. 208–213 (1994)

20. Jones, D.G., Malik, J.: Determining three-dimensional shape from orientation and spatial frequency disparities. In: Sandini, G. (ed.) ECCV 1992. LNCS, vol. 588, pp. 661–669. Springer, Heidelberg (1992)

21. Molnár, J., Chetverikov, D.: Quadratic transformation for planar mapping of implicit surfaces. J. Math. Imaging Vis. **23**(6), 1129–1139 (2012)

Deep Learning on Visual Data

Hybrid CNN-HMM Model for Street View House Number Recognition

Qiang Guo$^{(\boxtimes)}$, Dan Tu, Jun Lei, and Guohui Li

Department of Information System and Management,
National University of Defense Technology, Changsha, China
guoqiang05@nudt.edu.cn

Abstract. We present an integrated model for using deep neural networks to solve street view number recognition problem. We didn't follow the traditional way of first doing segmentation then perform recognition on isolated digits, but formulate the problem as a sequence recognition problem under probabilistic treatment. Our model leverage a deep Convolutional Neural Network(CNN) to represent the highly variable appearance of digits in natural images. Meanwhile, hidden Markov model(HMM) is used to deal with the dynamics of the sequence. They are combined in a hybrid fashion to form the hybrid CNN-HMM architecture. By using this model we can perform the training and recognition procedure both at word level. There is no explicit segmentation operation at all which save lots of labour of sophisticated segmentation algorithm design or finegrained character labeling. To the best of our knowledge, this is the first time using hybrid CNN-HMM model directly on the whole scene text images. Experiments show that deep CNN can dramaticly boost the performance compared with shallow Gausian Mixture Model(GMM)-HMM model. We obtaied competitive results on the street view house number(SVHN) dataset.

1 Introduction

Though the research of recognizing handwritten and machine printed characters have been last for several decades [1,2], recognizing text in natural scene still remains a difficult computer vision problem. Text captured in unconstrained natural scene show quite large appearance variability. They have different fonts, scales, rotations, lightening conditions etc.

Tradition methods for doing this task always fall in a separated pipeline of first segmenting the image to extract isolated characters then perform recognition on the extracted characters. Segmentation and recognition are long been considered to couple with each other which make it difficult to do either in a separated way. Meanwhile, hand designed segmentation algorithm often became fragile when facing natural scene images. Also, we need large amount of isolated characters to cover the characters variability. If the captured image is the whole text, that would need to label the bounding box of each character for training, which is a highly labor consuming work.

© Springer International Publishing Switzerland 2015
C.V. Jawahar and S. Shan (Eds.): ACCV 2014 Workshops, Part I, LNCS 9008, pp. 303–315, 2015.
DOI: 10.1007/978-3-319-16628-5_22

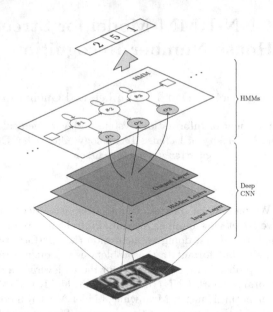

Fig. 1. Hybrid CNN-HMM Model. This image demonstrates the recognition process of our model. Input image is processed by sliding window to extract a sequence of frames. Each frame is normalised to the same scale and fed into the deep CNN which produce its posterior probability belonging to a category. This probability, after normalization, is used as the output probability of HMM. HMM is used to infer the most probable digits out of the frame sequence.

Our aim is to make the training and recognizing process performed directly on the whole text image. In this paper, we address the problem on a special natural scene text dataset of street view number images. Those images are collected from street view imagery contain only digits which make us focus on the image modeling problem other than language model.

Digits in street view number images are generally arranged in a sequential manner though sometimes in different vertical position. So we formulate the problem as a sequence recognition problem. To this end, the image is preprocessed to extract a sequence of frames which is seemed as observations at different time. In a generative way, the frames are seemed to be generated by an underlying state sequence which in this situation are digits from 0 to 9 and background clutter.

We propose a method in an unified framework integrating segmentation and recognition by hybriding hidden Markov model with deep convolutional neural networks. Figure 1 shows the architecture of our model. Sliding window is performed on the input image to extract a sequence of frames. Hiddem Markov model (HMM) is used to handle time variability and deep convolutional neural networks (CNN) deal with all kinds of character variations. Deep CNN is critical to the performance of the model for its great representation capability.

We conduct a bootstrap process to get initial labels of frames. In the bootstrap stage we train a Gaussian mixture model(GMM)-HMM model to get the initial frame assignments, then switch to train the hybrid CNN-HMM model. We show that starting with a not-so-well GMM-HMM model, we gain dramatic performance improvements after using CNN as the character model.

Similar methods have been used in speech and handwriting recognition communities, but to our knowledge, no one had investigated this model on scene text recognition task. Experiments show that deep CNN can dramaticly boost the performance compared with shallow Gausian Mixture Model(GMM)-HMM model. We obtaied competitive results on the street view house number(SVHN) dataset.

2   Related Work

Street view number recognition can fall in the category of natural scene text recognition problem. Research on scene text recognition already started in mid-90s [3], but still a not solved problem. Different from machine-printed character or handwriting recognition problem, recognizing text in natural scene has its special difficulties. Text captured in unconstrained natural scene show quite large appearance variability. They have different fonts, scales, rotations, lightening conditions and also organized in various kind of layouts.

Traditional methods deal with this problem are based on sequential character classification by either sliding window [4,5] or connected components [6,7], after which a word prediction is made by grouping character classifier predictions in a left-to-right manner. More recent works [8–10] make use of over-segmentation methods, guided by a supervised classifier to generate candidates. Words are recognized through a sequential beam search optimization over character candidates.

Often a small fixed lexicon as language model to constrain word recognition [4,5] in scene text recognition. However, the problem of recognizing street view numbers could not benefit from language model, which makes the problem more difficult.

In all these systems, character model is the most critical component. Nowadays, as Convolutional Neural Networks are showing more and more powerful capability in object recognition tasks [11–14]. Some works have used CNN to tackle scene text recognition problem [5,9,13,15]. Among these research, CNN shows great capability to represent all kinds of character variation in natural scene and still holding highly discrimitivity. Convolutional networks has successfully applied to recognition of handwritten numbers in 90s [16,17].

Another important issue is how to infer the characters from the whole image when it's difficult to isolate each single character from the image. Most of previous work on SVHN dataset recognize isolated digits, except Goodfellow etc. [15]. They solve the problem by directly using a deep large CNN to model the whole image and with a simple graphical model as the top inference layer. Inference of the position of digits in the number image is done by the network implicitly. Segmentation and recognition is performed in a unified way. However, as addressed by the authors [15], the method rest heavily on the assumption that the sequence is of bounded length, with a small number length.

Alsharif *et al.* [8] also used hybrid model, but only used HMM for segmentation. The whole process is still a combination of explicitly segmentation and character candidates recognition.

In this work, we proposed a model to combine segmentation and recognition. Our model is different from all other methods to this problem. Our character model is build upon the success of large CNN architecture on large scale image recognition task and with our tuning for the special problem. And, we use a effective graphical model – Hidden Markov Model – to inference the digits of in the image. The two parts are combined in a hybrid fashion [18].

Our model is motivated by recently successful application of deep neural networks on speech recognition [19,20]. In speech recognition community, deep learning methods have take over traditional shallow GMM-HMM model and achieve significant improvements [19–22].

3   Problem Formulation

We formulate the street view number recognition task as a sequence recognition problem.

Let I represent a street view number image which contains unknown amount of digits. We treat the whole number image as a concatenated sequence of frames arranged horizontally. The frame sequence is represented as $O = \{o_1, o_2, \cdots, o_T\}$, in which o_i corresponds to feature of the ith frame. The length of the sequence is T. Define $Y = \{y_1, y_2, \cdots, y_L\}$ as the label of the image. L is the amount of digits in the image, y_i is the ith digit's label.

We treat the frames as being generated sequentially from a Markov process that transit between states $S = \{s_1, s_2, \cdots, s_K\}$. That brings the use of hidden Markov model.

In our setting, the frames are categorized to 11 categories which contains 10 digits and a nul category, $\{0, \cdots, 9\} \cup \{\text{nul}\}$. The nul category represents non-digit frames which contain pre- or post-digit background, inter-digit interval and clutter frames.

Define $\mathcal{M}$ as the HMM model. The key parameters of $\mathcal{M}$ are the initial state probability distribution $\pi = \{p(q_0 = s_i)\}$, the transition probabilities $a_{ij} = p(q_t = s_j|q_{t_1} = s_i)$, and a model to estimate the observation probabilities $p(o_t|s_i)$. Training samples are given in the form of $\{O_i, S_i\}$, where $i \in \{1, \cdots, N\}$, N is the size of training set. Note there's no need of digits boundary information, but only their digit labels.

We first use GMM as the character model, then switch to CNN. The labels for training CNN are got by performing *forced alignment* on the frame sequence. The learning process involves learning the transition probability of $\mathcal{M}$, the mean m_i, variance σ_i of each GMM components and the parameters of CNN.

Recognition of the numbers is performed by *maximum a posteriori*(MAP) estimation. That is, given O we want to find S which satisfize

$$\hat{S} = \underset{S}{\operatorname{argmax}}\ \log P(S|O).$$

We use *Viterbi algorithm* [23] to get the most probable state sequence. After that, we eliminate all nul states to get the final recognition results.

There is no need of explicit segmentation in both training and recognition. Actually, during recognition and forced alignment, segmentation is implicitly performed within the inference process. To the user, the supplied data are only the whole number images and labels of all digits in the image, no position information is needed.

4   Hybrid CNN-HMM Model

4.1   Model Architecture

The whole architecture of the hybrid CNN-HMM model is showed in Fig. 1.

We use HMM to model the dynamics of frame sequences and deep Convolutional Neural Network to model the frame appearance. Specifically, CNN is used to approximate the emission probability $p(o_t|s_t)$ of each frame under a given state. These two components constitute the hybrid CNN-HMM model.

To train CNN we need to assign each frame a label to tell which state it belongs to. So we conduct a bootstrap process. A GMM-HMM model is trained to initialize the transition probabilities among HMM states and give an initial label assignment to each frame. This is achieved by doing forced alignment with the initial GMM-HMM model. The GMM-HMM model could give a not-perfect, but somehow, reasonable alignment.

Each frame category has a corresponding HMM, that means we have 11 HMMs in total. We use 3-state HMM to model each digit and 1-state HMM for background clutter frames. The 3-state HMM models pre-digit, mid-digit and post-digit frames. Figure 2 shows the structure of the HMM. Circle nodes represent hidden (without filling) and observation states (with filling). Note we

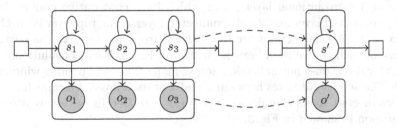

Fig. 2. HMM models. The left is 3-state HMM $\mathcal{M}$, the right is 1-state HMM $\mathcal{M}'$. Dashed arrows demonstrate the conversion of $\mathcal{M}$ to $\mathcal{M}'$.

also add non-emission states, which is represented as rectangle nodes, at the head and tail for the convenience of concatenating HMMs.

When using generative model, *e.g.* GMM, and training with MLE criterion, its better to divide the feature space into finer categories. That makes each GMM easier to represent its feature space. However, this treatment is not good for discrimitive models, for that would introduce ambiguities among inter- and intra class variations. Its properer to treat different variations of a digit as the same category. So, after bootstrap training we convert the 3-state HMMs to 1-state HMMs, denoted as $\mathcal{M}'$, to leverage the strong invariance representation capability of deep CNN. Frames belong to all states within a 3-state HMM are assigned to one category corresponding to the single state of $\mathcal{M}'$. Experiment demonstrates that CNN-HMM dramaticly improve the performance of GMM-HMM.

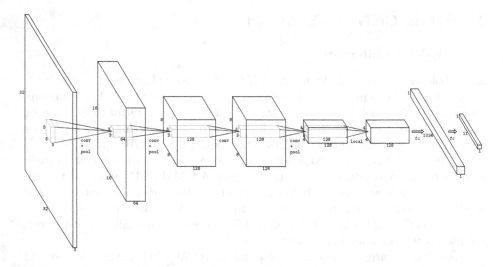

Fig. 3. Deep CNN architecture used in this work

Our CNN architecture is implemented upon Alex Krizhevsky's ConvNet [12]. We investigated different configurations of architecture. Our best architecture consists of 4 convolutional layers, 3 of which have consecutive pooling layers, 1 locally connected layer and 2 fully connected layers. The top layer is an 11-way softmax layer. All connections are feedforward from one layer to the next. The filter size of each convolution layer is showed in Fig. 3. Each convolutional layer includes local response normalization across maps. The max pooling window size is 2×2. The stride alternates between 2 and 1 at each layer. All convolution and locally connected layers contain rectifier units [24]. Details of the architecture's configuration is showed in Fig. 3.

4.2   Training Procedure

Feaure Extraction. We first extract frames of each image by sliding window with overlapping of consecutive windows. After that, we perform PCA on the frames data. Features are extracted by projecting each frame on the principal components. Then the feature is passed to next stage as observations.

Bootstrap Training of GMM-HMM. Then we use GMM-HMM as the bootstrap model to train the initial transition probability and get frame-state assignment. The extracted PCA features are fed to GMM-HMM model. We initialize the mixture components by using global data mean and variance with random noise added.

Each HMM corresponds to a digit category and contains three states to adapt pre-digit, mid-digit and post-digit frames. Background clutter frames are modeled with 1-state HMMs for there are not much obvious sequential difference among them.

The GMM-HMM model is trained with Baum-Welch algorithm [25,26] under maximum-likelihood (ML) criterion until the likelihood converges.

To use CNN, we need to supply the label of each frame for CNN is trained in a supervised way. After training of GMM-HMM, we perform forced alignment on all the images. That assigns each frame to a corresponding HMM state. This category is used as the label for CNN. To make CNN robust to intra category variations while keeping discrimitive to inter category variations, we merge the three states in a HMM to one state and observation frames corresponding to these states are fall in the same category of this state. The switching process is demonstrated in Fig. 2.

After the bootstrap training we obtain the label of each frame. CNN is then trained with these generated data. We split the dataset into training and validation set. The network is trained under cross entropy objective by stochastic gradient descent (SGD) until the validation error stops to drop.

Note that the output of softmax layer is an estimation of the state posterior probability $p(s_t|o_t)$ [27], while the observation probability in HMM is

$$p(o_t|s_t) = \frac{p(s_t|o_t)p(o_t)}{p(s_t)}$$

where $p(s_t)$ is the prior probability of each state estimated from the training set, and $p(o_t)$ is independent of the true digit label and thus can be ignored. Actually, the output probability used in hybrid model is the posterior probability divided by prior probability, $\frac{p(s_t|o_t)}{p(s_t)}$, which is called *scaled likelihood* [18].

Embedded Viterbi Training of CNN-HMM. After the bootstrap training process we get label frames as the training data for CNN and initial states transition probability of HMM. Then hybrid CNN-HMM model is trained with *embedded Viterbi* algorithm [28]. The main training procedure is summarized in Algorithm 1.

Algorithm 1. Embedded Viterbi Training Algorithm

---

1 $t \leftarrow 0$;
2 $\Delta AP \leftarrow$ inf;
3 Train a gmm-hmm model, assign each output HMM state a label-id;
4 Use *Viterbi-decoding* algorithm to recognize each number image and evaluate average precision AP_t;
5 **while** $\Delta AP > 0$ **do**
6 | Use *forced-alignmnet* algorithm to assign each frame a label-id as its category;
7 | Use labeled frames to train CNN as cnn_t;
8 | Get the prior probability $p(c_i)$ of category c_i, where $i = 1, 2, \cdots, N_c$;
9 | Feed each frame to cnn_t to get its posterior probability $p(c_i|x_j)$;
10 | Compute the scaled likelihood $p_{scaled}(x_j|c_i)$;
11 | Perform *embeded-training* to re-estimate the transition probability of HMM, denote the new hybrid model as cnn-hmm_t;
12 | Use *Viterbi-decoding* algorithm to recognize each number image and evaluate average precision AP_{t+1};
13 | $\Delta AP \leftarrow AP_{t+1} - AP_t$;
14 | $t \leftarrow t + 1$;
15 **end**
16 Output cnn_t as the final CNN model and cnn-hmm_t as the final CNN-HMM hybrid model;

The main idea of embedded Viterbi algorithm is to alternatively updating CNN and HMM until the average precision stops to improve. The procedure iteratively improve the best assignment of each frame, and can be proved to converge to a local optimum. This makes it possible to use supervised model, such as CNN, without the need of hand-labeled data.

5   Experiments

5.1   Dataset

The Street View House Numbers (SVHN) dataset [29] contains images captured from Google Street View. It includes two kinds of images. One have over $600,000$ cropped digit images, the other is constitute with more than $200,000$ images containing indeterminate number of digits. Our work deal with the later one.

Nearly all previous work are done on isolated digits, except the work by Goodfellow *et al.* [13] which use loosely cropped images. We solve this problem by integrating segmentation and recognition in a different way. Images in SVHN show quite large appearance variability, image blur and unnormalized layouts. Those make the recognition on the whole image quite difficult.

We preprocess the dataset similar with the way Goodfellow *et al.* [13] does. First find the smallest bounding box contain all individual digit bounding box, then expanding the box by 30 % then crop out the expanded bounding box. Note, we do not resize all cropped images to the same size.

We set the frame width to 10 pixel with 4 pixel overlap to do sliding window, and then randomly selected 100,000 frames to train the PCA feature. We choose 60 principal components which preserve 98 % percentage of the data information, then project each frame to the 60 eigenvectors to get the coefficients as the PCA feature.

We do not use any isolated digit images to train HMMs and CNN, training is totally performed in an embedded way at word level.

5.2 Results and Analysis

We investigate the relationship between recognition accuracy and the number of mixture components for GMM-HMM. For each setting the model is trained until accuracy convergence. The result is showed in Fig. 4. As shown in the figure, recognition accuracy increases with the number of mixture components but stops at 800, only achieve a not-so-well performance of 0.55 accuracy. Continually increasing the amount of mixture components makes the model tend to overfit.

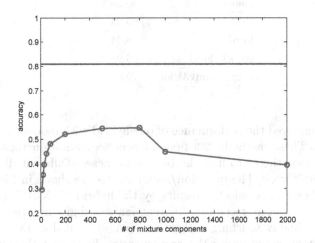

Fig. 4. Relationship between recognition accuracy and the number of mixture components for GMM-HMM (red line). The blue line is the final accuracy of CNN-HMM (Color figure online).

The training of hybrid CNN-HMM model starts with this model. After embedded Viterbi training we get an accuracy of **0.81**, which improve the performance of GMM-HMM model by 47.2 %. This demonstrates the represention superiority of deep model vs shallow model.

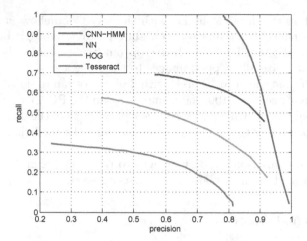

Fig. 5. Performance comparsion with traditional methods

Table 1. Performance after removing different layers

Ablation layers	Accuracy(%)
No ablation	**81.07**
conv3	80.26
full1	80.04
local	78.44
conv3&local	76.94
conv2&conv3&local	70.03

We also compared the performance of our hybrid with model with traditional methods [29]. These methods [29] first do over segmentation then search over segmentation hypotheses to find the best hypotheses. Different digit classifiers are tried in their work. The precision/recall curves are shows in Fig. 5.

Figure 6 shows some selected results by the hybrid CNN-HMM model, note that the model correctly recognize different fonts of digits in street view images when they are highly touching each other, blured, scaled or even their layout are slanted. Also, we visualize the segmentation lines over the images, which are automatically got by Viterbi decoding. The segmentation demonstrated the resonablity of our model and can be used to harvest large amount of digits directly from the images for training other character models.

We also conducted ablation study for the contribution of different layers of CNN. Table 1 summarized the accuracy after removing different layers. Through the ablation study we can see that depth of the network is important to the system performance. Dropping three layers (two convolutional layers and one locally connected layer) decreases the performance by degree of 13.6 %.

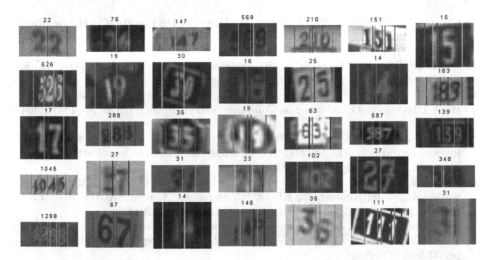

Fig. 6. Recognition results of hybrid CNN-HMM model. Segmentation line cuts generated by Viterbi algorithm are overlapped on the image. Two lines with the same color segment out a digit.

6   Conclusion

We proposed hybrid CNN-HMM model to recognize street view numbers. The method utilized CNN – a model recently performed promisingly well at many computer vision problems, and HMM – a classical graphical model which can effectively do inference on sequence data. In this work, they are integrated in a hybrid fashion. By using this model, we can perform training and recognition both directly at word level, not need any labeling labour or design sophisticated segmentation algorithm. We achieve promising result on SVHN dataset. Experiments shows that, by using deep CNN substitute GMM with HMM, the hybrid model can significantly increases the recognition performance. And, this model can easily extended to general scene text recognition by adopting a language model, whether a constraint lexicon or n-gram model.

Yet, HMM have a few shortcomings, for example, lack of context consideration, hypnosis and short at discrimination. Future work should be done by training HMM discrimitively under MMI [30] or MCE [31] criterion or using Conditional Random Field(CRF) [32]. On the other hand, adjacent frames should be considered when training CNN to incroperate context information. The architecture of CNN we used is pretty off-the-shelf. Better architecture need to be investigated.

References

1. Nagy, G.: Twenty years of document image analysis in PAMI. IEEE Trans. Pattern Anal. Mach. Intell. **22**, 38–62 (2000)

2. Cheriet, M., El Yacoubi, M., Fujisawa, H., Lopresti, D., Lorette, G.: Handwriting recognition research: twenty years of achievement and beyond. Pattern Recogn. **42**, 3131–3135 (2009)
3. Ohya, J., Shio, A., Akamatsu, S.: Recognizing characters in scene images. IEEE Trans. Pattern Anal. Mach. Intell. **16**, 214–220 (1994)
4. Wang, K., Babenko, B., Belongie, S.: End-to-end scene text recognition. In: 2011 IEEE International Conference on Computer Vision (ICCV), IEEE, pp. 1457–1464 (2011)
5. Wang, T., Wu, D.J., Coates, A., Ng, A.Y.: End-to-end text recognition with convolutional neural networks. In: 2012 21st International Conference on Pattern Recognition (ICPR), IEEE, pp. 3304–3308 (2012)
6. Neumann, L., Matas, J.: A method for text localization and recognition in real-world images. In: Kimmel, R., Klette, R., Sugimoto, A. (eds.) ACCV 2010, Part III. LNCS, vol. 6494, pp. 770–783. Springer, Heidelberg (2011)
7. Neumann, L., Matas, J.: Real-time scene text localization and recognition. In: 2012 IEEE Conference on Computer Vision and Pattern Recognition (CVPR), IEEE, pp. 3538–3545 (2012)
8. Alsharif, O., Pineau, J.: End-to-end text recognition with hybrid HMM maxout models (2013). arXiv preprint arXiv:1310.1811
9. Bissacco, A., Cummins, M., Netzer, Y., Neven, H.: PhotoOCR: reading text in uncontrolled conditions. In: ICCV (2013)
10. Neumann, L., Matas, J.: Scene text localization and recognition with oriented stroke detection. In: ICCV (2013)
11. Ciresan, D.C., Meier, U., Gambardella, L.M., Schmidhuber, J.: Convolutional neural network committees for handwritten character classification. In: ICDAR, pp. 1250–1254 (2011)
12. Krizhevsky, A., Sutskever, I., Hinton, G.E.: Imagenet classification with deep convolutional neural networks. In: NIPS, vol. 1, p. 4 (2012)
13. Goodfellow, I.J., Warde-Farley, D., Mirza, M., Courville, A., Bengio, Y.: Maxout networks (2013). arXiv preprint arXiv:1302.4389
14. Zeiler, M.D., Fergus, R.: Visualizing and understanding convolutional networks. CoRR **abs/1311.2901** (2013)
15. Goodfellow, I.J., Bulatov, Y., Ibarz, J., Arnoud, S., Shet, V.: Multi-digit number recognition from street view imagery using deep convolutional neural networks (2014). arXiv preprint arXiv:1312.6082
16. LeCun, Y., Bottou, L., Bengio, Y., Haffner, P.: Gradient-based learning applied to document recognition. Proc. IEEE **86**, 2278–2324 (1998)
17. Matan, O., Burges, C.J.C., LeCun, Y., Denker, J.S.: Multi-digit recognition using a space displacement neural network. In: NIPS, pp. 488–495 (1991)
18. Bourlard, H.A., Morgan, N.: Connectionist Speech Recognition: A Hybrid Approach. Kluwer Academic Publishers, Norwell (1993). ISBN: 0792393961
19. Dahl, G.E., Yu, D., Deng, L., Acero, A.: Context-dependent pre-trained deep neural networks for large-vocabulary speech recognition. IEEE Trans. Audio Speech Lang. Process. **20**, 30–42 (2012)
20. Hinton, G., Deng, L., Yu, D., Dahl, G.E., Mohamed, A.R., Jaitly, N., Senior, A., Vanhoucke, V., Nguyen, P., Sainath, T.N., et al.: Deep neural networks for acoustic modeling in speech recognition: the shared views of four research groups. IEEE Sig. Process. Mag. **29**, 82–97 (2012)
21. Graves, A., Mohamed, A.R., Hinton, G.: Speech recognition with deep recurrent neural networks. In: 2013 IEEE International Conference on Acoustics, Speech and Signal Processing (ICASSP), IEEE, pp. 6645–6649 (2013)

22. Sainath, T.N., Kingsbury, B., Ramabhadran, B., Fousek, P., Novak, P., Mohamed, A.R.: Making deep belief networks effective for large vocabulary continuous speech recognition. In: 2011 IEEE Workshop on Automatic Speech Recognition and Understanding (ASRU), IEEE, pp. 30–35 (2011)
23. Forney, G.D.J.: The viterbi algorithm. Proc. IEEE **61**, 268–278 (1973)
24. Jarrett, K., Kavukcuoglu, K., Ranzato, M., LeCun, Y.: What is the best multistage architecture for object recognition? In: ICCV, pp. 2146–2153 (2009)
25. Baum, L.E., Petrie, T., Soules, G., Weiss, N.: A maximization technique occurring in the statistical analysis of probabilistic functions of Markov chains. Ann. Math. Stat. **41**, 164–171 (1970)
26. Baum, L.E.: An inequality and associated maximization technique in statistical estimation for probabilistic functions of a Markov process. Inequalities **3**, 1–18 (1972)
27. Richard, M.D., Lippmann, R.P.: Neural network classifiers estimate Bayesian a posteriori probabilities. Neural Comput. **3**, 461–483 (1991)
28. Morgan, N., Bourlard, H.: Continuous speech recognition. IEEE Sig. Process. Mag. **12**, 24–42 (1995)
29. Netzer, Y., Wang, T., Coates, A., Bissacco, A., Wu, B., Ng, A.Y.: Reading digits in natural images with unsupervised feature learning. In: NIPS Workshop on Deep Learning and Unsupervised Feature Learning, vol. 2011 (2011)
30. Kapadia, S., Valtchev, V., Young, S.: Mmi training for continuous phoneme recognition on the timit database. In: 1993 IEEE International Conference on Acoustics, Speech, and Signal Processing, ICASSP 1993, vol. 2, pp. 491–494 (1993)
31. Juang, B.H., Hou, W., Lee, C.H.: Minimum classification error rate methods for speech recognition. IEEE Trans. Speech Audio Process. **5**, 257–265 (1997)
32. Lafferty, J.D., McCallum, A., Pereira, F.C.N.: Conditional random fields: probabilistic models for segmenting and labeling sequence data. In: Proceedings of the Eighteenth International Conference on Machine Learning, ICML 2001, pp. 282–289 (2001)

View and Illumination Invariant Object Classification Based on 3D Color Histogram Using Convolutional Neural Networks

Earnest Paul Ijjina[✉] and C. Krishna Mohan

Indian Institute of Technology Hyderabad,
Yeddumailaram 502205, Telangana, India
{cs12p1002,ckm}@iith.ac.in

Abstract. Object classification is an important step in visual recognition and semantic analysis of visual content. In this paper, we propose a method for classification of objects that is invariant to illumination color, illumination direction and viewpoint based on 3D color histogram. A 3D color histogram of an image is represented as a 2D image, to capture the color composition while preserving the neighborhood information of color bins, to realize the necessary visual cues for classification of objects. Also, the ability of convolutional neural network (CNN) to learn invariant visual patterns is exploited for object classification. The efficacy of the proposed method is demonstrated on Amsterdam Library of Object Images (ALOI) dataset captured under various illumination conditions and angles-of-view.

1 Introduction

Object recognition is an active area of research for the last five decades [1] and efficient recognition of objects under varying illumination conditions is a problem yet to be solved. Visual object classification is an important step in visual recognition and semantic analysis of visual content. Some of the challenges in object classification from 2D images is the loss of depth information, variations in the visual information captured, due to the change in view-angle, illumination color and illumination direction of the object. Techniques relying solely on shape and texture features are computationally expensive and inefficient to classify non-symmetric objects with complex shape. Thus, features incorporating color information should be considered for the design of an efficient object classification mechanism robust to variations in object viewpoint and illumination conditions.

Representing object images as graphs built over corner point, is proposed in [2] to classify objects using graph matching. The effectiveness of this approach depends upon the robustness of the graph representation against varying illumination conditions and angles of view. The dependence of this approach on corner points makes the representation less distinctive and thereby affects the efficiency. To compute similarity of images at multiple resolutions effectively, a new multi-resolution distance metric, Manhattan-pyramid distance is proposed in [3].

© Springer International Publishing Switzerland 2015
C.V. Jawahar and S. Shan (Eds.): ACCV 2014 Workshops, Part I, LNCS 9008, pp. 316–327, 2015.
DOI: 10.1007/978-3-319-16628-5_23

A multiset discriminant canonical correlation method namely multiple principle angle [4] that iteratively learns multiple subspaces and the global discriminant subspace to consider both local and global canonical correlations is used to classify images of objects captured from different view angles. For object classification, GMMs based on views of each object are built in [5] from global models using maximum likelihood estimation followed by an adaptation step to minimize the kNN classification error rate. The GMMs are combined to minimize the distance between objects of the same class and maximize the distance between objects of different classes. This method may not be effective for illumination invariant object recognition due to its dependence on shape information. An attention guided model for object recognition is proposed in [6] that learns the probability of an objects visual appearance having a range of values within a particular feature map. For a given test image, the possible candidate classes were identified along with their probabilities. As color is one of the critical feature, the model encounters more ambiguity when classifying images with variation in illumination conditions.

Label consistent K-SVD algorithm [7] associates label information with each dictionary item to learn a discriminative dictionary for object classification using spatial pyramid features. Multiple sets of features are combined using multiple kernel learning (MKL) for object recognition [8]. Representation learning algorithms [9] like deep learning, autoencoder and deep networks that learn from generic priors are used for object representations and classification. Convolutional neural network based feature extraction and classification models [10,11] trained on ImageNet dataset, produced competitive results for localization, detection and classification tasks for the last two years.

Some of the limitations of existing approaches is their lack of robustness to illumination variation and high computationally complexity. In this paper, we propose a view and illumination invariant object classification system using a convolutional neural network. The reminder of this paper is organized as follows: In Sect. 2, the proposed approach, image representation and the object classifier are discussed. Experimental results were discussed in Sect. 3. Section 4 gives conclusion of this work.

2 Proposed Approach

In this work, we propose a method for view and illumination invariant object classification based on 3D color histogram information using a convolutional neural network (CNN). The steps involved in the generation of 2D representation are detailed in the following section.

2.1 Image Representation

Due to the robustness of color distribution against changes in viewpoint and illumination conditions, a 2D representation preserving the neighborhood information of color bins in 3D color histogram of an image is used as the feature.

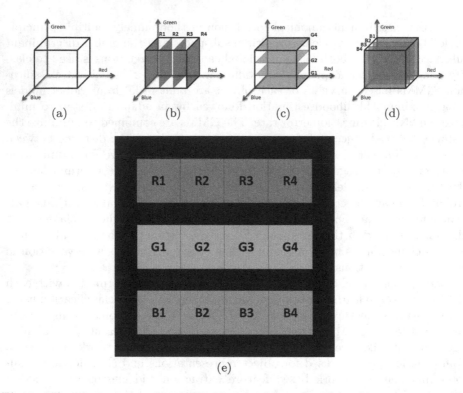

(a)   (b)   (c)   (d)

(e)

Fig. 1. 2D representation of an RGB 3D color histogram. (a) cube representing 3D color histogram (b) slices of cube along red-axis (c) slices of cube along green-axis (d) slices of cube along blue-axis (e) proposed 2D representation of a 3D color histogram (Color figure online)

This study considers RGB 3D histogram with n bins (considering 4 for illustration) along each axis, resulting in a $4 \times 4 \times 4$ cube with each axis representing a color in RGB as shown in Fig. 1(a). Figure 1(b) shows the slices of the cube R1, R2, R3, R4 along red-axis. Similarly, the slices along green axis G1, G2, G3, G4 and along blue axis B1, B2, B3, B4 are shown in Fig. 1(c) and (d). These 4×4 red, green and blue slices are arranged in a 20×20 matrix with a margin of 2 elements from each border and between slices of different color, to construct the 2D representation of an image, as shown in Fig. 1(e).

The proposed 2D representation preserves the neighborhood information of color bins along an axis in slices along the remaining axis. The 2D representation of some objects in ALOI dataset is shown in Fig. 2.

Fig. 2. The 2D representation of images of some ALOI objects

From the 2D representation of objects in Fig. 2, it can be observed that the local patterns provide discriminative information useful for classification. Different images of the same object may have different patterns due to variations in image capturing conditions like view-angle, illumination color and illumination direction of the object. The classifier employed to capture the variations in local patterns of each object for effective classification, is elaborated in the following section.

2.2 Object Classification Using CNN

A convolutional neural network (CNN) [12] is a feed-forward neural network capable of recognizing local patterns with some degree of shift and distortion. This characteristic is explored to classify objects from the local patterns in their color histogram. The convolutional neural networks (CNNs) have been shown to outperform the standard fully connected deep neural networks in various computer vision challenges. A typical CNN architecture used as a classifier [13] consists of an alternating sequence of convolution and subsampling layers followed by a neural network for classification. The architecture used in the proposed approach is shown in Fig. 3. If we consider a 20×20 image as input, 2×2 mask in subsampling layers S1 & S2, a 5×5 mask in convolution layers C1 & C2, 15 feature maps in F1 & F2 and 30 feature maps in F3 & F4, then (1) F1 represents 15 feature maps of size 16×16 (2) F2 represents 15 feature maps of size 8×8 (3) F3 represents 30 feature maps of size 4×4 and (4) F4 represents 30 feature maps of size 2×2. The 120 feature (corresponding to F4) generated at the output of second subsampling layer are given as input (I) to a fully connected neural network, to generate an output O to classify the N objects.

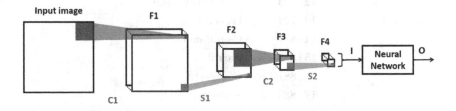

Fig. 3. CNN architecture in the proposed approach

The CNN is trained using back-propagation algorithm in batch-mode for 2500 epochs on training dataset and tested on test dataset to classify objects from the 2D representation of their images. Two-fold cross validation is used to evaluate the performance of the proposed approach. The train and test datasets are obtained by arranging the images of an object in order and assigning the odd image to the train dataset and even to the test dataset.

3 Experimental Results

The proposed approach is evaluated on vegetable and fruit objects in ALIO dataset [14], that consists of images of 1000 objects captured under 24 configurations of illumination direction, 12 configuration of illumination color and 72 directions of object view point. As two-fold cross validation is used for evaluation, the initial train and test datasets are obtained by considering all the odd images of an object as training images and even as testing images. The train and test datasets are interchanged during cross-validation. The vegetable and fruit objects in ALOI dataset considered for evaluation are listed in Tables 1 and 2.

Table 1. Vegetables

#	ALOI #	Object name
1	17	Red onion
2	281	Garlic
3	287	Red onion
4	324	Red pepper
5	709	Big mushroom
6	711	Small mushroom
7	714	Onion
8	717	Small onion
9	718	Garlic
10	719	Tomato
11	720	Red onion
12	723	Flat french bean
13	724	French bean
14	877	Cauliflower
15	880	Carrot
16	881	Courgette
17	883	Asperges
18	884	Rettig
19	885	Sweet potato
20	887	Witlof
21	889	Egg plant
22	948	Green capsicum
23	952	Artisjok
24	953	Yellow capsicum
25	954	Reddish

Table 2. Fruits

#	ALOI #	Object name
1	3	Apricot
2	52	Hairy ball
3	69	Tomato
4	82	Apple
5	102	Kiwi
6	273	Lemon2
7	446	Orange
8	567	Pear
9	649	Apple
10	650	Kiwi
11	651	Lemon
12	705	Green capsicum
13	706	Red capsicum
14	707	Mango
15	708	Kiwi
16	710	Apple
17	712	Mandarin
18	713	Lemon
19	715	Unknown fruit
20	716	Unknown fruit 2
21	721	Pear
22	722	Lemon
23	870	Pineapple
24	873	Mango
25	879	Cucumber
26	882	Orange
27	888	Melon
28	947	Sherry tomatos
29	950	Banana's

3.1 ALOI Vegetable Objects

Some examples of vegetable object images considered in this evaluation are shown in Fig. 4. The Fig. 4(a) and (b) shows the inter-class similarity of objects; Fig. 4(c) and (d) the intra-class dissimilarity of an object in terms of color profile. The Fig. 4(e) and (f) are images of the same object captured under different direction of illumination and Fig. 4(g) and (h) are images of the same object with different illumination temperate. This inter-class similarity and intra-class diversity makes vegetable categorization a challenging task.

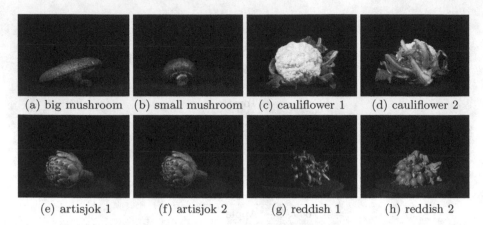

| (a) big mushroom | (b) small mushroom | (c) cauliflower 1 | (d) cauliflower 2 |
| (e) artisjok 1 | (f) artisjok 2 | (g) reddish 1 | (h) reddish 2 |

Fig. 4. Examples images of ALOI vegetable objects

Table 3. Average classification error of CNN classifier (in %) for vegetable objects with execution time/iteration in parenthesis

# of bins	3 × 3 mask			5 × 5 mask			7 × 7 mask		
	5FM	15FM	25FM	5FM	15FM	25FM	5FM	15FM	25FM
3	12.6	6.9	5.7	11.3	4.3	4.8			
	(3.4 s)	(22.2 s)	(58 s)	(4 s)	(25 s)	(64 s)			
4	4.2	3.4	1.3	1.8	1.2	1.4	2.2	1.4	1.3
	(4.4 s)	(29 s)	(78 s)	(4.9 s)	(31.4 s)	(86 s)	(5.9 s)	(34 s)	(87 s)
5	1.1	0.96	76	1.3	0.74	60	0.59	0.74	1.0
	(7 s)	(45.8 s)	(120 s)	(9.5 s)	(60 s)	(152 s)	(12.6 s)	(77 s)	(195 s)
6	0.74	80	83	0.81	48	92	0.74	1.0	92
	(8.9 s)	(55 s)	(142 s)	(13.6 s)	(89.5 s)	(230 s)	(16.2 s)	(102.5 s)	(276.5 s)

The 25 ALOI vegetable objects listed in Table 1 are used for evaluation. The CNN classifier is trained in batch mode with a batch size of 90 and evaluated on the test dataset. The following section presents the impact of the number of color bins used in 2D representation, the number of feature maps considered and the size of mask used in convolution layers.

Impact of Configuration Changes. We assume that same size of convolution mask is used in both convolution layers and that the number of feature maps in F3 & F4 is double the number of feature maps in F1 & F2. The impact of number of bins used in the generation of color histogram, the size of the mask used in convolution layers and the number of feature maps used in the first convolution layer on the performance of vegetable object classification is presented in Table 3.

From Table 3, it can be observed that when the number of color bins is 3 or 4, increase in the number of feature maps generally improves the performance. When the number of color bins is 5 or 6, increase in number of feature maps deteriorates the performance. This suggests that a solution with optimal performance and time-complexity can be identified by considering the right set of values for these parameters. The computation time per iteration (5 epochs) also increases with the increase in number of feature maps, due to the increase in number of free variables to be tuned by the back-propagation algorithm. The confusion matrix of the trained classifier for images of vegetable category is shown in Fig. 5.

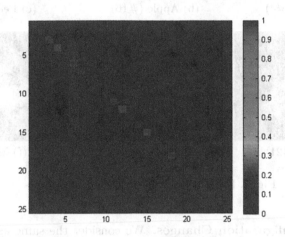

Fig. 5. Confusion matrix of vegetable object images from ALOI dataset

The class labels from top-left to top-right and bottom-left are in the order given in Table 1. The average classification error of our approach using 2-fold cross validation for vegetable objects is 0.74 %.

3.2   ALOI Fruit Objects

Among the 29 ALOI fruit objects considered, there are multiple instances of the same object class like for apple, pear, kiwi etc., capturing the raw and ripe variants of these fruits, thereby making this a fine-grained classification. Some example fruit images are shown in Fig. 6, where Fig. 6(a), (b) and (c), (d) are multiple instances of the same object i.e., apple and pear respectively, with different color profiles. The images in Fig. 6(e) and (f) shows the complex texture and color of fruits being recognized. Thus, fruit classification is relatively more challenging than vegetable classification due to their complex texture and variants.

The CNN classifier is trained on the 29 fruit objects with a batch size of 87. The following section presents the impact of the number of color bins used in 2D representation, the number of feature maps considered and the size of mask used in convolution layers.

(a) Apple (#4)   (b) Apple (#16)   (c) Pear (#8)

(d) Pear (#21)   (e) Pineapple   (f) Mango

Fig. 6. Examples images of ALIO fruit objects

Impact of Configuration Changes. We consider the same assumptions on the configuration of CNN architecture as we did for the classification of vegetable objects. The affect of number of bins used in the generation of color histogram, the size of the mask used in convolution layers and the number of feature maps used in the first convolution layer on the performance of fruit object classification is shown in Table 4.

Table 4. Average classification error of CNN classifier (in %) for fruit objects

# of bins	3 × 3 mask			5 × 5 mask			7 × 7 mask		
	5FM	15FM	25FM	5FM	15FM	25FM	5FM	15FM	25FM
3	7.21	4.85	4.34	6.25	3.57	3.7			
4	3.95	3	3	3.83	2.29	2.49	2.74	2.49	1.85
5	1.59	52.87	93.1	1.91	1.66	66.21	2.49	1.59	42.4
6	1.53	69.15	96.55	1.66	86.2	89.14	1.66	1.47	48

The confusion matrix of the trained classifier for images of vegetable objects is shown in Fig. 7.

The class labels are as listed in Table 2. The average classification error of the proposed approach using 2-fold cross validation is 2.17 %.

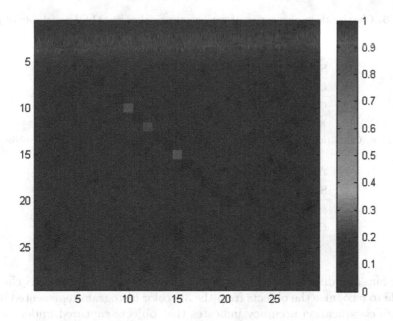

Fig. 7. Confusion matrix of fruit objects images from ALOI dataset

3.3 Analysis of Results and Comments

The low misclassification error shown in Figs. 5 and 7 is due to existence of similar color profile for some objects under certain image capturing conditions. Figure 8 shows some examples of objects Fig. 8(a) and (c) misclassified as objects of Fig. 8(b) and (d).

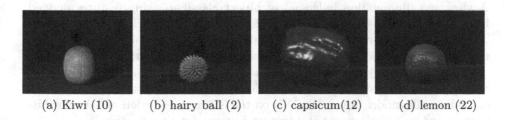

(a) Kiwi (10)   (b) hairy ball (2)   (c) capsicum(12)   (d) lemon (22)

Fig. 8. Some misclassified fruit objects

The classification accuracy of ALOI vegetable, fruit objects considering support vector machine (SVM) and k-nearest neighbor (kNN) classifiers using 3D color histogram features considered in this paper, for various values of number of color bins is shown in Tables 5 and 6 respectively.

From Tables 5 and 6, it can be observed that accuracy improves with the increase in the number of color bins. Experiments conducted on varying number

Table 5. Classification accuracies of various approaches on ALOI vegetable objects

Approach	3 bins	4 bins	5 bins	6 bins
SVM	46.74 %	60.37 %	66.81 %	72.07 %
kNN	79.85 %	85.11 %	89.18 %	91.18 %
CNN	95.7 %	98.8 %	99.26 %	99.26 %

Table 6. Classification accuracies of various approaches on ALOI fruit objects

Approach	3 bins	4 bins	5 bins	6 bins
SVM	47.41 %	58.56 %	64.01 %	63.54 %
kNN	80.74 %	85.39 %	87.51 %	90.50 %
CNN	96.3 %	98.15 %	98.41 %	98.53 %

of color bins and different configurations of CNN suggest that the CNN classifier was able to recognize the objects from the 3D color histogram represented in 2D. The high classification accuracy indicates that objects captured under varying illumination-color are also well classified suggesting that the CNN classifier was able to capture the illumination variations of objects. In contrast to the existing approaches like [15,16], we considered a class of ALOI objects to evaluate the performance of the proposed approach. The proposed approach could not be compared with the existing approaches due to the dissimilarity in the objects considered for evaluation.

4 Conclusion

A view and illumination independent object classifier, using features derived from 3D color histogram and CNN architecture for classification is presented. The low classification error on ALOI fruits and vegetables objects suggests that the CNN classifier was able to capture the view and illumination invariant characteristics of the objects. The limitation of this approach is its inability to discriminate objects with same color profile in images. The future work includes extending this model to use 3D CNN on the 3D representation of 3D color histogram and to include shape and texture features for classification.

References

1. Andreopoulos, A., Tsotsos, J.K.: 50 years of object recognition: directions forward. Comput. Vis. Image Underst. **117**, 827–891 (2013)
2. Albarelli, A., Bergamasco, F., Rossi, L., Vascon, S., Torsello, A.: A stable graph-based representation for object recognition through high-order matching. In: International Conference on Pattern Recognition (ICPR), IEEE, pp. 3341–3344 (2012)

3. Chauhan, A., Lopes, L.S.: Manhattan-pyramid distance: a solution to an anomaly in pyramid matching by minimisation. In: International Conference on Pattern Recognition (ICPR), IEEE, pp. 2668–2672 (2012)
4. Ya, S., Fu, Y., Gao, X., Tian, Q.: Discriminant learning through multiple principal angles for visual recognition. IEEE Trans. Image Process. **21**, 1381–1390 (2012)
5. Wang, M., Gao, Y., Ke, L., Rui, Y.: View-based discriminative probabilistic modeling for 3d object retrieval and recognition. IEEE Trans. Image Process. **22**, 1395–1407 (2013)
6. Elazary, L., Itti, L.: A Bayesian model for efficient visual search and recognition. Vis. Res. **50**, 1338–1352 (2010)
7. Jiang, Z., Lin, Z., Davis, L.S.: Label consistent K-SVD: learning a discriminative dictionary for recognition. IEEE Trans. Pattern Anal. Mach. Intell. **35**, 2651–2664 (2013)
8. Bucak, S.S., Jin, R., Jain, A.K.: Multiple kernel learning for visual object recognition: a review. IEEE Trans. Pattern Anal. Mach. Intell. **36**, 1354–1369 (2014)
9. Bengio, Y., Courville, A.C., Vincent, P.: Representation learning: a review and new perspectives. IEEE Trans. Pattern Anal. Mach. Intell. **35**, 1798–1828 (2013)
10. Krizhevsky, A., Sutskever, I., Hinton, G.E.: Imagenet classification with deep convolutional neural networks. In: Neural Information Processing Systems (NIPS), pp. 1106–1114 (2012)
11. Sermanet, P., Eigen, D., Zhang, X., Mathieu, M., Fergus, R., LeCun, Y.: Overfeat: integrated recognition, localization and detection using convolutional networks. Computer Research Repository (CoRR) (2013)
12. Bengio, Y.: Learning deep architectures for AI. Found. Trends Mach. Learn. **2**, 1–127 (2009)
13. Palm, R.B.: Prediction as a candidate for learning deep hierarchical models of data. Master's thesis, Technical University of Denmark, Asmussens Alle, Denmark (2012)
14. Geusebroek, J.M., Burghouts, G.J., Smeulders, A.W.M.: The Amsterdam library of object images. Int. J. Comput. Vis. **61**, 103–112 (2005)
15. Albarelli, A., Bergamasco, F., Rossi, L., Vascon, S., Torsello, A.: A stable graph-based representation for object recognition through high-order matching. In: ICPR, IEEE, pp. 3341–3344 (2012)
16. Smagghe, P., Buessler, J.L., Urban, J.P.: Déjà Vu object localization using IRF neural networks properties. In: IJCNN, IEEE, pp. 1–8 (2013)

Human Action Recognition Using Action Bank Features and Convolutional Neural Networks

Earnest Paul Ijjina[✉] and C. Krishna Mohan

Indian Institute of Technology Hyderabad,
Yeddumailaram 502205, Telangana, India
{cs12p1002,ckm}@iith.ac.in

Abstract. With the advancement in technology and availability of multimedia content, human action recognition has become a major area of research in computer vision that contributes to semantic analysis of videos. The representation and matching of spatio-temporal information in videos is a major factor affecting the design and performance of existing convolution neural network approaches for human action recognition. In this paper, in contrast to the traditional approach of using raw video as input, we derive attributes from action bank features to represent and match spatio-temporal information effectively. The derived features are arranged in a square matrix and used as input to the convolutional neural network for action recognition. The effectiveness of the proposed approach is demonstrated on KTH and UCF Sports datasets.

1 Introduction

Human action recognition is a complex computer vision task for which efficient techniques are yet to be proposed to address the problem thoroughly. Human actions based on the subjects and objects involved in the action, can be classified into (1) gestures performed by a single subject (2) interaction among subjects and (3) interaction of a subject with object. Human action recognition is generally accomplished by extracting discriminative features from video and processing them using pattern recognition techniques to classify the video into their corresponding action classes. Feature learning techniques like deep learning, that can learn the features directly from video data are also employed for action classification [1,2].

Some of the commonly used features for human action recognition are HOG [3], HOF, action bank [4] and dense trajectories [5]. Zhuolin Jiang et al. [6] proposed 'label consistent K-SVD' algorithm to learn discriminative dictionaries for action recognition using action bank features. Sadanand et al. [4] used SVM and random forest classifier to recognize actions using action bank features. Baumann et al. [7] trained random forest classifiers for motion information and static object appearance separately and combined their probabilities to classify a video. Heng Wang et al. [5] proposed the use of dense trajectories and motion boundaries descriptors for human action recognition. With local motion information being captured by trajectories, a dense representation covers motion in

C.V. Jawahar and S. Shan (Eds.): ACCV 2014 Workshops, Part I, LNCS 9008, pp. 328–339, 2015.
DOI: 10.1007/978-3-319-16628-5_24

both foreground and background and a descriptor based on motion boundary histograms is considered. Benjamin Z. Yao et al. [8] proposed the use of animated pose templates, that consists of a shape template and a motion template, to classify human actions.

Baccouche Moez et al. [1] proposed a neural-based deep model that learns spatio-temporal features from videos using 3D convolutional neural network and uses a recurrent neural network to classify a video from the temporal evolution of learned features. Experiments were conducted on KTH dataset considering the person-centered bounding box region as input to the system and by employing long short-term memory recurrent neural networks for classification of videos from the features extracted by 3D CNN over time. Shuiwang Ji et al. [2] proposed a 3D CNN model for action recognition that performs convolution and sub-sampling operations on multiple input channels extracted from adjacent input frames. The five different input channels considered are: gray value of pixels; the gradients along horizontal and vertical directions; and the optical flow along horizontal and vertical directions computed using hardwired layers. Majority voting is used to classify the videos from the prediction of individual frames. Experiments were conducted on KTH and TRECVid 2008 London Gatwick datasets.

In this paper, we propose an approach for human action recognition using action bank features. The use of template based action detectors to compute features that represent the similarity of an action with the corresponding action bank detector, is the motivation behind the use of action bank features in our proposed approach. As the size of the action bank features remains constant irrespective of the length of the video, the amount of data that needs to be processed by the system to classify a video remains constant. Thus, the system can be designed to classify a video from a single forward computation of the input data, thereby avoiding the need for a voting scheme for overall classification. The reminder of this paper is organized as follows: In Sect. 2, the proposed approach for human action recognition, feature extraction and convolutional neural network (CNN) classifier are discussed. Experimental results were discussed in Sect. 3. The last section gives conclusions of this work.

2 Proposed Approaches

In this paper, we propose convolutional neural network approaches for human action recognition using attributes derived from action bank features. An action bank consists of a predefined set of action detectors which are used to generate the corresponding action bank features for a video. An action bank feature, for a given input action is a measure of similarity of the input action with the corresponding action detector. Hence, identical actions will have similar action bank features as shown in Fig. 1.

The similarity of action bank features for identical actions is explored, to recognize actions from their local patterns using convolutional neural network. To reduce the size of input data, new attributes are derived from action bank

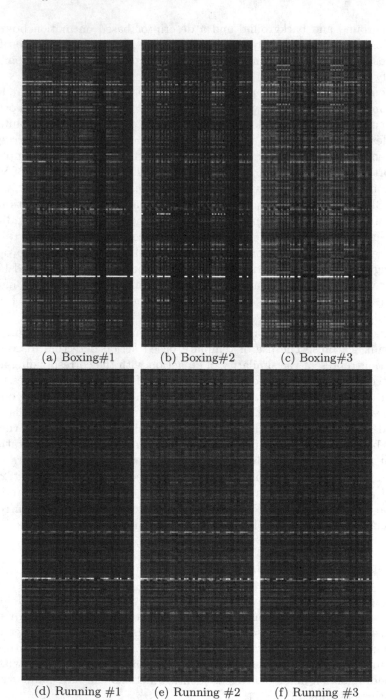

(a) Boxing#1   (b) Boxing#2   (c) Boxing#3

(d) Running #1   (e) Running #2   (f) Running #3

Fig. 1. Action bank representation of boxing and running videos.

features and are arranged in a square matrix. A convolutional neural network is trained to recognize actions from the local patterns in this matrix representation.

In this paper, we propose two approaches for human action recognition using convolutional neural networks with features derived from action bank representation of videos. The action bank proposed by Sadanand et al. [4] is used to generate the action bank representation of videos without considering the action associated with each action bank detector. The two approaches differ in computing the attributes from action bank features and also the way these derived features are organized. We use the same convolutional neural network architecture for classification in both approaches. The typical architecture of a CNN classifier [9] consists of an alternating sequence of convolution and subsampling layers followed by a neural network (NN) for classification. The common CNN architecture considered in the two approaches, $3C - 2S - 3C - 2S$ is shown in Fig. 2 whose configuration is mentioned in Table 1.

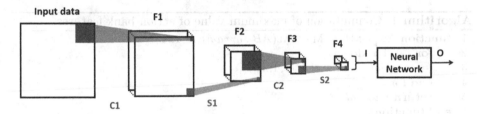

Fig. 2. CNN architecture considered in the proposed approaches

Table 1. CNN configuration considered in the proposed approaches

Layer: Template size	Feature map: #, size
C1: 3×3	F1: p, $(N - 2) \times (N - 2)$
S1: 2×2	F2: p, $\frac{N-2}{2} \times \frac{N-2}{2}$
C2: 3×3	F3: 2*p, $(\frac{N-2}{2} - 2) \times (\frac{N-2}{2} - 2)$
S2: 2×2	F4: 2*p, $\frac{\frac{N-2}{2} - 2}{2} \times \frac{\frac{N-2}{2} - 2}{2} = s \times s$

The CNN configuration used in the two approaches differ in terms of the size of input ($N \times N$), the # of feature maps considered (p) and the size of the output (O). Back-propagation algorithm in batch mode is used to train the CNN architecture.

The CNN architecture places an additional constraint on the size of the square matrix (N), that is given as input to the CNN. In addition to the requirement that, the size of the square matrix should be large enough to contain all the derived features, the side of this square matrix (N) must satisfy the formula $N = 6 + 4 \times s$ for some integral value of s. The two CNN approaches for human action recognition are elaborated in detail in the following subsections.

2.1 First Approach

In the first approach, the maximum value of each action bank feature is considered to provide discriminative information to recognize human actions. As the maximum value of an action bank feature indicate the extent of (partial) similarity of an action with the corresponding action detector, the maximum values of action bank features are used for classification of actions in KTH dataset. The KTH dataset consists of six types of actions and an action bank with 202 action detectors is used to generate the action bank features. The procedure described in Algorithm 1 is used to compute the maximum values of action bank features for a video, which are then arranged in a 34 × 34 matrix in row major order with a margin of 2 elements across the border and 1 element between the values as shown in Fig. 3. As 3 × 3 templates are used in the convolution layer of CNN, this arrangement of derived features is considered for better classification performance.

Algorithm 1. Computation of maximum value of action bank features

1: **function** ACTIONBANKMAXVAL($AB : array[1..n, 1..w]$)
2:     **for** $i \leftarrow 1, n$ **do**
3:         $maxVal[i] \leftarrow \mathbf{max}(AB[i, 1..w])$
4:     **end for**
5:     **return** $maxVal$
6: **end function**

A CNN configuration with $p = 8$, $O = 6$ and $N = 34$ is trained using backpropagation algorithm with a batch size of 18 elements on the training dataset for 500 epochs to obtain an accuracy of 96.75 %. The variation of misclassification error against iteration during training is shown in Fig. 5. The confusion matrix of the proposed approach is shown in Fig. 4.

The performance of existing approaches for human action recognition on KTH dataset is given in Table 2. It can be observed that the performance of the proposed approach is comparable with the current state of the art algorithms for human action recognition on KTH dataset. Even though the proposed approach utilizes all the action bank features to compute the corresponding maximum values in Algorithm 1, only 1.3 % ($\frac{1}{73} \times 100$) of the action bank data is used for action recognition. The proposed approach when applied to UCF sports dataset was not able to classify the 10 actions which may be due to inadequacy of discriminative information in the derived features. Experiments exploring other possible derived features led to the development of the second approach discussed in the next section.

2.2 Second Approach

Some of the discriminative information in action bank features may have been lost due to the computation and consideration of maximum values of action bank features as the derived feature in the first approach. The first approach

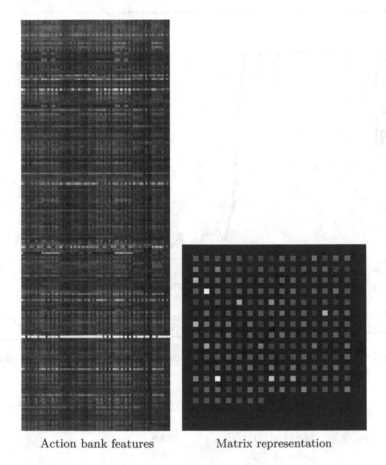

Action bank features   Matrix representation

Fig. 3. The action bank representation of KTH boxing #1 and the square matrix representation of maximum values of all action bank features.

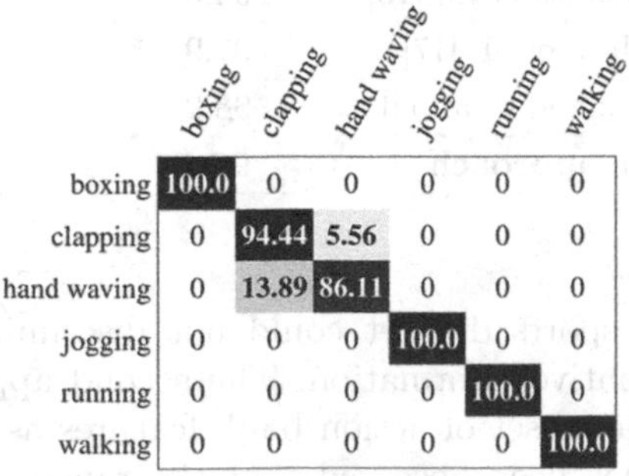

	boxing	clapping	hand waving	jogging	running	walking
boxing	100.0	0	0	0	0	0
clapping	0	94.44	5.56	0	0	0
hand waving	0	13.89	86.11	0	0	0
jogging	0	0	0	100.0	0	0
running	0	0	0	0	100.0	0
walking	0	0	0	0	0	100.0

Fig. 4. Confusion matrix of the proposed approach for human action recognition on KTH dataset

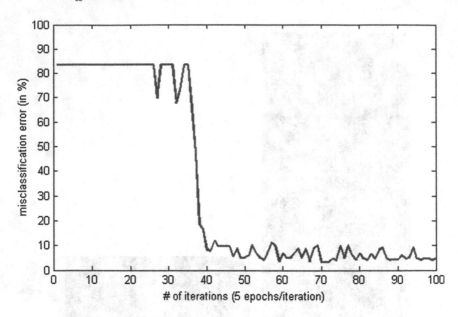

Fig. 5. Variation of misclassification error against training iteration for the proposed approach on KTH dataset

Table 2. Performance comparison of the proposed approach with existing techniques on KTH dataset

Approach	Accuracy (in %)
Liu et al. [10]	91.6
Liu et al. [11]	93.8
Le et al. [12]	93.9
Yimeng Zhang et al. [13]	94.0
Heng Wang et al. [14]	94.2
Wu et al. [15]	94.5
Kovashka et al. [16]	94.5
O'Hara et al. [17]	97.9
Sadanand et al. [4]	98.2
Our approach	**96.75**

when applied to UCF sports dataset could not discriminate the actions due to insufficient discriminative information. The second approach addresses this deficiency by utilizing a subset of action bank features as the derived features. From our analysis, it has been observed that the range of values in an action bank feature could be different in the index ranges [1 37] and [38 73] as shown in Fig. 6. Thus, instead of considering the action bank features, we split the action bank features into two vectors corresponding to the ranges [1 37] and [38 73]

Algorithm 2. Computation of split action bank features

```
1:  function SPLITACTIONBANKFEATURES(AB : array[1..n, 1..73])
2:      for i ← 0, n − 1 do
3:          sabIdx ← i × 2 + 1
4:          SplitAB[sabIdx, 1..37] ← max(AB[i + 1, 1..37])
5:      ▷ split action bank feature corresponding to the action bank feature range [1 37]
6:          SplitAB[sabIdx + 1, 1..38] ← max(AB[i + 1, 38..73])
7:      ▷ split action bank feature corresponding to the action bank feature range [38 73]
8:      end for
9:      return SplitAB                                           ▷ a 2n × 38 matrix
10: end function
```

---

using Algorithm 2, resulting in split action bank features. The split action bank features generated for the l videos in the dataset are then used by Algorithm 3 to identify the indexes of the r most significant split action bank features.

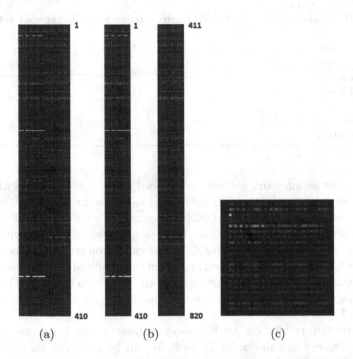

Fig. 6. Feature extraction from split action bank features and selected split action bank indexes. (a) the action bank representation of a video (b) the split action bank features and (c) the matrix representation of selected split action bank features

For our experiments on UCF sports dataset ($l = 140$), we considered the action bank features generated by an action bank of size 410 and computed the $r = 20$ most significant split action bank features for the entire dataset. These 20 most

---

Algorithm 3. Computation of indexes of r significant split action bank features for a dataset

---

```
 1: function SIGACTIONBANKFEAT(ABF : array[1..l, 1..m, 1..w], r : int)
 2:     for i ← 1, l do
 3:         AB ← ABF[i, 1..m, 1..w]                    ▷ consider i^{th} split action bank feature
 4:         ABFMaxVal[i, 1..m] ← ACTIONBANKMAXVAL(AB)       ▷ compute max.
    values of all split action bank features
 5:     end for
 6:     ABMaxVal[1..m] ← max(ABFMaxVal[1..l, 1..m])  ▷ compute the maximum
    across all l instances
 7:
 8:     SortABMaxVal[1..m] ← sort(ABMaxVal[1..m])
 9:               ▷ sort the max. value of all split action bank features in descending order
10:     threshold ← SortABMaxVal[r]       ▷ the cut-off value to select r split action
    bank features
11:
12:     selIter ← 1
13:     for i ← 1, n do
14:         if ABMaxVal[i] ≥ threshold then  ▷ select r significant split action bank
    features
15:             selABInd[selIter] ← i
16:             selIter ← selIter + 1
17:         end if
18:     end for
19:
20:     return selABInd
21: end function
```

---

significant split action bank features are placed in 42×42 matrix, with a margin of one element on the top and bottom of each feature and a left margin of 3 elements, as shown in Fig. 6. A CNN configuration with $p = 4$, $O = 10$ and $N = 42$ is considered and trained using back-propagation algorithm with a batch size of 10 elements. Leave-one-out(LOO) cross-validation strategy is used to evaluate the performance of the proposed approach that resulted in an average classification accuracy of 96.4 %, whose confusion matrix is shown in Fig. 7. The number of epochs the CNN is trained for each action during leave-one-out cross-validation is shown in Fig. 8.

The reported results on UCF sports dataset using leave-one-out cross-validation strategy are shown in Table 3. It can be observed that the performance of the proposed approach is better when compared with the existing algorithms using action bank features for human action recognition on UCF sports dataset. Even though the proposed approach analyzes the entire action bank features to find the most significant split action bank features, only 2.5 % ($\frac{20}{820} \times 100$) of the action bank feature data is used for action recognition.

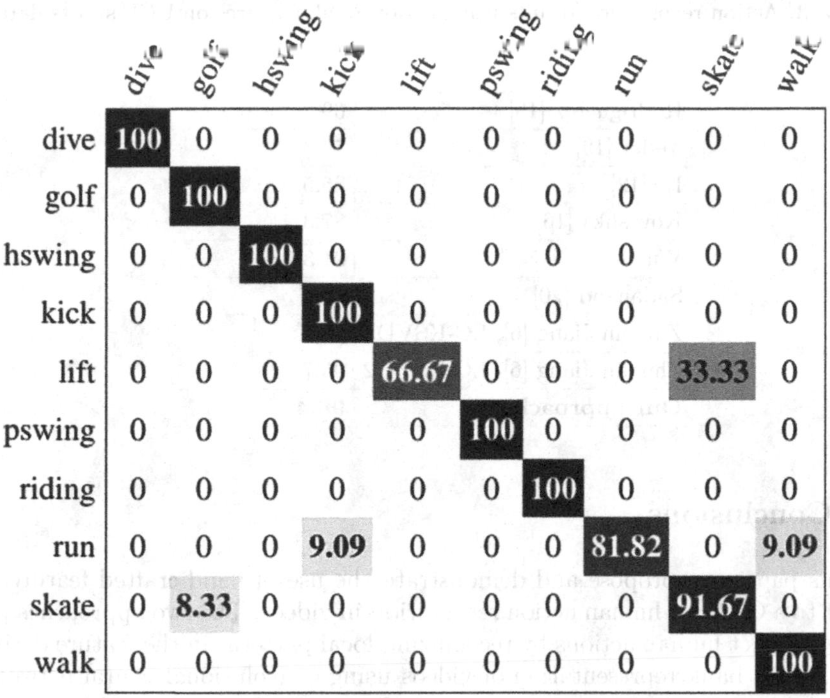

dive golf hswing kick lift pswing riding run skate walk
dive 100 0 0 0 0 0 0 0 0 0
golf 0 100 0 0 0 0 0 0 0 0
hswing 0 0 100 0 0 0 0 0 0 0
kick 0 0 0 100 0 0 0 0 0 0
lift 0 0 0 0 66.67 0 0 0 33.33 0
pswing 0 0 0 0 0 100 0 0 0 0
riding 0 0 0 0 0 0 100 0 0 0
run 0 0 0 9.09 0 0 0 81.82 0 9.09
skate 0 8.33 0 0 0 0 0 0 91.67 0
walk 0 0 0 0 0 0 0 0 0 100

Fig. 7. Confusion matrix of second approach for human action recognition on UCF sports dataset

# of iterations (5 epochs/iteration)
action #
0
50
100
150
200
250
300
350
400
20
40
60
80
100
120
140

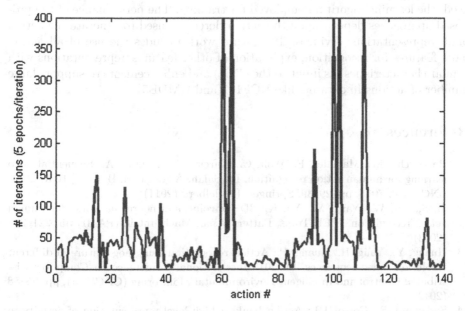

Fig. 8. Plot of action vs # of iterations for convergence of the proposed approach using Leave-one-out cross-validation strategy

Table 3. Action recognition results using action bank features on UCF sports dataset

Approach	Accuracy (in %)
Rodriguerez [18]	69.2
Yeffet [19]	79.3
Le [12]	86.5
Kovashka [16]	87.3
Wu [15]	91.3
Sadanand [20]	95.0
Zhuolin Jiang [6], LC-KSVD1	95.7
Zhuolin Jiang [6], LC-KSVD2	95.7
Our approach	**96.4**

3 Conclusions

In this paper, we propose and demonstrate the use of hand-crafted features as input to a CNN for human action recognition in videos. The two approaches presented, detect human actions by recognizing local patterns in the feature derived from action bank representation of videos using convolutional neural networks. Experimental studies suggests that the performance of the proposed approaches is better when compared with the current state of the art CNN approaches for action recognition and can be fine-tuned further in terms of the derived features used, the learning algorithm employed for training. The performance of the proposed approaches depend upon the action detectors used to generate the action bank representation of videos. The future work includes the use of all action bank features for recognition, exploration of other features/representations with similar characteristics (as input to the CNN) and enhancements to support large number of actions in datasets like UCF101 and HMDB51.

References

1. Baccouche, M., Mamalet, F., Wolf, C., Garcia, C., Baskurt, A.: Sequential deep learning for human action recognition. In: Salah, A.A., Lepri, B. (eds.) HBU 2011. LNCS, vol. 7065, pp. 29–39. Springer, Heidelberg (2011)
2. Ji, S., Xu, W., Yang, M., Yu, K.: 3D convolutional neural networks for human action recognition. IEEE Trans. Pattern Anal. Mach. Intell. (PAMI) **35**, 221–231 (2013)
3. Huang, Y., Yang, H., Huang, P.: Action recognition using hog feature in different resolution video sequences. In: 2012 International Conference on Computer Distributed Control and Intelligent Environmental Monitoring (CDCIEM), pp. 85–88 (2012)
4. Sadanand, S., Corso, J.J.: Action bank: a high-level representation of activity in video. In: IEEE Conference on Computer Vision and Pattern Recognition (CVPR), pp. 1234–1241 (2012)

5. Wang, H., Kläser, A., Schmid, C., Liu, C.L.: Dense trajectories and motion bounding descriptors for action recognition. Int. J. Comput. Vis. **103**, 60–79 (2013)

6. Jiang, Z., Lin, Z., Davis, L.: Label consistent K-SVD: learning a discriminative dictionary for recognition. IEEE Trans. Pattern Anal. Mach. Intell. (PAMI) **35**, 2651–2664 (2013)

7. Baumann, F.: Action recognition with HOG-OF features. In: Weickert, J., Hein, M., Schiele, B. (eds.) GCPR 2013. LNCS, vol. 8142, pp. 243–248. Springer, Heidelberg (2013)

8. Yao, B., Nie, B., Liu, Z., Zhu, S.C.: Animated pose templates for modeling and detecting human actions. IEEE Trans. Pattern Anal. Mach. Intell. (PAMI) **36**, 436–452 (2014)

9. Palm, R.B.: Prediction as a candidate for learning deep hierarchical models of data. Master's thesis, Technical University of Denmark, Asmussens Alle, Denmark (2012)

10. Liu, J., Kuipers, B., Savarese, S.: Recognizing human actions by attributes. In: IEEE Conference on Computer Vision and Pattern Recognition (CVPR), pp. 3337–3344 (2011)

11. Liu, J., Luo, J., Shah, M.: Recognizing realistic actions from videos 'in the wild'. In: IEEE Conference on Computer Vision and Pattern Recognition (CVPR), pp. 1996–2003 (2009)

12. Le, Q., Zou, W., Yeung, S., Ng, A.: Learning hierarchical invariant spatio-temporal features for action recognition with independent subspace analysis. In: IEEE Conference on Computer Vision and Pattern Recognition (CVPR), pp. 3361–3368 (2011)

13. Zhang, Y., Liu, X., Chang, M.-C., Ge, W., Chen, T.: Spatio-temporal phrases for activity recognition. In: Fitzgibbon, A., Lazebnik, S., Perona, P., Sato, Y., Schmid, C. (eds.) ECCV 2012, Part III. LNCS, vol. 7574, pp. 707–721. Springer, Heidelberg (2012)

14. Wang, H., Klaser, A., Schmid, C., Liu, C.L.: Action recognition by dense trajectories. In: IEEE Conference on Computer Vision and Pattern Recognition (CVPR), pp. 3169–3176 (2011)

15. Wu, X., Xu, D., Duan, L., Luo, J.: Action recognition using context and appearance distribution features. In: IEEE Conference on Computer Vision and Pattern Recognition (CVPR), pp. 489–496 (2011)

16. Kovashka, A., Grauman, K.: Learning a hierarchy of discriminative space-time neighborhood features for human action recognition. In: IEEE Conference on Computer Vision and Pattern Recognition (CVPR), pp. 2046–2053 (2010)

17. O'Hara, S., Draper, B.: Scalable action recognition with a subspace forest. In: IEEE Conference on Computer Vision and Pattern Recognition (CVPR), pp. 1210–1217 (2012)

18. Rodriguez, M., Ahmed, J., Shah, M.: Action mach a spatio-temporal maximum average correlation height filter for action recognition. In: IEEE Conference on Computer Vision and Pattern Recognition (CVPR), pp. 1–8 (2008)

19. Yeffet, L., Wolf, L.: Local trinary patterns for human action recognition. In: IEEE 12th International Conference on Computer Vision, pp. 492–497 (2009)

20. Sadanand, S., Corso, J.: Action bank: a high-level representation of activity in video. In: IEEE Conference on Computer Vision and Pattern Recognition (CVPR), pp. 1234–1241 (2012)

Deep Learning in the EEG Diagnosis
of Alzheimer's Disease

Yilu Zhao[⊠] and Lianghua He

Electronics and Information Engineering School,
Tongji University,
No. 4800, Cao'an Road, Jiading District, Shanghai, China
zhaoyilu-2008-19@163.com

Abstract. EEG (electroencephalogram) has a lot of advantages compared to other methods in the analysis of Alzheimer's disease such as diagnosing Alzheimer's disease in an early stage. Traditional EEG analysis
method needs a lot of artificial works such as calculating coherence
between different pair of electrodes. In our work we applied deep learning network in the analysis of EEG data of Alzheimer's disease to fully
use the advantage of the unsupervised feature learning. We studied EEG
based deep learning on 15 clinically diagnosed Alzheimer's disease patients
and 15 healthy people. Each person has 16 electrodes. The time domain
EEG data of each electrode is cut into 40 data units according to the data
size in a period. In our work we first train the deep learning network with
25 data units on each electrode separately and then test with 15 data units
to get the accuracy on each electrode. Finally we will combine the learning
results on 16 electrodes and train them with SVM and get a final result.
We report a 92 % accuracy after combining 16 electrodes of each person. In
order to improve the deep learning model on Alzheimer's disease with the
upcoming new data, we use incremental learning to make full use of the
existing data while decrease the expenses on memory space and computing time by replacing the exising data with new data. We report a 0.5 %
improvement in accuracy with incremental learning.

1 Introduction

1.1 Background of Study in Alzheimer's Disease (AD)

Alzheimer's disease (AD) is the most common cause of dementia all over the
world. It's not a familial disease, so it has more than 15 million people affected
worldwide sporadically. It not only takes a lot of health expenditures and costs
of care but influent caregivers' normal life and work as well. Alzheimer's disease
puts a heavy burden on society. A lot of work and research have been done on
diagnosing it as early as possible.

There are a lot of methods used in the analysis of Alzheimer's disease (AD).
Neuroimaging like CT and MRI plays an important part in the diagnosis of
Alzheimer's disease. But there are also some other causes of dementia, such as
subdural hematoma and brain tumor. Since the changes of early stage Alzheimer's

© Springer International Publishing Switzerland 2015
C.V. Jawahar and S. Shan (Eds.): ACCV 2014 Workshops, Part I, LNCS 9008, pp. 340–353, 2015.
DOI: 10.1007/978-3-319-16628-5_25

patient is subtle. It takes a long time to diagnosis it under clinical follow up testing and analyzing but the accuracy is still comparatively low, with sensitivity of around 80 % and specificity of 70 % according to Sarah Hulbert and Hojjat Adeli in [14].

Luckily, researches by computer aided methods such as Electroencephalogram (EEG) and magneto encephalogram (MEG) has been used in diagnosis of Alzheimer's disease (AD) these years. The advantage of using computer is that it will help clinicians diagnose the disease earlier and easier. This means we will have a higher chance of slowing down the development of the disease.

1.2 EEG Analysis of Alzheimer's Disease (AD)

The EEG analysis of Alzheimer's disease has been used as a tool for differential diagnosis and early detection of AD for decades since Hans Berger first observed pathological EEG sequences in a historically verified AD patient. EEG abnormalities which reflect anatomical and functional damages of the cerebral cortex are frequently shown in AD, that's the main reason why EEG has been intensively researched. Researchers have studied EEG diagnosis of AD with different methods.

One of the most important methods is to diagnose AD by analyzing the brain coherence with EEG. EEG coherence is supposed to give information about connections between different recording electrodes and is used as the evidence of structural and functional connections between cortical areas underlying the recording electrodes. A lot of studies have discovered a pattern of decrease in AD coherence. Sankari et al. in [13] studied on the intrahemispheric, interhemispheric and distal brain coherence in AD patients and discovered a pattern of decrease in AD coherence. Absolo et al. in [1] used sample entropy in the diagnosis of AD and got the accuracy of 77.27 %. Sankari and Adeli in [13] present a probabilistic neural network model which uses features extracted in coherence and wavelet coherence studies in AD and reported a classification accuracy of 100 %. T. Locatelli in [15], evaluate whether short-distance or long-distance coherence is more affected and the possible diagnostic value of coherence analysis in AD.

The second method for EEG analysis is Entropy method. It's helpful in the analysis of Alzheimer's disease because the brain is a nonlinear dynamical system. Entropy methods don't require large data sets. There are different entropy methods used for the analysis of EEG data such as approximate entropy(ApEn), sample entropy(SampEn), and multiscale entropy(MSE). Absolo et al. in [1] used sample entropy to differentiate between groups of AD patients (11) and HCs (11). They report an accuracy of 77.27 % on distinguishing AD patients and Healthy people at all four electrodes using receiver operating characteristic plots to test the accuracy of the SampEn method.

Another method is wavelets. Polikar et al. in [11] compare different types of wavelets for analysis of event-related potentials and get the highest average classification of 78.2 % for Daubechies wavelets. Ahmadlou et al. in [3] use fractal dimension (FD) to model the dynamical changes in the Alzheimer's disease patient's brain. They use the two global features and linear discriminant analysis (LDA) to classify ADs and HCs and report a high classification accuracy of 99.3 % with sensitivity of 100 % and specificity of 97.8 %. Sankari et al. in [12] investigate wavelet coherence of EEG records obtained from AD patients and HCs. Sankari and Adeli in [13] present a probabilistic neural network model for the classification of AD patients and HCs using features extracted in coherence and wavelet coherence.

The last but also important method is Graph theory. These methods use nodes, or vertices, to represent objects, edges to represent relationships, and weights assigned to edges to represent the strength of those relationships. Generally, three types of networks can be created using the graph theory: (1) small-world networks (Han et al. in [8], which are characterized by large clustering of nodes and small path lengths between nodes, (2) ordered networks, which have large clusters and large path lengths, and (3) random networks, which have small clusters and small path lengths. The underlying goal of this method is to measure the connectivity in various brain networks and the change in the connectivity transferring from HC to AD. Ahmadlou et al. in [2] present a chaos-wavelet methodology using a recently developed concept in graph theory called visibility graph (VG) by applying two classifiers to the selected features and discovering effective classification features and mathematical markers. The used classifiers include a radial basis function neural network (RBFNN) (Zhou et al. in [16]) and a two-stage classifier consisting of principal component analysis (PCA) (Al-Naser and Soderstrom in [4]; Meraoumia et al. in [10]) and the RBFNN. They obtained a diagnostic accuracy of 97.7 %using the discovered features and a two-stage classifier (PCA-RBFNN). McBride et al. in [9] compute the inter channel coherence for each pair of 30 EEG electrodes and perform graphical analysis using network features. They use LOO(leave-one-out) cross-validation method and the ML technique SVM as a classifier to differentiate AD from HC and MCI under various resting conditions.

The EEG methods we talked above are proved to be very helpful in the diagnosis of Alzheimer's disease. While those methods have good performance in the analysis, there are still several problems we should consider about. Firstly, how can we improve the accuracy when there are so much noise in the EEG data of a patient. We can reduce the influence of noise by decreasing the sensitivity of the analyzing model as well as by trying to find more features which can represent the data best. Secondly, most diagnosis with EEG data are operated after the whole set of data is totally collected and we can only use it once. How can we make full use of the previous data to help improve the diagnosis this time, in other words, how can we make the outdated data inheritable? We try to make full used of the experiment data by getting the features of it and initialize the learning model with features we used. Deep learning has the advantage of learning features in a unsupervised way. And training data of deep learning network can be flexible, which makes the EEG analysis easier and handy.

1.3 Deep Learning Networks

Deep learning has been researched for many years but there was not too much progress until 2006 when Geoffrey E. Hinton published an article called A Fast Learning Algorithm for Deep Belief Nets on the journal Science. The reason why deep learning is so popular among machine learning researchers is that it has shown great learning and classification performance in many areas, such as speech voice recognition, handwritten character recognition. Deep learning is helpful in machine learning because deep learning can exact useful features layer by layer automatically rather than artificially according to G.E. Hinton in [5]. In their work, they constructed a deep auto encoder by layer wise pretraining with pixel vectors of a picture and doing back propagation with the label of the picture through the whole auto encoder to fine-tune the weights for optimal reconstruction. On a widely used version of the MNIST handwritten digit recognition task they report the error rate is 1.2 %.

1.4 Our Contribution in EEG Diagnosis of Alzheimer's Disease

We applied deep learning algorithm in the EEG analysis of Alzheimer's disease to explore the possibility of making analyzing process easier by unsupervised feature learning. Firstly, by using the handy machine to collect patients' brain data without harming the patients or invading their bodies, we collected EEG data of 30 people from hospital. There are 15 Alzheimer's disease patient and 15 healthy people. After denoising under some thresholds and filters we got the processed data. Secondly, we built the deep learning structure for EEG analysis. The existing deep learning networks are often used to learn figures and we need to adapt it to EEG data by changing the input layer. G.E. Hinton in [5] used the input vector of $784(28 \times 28)$ data as the input layer but we will choose 500 or the data size which can be divided by 500 as the input vector according to the period of EEG data. The amount of hidden layers and amount of nodes in each layer also need to be considered as factors which influence the performance of deep learning network in EEG analysis. We carried out the experiment under different settings of hidden layer amount and nodes amount. And we choose the best setting as our experimental learning structure. Finally, we tried to combine 16 electrodes to get better analysis results. We used SVM to train the learning results on each electrode and got an accuracy of 92 % on the combined electrodes of a person. With the help of deep learning algorithm, we need less data because we can get more learning feature in a unsupervised way, what's more, we can do the learning process as long as we have an integrated data units instead of waiting for the whole EEG data be collected. By using incremental learning, we don't have to save all of the existing EEG data of Alzheimer's disease patients and update the learning model with new data. So the analysis of Alzheimer's disease's EEG data becomes easier to handle and will be used more commonly because all we need is the a machine to collect EEG data.

2   Deep Learning Used in EEG Diagnosis of Alzheimer's Disease

2.1   Theories of Deep Learning Network

Deep learning is a machine learning model which has many hidden layers except the input layer and output layer. It's built to simulate the human brain's working mechanism on learning. Deep learning theory originates from artificial neural network. The difference between deep learning and artificial network is as follows. The neural network initials the weights randomly and trains the whole network by back propagation while deep learning uses layer-wise greedy learning algorithm to initial the learning network and trains the network layer by layer wise network like Restricted Boltzmann Machine. And then use the supervised learning algorithm to optimize the parameters.

The structure of a deep learning network is a several-layer learning network. It includes the input and output layers as well as few hidden layers. The nodes in the same layer are independent and the connections between nodes from two layers are valued by weights. There are weights of two directions between every two layers except for the top layer. So actually the deep learning network is a graph model with a single layer neural network. We also can take the deep learning as a stack of encoding decoding layers. Here we can use RBM or auto encoder in each layer. Geoffrey E. Hinton promoted a framework of deep learning in [6]. In each single layer the learning process is greedily initialize the bias and weights between visible and hidden nodes. After the learning network is greedy initialized layer by layer, we need to fine-tune the network. Hinton promoted a supervised way called wake-sleep algorithm in [7]. It includes two parts called wake and sleep. In the wake stage, it recognizes the features and weights to produce the states of nodes in every layer. It will also adjust the weights in downward direction which are also called generated weights. In the sleep stage we use the note values from the top layer and generated weights which are learnt in wake stage to produce the lower layer's note values and adjust the upward weights which are also called recognize weights.

2.2   EEG Based Deep Learning Model of Alzheimer's Disease Analysis

The reason why we use deep learning network in the EEG analysis of Alzheimer's disease is that it can automatically generate helpful learning features which can make the analysis more efficient. The function of the EEG based deep learning model is like follows (Fig. 1):

The basic process of deep learning in EEG diagnosis of Alzheimer's disease includes two steps: the first step is the unsupervised training of a deep learning network with the EEG data of Alzheimer's disease patient and healthy people. The second step is to fine-tune the deep learning network with labels of the EEG data.

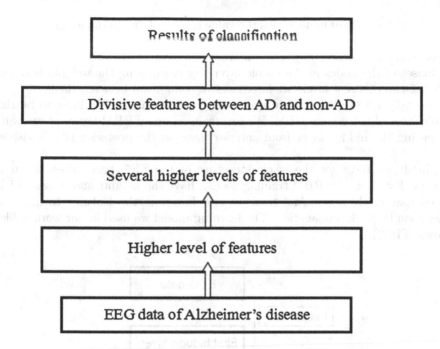

Fig. 1. EEG based deep learning network in the diagnosis of Alzheimer's disease. We use the several layer deep learning network to learn features that can distinguish Alzheimer's disease and healthy people. And then we use these features to do Alzheimers disease analysis.

The detailed learning process is as follows: We choose to use RBM as the basic structure in our work. By training a stack of n layers of RBM we model the deep learning network with weights generated from EEG data. We input the EEG data and train the first layer (bottom layer) of the DL network with random weight and biases, by minimizing the gap between decoded value and input value we can initialize the weight and bias and get the encoded value. We will carry on the learning process by taking the encoded data as the input of the next layer and repeat the first layer's initializing process. By doing the same layer by layer, we get the network with features which can represent the input better because of the restriction of the capacity and sparsity. Then we will use the parameters of every layer to initialize the whole network and then adjust the network in a supervised way with two kind of labels represent Alzheimer's disease or healthy people.

After we have got the data for training and testing from the raw EEG data, we start to input the data into the deep learning network. Firstly, we take the data units we have got earlier as the visible data of the first RBM layer. Take the node i from visible layer and the node j from the hidden layer for example.

$$p(h_j = 1|v) = \sigma \left(\Sigma_{i=1}^{m} w_{ij} \times v_i + c_j \right) \tag{1}$$

here, h_j is the value of node j, v_i is the value of node i, w_{ij} is the weight between i and j. c_j is the bias of hidden layer node j.

We can use the same way to reconstruct the value of the visible layer and get the biases of the nodes in the visible layer. By calculating the weights between nodes of two different layers we have got the connection between them.

Secondly we use the hidden layer data from the first RBM layer to be the visible layer of the second RBM. We can train as many RBM layers as we want by reusing the hidden layer from anterior layer as the posterior layer's visible layer.

Finally we have got the whole structure of the EEG based deep learning network. By layer wise RBM training we can have the weights and biases which are comparatively good. And then we can fine-tune the features to improve generation by back propagation. The learning model we used in our work is like follows (Fig. 2):

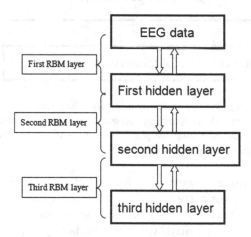

Fig. 2. The structure of the EEG based deep learning network. It's actually a stack of RBMs. We use the EEG data as the visible layer of the first RBM and then we use the hidden layer of the first RBM as the visible layer of the second RBM layer. We can set as many RBM layers as we want.

In our work, we explored the best structure of the EEG based deep learning network by experimenting with different amount of layers and different nodes in each layer. We finally find the layers and the nodes that can minimize the error of classify. We carried out the experiment on all sixteen electrodes with this optimal learning structure.

3   Experiment Design in Alzheimer's Disease (AD)

3.1   Experiment Data

In our work, we try to firstly build the EEG based auto-deep learning network and then evaluate it with 30 people's EEG data. We firstly collect the EEG data

from 15 Alzheimer's disease and 15 healthy people by qualified technicians under
standard procedures and were recorded at rest state, with eyes closed. The data
we used includes the EEG data in 16 electrodes. We can not used the raw data
directly. We need to denoise the raw EEG data by setting thresholds and using
filters.

Our data for training is the EEG data units we have got from the processed
EEG data earlier. The EEG data of Alzheimer's disease is different from the
image data. Compared to the two dimension data, the EEG data is one dimension data. So we should consider the data integrity of our input layer. Sixteen
electrodes were fixed by collodium and placed over the scalp according to the
1020 International System (Fp1, Fp2, Fz, F3, F4, F7, F8, C3, C4, Pz, P3, P4,
T5, T6, O1 and O2). Twenty minutes of EEG were collected with a band pass
of 0.530 Hz and digitized at a sample. By cutting the collected EEG data into
data units we can obtain more data for training and testing compared to using
the data of a single person. Here we have got 30 peoples EEG data with 16
electrodes per person. And by cutting each electrode data into 25 data units for
training and 15 data units for testing we can have 30 batches of data with the
training batch size being 25 and testing batch size 15. In each batch, we will
input into the deep learning network one data unit at a time which means we
have 750 training data units and 450 testing data units for 30 people on one
electrode. The size of the data unit is also called the dim of the data. We will
choose different dims of data such as 500, 1000, 1500, and 2000.

3.2   Experiment Structure of Learning Model

The structure of the net is represented by the amount of nodes in each layer like
2000-500-500-2000-2, here it means we put 2000 dim data into the visible layer
and set the amount of three hidden layers' nodes as 500,500,2000 separately,
and the last layer is the classify layer with two nodes representing Alzheimer's
disease and healthy people. Firstly we should determine how many layers we will
use as hidden layer in our deep learning net and then explore how many nodes
we should use in each layer.

3.3   Incremental Learning for Updating Deep Learning Analysis
    Model in Alzheimer's Disease

The existing EEG data of Alzheimer's disease takes huge memory space and it's
impossible for us to save all the EEG data after we have obtained the analysis
model. But an static analysis will soon become outdated with the upcoming
experimental data. So we use incremental learning in our experiment to update
the analysis model. We make a rule on how to choose data to abandon and how
to replace experimental data with new data to update the model. We make full
use of the existing data by saving learning features in the deep learning network
as parameters. Abandoning part of existing data saves a lot of memory space.
When the amount of new data has reached to our experimental standard we will
train the learning model again with data set which includes part of existing data
and the new data.

4   Experiment in Alzheimer's Disease (AD)

4.1   Experiment to Find the Best Amount of Layers

Firstly, we trained the AD patients data and normal people's data on 16 different electrodes in time domain. We set a deep auto-encoder network which has several hidden layers besides the input layer and output layer. By setting the input visible layer data as 500, we explore how many hidden layers we should use in the EEG based deep learning network by changing the amount of layers. The amount of hidden layer changing from 3 to 7. The experiment sample is with 10 people including 5 healthy people and 5 AD patients. The accuracy of each experiment is the average accuracy of 16 electrodes on the test data set with 150 data of 10 people. The results of the different networks is as follows (Table 1).

Table 1. The accuracy of EEG based deep learning network with different amount of layers.

Layer amount	Accuracy
3	86 %
4	83.6 %
5	78.1 %
6	80.4 %
7	75.33 %

Considering the different performance of different hidden layers in a deep learning network, we choose the amount of layers which has the best performance as our default amount of layers in the followed experiments. So we will continue our experiment with three hidden layers.

Secondly, we will do experiment to find the best amount of nodes in each layer.

We carry out the experiment on the influence of the amount of nodes in each layer and decide how many nodes we should set for each hidden layer. We will fix two of the three hidden layers and change one layer to see the difference. Here is the accuracy according to our experiment design on the same data set.

We will set the first two layers as 500 and 500 nodes. And change the third layer. The accuracy of each experiment is the average accuracy of 16 electrodes on the test data set with 450 data of 30 people. The results are shown as follows (Figs. 3, 4 and 5 and Tables 2, 3 and 4):

Secondly, we change the amount of the second layer and fix the first layer as 500 and the third layer as 2000 nodes.

Finally, we will change the first layer's nodes and fix the rest and fix the second layer as 500, the third as 2000.

With the results from the above experiments, we choose the best model of deep learning network in the analysis of Alzheimer's disease. We set the amount

Table 2. The accuracy of EEG based deep learning network with different amount of nodes in 3rd hidden layer.

Nodes	Accuracy	Nodes	Accuracy
500	80 %	2500	82 %
1000	82.9 %	3000	83.7 %
1500	78.2 %	3500	81.2 %
2000	85 %	4000	84 %

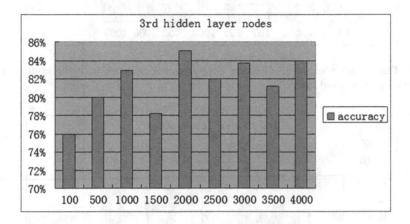

Fig. 3. The accuracy of the EEG based deep learning network with the node amount changing in the third layer.

Table 3. The accuracy of EEG based deep learning network with different amount of nodes in 2nd hidden layers.

Nodes	Accuracy	Nodes	Accuracy
100	77 %	500	82 %
200	79 %	1000	84.9 %
300	76 %	2000	80.33 %
400	80.7 %	3000	76 %

Table 4. The accuracy of EEG based deep learning network with different amount of nodes in 1st hidden layer.

Nodes	Accuracy	Nodes	Accuracy
100	75.1 %	500	85.1 %
200	80.5 %	1000	83.9 %
300	79.8 %	2000	82.3 %
400	83.4 %	3000	77 %

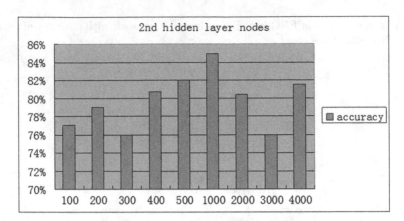

Fig. 4. The accuracy of the EEG based deep learning network with the node amount changing in the second layer.

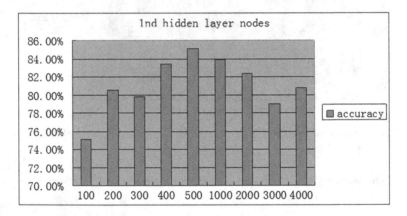

Fig. 5. The accuracy of the EEG based deep learning network with the node amount changing in the first layer.

of layers as 3 and the amount of nodes in each layer as 500, 1000, 2000 separately. Now we work on exploring the relation between the change of training data size and accuracy of classification. According the sampling frequency, the smallest experimental data size is 500, which is sampled in a complete period. We also tried different data sizes which are in integral multiple of 500. Based on the former learning model we choose. We have experimented with the input data size from 500 to 10000, when the data size is 500, the accuracy is under 50 %, but with the increasing of the data size, the accuracy also increased to be around 90 %, when the data size become 2000, the accuracy also reach the peak which is 92.5 %, then we got the decrease of the accuracy from 2000 to 10000. So we choose the training data size as 2000.

4.2 Experiment to Update Analysis Model with Incremental Learning

There are two steps in the incremental learning of the analysis model. Firstly, we define a rule on how to abandon useless data after the deep learning model on EEG data has been built. Secondly, we use the saved data and model to train with new data for updating the model. According to the different analysing result on different data, we calculate the average error rate of each data in the train set. Because the data which has a lower classification error can be better represented by the feature in the analysis model, we choose to abandon them. The data with low classification accuracy will be kept until the next round of training. By replacing 250 of 750 existing data with new data, we update the analysis model by making full use of the features in the deep learning model. The amount of misclassified data of 750 testing data in each electrode changes differently. But in 12 of the total 16 electrodes, we got less misclassified data than updating the model. The combination accuracy of 16 electrodes after incremental learning is 0.5 % higher than it was before incremental learning.

4.3 Comparison of Deep Learning Analysis with Other EEG Analysis Method on Alzheimer's Disease

There have been a lot of research in the EEG analysis of Alzheimer's disease. But the problem is that different methods are studied on different datasets. So it's a little bit hard for us to compare our method to others in a general dataset. Here are some methods we mentioned in the introduction and the different accuracies they have got. SampEn have got 72.3 % by testing 11 AD patient and 11 healthy people. Graph theory/SVM has got the result as 93.8 with 17 AD patients and 15 healthy people. Wavelets-chaos has the result 99.3 % with the sample 20&7(AD&HC). Since the analysis of Alzheimer's disease is a complicated process, we should consider both the accuracy of analysis and the practicability of the method. Using deep learning in the analysis of Alzheimer's disease can have both the high accuracy and the low cost of time and expenses.

5 Conclusions

In our study we applied the deep learning network into the EEG diagnosis of Alzheimer's disease. We modify the structure of deep learning network to adapt to the feature of EEG data and get the suitable structure of the deep learning network by changing the layers and the amount of nodes on each layer. We conducted experiment on 30 people with 15 Alzheimer's disease and 15 healthy people by imputing 16 electrodes' data of each people separately into the deep learning network. After all the deep learning on the 16 isolated electrodes, we combine them to get a final result with the help of SVM and finally get the accuracy of 92 %. Here we combine the advantages of the EEG analysis and deep learning network by analyzing the Alzheimer's disease in an early stage with flexible size of data. We have the possibility to do early stage analysis of Alzheimer's disease more commonly since we only need the machine to collect EEG data and can use deep learning network to analyze the data collected as long as the data size is larger than a period's data size. We also applied incremental learning in the update of the existing analysis model and have improved the analysis result with 0.5 % higher accuracy. In our future work, we will try to improve the deep learning network on EEG data of Alzheimer's disease by applying the algorithms of transfer learning to make full use of the existing features in other well studied deep learning models.

Acknowledgement. This work was supported by National Natural Sciences Foundation of China (No.61272267,61170220,51075306,61273261), Program for New Century Excellent Talents in University(NCET-11-0381), Fundamental Research Funds for the Central Universities, State Key Laboratory of Software Engineering.

References

1. Absolo, D., Hornero, R., Espino, P., Alvarez, D., Poza, J.: Entropy analysis of the EEG background activity in Alzheimer's disease patients. Physiol. Meas. **27**, 241–253 (2006)
2. Ahmadlou, M., Adeli, H., Adeli, A.: New diagnostic EEG markers of the Alzheimer's disease using visibility graph. J. Neural Transm. **117**, 1099–1109 (2010)
3. Ahmadlou, M., Adeli, H.: Fuzzy synchronization likelihood with application to attention-deficit/hyperactivity disorder. Clin. EEG Neurosci. **42**, 6–13 (2011)
4. Al-Naser, M., Soderstrom, U.: Reconstruction of occluded facial images using asymmetrical principal component analysis. Integr. Comput. Aided Eng. **19**, 273–283 (2012)
5. Hinton, G.E., Salakhutdinov, R.: Reducing the dimensionality of data with neural networks. Science **313**(5786), 504–507 (2006)
6. Hinton, G.E., Osindero, S., Teh, Y.: A fast learning algorithm for deep belief nets. Neural Comput. **18**, 1527–1554 (2006)
7. Hinton, G.E., Dayan, P., Frey, B.J., Neal, R.M.: The wake-sleep algorithm for unsupervised neural networks. Science **268**, 1558–1161 (1995)
8. Han, F., Wiercigroch, M., Fang, J.A., Wang, Z.: Excitement and synchronization of small-world neuronal networks with short-term synaptic plasticity. Int. J. Neural Syst. **21**, 415–425 (2011)
9. McBride, J., Zhao, X., Munro, N., Smith, C., Jicha, G., Jiang, Y.: Resting EEG discrimination of early stage Alzheimer's disease from normal aging using inter-channel coherence network graphs. Ann. Biomed. Eng. **41**, 1233–1242 (2013)
10. Meraoumia, A., Chitroub, S., Bouridane, A.: 2D and 3D palmprint information, PCA and HMM for an improved person recognition performance. Integr. Comput. Aided Eng. **20**, 303–319 (2013)
11. Polikar, R., Topalis, A., Green, D., Kounios, J., Clark, C.: Comparative multiresolution wavelet analysis of ERP spectral bands using an ensemble of classifiers approach for early diagnosis of Alzheimer's disease. Comput. Biol. Med. **37**, 542–558 (2007)
12. Sankari, Z., Adeli, H., Adeli, A.: Wavelet coherence model for diagnosis of Alzheimer's disease. Clin. EEG Neurosci. **43**, 268–278 (2012)

13. Sankari, Z., Adeli, H.: Probabilistic neural networks for EEG-based diagnosis of Alzheimer's disease using convolutional wavelet coherence. J. Neurosci. Methods **197**, 165–170 (2011)
14. Hulbert, S., Adeli, H.: EEG/MEG- and imaging-based diagnosis of Alzheimer's disease. Rev. Neurosci. **24**(6), 563–576 (2013)
15. Locatellia, T., Cursia, M., Liberatib, D., Franceschia, M., Comia, G.: EEG coherence in Alzheimers disease. Electroencephalogr. Clin. Neurophysiol. **106**, 229–237 (1998)
16. Zhou, L.R., Ou, J.P., Yan, G.R.: Response surface method based on radial basis functions for modeling large-scale structures in model updating. Comput. Aided Civil Infrastruct. Eng. **28**, 210–226 (2013)

Pedestrian Detection with Deep Convolutional Neural Network

Xiaogang Chen, Pengxu Wei, Wei Ke, Qixiang Ye, and Jianbin Jiao[✉]

School of Electronic, Electrical and Communication Engineering,
University of Chinese Academy of Science, Beijing, China
jiaojb@ucas.ac.cn

Abstract. The problem of pedestrian detection in image and video frames has been extensively investigated in the past decade. However, the low performance in complex scenes shows that it remains an open problem. In this paper, we propose to cascade simple Aggregated Channel Features (ACF) and rich Deep Convolutional Neural Network (DCNN) features for efficient and effective pedestrian detection in complex scenes. The ACF based detector is used to generate candidate pedestrian windows and the rich DCNN features are used for fine classification. Experiments show that the proposed approach achieved leading performance in the INRIA dataset and comparable performance to the state-of-the-art in the Caltech and ETH datasets.

1 Introduction

Pedestrian detection has been one of the most extensively studied problems in the past decade. One reason is that pedestrians are the most important objects in natural scenes, and detecting pedestrians could benefit numerous applications including video surveillance and advanced driving assistant systems. The other reason is that pedestrian detection has been the touchstone of various computer vision and pattern recognition methods. The improvement of pedestrian detection performance in complex scenes often indicates the advance of relevant methods.

Two representative works in pedestrian detection are the VJ [1] detector and HOG [2] detector. The VJ detector employed the framework of using simple Haar-like features and cascade of boosted classifiers, achieved a very fast detection speed. This framework is further developed by Dollár *et al.* who proposed the Integral Channel Features [3], including multiple types of features (grayscale, LUV color channels, gradient magnitude, etc.) that can be quickly computed using integral images. The Integral Channel Features is simple but effective, and widely used in many state-of-the-art pedestrian detectors [4–8]. On the other hand, the success of HOG detector encouraged the usage of complex features, like Local Binary Pattern [9], Dense SIFT [10] and Covariance Descriptor [11], etc. Also, based on HOG, Felzenszwalb *et al.* proposed the Deformable Part Based Model (DPM) [12] which made a breakthrough in pedestrian detection.

© Springer International Publishing Switzerland 2015
C.V. Jawahar and S. Shan (Eds.): ACCV 2014 Workshops, Part I, LNCS 9008, pp. 354–365, 2015.
DOI: 10.1007/978-3-319-16628-5_26

Since the feature extraction pipelines in above methods are designed manually, they can be categorized as hand-craft features. In recent researches, with the steady advance of deep learning [13] and unsupervised feature learning [14], learnable features gain significant attentions. Specially, the Deep Convolutional Neural Network (DCNN) proposed by Krizhevsky *et al.* [15] achieved record-breaking results in ImageNet Large Scale Visual Recognition Challenge 2012. Afterwards, its specific network structure has been widely used in image classification and object detection [16–19]. In [16], Donahue *et al.* showed that features generated from a classifying CNN perform excellently in related vision tasks, implying that DCNN can be used as a generic feature extractor.

In the field of pedestrian detection, many feature learning and deep learning methods have been introduced recently. In [20], Sermanet *et al.* proposed a two layers convolutional model and layers were pre-trained by convolutional sparse coding. In [21], Ouyang *et al.* conducted Restricted Boltzmann Machine (RBM) in modeling mutual visibility relationship for occlusion handling. And in [22] authors further cooperated with Convolutional Neural Network, and proposed a joint deep learning framework that jointly consider four key components in pedestrian detection: feature extraction, deformation model, occlusion model and classifier.

In [20], Convolutional Neural Network has been successfully applied in pedestrian detection, where the used network structure have only 2 layers. In contrast, Krizhevsky's CNN [15] that has 7 layers is much deeper. In this paper, we try to make it clear that whether the usage of a larger and deeper Convolutional Neural Network for feature extraction can further improve the performance of pedestrian detection or not. When using large CNN for feature extraction, the commonly used "sliding-window" detection paradigm is hard to work for the computational efficiency problem [17]. To improve detection efficiency, pre-localization approaches such as the selective search method [23] has been used to generate proposal regions, which is a "recognition-using-regions" paradigm [24]. In pedestrian detection task, however, it is observed in experiments that the "recognition-using-regions" paradigm based on "Selective Search" is infeasible. The reason is that pedestrian detection requires precise localization before it can obtain a good detection performance, but the selective search method cannot provide precise localization of pedestrians.

In this paper, we propose to cascade simple Aggregated Channel Features (ACF) and rich Deep Convolutional Neural Network (DCNN) features for efficient and effective pedestrian detection in complex scenes. To generate precisely localized candidate pedestrian windows, we employ a cascade of Adaboost classifiers on Aggregated Channel Features (ACF detector) [5]. We reduce the stage number of the original ACF detector to two so that most of pedestrian windows could be kept for fine classification. We propose to use the DCNN pre-trained on a large image set so that the network parameters are well learned. On the candidate pedestrian windows, DCNN features and a linear SVM classifier are used to perform pedestrian classification and fine detection. Our proposed approach is also an attempt to combine hand-craft features with learned features, which is seldom investigated in existing works. The flowchart of the proposed pedestrian detection approach is as Fig. 1.

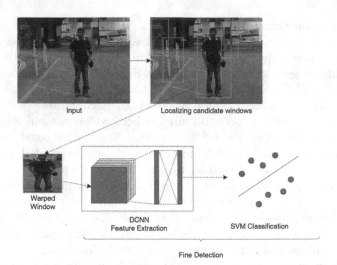

Fig. 1. Flowchart of the proposed pedestrian approach. A fast candidate window localization method is first applied to input image, then all candidate windows are warped to require size of the fine detection stage, where they will be classified. For details, see the following sections.

The rest of this paper is organized as follows: in Sect. 2 we introduce our pedestrian detection approach. In Sect. 3, we report the experiment results on three benchmark datasets, INRIA, Caltech, ETH. In Sect. 4, we conclude our works and have a discussion.

2   Pedestrian Detection Approach

The proposed pedestrian detection approach can be regarded as a two-stages system. In the first stage, we conduct simple channel features and boosted classifiers, in order to rapidly filter out as many negative windows as possible, while keeping all the positive windows. Then in next, the fine detection stage, we use DCNN to extract features from these windows, and SVM for classification. The flowchart of proposed approach is draw in Fig. 1.

2.1   Localizing Candidate Windows

In this stage, we employ the Aggregate Channel Features detector [5] for localizing candidate windows. Following the "channel features + boosted classifier" schema, three types of channel features are used: normalized gradient magnitude, histogram of oriented gradients and LUV color channels.

These channels are first generated from input images, then summed up in 4×4 pixel grid and smoothed, yields a 5120 dimensions feature pool. Next, a bootstrapping iteration is conducted over the feature pool to construct a cascade of classifiers. Here we built a cascade of 2 stages, that combined 32 and 128 classifiers in each.

Since our purpose is to quickly filter out as many negative windows as possible, while keeping all positive windows, it is important to measure performance with ground truth cover rate and runtime. The ground truth cover rate measures how many positive windows can be detected by proposal windows. A ground truth window is consider detected by a proposal given that their area of overlap exceed 50 % [25]. In Table 1, we compare the proposed method with Selective Search [23] and Objectness [26] on INRIA pedestiran dataset, using the measures indicated above.

Table 1. Comparison of region proposal methods.

	Selective search	Objectness	ACF	Selective search on Caltech dataset
Cover rate	97.62 %	93.55 %	**98.13 %**	1.9 %
Runtime	~4 s	~4 s	<**0.5 s**	~4 s

It can be seen that although Selective Search achieves comparable cover rate on INRIA dataset, the processing time is much slower than ACF. Notice that Selective Search is a segmentation based method that performance is strongly affected by image quality. So we further conduct Selective Search on Caltech dataset, where the cover rate dramatically down to 1.9 percents because the image quality of Caltech dataset is much worse than INRIA. The Objectness method did not show advantages in either cover rate or runtime.

On the other hand, pedestrian detection requires the bounding box tightly surround pedestrians, where the general region proposal methods might fail because most of them are designed to capture object in any aspect ratio, ignoring the fact that pedestrians are more like rigid object. This inaccuracy will affect applications built on pedestrian detection results. As in Fig. 2, the blue box indicates ground truth annotation, and the red box is the candidate window overlapping mostly with ground truth. Due to the region grouping setting, Selective Search methods generate candidate windows that cover only half of the pedestrian body. Since ACF detector is designed for pedestrian detection, it won't suffer from this problem.

2.2 Fine Detection

In the fine detection stage, to classify the candidate windows passed the previous stage, DCNN is employed. Following the network architecture proposed by Krizhevsky et al. [15], we used the RCNN package [17] which utilize the Caffe [27] to implement DCNN.

The architecture of used DCNN is presented in Fig. 3, which has 7 layers. Notice that the DCNN requires input images of 227 × 227 pixels size, so first we

Fig. 2. Fail cases of using selective search. The candidate windows (red rectangles) that mostly overlapping with ground truth (blue rectangles) only cover half of the pedestrian (Color figure online).

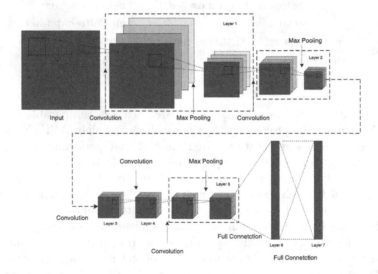

Fig. 3. Architecture of DCNN

simply warp candidate windows to the required size. Notice that the warping causes a distortion of images which will affect the information carried within, however it is observed from experiments that warping works well. In the first layer, the warped images are filtered with 96 kernels of size $11 \times 11 \times 3$ pixels with a stride of 4 pixel, then max-pooling is applied in 3×3 grid. The second layer has the same pipeline as first layer, with 256 kernels of size $5 \times 5 \times 48$, and max-pooling in 3×3 grid. Afterwards, there are two convolution layers without pooling, which both contains 384 kernels. In the fifth layer, again, the output of previous layer is first convoluted with 256 kernels then applied spatial max-pooling in 3×3 pixel grid. The last two layers of the network are fully connected layer, which both contains 4096 nodes respectively. The DCNN eventually output features of 4096 dimensions from the last layer. The activation function used in the convolution and full connected layer is Rectified Linear function $f(x) = \max(0, x)$. For more details about network parameters and training protocol, we refer reader to [15].

After obtaining the training features, we train a linear SVM for classification. As common practice [12], a bootstrapping process is conducted to improve classification. We mine hard negative samples from training dataset and retrain SVM with it. It is worth nothing that, the bootstrapping converges quickly in single iteration, compare with DPM [12] that runs in multiple iterations, indicating the capacity of DCNN in modeling complex images.

3   Experiments

In this section, we evaluate the proposed pedestrian detection approach on three well-known benchmarks: INRIA, Caltech and ETH datasets. Before getting into specific experiments, there are some issues in using DCNN, which are the usage of pre-trained model and the feature layers. In first subsection, we discuss these problems and show the comparison experiment results. The results on INRIA, Caltech and ETH are presented in second and third subsections. All evaluations follow the protocols proposed by Dollár *et al.* [28].

3.1   Model Setting

As common practice, we train the detector with INRIA dataset. However, it is obviously insufficient to train the DCNN model with INRIA dataset, since the model contains millions of parameters, that will easily leads to over-fitting. In [16], Donahue *et al.* generated features from a network pre-trained with the ImageNet dataset, and successfully apply them to other vision tasks, shows generalization capacity of DCNN. Here we follow the same strategy that use a pre-trained CNN as a blackbox feature extractor. We use the pre-trained models provided in the RCNN package, which were trained on the PASCAL VOC 2007, 2012 and ImageNet dataset, respectively.

On the other hand, another issue concerned is the usage of feature layers, recent DCNN based research [17] suggested that the FC6 (Short for Fully Connected layer 6) features usually outperforms the FC7 and POOL5 (Short for Pooling layer 5) layer features. To achieve comprehensive understanding, in our experiments, we compare the performance of different model-layer combinations. The result are presented in Fig. 4.

It can be seen that the DCNN pre-trained from PASCAL VOC 2007 dataset generally outperforms models trained from VOC 2012 and ImageNet datasets. The model pre-trained in ImageNet dataset perform unexpectedly poor, considering that both PASCAL VOC and ImageNet datasets contain the category of person, which is no exactly pedestrian but share most characteristics with it, so all models are expected to have similar performances. We consider the performance gap might be that the PASCAL VOC dataset contain less categories of objects than ImageNet dataset, which is 20–200, then the DCNN model trained

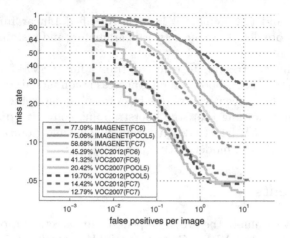

Fig. 4. Comparison of different model settings on INRIA dataset

from VOC can have more parameters to characterize persons that leads to better performance.

For feature layer selection, with different pre-trained model, the FC7 features performed best. These experiments gave the hints for constructing better DCNN model for pedestrian detection, so in this work we choose to use the pre-trained model from VOC 2007 and FC7 layer feature for further study.

3.2   INRIA Dataset

We first evaluate the proposed approach in INRIA dataset. In this experiment, we mirror the positive samples for augmentation, then generate features from the DCNN for SVM training, and run single iteration of bootstrapping.

Evaluation is based on the fixed INRIA annotations provided by [20], which include additional "ignore" labels for pedestrians miss labelled in original annotations [2]. To compare with major state-of-the-art pedestrian detectors, the log-average miss rate is used and is computed by averaging the area under curve (AUC) from 9 discrete false positive per image (FPPI) rates [25]. We plot all comparing DET curves (miss rate versus FPPI) in Fig. 5. The abbreviation DCNN short for Deep Convolutional Neural Network indicates our results.

The proposed system outperforms most state-of-the art pedestrian detectors with an average 12.79 % miss rate. We can see that about 30 % performance improvement is obtained compare with ConvNet [20], the substantial gain proved that the larger and deeper Convolutional Neural Network indeed improve pedestrian detection. In the other hand, 21 % improvements is gained from ACF detector [5], which is used for region proposal in our approach. Notice that original ACF detector, 4 stages of boost classifiers are used while 2 stages in our approach, indicating that the DCNN outperforms the higher stages boost classifiers.

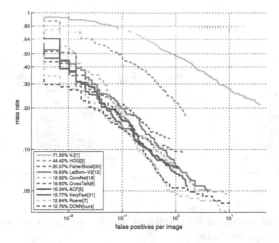

Fig. 5. Comparison of different methods on INRIA dataset. DCNN refers to our proposed detection approach.

3.3 Caltech and ETH Datasets

In this section, we train the proposed system with INRIA dataset, then apply it to Caltech and ETH dataset. We do not bootstrap our system on these datasets, in order to discover if the features extracted from DCNN can generate to other datasets. Comparison results are plot in Figs. 6 and 7 respectively.

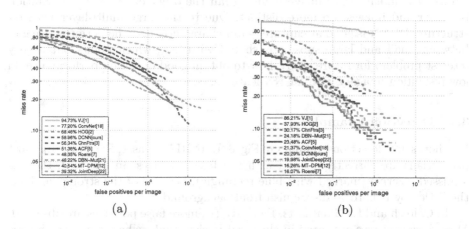

Fig. 6. Experimental results on Caltech dataset. (a) Results on '*Reasonable*' pedestrians. (b) Results on '*Large*' pedestrians

We can observe that, although the proposed system achieve excellent performance on the INRIA dataset, results on Caltech and ETH datasets are less

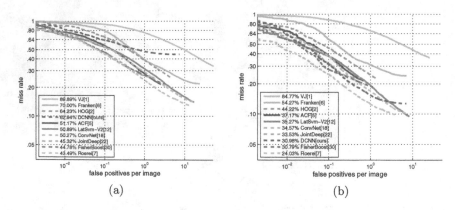

Fig. 7. Experimental results on ETH dataset. (a) Result on '*Reasonable*' pedestrians. (b) Result on '*Large*' pedestrians

impressive. The proposed approach performs poorly in the "Reasonable" subset, however, in the "Large" subset it is better than ConvNet as in the INRIA experiments. Notice that the performance of DCNN is comparable with another CNN based method JointDeep [22] which employed deformation and occlusion handling pipelines in their method.

We conclude the decrease of performance might have several reasons. First is that we do not conduct any specific fine tuning on both dataset which might affect the performance. Another reason might relate to image quality, the INRIA dataset has much better image quality than the other datasets, the resolution is higher and images are more distinct. Due to the large multi-layer network structure, DCNN is good at capture detail characteristics. The performance gap between small and large subsets is shared with [20], which points out a very interest problem for future study, how to obtain good detection performance on low resolution imagery with DCNN features.

3.4 Detection Examples

We show some detection examples in Fig. 8. In INRIA dataset, Fig. 8(a)–(c), most pedestrians are correctly located with few false positives. In Fig. 8(a), there is a missed positive (marked with blue rectangle), because of the strong sunlight, the little boy is hard to distinguish from background.

In Caltech and ETH datasets, Fig. 8(d)–(i), more false positives are observed. Most false positives appeared in clustered backgrounds, where trees, trash cans and billboards are prone to be recognized as pedestrians. As we analyzed in previous section, a specific hard samples mining procedure will help to reduce these types of false alarms. Also, conducting deformation and occlusion handling pipeline in pedestrian detection will boost the performance in crowd scenes.

(a) (b) (c)

(d) (e) (f)

(g) (h) (i)

Fig. 8. Detection examples (Color figure online)

4 Conclusions

We proposed a state-of-the-art pedestrian detection method, which combines the successful Aggregated Channel Features detector and Deep Convolutional Neural Network. An ACF detector is used to generate candidate pedestrian windows, and a DCNN based detector is used to extract features for classification. Benefitting from the large network structure, the proposed method gains substantial improvement over previous CNN based methods, and achieves leading performance in INRIA dataset and comparable performance in Caltech and ETH datasets.

The proposed method does not conduct fine tuning on the experiment datasets and does not include a specified pipeline for occlusion handling, leaving room for further improvements. In addition, improving the performance of DCNN in low resolution images is worth working.

Acknowledgement. This work was supported in Part by National Basic Research Program of China (973 Program) with Nos. 2011CB706900, 2010CB731800, and National Science Foundation of China with Nos. 61039003, 61271433 and 61202323.

References

1. Viola, P., Jones, M.J.: Robust real-time face detection. Int. J. Comput. Vis. **57**, 137–154 (2004)
2. Dalal, N., Triggs, B.: Histograms of oriented gradients for human detection. In: Proceedings of IEEE Conference on Computer Vision and Pattern Recognition, pp. 886–893 (2005)
3. Dollár, P., Zhuowen, T., Perona, P., Belongie, S.: Integral channel features. In: British Machine Vision Conference, vol. 2 (2009)
4. Dollár, P., Belongie, S., Perona, P.: The fastest pedestrian detector in the west. In: British Machine Vision Conference, vol. 2 (2010)
5. Dollár, P., Appel, R., Belongie, S., Perona, P.: Fast feature pyramids for object detection. IEEE Trans. Pattern Anal. Mach. Intell. **36**, 1532–1545 (2014)
6. Mathias, M., Benenson, R., Timofte, R., Gool, L.V.: Handling occlusions with franken-classifiers. In: Proceedings of IEEE International Conference on Computer Vision, pp. 1505–1512 (2013)
7. Benenson, R., Mathias, M., Tuytelaars, T., Gool, L.V.: Seeking the strongest rigid detector. In: Proceedings of IEEE Conference on Computer Vision and Pattern Recognition, pp. 3666–3673 (2013)
8. Dollár, P., Appel, R., Kienzle, W.: Crosstalk cascades for frame-rate pedestrian detection. In: IEEE European Conference on Computer Vision, pp. 645–659 (2012)
9. Xiaoyu, W., Han, T., Shuicheng, Y.: An HOG-LBP human detector with partial occlusion handling. In: Proceedings of IEEE International Conference on Computer Vision, pp. 32–39 (2009)
10. Vedaldi, A., Gulshan, V., Varma, M., Zisserman, A.: Multiple kernels for object detection. In: Proceedings of IEEE International Conference on Computer Vision, pp. 606–613 (2009)
11. Tuzel, O., Porikli, F., Meer, P.: Pedestrian detection via classification on riemannian manifolds. IEEE Trans. Pattern Anal. Mach. Intell. **30**, 1713–1727 (2008)
12. Felzenszwalb, P.F., Girshick, R.B., McAllester, D., Ramanan, D.: Object detection with discriminatively trained part-based models. IEEE Trans. Pattern Anal. Mach. Intell. **32**, 1627–1645 (2010)
13. Bengio, Y.: Learning deep architectures for AI. Found. Trends Mach. Learn. **2**, 1–127 (2009)
14. Bengio, Y., Courville, A., Vincent, P.: Representation learning: a review and new perspectives. IEEE Trans. Pattern Anal. Mach. Intell. **35**, 1798–1828 (2013)
15. Krizhevsky, A., Sutskever, I., Hinton, G.E.: ImageNet Classification with Deep Convolutional Neural Networks. In: Neural Information Processing Systems (2012)
16. Donahue, J., Yangqing, J., Vinyals, O., Hoffman, J., Ning, Z., Tzeng, E., Darrell, T.: DeCAF: a deep convolutional activation feature for generic visual recognition. In: International Conference on Machine Learning (2014)
17. Girshick, R., Donahue, J., Darrell, T., Malik, J.: Rich feature hierarchies for accurate object detection and semantic segmentation. In: IEEE Conference on Computer Vision and Pattern Recognition (2014)
18. Sermanet, P., Eigen, D., Zhang, X., Mathieu, M., Fergus, R., LeCun, Y.: OverFeat: integrated recognition, localization and detection using convolutional networks. In: International Conference on Learning Representations (2014)
19. Zeiler, M.D., Fergus, R.: Visualizing and understanding convolutional neural networks. In: IEEE European Conference on Computer Vision, pp. 818–833 (2014)

20. Sermanet, P., Kavukcuoglu, K., Chintala, S., LeCun, Y.: Pedestrian detection with unsupervised multi-stage feature learning. In: Proceedings of IEEE Conference on Computer Vision and Pattern Recognition, pp. 3626–3633 (2013)
21. Ouyang, W., Zeng, X., Wang, X.: Modeling mutual visibility relationship in pedestrian detection. In: Proceedings of IEEE Conference on Computer Vision and Pattern Recognition, pp. 3222–3229 (2013)
22. Ouyang, W., Wang, X.: Joint deep learning for pedestrian detection. In: Proceedings of IEEE Conference on Computer Vision and Pattern Recognition, pp. 2056–2063 (2013)
23. Van de Sande, K., Uijlings, J., Gevers, T., Smeulders, A.: Segmentation as selective search for object recognition. In: Proceedings of IEEE International Conference on Computer Vision, pp. 1879–1886 (2011)
24. Chunhui, G., Lim, J., Arbeláez, P., Malik, J.: Recognition using regions. In: Proceedings of IEEE Conference on Computer Vision and Pattern Recognition, pp. 1030–1037 (2009)
25. Dollar, P., Wojek, C., Schiele, B., Perona, P.: Pedestrian detection: an evaluation of the state-of-the-art. IEEE Trans. Pattern Anal. Mach. Intell. **34**, 743–761 (2012)
26. Alexe, B., Deselaers, T., Ferrari, V.: Measuring the objectness of image windows. IEEE Trans. Pattern Anal. Mach. Intell. **34**, 2189–2202 (2012)
27. Yangqing, J.: Caffe: An Open Source Convolutional Architecture for Fast Feature Embedding (2012). http://caffe.berkeleyvision.org/
28. Dollár, P., Wojek, C., Schiele, B., Perona, P.: Pedestrian detection: a benchmark. In: Proceedings of IEEE Conference on Computer Vision and Pattern Recognition, pp. 304–311 (2009)
29. Luo, P., Tian, Y., Wang, X., Tang, X.: Switchable deep network for pedestrian detection. In: Proceedings of IEEE Conference on Computer Vision and Pattern Recognition, pp. 899–906 (2013)
30. Shen, C., Wang, P., Paisitkriangkrai, S., van den Hengel, A.: Training effective node classifiers for cascade classification. Int. J. Comput. Vis. **103**, 326–347 (2013)
31. Gool, L.V., Mathias, M., Timofte, R., Benenson, R.: Pedestrian detection at 100 Frames Per Second. In: Proceedings of IEEE Conference on Computer Vision and Pattern Recognition, pp. 2903–2910 (2012)

Workshop on Scene Understanding for Autonomous Systems

Surface Prediction for a Single Image of Urban Scenes

Foat Akhmadeev[✉]

Institute of Computer Mathematics and Information Technologies,
Kazan Federal University, Kazan, Russian Federation
foat.akhmadeev@gmail.com

Abstract. In the paper we present a novel method for three-dimensional scene recovering from one image of a man-made environment. We use image segmentation and perspective cues such as parallel lines in space. The algorithm models a scene as a composition of surfaces (or planes) which belong to their vanishing points. The main idea is that we exploit obtained planes to recover neighbor surfaces. Unlike previous approaches which use one base plane to place reconstructed objects on it, we show that our method recovers objects that lie on different levels of a scene. Furthermore, we show that our technique improves results of other methods. For evaluation we have manually labeled two publicly available datasets. On those datasets we demonstrate the ability of our algorithm to recover scene surfaces in different conditions and show several examples of plausible scene reconstruction.

1 Introduction

Generally, some computer vision and image processing applications may benefit from scene reconstruction. E.g. some reconstructed objects like pillars can be used to avoid them by robots [13]. This knowledge may greatly improve object detection and their understanding [12,13,18,23]. Moreover, full scene recovering can be applied in rendering synthetic objects into photos [13,20,22]. Image surfaces can be used in image segmentation, e.g. road, buildings [8]. In addition, spatial layout and spatial understanding can be improved by scene recovering [12,23].

A plausible recovery of a 3D scene is a major problem for most single view 3D reconstruction algorithms. Look at the image in Fig. 1a. From this single image you can see the basic structure of the scene, which consists of orientation of main surfaces like walls, floor and ceiling. Nevertheless, automatic reconstruction of a scene from a monocular image is challenging due to the ambiguity of the perspective projection. Often an image of interests comes from urban or indoor scenes where certain structural regularities are presented, this can help us to disambiguate this problem. Those scenes usually contain a lot of straight lines (see Fig. 1b) which can be grouped as parallel lines in space. Most viewers can recognize structure of a scene even from that image. However, there are a lot of missing lines, because not all of them are perfectly detected by low level

© Springer International Publishing Switzerland 2015
C.V. Jawahar and S. Shan (Eds.): ACCV 2014 Workshops, Part I, LNCS 9008, pp. 369–382, 2015.
DOI: 10.1007/978-3-319-16628-5_27

(a)   (b)

Fig. 1. Most people can infer the 3D structure of the scene from the image above. It is a little harder from (b), but it can be done too.

algorithms. People can still understand that scene because (a) we see all objects in a case of perspective, so we can recognize objects if we see their projections on a plane, (b) we can show where objects boundaries are, because we see how lines are grouped on that image.

In the paper we present an algorithm whose purpose is to reconstruct a 3D model from a single image. The basic assumption which we chose is that the main characteristics of a scene should have an agreement with "Manhattan" worlds [3]. This assumption states that most of horizontal lines in space are divided into two orthogonal groups.

The first step of our method includes the use of parallel lines in space for recovering scene planes. Then, those planes are used to predict other scene surfaces. The information extracted by the algorithm contains main planes of a scene like floors, ceilings, walls and facets of big objects. This can be done only through getting main camera parameters from an image. Camera characteristics like the focal length are gotten automatically. To get the focal length we use vanishing points. Vanishing point is a single point on the image plane which is created by intersection of projections of parallel lines on an image.

The novelty of the algorithm is that it recovers objects from different levels of a scene and it does not use only one bearing surface, e.g. ground. In addition, our method works on both indoor and outdoor man-made scenes. Next, we propose the formulation of *surface prediction* approach. Eventually, we show combinations of our algorithm with other approaches that outperform state-of-the-art methods.

2   Prior Work

Recent years have seen a large progress in scene understanding. Many works state that the Manhattan world assumption can be enough to generate plausible interpretation of man-made scenes. E.g. In [2] Boulanger et al. have extracted camera parameters and generated a simple 3D model from an image. They have used straight lines and three vanishing points to produce this result.

Spatial layout reconstruction was shown in [10–12,23,24]. The main base of those works is the Manhattan world assumption. Gradient features have been used to produce 3D boxes for big objects like beds in [11,12]. In [23,24] *orientation map* was used for this purpose. We applied this orientation map as a starting point of our algorithm.

It is important to get vanishing points for scenes of Manhattan type. In [6] Denis et al. have shown that a better way for extracting vanishing points is to use localized edges of objects rather than gradient maps. In [31] a fast algorithm was presented to extract three vanishing points. To improve this result an Expectation–maximization algorithm was used. In [33] vanishing points recovering was expanded to high-level geometric primitives like horizon and zenith. In addition, they have presented framework which can be used for images that are not in agreement with the Manhattan world assumption.

Recovering 3D structure of urban scenes can be done by using 3D model which consists of vertical walls and ground plane, where ground-vertical boundary is a continuous polyline. This was shown in [1].

Some works have shown that understanding the type of a surface can produce a good 3D model. Labeling geometric classes for image surfaces was shown in [15–17]. In [13,18] those works were combined to produce 3D models of scenes. They have used various cues like color, texture gradient and projection information (vanishing points). In [28,29] surface connectivity and image depth was used to generate plausible 3D model. All those works use this information to train a surface classifier on a training set of images. Moreover, they have shown that the use of superpixels instead of pixels significantly improves algorithm speed and surface recovering. We combined our approach with [17] to improve our result and to show that the proposed algorithm can be easily combined with other methods.

The significance of occlusion boundaries was shown in [13,18,19]. In [8] occluded surfaces were restored from both indoor and outdoor images. Those results can improve 3D scene reconstruction.

In [30] depth ordered planes were used. Those planes are generated using vanishing points and relative depth cues. Spatial layout of cluttered rooms was successfully applied for rendering synthetic objects into photographs in [20]. An interesting observation has been made in [9] that the world can be presented as a set of 3D blocks. Besides, it uses density classes to estimate each block, such as "light" for trees and "high-density" for buildings.

As we see, there are various methods to get a three-dimensional model from one image. For a wide range of image types training algorithms have shown the highest percentage of getting correct surfaces, but we focus only on a man-made environment which is mostly in agreement with the Manhattan world assumption. As a result, we decided to use this assumption as a base of our work.

3   Overview of Approach

In the paper we introduce *surface prediction* approach. By surface prediction here, we understand how recovered surfaces can be used to predict their neighbor planes. As we can see on Fig. 2 there can be a cube, but one surface is missing (top). Moreover, there are can be two missing surfaces instead of one. What can we do with that? Assume, that if we know one object surface or surface of neighbor object, then we can find other surfaces. We say that each plane or surface votes for neighbor planes. On Fig. 2 top surface receives 2 votes. As we will see, this simple assumption helps us to recover main surfaces of a scene.

The main difference of our method is that it restores most of the planes of a scene without building hypothesis for objects. We do not use only one bearing surface (like ground in [1]). Furthermore, we recover objects from different levels of a scene, because all surfaces are generated iteratively bearing on each other. Moreover, we use the Manhattan world assumption as the base for recovering scene surfaces. Consequently, our approach works on both indoor and outdoor man-made scenes. We do not generate object fitting, but we think that it can be easily obtained if blocks world assumption is used.

First, in our work we automatically find vanishing points and the focal length. Vanishing points are computed using straight lines of objects. To make computation easier we state that the vertical vanishing point is the point at infinity. For the focal length we assume that the principal point is located in the center of an image. Then, we calculate the focal length from vanishing points. This information is needed to recover 3D surfaces. In the second part of this paper we present our new method for surface recovering. Its origin is based on *orientation map* (*omap*) [24]. We collect votes (predicted surfaces) from each surface from *omap* and choose the most appropriate result. To improve our result we use surface approximation based on the Manhattan world assumption. Last, we present experimental validation on two datasets. To evaluate our algorithm we have used Delage et al. [5] and York Urban [6] datasets. Most of the images here are in agreement with the Manhattan world assumption. For surface testing those images were manually labeled with LabelMe tool [27].

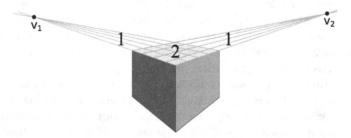

Fig. 2. We recovered two planes of a figure, but we do not have the third (top surface). Numbers indicate votes for surfaces. v_1 and v_2 are two horizontal vanishing points.

4   Finding Camera Parameters

For finding vanishing points we follow [26,31] approaches with some modifications. Initially, we extract object edges with Canny edge detector. Next, we use the following steps to extract line segments from obtained edge map. First, to speed up computation, we remove all junctions on received binary image. Then, we extract connected components using flood fill algorithm. As the next step we get line segments with Kovesi Matlab toolbox [21]. Each edge of an edge map is divided into straight line segments by splitting it when standard deviation of pixels is larger than one pixel.

Next, we use j-linkage algorithm [32] to get vanishing points from obtained lines. We need a consistency measure to link lines to theirs vanishing points. If we use the distance between a vanishing point and a line, for only a small deviation of a line from a point, we can get large distances, e.g. vanishing points at infinite. To avoid that, we use approximation from [31], it represents the distance between edge ending and a line through a vanishing point and the edge center. You can see obtained lines on Fig. 3a.

After this step, to get three orthogonal vanishing points we use the Manhattan world assumption and the guess that the vertical vanishing point is the point at infinite. Once vanishing points are found on an image, we get the focal length of the camera by finding a focal length that makes angles 90 degrees with vectors to vanishing points and assuming that principal point location is in the center of an image [4].

4.1   Finding Surface Coordinates

Now, to compute object surfaces we need only a complete orientation map which we obtain in the next part of this article. When we have it, we reconstruct a 3D scene only to the scale factor, because we do not know original metric of a scene. We iteratively compute coordinates for each surface and its neighbors. To minimize an error, we choose surfaces with maximum area and neighbors, then compute them first. Basing on their received coordinates we compute others. Due to the iterative process of 3D model recovering we restore objects that lie on different levels of a scene.

5   Orientation Map

First, to recover the orientation of surfaces we need a starting point. We get it from [23,24]. We use an orientation map which is described in those articles. We give here only a brief description. An example of orientation map can be seen on Fig. 3b. This map indicates the orientation of some region which is produced by two vanishing points. The main idea is that if you have a region which consists of two groups of lines referred to two different vanishing points, then those lines can produce a surface. On Fig. 3a we can see that the horizontal planes consist of green and blue lines (marked as 2 and 3 respectively), while vertical are from

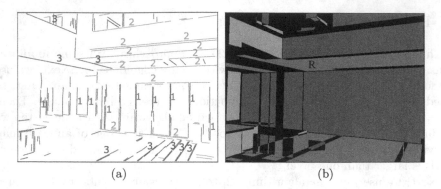

Fig. 3. (a) Line segments from Sect. 4 which are labeled with respect to their vanishing points (only long lines have labels). (b) Initial orientation map from [24]. Color represents orientation of regions and lines.

red and blue (1 and 3) or red and green (1 and 2) lines. Each line segment is extended until it abuts against a line (stopping parameter) which is orthogonal to the surface this line segment generates. This assumption is correct, because a line cannot lie on a plane which is orthogonal to that line. Detailed explanation and formal description can be seen in [24].

6   Surface Recovering

There are some problems with original orientation map (*omap*) from Sect. 5: (1) incorrect surfaces, which are formed due to the wrong lines, (2) *omap* approach cannot restore a surface without lines, as you can see there are a lot of free space on Fig. 3b, (3) in addition to the previous point, *omap* method does not restore occluded surfaces, e.g. walls of a building behind a tree.

To overcome problems that were mentioned in the previous paragraph we propose our new *surface prediction* (*SP*) approach. First, we remove all surfaces whose area is lower than the threshold. In that case we do not loose much information, because we will fill this free space during next steps. Second, our algorithm recovers missing planes by predicting them using surfaces that were restored in Sect. 5. Finally, our method reconstruct occluded surfaces. For example, if one part of a building wall was occluded, then a recovered segment of the wall will vote for the occluded part.

Our method extends the idea of line continuation from [24] to plane continuation. Moreover, we introduce a voting scheme to get final orientation of surfaces. Next, we propose a surface approximation step to further improve the result. We have found that it is especially useful for building 3D models.

6.1   Surface Prediction

Suppose, we know one surface of an object, in our case it is an initial orientation map from Sect. 5, then we can recover others. For example, if we reconstruct

(a)   (b)   (c)

Fig. 4. The first two figures show predicted surfaces for one red plane from Fig. 3b (marked as R). (a) shows surfaces without a stopping parameter, on (b) for the stopping parameter we use neighbor planes from Sect. 5, on (c) you can see the final result which is obtained after adding together all predicted surfaces.

only one cube facet, then we can say that the cube can have other facets or it lies on a different object (there are no flying objects on an image or they are supposed to be a noise). Following this assumption we define a position and an orientation of other surfaces, see Fig. 4 for an example of predicted surfaces for a red plane. Basically, we continue boundaries of planes from Sect. 5 to vanishing points. Then, we calculate votes for each orientation from all predicted surfaces and receive full orientation map.

The formal description of the method: let $S_x = \{s_{x,1}, s_{x,2}, ..., s_{x,n_x}\}$ be the set of surfaces (or planes), which are orthogonal to the x, where $x \in \{1, 2, 3\}$ denotes one of the three orientations and n_x is the number of planes with corresponding orientation. Remember that in Manhattan world, we have only 3 vanishing points (2 horizontal, 1 vertical) and each surface is produced by 2 of them. Consequently, we have only 3 orientations for surfaces. Next, the set of points which are produced by Ramer–Douglas–Peucker algorithm [7,25] for each surface boundary is:

$$P_{x,n} = \{p_{x,1,1}, \ldots, p_{x,n_x,m_x}\} \tag{1}$$

where m_x is a number of points which are obtained by this approximation and $n \in \{1 : n_x\}, m \in \{1 : m_x\}$.

Algorithm 1. *Predict neighbors for surface*

Require: Points from approximation $P_{x,n}$, vanishing points v_d, where $d \in \{1, 2, 3\}$.
Ensure: Orientation map for each surface $O_{x,n}$.
1: $p \leftarrow P_{x,n}[1]$
2: **for** $i \leftarrow 2$ **to** m_x **do**
3:     $g \leftarrow linesegment(p, P_{x,n}[i])$, a line segment between two points.
4:     $O_{x,n} \leftarrow O_{x,n} + getomap(g, v_d)$ (*getomap* returns orientation map $\widehat{O}$ from Sect. 5 for g and v_d)
5:     $p \leftarrow P_{x,n}[i]$
6: **end for**

After that, we get orientation maps $O_{x,n}$ by Algorithm 1, you can see the example for the vertical vanishing point on Fig. 4. It is important that Fig. 4b presents the result which uses the same idea for a line segment continuation as the one in Sect. 5, the only difference is that continuation stops when a line segment abuts a surface, not a line. Then, we sum up all surfaces:

$$O_x = \sum_{1 \le n \le n_x} O_{x,n} \tag{2}$$

This addition represents voting for surface prediction. During algorithm evaluation we realized that horizontal surfaces performed poorly compared to vertical, due to the fact that ground and floor parts of an image usually contain many incorrect surfaces from original *omap*. Therefore, we decided to use surfaces from [17] for horizontal surfaces in Eq. (2). In the testing section we show results with and without this information. The final equation for the whole orientation map O calculates as:

$$O = \widehat{O} \vee max_x(O_x) \tag{3}$$

where $x \in \{1, 2, 3\}$ and $\widehat{O}$ is an orientation map from Sect. 5, the function max_x denotes orientation of a pixel on the base of votes for each orientation, which we obtain when we get the sum from Eq. (2). The result can be seen in Fig. 4c.

6.2 Approximation of Surface Edges

Orientation map which is obtained in the previous part, see Fig. 4c, is not good. It has a lot of noise among edges of different channels. Due to the iterative process of building a 3D model, see Sect. 4.1, this may lead to gross errors at those surfaces that are computed in the last instance. Thus, we need smooth boundaries for surfaces. To get a better result we must use lines which belong to vanishing points for those boundaries. We propose the following method to do that.

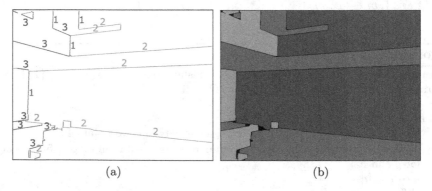

(a)   (b)

Fig. 5. Smoothed boundaries and a final orientation map. Boundaries (a) divided into three groups of line segments. The final orientation map (b).

First of all, we use erosion for each channel of orientation map. Next, we add all channels together and get a binary image (after voting, channels do not intersect)

$$B = \sum_{1 \le x \le 3} O_x \tag{4}$$

Then, we reverse image $\neg B$ and use thinning operation on it. We break obtained edge map into line segments as we did in Sect. 4. After that, we link all line segments to their vanishing points through consistency measure, see Sect. 4. We need to align each line segment to its vanishing point. For $g = linesegment(t_1, t_2)$, where t_1 and t_2 are endpoints of g, let m be midpoint of g and l be the line which passes through m and v (vanishing point), then aligned line segment is:

$$r = (proj(t_1, l), proj(t_2, l)) \tag{5}$$

where $proj(t, l)$ function returns the projection point of a point t on a line l. The result can be seen on Fig. 5a.

Obtained line segments are added together and imposed to an orientation map dividing it to distinct components. After noise removing we get smoothed orientation map, see Fig. 5b.

Fig. 6. From left to right: orientation map after multiscaling, two projections of the reconstructed 3D model.

To get a better result we can use different scales of an image or *multiscaling*. However, the algorithm execution time grows a lot. The idea is pretty simple: we choose several scales of an image, compute an orientation map for each scale, then we add all orientation maps from all scales together and use voting to choose the best orientation for each pixel. The result can be seen in Fig. 6.

7 Experiments and Results

We have tested our algorithm on Delage et al. [5] dataset (48 images) of indoor scenes and we have also labeled its images with LabelMe tool [27]. In particular, we have manually marked ground truth orientation for each surface formed by vanishing points. You can see the percentage of pixels that have the correct orientation on Fig. 7.

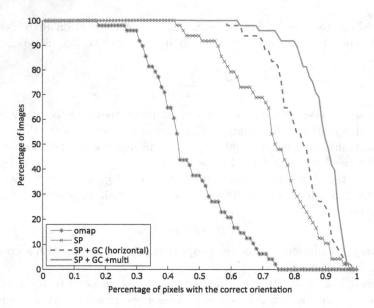

Fig. 7. Percentage of pixels with the correct orientation. Compared methods: *omap* — raw orientation map from [24], *SP* — Surface Prediction approach (Sect. 6), *GC* (Geometric Context) — [17] for all channels, *GC(horizontal)* — [17] to vote for horizontal surfaces (red channel on images), *multi* — *multiscaling*, see Sect. 6.2.

Here we present a comparison of the average percentage of correct pixels among all images in Delage et al. dataset. First, we have tested our *SP* method and got 73.2 %. With multiscaling this result was improved to 73.6 %. Multiscaling for *SP* added less than a percent for an output result, so it is not so useful due to the additional time cost. We applied it only to the final result to slightly improve it and to obtain better 3D models (boundaries which are produced after multiscaling are much smoother). After that, to improve the result we combined *SP* with [17] (*Geometric Context*) approach for horizontal surfaces *GC(horizontal)* and in that case this combination outperformed method that

Table 1. Results that were obtained on Delage et al. [5] (second column) and York Urban [6] (third column) datasets. The first column shows used methods for surface recovering, two last columns denote an average percentage of correct pixels for surface orientation.

Method components	Avg Delage	Avg York
omap	46.7	51.8
SP	73.2	73
Lee et al. [24]	80	–
SP + GC(horizontal)	81.9	81.1
GC [14,17]	87	–
SP + GC + multi	88.9	–

was shown in [24] (81.9 % versus 80 %). Next, as reported in [24], they got 87 % for [14] approach on Delage et al. dataset, we have tested this with *GC* for all orientations and got nearly the same result. Interestingly, *GC + multiscaling* produced worse result (86 %) than *GC* itself. Finally, we combined *GC* for all surfaces with our Surface Prediction method and got 88.9 %, that outperformed *GC* for indoor scenes. The average percentage for each method can be seen in Table 1.

For Delage et al. dataset the number of images with ≥ 50 % of correct surfaces is near 94 % for *SP* and 100 % for its combinations with *GC*.

We have also tested our approach on York Urban [6] dataset (102 images), which consists of indoor and outdoor urban scenes. On this dataset we got 81.1 % correct surfaces in average with *SP+GC*(*horizontal*) params, which corresponds

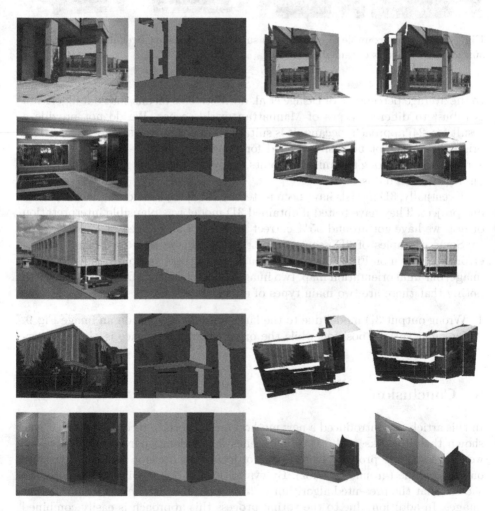

Fig. 8. Scene reconstruction examples. From left to right: original image, obtained orientation map, two views of a 3D model.

Fig. 9. An example of failure in the presence of significant noise (trees).

Fig. 10. Another example of errors when surface positions are irregular while most of surface orientation is obtained correctly.

to the average percentage of Delage et al. dataset. This means that our algorithm is robust to different types of Manhattan world scenes. We do not provide a result for [24] approach, because it is suitable only for indoor scenes. In addition, original *GC* cannot be easily applied for images of outdoor scenes for vertical surfaces which belong to vanishing points, so we did not use it in our experiments on York Urban dataset.

Eventually, 3D models have been tested by four people who were not a part of our project. They have tested if obtained 3D model was plausible interpretation or not, we have got around 55 % correct models for all images.

Some examples of 3D scene reconstruction can be seen on Fig. 8. Typical errors are shown on Figs. 9 and 10. For each photo there are four images: original image, our final orientation map, two images of an obtained 3D model. It is worth noting that there are two main types of errors:

1. Wrong output 3D model due to the large amount of noise on an image Fig. 9.
2. Irregular surface position while the orientation of a surface is obtained correctly Fig. 10.

8   Conclusion

In this article, we introduced a new method for 3D surface reconstruction. It was shown that it works on both urban scenes and indoor apartments. Moreover, we addressed the problem of restoring objects that are not connected to only one base plane (ground or floor). By evaluation on popular image datasets, we showed that the presented algorithm restores plausible 3D models from different images. In addition, due to the voting process this approach is easily combined with other algorithms and obtained results outperformed several state-of-the-art techniques. The proposed concept of *surface prediction* can be useful for many computer vision applications as well.

Acknowledgements. To Jiri Matas for valuable comments on the paper and to Evgeny Stolov for the help during the research.

References

1. Barinova, O., Konushin, V., Yakubenko, A., Lee, K.C., Lim, H., Konushin, A.: Fast automatic single-view 3-d reconstruction of urban scenes. In: Forsyth, D., Torr, P., Zisserman, A. (eds.) ECCV 2008, Part II. LNCS, vol. 5303, pp. 100–113. Springer, Heidelberg (2008)
2. Boulanger, K., Bouatouch, K., Pattanaik, S.: Atip: A tool for 3d navigation inside a single image with automatic camera calibration. In: Proceedings of the EG UK Theory and Practice of Computer Graphics 15 (2006)
3. Coughlan, J.M., Yuille, A.L.: Manhattan world: compass direction from a single image by bayesian inference. In: The Proceedings of the Seventh IEEE International Conference on Computer Vision, vol. 2, pp. 941–947 (1999)
4. Criminisi, A., Reid, I., Zisserman, A.: Single view metrology. Int. J. Comput. Vis. **40**, 123–148 (2000)
5. Delage, E., Lee, H., Ng, A.: A dynamic bayesian network model for autonomous 3d reconstruction from a single indoor image. Comput. Vis. Pattern Recogn. **2**, 2418–2428 (2006)
6. Denis, P., Elder, J.H., Estrada, F.J.: Efficient edge-based methods for estimating manhattan frames in urban imagery. In: Forsyth, D., Torr, P., Zisserman, A. (eds.) ECCV 2008, Part II. LNCS, vol. 5303, pp. 197–210. Springer, Heidelberg (2008)
7. Douglas, D.H., Peucker, T.K.: Algorithms for the reduction of the number of points required to represent a digitized line or its caricature. Cartographica Int. J. Geog. Inf. Geovisualization **10**, 112–122 (1973)
8. Guo, R., Hoiem, D.: Beyond the line of sight: labeling the underlying surfaces. In: Fitzgibbon, A., Lazebnik, S., Perona, P., Sato, Y., Schmid, C. (eds.) ECCV 2012, Part V. LNCS, vol. 7576, pp. 761–774. Springer, Heidelberg (2012)
9. Gupta, A., Efros, A.A., Hebert, M.: Blocks world revisited: image understanding using qualitative geometry and mechanics. In: Daniilidis, K., Maragos, P., Paragios, N. (eds.) ECCV 2010, Part IV. LNCS, vol. 6314, pp. 482–496. Springer, Heidelberg (2010)
10. Hedau, V., Hoiem, D., Forsyth, D.: Recovering the spatial layout of cluttered rooms. In: IEEE 12th International Conference on Computer Vision, pp. 1849–1856 (2009)
11. Hedau, V., Hoiem, D., Forsyth, D.: Thinking inside the box: using appearance models and context based on room geometry. In: Daniilidis, K., Maragos, P., Paragios, N. (eds.) ECCV 2010, Part VI. LNCS, vol. 6316, pp. 224–237. Springer, Heidelberg (2010)
12. Hedau, V., Hoiem, D., Forsyth, D.: Recovering free space of indoor scenes from a single image. In: IEEE Conference on Computer Vision and Pattern Recognition (CVPR), pp. 2807–2814 (2012)
13. Hoiem, D.: Seeing the world behind the image: Spatial layout for three-dimensional scene understanding (2007)

14. Hoiem, D., Efros, A., Hebert, M.: Putting objects in perspective. Int. J. Comput. Vis. **80**, 3–15 (2008)
15. Hoiem, D., Efros, A.A., Hebert, M.: Automatic photo pop-up. ACM Trans. Graph. TOG) **24**, 577–584 (2005). ACM
16. Hoiem, D., Efros, A.A., Hebert, M.: Geometric context from a single image. In: Tenth IEEE International Conference on Computer Vision, ICCV 2005, vol. 1, pp. 654–661. IEEE (2005)
17. Hoiem, D., Efros, A.A., Hebert, M.: Recovering surface layout from an image. Int. J. Comput. Vis. **75**, 151–172 (2007)
18. Hoiem, D., Efros, A.A., Hebert, M.: Closing the loop in scene interpretation. In: IEEE Conference on Computer Vision and Pattern Recognition, CVPR 2008, pp. 1–8. IEEE (2008)
19. Hoiem, D., Efros, A.A., Hebert, M.: Recovering occlusion boundaries from an image. Int. J. Comput. Vis. **91**, 328–346 (2011)
20. Karsch, K., Hedau, V., Forsyth, D., Hoiem, D.: Rendering synthetic objects into legacy photographs. ACM Trans. Graph. TOG **30**, 157 (2011)
21. Kovesi, P.D.: Matlab and octave functions for computer vision and image processing (2000). http://www.csse.uwa.edu.au/pk/research/matlabfns/
22. Lalonde, J.F., Hoiem, D., Efros, A.A., Rother, C., Winn, J., Criminisi, A.: Photo clip art. ACM Trans. Graph. **26**, 3 (2007). ACM
23. Lee, D.C., Gupta, A., Hebert, M., Kanade, T.: Estimating spatial layout of rooms using volumetric reasoning about objects and surfaces. In: Proceedings of the NIPS, vol. 1, p. 3. Vancouver, BC (2010)
24. Lee, D.C., Hebert, M., Kanade, T.: Geometric reasoning for single image structure recovery. In: IEEE Conference on Computer Vision and Pattern Recognition, CVPR 2009, pp. 2136–2143. IEEE (2009)
25. Ramer, U.: An iterative procedure for the polygonal approximation of plane curves. Comput. Graph. Image Proc. **1**, 244–256 (1972)
26. Rother, C.: A new approach to vanishing point detection in architectural environments. Image Vis. Comput. **20**, 647–655 (2002)
27. Russell, B.C., Torralba, A., Murphy, K.P., Freeman, W.T.: Labelme: a database and web-based tool for image annotation. Int. J. Comput. Vis. **77**, 157–173 (2008)
28. Saxena, A., Chung, S.H., Ng, A.Y.: 3-d depth reconstruction from a single still image. Int. J. Comput. Vis. **76**, 53–69 (2008)
29. Saxena, A., Sun, M., Ng, A.Y.: Make3d: learning 3d scene structure from a single still image. IEEE Trans. Pattern Anal. Mach. Intell. **31**, 824–840 (2009)
30. Stella, X.Y., Zhang, H., Malik, J.: Inferring spatial layout from a single image via depth-ordered grouping. In: CVPR Workshop (2008)
31. Tardif, J.P.: Non-iterative approach for fast and accurate vanishing point detection. In: 2009 IEEE 12th International Conference on Computer Vision, pp. 1250–1257. IEEE (2009)
32. Toldo, R., Fusiello, A.: Robust Multiple structures estimation with J-linkage. In: Forsyth, D., Torr, P., Zisserman, A. (eds.) ECCV 2008, Part I. LNCS, vol. 5302, pp. 537–547. Springer, Heidelberg (2008)
33. Tretyak, E., Barinova, O., Kohli, P., Lempitsky, V.: Geometric image parsing in man-made environments. Int. J. Comput. Vis. **97**, 305–321 (2012)

Scene Parsing and Fusion-Based Continuous Traversable Region Formation

Xuhong Xiao[⊠], Gee Wah Ng, Yuan Sin Tan, and Yeo Ye Chuan

DSO National Laboratories, 20 Science Park Drive,
Singapore 118230, Singapore
xxuhong@dso.org.sg

Abstract. Determining the categories of different parts of a scene and generating a continuous traversable region map in the physical coordinate system are crucial for autonomous vehicle navigation. This paper presents our efforts in these two aspects for an autonomous vehicle operating in open terrain environment. Driven by the ideas that have been proposed in our Cognitive Architecture, we have designed novel strategies for the top-down facilitation process to explicitly interpret spatial relationship between objects in the scene, and have incorporated a visual attention mechanism into the image-based scene parsing module. The scene parsing module is able to process images fast enough for real-time vehicle navigation applications. To alleviate the challenges in using sparse 3D occupancy grids for path planning, we are proposing an approach to interpolate the category of occupancy grids not hit by 3D LIDAR, with reference to the aligned image-based scene parsing result, so that a continuous $2\frac{1}{2}D$ traversable region map can be formed.

1 Introduction

It is widely accepted that humans possess two distinct visual pathways, the ventral stream (known as "what pathway") involved with object identification and recognition, and the dorsal stream (or, "where pathway") involved with processing the object spatial location relevant to the viewer [1,2]. Similarly, for autonomous vehicle to successfully drive on road, our scene understanding systems must tell the vehicles not only what are in the environment, but also where the roads and obstacles are in a physical coordinate system, other than in the image plane.

In the last two decades, the study on computer vision systems mainly focus on solving the problem about what are in a scene, taking images as input. Many prototypes detect/recognize individual objects in the images. Effective features, such as SIFT [3] and HOG [4], have been proposed and widely used in these systems. The HMAX [5,6] tried to model objects via a hierarchical representation involving increasingly complex features. In [7], Latent SVM was proposed to learn deformable part models. One common feature of these systems is that they all apply sliding windows to exhaustively search for object locations. To reduce

© Springer International Publishing Switzerland 2015
C.V. Jawahar and S. Shan (Eds.): ACCV 2014 Workshops, Part I, LNCS 9008, pp. 383–398, 2015.
DOI: 10.1007/978-3-319-16628-5_28

unnecessary computation and speed-up the detection process, strategies like combining multiple, increasingly complex classifiers in a "cascade" have been proposed [8] and adapted in [9]. Some other systems recognize overall scene categories (e.g., beach) from the image, usually making use of features reflecting the "gist" of scenes [10–14]. To facilitate autonomous systems' response to environment, it is insufficient to recognize only a certain category of objects. Instead, scene parsing, which conducts classification for all parts in the scene including road and vegetation, is required to form a continuous traversable map for navigation. While early scene parsing systems tried to classify each pixel separately, recent scene parsing systems perform the task on the superpixel level. In the superParsing system [15], features like shape, SIFT, color and appearance are extracted for each superpixel. Total scene understanding, which conducts scene parsing and object detection simultaneously, has also been studied, as in [16,17]. Inspired by the findings in cognitive science, the strategies taken for object recognition and scene understanding also evolves from pure bottom-up processing to combination of both bottom-up processing and top-down facilitation [18,19]. Graph models have been proposed to control the top-down process, among which, CRF (Conditional Random Field) has been the most popular framework [20–22].

Another line of research in perception for autonomous vehicles is to understand "where" the objects are in a physical coordinate system, which is necessary for autonomous vehicles to navigate or to response to the environment. 3D LIDAR sensors have been widely used to achieve it, as demonstrated in DARPA Grand Challenges [23]. 3D LIDAR-based scene classification is usually conducted on occupancy grids, whereby the physical region is divided into 3D cubes or 2D grids of pre-defined size. Authors in [24] have made attempt to classify, from LIDAR responses, classes including surface (ground bare terrain surface, solid object, large tree trunk), linear structures (wires, thin branches) and scatter (tree canopy, grass), based on eigen-values of the covariance matrix of the local pointcloud. In [25], an SVM classifier is trained using features extracted from LIDAR points, such as intensities of LIDAR points, scatterness, linearness, and surfaceness. One disadvantage of 3D LIDAR based classification is that the classified grids are not continuous due to the sparseness of pointcloud. Furthermore, the classified grids will get even sparser with the increase of distance from the vehicle, resulting in extra challenges to the path planning and navigation modules.

With the progress in the above two directions, it is natural to compensate LIDAR with image based scene understanding. For example, Stanley [26], the 1st winner of DARPA Grand Challenge 2004, made use of 3D LIDAR data to conduct terrain labelling, and image-based vision analysis for early warning of obstacles in the distance beyond the range of LIDAR. Little fusion of image and LIDAR was involved in the Stanley system. There are also work to fuse LIDAR and camera for road boundary detection [27,28], but such tracking based approaches are constrained to roads with nearly parallel borders or when the road model is known *a prior*. They do not work well on complex terrains, e.g., open fields without clear road boundary.

Our work follows the line of fusion of image and LIDAR for scene parsing. The scene parsing module is an integral part of the Cognitive Architecture, for which we have proposed a computational infrastructure that defines the various regions and functions working as a whole to produce human-like intelligence [2,31]. In particular, for scene parsing, the architecture has specified sub-functions including initial scene classification based on bottom-up, low-level features, top-down facilitation, visual attention-based priming and object-level fine grained classification. In this paper, we will first give some details about our strategy for initial classification and context-based top-down facilitation. We will then focus on reporting our effort to combine the LIDAR and scene parsing result to create a continuous traversable region map for vehicle navigation in outdoor, off-road environment. In the same time, we will also exemplify how the visual attention mechanism is implemented and applied to enhance the capability of man-made obstacle detection from images.

Compared with existing work, our system has some special features. First, we have incorporated not only the contextual-based top-down processing in scene parsing, but also the visual attention mechanism to enhance the capability of obstacle (man-made objects) detection. Although strategies of applying different features to detect various objects have been widely used [33], traditional superpixel level scene parsing techniques treat all parts of an image equally, ignoring the function of visual attention in finding objects of interest [34]. Secondly, although it is not new anymore to utilize the top-down process to enhance scene parsing, most existing systems only make use of the co-occurrence or neighboring relations between objects in the process, while our approach explicitly interpret the relative spatial relations between irregular-shape components in a semantic way (for example, in a front view image, road regions cannot be above the sky, but can be below the sky). It is no doubt that the interpretation of specific spatial relation will solve uncertainties in the initial classification more efficiently than simple neighboring relations. Thirdly, to our knowledge, this is the first work to generate continuous road regions from combination of scene classification and LIDAR detections, making it possible to provide continuous traversable region map under complex situations, e.g., when there are obstacles on road and the road boundary is cluttered or in irregular shapes.

2   Image-Based Scene Understanding in the Cognitive Architecture

Figure 1 presents the high-level structure of the visual perception module, showing the biological regions where the sub-functions are accomplished in human brains and the interaction between the perception module and other modules, such as the Reasoner module, in our Cognitive Architecture. Total scene understanding is achieved via two processes:

1. An interactive process among initial classification and top-down facilitation, which emulates the typical bottom-up and top-down interaction [31,32], and

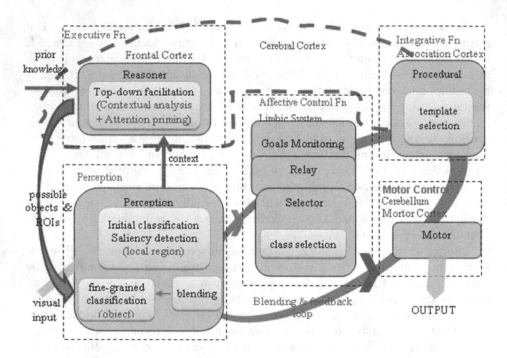

Fig. 1. Building block of the visual perception module in DSO-CA

the visual attention process [34]. Initial classification refers to the early, coarse-level classification (e.g., road, vegetation) of scene parts based on low-level features. The visual attention mechanism is responsible for identifying potential objects of interest, some of which may have been successfully detected by the coarse-level initial classification sub-module. Furthermore, it will also pick up some new objects of interest which are unknown to the learning-based initial classification sub-module. Top-down facilitation is not only responsible for making use of contextual knowledge to resolve uncertainties in initial classification through a contextual analysis process, but also responsible for suggesting regions of interest worthy of further attention via attention priming.

2. Fine-grained object classification which determines more specific categories of objects of interest (e.g., pedestrians, bus, cars, etc.), identified by attention priming. Known object classification techniques, such as HOG-based classification [4], and the deformable part-based models [7], can be adapted for the purpose. In particular, a special blending mechanism, as proposed in [29] has been adapted in the Cognitive Architecture, as shown in Fig. 1.

In comparison, most state-of-the-art scene parsing systems treat every part of the image equally, in that they classify each part (an object may be segmented into several parts), while individual object detection systems search for specific categories of objects via sliding window. There is barely a system that conducts both functions simultaneously. Furthermore, although sliding window based search of objects is effective in engineering systems, it is not consistent with the biological way of object detection, and it takes unnecessary longer time to process irrelevant information. According to cognitive findings, visual search is conducted via the visual attention mechanism: individual objects pop-out due to its saliency, and task-based intention help to achieve selective attention [34], resulting in detections of potential objects of interest. Fine-grained object classification is conducted only to verify these potential object in the identified regions. The integration of attention mechanism in our framework makes it more similar to biological system in quickly switching the focus of process to the regions potentially containing objects of interest.

A detailed breakdown of the scene understanding module is illustrated in Fig. 2. The scene part parsing mechanism will classify each superpixel via a bottom-up initial classification process and a top-down contextual analysis process. The visual attention mechanism involves bottom up saliency map generation and binding, as well as top-down attention priming.

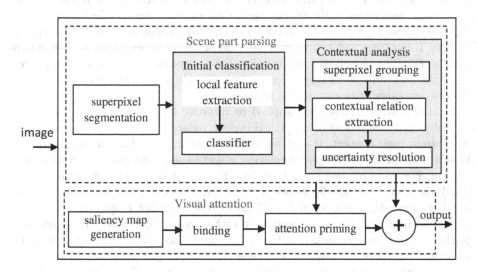

Fig. 2. Detailed diagram of the implementation of scene understanding

2.1 Initial Classification

The initial classification sub-module consists of algorithms to achieve early, coarse-level classification of local image regions. Each image is over-segmented into superpixels as described in [35]. Color and texture features are extracted to describe each superpxiel. In detail, histograms of RGB/HSV color features, anisotropic Gauss filtering responses [36], Gabor filter responses [37] and Local Binary Patterns [38], can be extracted for each superpixel.

Popular classifiers, such as Support Vector Machine (SVM) and Multiple Layer Perceptron (MLP) are included in the Perception module of the Cognitive Architecture. According to our experience, the MLP classifier can consistently achieve similar accuracy to that achieved by SVM with RBF kernel, which is much better than linear SVM for natural scene parsing. Besides, MLP can run much faster than SVM with RBF kernel when there are thousands of support vectors generated, which is always the case for natural scene parsing.

2.2   Top-Down Facilitation via Contextual Analysis

The design of the top-down facilitation strategy in the Cognitive Architecture is mainly motivated by the cognitive findings as reported in [39]. According to [39], there are two factors contributing to top-down facilitation: the object-based and context-based facilitation. The object-based mechanism refers to the case that "initial guess" of object type based on low level information triggers a more fine-grained object classification process, which is emulated by the attention priming mechanism in the framework. The context-based mechanism triggers top down facilitation through contextual association between objects in scenes. The contextual association activates predictive information about which objects are likely to appear together, and can influence the "initial guesses" about an object's identity. It has been widely accepted [40] that contexts affects classification in two aspects: (1). The presence of objects that have a unique interpretation improves recognition of ambiguous object in a scene; (2). Proper spatial relations among objects decreases error rate in the recognition.

Spatial relations have been applied to improve object detection, which confines objects by bounding boxes [41]. However, most superpixel-based scene parsing systems limit contextual analysis to the co-occurrence of neighboring objects. As a matter of fact, the spatial relation is very useful for natural scene classification. For example, we know that "road" cannot be on top of a "tree", but "road" is possible to appear beside a "tree". In this case, simple co-occurrence, or adjacency relation between "road" and "tree" will not help much in resolving uncertainties in the classification of either "road" or "tree". However, the spatial relation, one is on top of the other, will clearly verify the situation. To efficiently interpret such semantic relations, we first group the superpixels into connected components based on their initial classification. For example, as shown in Fig. 3(b), all superpixels falling on the circled component, which are initially classified as "sky", are grouped together. The contexts about whether its top, bottom and sides are tree, sky, etc., are extracted and represented as soft evidence passed to a learned Nave Bayes structure illustrated in Fig. 4. Each node in the structure, except the node "category", represents a spatial relation. For example, the node "isTopSky" has two values: 0 if the top of the component under consideration is not "sky", 1 otherwise. Inference based on soft evidence updates the probability of the values of node "category", i.e., the classification of the component under consideration. For example, based on the evidence that the top and two sides of the circled component in Fig. 3(b) are all "road", the classification of the component will be updated to "road", as shown in Fig. 3(c).

(a)   (b)   (c)

Fig. 3. Illustration of the contextual analysis process. (a). Original image; (b). Result of initial classification, with misclassifications for some parts; (c). Result after contextual analysis.

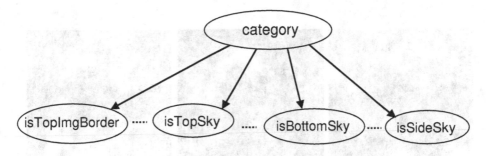

Fig. 4. Nave Bayes structure for contextual analysis.

2.3   Visual Attention

In a computer vision system, the purpose of visual attention module is to determine the region of objects of interest. The visual attention mechanism in the framework consists of a saliency map generation submodule, a binding submodule and a top-down priming submodule. In the following, we will use the image in Fig. 5(a) to illustrate how the mechanism works.

First, an initial saliency map is computed for the image based on the algorithm in [42]. This initial saliency map is further refined by suppressing the effect of large patches of background, whereby the background region is determined based on the consistency of color/intensity of the image. The refined saliency map is shown in Fig. 5(b). The task of binding is to group the saliency regions into proto-objects, which are believed to be the form of output in visual attention [43]. The MSER algorithm [44] is modified to accomplish the binding task. Major modifications to the MSER algorithm lie in the change of maximally stable criteria. Besides the ratio of region change as defined in [45], we also consider the orientation consistency when two components are to be merged, as well as the contour completeness. The bounding boxes of the initial proto-objects are shown in Fig. 5(c). There are false alarms of proto-objects in Fig. 5(c) because only information from the bottom-up saliency map and the contour is made use of at this stage. The top-down process - attention priming, steps in to reduce false alarms. It works as follows: First, it makes use of domain knowledge about the task requirement to determine potential object size and object types. For example, for a driving vehicle, the potential objects of interest may be cars and other obstacles on the road that the vehicle should avoid colliding. This inference will further activate the use of other knowledge such as "obstacles should be on the road, not on the top of trees". Combining such domain knowledge with the image context derived from the scene parsing process (as shown in Fig. 5(e)), the priming process will switch the attention to the true object of interest, removing the false alarms that does not fit for our domain knowledge, leading to the result as in Fig. 5(d), with only one potential obstacle. This visual attention result is then applied to update the scene part parsing result on superpixel level, and the final scene understanding result is shown in Fig. 5(f).

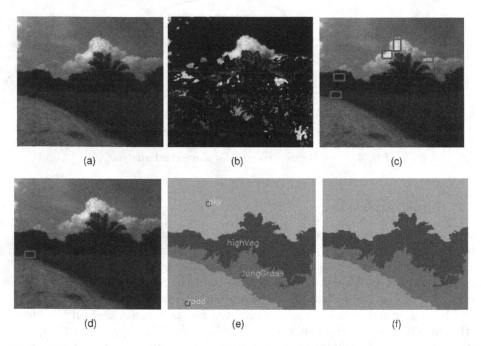

Fig. 5. Illustration of the Visual attention process. (a). Original image; (b). Saliency map; (c). Bounding boxes for initial proto-objects output by binding process, (d). Final bounding box for a proto-object after attention priming; (e). Output of the scene part parsing mechanism; (f). Final output combing scene part parsing and visual attention.

The example shown in Fig. 5 clearly demonstrates one function of the visual attention mechanism - it can discover some potential objects of interest that the scene part parsing mechanism misses. As shown in Fig. 5(e), the obstacle (in this case, the truck) is misclassified as "longGrass" by the scene part parsing mechanism. There are two possible reasons for the misclassification. It is possible that the superpixel level classification, which is based on only partial structure of objects, is not effective in detection of obstacles. It is possible too that the object is "new" to the training-based scene part parsing process. However, the visual attention mechanism, which is based on contrast and contours, successfully makes the obstacle "pop out", and complement the scene part parsing mechanism in detection of obstacles. On the other hand, the information from scene part classification provides scene contexts for attention priming.

The scene part parsing mechanism has been tested with many scenarios of different terrains. In a recent exercise, we have collected a variety of data covering diversified terrains like cluttered unstructured fields, narrow tracks with water bodies and wide open areas with ponds. From the data collected, we selectively labelled 356 images, with effort to ensure that the selected images are sufficient to reflect the diversities in terrain and illumination changes. From these labelled images, we have randomly chosen 178 images to train the classifier for initial classification and the Naive-Bayes model for contextual analysis, while the remaining 178 images for test. The F-measure (measured on superpixel level) for the major categories is shown in Table 1.

Table 1. F-measure for different categories

Processing stage	F-measure (%)					
	Road	HighVeg	LongGrass	Sky	Water	Obstacle
Initial classification	0.961	0.910	0.767	0.991	0.726	0.624
After contextual analysis	0.965	0.922	0.784	1.0	0.731	0.622

As can be easily observed from Table 1, the top-down contextual analysis improves the classification performance for most categories, and the scene part parsing submodule performs constantly well for categories like "road", "highVeg" (representing high vegetation) and "sky", which are the major categories in off-road natural scenes. Considering that the vehicle has to traverse over long grass on narrow tracks, a category "longGrass" is added to the classification module. Although "longGrass" is quite confusing with "highVeg", the scene part parsing mechanism works reasonably well for the category.

Another feature of the module is its high efficiency in computation. It takes only about 0.08 s to process an image of 400 × 300 in a 64-bit Windows system with Inter i7-3520 Dural Core and 8GB RAM.

The scene part parsing submodule is not very successful in classification of "obstacle", which includes all kinds of man-made structures/objects in our case. It is reasonable considering the diversity of obstacles and that the scene parsing is conducted on superpixel level, which may not extract complete topological features to interpret the structure of obstacles as a whole. Its weakness in obstacle detection can be partially complemented by the visual attention mechanism, as shown in Fig. 5. It will also be complemented by the LIDAR obstacle detector, which is known to be good for obstacle detection.

3   Fusion-Based Traversable Region Formation

Figure 6 presents our framework to fuse the image-based scene parsing module and LIDAR detections. Both monocular camera and LIDAR are installed on top of an autonomous vehicle. The LIDAR applied is Velodyne 64-HDL with 64 sensors to generate pointcloud. A LIDAR based detector classifies the 3D points into ground points or obstacle points mainly based on their z-coordinates (the height of objects). It futher groups the obstacle points into clusters based on the distance between the points. A data registration process is conducted to align the image frames and LIDAR scans in time and estimate the homography transform so as to acquire the point-level correspondence between image pixels and 3D LIDAR points.

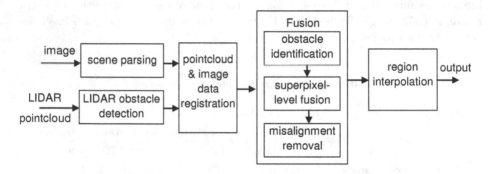

Fig. 6. Framework for image and LIDAR fusion.

Figure 7 illustrates an example of data registration. The original image is shown in Fig. 7.(a), and the aligned LIDAR occupancy grid with cell size 20 cm × 20 cm) is shown in Fig. 7.(e). Figure 7(f) shows the clusters (on occupancy grid) provided by the LIDAR detector, while Fig. 7(b) shows the correspondence between the LIDAR points and the image after data registration. In both Fig. 7(b) and Fig. 7(f), ground points are represented in gray, while each other color corresponds to an individual cluster output by the LIDAR detector.

3.1   Image and LIDAR Fusion

The fusion module first identifies the LIDAR clusters belonging to man-made "obstacle" and "longGrass" based on features including the height of the cluster, its position relative to the ground plane, depth difference within the cluster, and the image-based scene parsing result. It then makes use of LIDAR information to achieve more reliable scene parsing. There are several cases that LIDAR information will help to resolve ambiguities in image-based scene parsing. For example, if all 3D points projected to a superpixel are on ground level, the

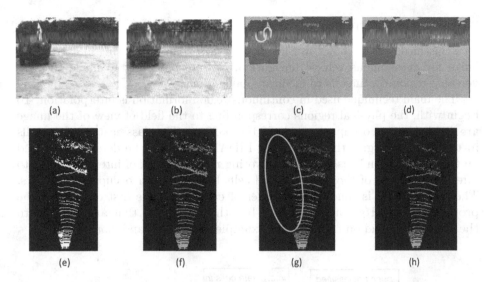

Fig. 7. Illustration of the fusion process. (a). Original image; (b). LIDAR clusters projected to the image; (c). scene classification without fusion, with errors in regions A and B; (d). scene classification after fusion; (e).Original LIDAR pointcloud (on occupancy grid); (f). LIDAR clusters output by the LIDAR detector; (g). Classification of LIDAR cells with presence of misalignment; (h). Classification of LIDAR cells with misalignment removed.

superpixel is unlikely to be part of an obstacle. On the other hand, if most of the 3D points projected to the superpixel belong to obstacle, the superpixel is unlikely to be classified as "road". The fusion strategy is similar to that applied in [29]. Once the classification of superpixels is updated, the categories of the individual 3D points are set to be the same as the corresponding image point. Accordingly, the classification of the cells in the occupancy grid that are occupied by LIDAR points can be determined based on that of points in it using majority voting rule. As shown in Fig. 7(g), gray points correspond to ground cells, red for obstacle cells, green for long grass and dark green for high vegetation which is non-traversable.

Another issue in the fusion process is the imperfectness of data registration. For example, in Fig. 7(g), there are some scattered false alarms of "obstacle" cells further away from the real obstacle. This is because, due to misalignment, some 3D points of road and vegetation in the distance are projected to the obstacle in the image, resulting in misclassification to these 3D points. The misalignment removal mechanism removes such misclassified points by referring to the classification of 3D clusters, utilizing ad-hoc rules. For example, If LIDAR points of a non-obstacle cluster is projected to an "obstacle" in the image, they are removed from the occupancy grid. After misalignment removal, we will get the updated, more reliable classification of occupancy grid , as shown in Fig. 7(h).

3.2   Region Interpolation

The classified cells as shown in Fig. 7(h) are very sparse, resulting in extra challenge for autonomous vehicle navigation. We are intending to make the traversable region continuous to facilitate navigation.

The main technique used in continuous region formation is interpolation. To begin with, the physical regions corresponding to the field of view of the image are divided into an occupancy grid of 20 cm × 20 cm. The classification of the cells in the occupancy grid that are hit by LIDAR points can be determined based on that of points in it using majority voting rule. The task of interpolation is to infer the category of those unoccupied cells based on their occupied neighbors. The interpolation is conducted in order of categories. The first category to be processed is "road/ground", followed by other categories that are adjacent to the road or overlaid on the road, for example, "obstacle" and "longGrass".

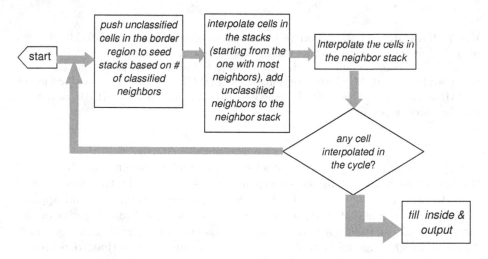

Fig. 8. Flowchart of the interpolation process along a border.

For each category of interest (e.g., "road", "obstacles" and "longGrass"), the connected components are first identified in the image plane. For each connected component, the classified cells on the LIDAR occupancy grid are identified along the left, right, top and bottom borders of the component. The bordering cells along each direction are discontinuous, and they may not reflect the actual border position due to the sparseness of the classified cells. However, they form a proper estimation about the border regions of the component. For each direction, the search and interpolation of actual borders will be around the initial bordering region. As shown in Fig. 8, along each direction, the non-classified cells which have at least one neighboring cell classified, are pushed to respective seed stacks based on the number of classified neighbors. A classified neighbor is a cell on the occupancy grid, the category of which has been either inferred in the fusion process, or interpolated before. The interpolation will start from the cells with the most classified neighbors: the z-coordinate of the cell is set to the averaged z-coordinates of its classified neighbors, the coordinates of the four corners of the cell (the x and y coordinates of the cell is known based on the location of the cell on the occupancy grid) is then projected to the image plane. If all the four corners fall in the connected component under consideration, the cell is considered to belong to the component. Therefore, its category is set to be the same category as the component. In the meantime, its unclassified neighbors are pushed to the neighbor stack. After all the cells in the seed stacks are interpolated, the cells in the neighbor stack will be interpolated. This interpolation process will be repeated until there is no new cell processed in the iteration.

Figure 9(a)–(c) illustrate the intermediate results of the 1st, 3rd and 10th iterations to interpolate the large road region in Fig. 7(a), along the left border. The points in cyan correspond to the cells in the seed stacks in the iteration.

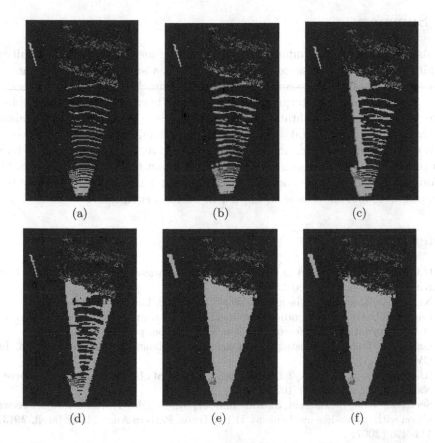

Fig. 9. Illustration of the interpolation process.(a). 1st iteration along left (seed cells are in purple); (b).3rd iteration along left; (c).10th iteration along left; (d). 4 borders of road are interpolated; (e). internal area of road are filled; (f). obstacle and long grass regions are interpolated (Colour figure online)

Figure 9(d) illustrates the result after interpolation is conducted along left, right, top and bottom sides of the road region. As can be observed, most of the interpolated borders are continuous. The border regions are then smoothed and the internal holes are filled to achieve a continuous region as shown in Fig. 9(e). Figure 9(f) presents the result after the interpolation process are also applied to category "longGrass" and "obstacle". The region for "longGrass" is not fully continuous even after the interpolation. One of the reasons is that due to irregular height of the long grass, the estimated coordinates of the four corners of unclassified cells covered by long grass are not accurate enough to get the correct projection to image pixels. The other reason is that there is no clear-cut boundary between "longGrass" regions and "highVeg" regions, as shown in Fig. 9(a). Likewise, there are gaps between the road boundary and "longGrass" region. However, the continuous road region and the approximation of borders between "road" and other categories form a good 2 1/2 D map for navigation purpose.

4   Summary

In this paper, we have introduced the scene parsing mechanism in our Cognitive Architecture, which has successfully integrated a visual attention mechanism with the super-pixel level scene parsing mechanism. We have also proposed a novel way to explicitly interpret spatial relations between objects and applied them to the top-down facilitation process of the scene part parsing mechanism. The scene part parsing and visual attention mechanisms have been tested in many experiments and trials and proved to be effective. In the meantime, we have also proposed a new approach to acquire a continuous $2\frac{1}{2}D$ map of traversable regions via fusion of image and LIDAR detections. This algorithm is under test involving autonomous vehicle navigation in off-road environment.

References

1. Goodale, M.A., Milner, A.D.: Separate visual pathways for perception and action. trends Neurosci. **15**(1), 20–25 (1992)
2. Ng, G.W.: Brain-Mind Machinery. World Scientific, London (2009)
3. Lowe, D.G.: Object recognition from local scale-invariant features. In: Proceedings of the International Conference on Computer Vision, pp. 1150–1157 (1999)
4. Dalal, N., Triggs, B.: Histograms of oriented gradients for human detection. In: CVPR (2005)
5. Riesenhuber, M., Poggio, T.: Hierarchical models of object recognition in Cortex. Nature Neurosci. **2**, 1019–1025 (1999)
6. Serre, T., Wolf, L., Bileschi, S., Riesenhuber, M., Poggio, T.: Robust object recognition with cortex-like mechanisms. IEEE Trans. Pattern Anal. Mach. Intell. **29**(3), 411–426 (2007)
7. Felzenszwalb, P., McAllester, D., Ramanan, D.: A discriminatively trained multi-scale deformable part model. In: CVPR (2008)
8. Viola, P., Michael J.J.: Rapid object detection using a boosted cascade of simple features. In: CVPR (2001)

9. Felzenszwalb, P., Girshick, R. McAllester, D.: Cascade object detection with deformable part models. In: CVPR (2010)
10. Laxebnik, S., Schmid, C., Ponce, J.: Beyond bags of features: spatial pyramid matching for recognizing natural scene categories. In: CVPR (2006)
11. Oliva, A., Torralba, A.: Modeling the shape of the scene: a holistic representation of the spatial envelope. Int. Comput. Vis. **42**(3), 145–175 (2001)
12. Torralba, A., Murphy, K., P., Freeman, W.T., Rubin, M. A.: Context-based vision system for place and object recognition. In: ICCV, pp. 1023–1029 (2003)
13. Siagian, C., Itti, L.: Rapid biologically-inspired scene classication using features shared with visual attention. PAMI **29**(2), 300–312 (2007)
14. Renniger, L., Malik, J.: When is scene identification just texture recognition? Vis. Res. **44**, 2301–2311 (2004)
15. Tighe, J., Lazebnik, S.: SuperParsing: scalable nonparametric image parsing with superpixels. In: Daniilidis, K., Maragos, P., Paragios, N. (eds.) ECCV 2010, Part V. LNCS, vol. 6315, pp. 352–365. Springer, Heidelberg (2010)
16. Li, L.J., Socher, R., Li, F.F.: Towards total scene understanding: classification, annotation and segmentation in an automatic framework. In: CVPR (2009)
17. Du, L., Ren, L., Dunson, D., B., Carin, L.: A Bayesian model for simultaneous image clustering, annotation and object segmentation. In: NIPS (2009)
18. Rabinovich, A., Vedaldi, A., Galleguillos, C.: Object in context. In: ICCV (2007)
19. Galleguillos, C., Belongie, S.: Context-based object categorization: a critical survey. J. Comput. Vis. Image Underst. **114**(6), 712–722 (2010)
20. He, X., Zemel, R., Carreira-Perpindn, M.A.: Multiscale conditional random fields for image labelling. In: CVPR, pp. 695–702 (2004)
21. Kumar, S., Hebert, M.: A hierarchical field framework for unified context-based classification. In: ICCV, pp. 1284–1291 (2005)
22. Verbeek, J., Triggs, B.: Scene segmentation with conditional random fields learned from partially labeled images. In: NIPS (2008)
23. http://en.wikipedia.org/wiki/DARPA_Grand_Challenge
24. Vandapel, N., Huber, D.F., Kapuria, A., Hebert, M.: Natural terrain classification using three-dimensional Ladar data for ground robot mobility. J. Field Robot. **23**(10), 839–861 (2006)
25. Himmelsbach, M., Luettel, T., Wuensche, H.J.: Real-time object classification in 3D point clouds using point feature histograms. In: Proceedings of IEEE/RSJ International Conference on Intelligent Robots and Systems, USA (2009)
26. Thrun, S., et al.: Stanley: the robot that won the DARPA grand challenge. J. Robot. Syst. **23**(9), 661–692 (2006)
27. Rasmussen, C.: A hybrid vision+Ladar rural road follower. In: Proceedings of the IEEE Conference on Robotics and Automation, pp. 156–161 (2006)
28. Manz, M., Himmelsbach, M., Luettel, T., Wuensche, H.: Detection and tracking of road networks in rural terrain by fusing vision and LIDAR. In: Proceedings IEEE/RSJ International Conference on Intelligent Robots and Systems, pp. 4562–4568 (2011)
29. Ng, G.W., Xiao, X., Chan, R.Z., Tan, Y.S.: Scene understanding using DSO cognitive architecture. In: Proceedings of the 15th International Conference on Information Fusion (2012)
30. Zhao, G., Xiao, X., Yuan, J., Ng, G.W.: Fusion of 3D-LIDAR and camera data for scene parsing. J. Vis. Commun. Image Represent. **25**(1), 165–183 (2013)
31. Hochstein, S., Ahissar, M.: View from the top: hierarchies and reverse hierarchies in the visual system. Neuron **36**, 791–804 (2002)

32. Bar, M.: A cortical mechanism for triggering top-down facilitation in visual object recognition. J. Cogn. Neurosci. **15**(4), 600–609 (2003)
33. Yao, J., Fidler, S., and Urtasun, R.: Describing the scene as a whole: joint object detection, scene classfication and semantic segmentation. In: CVPR (2012)
34. Kasther, S., Ungerleider, G.: Mechanisms of visual attention in the human cortex. Annu. Rev. Neural Sci. **23**, 315–341 (2000)
35. Felzenszwalb, P., Huttenlocker, D.: Efficient graph-Based imagesegmentation. IJCV **2**, 167–181 (2004)
36. http://www.robots.ox.ac.uk/vgg/research/textclass/filters.html
37. http://www.mit.edu/jmutch/fhlib
38. Ojala, T., Pietikainen, M., Maenpaa, T.: Multi-resolution gray-scaleand rotation invariant texture classification with local binary patterns. PAMI **24**(7), 971–986 (2002)
39. Fenske, M.J., Aminoff, E., Gronau, N., Bar, M.: Top-down facilitation of visual object recognition: object-based and context-based contributions. Prog. Brain Res. **155**, 3–21 (2006)
40. Oliva, A., Torralba, A.: The role of context in object recognition. Trends Cogn. Sci. **11**(2), 520–527 (2007)
41. Desai, C., Ramanan, D., Fowlkes, C.C.: Discriminative models for multi-class object layout. IJCV **2**, 169–176 (2012)
42. Achanta, R., Hemami, S., Estrada, F., Susstrunk, S.: Frequency-tuned Salient Region Detection. In: CVPR (2009)
43. Rensink, R.A.: The dynamic representation of scenes. Visual Cognition **7**(1/2/3), 17–42 (2000)
44. Nistér, D., Stewénius, H.: Linear time maximally stable extremal regions. In: Forsyth, D., Torr, P., Zisserman, A. (eds.) ECCV 2008, Part II. LNCS, vol. 5303, pp. 183–196. Springer, Heidelberg (2008)
45. Matas, J., Chum, O., Urban, M., Pajdla, T: Robust wide baseline stereo from maximally stable extremal regions. In: BMVC (2002)

Combining Multiple Shape Matching Techniques with Application to Place Recognition Task

Karel Košnar[(⊠)], Vojtěch Vonásek, Miroslav Kulich, and Libor Přeučil

Department of Cybernetics, Faculty of Electrical Engineering,
Czech Technical University in Prague, Technicka 2,
166 27 Prague 6, Czech Republic
{kosnar,kulich}@labe.felk.cvut.cz

Abstract. Many methods have been proposed to solve the problem of shape matching, where the task is to determine similarity between given shapes. In this paper, we propose a novel method to combine many shape matching methods using procedural knowledge to increase the precision of the shape matching process in retrieval problems like place recognition task. The idea of our approach is to assign the best matching method to each template shape providing the best classification for this template. The new incoming shape is compared against all templates using their assigned method. The proposed method increases the accuracy of the classification and decreases the time complexity in comparison to generic classifier combination methods.

1 Introduction

The shape matching problem is studied in various form: (1) the computation problem is to compute the dissimilarity measure between two shapes, (2) decision problem is to decide if two shapes are similar enough to represent the same object and (3) the retrieval problem is to chose the most similar shape from the set of templates. The shape matching is a problem that is solved in context of computer vision, pattern recognition and robotics. Shape matching is one of the key tasks in shape retrieval, object recognition, visual scene understanding or place recognition.

There are many shape matching techniques solving the computation problem and providing dissimilarity measure between patterns. This dissimilarity measure is a function defined on pairs of shapes indicating the degree of resemblance of patterns. A dissimilarity measure should be invariant for the geometrical transformation group and ideally has the properties of metric. The decision problem or retrieval problem is then mostly solved by applying the thresholding or nearest neighbor method respectively making use of the pairwise dissimilarity measure.

In robotics, the place recognition problem is equivalent to the retrieval problem. The task of place recognition is to decide whether a robot is revisiting an already known location or is visiting unknown location using only sensor information. This is crucial for applications like pose initialization and localization using prior maps, closing large loops, simultaneous localization and mapping

© Springer International Publishing Switzerland 2015
C.V. Jawahar and S. Shan (Eds.): ACCV 2014 Workshops, Part I, LNCS 9008, pp. 399–412, 2015.
DOI: 10.1007/978-3-319-16628-5_29

(SLAM), localization in topological maps or merging maps collected at different time or by multiple robots.

Different types of sensors like cameras, laser range-finders or RGB-D sensors can be used to obtain the information about the places. As the sensor information is limited, two main problems arise: (1) Perceptual aliasing when two different places can be perceived as the same and (2) perceptual variability when the same place provides different sensor readings under different circumstances. The perceptual aliasing can be minimized by using as detailed place description as possible. On the other hand, the detailed description increases the perceptual variability.

Usually the comparison is not performed on raw sensor data, but these data are processed into form of descriptors to decrease memory consumption, increase speed of comparison and robustness. One of the possible representation of the raw sensor data is a shape of the place. In this case, the shape matching techniques is possible to use.

Each shape matching method works for certain class of shapes better and for other type worse. Therefore it is not easy to choose the proper method for given set of templates or set a generally suitable thresholds. In robotics, if the environment consists many subareas of different types (e.g. indoor and outdoor if is considered only the rawest division) it can be impossible to recognize places using only one method. The solution is to use more methods and combine their outputs.

As the retrieval problem or place recognition task can be seen as a multi-class classification problem, it is possible to use the methods for combining classifier, which are widely studied. Classifier combination techniques operate on the outputs of individual classifiers and usually fall into one of two categories. In the first approach the outputs are treated as inputs to a generic classifier, and the combination algorithm is created by training this, sometimes called secondary, classifier. The advantage of using such a generic combinator is that it can learn the combination algorithm and it can automatically account for the strengths and score ranges of the individual classifiers. In the second approach, a function or a rule combines the classifier scores in a predetermined manner. For the review see [1].

In the field of mobile robotics and specifically in place recognition, different classifier combination techniques are used. The voting scheme and nearest neighbor is used in [2]. The AdaBoost is widely used in place recognition and semantic classification of places [3–5]. Another method used in robotics as a generic classifier is Support Vector Machine (SVM) [6].

This paper introduces novel method for combining the methods for retrieval problem based on the knowledge of the template set. As the place recognition methods always compare sensor information representing actual place with data stored in the database of known places, the properties of the reference place from database is known. This information is used to decide which method should be used.

The proposed method finds the best classifier for each known place, which distinguishes this place from all other known places. If a new place is visited, it is compared with all known places. In our method, when the new place is compared with a given known place, the knowledge of corresponding best method is employed. It allows to improve the precision of the classification as this knowledge is utilized. It is called procedural knowledge as defined by [7] because it describes *how to proceed* the comparison involving content specific rules, strategies and actions.

As only one method is used when a new incoming place is compared to stored place, the time complexity is significantly lower than the generic classifier methods combining all the available methods together.

The rest of the paper is organized as follows: next section describes the proposed procedural knowledge-based method combining multiple classifiers. The Sect. 3 describes experimental setup, used datasets and classifiers for combining. The Sect. 4 displays resulting accuracy and time consumption. The paper concludes in Sect. 5.

2   Method

The proposed procedural-knowledge based method utilizes specific property of the retrieval problem or place recognition task in robotics. The input to the method is an unknown shape and the set of known template shapes. The template set is known in advance, therefore it is possible to utilize this knowledge to improve performance of the classification.

Let the template set or database $D = \{d_i^j; i = 1 \ldots n, j = 1 \ldots k\}$ consists of k classes of shapes, where every class is represented by n instances of shapes. Each shape is described by the polygon $d_j^i = \{(x, y) \in \mathbb{R}^2\}$. The notation d_i^j means that the shape d is i^{th} instance of the class j.

The shape matching method, which correctly distinguishes the particular class of shapes from the most other shapes in the template set is used always to measure dissimilarity of this class of shapes. This best method is called procedural knowledge and each class has exactly one shape matching method assigned as a procedural knowledge.

In robotics, the database is called a map, a class is equivalent to the physical location in the environment and instances are the measurements or sensor readings taken in this place and stored in the map. So each place in the environment is represented by n shape descriptors stored in the map D. It is assumed, that each place or class is described by exactly n shapes, without loss of generality.

Let there is set of shape matching methods F, where each method $f \in F$ computes the dissimilarity of two shapes

$$f_i(d_l^x, d_k^y) : D \times D \mapsto \langle 0, 1 \rangle,$$

where the dissimilarity takes value from an interval $\langle 0, 1 \rangle$. The dissimilarity is 0 if the two shapes are identical and takes maximal value 1 if they are totally different.

The shape matching method can provides the dissimilarity measure in range $\langle 0, m_i \rangle$, but all the results can normalized by dividing dissimilarity measure by maximal value m_i. The normalized dissimilarity measure is assumed in the rest of paper without loss of generality.

The classification is then performed using the threshold ϑ_i as

$$c(f_i, (d_k^x, d_l^y), \vartheta_i) = \begin{cases} 1 & \text{if } f_i(d_k^x, d_l^y) < \vartheta_i \\ 0 & \text{if } f_i(d_k^x, d_l^y) \geq \vartheta_i \end{cases},$$

where 1 means that d_i^x and d_k^y are from the same class of shapes or represent the same place and classifier declares that $x = y$ and 0 means that they represent different classes or places and $x \neq y$.

The method works in two phases: learning phase and classification phase. In the learning phase, the best classifier is determined for each place in a database. The classification phase takes these best classifiers and compare the novel place with each place in database making use of the procedural knowledge to determine the best classifier.

2.1 Learning Phase

The proposed procedural knowledge-based method divides the dataset D into k disjunctive parts D_p such $\bigcup_{p=1}^{k} D_p = D$ and $\bigcap_{p=1}^{k} D_p = \emptyset$, where each part $D_p = \{d_i^p : i = 1 \dots n\}, p = 1, \dots, k$ contains all shapes of the same class p.

Then, the set of shape pairs $R_p = \{(x, y); x \in D_p, y \in D\}$ are created for each database part D_p, where each element from D_p is paired with each element from the full database D. Any set of pairs $R = R^+ \cup R^-$ is a union of positive examples $R^+ = \{(x, y) : x, y \in D_p\}$, where all shapes to same class and negative examples $R^- = \{(x, y) : x \in D_p, y \in D \setminus D_p\}$, where shapes to different classes.

For each class $p = 1, \dots, k$ represented by shapes D_p, a best-matching method f_p^* and corresponding threshold ϑ_p^* is chosen to maximize the F-Score:

$$(f_p^*, \vartheta_p^*) = \arg \max_{f \in F, \vartheta_p^f = <0,1>} \text{FScore}(c(f, R_p, \vartheta_p^f)),$$

where F is a set of all available classificatiors. The F-Score is computed on the pairs R_p. The F-Score represents the quality of the classifier and is computed as a harmonic mean of the precision and the recall using

$$\text{FScore}(c(f, R, \vartheta)) = (1 + \beta^2) \frac{pr(c(f, R, \vartheta)) \cdot re(c(f, R, \vartheta))}{(\beta^2 \cdot pr(c(f, R, \vartheta))) + re(c(f, R, \vartheta))}, \quad (1)$$

where the precision

$$pr(c(f, R, \vartheta)) = \frac{\sum_{r \in R^+} c(f, r, \vartheta)}{\sum_{r \in R} c(f, r, \vartheta)}$$

is a fraction of true positive hits out of all instances classified as true (true positive and false positive hits) and the recall (or true positive rate)

$$re(c(, f, R, \vartheta)) = \frac{\sum_{r \in R^+} c(f, r, \vartheta)}{|R^+|}$$

is a fraction of true positive hits of a classifier out of all positive cases in a dataset with a given threshold ϑ_i. As there is no preference for precision or recall, we set $\beta = 1$. The perfect classifier has the FScore $= 1$. The FScore $= 0$ means that classifier classifies all the positive example wrongly.

We use the F-Score quality measure because the numbers of positive and negative examples are unbalanced. As the number of the negative examples are always significantly higher due to one-to-rest training set, the F-Score provides more relevant results than the widely used accuracy measure.

2.2 Classification Phase

The classification phase is very easy. Let there is a unknown shape x. This shape can be acquired, when the robot visit a novel place and processes the sensor data to acquire the shape x describing this place. The aim of the classification phase is to decide, to which class the unknown shape x belongs.

The set of pairs

$$\forall p = 1 \ldots k; R_p^x = \{(x, d_i^p) : d_i^p \in D_p\}$$

is created for each known class p, where unknown shape x is paired with all shapes d_i^p belonging to given class p.

The result class for the unknown shape x is determined using the nearest neighbor method. The class $cl(x)$ of the unknown shape x according the given database is then computed as

$$cl(x) = \arg \min_{r_p^x \in R_p^x} (f_p^*(r_p^x) \cdot c(f_p^*, r_p^x, \vartheta_p^*)),$$

where pairwise dissimilarity is computed using the class p best method normalized method f_p^*. There are used only the dissimilarity measures with shapes that are classified as similar by classifier and the best threshold ϑ_p^*. The pair with the lowest normalized dissimilarity measure is taken from all the pairs of the shape x and each shape from the database. The template shape from this pair denotes the resulting class.

Main advantage of this method is, that during the classification phase only one dissimilarity measure is computed for each pair. This significantly increases speed of the matching process in comparison to other combining methods, as they usually require to compute all the dissimilarity measures for each pair, as the computation of the dissimilarity measure can be very computational intensive.

3 Experiments

Experiments are mainly conducted in context of place recognition, where the places are describe by shapes making use of laser range-finder sensors.

All shapes are represented by closed polygons. Four different datasets used in the experiments are described in following section. Examples of the shapes taken from used datasets are depicted on Fig. 1.

Dissimilarity of the places is computed by different methods listed below. The methods providing dissimilarity of the shapes are selected to cover wide range of different approaches selected.

The proposed procedural knowledge-based method is compared to the AdaBoost method, as it is popular in robotics. The implementation of the AdaBoost.M1 from Matlab is used.

3.1 Datasets

Four datasets are used to show performance of the proposed method. The first dataset is a MPEG7 part B [8], which is commonly used for shape matching method comparison in pattern recognition and computer vision. This dataset is chosen as an ethalon, showing the performance of the presented method in context of shape retrieval problem. MPEG7 dataset contains binary images of different shape silhouettes. The MPEG7 dataset contains 70 different shapes each in 20 variants. To get the same format as in other datasets, the images are converted to polygons. Then the same set of methods is applicable to the all datasets.

The second dataset (called Robotic) is collected from a real environment by a mobile robot equipped with two real laser range finders Sick LMS200, in configuration providing together full 360 degree range scan. The robot is placed to 15 different places in an office building and the robot took 12 different scans around each place. These scans are taken equidistantly on the circular trajectory with 0.6 m diameter, which ensures various orientations of the scans.

Two other datasets are generated synthetically in a robotic simulator. The third dataset called Box is generated from a planar environment, where boxes of various orientation and size are placed randomly. Then, 11 places are chosen and for each place 21 different range scans with varying orientation and displacement (limited to 1 m) are generated. This dataset simulates a cluttered unstructured environment, but with lot of significant and detectable points (like corners).

The fourth dataset called Surface is generated from a 3D undulated surface. The range scans are generated in the same manner as in the Box dataset. This dataset simulated another type of unstructured environment without significant points.

3.2 Methods

Various methods suitable for place recognition from laser scans are used. As an initial source of methods is used work [9], where methods used in computer vision, shape matching and robotic mapping are compared. Selected methods are outlined in the following lines to provide main ideas of each method. For detailed description, see the original papers cited in each section.

Fourier Transformation (Fft). The shape is treated as a function in a polar coordinates system and is described by the coefficients of the Fourier

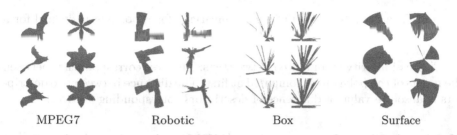

| MPEG7 | Robotic | Box | Surface |

Fig. 1. Example of places from the used datasets.

transformation [10]. Only amplitudes of the Fourier coefficients are considered for the descriptor to assure a rotation invariance. To minimize influence of noise in the data, the only first 20 harmonic functions are considered. The dissimilarity of two descriptors is computed as the Euclidean distance in 20 dimensional space.

Tangent Space. Traditionally, the closed polygon can be represented as a list of vertices or by giving a list of line segments. Alternatively, a polygon can be represented using a tangent space [11] - a list of angle-length pairs, whereby the angle at a vertex is an accumulated tangent angle at this point while the length is the normalized accumulated length of polygon sides up to this point. As the tangent space representation depends on the starting vertex, the dissimilarity of two polygons is a minimal difference between all possible variants of the tangent space representation.

Scan Line. The scan line matching algorithm [12] computes a shape descriptor from the intersection of randomly placed lines with the polygon. All the intersecting points are ordered and form n compact intervals. The descriptor is a vector of values computed for different lengths of interval, where only intervals greater than given threshold are used. If an interval lies strictly on an interior or exterior of the polygon, the descriptor value is incremented. If the interval represents a collection of intervals both interior and exterior, the descriptor value is decremented. The dissimilarity of two polygons is a sum of absolute values of descriptors components difference.

Ring Projection. The ring projection [13] algorithm computes the intersection of the polygon with the growing circle placed in the center of the polygon analogically to *Scan line* method. The value of ring projection is a fraction of circle part inside the polygon to circle circumference. The dissimilarity of two polygons is one minus a normalized correlation of polygon descriptors.

Integral Invariant. The integral invariant method [14] relies on measurement of dissimilarity between two curves that represents an integral invariant of the polygon. The discrete version of integer invariant for a polygon is defined for a given vertex as a logarithm of sum of Euclidean distances from given vertex to all others.

The dissimilarity is computed in two steps: The best correspondence between the points of the polygons is computed at first. The distance between two descriptors is absolute value of difference of descriptors corresponding vertices.

Multi-scale Shape Representation (MRM). The multi-scale shape representation [15] stores convexity/concavity of the polygon at different scale levels for each point. Different scale levels for polygon are computed by the convolution with the Gaussian kernel. The convexity and concavity of the curve is measured as a displacement of the contour between two consecutive scale levels, which is measured as the Euclidean distance of the corresponding contour points from two consecutive scale levels. Dissimilarity is based on the cost of the optimal path found by the dynamic programming in the matrix of mutual points distances and normalized by complexity of the compared curves.

Fast Laser Interest Region Transform. Fast laser interest region transform (FLIRT) method [16] is a multi-scale interest region operator for a 2D range data which combines the curvature-based detector with the β-grid descriptor. Interest points at the given scale correspond to points, where scale level equals the inverse of the local curvature of the smoothed signal. These interest points are described by β-grids. Place recognition using FLIRT is done by applying the RANSAC algorithm between two shapes. The re-projection error is used as the measure. The FLIRTLib [17] implementation is used in this paper with default parameters.

Shape Context. This method is based on assignment of a shape context [18] to each polygon vertex, which describes a near neighborhood of the vertex in question. The shape context is defined as a two-dimensional histogram of logarithmic polar distances from a particular vertex to other vertices in the polygon. The dissimilarity measure between two polygons can be then computed as follows: the shape context is computed for each vertex first and the shape distance between vertices of the two polygons are computed consequently. The distance between two polygons is obtained as the sum of distances between resulting matched vertex pairs.

Inner Distance. Inner distance method improves *Shape context* method described above. The inner distance of two vertices of the polygon is defined as a length of shortest path connecting the vertices under the condition that whole connecting path lies inside the polygon. The inner distance captures better the shape structure and is insensitive to articulation. Rest of the computation is same as in the *Shape context* method.

Geometric Moments. These methods compute the descriptors using geometric moments of the polygon, that are invariant to translation and scale [19]. Computation of the moments for polygon is well described in [20]. The dissimilarity of two polygons is computed as difference of corresponding moments normalized by the number of moments. In the experiments, the maximal order of polygon is set to 3, as the moments of higher order are sensitive to noise.

Zernike Moments. Contrary to the geometric moments, the Zernike moments are computed using complex polynomials that form an orthogonal basis [21]. The orthogonal moments allows computing arbitrary high moments of the input images. Similarily to the geometric moments or harmonics in Fourier transform, the low-order Zernike moments describe rough image properties, while the moments of higher orders are mainly influenced by detail in the image and therefore, they are more sensitive to noise. In this paper, Zernike moments of 15th order are utilized. For the details, see [22].

4   Results

The performance of proposed method is evaluated on all four datasets. The speed and accuracy of the proposed procedural knowledge-based method is compared with AdaBoost.M1 method. The results of single shape matching methods are also depicted for comparison.

As place recognition methods are expected to work with a database, where more than one sensor reading is stored for each place, the cross-validation method is used to obtain results. There are n instances for each place in the dataset. The $n-1$ instances of each place are taken as the training set used for learning classifiers. Then the remaining instances of each place is taken as an unknown inputs and localized against the training set. This is performed n times for each place instance index.

The accuracy of the place recognition is measured using F-Score (Eq. 1). Results are depicted on Fig. 2 as boxplots, the bottom and top of the box are the first and third quartiles, and the band inside the box represents the median. The cross inside the box is a mean value of measured F-Scores. The ends of the whiskers represent the 9th percentile and the 91st percentile or the worst and the best cases respectively.

The performance of solving the retrieval problem is measured by so called Bull's eye score. This score expects that the dataset D can be divided into a set of k disjunctive classes $C = \{c_1, \ldots, c_k\}$; $\bigcup_{c \in C} c = D$; $\bigcap_{c \in C} c = \emptyset$ and each class $\forall c \in C, |c| = n$ has the same number n of shapes. Every shape is compared to all other shapes and the $m = 2n$ best matches are considered. The number of true positive hits h_i (both shapes are from same class) from the m best matches is computed for each shape $i \in D$. The Bull eye score B is then the ratio of the total number of shapes from the same class to the highest possible number $B = \frac{\sum_{i \in D} h_i}{|D| n}$. Thus, the best possible rate is 100 %.

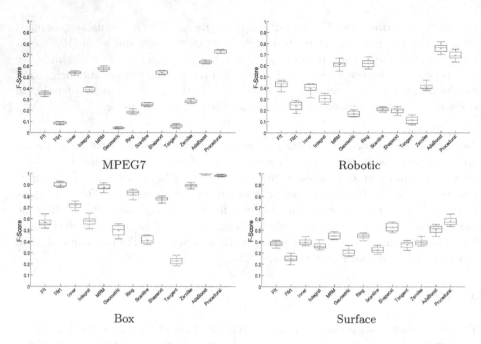

Fig. 2. F-Scores for datasets.

Table 1. Processing time [ms] of the classification phase.

Method	Datasets			
	MPEG7	Robot	Box	Surface
FFT	0.52	0.52	0.58	0.52
FLIRT	123.59	191.23	993.00	657.43
Inner	42.49	17.81	37.46	52.88
Integral	0.65	1.60	0.88	0.97
MRM	162.29	87.63	140.86	139.37
Moments	0.69	1.46	3.24	4.37
Ring	2.20	4.46	25.53	11.49
Shapecontext	25.26	10.47	11.74	38.23
Scanline	47.51	70.79	125.97	195.46
Tangent	3.20	15.33	131.61	129.74
Zernike	164.52	65.30	103.60	129.47
AdaBoost	572.90	466.60	1574.50	1360.00
Procedural Knowledge	72.50	32.60	242.10	166.30

The results for the selected methods and datasets are summarized in the Table 3. The best score is displayed in a bold font for each dataset, considering only the single shape matching techniques.

Table 2. Learning time [s].

Method	Datasets			
	MPEG7	Robot	Box	Surface
AdaBoost	475.19	7.16	11.57	6.61
Procedural Knowledge	102.78	3.65	4.17	4.17

Table 3. Bull's eye score.

Method	Datasets			
	Mpeg7	Robot	Box	Surface
Inner	76.34	59.07	98.04	63.55
Fft I	65.40	72.22	78.54	57.43
Integral	48.21	51.99	83.24	50.38
Ring	34.84	81.20	**98.80**	52.34
Shapecontext	63.72	35.37	92.35	**70.95**
Zernike	47.98	64.26	97.67	55.62
Moments	16.97	36.39	74.81	51.27
Scanline	41.42	38.98	67.59	50.48
MRM	**77.95**	**81.76**	97.20	60.52
Tangent	18.38	30.37	39.08	58.94
Flirt	16.12	43.01	97.84	31.99
AdaBoost	91.32	94.19	99.99	56.56
Procedural Knowledge	92.89	92.68	99.46	80.11

The accuracy of the proposed procedural-knowledge based combining method is significantly higher than all the single shape matching methods. The average performance of the proposed method is always better than the best case performance of the best single method. This is a important property of the proposed method, addition of any even a very bad shape matching method cannot make an accuracy lower. This is not surprising as always the best possible method is used for particular place. Moreover, procedural knowledge-based method has no requirements on the properties of the partial methods in contrast to AdaBoost.

In comparison, a not suitable shape matching method can negatively influence the resulting accuracy of the AdaBoost method. AdaBoost requires that used partial methods performs better than random guess. This condition is not fulfilled when the Surface dataset is used, and AdaBoost fails to find good classifier for this dataset. As can be seen on the Surface dataset, the accuracy of the AdaBoost is lower than the best shape matching method. The accuracy of proposed method is comparable with the AdaBoost method on the other datasets.

A low computation time of the classification phase is a big advantage of the proposed procedural knowledge-based method. The computation time of the classification phase is crucial in the retrieval problem and place recognition task, as many partial comparisons are necessary to retrieve the shape class or localize the robot inside the known map. As the proposed procedural-knowledge based method utilizes only one dissimilarity computed by single method selected according the procedural knowledge, the computation time of procedural-knowledge based method is always lower (or equal), than the computation time of the slowest shape matching method. The number of used shape matching method does not influence the time consumption of the classification phase.

In comparison, the complex methods are typically worse than single method in term of time consumption, as they combine outputs from more than one shape matching method. The AdaBoost method needs the dissimilarity values computed by all methods, therefore the time consumption is increasing with every added shape matching method. The mean time necessary to compare one pair of places are summarized in Table 1. The computation is performed in the Matlab on the computer with Intel Xeon 3.19 Ghz computer with 8 GB RAM.

The proposed procedural knowledge-based method is less time consuming than AdaBoost method, even in the learning phase. The time consumption to learn the AdaBoost classifier and the procedural knowledge-based classifier is summarized in the Table 2. The size of learning set significantly influence the learning time as well as the number of used shape matching methods.

5   Conclusion

This paper presented a novel retrieval method based on the procedural knowledge. This method utilizes that the database is known in advance and chooses the best shape matching method for each class of shapes in database that distinguishes it from others. These chosen methods are called the procedural knowledge.

The utilization of the procedural knowledge significantly increase the accuracy of the retrieval problem and the place recognition task without additional computational costs. The performance of the novel method is always better than the best single shape matching method.

The procedural knowledge-based method require lower computation time in learning as well as in classification phase, in comparison to the AdaBoost method. This is caused by selecting only the one best classifier for each class of shapes, therefore during the classification phase, only one single dissimilarity measure is computed for each shape. Contrary, the AdaBoost requires to compute all the dissimilarity measures for each shape, which is time consuming.

The proposed method is suitable in all retrieval tasks, where the dissimilarity of unknown object with a database is required to compute, especially when the database is not homogeneous. In context of autonomous systems is it mainly the place recognition task and robot global localization. In these cases, the proposed procedural knowledge-based method provides the best results.

Acknowledgement. This work has been supported by the Czech Science Foundation under research project No. 13-30155P and by the Technology Agency of the Czech Republic under the project no. TE01020197 "Centre for Applied Cybernetics". The experiments have been run using Grid Infrastructure Metacentrum (project No. LM2010005).

References

1. Tulyakov, S., Jaeger, S., Govindaraju, V., Doermann, D.: Review of classifier combination method. In: Machine Learning in Document Analysis and Recognition, pp. 361–386 (2008)
2. Ulrich, I., Nourbakhsh, I.R.: Appearance-based place recognition for topological localization. In: ICRA, pp. 1023–1029. IEEE (2000)
3. Mozos, O., Stachniss, C., Burgard, W.: Supervised learning of places from range data using AdaBoost. In: IEEE International Conference on Robotics and Automation, pp. 1730–1735 (2005)
4. Stachniss, C., Mozos, O.M., Burgard, W.: Speeding-up multi-robot exploration by considering semantic place information. In: IEEE International Conference on Robotics and Automation (2006)
5. Soares, S.G., Arajo, R.: Semantic place labeling using a probabilistic decision list of AdaBoost classifiers. Int. J. Comput. Inf. Syst. ind. Manag. Appl. **6**, 548–559 (2014)
6. Pronobis, A., Martinez Mozos, O., Caputo, B.: SVM-based discriminative accumulation scheme for place recognition. In: IEEE International Conference on Robotics and Automation (2008)
7. Cauley, K.M.: Studying knowledge acquisition: distinctions among procedural, conceptual and logical knowledge. In: Annual Meeting of the American Educational Research Association, 67th, Office of Educational Research and Improvement, Educational Resources Information Center (ERIC) (1986)
8. Sikora, T.: The MPEG-7 visual standard for content description-an overview. IEEE Trans. Circuits Syst. Video Technol. **11**, 696–702 (2001)
9. Kosnar, K., Vonasek, V., Kulich, M., Preucil, L.: Comparison of shape matching techniques for place recognition. In: European Conference on Mobile Robots (ECMR) (2013)
10. Kauppinen, H., Seppänen, T., Pietikäinen, M.: An experimental comparison of autoregressive and fourier-based descriptors in 2d shape classification. IEEE Trans. Pattern Anal. Mach. Intell. **17**, 201–207 (1995)
11. Van Otterloo, P.: A Contour-Oriented Approach to Shape Analysis. Prentice Hall International, New York (1991)
12. Cannon, M., Warnock, T.: A shape descriptor based on the line scan transform (2004)
13. Tang, Y., Li, B., Ma, H., Lin, J.: Ring-projection-wavelet-fractal signatures: a novel approach to feature extraction. IEEE Trans. Circuits Syst. II Analog Digital Signal Proc. **45**, 1130–1134 (1998)
14. Manay, S., Hong, B.W.: Integral invariants for shape matching. IEEE Trans. Pattern Anal. Mach. Intell. **28**, 1602–1618 (2006)
15. Adamek, T., O"Connor, N.: A multiscale representation method for nonrigid shapes with a single closed contour. IEEE Trans. Circuits Syst. Video Technol. **14**, 742–753 (2004)

16. Tipaldi, G.D., Braun, M., Arras, K.O.: FLIRT: Interest regions for 2d range data with applications to robot navigation. In: Proceedings of the International Symposium on Experimental Robotics (ISER), Dehli, India (2010)

17. Tipaldi, G.D., Arras, K.O.: Flirtlib (2010). http://srl.informatik.uni-freiburg.de/~tipaldi/FLIRTLib/. Accessed 20 March 2013

18. Belongie, S., Malik, J., Puzicha, J.: Shape matching and object recognition using shape contexts. IEEE Trans. Pattern Anal. Mach. Intell. **24**, 509–522 (2002)

19. Hu, M.K.: Visual pattern recognition by moment invariants. IRE Trans. Inf. Theory **8**, 179–187 (1962)

20. Steger, C.: On the calculation of arbitrary moments of polygons. Technical report FGBV-96-05, Munchen University, Munchen, Germany (1996)

21. Teague, M.R.: Image analysis via the general theory of moments*. J. Opt. Soc. Am. **70**, 920–930 (1980)

22. Li, S., Lee, M.C., Pun, C.M.: Complex zernike moments features for shape-based image retrieval. IEEE Trans. Syst. Man Cybern. Part A Syst. Hum. **39**, 227–237 (2009)

A Model-Based Approach for Fast Vehicle Detection in Continuously Streamed Urban LIDAR Point Clouds

Attila Börcs[⊠], Balázs Nagy, Milán Baticz, and Csaba Benedek

Distributed Events Analysis Research Laboratory,
Institute for Computer Science and Control of the Hungarian
Academy of Sciences, Kende utca 13-17, Budapest 1111, Hungary
{attila.borcs,balazs.nagy,milan.baticz,csaba.benedek}@sztaki.mta.hu

Abstract. Detection of vehicles in crowded 3-D urban scenes is a challenging problem in many computer vision related research fields, such as robot perception, autonomous driving, self-localization, and mapping. In this paper we present a model-based approach to solve the recognition problem from 3-D range data. In particular, we aim to detect and recognize vehicles from continuously streamed LIDAR point cloud sequences of a rotating multi-beam laser scanner. The end-to-end pipeline of our framework working on the raw streams of 3-D urban laser data consists of three steps (1) producing distinct groups of points which represent different urban objects (2) extracting reliable 3-D shape descriptors specifically designed for vehicles, considering the need for fast processing speed (3) executing binary classification on the extracted descriptors in order to perform vehicle detection. The extraction of our efficient shape descriptors provides a significant speedup with and increased detection accuracy compared to a PCA based 3-D bounding box fitting method used as baseline.

1 Introduction

1.1 Problem Statement

Efficient and fast perception of the surrounding environment has a major impact in mobile robotics research with many prominent application areas, such as autonomous driving, driving assistance systems, self localization and mapping, and obstacle avoidance [1,2]. Future mobile vision systems promise a number of benefits for the society, including prevention of road accidents by constantly monitoring the surrounding vehicles or ensuring more comfort and convenience for the drivers. Outdoor laser scanners, such as LIDAR mapping systems particularly have become an important tools for gathering data flow for these tasks since

---

Csaba Benedek—This work was partially funded by the Government of Hungary through a European Space Agency (ESA) Contract under the Plan for European Cooperating States (PECS), and by the Hungarian Research Fund (OTKA #101598).

© Springer International Publishing Switzerland 2015
C.V. Jawahar and S. Shan (Eds.): ACCV 2014 Workshops, Part I, LNCS 9008, pp. 413–425, 2015.
DOI: 10.1007/978-3-319-16628-5_30

they are able to rapidly acquire large-scale 3-D point cloud data for real-time vision, with jointly providing accurate 3-D geometrical information of the scene, and additional features about the reflection properties and compactness of the surfaces. Moreover, LIDAR sensors have a number of benefits in contrast to conventional camera systems, *e.g.* they are highly robust against daily illumination changes, and they may provide a larger field of view. Robust detection and recognition of vehicles in 3-D urban scenarios is one of the major challenges in any robot perception related task. In this paper we focus on the vehicle detection problem relying on large-scale terrestrial point clouds recorded in different crowded urban scenarios, such as main roads, narrows streets and wide intersections. More specifically we use as input a raw point cloud stream of a rotating multi-beam (RMB) laser acquisition system. The problem of detection and recognition of certain types of object characteristics on streaming point clouds is challenging for various reasons. First, the raw measurements are noisy and contain several different objects in cluttered regions. Second, in crowded scenes the vehicles, pedestrians, trees and street furnitures often occlude each other causing missing or broken object parts in the visible measurement streams. Third, typically by terrestrial laser scanning the point cloud density rapidly decreases as a function of the distance from the sensor [3], which fact may cause strongly corrupted geometric properties of the object appearances, misleading the recognition modules. Further requirements arise for navigation or autonomous driving systems, where the data is continuously streamed from a laser sensor mounted onto a moving platform, and we are forced to complete the object detection and recognition tasks within a very limited time frame.

1.2   Related Works

Significant research efforts are expended nowadays for solving object recognition problems in point clouds obtained by 3-D laser scanners. Extracting efficient object descriptors (i.e. *features*) is an essential part in each existing technique, which step usually implements one of the two following strategies.

According to the *first* strategy, the shape and the size of the objects are approximated by 3-D bounding boxes. In [4] a framework has been proposed for object classification and tracking. The basic idea is to use an octree based Occupancy Grid representation to model the surrounding environment, and simple features for object classification, such as the length ratios of object bounding boxes. In that method three different object classes are considered: pedestrians, bicycles and vehicles. In our case, however, the observed environment consists of complex urban scenarios with many object types such as trees, poles, traffic signs, and occluded wall regions. Here simple features may not be robust enough for efficient object classification, due to the largely diverse appearances of the considered object shapes throughout an entire city. Other approaches derive 3-D bounding boxes for recognition via Principal Component Analysis (PCA) techniques. The authors of [5] and [6] calculate statistical point cloud descriptors: they compute saliency features which capture the spatial distribution of points in a local neighborhood by covariance analysis. The main orientation of an object is derived from the principal components (eigenvalues and corresponding eigenvectors), considering the covariance matrix of the 3-D point positions. Object classification is achieved by three saliency features, namely *scatter*ness, *linear*ness and *surface*ness, which are calculated as linear combinations of the eigenvalues.

Following the *second* strategy, a group of existing object classification techniques use different features, based on shape and contextual descriptors [7–9]. Reference [7] propose a system for object recognition, by clustering nearby points from a set of potential object locations. Thereafter, they assign the points near the estimated object locations to foreground and background sets using a graph-cut algorithm. Finally a feature vector is built for each point cluster, and the feature vectors are labeled by a classifier, trained on a manually collected object set. In [8] an algorithm is presented for fast segmentation of point cloud regions, followed by a 3-D segment classification step which is based on various 3-D features, like the spin image or the spherical harmonic descriptor. Reference [9] introduce an approach for detecting and classifying different urban objects from a raw stream of 3-D laser data such as cars, pedestrians and bicyclists. For this purpose a graph-based clustering algorithm, different shape descriptors and shape functions are exploited, such as the spin image and PCA based eigenvectors. Although these general shape and context based methods may provide more precise recognition rates for urban objects than the 3-D bounding box based techniques, they are computationally more expensive, and often do not perform in real time.

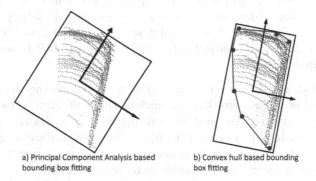

a) Principal Component Analysis based bounding box fitting   b) Convex hull based bounding box fitting

Fig. 1. Demonstrating the limitations of PCA based bounding box approximation, and the advantages of the proposed convex hull based bounding box fitting technique on the top-view projection of a selected vehicle in the point cloud

2 Contributions of the Proposed Approach

In this work we present a real-time model-based system for vehicle detection and extraction from continuously streamed LIDAR point clouds, which are captured in challenging urban scenarios. By constructing the proposed vehicle model, we combine three novel feature extraction steps. *First* we propose a new convex hull based 2-D bounding box fitting method, which is used for fast and precise estimation of the location, size and orientation parameters of the vehicle candidates. *Second*, we develop a 3-D sphere based feature, which is used for approximating the principal curvatures of the objects in 3-D. *Third*, we extract the object contours from the side-view, in order to obtain a representative shape characteristics of vehicles in 2-D. Our model gives two major contributions over existing approaches:

◊ *Fast 2-D bounding box fitting for cluttered and partially incomplete objects:* It is highly challenging to fit precise bounding boxes around the objects in RMB LIDAR range data streams, since we should expect various artifacts of self-occlusion, occlusion by other object, measurement noise, inhomogeneous point density and mirroring effects. These factors drastically change the appearances of the 3-D objects, and the conventional principal component analysis (PCA) based techniques [4,5] may not give sufficient results. Especially, in the RMB point cloud streams, only the object side facing the sensor is clearly visible, and the opposite side of the object is usually completely or partially missing. For this reason, if we calculate by PCA covariance analysis the principal directions of a point cloud segment identified as a vehicle candidate, the eigenvectors usually do not point towards the main axes of the object, yielding inaccurately oriented bounding boxes, as demonstrated in Fig. 1. In contrast to PCA solutions in 3-D, we calculate the 2-D convex hull of the top-view projection of the objects, and we derive the 2-D bounding boxes directly from the convex hull. As shown later, this strategy is less sensitive to the inhomogeneous point density and the presence of missing/occluded object segments, since instead of calculating spatial point distributions for the entire object's point set, we capture here the local shape characteristics of the visible object parts, and fit appropriate 2-D bounding boxes with partial matching.

◊ *Lightweight shape analysis for streaming data:* For enhancing the classification performance of object recognition, an efficient shape descriptor, called the spin image, has been adopted in several previous methods [7–9]. Spin image based features can be used to approximate the object shapes by surface meshes, yielding robust solutions for object classification and recognition. However, the demand of real-time performance is not feasible here, since estimating different surface models for 3-D data is a computationally expensive task. Therefore, these mesh based models are not directly designed for continuously streamed range data. In our solution, we propose two new features for approximating the principal curvatures of the objects in 3-D. The extracted contour of the detected object's side-view profile can be compared to a reference vehicle contour model, which is obtained by supervised training. To train our classifier, we use vehicle samples from a manually annotated point cloud database. Our proposed features are also able to sufficiently model the shape characteristics of vehicle objects, while they can be calculated very quickly, offering a decent trade-off with respect to speed and accuracy.

The description of the proposed model-based recognition framework is
ᵒᵗʳᵘⁱ ᶜᵒⁿᵗⁱⁿᵘᵉᵈ ᵃˢ follows. In Sect. 3 we briefly present the preprocessing steps. First,
a hierarchical grid based data structure is introduced, which will allow us to
perform fast retrieval of 3-D point cloud features for segmentation and detection
purposes. Second, we propose a point cloud segmentation algorithm, to distinguish the *foreground* regions of the captured scene containing the moving or
static field objects, from the *background* composed of the roads and other terrain parts. Third, a fast connected component analysis algorithm is presented
for separating individual objects within the foreground regions. The main new
contributions of this paper are related to vehicle detection and localization. In
Sect. 4, we introduce our *vehicle model* with defining an efficient set of features,
which characterize vehicle candidates in the point clouds. In Sect. 5, a Support
Vector Machine (SVM) based object classification process is described based
on the previous features. Finally we report on the experiments and present the
evaluation results in Sect. 6.

3   Point Cloud Segmentation and Object Separation

In this section, we introduce the point cloud preprocessing module of the proposed system to prepare the data for the vehicle detection step. An efficient
grid based method was presented [10,11], which will be adopted here for robust
foreground extraction and 3-D object separation, in challenging dense urban
environments where several nearby object may be located close to each other.

◇ *The Hierarchical Grid Model data structure*: We fit a regular 2-D grid S with
W_S rectangle side length onto the $P_{z=0}$ plane (using the RMB sensor's vertical
axis as the z direction and the sensor height as a reference coordinate), where
$s \in S$ denotes a single cell. We assign each $p \in P$ point of the point cloud to
the corresponding cell s_p, which contains the projection of p to $P_{z=0}$. Let us
denote by $P_s = \{p \in P : s = s_p\}$ the point set projected to cell s. Moreover, we
store the height coordinate and different height properties such as, maximum
$z_{\max}(s)$, minimum $z_{\min}(s)$ and average $\hat{z}(s)$ of the elevation values within cell s,
which quantities will be used later for foreground separation.

For robust object detection a denser grid resolution is also required in this
grid data structure, therefore the cell s of the coarse grid level is subdivided into
smaller cells $s'_d | d \in \{1, 2, \dots, \xi^2\}$, with cell side length $W_{s'_d} = W_s / \xi$, where ξ is a
scaling factor (used $\xi = 3$). We store references for each 3-D point in the coarse
and dense grid levels as well (Fig. 2).

◇ *Foreground separation and object detection*: The foreground separation is
achieved on the coarse grid level of the above presented grid data structure.
Our goal is to discriminate foreground regions *e.g.* street furnitures, pedestrians,
vehicles, walls and other street objects, and background regions composed of the
roads, sidewalks and grassy ground. For ground segmentation we apply a locally
adaptive terrain modeling approach similarly to [10], which is able to accurately
extract the road regions, even if their surfaces are not perfectly planar. We use

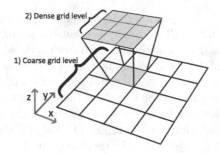

Fig. 2. Visualization of our *hierarchical grid model* data structure - *(bottom)* the coarse grid level: the 3-D space coarsely quantized into 2-D grid cells, *(top)* the dense grid level: each grid cell on the coarse level subdivided into smaller cells.

point height information for assigning each grid cell to an appropriate cell class. Before that, we detect and remove grid cells that belong to irrelevant *clutter* regions, thus we will not visit these cells later and save processing time. We classify each cell to *clutter*, which contains less points than a predefined threshold (typically 4–8 points). After clutter removal all the points in a cell are classified as *ground*, if the difference of the minimal and maximal point elevations in the cell is smaller than a threshold (used 25 cm), moreover the average of the elevations in neighboring cells does not exceeds an allowed height range based on a globally estimated digital terrain map. The first criterion ensures the flatness or homogeneity of the points. Given a cell with 60 cm of width, this allows 22.6° of elevation within a cell; higher elevations are rarely expected in an urban scene. The rest of the points in the cloud are assigned to class *foreground* belonging to vehicles, pedestrians, mail boxes, billboards etc.

After the foreground separation step, our aim is to find distinct groups of points which belong to different urban objects on the foreground. For this task we use the *hierarchical grid model*: On one hand, the coarse grid resolution is appropriate for a rough estimation of the 3-D blobs in the scene, in this way we can also roughly estimate the size and the location of possible object candidates. On the other hand, using a dense grid resolution beside a coarse grid level, is efficient for calculate point cloud features from a smaller subvolume of space, therefore we can refine the detection result derived from the coarse grid resolution.

The proposed object detection algorithm consists of three main steps: *First*, we visit every cell of the coarse grid and for each cell s we consider the cells in its 3×3 neighborhood (see Fig. 3(a, b)). We visit the neighbor cells one after the other in order to calculate two different point cloud features: (i) the maximal elevation value $Z_{max}(s)$ within a coarse grid cell and (ii) the point cloud density (*i.e.* point cardinality) of a dense grid cell. *Second* our intention is to find connected 3-D blobs within the foreground regions, by merging the coarse level grid cells together. We use and elevation-based cell merging criterion for perform this step. $\psi(s, s_r) = |Z_{max}(s) - Z_{max}(s_r)|$ is a merging indicator, which

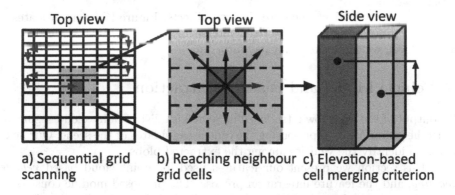

a) Sequential grid scanning   **b) Reaching neighbour grid cells**   **c) Elevation-based cell merging criterion**

Fig. 3. The step by step demonstration of the object detection algorithm

measures the difference between the maximal point elevation within cell s and its neighboring cell s_r. If the ψ indicator is smaller than a predefined value, we assume that s and s_r belong to the same 3-D object (see Fig. 3(c)). *Third*, we perform a detection refinement step on the dense grid level. The elevation based cell merging criterion on the coarse grid level often yields that nearby and self-occluded objects are merged into a same blob. We handle this issue by measuring the point density in each sub-cell s'_d at the dense grid level. Our assumption is here that the nearby objects, which were erroneously merged at the coarse level, could be appropriately separated at the fine level, as the examples in Fig. 4 show. Let us present three typical urban scenarios when the *simple* coarse grid model merges the close objects to the same extracted component, while using a *hierarchical* grid model with coarse and dense grid level, the objects can be appropriately separated. We consider two neighboring super-cell pairs -marked by red - in Fig. 4(a, b), respectively. In both cases the cells contain points from different objects, which fact cannot be justified at the coarse cell level. However, at the dense level, we can identify connected regions of near-empty sub-cells

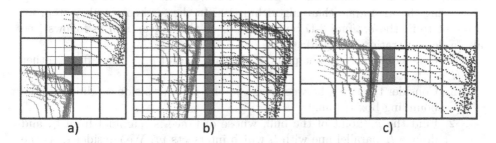

a)   b)   c)

Fig. 4. Separation of close objects at the dense grid level. [color codes: green lines = coarse grid level, black lines = dense grid level, grey cells = examined regions for object separation] (Color figure online) (denoted by gray), which separate the two objects. Figure 4(c) demonstrates a third configuration, when a super-cell intersects with two objects, but at the sub-cell level, we can even find a separator line.

4   Vehicle Model and Feature Extraction

The output of the workflow introduced in Sect. 3 is a list of point cloud segments (called hereafter *blobs*) representing the object candidates of the scene. Our next goal is to identify the vehicles among the extracted blobs.

In this section we present our features used for point cloud based *vehicle modeling*, and the feature integration process. The proposed module consists of the combination of three descriptors in order the find the optimal trade-off w.r.t. speed and accuracy. First, we approximate the position, size, and orientation parameters of the obtained objects by fitting a 2-D bounding box to the point cloud blobs. In contrast to previous works [4–6] we do not calculate spatial point distributions over the entire blobs, due to reasons detailed in Sec. 2. Instead, we determine the vehicle's 2-D convex hull with a matching step considering only the visible structure elements. In this way our solution is independent of the local point cloud density at the given object position, and performs efficiently even if a significant part of the object is occluded. On the other hand, we have observed that the special curvature of the vehicle shape, especially around the windshields, is a characteristic feature for visual recognition, which can be quickly estimated and efficiently used for identification. The detailed explanation of the proposed feature extraction strategy is presented as follows:

◇ $2 - D\ bounding\ box\ fitting\ via\ convex\ hulls$: Let us consider the output of the preprocessing module (Sect. 3) on the dense grid level of the *Hierarchical Grid Model*. At this step, we only use the width and a depth (X,Y) coordinates of the points, and the height coordinates (Z) are ignored. We mark *first* which cells are occupied by several points, and which ones are empty. Next we visit the 3×3 neighborhood of each occupied cell, and filter out the cells, where all neighbors are occupied as well. In this way we can roughly estimate the boundary cells of the object. Then, we construct the convex hull from the points of the boundary cells using the monotone chain algorithm [12]. In the forthcoming key step, we attempt to fit the *optimal* 2-D bounding box to the convex hull as follows (see also a demonstration in Fig. 5):

- Visit the consecutive point pairs of the hull p_i and p_{i+1}, one after another ($i = 1, 2, \ldots, i_{\max}$):
    1. Consider the line l_i between point p_i and p_{i+1}, as a side candidate of the bounding box rectangle.
    2. Find the $p_\star$ point of the hull, whose distance is maximal from l_i, and draw a $l_\star$ parallel line with l_i which intersects $p_\star$. We consider $l_\star$ as the second side candidate of the bounding box.
    3. Project all the points of the convex hull to the line l_i, and find the two extreme ones p' and p''. The remaining two sides of the bounding box candidate will be constructed by taking perpendicular lines to l_i, which intersect p' and p'' respectively.

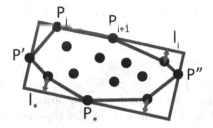

Fig. 5. Demonstration of the fast 2-D bounding box fitting algorithm for the convex hull of the top-view object projection (the bounding box is shown marked by gray color)

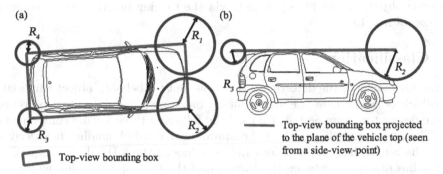

Fig. 6. Demonstration of the principal curvature feature with 3-D spheres

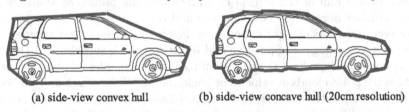

(a) side-view convex hull   (b) side-view concave hull (20cm resolution)

Fig. 7. Demonstration of side-view convex and concave hulls

- Chose the optimal bounding box from the above generated rectangle set by minimizing the average distance between the points of the convex hull and the fitted rectangle.

◇ *Principal curvature estimation with 3 − D spheres*: Our aim here is to place four spheres near to the four top corners of the vehicle's roof, in order to examine the typical curvatures around this regions, especially close to the windshields. First we vertically shift the 2-D bounding box obtained by the previous feature extraction step to the maximal elevation within the vehicle's point set. This configuration is demonstrated in Fig. 6 both from top-view and from side-view. Then we set the center points of the spheres to the corner points of the shifted 2-D bounding rectangle. We start to increase the radius of the spheres as long as they hit a 3-D point from the vehicle. Our assumption is that due to the typical slope around the windshields, we should experience significant differences between the radiuses of the four spheres. We can also observe in Fig. 6 that the radiuses of the frontal spheres (R_1 and R_2) are significantly larger than the radiuses of the spheres at the back side (R_3 and R_4). We use in the following the four radius values in the object's feature vector.

◇ *Shape approximation of side-view profile using convex and concave hulls:* at this step, we project the point clouds of the object candidates to a vertical plane which is parallel to the main axis of the top-view bounding box. Thereafter, we fit to the 2-D side-view object silhouettes a convex hull, and a concave hull with 20 cm resolution. Here the shape features are the contour vectors of the convex and concave hulls themselves, so that we store the contours of sample vehicles with various prototypes in a library, and we compare the contours of the detected objects to the library objects via the turning function based polygon representation [13].

5   Classification

In the final stage of the detection process, we must label each object candidate as *vehicle* or *background* (*i.e.* non-vehicle), based on a feature vector composed of the descriptors from Sect. 4. We used the following three feature components: (1) The length and the width of the approximated 2-D bounding box derived from the convex hull. (2) The four radius values of the 3-D spheres, as well as the radius difference between the frontal and the back sphere pairs, which are proposed for principal curvature estimation. (3) The difference between the concave side profile hull of the vehicle candidate and the prototype shape, which is a real number normalized between zero and one. Consequently, the resulting feature vector consists of eight dimensions. Following a supervised approach, we created first a training set of positive and negative vehicle samples. For this purpose, we have developed a 3-D annotation tool, which enables labeling the urban objects in the point clouds as vehicles or background. We have manually collected more than 1600 positive samples (*i.e.* vehicles), and also generated 4000 negative samples from different scenarios recorded in the streets of Budapest, Hungary. The negative samples were created by a semi-automatic process, cutting random regions from the point clouds, which were manually verified. In addition, 12715 positive vehicle samples, and 3396 negative samples (different street furniture and other urban objects) have been selected from the KITTI Vision Benchmark Suite [14] and used for the training of the classifier. We have performed a binary classification between the *vehicles* and the *background objects* by a Support Vector Machine (SVM) using the toolkit from [15]. Here we used $C = 0.1078$ as the kernel function parameter, and $\nu = 0.0377$ as the upper bound on the fraction of margin errors.

6   Experiments

We evaluated our method on four LIDAR point cloud sequences, concerning different types of urban scenarios, such as main roads, narrow streets and inter-

Table 1. Numerical comparison of the detection results obtained by the Principal Component Analysis based technique [5] and the proposed Model-based framework. The number of objects (NO) are listed for each data set, and also in aggregate.

Point cloud dataset	NO	PCA based approach [5]		Prop. Model-based approach	
		F-rate(%)	Avg. Processing Speed (fps)	F-rate(%)	Avg. Processing Speed (fps)
Budapest dataset #1	567	73	15	89	24
Budapest dataset #2	1141	71	12	90	21
Budapest dataset #3	368	57	13	80	22
KITTI dataset [14]	614	62	14	78	25
Overall	2690	68	13.5	86	23

sections. Three scenarios have been recorded in the streets of Budapest, Hungary, and the fourth scenario has been selected from the KITTI Vision Benchmark Suite [14]. All the test sequences have been recorded by a Velodyne HDL-64E S2 rotating multi-beam LIDAR sensor, with a 10 Hz rotation speed. We have compared our *Model-based approach* to a reference method, which uses a simple occupancy grid representation for foreground separation, and applies Principal Component Analysis (PCA) based features for object classification [5].

Qualitative results of our proposed model on four sample frames are shown in Fig. 8.[1] During the quantitative evaluation, we verified the proposed method and the reference PCA based technique on 2690 vehicles, using the Ground Truth (GT) information. To enable fully automated evaluation, we needed to make first a non-ambiguous assignment between the detected objects and GT object samples, where we used the Hungarian algorithm [16] to find optimal matching. Thereafter, we counted the number of Missing Objects (MO), and the Falsely detected Objects (FO). These values were compared to the Number of real Objects (NO), and the F-rate of the detection (harmonic mean of precision and recall) was also calculated. We have also measured the processing speed of the two methods in frames per seconds (fps). The numerical performance analysis is given in Table 1. The processing speed of the individual modules of the framework on the entire test dataset, which consist of 2690 vehicles is the following: (1) Building hierarchical grid data structure - *13* ms (2) Point cloud segmentation - *4* ms (3) Object separation step - *6* ms (4) Feature extraction step - *18* ms (5) SVM decision *2* ms. The results confirm that the proposed model surpasses the PCA based method in F-rate for all the scenes. Moreover, the proposed *Model-based approach* is significantly faster on the streaming data, and in particularly, it gives more reliable results in the challenging crowded urban scenarios (#2 and #3), where several vehicles are occluded by each other, and the scene contains various types of other objects and street furnitures such as

---

[1] Demonstration videos are also available at the following url: https://vimeo.com/ pointcloudprocessing.

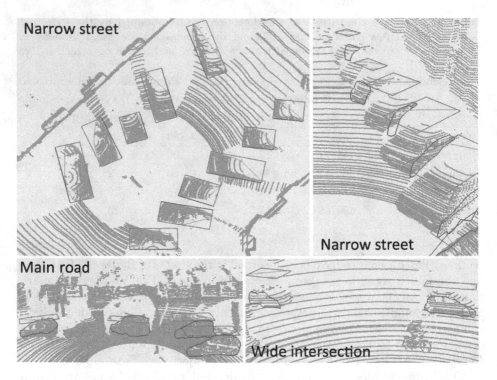

Fig. 8. Qualitative results of vehicle detection with displaying the top-view bounding boxes (by red) and the side view concave hulls (blue) extracted by the algorithm (Color figure online)

walls, traffic signs, billboards, pedestrians etc. The proposed method only fails on highly occluded vehicles, where the objects are broken into many parts or the majority of the vehicle's point cloud is missing. As for the computational speed, we measured 13.5 fps in average with the Principal Component Analysis based technique [5] and 23 fps with the proposed *Model-based approach*.

7   Conclusion

In this work we have proposed a novel *Model-based framework*, which uses three new descriptors for robust detection of vehicles in continuously streamed point cloud sequences of a rotating multi-beam LIDAR sensor. Due to the presented features we have observed a reliable performance in challenging dense urban scenarios with multiple occlusions and the presence of various types of scene objects. The model has been quantitatively validated based on Ground Truth data, and the advantages of the proposed solution versus a state-of-the-art technique have been demonstrated.

References

1. McNaughton, M., Urmson, C., Dolan, J.M., Lee, J.W.: Motion planning for autonomous driving with a conformal spatiotemporal lattice. In: ICRA, pp. 4889–4895 (2011)
2. Levinson, J., Montemerlo, M., Thrun, S.: Map-based precision vehicle localization in urban environments In: Proceedings of Robotics: Science and Systems, Atlanta, GA, USA (2007)
3. Behley, J., Steinhage, V., Cremers, A.B.: Performance of histogram descriptors for the classification of 3d laser range data in urban environments. In: ICRA, pp. 4391–4398 (IEEE)
4. Azim, A., Aycard, O.: Detection, classification and tracking of moving objects in a 3D environment. In: IEEE Intelligent Vehicles Symposium (IV), Alcalá de Henares, Spain (2012)
5. Himmelsbach, M., Müller, A., Luettel, T., Wuensche, H.J.: LIDAR-based 3D object perception. In: Workshop on Cognition for Technical Systems, Munich (2008)
6. Lalonde, J.F., Vandapel, N., Huber, D., Hebert, M.: Natural terrain classification using three-dimensional ladar data for ground robot mobility. J. Field Robot. **23**, 839–861 (2006)
7. Golovinskiy, A., Kim, V.G., Funkhouser, T.: Shape-based recognition of 3D point clouds in urban environments, Kyoto, Japan (2009)
8. Douillard, B., Underwood, J., Vlaskine, V., Quadros, A., Singh, S.: A pipeline for the segmentation and classification of 3d point clouds. In: ISER (2010)
9. Wang, D.Z., Posner, I., Newman, P.: What could move? finding cars, pedestrians and bicyclists in 3d laser data. In: Proceedings of IEEE International Conference on Robotics and Automation (ICRA), Minnesota, USA (2012)
10. Józsa, O., Börcs, A., Benedek, C.: Towards 4D virtual city reconstruction from Lidar point cloud sequences. In: ISPRS Workshop on 3D Virtual City Modeling. ISPRS Annals Photogram. Rem. Sens. and Spat. Inf. Sci., Regina, Canada, vol. II-3/W1, pp. 15–20 (2013)
11. Börcs, A., Nagy, B., Benedek, C.: Fast 3-D urban object detection on streaming point clouds. In: Bronstein, M., Agapito, L., Rother, C. (eds.) ECCV 2014 Workshops, Part II. LNCS, vol. 8926, pp. 628–639. Springer, Heidelberg (2015)
12. Andrew, A.: Another efficient algorithm for convex hulls in two dimensions. Inf. Process. Lett. **9**, 216–219 (1979)
13. Kovács, L., Kovács, A., Utasi, A., Szirányi, T.: Flying target detection and recognition by feature fusion. SPIE Opt. Eng. **51**, 117002 (2012)
14. Geiger, A., Lenz, P., Urtasun, R.: Are we ready for autonomous driving? The kitti vision benchmark suite. In: Conference on Computer Vision and Pattern Recognition (CVPR) (2012)
15. King, D.E.: Dlib-ml: A machine learning toolkit. J. Mach. Learn. Res. **10**, 1755–1758 (2009)
16. Kuhn, H.: The Hungarian method for the assignment problem. Naval Res. Logist. Q. **2**, 83–97 (1955)

Large-Scale Indoor/Outdoor Image Classification via Expert Decision Fusion (EDF)

Chen Chen[(✉)], Yuzhuo Ren[(✉)], and C.-C. Jay Kuo[(✉)]

Department of Electrical Engineering,
University of Southern California, Los Angeles, CA 90089, USA
{chen80,yuzhuore}@usc.edu, cckuo@sipi.usc.edu

Abstract. In this work, we propose an Expert Decision Fusion (EDF) system to tackle the large-scale indoor/outdoor image classification problem using two key ideas, namely, data grouping and decision stacking. By data grouping, we partition the entire data space into multiple disjoint sub-spaces so that a more accurate prediction model can be trained in each sub-space. After data grouping, the EDF system integrates soft decisions from multiple classifiers (called experts here) through stacking so that multiple experts can compensate each other's weakness. The EDF system offers more accurate and robust classification performance since it can handle data diversity effectively while benefiting from data abundance in large-scale datasets. The advantages of data grouping and decision stacking are explained and demonstrated in detail. We conduct experiments on the SUN dataset and show that the EDF system outperforms all existing methods by a significant margin with a correct classification rate of 91 %.

1 Introduction

Indoor/outdoor scene classification is one of the basic scene classification problems in computer vision. Its solutions contribute to general scene classification [1–6], image tagging [7–9], and many other applications [10–13]. As compared to general scene classification problems, the indoor/outdoor scene classification problem has a clearer definition, namely, whether the scene is inside or outside a man-made structure with enclosed roofs and walls. Since the man-made structure is well-defined, the decision is unambiguous under various circumstances.

Indoor/outdoor classification allows a precise characterization of a wide range of images with diversified semantic meanings. For example, images from ① kitchen to ⑨ green house in the left column of Fig. 1 should all be classified as indoor images. In contrast with other scene classification problems [14–16], semantic objects in the scene may not help much in the decision. For example, indoor ⑤ swimming pool and outdoor ⑤ swimming pool in Fig. 1 share the same salient semantic object (*i.e.*, the pool), yet they should be classified differently from the aspect of indoor/outdoor scene classification. The same observation occurs in quite a few real-world images.

© Springer International Publishing Switzerland 2015
C.V. Jawahar and S. Shan (Eds.): ACCV 2014 Workshops, Part I, LNCS 9008, pp. 426–442, 2015.
DOI: 10.1007/978-3-319-16628-5_31

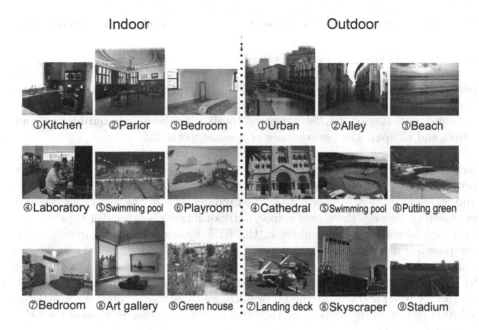

Fig. 1. Exemplary indoor and outdoor scene images from the test dataset are given in the left and right of the dash line, respectively (Color figure online).

Millions of images have been created every day due to the popularity of smart phones. Due to the huge size and great diversity of image data, applications such as large-scale image search [17] and tagging [18] will benefit from accurate indoor/outdoor classification results. Several methods, including SP [19], VFJ [20], SSL [21], PY [22], KPK [23] and XHE [24], were proposed to tackle this problem based on image datasets consisting of about 1,000 images. It is not clear whether the reported performance of these methods is scalable to large-scale datasets consisting of more than 100,000 images. This is the main focus of our current research.

To address the large-scale indoor/outdoor scene classification problem, we propose an Expert Decision Fusion (EDF) system that consists of two key ideas – data grouping and decision stacking. In contrast with prior art, the proposed EDF system is less concerned with the search of new features but on a meaningful way to partition the dataset and organize basic indoor/outdoor classifiers in an effective way to lead to a more accurate and robust classification system. For convenience, each basic indoor/outdoor classifier is called an "expert" in this paper.

The design of the EDF system can be described as follows. We select a set of experts as the constituent members of the EDF system. After evaluating a few existing indoor/outdoor image classifiers [19–33], we choose 6 experts. They are SP [19], VFJ [20], SSL [21], PY [22], KPK [23] and XHE [24]. Furthermore, we developed three new experts (namely, HSH, TN and HDH) on our own.

To handle the problem of data diversity, we propose an effective way to partition data samples into multiple groups, where data in one group are more homogeneous to model and predict. Furthermore, the EDF system integrates soft decisions of constituent experts via stacking [34–36] to offer a better classification performance than each individual expert in each partitioned sub-space. To illustrate the advantage of the EDF system, we label all images in the SUN [24] dataset (consisting of 108,754 images in total) with the indoor/outdoor ground truth, and compare the performance of a set of methods.

There are several contributions of this work. First, to the best of our knowledge, this is the first study on the large-scale indoor/outdoor scene classification problem with a dataset exceeding 100,000 images. The developed methodology and learned experience contribute to the fundamentals of "big data" science and engineering. Second, three new indoor/outdoor image classifiers (or experts) are proposed as constituent members in the EDF system. Third, we demonstrate the power of data grouping and decision stacking in the design of the EDF system. Finally, the proposed EDF system reaches a correct classification rate of 91 % against the SUN dataset, which offers 6–26 % performance improvement over other benchmarking methods. Besides, we show that it provides a scalable solution by examining its performance as a function of different sizes of the SUN dataset.

One important lesson learned from our current study is that, as the data size becomes larger, there are two competing factors that have a high impact on the performance of a classification system – data diversity and data abundance. The former demands a better classifier design. In this work, we propose the use of data grouping and decision stacking to achieve this goal. Once the data diversity problem is addressed, data abundance actually helps improve the performance of a robust classifier.

The rest of this paper is organized as follows. We describe constituent experts in the EDF system in Sect. 2, which include existing indoor/outdoor scene classifiers as well as three newly developed classifiers. Then, the design of the EDF system is detailed in Sect. 3. Experimental results are reported and discussion is given in Sect. 4. Finally, concluding remarks and possible future extensions are presented in Sect. 5.

2   Description of Constituent Experts

2.1   Six Experts from Existing Work

Many indoor/outdoor scene classification solutions have been proposed in the past 15 years. The main focus has been on the selection of discriminant features. Low-level features such as color, texture and shape have been examined. For example, color histograms [19,21,23] and color moments [20] are two popular features. The global color pattern of tiny images [24] offers another color descriptor. Besides the RGB color space, other color spaces such as Ohta [37], LST and HSV, were studied [19,21,23]. Texture features were applied to indoor/outdoor scene classification [19,21]. The MSAR [38] and the multi-scale wavelet [39] are used as local texture descriptors.

Features such as edge angle histograms [23] and responses of Gabor filters (GIST, [22,29]) were used in recent works. KPK [23] partitions an image into one horizontal block in the top portion and four vertical blocks in the middle and lower portions and assigns different weights to features in these five blocks for further processing. Rather than partitioning an image into blocks, PY [22] computes the GIST [40] features from the original image and its edge map separately and cascades the two responses into a feature vector.

The performance of these classifiers approaches to their limits quickly as the image dataset becomes larger. We choose six of them as constituent experts of the EDF system, denoted by SP [19], VFJ [20], SSL [21], PY [22], KPK [23] and XHE [24]. We implement all of them by ourselves in the experimental section since none of the source codes is available.

Feature extraction and classifier training are two basic steps in developing an expert. Machine learning has been widely used in classifier training. The K-Nearest Neighbor (KNN) algorithm and the Learning Vector Quantization (LVQ) [41]) were considered in [19,20,29], respectively, where the choice of a good distance measure was the main issue. Later, the Support Vector Machine (SVM) [42] was used in [21–23] and the Probability Neural Network (PNN) [43] became popular due to their good performance and the availability of open source codes.

2.2 Three New Experts

We propose three new experts based on the features of Thermal Noise (TN), the Hue-Saturation Histogram (HSH) and the Hue-Dark Histogram (HDH). Their justification and implementation are detailed below.

The TN Expert. Thermal noise [44] arises in the image acquisition process due to poor illumination, high temperature, etc. Typically, indoor scenes have weak lighting sources and lower temperatures while outdoor scenes have stronger natural light and higher temperatures. For this reason, we propose to use TN to differentiate indoor/outdoor scenes. In the feature extraction step, noise levels in different color channels are calculated as a descriptor. First, we adopt a bilateral filter approach [45] to denoise each channel of the RGB, HSV and YUV color representations of an input image. Then, absolute differences between the original and the denoised image channels are computed to yield 9 noise maps for a single color image. Finally, standard deviations of all noise maps are concatenated to form a feature vector. In the model training step, we adopt linear SVM in the package [42] and the 5-fold cross validation process for performance evaluation.

The HSH Expert. The HSV color space is strongly linked to human visual perception. Here, we use a "modified" Hue-Saturation Histogram (HSH) to characterize the global color distribution of an image based on the following observation. Image pixels with low value (V) and low saturation (S) components do not contribute to the discrimination of indoor/outdoor scenes since too dark or bright pixels are not reliable in the decision. Hence, the hue-saturation histogram is only calculated in a partial volume of the HSV color space by excluding dark and bright pixels in our implementation. That is, for a pixel with its HSV color coordinates (h, s, v) we will include this pixel in the histogram calculation only if $v \geq T_v$, $s \geq T_s$. Otherwise, it is abandoned. T_v and T_s are empirically set to 0.2 and 0.1, respectively. We quantize the hue values into 16 bins and adopt a 5-bin saturation histogram for each hue bin. Consequently, we obtain an 80-bin hue-saturation histogram of an image. This HSH descriptor is used to train a linear SVM classifier to yield the HSH expert.

The HDH Expert. The dark channel was introduced in [46]. For a given pixel, its dark channel value, denoted by D, is the lowest one among its R, G, B three channel values. We have an interesting observation, namely, the dark channel patterns are different for indoor and outdoor scenes. Bright red objects in indoor scenes, such as the carpet in ②parlor and sheet in ③bedroom in Fig. 1, usually have small dark channel values since they have small values in green and blue channels. In contrast, due to lighting conditions, red objects in outdoor scenes, such as walls in ①urban and the sunset halo in ③beach in Fig. 1 have larger dark channel values. Furthermore, we observe different relations between certain colors and their dark channel values in indoor and outdoor images. To model this relationship and design a suitable feature, we partition an image into $4 \times 4 = 16$ sub-images and calculate the hue-dark histogram in each sub-image. For the hue-dark histogram, we quantize the hue channel values into 16 bins and compute a 5-bin dark value histogram in each hue bin to result in a 80-bin hue-dark histogram (HDH). Then, we concatenate the HDH descriptors of 16 sub-images to yield the final HDH descriptor, which is a 1280-dimensional feature vector. Again, a SVM model is used to train the HDH expert.

2.3   Feature Selection via ANOVA

A correct classification rate of 90 % for several experts such as SP, VFJ and SSL was reported before on experiments with around 1,000 images. However, when the data size becomes much larger, we see a significant performance drop between the training stage and the testing stage for experts with a high dimensional feature vector. This phenomenon is attributed to the over-fitting of high-dimensional feature vector in the training stage. We list the feature vector dimension of each expert in Table 1. To avoid the overfitting problem in classifier training and reduce

Table 1. The feature dimension numbers of experts before (labeled as "original") and after (labeled as "selected") the ANOVA feature selection process.

Experts	SP	VFJ	SSL	PY	KPK	XHE	TN	HSH	HDH
Original	1536	600	880	1024	80	768	9	80	1280
Selected	360	40	280	300	80	500	9	80	240

the training-testing performance gap, one solution is to select a smaller set of discriminant features. In the following, we use the well-known analysis of variance (ANOVA [47]) method for feature selection.

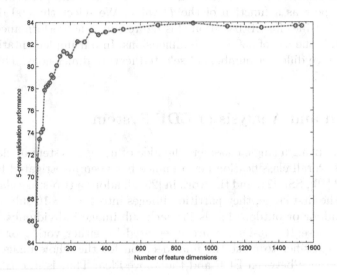

Fig. 2. The correct classification rate curve of SP with top D dimensions of the SP feature vector used as the training features and being evaluated by 5-fold cross validation.

One of the most widely used tools in ANOVA is the F-test [48]. For a single feature dimension, its F-value is defined as the ratio of the between-group variance and the within-group variance. Let $\bar{Y}$ be the mean of all data in this feature dimension, and K is the number of groups. We have indoor/outdoor two groups so that $K = 2$. We use $\bar{Y}_i$ and n_i to denote the sample mean and the observation number over this dimension in the ith group. Then, the between-group variance of a single feature dimension can be written as

$$Var_{bg} = \sum_{i=1}^{K} n_i(\bar{Y}_i - \bar{Y})^2/(K - 1). \tag{1}$$

Furthermore, its within-group variance can be expressed as

$$Var_{wg} = \sum_{ij} n_i(Y_{ij} - \bar{Y}_i)^2/(N - K), \tag{2}$$

where Y_{ij} is the jth observation in the ith out of K groups and N is the overall sample size. Finally, the F-value can be written mathematically as

$$F = \frac{Var_{bg}}{Var_{wg}}. \tag{3}$$

In classification, the larger F value is, the more discriminative this feature dimension is. After computing the F values of all feature dimensions of an expert's feature vector, we rank them from top to bottom and select the top D as desired features for classification. Figure 2 shows the performance curve of the SP expert as a function of the D value. We select the 360 dimensions with larger F values for SP based on this figure since the performance becomes saturated after the use of 360 feature dimensions. In our implementation, different experts have different numbers of selected feature dimensions. This result is listed in Table 1.

3   Design and Analysis of EDF System

Instead of adopting a single classifier, the idea of using a system of classifiers to improve the overall classification performance has been investigated before. For example, SP [19], SSL [21] and the work in [29] all adopt a two-stage classification system. At the first stage, they partition images into $4 \times 4 = 16$ sub-images and determine indoor or outdoor labels for each sub-image individually. Then, the decisions of these 16 sub-images are integrated by either voting or training a second-level classifier to make the final decision for the whole image. There is a major difference between EDF and the above idea. That is, the EDF system does not integrate decisions from sub-images but decisions from multiple experts made for the whole image. In the following, we first conduct the analysis on a single expert's decision in Sect. 3.1. Then, we explain the "diversity gain" of any two experts in Sect. 3.2. Finally, we discuss the structure of the EDF system in Sect. 3.3.

3.1   Analysis of Single Expert Decision

Before considering the collaboration of experts, we first analyze the decision behavior of a single expert. Without loss of generality, we use expert KPK as an illustrative example. For the jth image sample, denoted by I_j, KPK can generate a soft decision score, d_j^{kpk}, for it using its sample-to-boundary distance normalized to the range $[0, 1]$, where 0 and 1 indicate the indoor and outdoor scenes with complete confidence, respectively. When there is only one expert, we

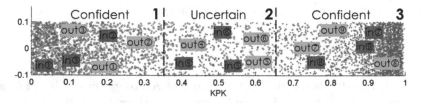

Fig. 3. The distribution of soft KPK decision scores d^{kpk} from 5,000 random samples (Color figure online).

need to quantize the soft decision score into a binary decision. That is, we divide interval $[0, 1]$ into two subintervals $S_1 = [0, T)$ and $S_0 = [T, 1]$, where $0 < T < 1$ is a proper threshold value (typically, $T = 0.5$). If $d_j^{kpk} \in S_1$, I_j is classified to an indoor image. Otherwise, $d_j^{kpk} \in S_2$ and I_j is classified to an outdoor image.

When soft decision score d_j^{kpk} is closer to threshold T, expert KPK is less confident about its decision. To take this into account, we may partition the entire decision interval into 3 subintervals $S_1 = [0, T_1)$, $S_2 = [T_1, T_2)$, and $S_3 = [T_2, 1]$, where $0 < T_1 < T_2 < 1$ are two thresholds. Parameters T_1 and T_2 are set to 0.35 and 0.65 in our implementation. Subintervals S_1 and S_3 are called the confident regions while subinterval S_2 is called the uncertain region.

We show the distribution of soft KPK decision scores collected from 5000 sample images randomly selected from the SUN database in Fig. 3, where red circles and green crosses denote indoor and outdoor image samples, respectively. To avoid the overlap of cluttered samples along the x-axis, we generate a vertical random shift between -0.1 to 0.1 for each sample, which is purely for the visualization purpose and has no practical meaning. We see that most red circles are in S_1 while most green crosses are in S_3. They can be correctly classified by KPK. On the other hand, there are few red circles in S_3 and few green crosses in S_1, and they will be misclassified by KPK. There are some red circles and green crosses in S_2, which are difficult to set apart using the soft KPK decision scores. It is apparent that the criteria of a good expert can be stated as:

1. it has a larger ratio of correct versus incorrect decision samples in S_1 and S_3; and
2. it has a smaller percentage of samples in S_2.

We will discuss ways to achieve the above goal by inviting the second expert to join the decision-making process in Sect. 3.2.

To gain more sights, we show the KPK soft scores of all 18 images in Figs. 1 and 3. Note that we select the images in Fig. 1 carefully so that there are three representative indoor and outdoor images in each sub-interval in Fig. 3. Visual inspection of these sample images will help us understand the strength and weakness of KPK.

Images in the Uncertain Region. For images in S_2, KPK cannot make a firm decision. Indoor images ④-⑥ and outdoor images ④-⑥ lie in this region. The two swimming pool images, images ⑤ in both indoor and outdoor categories, have similar elements such as blue water and dark tops. In addition, indoor and outdoor images ⑥ also share similar color patterns.

For images in S_1 and S_3, KPK has confident soft scores. They can be further divided into two cases.

Correctly Classified Images. Indoor images ①-③ and outdoor images ⑦-⑨ are correctly classified. Recall that KPK partitions an image into 5 blocks (namely, one horizon block in the top and four parallel vertical blocks in the lower portion). Indoor images ①-③ have small d^{kpk} values since they all have red and wooden objects at the bottom part of the images and shell-white ceilings or walls, which are easy to classify with KPK's block-based color and edge descriptors. Similarly, the top horizontal block carries the valuable sky information for outdoor images ⑦-⑨.

Misclassified Images. Outdoor images ①-③ and indoor images ⑦-⑨ are misclassified although their scores fall in the confident regions. They are called outliers. Outdoor images ①-③ all have dark colors and clear edge structures over the entire image, which misleads KPK. The blue top part of indoor image ⑦ is also misleading. Indoor image ⑧ is difficult since its wall contains the outdoor view and painting. Indoor image ⑨ can be even challenging to human being since one may make a different decision depending on the existence of the ceiling and the wall.

For the outlying images, low-level features mislead KPK to draw a confident yet wrong conclusion. Human can make a correct decision by understanding the semantic meaning of the scenes such as the river, the street and the ocean in outdoor images ①-③, respectively. Furthermore, indoor images ③ and ⑦ have the same semantic theme (bedroom) but different low-level features (color and texture patterns).

It is well known that there exists a gap between low-level features and high-level semantics of an image, which explains the fundamental limits of experts that rely purely on features in decision-making. Despite the semantic gap, a well-designed feature-based classifier can offer a reasonable classification performance due to the strong correlation between good low-level features and high-level semantics in a great majority of images.

3.2   Diversity Gain of Two Experts

As the size of image data becomes larger and their contents become more diversified, it is challenging to design a single expert that can handle all image types effectively. It is a natural idea to get the opinions of multiple experts and combine their opinions to form one final decision. In this subsection, we consider the simplest two-expert case. Intuitively, such a system may not work well under the following two scenarios: (1) if the opinions of two experts are too similar to each other; or (2) if one expert is significantly better than the other. In both scenarios, we do not benefit much by inviting the second expert in the decision process. Scenario (2) is self-evident. We will focus on scenario (1) by investigating the diversity gain of the two-expert system.

Without loss of generality, we choose KPK and PY as the two experts. The soft decision scores of expert PY, denoted by d^{py}, for the same 5,000 samples are plotted along the vertical axis in Fig. 4. The jth sample image represented by a red circle (indoor) or a green cross (outdoor) has a 2-D coordinate, (d_j^{kpk}, d_j^{py}), whose coordinate domain is called the KPK-PY soft decision map. With different combinations of soft decisions from the two experts, we can divide the 2-D decision space into 9 regions. KPK and PY have consistent opinions in their soft decisions in regions 1, 5 and 9, complementary decisions in regions 2, 4, 6 and 8, and contradictory opinions in regions 3 and 7.

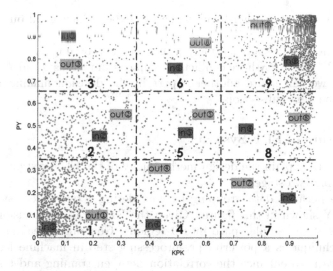

Fig. 4. The distribution of KPK-PY soft decision scores for the same 5,000 randomly picked samples shown in Fig. 3 (Color figure online).

By comparing Figs. 3 and 4, we see that PY can help KPK in resolving some decision ambiguities in regions 4–6. That is, PY can offer more confident scores for images in regions 4 and 5 than KPK. Similarly, KPK can help PY in resolving some decision ambiguity in regions 2, 5 and 8 since KPK is confident for images in regions 2 and 8. KPK and PY offer complementary strength since they examine different low-level features in evaluating an input image. KPK focuses on the local color and edge distributions while PY focuses on global scene structures. Indoor/outdoor images ②, ④, ⑥ and ⑧ are exemplary images in regions 2, 4, 6 and 8, respectively.

We focus on indoor/outdoor images ④ and ⑥, for which KPK does not have a confident score. Recall that PY [22] does not partition an image into multiple sub-images but computes the GIST features from the original image and its edge map separately and cascades the two responses into a feature vector. As a result, PY can make a more confident decision. PY's decisions on indoor image ④ (complicated scene structure for the whole image) and outdoor image ⑥ (textures of grass and leaves) are correct, yet PY's decisions on indoor image ⑥ (similar to outdoor image ⑥) and outdoor image ④ (consisting of many straight vertical lines similar to the view observed inside a church building) are not accurate. Since there are more indoor images than outdoor images in region 4 and more outdoor images than indoor images in region 6, PY does contribute to the correct classification rate in regions 4 and 6.

The same discussion applies to regions 2 and 8, where KPK helps PY in resolving ambiguity in a positive way. Region 5 remains to be ambiguous in the two-expert system. If the two experts share very similar opinions, most samples will fall in regions 1, 5 and 9 so that the two-expert system does not offer a clear advantage. On the other hand, if the two experts have good but diversified opinions, we will observe more samples in the four complementary regions and, as a result, the overall classification performance can be improved.

Finally, PY and KPK have conflicting opinions in regions 3 and 7. To resolve the conflict, we can invite another expert in the decision making process as detailed in the next subsection.

3.3   Structure of EDF System

We use Fig. 5 to explain the design methodology of the EDF system. It consists of the following two stages.

First, we perform data grouping by considering a two-expert system (say, KPK and PY shown in this figure). Given the KPK-PY soft decision map, we partition the data sample space into 9 regions. Generally speaking, the data grouping technique is a powerful pre-processing step in machine learning. Its main purpose is to enhance the correlation between training and testing samples. A good grouping strategy can contribute to the overall performance of the learning-based system significantly. We have tried different combinations of two experts from nine experts introduced in Sect. 2, and found that KPK and PY provide the best results due to their excellent individual performance and good complementary property. After grouping, the diversity of data samples in each region is reduced.

Second, we fuse the soft decisions of all nine experts in each region. We compare two methods in Sect. 4 – voting and stacking. For voting, we binarize the soft decision of each expert and use the simple majority voting rule to fuse expert's

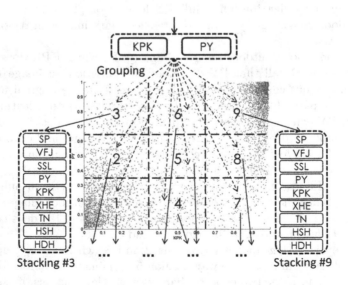

Fig. 5. The structure of the EDF system, where Stacking #3 indicates the stacking method in Region 3 of the joint KPK-PY soft decision map.

decisions. For stacking, we build a meta-level classification model that takes soft scores of all experts as the input features and make a final binary system decision. Since the training data in each region are different, different meta-level data models are built for different regions. The meta-level classifier is trained by linear SVM using samples with known binary outputs and, then, the trained model is used to predict samples with unknown binary outputs in the test. The correct classification rates of voting-based and stacking-based EDF systems are compared in Sect. 4. We will show that the stacking approach provides a better result.

Both grouping and stacking provide powerful tools to handle the problem of data diversity. Through grouping, we have more and smaller homogeneous datasets rather than one large highly heterogenous dataset. Through stacking, we can improve the robustness of the final decision in each region by leveraging the complementary strength of multiple experts.

4 Experimental Results

In the indoor/outdoor scene image classification literature, datasets used by other research groups are either too small or not available to the public. For example, the Kodak consumer image dataset, tested by SP [19] and SSL [21], contains 1343 images. Coral, used by VFJ [20], is not available to the public. A benchmark of 1000 images used in [29] is also not available. KPK [23] collects around 1200 images from the Internet, yet they are not released to the public. Two datasets consisting of 390 and 968 images, respectively, and used in [43] are accessible from their websites. Recently, a very large dataset, SUN, was published by [24] for the general scene classification benchmark. It consists of 397 well-sampled scene category indexes and 108,754 images. We labeled the whole SUN dataset into 47,260 indoor images and 61,494 outdoor images. Our experiments are conducted with respect to this dataset.

First, we show the correct classification rates of nine experts against the full SUN dataset in Table 2 without data grouping. The 5-fold cross validation is adopted in the experiment and the averaged performance is listed. There are one, three and five experts with correct classification rates between 60–69 %, 70–79 % and 80–89 %, respectively. Expert KPK has the best performance with a correct classification rate of 85.30 %.

Next, we show the correct classification rate achieved by nine experts and the EDF system in Regions 1–9 in Table 3, where the 5-fold cross validation is conducted in each region and the averaged performance is reported. For the last

Table 2. Correct classification rate of nine experts conducted on the full dataset (in the unit of %).

SP	VFJ	SSL	PY	KPK	XHE	TN	HSH	HDH
82.80	79.01	84.69	82.28	85.30	78.68	64.52	77.84	82.23

Table 3. Classification performance comparison of nine experts and EDF in Regions 1–9 on the full dataset (in the unit of %).

	SP	VFJ	SSL	PY	KPK	XHE	TN	HSH	HDH	EDF$_V$	EDF$_S$	EDF
1	92.22	92.22	92.22	92.22	92.22	92.22	92.22	92.22	92.22	92.22	92.43	92.43
2	76.91	75.99	78.24	80.07	75.96	75.96	75.92	75.94	77.34	75.96	86.32	86.32
3	71.63	61.18	73.18	70.07	66.17	73.18	61.47	66.75	70.49	77.48	81.36	81.36
4	83.46	83.22	83.67	83.22	83.22	83.22	83.22	83.32	83.41	83.22	85.58	85.58
5	68.96	56.87	70.50	65.21	60.23	70.16	61.57	66.14	67.79	68.23	79.84	79.84
6	81.17	81.07	81.72	80.90	81.09	80.99	80.87	81.06	81.70	80.90	86.32	86.32
7	70.04	63.06	71.72	68.68	66.14	71.06	63.06	64.62	68.80	65.60	80.12	80.12
8	72.51	70.56	72.58	75.35	70.37	72.44	70.35	70.58	72.05	70.51	81.54	81.54
9	96.70	96.74	96.74	96.75	96.74	96.74	96.74	96.74	96.74	96.74	96.74	96.74
All	88.64	87.44	88.96	88.64	87.87	88.73	87.56	88.00	88.56	88.52	91.15	91.15

row, we perform the weighted sum of correct rates in nine regions based on their sample population to derive the results with respect to "all" data samples. For EDF$_V$, we binarize the soft decision of each expert and use the voting scheme to fuse expert's decisions. The majority rule is used to select the final system decision. For EDF$_S$, we adopt the stacking scheme to fuse experts' decisions. That is, we build a meta-level on top of all soft decisions which learns the fusion rule with an SVM classifier that treats experts' soft decisions as features. We see that the performance of EDF$_S$ is no worse than EDF$_V$ in all regions. Thus, it is chosen to be the final EDF solution.

By comparing results in Tables 2 and 3, we see clearly that the performance of each expert has improved a lot (ranging from 4–11 %) due to data grouping. After data grouping, the performance gap among different experts narrows down significantly. Their correct classification rates now are in the range of 87.44–88.96 %. The EDF system can achieve a correct classification rate of 91.15 % by stacking all experts in each region. With the combination of grouping and stacking, the EDF system can outperform traditional experts (without grouping) by a margin of 6–26 %.

Finally, we plot the performance of each individual expert (without data grouping) and the EDF system as a function of the size of the dataset in Fig. 6. We select subsets of increasing sizes randomly from the SUN dataset and list the size in the x-axis while the averaged correct classification rate using the 5-fold cross validation is shown in the y-axis. The vertical segment on each marker indicates the standard deviation of a particular test. We see that the performance of some individual experts stay flat while others drop as the data size becomes large. In contrast, the performance of the EDF system improves as the data size becomes larger.

When the data size becomes larger, there are two competing factors that influence the performance of a classification system in two opposite directions. On one hand, the data become more diversified and the performance of an expert

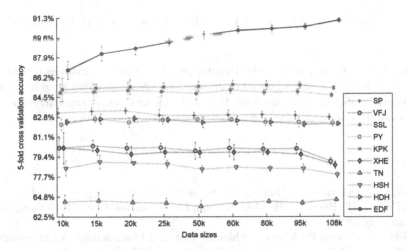

Fig. 6. Comparison of classification performance of nine experts and EDF as a function of the size of dataset.

may go down if its model cannot handle a diversified data set. On the other hand, the number of similar data (*i.e.*, belonging to the same data type) becomes more. Data abundance helps improve the performance of learning-based classifiers. Figure 6 implies that the data diversity problem is under control by the robust EDF system so that the EDF system benefits more from data abundance. We expect the EDF system performance to be level-off at a certain data size although we have not yet observed such a phenomenon in Fig. 6. This is because that the EDF system does not address the semantic gap issue.

5 Conclusion and Future Work

An Expert Decision Fusion (EDF) system was developed to address the large-scale indoor/outdoor image classification problem in this work. As compared with the traditional classifiers (or experts), the EDF system consists of two key ideas: (1) grouping of data samples based on the soft decisions of two experts into 9 regions; and (2) stacking of soft decisions from all constituent experts to enhance the classification performance in each region. It was shown by experimental results that the proposed EDF system outperforms all traditional classifiers in the classification accuracy by a margin of 6–26 % on the large-scale SUN image dataset. The classification performance of EDF improves as the size of the dataset grows, which can be explained by its capability of handling data diversity. With this capability in place, as the dataset grows to a very large size, data abundance becomes a more dominant factor than data diversity. Thus, the EDF system offers a robust and scalable solution.

As discussed in Sect. 3, there is some fundamental limits in the feature-based classifiers since they do not take the image semantics into account. We expect to see a point where the performance of EDF becomes saturated, which will be the true upper performance bound of EDF. To achieve this goal, we need to look for some dataset even larger than SUN. To improve the performance of EDF furthermore beyond the saturation point, we need to look for semantic-based experts. This is clearly a very challenging problem since it involves object and scene recognition. Finally, a good indoor/outdoor image classifier is an important pre-processing step to scene analysis. It is desirable to leverage our current results to obtain better methods for scene classification and recognition.

References

1. Bosch, A., Zisserman, A., Muñoz, X.: Scene classification via pLSA. In: Leonardis, A., Bischof, H., Pinz, A. (eds.) ECCV 2006. LNCS, vol. 3954, pp. 517–530. Springer, Heidelberg (2006)
2. Fei-Fei, L., Perona, P.: A bayesian hierarchical model for learning natural scene categories. In: IEEE Computer Society Conference on Computer Vision and Pattern Recognition, 2005, CVPR 2005, vol. 2, pp. 524–531. IEEE (2005)
3. Quattoni, A., Torralba, A.: Recognizing indoor scenes. In: 2009 IEEE Conference Computer Vision and Pattern Recognition (CVPR) (2009)
4. Li, L.J., Su, H., Xing, E.P., Li, F.F.: Object bank: A high-level image representation for scene classification & semantic feature sparsification. In: NIPS, vol. 2, p. 5 (2010)
5. Wu, J., Rehg, J.M.: Centrist: A visual descriptor for scene categorization. IEEE Trans. Pattern Anal. Mach. Intell. **33**, 1489–1501 (2011)
6. Vailaya, A., Figueiredo, M., Jain, A., Zhang, H.: Content-based hierarchical classification of vacation images. In: IEEE International Conference on Multimedia Computing and Systems, 1999, vol. 1, pp. 518–523. IEEE (1999)
7. Lim, J.H., Jin, J.S.: A structured learning framework for content-based image indexing and visual query. Multimed. Syst. **10**, 317–331 (2005)
8. Chatzichristofis, S.A., Boutalis, Y.S.: CEDD: Color and edge directivity descriptor: A compact descriptor for image indexing and retrieval. In: Gasteratos, A., Vincze, M., Tsotsos, J.K. (eds.) ICVS 2008. LNCS, vol. 5008, pp. 312–322. Springer, Heidelberg (2008)
9. Vailaya, A., Jain, A., Zhang, H.J.: On image classification: City images vs. landscapes. Pattern Recogn. **31**, 1921–1935 (1998)
10. Zhang, L., Li, M., Zhang, H.J.: Boosting image orientation detection with indoor vs. outdoor classification. In: Sixth IEEE Workshop on Applications of Computer Vision, 2002, (WACV 2002), Proceedings, pp. 95–99. IEEE (2002)
11. Battiato, S., Curti, S., La Cascia, M., Tortora, M., Scordato, E.: Depth map generation by image classification. Proc. SPIE **5302**, 95–104 (2004)
12. Bianco, S., Ciocca, G., Cusano, C., Schettini, R.: Improving color constancy using indoor-outdoor image classification. IEEE Trans. Image Process. **17**, 2381–2392 (2008)
13. Boutell, M.R., Luo, J., Shen, X., Brown, C.M.: Learning multi-label scene classification. Pattern Recogn. **37**, 1757–1771 (2004)
14. Vailaya, A., Figueiredo, M., Jain, A., Zhang, H.J.: Content-based hierarchical classification of vacation images. IEEE Int. Conf. Multimed. Comput. Syst. **1**, 518–523 (1999)

15. Lazebnik, S., Schmid, C., Ponce, J.: Beyond bags of features: Spatial pyramid matching for recognizing natural scene categories. In: IEEE Computer Society Conference on Computer Vision and Pattern Recognition, 2006, vol. 2, pp. 2169–2178 (2006)

16. Vogel, J., Schiele, B.: Semantic modeling of natural scenes for content-based image retrieval. Int. J. Comput. Vis. **72**, 133–157 (2007)

17. Jégou, H., Douze, M., Schmid, C.: Improving bag-of-features for large scale image search. Int. J. Comput. Vis. **87**, 316–336 (2010)

18. Li, Y., Crandall, D.J., Huttenlocher, D.P.: Landmark classification in large-scale image collections. In: 2009 IEEE 12th International Conference on Computer Vision, pp. 1957–1964. IEEE (2009)

19. Szummer, M., Picard, R.W.: Indoor-outdoor image classification. In: Content-Based Access of Image and Video Database, pp. 42–51. IEEE (1998)

20. Vailaya, A., Figueiredo, M.A., Jain, A.K., Zhang, H.J.: Image classification for content-based indexing. IEEE Trans. Image Process. **10**, 117–130 (2001)

21. Serrano, N., Savakis, A., Luo, A.: A computationally efficient approach to indoor/outdoor scene classification. In: 16th International Conference on Pattern Recognition, Proceedings, vol. 4, pp. 146–149 (2002)

22. Pavlopoulou, C., Yu, S.: Indoor-outdoor classification with human accuracies: Image or edge gist? In: 2010 IEEE Computer Society Conference on Computer Vision and Pattern Recognition Workshops (CVPRW), pp. 41–47 (2010)

23. Kim, W., Park, J., Kim, C.: A novel method for efficient indoor-outdoor image classification. J. Sig. Process. Syst. **61**, 251–258 (2010)

24. Xiao, J., Hays, J., Ehinger, K.A., Oliva, A., Torralba, A.: Sun database: Large-scale scene recognition from abbey to zoo. In: 2010 IEEE Conference on Computer Vision and Pattern Recognition (CVPR), pp. 3485–3492. IEEE (2010)

25. Luo, J., Savakis, A.: Indoor vs outdoor classification of consumer photographs using low-level and semantic features. In: 2001 International Conference on Image Processing, 2001, Proceedings, vol. 2, pp. 745–748 (2001)

26. Kane, M.J., Savakis, A.: Bayesian network structure learning and inference in indoor vs. outdoor image classification. In: Proceedings of the 17th International Conference on Pattern Recognition, 2004, ICPR 2004, vol. 2, pp. 479–482. IEEE (2004)

27. Traherne, M., Singh, S.: An integrated approach to automatic indoor outdoor scene classification in digital images. In: Yang, Z.R., Yin, H., Everson, R.M. (eds.) IDEAL 2004. LNCS, vol. 3177, pp. 511–516. Springer, Heidelberg (2004)

28. Radu, V.: Application. In: Radu, V. (ed.) Stochastic Modeling of Thermal Fatigue Crack Growth. ACM, vol. 1, pp. 63–70. Springer, Heidelberg (2015)

29. Payne, A., Singh, S.: Indoor vs. outdoor scene classification in digital photographs. Pattern Recogn. **38**, 1533–1545 (2005)

30. Payne, A., Singh, S.: A benchmark for indoor/outdoor scene classification. In: Singh, S., Singh, M., Apte, C., Perner, P. (eds.) ICAPR 2005. LNCS, vol. 3687, pp. 711–718. Springer, Heidelberg (2005)

31. Hu, G.H., Bu, J.J., Chen, C.: A novel bayesian framework for indoor-outdoor image classification. In: 2003 International Conference on Machine Learning and Cybernetics, vol. 5, pp. 3028–3032. IEEE (2003)

32. Efimov, S., Nefyodov, A., Rychagov, M.: Block-based image exposure assessment and indoor/outdoor classification. In: Proceedings of 17th Conference on Computer Graphics GraphiCon (2007)

33. Tao, L., Kim, Y.H., Kim, Y.T.: An efficient neural network based indoor-outdoor scene classification algorithm. In: 2010 Digest of Technical Papers International Conference Consumer Electronics (ICCE), pp. 317–318. IEEE (2010)
34. Wolpert, D.H.: Stacked generalization. Neural Netw. **5**, 241–259 (1992)
35. Deng, L., Yu, D., Platt, J.: Scalable stacking and learning for building deep architectures. In: 2012 IEEE International Conference on Acoustics, Speech and Signal Processing (ICASSP), pp. 2133–2136. IEEE (2012)
36. Džeroski, S., Ženko, B.: Is combining classifiers with stacking better than selecting the best one? Mach. Learn¿ **54**, 255–273 (2004)
37. Ohta, Y.I., Kanade, T., Sakai, T.: Color information for region segmentation. Comput. Graph. Image Process. **13**, 222–241 (1980)
38. Mao, J., Jain, A.K.: Texture classification and segmentation using multiresolution simultaneous autoregressive models. Pattern Recogn. **25**, 173–188 (1992)
39. Daubechies, I., et al.: Ten Lectures on Wavelets, vol. 61. SIAM, Philadelphia (1992)
40. Oliva, A., Torralba, A.: Modeling the shape of the scene: A holistic representation of the spatial envelope. Int. J. Comput. Vis. **42**, 145–175 (2001)
41. Radu, V.: Application. In: Radu, V. (ed.) Stochastic Modeling of Thermal Fatigue Crack Growth. ACM, vol. 1, pp. 63–70. Springer, Heidelberg (2015)
42. Chang, C.C., Lin, C.J.: Libsvm: a library for support vector machines. ACM Trans. Intell. Syst. Technol. (TIST) **2**, 27 (2011)
43. Gupta, L., Pathangay, V., Patra, A., Dyana, A., Das, S.: Indoor versus outdoor scene classification using probabilistic neural network. EURASIP J. Appl. Sig. Process. **2007**, 123–123 (2007)
44. Johnson, J.B.: Thermal agitation of electricity in conductors. Phys. Rev. **32**, 97 (1928)
45. Tomasi, C., Manduchi, R.: Bilateral filtering for gray and color images. In: Sixth International Conference on Computer Vision, 1998, pp. 839–846. IEEE (1998)
46. He, K., Sun, J., Tang, X.: Single image haze removal using dark channel prior. IEEE Trans. Pattern Anal. Mach. Intell. **33**, 2341–2353 (2011)
47. Iversen, G.R., Norpoth, H.: Analysis of Variance. Sage, Newbury Park (1987)
48. Lomax, R.G., Hahs-Vaughn, D.L.: Statistical Concepts: A Second Course. Routledge, New York (2013)

Search Guided Saliency

Shijian Lu[1]([✉]), Byung-Uck Kim[1], Nicolas Lomenie[2],
Joo-Hwee Lim[1], and Jianfei Cai[3]

[1] Institute for Infocomm Research, A*STAR, 1 Fusionopolis Way,
#21-01 Connexis, 138632 Singapore, Singapore
`slu@i2r.a-star.edu.sg`
[2] The Laboratory of Informatics Paris Descartes, University Paris Descartes,
12 Rue de L'Ecole de Mdecine, 75006 Paris, France
[3] School of Computing Engineering, Nanyang Technological University,
Nanyang Avenue, 639798 Singapore, Singapore

Abstract. We propose a new type of saliency as inspired by findings
from visual search studies - the searching difficulty is correlated with
the target-distractor contrast, the distractor homogeneity, as well as the
target uniqueness. By putting an image pixel as the target and the surrounding pixels as distractors, a search guided saliency model is designed
in accordance with these findings. In particular, three saliency measures
in correspondence to the three searching factors are simultaneously computed and integrated by using a series of contextual histograms. The
proposed model has been evaluated over three public datasets and experiments show superior prediction of the human fixations when compared
to the state-of-the-art models.

1 Introduction

Visual saliency [1] characterizes the distinct perceptual quality of an object or
image region with respect to its surrounding. It helps to serialize the attending
of objects in scenes which the human vision system cannot process in parallel
due to the tremendous amount of visual information involved. Computational
modeling of visual saliency aims to build an attention model that is capable of
predicting where people will look at given an image or scene. It has increasingly
attracted research interest in recent years due to its importance in both human
visual attention study and a wide range of applications in object detection, object
segmentation, visual search, etc. [2–5].

Quite a number of saliency models [9] have been reported in recent years that
exploit the local contrast and global image contrast. The local contrast based models make use of a center surround difference to compute the contrast of an object or
image region with respect to its surrounding [10–13]. Itti and Koch's model [11,12]
is probably one of the earliest that exploit the center surround difference which is
computed using a set of spatial filters. Other approaches have also been proposed
that exploit decision-theoretic discrimination [10], object-level segmentation [13],
etc. for center surround difference computation. The major limitation of the local

© Springer International Publishing Switzerland 2015
C.V. Jawahar and S. Shan (Eds.): ACCV 2014 Workshops, Part I, LNCS 9008, pp. 443–456, 2015.
DOI: 10.1007/978-3-319-16628-5_32

Fig. 1. Illustration of the Search Guided Model: The search guided model is tolerant to image edges and dynamic structures and capable of predicting the human fixations accurately. For the sample images in the first row (from AIM dataset [6], SR dataset [7], and MIT300 dataset [8]), rows 2–3 show the searched guided saliency maps and the corresponding fixational maps, respectively.

contrast based models is that certain global features which are closely related to the perceptual saliency is not captured.

A number of models have been reported to incorporate the global contrast that is often pertaining to the perceptual rarity, uniqueness, unusualness. In particular, image histograms have been extensively used to capture the low-frequency global features. For example, Cheng et al. [14] employ color histograms to capture the global color contrast and combine it with a local region contrast for saliency computation. Lu et al. [15,16] exploit a 2D co-occurrence histogram that captures both local contrast and global unusualness simultaneously. Contextual information has also been exploited [17–19] to capture both local and global contrast in different ways. In addition, several frequency space models [7,20–22] are reported that compute saliency based on global unusual amplitude or phase spectrum of the Fourier transform of an image.

Though local and global contrast has been exploited in various ways, the contrast-based models are often over-responsive to inconspicuous image edges which are associated with the local contrast and often represent certain globally abnormal features. Several learning based models [6,23–27] have been proposed to learn the statistics or eye fixation data directly. These learning based models are not sensitive to inconspicuous image edges but the computed saliency often lacks discrimination between salient and inconspicuous objects.

We propose a novel saliency model as inspired by findings from the visual search studies, i.e., searching difficulty is determined by the target-distractor

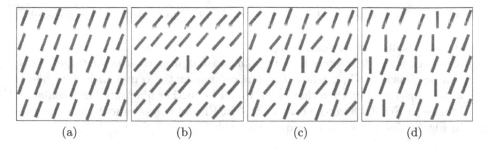

(a)   (b)   (c)   (d)

Fig. 2. Visual Search Principle: Compared with the target (the vertical bar at the center) in (a), the target in (b) has a higher pop-out effect due to its higher target-distractor difference, the target in (c) has a lower pop-out effect due to its lower distractor homogeneity (the distractors have the same average slant angle as those in (a)), and the target in (d) has a lower pop-out effect due to its lower uniqueness level.

difference, the center uniqueness, as well as the distractor homogeneity as illustrated in Fig. 1. In the proposed model, the three saliency features are computed and integrated by using a series of *contextual histograms* that can be directly determined within a local neighborhood window. The proposed model has a number of novelties. First, it computes and integrates several search-guided saliency features and obtains superior human fixation prediction performance compared with state-of-the-art models. Second, it makes use of contextual histograms and overcomes one typical limitation of many existing models, i.e., high saliency response around inconspicuous image edges or other dynamic structures. Third, it is simple and easy for implementation.

2   Principle of Search Guided Saliency

The visual search guided saliency is inspired by the feature integration and stimulus similarity theory [28,29] as illustrated in Fig. 2 - the target will have a shorter searching time and stronger pop-out effect when the target-distractor contrast is stronger, the surrounding distractors have a higher homogeneity level and the target has a high uniqueness level. It integrates three principles from the visual search studies [28–30] including:

1. A target will have a stronger pop-out effect when it has a larger contrast to the surrounding distractors. As shown in Fig. 2, the target - the vertical bar at the center in Fig. 2b has a stronger pop-out effect than that in Fig. 2a. This can be further illustrated by simulated images in Fig. 3 where the target pixel at the center in Fig. 3b is more salient than the one in Fig. 3a due to its stronger contrast to the surrounding distractor pixels.
2. A target will have a stronger pop-out effect when the surrounding distractors have a higher homogeneity level. As shown in Fig. 2, the vertical bar at the center in Fig. 2c has a lower pop-out effect than that in Fig. 2a. This can be further illustrated by simulated images in Fig. 3 where the target pixel

in Fig. 3c is less salient than the one in Fig. 3a due to its lower homogeneity level.

3. A target will have a stronger pop-out effect when it has a higher uniqueness level, i.e., fewer distractors have the same visual properties as the target. This can be shown in Fig. 2 where the vertical bar at the center in Fig. 2d has a lower pop-out effect than that in 2a. It can be further illustrated by simulated images in Fig. 3 where the target pixel in Fig. 3d is less salient than the one in Fig. 3a due to its lower uniqueness level.

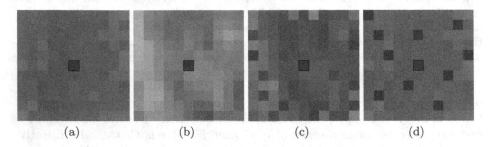

(a)	(b)	(c)	(d)

Fig. 3. Visual Search Principle: Compared with the target pixel at the center of the image in (a), the target pixel in (b) has higher saliency due to its higher center-surround contrast, the target pixel in (c) has lower saliency due to its lower surround homogeneity (the distractor pixels have the same mean but much smaller variance than those in (a)), the target pixel in (d) has lower saliency due to its lower uniqueness.

Under the same target-distractor paradigm, the proposed model constructs a contextual histogram based on the "distractor" pixels that surround each "target" pixel at the center. The saliency of the target pixel is computed by integrating three search-inspired saliency measures including the surround homogeneity, the center uniqueness, and the center surround contrast, all of which can be computed from a series of contextual histogram simultaneously. Due to the integration of the three saliency features by using the contextual histograms, the proposed model is tolerant to the image edges and demonstrates better prediction of the human fixations when compared with those local and global contrast based models as illustrated in Fig. 4.

Related models also exploit the image histogram and contextual information to capture the local and global image contrast [15–17,22,31]. The frequency space model in [22] makes use of the global contrast which is tolerant to image edges but often detects only the boundary of salient objects as illustrated in Fig. 4b. The context models [17,31] integrate the local and global contrast and but are over-responsive to inconspicuous image edges and corners as shown in Fig. 4c. The histogram based models [15,16] exploit occurrence and co-occurrence of image intensity and color to capture the local and global contrast concurrently. They are capable of detecting multiple salient objects with a complex background but are also over-responsive to image edges as shown in Fig. 4d.

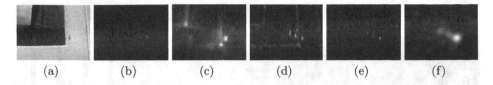

(a)   (b)   (c)   (d)   (e)   (f)

Fig. 4. Search Guided Saliency: For the sample image in Fig. 3a [6], the images in Figs. 3b–e show the saliency that is computed using the spectral residual [22], the context [17], the co-occurrence histogram [16], the search guided principles, and the corresponding fixational map (Gaussian smoothing of eye fixations of 20 subjects), respectively.

3 Visual Search Guided Saliency Modeling

This section describes the visual search guided saliency modeling technique. For each "target" pixel, several contextual histograms are first constructed based on the "distractor" pixels that surround the target pixel. A saliency level is then determined by integrating the three saliency measures that are computed from each contextual histogram. The overall saliency is finally determined by integrating the saliency that is computed across multiple contextual histograms and multiple image channels.

3.1 Contextual Histogram

A contextual histogram is constructed to emulate the target-distractor paradigm. For each target pixel, a contextual histogram H is constructed by using a number of distractor pixels that surround the target pixel at the center. Neighborhoods of different shapes such as circular or square-shaped can be used to pick the distractor pixels. At the same time, neighborhoods of different sizes can be used to capture contexts of different distances to the target pixel (to be described in Sect. 3.4). Note that the distractor pixels are picked along the neighborhood boundary instead of from within the neighborhood.

Figure 5b shows the contextual histogram of three typical image pixels as labeled in the image in Fig. 5a, where histogram graphs have the same color as the corresponding labeling neighborhood squares. In particular, the three example pixels have the same intensity (80 as indicated by the arrow) but are picked from a homogeneous image region (blue square), an inconspicuous image edge (red square), and a salient image region (brown square), respectively. The corresponding three contextual histograms are distinctive as illustrated in Fig. 5b. For the pixel in the homogeneous region, its contextual histogram (blue graph) has a large global peak and its intensity lies close to the global peak. For the pixel along the image edge, its contextual histogram (red graph) usually has two major peaks and its intensity lies somewhere between the two major peaks. For the pixel in the salient image region, its contextual histogram (brown graph) often has a large global peak and its intensity lies far away from the global peak.

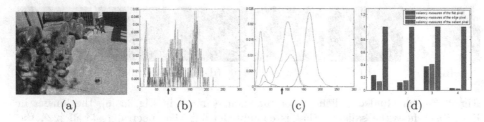

Fig. 5. Search Guided Saliency Measures: (a) shows a sample image with three typical pixels from a homogeneous region (square neighborhood of blue color), an image edge (square neighborhood of red color) and a salient region (square neighborhood of brown color). (b) and (c) show the contextual histogram H and the smoothed contextual histogram H' of the three pixels, respectively (with the same coloring as the square neighborhood in Fig. 5a). (d) shows the three saliency measures of the three pixels as well as their integrated saliency (Colour figure online).

The contextual histogram can be smoothed to suppress the undesired saliency response (for the three saliency measures) around the image edge. For the three contextual histograms shown in Fig. 5b, Fig. 5c shows the smoothed contextual histograms by using a Gaussian filter. As Fig. 5c shows, the salient pixel and the one in the homogeneous region still have a small and large histogram values, respectively, after the smoothing. As a comparison, the edge pixel has a much higher value after the smoothing because the smoothing raises its histogram values due to the two major peaks at both sides as illustrated in Fig. 5c. The three saliency measures can be computed from the smoothed contextual histogram as shown in Fig. 5d (to be described in the next subsection).

3.2   Search Guided Saliency Measures

With respect to the three visual search principles, three saliency measures can be defined and computed from the contextual histogram simultaneously. The first saliency measure is center surround difference that has been widely exploited in the literature. It is defined as follows:

$$S_c(x,y) = \|I(x,y) - H_c\| \tag{1}$$

where $I(x,y)$ denotes the intensity of the target pixel at (x,y). H_c is the centroid of the contextual histogram which is equal to the mean of the surrounding distractor pixels that are picked to construct the histogram. Note that the contextual histogram here is the original before smoothing.

Image pixels in a salient region usually have much larger center-surround difference S_c than those in a homogeneous region whose intensity is usually close to the histogram centroid H_c. Edge pixels often have a small S_c because the intensity of edge pixels usually lies between the two major histogram peaks and will be close to the histogram centroid H_c as illustrated in Fig. 5c (red graph). This can be further illustrated in Fig. 5d where the first bar group (with the same labeling colors as in Figs. 5a–c) shows the S_c of the three pixels which is normalized by their maximum. As Fig. 5d shows, the pixel in a salient region has a much higher S_c than the edge pixel and the pixel in a homogeneous region.

The second saliency measure is the surround homogeneity which is closely correlated with the perceptual visual saliency. Within a smoothed contextual histogram H', the surround homogeneity is mainly demonstrated by a large global peak. This saliency measure is defined as follows:

$$S_h(x,y) = \left(H'_x - H'\big(I(x,y)\big)\right)^p \tag{2}$$

where H'_x denotes the global peak of the smoothed contextual histogram H'. Parameter p is a number larger than 1 which controls the weight of this saliency measure.

For image pixels having the same center-surround difference, those with a more homogeneous surrounding should have a larger S_h. In particular, target pixels with a more homogeneous surrounding should have a larger histogram peak H'_x but a smaller $H'(I(x,y))$, where the smaller $H'(I(x,y))$ is largely due to fewer distractor pixels whose intensity lies between the global peak intensity and the target pixel's intensity $I(x,y)$. This can be illustrated in Fig. 5d where the second bar group shows the S_h of the three sample pixels that is normalized by their maximum. It should be noted that image pixels in a homogeneous region usually have a large H'_x but a small S_h because the histogram value of these pixels, i.e., $H'(I(x,y))$, is usually large and close to H'_x. As Fig. 5d shows, the pixel in a salient region has a much higher S_h than the edge pixel and the pixel in the homogeneous region.

The third saliency measure captures the uniqueness level of the target pixel. With a smoothed contextual histogram H', this measure is defined as follows:

$$S_u(x,y) = 1 - H'(I(x,y))^q \tag{3}$$

where $I(x,y)$ denotes the intensity of the target pixel and q is a number lying between 0 and 1 which controls the weight of this saliency measure in the integrated saliency.

Image pixels with a higher saliency level usually have a larger S_u. In particular, a higher center uniqueness level means fewer distractor pixels with the same intensity as the target pixel, i.e., a smaller $H'(I(x,y))$ that leads to a larger S_u. Note that S_u is related to the S_c and S_h as a larger center-surround difference and surround homogeneity usually lead to a higher center uniqueness. On the other hand, S_u captures certain specific saliency information, i.e., the center uniqueness, that is not captured in either S_c or S_h (as illustrated in Figs. 2d and 3d). This can be illustrated in Fig. 5d where the third bar group shows the S_u of the three sample pixels that is normalized by their maximum. As Fig. 5d shows, the pixel in a salient region has a much higher S_u than the edge pixel and the pixel in a homogeneous region.

3.3 Saliency Modeling

The saliency of an image can be determined by integrating the three saliency measures that are computed for different channel images of different scales. We use the **Lab** color space where channel **L** encodes the image lightness and contrast information and channels **a** and **b** encode the image color information. In addition, each channel image is down-sampled to n image scales to capture contexts of different sizes (to be described in Sect. 3.4). Note that image values in the three image channels are first mapped to 0~255 (first subtracted by the minimum intensity, then divided by the maximum value, and finally multiplied by 255) and then rounded to integers for the contextual histogram construction.

With the three saliency measures as defined in Eqs. 1–3 in the previous subsections, the saliency level of a target image pixel is determined as follows:

$$S(x, y) = S_c(x, y) * S_h(x, y) * S_u(x, y) \qquad (4)$$

where $S_c(x, y), S_h(x, y)$, and $S_u(x, y)$ denote the three saliency measures that are computed for the target image pixel at (x, y), respectively. A multiplication strategy is adopted because all the three saliency measures change in the same direction as the overall perceptual saliency. For the three sample pixels labeled in Fig. 5a, the fourth bar group in Fig. 5d shows the corresponding integrated saliency, where the pixel from a salient region has much higher saliency compared with the other two sample pixels.

The overall saliency of a target pixel can be finally computed as follows:

$$S_o(x, y) = \sum_{Lab} \max\big(S_1(x, y), \cdots, S_n(x, y)\big) \qquad (5)$$

where $S_1(x, y), \cdots, S_n(x, y)$ refer to the integrated saliency in Eq. 4 that is computed for one image channel of different scales. Two strategies are adopted to integrate the computed saliency. First, max-pooling is employed to take the maximum of the saliency that is computed across n image scales, i.e., $S_1(x, y), \cdots, S_n(x, y)$. Note that saliency computed at different image scales is first scaled back to the original image scale before the max-pooling. Second, average-pooling is implemented to determine the overall saliency level by averaging the saliency that is computed over the three image channels.

3.4 Discussion

The proposed model involves several parameters. In particular, p and q are used to control the weights of the surround homogeneity and the center uniqueness as described in Sect. 3.2. Evaluation over the eye fixational map shows that saliency can be detected properly when p and q are set around 0.05–0.2 and 2–4, respectively. Neighborhoods of different shapes and sizes can be set to pick the distractor pixels as described in Sect. 3.1. In our implementation, a square-shaped neighborhood is used and the neighborhood radius is set at 10 pixels. In addition, each channel image is down-sampled to n image scales for saliency

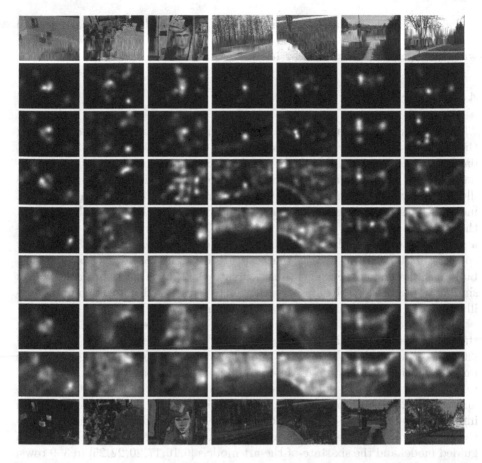

Fig. 6. Comparison of the search guided model with six state-of-the-art models over the AIM dataset: For the sample images in the first row, row 2 shows the corresponding fixational maps as described in Sect. 4, row 3–9 show the corresponding saliency maps by the search guided model and the six state-of-the-art models in [6,16,17,20,22,25], respectively.

computation as described in Sect. 3.3. In our implementation, five image scales are used where each image is down-sampled to 0.6, 0.5, 0.4, 0.3, and 0.2 of the original image scale. Last, the contextual histogram is smoothed by using a Gaussian filter as described in Sect. 3.1. The width of the filter window can be around 20–40 based on the humans' perceptible visual contrast.

The three saliency measures help to suppress the response at the inconspicuous edges and other dynamic structures such as tree branches effectively. First, edge pixels usually have a small S_c because their intensity lying between the two major histogram peaks is often close to the H_c, the mean of the distractor pixels that are counted into the contextual histogram. Second, edge pixels usually have a small S_h because they have a smaller global histogram peak H_x' but a larger $H'(I(x, y))$, largely due to the smoothing of the contextual histogram with two major peaks at the left and right sides. Third, edge pixels usually have a small S_u because a small q (e.g. $q = 0.1$) will left $H'(I(x,y))^q$ to be close to 1 even when $H'(I(x,y))$ is very small by itself.

4   Results

The proposed model has been evaluated over three public datasets including the SR dataset [7], the AIM dataset [6], and the MIT300 dataset [8]. The SR dataset consists of 62 static images and for each image, salient regions are manually labeled by four subjects which are further averaged to form a hit map as illustrated by the first three graphs in the second row of Fig. 7. The AIM dataset includes 120 static images and the corresponding fixational maps as illustrated in the second row in Fig. 6, which are created by Gaussian smoothing of the eye fixations that are collected from 20 subjects for each image. The MIT300 dataset consists of 300 natural images that capture different scenes such as humans, buildings, flowers, etc. For each image, fixations of 39 subjects are collected in similar way as the AIM dataset with which a fixational map is computed as illustrated in the last three graphs in the second row of Fig. 7.

The proposed model is compared with six state-of-the-art models including the context model [17], the signature model [22], the frequency tuned (FT) model [20], the AIM model [6], the SUN model [25], and the CCH model [16]. The implementations of the state-of-the-art models are downloaded from the authors' websites. For the search guided model, parameters p and q are set at 0.1 and 3, and the window width of the histogram filter is set at 30. Figure 6 show several images of the AIM dataset in the first row, the corresponding fixational maps in the second row, and the saliency maps that are computed by using the search guided model and the six state-of-the-art models [6,16,17,20,22,25] in 3–9 rows, respectively. Figure 7 show the saliency maps of the search guided model and the six compared models for the SR dataset (the first 3 images) and the MIT300 dataset (the last 3 images). As Figs. 6 and 7 show, the search guided model predicts the human fixations accurately.

In particular, the contrast-based models [16,17,22]are often over-responsive to the inconspicuous image edges as illustrated in rows 4, 5, and 7. As a comparison, the search guided models helps to suppress such "false alarms" effectively as shown in row 3. For example, most contrast-based models are over-responsive to the inconspicuous image edges and dynamic tree branches and grasses as shown in the second, fourth, fifth and seventh images in Fig. 6 where the search guided model has little responses as shown in row 3. The learning based models [6,25] are instead blurry where salient and non-salient regions both have certain saliency as illustrated in rows 6 and 8. In addition, the search guided model is capable of detecting salient objects or image regions of small scale such as the dark object in the second image and the red flowers in the fifth image in Fig. 6, largely due to the incorporated context homogeneity information. As a comparison, the contrast-based models often fail to detect such salient objects because other inconspicuous objects often have much higher contrast.

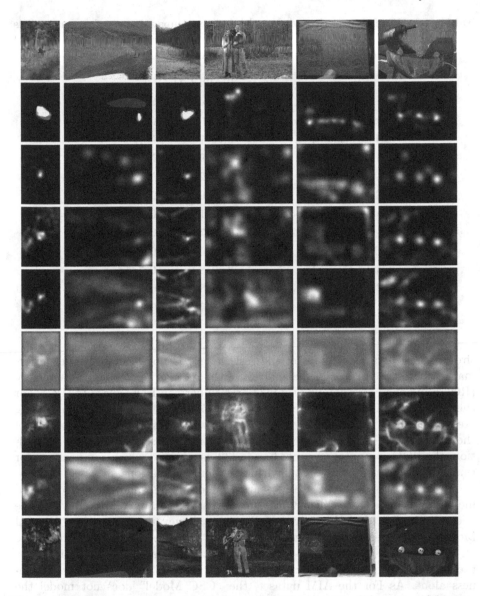

Fig. 7. Comparison of the search guided model with six state-of-the-art models over the SR and MIT300 datasets: For the sample images in the first row, row 2 shows the corresponding fixational maps as described in Sect. 4, row 3–9 show the corresponding saliency maps by the search guided model and the six state-of-the-art models in [6, 16, 17, 20, 22, 25], respectively.

Quantitative experiments have also been conducted based on the AIM dataset. The MIT300 dataset is not evaluated as only nine images have fixational maps available (used for comparison of different models on the authors' web-

Table 1. sAUC of the search guided model and six compared state-of-the-art models based on the AIM dataset [6] (CSC: center-surround contrast in Eq. 1; SH: surround homogeneity in Eq. 2; CU: center uniqueness in Eq. 3).

Models	Shuffled AUC	
	AIM dataset	SR dataset
Search guided model	**0.7311**	0.7224
SH+CU model	0.7217	N.A
SH model	0.7039	N.A
CU model	0.6942	N.A
CSC model	0.6899	N.A
Co-occurrence model in [16]	0.7221	0.7291
Signature model [22]	0.7147	0.6881
AIM model [6]	0.6990	0.7149
Context model in [17]	0.6958	**0.7458**
SUN model [25]	0.6813	0.6668
FT model [20]	0.5885	0.6108

site) whereas fixational maps of the rest images is not available. The performance is evaluated through the analysis of the receiver operating characteristic (ROC) and the corresponding shuffled area under the ROC curve (sAUC). For the saliency computed by different models, 25 rounds of Gaussian smoothing are implemented by changing the smoothing window size from 0.01 to 0.13 of the image width with an increase step of 0.005 as described in [22]. In addition, the ROC computation procedure in [32] is adopted which compensates for center-bias that commonly exists within the human fixations.

Table 1 shows the sAUC of the search guided models and the six compared models. With the three saliency measures, five sAUCs are computed where the "Search Guided Model" integrates all three saliency measures, the "SH+CU Model" integrates the surround homogeneity and the center uniqueness, the "CSC Model" uses the center-surround contrast alone, the "SH Model" uses the surround homogeneity alone, and the "CU Model" uses the center uniqueness alone. As For the AIM dataset, the "CSC Model" does not model the visual saliency well, with a sAUC at 0.6899. The "SH+CU Model" integrates the novel surround homogeneity and center uniqueness, which greatly outperforms the "CSC Model". In addition, the "SH Model" clearly outperforms the "CU Model", meaning that the surround homogeneity plays a heavier role in perceptual saliency compared with the center uniqueness. Furthermore, the "Search Guided Model" obtains a sAUC of 0.7311 which outperforms all sub-component models as well as the six contrast-based models. For the SR dataset, the search guided model obtains a sAUC of 72.24 % which is close to that of the AIM dataset (sAUC of the sub-component models are not computed). Note that the context based model [17] obtains a clearly higher sAUC, largely due to a face detector it incorporates that helps to predict the high saliency of human and animal faces within a number of images of the 3R dataset.

The proposed search guided model exploits only the low-level features. On the other hand, the human eyes are often attracted by familiar objects with semantic meaning such as human bodies, animals, human faces, vehicles, texts in scenes, etc. Relevant visual search models such as face detector, text detector, vehicle detector, etc. will be investigated and combined with the search guided model for better prediction of the human fixations.

5 Conclusions

This paper presents a novel saliency model that is inspired by visual search studies. Three saliency measures including the widely used center-surround contrast, the surround homogeneity, and the center uniqueness are defined and integrated for saliency modeling. A series of contextual histograms are constructed for each image pixel from which all the three saliency measures can be computed simultaneously. Experiments over three widely used public benchmarking datasets show that the proposed model predicts the human fixations accurately.

References

1. Tsotsos, J.: Analyzing vision at the complexity level. Behav. Brain Sci. **13**, 423–445 (1990)
2. Feng, J., Wei, Y., Tao, L., Zhang, C., Sun, J.: Salient object detection by composition. In: IEEE ICCV, pp. 1028–1035 (2011)
3. Marchesotti, L., Cifarelli, C., Csurka, G.: A framework for visual saliency detection with applications to image thumbnailing. In: IEEE ICCV, pp. 2232–2239 (2009)
4. Sharma, G., Jurie, F., Schmid, C.: Discriminative spatial saliency for image classification. In: IEEE CVPR, pp. 3506–3513 (2012)
5. Wang, P., Wang, J., Zeng, G., Feng, J., Zha, H., Li, S.: Salient object detection for searched web images via global saliency. In: IEEE CVPR, pp. 3194–3201 (2012)
6. Bruce, N., Tsotsos, J.: Saliency, attention, and visual search: an information theoretic approach. J. Vis. **9**, 1–24 (2009)
7. Hou, X., Zhang, L.: Saliency detection: a spectral residual approach. In: IEEE CVPR, pp. 1–8 (2007)
8. Judd, T., Durand, F., Torralba, A.: A benchmark of computational models of saliency to predict human fixations. MIT Computer Science and Artificial Intelligence Laboratory Technical Report, MIT-CSAIL-TR-2012-001 (2012)
9. Borji, A., Sihite, D., Itti, L.: Quantitative analysis of human-model agreement in visual saliency modeling: a comparative study. IEEE TIP **22**, 55–69 (2012)
10. Gao, D., Vasconcelos, N.: Bottom-up saliency is a discriminant process. In: IEEE ICCV, pp. 1–6 (2007)
11. Itti, L., Koch, C.: Computational modeling of visual attention. Nat. Rev. Neurosci. **2**, 194–203 (2001)
12. Itti, L., Koch, C., Niebur, E.: A model of saliency-based visual attention for rapid scene analysis. IEEE TPAMI **20**, 1254–1259 (1998)

13. Liu, T., Sun, J., Zheng, N., Tang, X., Shum, H.: Learning to detect a salient object. In: IEEE CVPR, pp. 1–8 (2007)
14. Cheng, M., Zhang, G., Mitra, N., Huang, X., Hu, S.: Global contrast based salient region detection. In: IEEE CVPR, pp. 409–16 (2011)
15. Lu, S., Lim, J.-H.: Saliency modeling from image histograms. In: Fitzgibbon, A., Lazebnik, S., Perona, P., Sato, Y., Schmid, C. (eds.) ECCV 2012, Part VII. LNCS, vol. 7578, pp. 321–332. Springer, Heidelberg (2012)
16. Lu, S., Tan, C., Lim, J.: Robust and efficient saliency modeling from image co-occurrence histograms. IEEE TPAMI 36, 195–201 (2013)
17. Goferman, S., Zelnik-Manor, Tal, A.: Context-aware saliency detection. In: IEEE CVPR, pp. 2376–2383 (2010)
18. Wang, L., Xue, J., Zheng, N., Hua, G.: Automatic salient object extraction with contextual cue. In: IEEE ICCV, pp. 105–112 (2011)
19. Margolin, R., Tal, A., Zelnik-Manor, L.: What makes a patch distinct? In: IEEE CVPR, pp. 1139–1146 (2013)
20. Achanta, R., Hemami, S., Estrada, F., Susstrunk, S.: Frequency-tuned salient region detection. In: IEEE CVPR, pp. 1597–1604 (2009)
21. Guo, C., Ma, Q., Zhang, L.: Spatio-temporal saliency detection using phase spectrum of quaternion fourier transform. In: IEEE CVPR, pp. 1–8 (2008)
22. Hou, X., Harel, J., Koch, C.: Image signature: highlighting sparse salient regions. IEEE TPAMI 34, 194–201 (2012)
23. Bruce, N., Tsotsos, J.: Saliency based on information maximization. NIPS 18, 155–162 (2006)
24. Judd, T., Ehinger, K., Durand, F., Torralba, A.: Learning to predict where humans look. In: IEEE ICCV, pp. 2106–2113 (2009)
25. Zhang, L., Tong, M.H., Marks, T.K., Cottrell, G.W.: Sun: a bayesian framework for saliency using natural statistics. J. Vis. 8, 1–20 (2008)
26. Zhao, Q., Koch, C.: Learning a saliency map using fixated locations in natural scenes. J. Vis. 3, 1–15 (2011)
27. Rudoy, D., Goldman, D., Shechtman, E., Zelnik-Manor, L.: Learning video saliency from human gaze using candidate selection. In: IEEE CVPR, pp. 1147–1154 (2013)
28. Treisman, A., Gelade, G.: A feature-integration theory of attention. Cogn. Psychol. 12, 97–136 (1980)
29. Duncan, J., Humphreys, G.: Visual search and stimulus similarity. Psychol. Rev. 96, 433–458 (1989)
30. Wolfe, J.: Guided search 2.0 a revised model of visual search. Psychon. Bull. Rev. 1, 202–238 (1994)
31. Borji, A., Itti, L.: Exploiting local and global patch rarities for saliency detection. In: IEEE CVPR, pp. 478–485 (2012)
32. Tatler, B., Baddeley, R., Gilchrist, I.: Visual correlates of fixation selection effects of scale and time. Vis. Res. 45, 643–659 (2005)

Salient Object Detection via Saliency Spread

Dao Xiang and Zilei Wang[✉]

Department of Automation, University of Science and Technology of China,
Hefei 230027, Anhui, China
zlwang@ustc.edu.cn

Abstract. Salient object detection aims to localize the most attractive objects within an image. For such a goal, accurately determining the saliency values of image regions and keeping the saliency consistency of interested objects are two key challenges. To tackle the issues, we first propose an adaptive combination method of incorporating texture with the dominant color, for enriching the informativeness and discrimination of features, and then propose saliency spread to encourage the image regions of the same object producing equal saliency values. In particular, saliency spread propagates the saliency values of the most salient regions to their similar regions, where the similarity serves for measuring the degree of belonging to the same object of different regions. Experimental results on the benchmark database MSRA-1000 show that our proposed method can produce more consistent saliency maps, which is beneficial to accurately segment salient objects, and is quite competitive compared with the advanced methods in previous literatures.

1 Introduction

Cognitive psychology research [1] indicates that given a visual scene, human vision is guided to particular parts by selective attention mechanism. These parts are called salient regions, and their saliency degree mainly depends on the state or quality of standing out from their neighbors. In computer vision, visual saliency simulates the functionality of selective attention, and concretely localizes the most salient and attention-grabbing regions or pixels in a digital image. Specifically, the saliency map represents the likelihood of each pixel belonging to salient regions with different values. Visual saliency estimation is much helpful to various vision tasks, such as object detection and recognition [2,3], adaptive image display [4], content-aware image editing [5], and image segmentation [6–8]. Recently, besides the eye-fixation prediction, visual saliency begins to serve object detection with the aim of segmenting salient objects from images. Particularly, this work focuses on such a detection goal.

Inspired by the pioneering work in [9], different saliency models for detecting salient objects were proposed. Most of them [7,8,10,11] use the superpixel-level color contrast to compute saliency map, due to the special attention of human vision to color and the robustness of superpixels compared with raw pixels [12]. However, these methods unavoidably suffer from unsatisfied segmentation results, *i.e.*, either producing incomplete objects or being contaminated

© Springer International Publishing Switzerland 2015
C.V. Jawahar and S. Shan (Eds.): ACCV 2014 Workshops, Part I, LNCS 9008, pp. 457–472, 2015.
DOI: 10.1007/978-3-319-16628-5_33

Fig. 1. Illustration of the effectiveness of our method in detecting salient objects. From top to bottom: input images, saliency maps obtained by our method, and our segmentation results.

by background. In our opinion, the reasons of leading to such unexpected results are two fold. The first is the insufficiency of color feature. Only adopting color feature works well for most natural images with considerable color variance between foreground and background, but not for the images without dominant color yet (*e.g.*, artificial images or gray-scale images). Consequently, the poor segmentations are produced (see Fig. 2) as little information is provided. Thus more visual cues need to be incorporated. Along this routine, some improved methods [13–15] have been proposed with different combination means of multiple features. The second is the inconsistency of the saliency of object regions. Under a certain saliency model, different parts of the ground truth salient object are likely not to produce uniform saliency, due to the object internal incoherence and the model sensitivity [16]. So different pixels or superpixels of the same object would have inconsistent saliency values. And such saliency map would result in the failure of exactly keeping the completeness of the segmented objects without absence of object parts or contamination of background (see Fig. 3). This fact is actually an important challenge of detecting salient objects.

To alleviate the aforementioned issues, we propose two concrete approaches in this paper to improve the performance of saliency detection. Firstly, we propose an adaptive feature fusion strategy for incorporating the texture with the main color feature.

Secondly, we propose a **saliency spread** mechanism to tackle the saliency inconsistency of object regions. The main idea of saliency spread is to spread saliency values of the most salient regions with high confidence to the similar regions by exploring the feature correlation of regions (probably belonging to the same object).

Figure 1 gives some examples of our proposed method to segment objects. Before elaborating on the details of our method, we review the related works

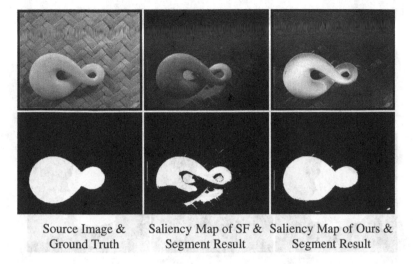

| Source Image & | Saliency Map of SF & | Saliency Map of Ours & |
| Ground Truth | Segment Result | Segment Result |

Fig. 2. Exemplar of the insufficiency of color features. For images whose foreground and background have similar color distributions, the method (SF [7] here) only using color leads to pool performance (middle column), while our method can achieve much better segmentation result due to incorporating texture (right column).

on detecting the salient objects. More detailed investigation and comparing can also be found in [16].

The rest of this paper is organized as follows. We first review the related works of saliency object detection in Sect. 2. Then we give an overview of our method in Sect. 3, and the detailed description of key models in Sect. 4. Finally, in Sect. 5, we report experimental results of the proposed method on public benchmark. The conclusions are provided in Sect. 6.

2   Related Works

In this paper, we focus on the data-driven bottom-up saliency detection. This kind of saliency is usually derived by primitive image features, such as color, texture, and edges. Based on the design ideology, the bottom-up saliency detection methods can roughly be classified into three categories: (1) frequency domain analysis based methods: the saliency is determined by the amplitude or phase spectrum [17,18]; (2) information theory based methods: Shannon's self-information [19] or the entropy of the sampled visual features [20] is maximized for achieving attention selectivity; (3) contrast based methods: the saliency map is computed by exploring the contrast of image pixels or regions. Now we briefly review the contrast based methods since this work falls into this one.

Actually, the contrast based methods have been proved to achieve the state-of-the-art performance [7,8,10,11,21–23]. Perceptual research results [24, 25] indicate that contrast is the most influential factor in low-level stimuli-driven attention. Itti *et al.* proposed the fundamental framework of the contrast

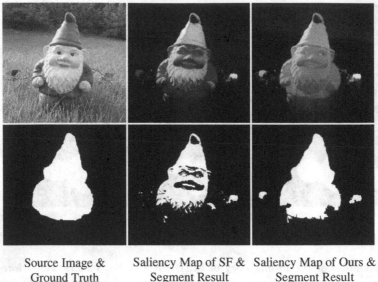

Source Image &   Saliency Map of SF &   Saliency Map of Ours &
Ground Truth   Segment Result   Segment Result

Fig. 3. Exemplar of the saliency inconsistency of object regions. For a salient object without uniform color distributions, the traditional methods (SF [7] here) fail to exactly segment the complete object (middle column), while our method with saliency spread can significantly improve the quality of segmentation (right column).

model [9], which particularly uses center-surrounded differences across multi-scale low-level features to detect saliency. A typical workflow of such methods includes extracting multiple low-level features (color, intensity, orientation, etc.) to construct prominent maps by determining the contrast of image regions to their surroundings, and combining these maps to form a final saliency map via a predefined fusion strategy.

The contrast based methods can use local or global information. The local contrast based methods utilize the neighborhoods to estimate the saliency of a certain image region. For example, Liu *et al.* [10] defines multi-scale contrast as a linear combination of contrasts in a Gaussian image pyramid. Ma *et al.* [23] generates a saliency map based on dissimilarities at the pixel-level, and extracts attended areas or objects using a fuzzy growing method. These local contrast based methods tend to highlight the object boundaries rather than the entire area, which limits the segmentation-like applications. In contrary, the global contrast based methods consider the contrast relations within the whole image to evaluate saliency of an image region. Zhai *et al.* [26] defines pixel-level saliency based on a pixel's contrast to all other pixels. Chen *et al.* [8] simultaneously evaluates global contrast differences and spatial coherence. Perazzi *et al.* [7] computes two kinds of contrasts (*i.e.*, uniqueness and the spatial distribution) of perceptually homogeneous regions with weighting parameters to compromise local and global contrast. Though these global models achieve more consistent results, they may fail to highlight the entire target objects, or get rid of background. In this work, we impute those inferiors to the insufficiency of color feature and the inconsistency of the saliency of object regions. Specifically, we propose two strategies to improve the saliency detection performance from the feature fusion and the saliency consistency.

For enriching the informativeness of features, we consider the texture to serve as a supplementation of color feature. In the previous literatures, the texture has actually been used for providing the information of spatial arrangement of color or intensities. Tang et al. [13] incorporates the LBP texture into color for providing diverse information, and the combined features can achieve a better saliency detection performance. Gopalakrishnan et al. [14] simultaneously computes the color saliency map and the orientation saliency map, then chooses the one of higher connectivity and less spatial variance as the final saliency map. However, these methods suffer from either model complexity or failing to find the accurate object boundary. In this work, we specifically use the LM filter bank [27] to produce the texture feature, and combine it with the color in an adaptive manner, which depends on the image content.

As for the inconsistency of salient object parts incurred by the model sensitivity [16], we propose saliency spread to alleviate it. Here we assume that different regions of the same object have similar color or texture distribution. So we can utilize the correlation of object parts to encourage the similar parts (likely belonging to the same object) producing equal saliency value. Specifically, we first pick out the most salient regions, and then use the relationship with these regions to enhance the saliency of similar regions, where the similarity of regions is determined by their color, texture, and position in practice. To the best of our knowledge, no similar works have been proposed yet.

3 Overview

In this section, we briefly introduce the framework of our proposed method. We follow the classical pipeline of the contrast based methods except that the proposed saliency spread is embedded. Therefore, as shown in Fig. 4, our method is composited of four key stages: (1) generating the superpixels of images as homogeneous regions, (2) computing the saliency values of image regions, (3) conducting saliency spread to highlight the saliency object, and (4) assigning each pixel a saliency value to produce the final saliency map.

3.1 SuperPixel Generation

This step is used to decompose an image into superpixels [28], which are small regions grouped by homogenous neighboring pixels with similar properties (color, brightness, texture, etc.). The superpixel-level saliency estimation is more robust and efficient than the pixel-level one in practice [7,8]. In fact, superpixels could capture image redundancy and abstract unnecessary details, which conforms with the regional perception mechanism of human vision. Moreover, superpixels

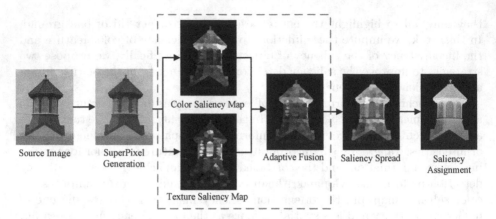

Fig. 4. The framework of our proposed saliency spread method, which includes super-pixel generation, regional saliency computation, saliency spread, and pixel-level saliency assignment. Particularly, saliency spread encourages the saliency values of regions belonging to the same objects to be consistent.

could significantly decrease the number of involved elements, which will reduce the computational complexity.

In this work, we adopt the SLIC method [29] to decompose an image into superpixels, which are denoted by $R = R_1, R_2, \ldots, R_M$. Specifically, SLIC employs K-means clustering to segment images in the CIELab color space, and consequently the compact, memory efficient and edge-preserving superpixels can be yielded.

3.2   Regional Saliency Computation

This stage is used for computing the saliency value of each region produced in the first step. Generally, the saliency of one region is determined by the properties of itself and the contrast relationship with its neighbors. From such considerations, we use two kinds of features (color and texture) for enriching the description, and define two contrast metrics, *i.e.*, uniqueness and distribution, to measure the saliency.

Here *uniqueness* represents the rarity or surprise of a region, which has actually been used for saliency detection in the previous works [7,8]. Such definition is natural to saliency computation since the regions with unusual surroundings are more attractive for human vision. In this work, we propose a revised version of such uniqueness by considering more information. *Distribution* denotes the spatial variances of features within a certain region. Roughly speaking, the distribution of features belonging to the foreground is probably more centralizing, while for the background it may exhibit more diverse with high spatial variance [7,10,30].

3.3 Saliency Spread

Most existing saliency estimation models directly obtain the final saliency map from the fused saliencies of multiple features computed in the previous stage. Different from them, we propose *saliency spread* to enhance the consistency of different regions of salient objects. In practice, it is observed that the saliency values of object regions can vary seriously due to the model sensitivity [16], even they have similar color and texture (see Fig. 4), or the feature variation of object parts. Saliency spread tried to tackle the issue by utilizing the similarity of image regions (probably belonging to the same object), and can be regarded as a special smoothing technique on regional saliency. Specifically, we first pick out the n most salient regions as a pseudo-object, and then enhance similar regions via propagating the saliency of the selected pseudo-object, where the color, texture, spatial position are comprehensively leveraged.

3.4 Saliency Assignment

The role of this step is to assign each pixel a saliency value using regional saliencies. Directly assigning each pixel the same value as the belonged region would lose detailed information within superpixels(*e.g.*, strong edges or small feature variations), and thus much error is caused. So we adopt the upsampling method used in [7], which works well due to the ability of capturing details and preserving edges.

4 Algorithm

In the following section we will give a detailed description of regional saliency computation and saliency spread, which form the main parts of our method.

4.1 Regional Saliency Computation

In this section, we show in detail how to measure the two kinds of contrast, *i.e.*, uniqueness and distribution, for color and texture respectively, and combine the power of them to generate the final regional saliency map.

Uniqueness. As mentioned before, uniqueness generally stands for the rarity of a region with its surroundings. Hence the key issues in uniqueness are to determine *surroundings* and characterize *rarity*. *Surroundings* represents the regions involved in computing *rarity*, which should have nonuniform significance due to their spatial positions. And *rarity* denotes the regional feature difference. Intuitively, distance is a proper choice to measure both of them. So we naturally give the definition of *color uniqueness* for R_i:

$$U_i^c = \sum_{j=1}^{M} r_j \cdot d_{i,j}^c \cdot d_{i,j}^p$$

$$d_{i,j}^c = \chi^2(c_i, c_j) = \sum_{k=1}^{t} \frac{(h_{1k} - h_{2k})^2}{h_{1k} + h_{2k}} \qquad (1)$$

$$d_{i,j}^p = \exp(-\frac{1}{2\sigma_u^2} \cdot \|p_i - p_j\|_2^2)$$

where r_j is the number of pixels in R_j, and emphasizes the contrast to bigger regions. $d_{i,j}^c$ is the chi-square distance between color histograms of R_i and R_j. c_i is the color histogram of R_i in Lab colorspace with $t = 60$ bins. A small variation of a or b channel could cause a remarkable change of color perception when they are close to 0, so we non-uniformly quantize a and b to 22 bins respectively with well chosen quantization near 0. To be more specific, the quantization intervals of a and b below 0 are set as follows: $[-127, -70], (-70, -60], (-60, -50], (-50, -40], (-40, -30], (-30, -25], (-25, -20], (-20, -15], (-15, -10], (-10, -5], (-5, 0]$. Symmetrically, the quantization density of a and b above 0 stays the same with the one below 0. We choose color histogram to alleviate the information loss of using mean color [7] or algorithm complexity caused by exhaustively computing distances among all the colors in R_i and R_j [8].

$d_{i,j}^p$ represents the spatial relationship between R_i and R_j, and renders R_j as more important when they're close. p_i is the mean position of R_i. The introduction of $d_{i,j}^p$ can effectively compromise the global and local contrast, allowing for a sensitivity to local color variation and meanwhile avoiding overemphasizing object edges. Actually in extreme cases, where $d_{i,j}^p = 1$, (1) is equivalent to a completely global uniqueness estimation [8], whereas $d_{i,j}^p \approx 0$ if R_i and R_j are not direct neighbors will yield a local contrast estimation [10]. Parameter σ_u tunes the range of the uniqueness operator. In practice, we find that $\sigma_u = 0.15$ is a well tuned value.

Similar to (1), we define *texture uniqueness* as:

$$U_i^t = \sum_{j=1}^{M} r_j \cdot d_{i,j}^t \cdot d_{i,j}^p$$

$$d_{i,j}^t = \|t_i - t_j\|_2^2 \qquad (2)$$

where t_i is the texture feature of R_i. Here we use the max response among the LM filter bank [27] to represent t_i. The LM set is a multi-scale, multi-orientation filter bank with 48 filters, which consists of first and second derivatives of Gaussians at 6 orientations and 3 scales making a total of 36, 8 Laplacian of Gaussian (LOG) filters, and 4 Gaussians.

With the above definitions, we combine the power of color and texture to get a enhanced *uniqueness* of R_i:

$$U_i = w \cdot U_i^c + (1 - w) \cdot U_i^t \qquad (3)$$

where w depends on the image. The contribution of color and texture differs across images. Hence it's not suitable to use a fixed value as the weight. Noticing that the more information the color or texture provides, the greater is its uniqueness variation, we use uniqueness variation to represent the contribution of color and texture. To be more specific, we set $w = \xi \cdot var(U_c)/(\xi \cdot var(U_c) + var(U_t))$, where ξ is a tuning parameter to highlight the importance of color, and $var(*)$ represents the variation. A similar idea can be found in [31], where the weights of color and texture are determined by computing the overlapping degree of their distributions given the foreground and background sample. In all our experiments, we set $\xi = 5$.

Distribution. Features belonging to the foreground are generally compact and exhibit low spatial variances. So we define regional distribution using the spatial variances of its features. The spatial variance of a feature corresponds to its occurrence elsewhere in the image, which can be measured by its spatial distance to the mean position. Thus we define *color distribution* for R_i as:

$$D_i^c = \sum_{j=1}^{M} r_j \cdot \|p_j - p_i^c\|_2^2 \cdot \tilde{d}_{i,j}^c$$

$$p_i^c = \sum_{j=1}^{M} r_j \cdot \tilde{d}_{i,j}^c \cdot p_j \qquad (4)$$

$$\tilde{d}_{i,j}^c = \exp(-\frac{1}{2\sigma_d^2}\chi^2(c_i, c_j))$$

where p_i^c is the weighted mean position of R_i in terms of color. $\tilde{d}_{i,j}^c$ denotes color similarity between R_i and R_j, which is defined with color distance. The parameter σ_d controls the role that color similarity plays, since a big σ_d tends to decrease the significance of regions with similar color, while a small one yields more sensitivity to color variation. In our experiments, we set $\sigma_d = 10$.

The *texture distribution* is defined in a similar way to (4):

$$D_i^t = \sum_{j=1}^{M} r_j \cdot \|p_j - p_i^t\|_2^2 \cdot \tilde{d}_{i,j}^t$$

$$p_i^t = \sum_{j=1}^{M} r_j \cdot \tilde{d}_{i,j}^t \cdot p_j \qquad (5)$$

$$\tilde{d}_{i,j}^t = \exp(-\frac{1}{2\sigma_d^2}\|t_i - t_j\|_2^2)$$

where t_i is again the texture feature. We combine color and texture distribution with adaptive weighting to obtain the *distribution* of region R_i:

$$D_i = w \cdot D_i^c + (1 - w) \cdot D_i^t \qquad (6)$$

Saliency Fusion. After obtaining uniqueness U_i and distribution D_i for region R_i, we now combine them to obtain a regional saliency map. Assuming that U_i and D_i are independent, we define the saliency value S_i^f of region R_i similar to [7]:

$$S_i^f = U_i \cdot \exp(-\lambda \cdot D_i) \tag{7}$$

The form of exponential function is chosen to emphasize D_i, which is more powerful to highlight salient regions. The scaling factor λ is empirically set to 3 in our experiments.

4.2   Saliency Spread

This step is to deal with the inconsistent saliencies of object parts. We assume that regions belonging to the same objects have similar properties, and choose n most salient regions as pseudo-objects, then spread saliency to the regions that are likely to belong to the selected objects. This can be formulated as:

$$S_i = S_i^f + \sum_{j=1}^{n} r_j \cdot S_j^f \cdot \exp(-\frac{\chi^2(c_i, c_j)}{2\alpha^2} - \frac{\|t_i - t_j\|_2^2}{2\beta^2} - \frac{\|p_i - p_j\|_2^2}{2\delta^2}) \tag{8}$$

where S_i^f is the saliency value of region S_i obtained from (7). α, β, δ are tuning parameters that adjust the significance of color, texture and spatial relations with selected pseudo-object regions, respectively. In our experiments, we set $\alpha = \beta = \delta = 2$, which is strong enough to guarantee that only the nearby regions with similar color and texture are enhanced. From Fig. 4 we can see that saliency spread could significantly increase the saliency value of the object regions and highlight the object as a whole. In our experiment, we empirically set $n = 30$.

Saliency spread can bring another benefit. Many methods are based on the assumption that the containing objects in an image are in a position near the center of the image. [10] use the distance from pixel x to the image center as a weight to assign less importance to colors nearby image boundaries, [11] treat the 15-pixel wide narrow border region of the image as pseudo-background region and extract the backgroundness descriptor. But such assumption is not always true, and the methods will have a poor performance on images in which objects reside near the boundaries. Our saliency spread could roughly determine the location of objects without the assumption, and this will help to relatively decrease the backgroundness saliency via (8).

The last step is a per-pixel saliency assignment. For pixel i:

$$Sal_i = \sum_{j=1}^{M} r_j \cdot S_j \cdot \exp(-\frac{1}{2\sigma_c^2} \cdot \chi^2(c_i, c_j) - \frac{1}{2\sigma_p^2} \cdot \|p_i - p_j\|_2^2) \tag{9}$$

where S_j are regional saliency surrounding pixel i. σ_c and σ_p are parameters controlling the sensitivity to color and position respectively, we set $\sigma_c = \sigma_p = \frac{1}{30}$ in the experiments. Finally, the resulted pixel-level saliency map is rescaled to the range $[0-255]$ for the purpose of exhibiting and comparing with the groundtruth.

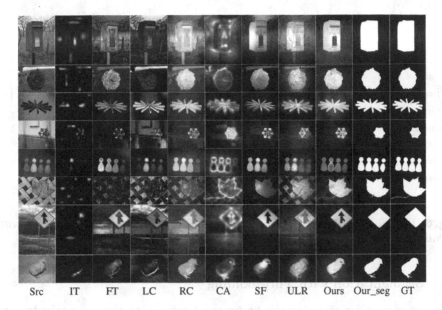

Src   IT   FT   LC   RC   CA   SF   ULR   Ours   Our_seg   GT

Fig. 5. Visual comparison of previous approaches to our method. Due to space limitation, only a part of the results are exhibited. Our method generates consistent and uniform salient regions. The segment results (Ours-Seg), which are obtained using adaptive threshold (Eq. 10), are also close to ground truth (GT).

5   Experiments

We evaluate the results of our approach on the commonly used MSRA-1000 databset provided by [32], which is a subset of MSRA [10]. MSRA-1000 is the largest of its kind [8] for saliency detection with accurate human-marked labels as binary ground truth rather than rectangle bounding boxes used in MSRA. We provide a comprehensive comparison of our method to 13 state-of-the-art saliency detection methods, including biologically-motivated saliency (IT [9]), purely computational fuzzy growing (MZ [23]), frequency domain based saliency (FT [32], SR [18]), spatiotemporal cues (LC [26]), graphed-based saliency (GB [33]), context-aware saliency (CA [30]), salient region detection (AC [21]), low-rank matrix recovery theory inspired saliency (LSMD [34], ULR [35]), and works related to our method (SF [7], HC [8], RC [8]). To evaluate these methods, we use author's implementation (when available) or the resulting saliency maps provided in [8]. A visual comparison of saliency maps obtained by these methods can be seen in Fig. 5.

In order to comprehensively evaluate the performance of our method, we conduct two experiments following the standard evaluation measures in [7,8,34]. In the first experiment, we segment saliency maps using fixed or adaptive threshold, and calculate precision and recall curves. In the second experiment, we use mean absolute error to evaluate how well the continuous saliency map match the binary ground truth.

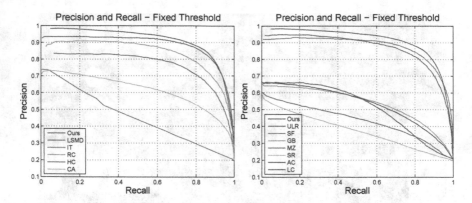

Fig. 6. Precision-Recall curves for fixed threshold of saliency maps. Compared with various methods, our approach achieves the best performance.

5.1   Segmentation with Thresholding

A common way for assessing the accuracy of saliency detection methods is to binarize each saliency map with fixed threshold or adaptive threshold, and compute its precision and recall rate. Precision (also called positive predictive value) represents the fraction of retrieved pixels that are relevant, while recall (also known as sensitivity) corresponds to the percentage of relevant pixels that are retrieved. They are often evaluated simultaneously, since a high precision can be obtained at the cost of a low recall and vice-versa.

Fixed Threshold. We first segment a saliency map with a fixed threshold $t \in [0, 255]$. After the segmentation, we compare the binarized image with ground truth to obtain its precision and recall. To reliably measure the capability of various methods highlighting salient regions in images, we vary the threshold t from 0 to 255 to generate a sequence of precision-recall pairs. After averaging over all the results of images in the dataset, we obtain the precision-recall curves, as Fig. 6 shows. As we can see, compared to other approaches, the saliency maps generated by our method with fixed threshold are more accurate, and closer to the ground truth on the whole.

Adaptive Threshold. Similar to [7, 35], we adopt the image dependent adaptive threshold, which is defined as twice the mean saliency value of the entire image [32]:

$$T_a = \frac{2}{W \times H} \sum_{x=1}^{W} \sum_{y=1}^{H} S(x, y) \tag{10}$$

where W and H are the width and height of the image, respectively. S is the obtained saliency map. Adaptive threshold is a simple but practical indicator for comparing quality among approaches, as the resulting segmentation could

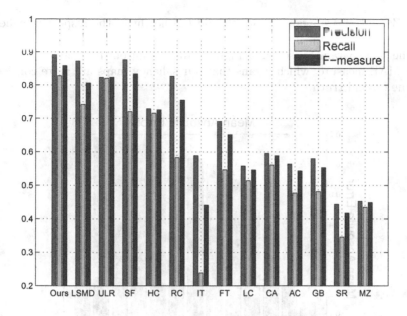

Fig. 7. Precision, recall, and F-measure for adaptive thresholds.

be directly utilized in other literatures. In addition to precision and recall, we also compute their weighted harmonic mean measure (or F-measure), which is defined as:

$$F_\beta = \frac{(1 + \beta^2) \cdot \text{Precision} \cdot \text{Recall}}{\beta^2 \cdot \text{Precision} + \text{Recall}} \quad (11)$$

Similar to previous works [7,32], we also set $\beta^2 = 0.3$. The result is given in Fig. 7. Our method achieves the best precision, recall and F-measure among all the approaches. Compared to SF, which is closest to our method, we have a significant improvement of recall (9 %), which means our method are likely to detect more salient regions, while keeping a high accuracy.

5.2   Mean Absolute Error

Ideally a saliency map should be equal to the ground truth, and each thresholding in $(0, 255)$ results in the same segmentation, *i.e.* the true object. Hence the more similar with the ground truth, the better is the saliency map and the algorithm generating it. Yet neither the precision nor recall measure consider such performance indicator. We adopt MAE (Mean Absolute Error) to measure the similarity between the continuous saliency map S and the binary ground truth GT, which is defined in [7]:

$$\text{MAE} = \frac{1}{W \times H} \sum_{x=1}^{W} \sum_{y=1}^{H} |S(x, y) - GT(x, y)| \quad (12)$$

where W and H are again the width and the height of the respective saliency map and ground truth image. We compute MAE by averaging over all images with the same parameter settings. Figure 8 shows that our method generates the lowest MAE measure, which means that our saliency maps are more consistent with the ground truth.

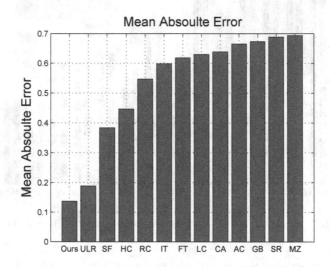

Fig. 8. Mean absolute error of the different saliency methods to ground truth.

6   Conclusions

In this work, we present a contrast based method for salient object detection, which follows the typical pipeline of contrast measures estimating and fusing. On this basis, we analysis the weakness of existing models, and attribute it to the insufficiency of color feature and the inconsistency of the saliency of object regions. Contrapose these deficiencies we present two improvements. Firstly, we incorporate texture as a complementary feature of color, to deal with images without dominant color. Secondly, we propose saliency spread, which propagates saliencies to regions that are likely to belong to the same objects and achieves more consistent saliency maps. Experiments show the superiority of our proposed schemes in terms of serval widely accepted indicators.

References

1. Mangun, G.R.: Neural mechanisms of visual selective attention. Psychophysiology **32**, 4–18 (1995)
2. Kanan, C., Cottrell, G.: Robust classification of objects, faces, and flowers using natural image statistics. In: CVPR, pp. 2472–2479. IEEE (2010)
3. Rutishauser, U., Walther, D., Koch, C., Perona, P.: Is bottom-up attention useful for object recognition? In: CVPR, vol. 2, p. II-37 (2004)

4. Chen, L.Q., Xie, X., Fan, X., Ma, W.Y., Zhang, H.J., Zhou, H.Q.: A visual attention model for adapting images on small displays. Multimedia Syst. 9, 353–364 (2003)
5. Ding, M., Tong, R.F.: Content-aware copying and pasting in images. Visual Comput. 26, 721–729 (2010)
6. Ko, B.C., Nam, J.Y.: Object-of-interest image segmentation based on human attention and semantic region clustering. JOSA A 23, 2462–2470 (2006)
7. Perazzi, F., Krahenbuhl, P., Pritch, Y., Hornung, A.: Saliency filters: contrast based filtering for salient region detection. In: CVPR, pp. 733–740 (2012)
8. Cheng, M.M., Zhang, G.X., Mitra, N.J., Huang, X., Hu, S.M.: Global contrast based salient region detection. In: CVPR, pp. 409–416 (2011)
9. Itti, L., Koch, C., Niebur, E., et al.: A model of saliency-based visual attention for rapid scene analysis. PAMI 20, 1254–1259 (1998)
10. Liu, T., Yuan, Z., Sun, J., Wang, J., Zheng, N., Tang, X., Shum, H.Y.: Learning to detect a salient object. PAMI 33, 353–367 (2011)
11. Jiang, H., Wang, J., Yuan, Z., Wu, Y., Zheng, N., Li, S.: Salient object detection: a discriminative regional feature integration approach. In: CVPR, pp. 2083–2090. IEEE (2013)
12. Felzenszwalb, P.F., Huttenlocher, D.P.: Efficient graph-based image segmentation. IJCV 59, 167–181 (2004)
13. Tang, K., Au, O.C., Fang, L., Yu, Z., Guo, Y.: Multi-scale analysis of color and texture for salient object detection. In: ICIP, pp. 2401–2404 (2011)
14. Gopalakrishnan, V., Hu, Y., Rajan, D.: Salient region detection by modeling distributions of color and orientation. Multimedia 11, 892–905 (2009)
15. Hu, Y., Xie, X., Ma, W.-Y., Chia, L.-T., Rajan, D.: Salient region detection using weighted feature maps based on the human visual attention model. In: Aizawa, K., Nakamura, Y., Satoh, S. (eds.) PCM 2004. LNCS, vol. 3332, pp. 993–1000. Springer, Heidelberg (2004)
16. Borji, A., Sihite, D.N., Itti, L.: Salient object detection: a benchmark. In: Fitzgibbon, A., Lazebnik, S., Perona, P., Sato, Y., Schmid, C. (eds.) ECCV 2012, Part II. LNCS, vol. 7573, pp. 414–429. Springer, Heidelberg (2012)
17. Guo, C., Ma, Q., Zhang, L.: Spatio-temporal saliency detection using phase spectrum of quaternion fourier transform. In: CVPR, pp. 1–8 (2008)
18. Hou, X., Zhang, L.: Saliency detection: a spectral residual approach. In: CVPR, pp. 1–8 (2007)
19. Bruce, N., Tsotsos, J.: Saliency based on information maximization. Adv. Neural Inf. Process. Syst. 18, 155 (2006)
20. Hou, X., Zhang, L.: Dynamic visual attention: searching for coding length increments. In: NIPS, vol. 5, p. 7 (2008)
21. Achanta, R., Estrada, F.J., Wils, P., Süsstrunk, S.: Salient region detection and segmentation. In: Gasteratos, A., Vincze, M., Tsotsos, J.K. (eds.) ICVS 2008. LNCS, vol. 5008, pp. 66–75. Springer, Heidelberg (2008)
22. Duan, L., Wu, C., Miao, J., Qing, L., Fu, Y.: Visual saliency detection by spatially weighted dissimilarity. In: CVPR, pp. 473–480. IEEE (2011)
23. Ma, Y.F., Zhang, H.J.: Contrast-based image attention analysis by using fuzzy growing. In: Proceedings of the Eleventh ACM International Conference on Multimedia, pp. 374–381. ACM (2003)
24. Einhäuser, W., König, P.: Does luminance-contrast contribute to a saliency map for overt visual attention? Eur. J. Neurosci. 17, 1089–1097 (2003)
25. Parkhurst, D., Law, K., Niebur, E.: Modeling the role of salience in the allocation of overt visual attention. Vision Res. 42, 107–123 (2002)

26. Zhai, Y., Shah, M.: Visual attention detection in video sequences using spatiotemporal cues. In: ACM Multimedia, pp. 815–824. ACM (2006)
27. Leung, T., Malik, J.: Representing and recognizing the visual appearance of materials using three-dimensional textons. IJCV **43**, 29–44 (2001)
28. Ren, X., Malik, J.: Learning a classification model for segmentation. In: Computer Vision, pp. 10–17. IEEE (2003)
29. Achanta, R., Shaji, A., Smith, K., Lucchi, A., Fua, P., Süsstrunk, S.: Slic superpixels. EPFL, Technical report **2**, 3 (2010)
30. Goferman, S., Zelnik-Manor, L., Tal, A.: Context-aware saliency detection. PAMI **34**, 1915–1926 (2012)
31. Shahrian, E., Rajan, D.: Weighted color and texture sample selection for image matting. In: CVPR, pp. 718–725. IEEE (2012)
32. Achanta, R., Hemami, S., Estrada, F., Susstrunk, S.: Frequency-tuned salient region detection. In: CVPR, pp. 1597–1604 (2009)
33. Harel, J., Koch, C., Perona, P., et al.: Graph-based visual saliency. Adv. Neural Inf. Process. Syst. **19**, 545 (2007)
34. Peng, H., Li, B., Ji, R., Hu, W., Xiong, W., Lang, C.: Salient object detection via low-rank and structured sparse matrix decomposition. In: AAAI (2013)
35. Shen, X., Wu, Y.: A unified approach to salient object detection via low rank matrix recovery. In: CVPR, pp. 853–860. IEEE (2012)

Biologically Inspired Composite Vision System for Multiple Depth-of-field Vehicle Tracking and Speed Detection

Lin Lin, Bharath Ramesh[✉], and Cheng Xiang

Department of Electrical and Computer Engineering,
National University of Singapore, Singapore 117576, Singapore
bharath.ramesh03@u.nus.edu

Abstract. This paper presents a new vision-based traffic monitoring system, which is inspired by the visual structure found in raptors, to provide multiple depth-of-field vision information for vehicle tracking and speed detection. The novelty of this design is the usage of multiple depth-of-field information for tracking expressway vehicles over a longer range, and thus provide accurate speed information for overspeed vehicle detection. A novel speed calculation algorithm was designed for the composite vision information acquired by the system. The calculated speed of the vehicles was found to conform with the real-world driving speed.

1 Introduction

Object tracking has been one of the most attractive topics in computer vision, and it has various practical applications so far, such as human-computer interaction [1,2], video surveillance [3–5], vehicle navigation [6], traffic monitoring and control [7–9], and motion analysis [10]. In this paper, the focus is on traffic monitoring in expressways for automatic overspeed vehicle detection. In the past few years, various vision-based methods have been designed to solve the traffic monitoring problems using a single [11–14] or stereo camera [15,16]; however, the performance of the above-mentioned systems is limited by a small tracking range due to the fixed depth-of-field of the cameras [17,18]. These systems perform vehicle tracking near the installed location, and hence they can only track and calculate vehicle speed within a small distance. Therefore, these systems are better suited for traffic monitoring situations such as congestion control and intersection monitoring. In practice, a long tracking range is crucial when high-speed vehicle monitoring is needed.

In contrast, the traditional speed detection approaches using sensors such as LIDAR/RADAR have several drawbacks [19,20]. These approaches generally work in this way: the sensors detect the presence of a possible overspeeding vehicle and trigger a camera to capture the image of the overspeeding vehicle. However, with a large amount of vehicles present in the scene, the detector might know one of the vehicles is overspeeding but not be able to single it out. Furthermore, the interference caused by big vehicles leads to unreliable results for speed detection.

© Springer International Publishing Switzerland 2015
C.V. Jawahar and S. Shan (Eds.): ACCV 2014 Workshops, Part I, LNCS 9008, pp. 473–486, 2015.
DOI: 10.1007/978-3-319-16628-5_34

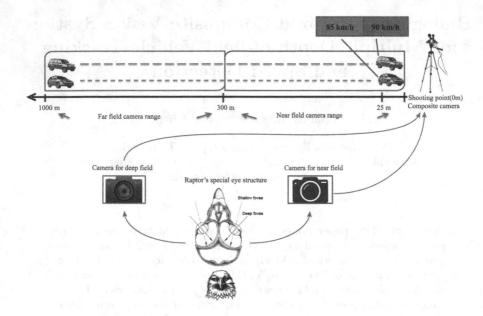

Fig. 1. Overview of the Composite Vision System.

Finally, the lack of coherence between sensor data and visual vehicle identity data seriously cripples law enforcement. Therefore, there is a need for a vision-based traffic monitoring system that can address all the above problems.

Reference [21] proposed a composite image sensor inspired by raptors' vision for deep-field object tracking, which uses multiple depth-of-field information to detect the presence of vehicles over a longer range. Although the object tracking range was increased, only a limited portion of the tracking result was used as a switch to activate another camera for license plate detection. In fact, this activation can be achieved by a single camera or other sensors. In other words, the visual data was not fully utilized, and the object tracking effort over a longer range did not serve a suitable application. In this paper, we propose a speed detection application using the multiple depth-of-field vision information, which utilizes the full tracking range.

As shown in Fig. 1, the proposed system uses a composite camera that can view both near field and far field synchronously, and monitor vehicle movements up to 1000 m away from the shooting point. The proposed system defines the tracking result, while vehicles are present in the composite camera's field of view. Later on, each vehicle's speed and identity information are output as a snapshot, which can be easily used for law enforcement. As a result, the drawbacks of present traffic monitoring systems are overcome by providing a long tracking range and coherence between speed information and vehicle identity. In summary, the proposed system provides a new solution to improve the capabilities of vision-based traffic monitoring systems.

The rest of the paper is organized as follows. Section 2 presents the composite camera design with implementation details; Sect. 3 presents the vehicle tracking

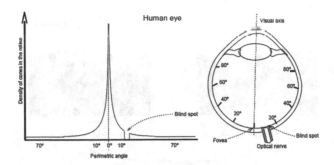

Fig. 2. Primates' retina photoreceptor distribution.

algorithm, followed by Sect. 4 with the proposed vehicle speed calculation algorithm, and Sect. 5 with experimental results and discussion. Finally, conclusions and future works are presented in Sect. 6.

2   Composite Camera Design

According to Tucker and Snyder [22,23], raptors such as falcons and eagles have two sets of foveae, which enables them to simultaneously observe objects from near and far distances, as shown in Fig. 1. The visual data obtained by the two sets of foveae are combined into a single nerve and sent to raptor's brain for processing. However, since the study about raptor's foveae to brain information mapping is limited, we adopt the primate retina model and log-polar transform [24] to simulate raptor's internal mapping.

By observing the non-uniform distribution of the cones around primates' fovea, as shown in Fig. 2, a logarithmic relationship for information around the fovea structure [25] can be established. Using log-polar mapping, the vision information received in each camera can be transformed into log-polar space regardless of the depth-of-field. Therefore, the problem of combining information from multiple depth-of-field is neatly solved as shown in Fig. 3. Furthermore, log-polar transformation (LPT) with ideal center point provides scale and rotation invariance. As a result, the scale change of vehicles caused by forward vehicle movement will be converted to horizontal shifting in the log-polar space with a fixed shape. Hence, the transformed LPT image could possibly provide relatively unchanged vehicle shape during the tracking process.

To have seamless stitching, the composite camera requires two cameras with different view angles. In addition, the two cameras should be placed as close as possible to reduce errors. As shown in Fig. 4, these nested cameras should have a special relation to achieve seamless stitching of different depth-of-field information. For ideal log-polar mapping and stitching, the camera relation factor K_b (Eq. 1) is set to be 10. In other words, the two camera lenses should have roughly about 10 times difference in their view angle.

$$K_b = \frac{\rho_b}{\rho_{max}} = \frac{tan(\theta_{far}/2)}{tan(\theta_{near}/2)} \tag{1}$$

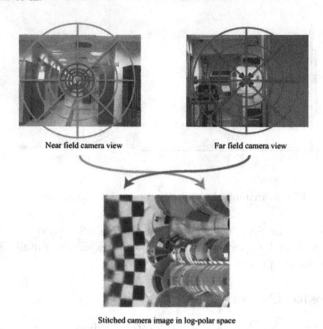

Fig. 3. Composite image stitching example. Best viewed in color.

$$K_b \approx \frac{\theta_{far}}{\theta_{near}}, if \theta_{far} \to 0, \theta_{near} \to 0 \tag{2}$$

2.1   Composite Camera Implementation

On the basis of the theoretical design introduced earlier, a market research showed that there is no readily available camera for synchronous viewing with different depth-of-fields and view angles. Therefore, the composite camera was implemented by choosing individual cameras and integrating them in a flexible hardware mount. As indicated in Fig. 5, USB 3.0 technology was selected after comparing various industrial standards. Further market survey about USB 3.0 cameras led us to choose Basler cameras, shown in Fig. 6(a).

This composite camera design is different from the work of [21], because USB 3.0 cameras provide high frame rate with minimal frame dropping and also support synchronous video acquisition with easy plug-and-play ability with laptops. This provides future development possibilities to easily install the vehicle monitoring system on mobile platforms such as police cars. On the contrary, Ref. [21] used ad hoc surveillance cameras that have several disadvantages. Surveillance camera standards such as Gigabit Ethernet/Fire Wire/Camera Link are not so suitable for outdoor vehicle monitoring, because they require a special connection interface, or separate power supply, or separate communication devices for data acquisition. These additional requirements limit the mobility and flexibility

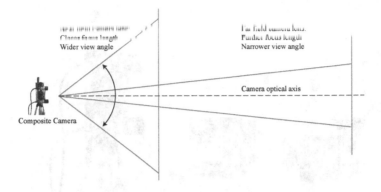

Fig. 4. Dual lens composite camera relationship.

	FireWire	Gigabit Ethernet	USB 2.0	USB 3.0	Camera Link
Bandwidth	80MB/s	100MB/s	40MB/s	440MB/s	680MB/s
Cable length	10m	100m	5m	3m	7m
Consumer acceptance	Declining	Excellent	Excellent	Excellent	None
Multiple cameras	Excellent	Good	Fair	Excellent	Fair
Power delivery	Excellent	Excellent (POE)	Fair	Good	None
Vision Standard	IIDC DCAM	GigE Vision	No	USB3 Vision	Camera Link

Fig. 5. Industrial Camera Standards [26].

of traffic monitoring system using surveillance cameras. Besides, our composite camera design includes a flexible hardware mount, which was designed to adjust the cameras' positional relationship arbitrarily (Fig. 6(c)).

Two lenses were carefully selected for the implementation of the composite camera to simulate raptor's vision. One of them is a wide angle camera which provides a 76.7 degree view angle, and the other has a narrow view angle about 7.9 degrees. Therefore, the two of them have approximately 10 times difference in viewing angle. The successfully built composite camera is shown in Fig. 6(b). The next objective was to synchronously capture videos from both these cameras.

Our initial trial in using MATLAB to synchronously capture videos failed, because it could not differentiate the two identical cameras with different lenses. Hence, some alternative software solutions were tested. Finally, a C# program with AForge.Net library was implemented to handle synchronized video acquisition from both cameras. After implementing the software, two plausible locations in Singapore - Tampines Expressway and Ayer Rajah Expressway - were selected to capture high-speed vehicle movements. Figure 7 shows the field test conducted on an overhead bridge 4.5 m above the expressway lanes.

(a) Basler cameras.

(b) Composite camera system.

(c) Hardware platform (without cameras).

Fig. 6. Composite Camera built using USB 3.0 industrial cameras.

2.2 Multiple Depth-of-field Data Processing

The stitching of the multiple depth-of-field images acquired from the composite camera is shown in Fig. 8. For the two Cartesian images acquired synchronously from the composite camera, log-polar transformation is applied individually. To achieve the scale and rotation invariant property in log-polar space, the road vanishing point has to be carefully determined, which serves as the central point for the log-polar transformation. After the transformation, there appears an over-sampled region around the inner part of the near view and an under-sampled region around the outer part of the far vision, shown in Fig. 8 as the blue areas. To stitch the two views, some duplicated regions in both views are discarded. The final stitched result has the scale and rotation invariant property of log-polar transformation [27], which helps to maintain the size of each vehicle from the shooting point. Moreover, well-separated road lanes assist tracking in particular lane(s) of interest. Notice that the stitched multiple depth-of-field view has a significant reduction in image content, which helps to fasten the object tracking process and therefore increase the computational efficiency. As shown in Fig. 8, the Cartesian image size of both views is 600×900, whereas the stitched log-polar space composite view result is 136×509 only.

Fig. 7. Composite Camera Field Test.

3   Vehicle Tracking

The seamless stitching from different depth-of-field cameras extended the object tracking range from about 300 m in the traditional vision-based vehicle tracking practice to up to 1000 m. To counter the complicated outdoor conditions, such as illumination change (shadow of cloud/tree/vehicles, brightness change, and reflections) and mechanical vibration, Gaussian mixture model (GMM) was adopted to extract and separate the moving vehicles from the background. Figure 9 shows the vehicle extraction result using GMM algorithm.

With successful object extraction by GMM, Kalman filter [28] was adopted as the object tracking method for multiple vehicles in the scene. The main advantage of using Kalman filter is that it is able to model the vehicle's acceleration in the video due to the prospective projection. Furthermore, it provides tolerance to a certain degree of occlusion by predicting the vehicle position based on previous vehicle states. Finally, Kalman filter provides a distance parameter to tolerate distortion and noise of object movements. Briefly, Kalman filter allows the system to track multiple vehicles while maintaining some prediction and tolerance to the complex vehicle motions, such as lane switching and variable acceleration.

4   Proposed Vehicle Speed Calculation

The most commonly adopted solution for vehicle speed detection is using LIDAR or RADAR signal along with surveillance cameras. One significant drawback is the lack of ability to determine the correct overspeed vehicle, because of insufficient coherence between speed information and the vehicle image saved. For

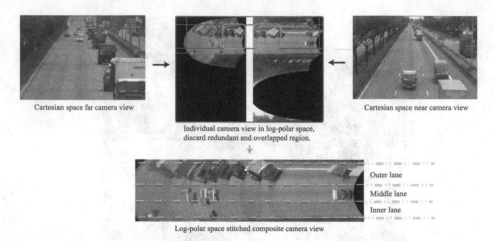

Fig. 8. A typical stitching process of the multiple depth-of-field images.

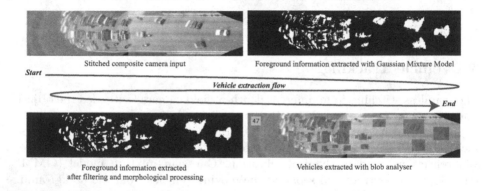

Fig. 9. Vehicle extraction with Gaussian mixture model.

instance, upon activation by the RADAR signal, the surveillance camera could capture an image with more than one vehicle in the scene including the overspeed vehicle. Another defect is that the RADAR/LIDAR accuracy is highly affected by interference from large vehicles. The proposed composite vision system aims to provide a vision-based solution that is able to gather speed information of a vehicle with its corresponding image, and thus overcome the drawbacks in the current speed monitoring systems.

4.1   Vehicle Speed Calculation

Because the vehicles move faster from deep to near field due to perspective projection, direct speed calculation from pixel coordinates is impractical. Therefore, an algorithm is proposed to transform the composite image location to the real-world location, and determine the vehicle speed in kilometer per hour. The vehicle speed calculation involves four main steps that are as follows:

1. Transform stitched log-polar space coordinates to single individual log-polar space coordinates.
2. Transform individual log-polar space coordinates to camera Cartesian space coordinates.
3. Transform camera Cartesian space coordinates to real world coordinates.
4. Use tracking time information and real world location to calculate vehicle speed

Step 1: An arbitrary position (u, v) in the stitched log-polar coordinates can be transformed to the single individual log-polar space coordinate (U,V) by the following relation, where $u_{InnerRingCrop}$ is the number of rings cropped out, $v_{LowerWedgeCrop}$ is the number of wedges cropped out, and $u_{stitchline}$ is the position of the stitch line.

1. When u,v falls in the far-field view range, that is, on the left of the stitching line
$$U = u + u_{InnerRingCrop} \tag{3}$$
$$V = v + v_{LowerWedgeCrop} \tag{4}$$
2. When u,v falls in the near-field view range, that is, on the right of the stitching line
$$U = u + u_{stitchline} + u_{InnerRingCrop} \tag{5}$$
$$V = v + v_{LowerWedgeCrop} \tag{6}$$

Step 2: The method to transform individual log-polar space coordinate (U,V) to their corresponding camera Catesian space coordinate (x,y) is defined as follows:

$$Distance = r_{min} \times e^{\frac{U \times log(\frac{r_{max}}{r_{min}})}{n_r - 1}} \tag{7}$$

$$Angle = V \times \frac{2\pi}{n_w} \tag{8}$$

$$x = Distance \times \cos(Angle) + x_c \tag{9}$$

$$y = Distance \times \sin(Angle) + y_c \tag{10}$$

where n_r represents the number of rings, n_w is number of wedges, x_c and y_c are chosen road vanishing point position, r_{max} and r_{min} are maximum and minimum radii used in stitching process. The dots in Fig. 10 demonstrate successful transformation between the two coordinates.

Step 3: The transformation from camera Cartesian space coordinates to real-world coordinates (x,z) follows the method proposed by Wu [11]. Taking into consideration the composite cameras height above the road and tilt angle θ

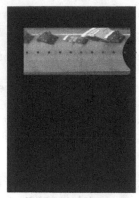 

Single log-polar space image   Cartesian space image with transformed locations
with target locations

Fig. 10. Speed calculation step 2 demonstration.

Fig. 11. Speed calculation step 3 demonstration.

from the road's forward direction, the real-world locations x (transverse direction on the road surface) and z (longitudinal or forward direction on the road surface) can be obtained. To validate the transformation result, the lane markers separation distance (12 m) defined by international traffic standard was used. As shown in Fig. 11, the estimated distance roughly matches with the real-world location.

Step 4: Finally, with the correct location information, the distance traveled by a vehicle in the real world can be calculated based on the Euclidean distance between the two points of interest. The speed calculation is effectively achieved by using the timing information provided during vehicle tracking process. About 20 calculation windows were used to average out the calculated speed of a vehicle, which is tracked from approximately 1000 m to 25 m away from the shooting point.

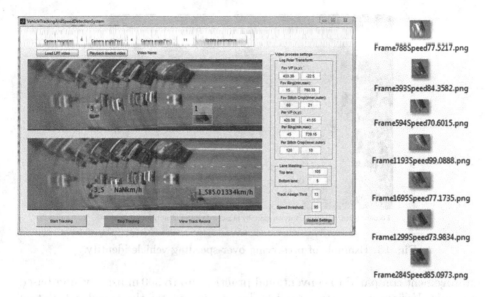

Frame788Speed77.5217.png

Frame393Speed84.3582.png

Frame594Speed70.6015.png

Frame1193Speed99.0888.png

Frame1695Speed77.1735.png

Frame1299Speed73.9834.png

Frame284Speed85.0973.png

Fig. 12. Results of the composite vision system.

5   Experimental Results and Discussion

5.1   Vehicle Speed Calculation Based on Multiple Depth-of-field Vision

To test the performance of the composite vision system, we arranged a car to drive with a known speed around 75 km/h and then recorded the video using our composite camera setup. With the recorded video, the composite vision system tracked the speed information with 20 calculation windows and calculated the average vehicle speed to be 78 km/h. This is a reasonable speed detection result, because the car's original speed was slightly varied around 5 km/h due to driving conditions on the expressway.

After establishing the baseline result, we tested the system on a variety of traffic conditions. Figure 12 shows the results from a video captured at Ayer Rajah Expressway in Singapore; the upper half of the MATLAB GUI displays the tracking result while the lower half shows the calculated speed. After each vehicle exits the scene, a snapshot of it is stored along with the time stamp and average speed. Notice that NaN appears as one of the vehicles' speed in Fig. 12. This is because the speed calculation algorithm waits for tracking to stabilize, which typically takes a few seconds.

Because the vehicle motion is tracked from far field to near field, the composite camera provides adequate information for the system to perform long distance tracking and speed calculation. This benefits the accuracy and reliability of the speed calculation. With the multiple depth-of-field viewing, the system is able to track the vehicle from approximately 1000 m away to 25 m nearby, which is a big

Locate desired vehicle in Cartesian space by preserved vehicle identity information

Fig. 13. Example of preserving over-speeding vehicle identity.

improvement compared to conventional practices up to 300 m using vision-based methods. Various parameters need to be considered for the speed calculation procedure, such as road lane vanishing point setting, log-polar transform parameters, and individual camera settings. However, the speed calculation is most affected by the precision of object tracking. From experiments in other parts of the island, we found that the most challenging aspect was low lighting conditions. In these cases, the speed detection result was not reliable due to tracking difficulties.

5.2   Benefits of Vehicle Identity Preservation

As discussed earlier, the main drawback of the present traffic monitoring systems is the lack of vehicle identity information. The composite vision system solves this problem by providing a snapshot of the exact overspeeding vehicle. Figure 12 shows the saving of snapshots during the tracking process. To augment evidence of the overspeeding vehicle, license plate detection with the aid of a high-resolution camera triggered by the composite vision system can be implemented. As a proof-of-concept, we conducted some experiments with the snapshot information to locate the vehicle in the Cartesian image. As shown in Fig. 13, the overspeeding vehicle can be precisely located using SIFT keypoint matching, which can possibly help the high-resolution camera take a close-up picture.

6   Conclusion

We proposed a composite vision system with multiple depth-of-field viewing ability that largely extended the tracking range of traditional traffic monitoring systems. By defining the overspeeding vehicle using the tracking result, strong relationship between identity and speed information was established.

Having deep field object tracking ability, the composite vision system can handle high-speed vehicle tracking and can compensate the drawbacks of present speed monitoring systems. Moreover, the system has the potential to do real-time tracking in complex road conditions and multiple lanes. It also has the potential for airborne unmanned vehicle navigation due to its versatility in being autonomous.

The future work of the proposed system includes the addition of a third camera for license plate detection. Besides, more cameras could be added to extend the tracking distance. Last but not least, the performance and reliability will be further investigated in the future work as well.

Acknowledgement. The authors would like to thank Mr. Jack Chin from Sodavision for helping us gather detailed information about camera devices and their accessories. We would also like to thank the software engineers from Basler Asia service for giving advice and guidance on using the Basler camera SDK and AForge.Net library to develop the programs.

References

1. Rehg, J.M., Kanade, T.: Visual tracking of high dof articulated structures: an application to human hand tracking. In: Eklundh, J.-O. (ed.) ECCV '94. LNCS, pp. 35–46. Springer, Heidelberg (1994)
2. Jacob, R.J., Karn, K.S.: Eye tracking in human-computer interaction and usability research: ready to deliver the promises. Mind **2**, 4 (2003)
3. Foresti, G.: Object recognition and tracking for remote video surveillance. IEEE Trans. Circuits Syst. Video Technol. **9**, 1045–1062 (1999)
4. Cohen, I., Medioni, G.: Detecting and tracking moving objects for video surveillance. In: IEEE Computer Society Conference on Computer Vision and Pattern Recognition, vol. 2, p. 325 (1999)
5. Javed, O., Shah, M.: Tracking and object classification for automated surveillance. In: Heyden, A., Sparr, G., Nielsen, M., Johansen, P. (eds.) ECCV 2002, Part IV. LNCS, vol. 2353, pp. 343–357. Springer, Heidelberg (2002)
6. Saripalli, S., Montgomery, J., Sukhatme, G.: Visually guided landing of an unmanned aerial vehicle. IEEE Trans. Robot. Autom. **19**, 371–380 (2003)
7. Coifman, B., Beymer, D., McLauchlan, P., Malik, J.: A real-time computer vision system for vehicle tracking and traffic surveillance. Transp. Res. Part C: Emerg. Technol. **6**, 271–288 (1998)
8. Kamijo, S., Matsushita, Y., Ikeuchi, K., Sakauchi, M.: Traffic monitoring and accident detection at intersections. IEEE Trans. Intell. Transp. Syst. **1**, 108–118 (2000)
9. Tai, J.C., Tseng, S.T., Lin, C.P., Song, K.T.: Real-time image tracking for automatic traffic monitoring and enforcement applications. Image Vis. Comput. **22**, 485–501 (2004)
10. Wang, L., Hu, W., Tan, T.: Recent developments in human motion analysis. Pattern Recogn. **36**, 585–601 (2003)
11. Wu, J., Liu, Z., Li, J., Caidong, G., Si, M., Tan, F.: An algorithm for automatic vehicle speed detection using video camera. In: 4th International Conference on Computer Science Education, ICCSE '09, pp. 193–196 (2009)

12. Clady, X., Collange, F., Jurie, F., Martinet, P.: Cars detection and tracking with a vision sensor. In: Proceedings of the Intelligent Vehicles Symposium, pp. 593–598. IEEE (2003)

13. Roessing, C., Reker, A., Gabb, M., Dietmayer, K., Lensch, H.: Intuitive visualization of vehicle distance, velocity and risk potential in rear-view camera applications. In: Intelligent Vehicles Symposium (IV), pp. 579–585. IEEE (2013)

14. Wang, C.C., Thorpe, C., Suppe, A.: Ladar-based detection and tracking of moving objects from a ground vehicle at high speeds. In: Proceedings of the Intelligent Vehicles Symposium, pp. 416–421. IEEE (2003)

15. Kormann, B., Neve, A., Klinker, G., Stechele, W., et al.: Stereo vision based vehicle detection. In: VISAPP (2), pp. 431–438 (2010)

16. Toulminet, G., Bertozzi, M., Mousset, S., Bensrhair, A., Broggi, A.: Vehicle detection by means of stereo vision-based obstacles features extraction and monocular pattern analysis. IEEE Trans. Image Process. **15**, 2364–2375 (2006)

17. Messner, R.A., Melnyk, P.B.: Mobile digital video system for law enforcement. In: VTC Spring, pp. 468–472 (2002)

18. Schoepflin, T.N., Dailey, D.J.: Dynamic camera calibration of roadside traffic management cameras for vehicle speed estimation. IEEE Trans. Intell. Transp. Syst. **4**, 90–98 (2003)

19. Simmoneau, B.: I-team: Controversy over speed cameras. News (2013)

20. Solicitors, M.D.: Guide to speed detection devices (2014)

21. Melnyk, P.: Biologically inspired composite image sensor for deep field target tracking. Ph.D. thesis, University of New Hampshire (2008)

22. Tucker, V.A.: The deep fovea, sideways vision and spiral flight paths in raptors. J. Exp. Biol. **203**, 3745–3754 (2000)

23. Snyder, A.W., Miller, W.H.: Telephoto lens system of falconiform eyes. Nature **275**, 127–129 (1978)

24. Messner, R.A., Szu, H.H.: An image processing architecture for real time generation of scale and rotation invariant patterns. Comput. Vis. Graph. Image Process. **31**, 50–66 (1985)

25. Schwartz, E.L.: Spatial mapping in the primate sensory projection: analytic structure and relevance to perception. Biol. Cybern. **25**, 181–194 (1977)

26. Richmond, B.: A Practical Guide to USB 3.0 for Vision Applications. Point Grey Research Inc, 12051 Riverside Way (2013)

27. Bailey, J.G., Messner, R.A.: Log-polar mapping as a preprocessing stage for an image tracking system. In: Robotics Conferences, International Society for Optics and Photonics, pp. 15–22 (1989)

28. Brown, R.G., Hwang, P.Y., et al.: Introduction to Random Signals and Applied Kalman Filtering, vol. 3. Wiley, New York (1992)

Robust Maximum Margin Correlation Tracking

Han Wang[✉], Yancheng Bai, and Ming Tang

National Lab of Pattern Recognition,
Institute of Automation Chinese Academy of Sciences, Beijing 100190, China
{han.wang,ycbai,tangm}@nlpr.ia.ac.cn

Abstract. Recent decade has seen great interest in the use of discriminative classifiers for tracking. Most trackers, however, focus on correct classification between the target and background. Though it achieves good generalization performance, the highest score of the classifier may not correspond to the correct location of the object. And this will produce localization error. In this paper, we propose an online Maximum Margin Correlation Tracker (MMCT) which combines the design principle of Support Vector Machine (SVM) and the adaptive Correlation Filter (CF). In principle, bipartite classifier SVM is designed to offer good generalization, rather than accurate localization. In contrast, CF can provide accurate target location, but it is not explicitly designed to offer good generalization. Through incorporating SVM with CF, MMCT demonstrates good generalization as well as accurate localization. And because the appearance can be learned in Fourier domain, the computational burden is reduced significantly. Extensive experiments on public benchmark sequences have proven the superior performance of MMCT over many state-of-the-art tracking algorithms.

1 Introduction

Visual tracking is a significant problem in computer vision and it has been used in various applications such as automatic object identification, automated surveillance, vehicle navigation *et al.* Visual tracking has made great progress in the last decades and there are many different tracking approaches, such as kernel based tracking [1], particle filter based tracking [2], and tracking by detection [3]. However, designing a robust tracker is still a challenging problem, as the tracking results can be greatly influenced by moving out of plane, illumination changes, occlusion [4] *et al.*

Recently, tracking by detection has become a hot topic in single object tracking [3]. It stems directly from the offline training object detection methods, and it turned the offline training to online training to solve tracking problems.

Avidan [3] uses SVM to build a classifier separating the object from the background. The classifier uses offline training SVM integrated with optical flow algorithm to locate the object. But as the classifier is offline trained, the tracker can not adapt to the appearance changes of the object. In order to solve this problem, ensemble tracking [5] algorithm has been proposed. The algorithm collected positive and negative samples from the object and background regions

© Springer International Publishing Switzerland 2015
C.V. Jawahar and S. Shan (Eds.): ACCV 2014 Workshops, Part I, LNCS 9008, pp. 487–500, 2015.
DOI: 10.1007/978-3-319-16628-5_35

to train the weak classifiers, and used adaboost to select the most effective weak classifiers. A weighted sum of the selected classifiers presents the final strong tracker. As selecting appropriate positive and negative samples can influence the tracking results a lot, Babenko [6] propose a more robust algorithm based on multiple instance learning. The algorithm is more robust and have more fault tolerance as instead of receiving a set of instances which are labeled positive or negative, the learner receives a set of bags that are labeled positive or negative. As wrong labeling can be always occurred in tracking, Z. Lalal proposed [7] using 'P-N learning' to estimate the samples that are wrong labeled. The tracker utilizes P-*expert* to find the wrong labeled positive samples and N-*expert* to find the wrong labeled negative samples.

All of the aforementioned algorithms have one thing in common, in training process, they all regarded the tracking problem as a bipartite classification problem. This can severely influence the localization performance of the tracker. Assume in frame t, if we have a d-dimensional solution vector $\mathbf{w}$, correlating it with the image search patch, the peak of the response map can represent the object center. The ideal response map obtains a sharp correlation peak, which is centered at the object center. However, the response map of these trackers usually exhibits very broad peaks as they use binary labels for training. Broad peak will cause poor localization performance, as the top of the peak may be spread over several pixels thus can not correspond to the target center. Hare proposed structured SVM tracker [8] labeling every sample differently to improve the localization performance. The training process of all aforementioned trackers is calculated in spatial domain. Thus these trackers can not choose dense sampling strategy which will becomes computational burdens. Instead, they choose sparse sampling as shown in Fig. 1(a): positive samples are usually randomly collected in the target's neighbour, which can make the results severely influenced by the selection of samples.

Bolme [9] proposed the MOSSE tracker using the adaptive Correlation Filter for tracking. It uses dense sampling strategy, shown in Fig. 1(b). And as the center patch labels 1 and the value of labels degrades as the distance between the sample and the target center increases. This strategy can keep the structure of the target and localize accurately. As the model is computed in its Fourier domain, the computational burden can be reduced a lot. CF can generate sharp peaks and thus provide good localization performance, but they are not explicitly designed to offer good generalization.

While SVM are designed to maximize the margin of different classes, it usually has good generalization performance. In principle, combining the design of CF and SVM, Andres [10] proposed the MMCF, an offline training algorithm for object detection. The classifier has good generalization and localization performance rather than SVM and CF. And it can be processed in Fourier domain for fast training. But the MMCF is an offline training method, in tracking, it can not adapt to the appearance changes of the target.

In this paper, however, we propose the MMCT. This tracker integrates the design of CF and SVM, using the two criteria to build the objective function.

The tracker uses dense sampling strategy around the target, and away from the target, it randomly sample the negative patches. By using the SVM to constrain the coefficients of CF, the response map of the tracker can produce discriminative sharp peaks around the target center and small values away from the target center.

Different from MMCF, we achieve an online learning algorithm. Instead of using a weighted sum of models to update the tracker as many traditional tracker do, we import the previous model into the objective function. Through incorporating the last model into the SVM constraint, the tracker of consecutive frames can maintain continuity and at the same time achieves good generalization. This makes the tracking model more robust, and as the objective function can be processed into Fourier domain, the computational burden can be dramatically reduced.

The rest of the paper is organized as follows. In Sect. 1, we give a brief introduction for the MMCF and then present our algorithm in details. Experiments and the results of comparing with other state-of-art algorithms are shown in Sect. 2. In Sect. 3, we will summarize our work.

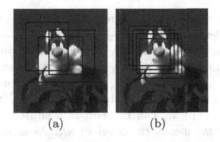

(a)   (b)

Fig. 1. This figure illustrates the sparse sampling strategy and dense sampling strategy. (a) illustrates the sparse sampling, it randomly samples p windows and save them; (b) illustrates the dense sampling, it samples all subwindows together and save one image.

1.1   Tracking Model

In this section, we will introduce our online tracking algorithm. In Sect. 2.1, we give a brief introduction of how CF works in tracking, and in Sect. 2.2, the offline training model used in object detection is represented. In Sect. 2.3, we introduce our online updating model and in Sect. 2.4, a detailed tracking strategy will be given.

1.2   The Adaptive Correlation Filter

As mentioned above, many traditional trackers use sparse sampling strategy, it means that several positive patches are randomly sampled around the object and all labeled 1. Obviously, there is a lot of redundancy because of the overlap between samples. Besides, as the labels of positive samples are all ones, it ignores the structure of the target, which can cause poor localization performance.

The Adaptive Correlation Filter [11] is firstly rooted on classical signal processing, and now widely used in localization and classification. It realizes dense sampling strategy around the object and at the same time, as it labels each sample differently, the model can present the structure of the target.

We start a general formulation to introduce the notation. First, we introduce the notation of circulant matrix [12]. If a matrix is circulant, means that if a n*n matrix $C(\mathbf{u})$ is extracted from the $n * 1$ vector $\mathbf{u}$ by concatenating all possible cyclic shifts of $\mathbf{u}$,

$$C(\mathbf{u}) = \begin{pmatrix} u_0 & u_1 & u_2 & \cdots & u_{n-1} \\ u_{n-1} & u_0 & u_1 & \cdots & u_{n-2} \\ u_{n-2} & u_{n-1} & u_0 & \cdots & u_{n-3} \\ \vdots & \vdots & \vdots & \ddots & \vdots \\ u_1 & u_2 & u_3 & \cdots & u_0 \end{pmatrix}. \tag{1}$$

Since the product $C(\mathbf{u})\mathbf{v}$ can be seen as the convolution of the two vectors $\mathbf{u}, \mathbf{v}$, we can compute it in Fourier domain, using

$$\widehat{C(\mathbf{u})\mathbf{v}} = \hat{\mathbf{u}}^* \odot \hat{\mathbf{v}} \tag{2}$$

where $\odot$ denotes the element-wise product, and $\hat{}$ denote Fourier transform, and * represents the complex-conjugate.

The dense sampling strategy at many subwindows in our paper is conceptually close to circulant matrix. In frame t, there are N target image patches from the last N frames $P_{t-N+1}, P_{t-1}, \ldots, P_{t-1}, P_t \in \mathbf{R}^{m*k}$. For each patch P_i, The dense sampling subwindows and their labels are $(\mathbf{x}_{i1}, y_{i1}), (\mathbf{x}_{i2}, y_{i2}), (\mathbf{x}_{ij}, y_{ij}) \cdots (\mathbf{x}_{id}, y_{id}), d = m * k$, where $\mathbf{x}_{ij}$ can be seen as a shifted vectorized version of image patch P_i, while y_{ij} means the label of $\mathbf{x}_{ij}$. As a linear classifier can be seen as $f(x) = \mathbf{w}^T * \mathbf{x} + b$, ignore the bias term b, just as [12] do, with quadratic loss, the objective minimization problem can be simply seen as

$$\min_{\mathbf{w}} \sum_{i=1}^{N} \|\mathbf{w}^T B_i - \mathbf{g}_i\|^2 \tag{3}$$

where $B_i = [\mathbf{x}_{i1}, \mathbf{x}_{i2}, ..., \mathbf{x}_{id}]$; $\mathbf{g}_i = [y_{i1}, y_{i2}, ..., y_{id}]^T$. Unlike traditional labeling strategy, in order to output sharp peaks, instead of using binary labels, the model uses a Gaussian function-like to represent $\mathbf{g}_i$ whose peak is at the object center. As the structure of B_i, the sampling subwindows $\mathbf{x}_{i1}, \mathbf{x}_{i2}, \ldots, \mathbf{x}_{id}$, is close to circulant matrix. So the Fourier transform of the Eq. 3 is as follows,

$$\min_{\hat{\mathbf{w}}} \sum_{i=1}^{N} \|\hat{\mathbf{w}} \odot \hat{\mathbf{x}}_i - \hat{\mathbf{g}}_i\|^2 \tag{4}$$

where $\hat{\mathbf{x}}_i$ is the vectorized version of 2-D Fourier transform of the image patch P_i. In tracking process, when the tracker $\mathbf{w}$ is correlated with the test image, the ideal response map $\mathbf{g}_t$ can obviously produce sharp peak to localize correctly. But as the tracker is not designed for classification, when the background clutters, it may not track well.

1.3   Offline Training Model

Trackers related to adaptive Correlation Filter are MOSSE [9] and Circulant [12] trackers, they have fast speed in tracking.

While in object detection, the SVM classifier is designed to maximize the margin and can always produce robust classifiers to classify the positive and negative samples. As the training samples are binary labeled, the output, which is resulting from cross-correlation of SVM templates with testing images, can not produce sharp peaks. As mentioned above, this will cause poor localization performance. Andres [10] propose an offline training object detection algorithm, MMCF. The MMCF uses two criteria combining the design of the SVM and CF. We first follow the notation in [10] to introduce the model.

The MMCF classifier is a multi-criteria classifier. The first criterion is SVM. Given N of training column vectors $\mathbf{x}_i \in \mathbb{R}^d$ and the class labels $t_i \in \{-1, 1\}$ $\forall i \in 1, \ldots, N$, the objective function of SVM can be expressed as follows,

$$\min_{\mathbf{w}, b} \mathbf{w}^T \mathbf{w} + C \sum_{i=1}^{N} \xi_i$$

$$s.t. \ t_i(\mathbf{w}^T \mathbf{x}_i + b) \geq c_i - \xi_i \qquad (5)$$

The second criterion is the CF, just as mentioned above, the objective function is $\min_{\mathbf{w}} \sum_{i=1}^{N} \|\mathbf{w}^T B_i - \mathbf{g}_i\|^2$, where $\mathbf{g}_i = [0, \ldots, 0, \mathbf{w}^T \mathbf{x}_i, 0, \ldots, 0]$, we prefer the center of the object is $\mathbf{w}^T \mathbf{x}_i$, while others close to 0. Combined with the SVM, the objective function can be seen as follows,

$$\min_{\mathbf{w}, b}(\mathbf{w}^T \mathbf{w} + C \sum_{i=1}^{N} \xi_i, \sum_{i=1}^{N} \|\mathbf{w}^T B_i - \mathbf{g}_i\|^2)$$

$$s.t. \ t_i(\mathbf{w}^T \mathbf{x}_i + b) \geq c_i - \xi_i \qquad (6)$$

where $c_i = 1$ for positive image patches and $c_i = \varepsilon$ for negative image patches, where ε is a small value constant. That means for positive image patchs, we expect a value above 1, while for negative patches, the expected value is close to 0. The large margin of SVM means good generalization performance, while the CF criterion makes sharper correlation peak. The objective function suggests a correlation response map, which has a sharp peak at the target center and small values everywhere else.

1.4   Online Tracking Model and Optimization

In tracking problems, we'd like to have dense sampling in the target center's neighbourhood to ensure good localization performance, at the same time, target should be separated from the background. We also need an online training strategy to adapt to the appearance changes of the object. Above these, we propose an online tracking model which can produce discriminative sharp peaks.

In our approach, suppose in frame t, after locating the object in $\mathbf{p}_t$, we extract the positive image patch P_t centered at $\mathbf{p}_t$ which has the same size with the target, and with the last k frames's k positive image patches, we have a positive training set $(P_t, P_{t-1}, \ldots, P_{t-k})$. To get the negative training set, we simply collected m patches away from the target in frame t, $(P_1, P_2, \ldots, P_m)$. We then train the online model w_{t+1} using the sample sets. Instead of using the simply weighted sum $\mathbf{w}_{t+1} = \mathbf{w}_t + \eta\mathbf{w}$ to update the model, where $\mathbf{w}$ is the trained model using current sample sets, we optimize $\|\mathbf{w}_{t+1} - \mathbf{w}_t\|^2$ in SVM criterion instead of $\|\mathbf{w}\|^2$ to keep the continuity between frames. Given $N = k + 2 + m$ of training column vectors $\mathbf{x}_i \in \mathbb{R}^d$ which is the vectorized version of P_i, and the class labels $t_i \in \{-1, 1\}$ $\forall i \in 1, \ldots, N$, the online tracking model can be expressed as follows,

$$\min_{\mathbf{w}_{t+1},b}(\|\mathbf{w}_{t+1} - \mathbf{w}_t\|^2 + C\sum_{i=1}^{N}\xi_i, \sum_{i=1}^{N}\|\mathbf{w}^T B_i - \mathbf{g}_i\|^2)$$
$$s.t.\ t_i(\mathbf{w}_{t+1}^T\mathbf{x}_i + b) \geq c_i - \xi_i \tag{7}$$

where $\mathbf{g}_i = [0, \ldots, 0, \mathbf{w}_{t+1}^T\mathbf{x}_i, 0, \ldots, 0]$, the nonzero value $\mathbf{w}_{t+1}^T\mathbf{x}_i$ is at the target center, and the other elements are all zeros. Just the same as the offline model, B_i represents the circulant matrix of $\mathbf{x}_i$, $c_i = 1$ for positive training set and $c_i = \varepsilon$ for negative training set. The objective function shows that in target center, we prefer a value of above 1, and the value decays to small values as the distance increases. The tracker uses dense sampling strategy around the target, so it can produce sharp peaks of the correlation output, and at the same time, using maximum margin to constrain the CF, the generalization performance improves a lot. With the $\|\mathbf{w}_{t+1} - \mathbf{w}_t\|^2$ constraint, the trackers of consecutive frames can maintain continuity.

In order to make use of the property that cross-correlation in the spatial domain is equivalent to multiplication in frequency domain, we transform Eq. 7 to its Fourier domain. We turn the SVM to the frequency domain by using the Parseval theorem. While the correlated part can be easily transformed to the Fourier domain as shown in Sect. 2.1. Then, Eq. 7 can be transformed as follows,

$$\min_{\hat{\mathbf{w}}_{t+1},b}(\|\hat{\mathbf{w}}_{t+1} - \hat{\mathbf{w}}_t\|^2 + C\sum_{i=1}^{N}\xi_i, \sum_{i=1}^{N}\|\hat{\mathbf{w}}_{t+1}^* \odot \hat{\mathbf{x}}_i - \hat{\mathbf{g}}_i\|^2)$$
$$s.t.\ t_i(\hat{\mathbf{w}}_{t+1}^\dagger\hat{\mathbf{x}}_i + b') \geq c_i - \xi_i \tag{8}$$

where $\dagger$ is the conjugate transpose. The multi-criteria function shown in Eq. 8 is formulated by two quadratic function, Refregier [13] showed that this can be optimized by minimizing a weighted sum of the two criteria, so it can be expressed as,

$$\min_{\hat{\mathbf{w}}_{t+1},b}\lambda\|\hat{\mathbf{w}}_{t+1} - \hat{\mathbf{w}}_t\|^2 + \lambda C\sum_{i=1}^{N}\xi_i + (1-\lambda)\sum_{i=1}^{N}\|\hat{\mathbf{w}}_{t+1} \odot \hat{\mathbf{x}}_i - \hat{\mathbf{g}}_i\|^2$$
$$s.t.\ t_i(\hat{\mathbf{w}}_{t+1}^\dagger\hat{\mathbf{x}}_i + b') \geq c_i - \xi_i \tag{9}$$

where λ represents the trades-off parameter between the margin criterion and localization criterion. When $\lambda = 1$ it equals to SVM tracker and vice versa.

For the second part, as $\mathbf{w}_{t+1}^T \mathbf{x}_i$ is the same as $\frac{1}{d}\hat{\mathbf{w}}_{t+1}^\dagger \hat{\mathbf{x}}_i$, using Pascal's theorem. The Fourier transform of $\mathbf{g}_i$ is as follows,

$$\hat{\mathbf{g}}_i = \mathbf{1} * (\frac{1}{d}\hat{\mathbf{x}}_i^\dagger \hat{\mathbf{w}}_{t+1}) \tag{10}$$

where $\mathbf{1}$ represents a column vector whose elements are all 1. using the diagonal matrix $\hat{X}_i \mathbf{1} = \hat{\mathbf{x}}_i$, then the right part of Eq. 8 can be expressed as follows,

$$\sum_{i=1}^{N} \|\hat{\mathbf{w}}_{t+1} \odot \hat{\mathbf{x}}_i - \hat{\mathbf{g}}_i\|^2 = \sum_{i=1}^{N} \hat{\mathbf{w}}_{t+1}^\dagger \hat{X}_i \hat{X}_i^* \hat{\mathbf{w}}_{t+1} - \frac{2}{d}\hat{\mathbf{w}}_{t+1}^\dagger \hat{X}_i \hat{\mathbf{g}}_i + \frac{1}{d^2}\hat{\mathbf{g}}_i^\dagger \hat{\mathbf{g}}_i$$

$$= \sum_{i=1}^{N} \hat{\mathbf{w}}_{t+1}^\dagger \hat{X}_i \hat{X}_i^* \hat{\mathbf{w}}_{t+1} - \frac{2}{d}\hat{\mathbf{w}}_{t+1}^\dagger \hat{X}_i \mathbf{1}\hat{\mathbf{x}}_i^\dagger \hat{\mathbf{w}}_{t+1} + \frac{1}{d^2}\hat{\mathbf{w}}_{t+1}^\dagger \hat{\mathbf{x}}_i \mathbf{1}^\dagger \mathbf{1}\hat{\mathbf{x}}_i^\dagger \hat{\mathbf{w}}_{t+1})$$

$$= \hat{\mathbf{w}}_{t+1}^\dagger \hat{Z}\mathbf{w}_{t+1} \tag{11}$$

where

$$\hat{Z} = \sum_{i=1}^{N}(\hat{X}_i \hat{X}_i^* - \frac{1}{d}\hat{\mathbf{x}}_i \hat{\mathbf{x}}_i^\dagger) \tag{12}$$

Subsume Eq. 11 into Eq. 9, we can rewrite Eq. 9 as follows,

$$min_{\hat{\mathbf{w}}_{t+1},b}\lambda\|\hat{\mathbf{w}}_{t+1} - \hat{\mathbf{w}}_t\|^2 + \lambda C \sum_{i=1}^{N} \xi_i + (1-\lambda)\hat{\mathbf{w}}_{t+1}^\dagger \hat{Z}\hat{\mathbf{w}}_{t+1}$$

$$s.t.\ t_i(\hat{\mathbf{w}}_{t+1}^\dagger \hat{\mathbf{x}}_i + b') \geq c_i - \xi_i \tag{13}$$

With one quadratic term subsumed into the other quadratic term, Eq. 13 can be rewritten as follows,

$$min_{\hat{\mathbf{w}}_{t+1},b}\hat{\mathbf{w}}_{t+1}^\dagger \hat{S}\hat{\mathbf{w}}_{t+1} + \lambda C \sum_{i=1}^{N} \xi_i - 2\lambda\hat{\mathbf{w}}_{t+1}^\dagger \hat{\mathbf{w}}_t$$

$$s.t.\ t_i(\hat{\mathbf{w}}_{t+1}^\dagger \hat{\mathbf{x}}_i + b') \geq c_i - \xi_i \tag{14}$$

where $\hat{S} = \lambda I + (1-\lambda)\hat{Z}$, as $0 < \lambda < 1$, $\hat{S}$ is positive definite matrix. And we can transform the data that $\tilde{\mathbf{w}} = \hat{S}^{\frac{1}{2}}\hat{\mathbf{w}}$ and $\tilde{\mathbf{x}}_i = \hat{S}^{-\frac{1}{2}}\hat{\mathbf{x}}_i$. So we can easily compute the dual form of Eq. 14,

$$\min_{\mathbf{a}} \mathbf{a}^T T \tilde{X}^\dagger \tilde{X} T \mathbf{a} + (\mathbf{c}^T - 2\tilde{X}^\dagger T \tilde{\mathbf{w}}_t)\mathbf{a}$$

$$s.t.\ \mathbf{0} \leq \mathbf{a} \leq \mathbf{1}C',\ \mathbf{a}^T \mathbf{t} = 0 \tag{15}$$

where $\tilde{X} = [\tilde{\mathbf{x}}_1,\ldots,\tilde{\mathbf{x}}_N]$, $\mathbf{t} = [t_1,\ldots,t_N]^T$, $\mathbf{c} = [c_1,\ldots,c_N]^T$, $C' = \lambda C$, and T is the diagonal matrix with $\mathbf{t}$ along the diagonal. With the dual form, we can optimize $\mathbf{a}$ using Sequential minimal optimization (SMO) [14]. SMO breaks this problem into many subproblems that each problem solve for one nonoverlap pair of $\mathbf{a} = [a_1, \ldots, a_N]^T$. It recursively solves for $\mathbf{a}$ until convergence, and after solving for $\mathbf{a}$, the tracking model $\hat{w}_{t+1}$ can be computed as follows,

$$\hat{\mathbf{w}}_{t+1} = \hat{S}^{-\frac{1}{2}}\widetilde{X}\mathbf{a} \tag{16}$$

Here as $\hat{S}$ is not a diagonal matrix, so when computing the inverse of the matrix, it is very computationally expensive. As the target patch dimension d is always very large, so we can approximate $\hat{S}$ as,

$$\hat{S} = \lambda I + (1 - \lambda)\hat{Z} = \lambda I + (1 - \lambda)\sum_{i=1}^{N}(\hat{X}_i\hat{X}_i^* - \frac{1}{d}\hat{\mathbf{x}}_i\hat{\mathbf{x}}_i^\dagger)$$

$$\approx \lambda I + (1 - \lambda)\sum_{i=1}^{N}\hat{X}_i\hat{X}_i^* \tag{17}$$

As the objective function can be processed in Fourier domain, the computational burden can be significantly reduced. And with the larger λ, the stronger generalization performance and smaller λ can make the model output sharper peak.

1.5   Tracking Process

In frame t, given the model $\mathbf{w}_t$ and $\mathbf{p}_{t-1}$ the center of frame t-1, the prediction process is to find the new target center $\mathbf{p}_t$. We cropped the search patch 1.5 times as big as the target, centered at $\mathbf{p}_{t-1}$ in frame t, and correlate it with w_t, get the correlation response map $\mathbf{g}_t$. The strength of the peak of $\mathbf{g}_t$ can be measured by the Peak to Sidelobe Ratio(PSR) [9]. To compute the PSR, we first divide the response map $\mathbf{g}_t$ into two portions. The peak represents the maximum value of the response map and the sidelobe is the rest of the pixels excluding an $11 * 11$ window around the peak. The PSR can be computed as $\frac{g_{max} - \mu_s}{\sigma_s}$, where g_{max} is the peak value, and μ_s, σ_s are the average and standard deviation of the sidelobe. The PSR can be used to detect the object occlusion or tracking failure. If PSR is smaller than 6 (experience in our experiment), the target is supposed to be missing, and we will search the whole image and stop updating the model. Algorithm 1 summarizes our tracking algorithm.

2   Experiments

We evaluated our tracking system on twelve challenging videos, all of the videos come from the benchmark. These videos contain many kinds of objects (car, pedestrian, human body, faces animals et al).

The proposed algorithm is implemented in MATLAB on a workstation with an Intel core i5 3.2 GHz processor and 16 G RAM. The average pruning time is 3–4 frames per second.

Algorithm 1. Maximum Margin Correlation Tracker.

Require:
    The current frame image, F_t;
    The current model, $\mathbf{w}_t$;
    The object center of frame t-1, $\mathbf{p}_{t-1}$;

Ensure:
    The new tracker $\mathbf{w}_{t+1}$ and the object center of frame t $\mathbf{p}_t$;
1: Cropped the current search image patch S_t centered at $\mathbf{p}_{t-1}$;
2: Getting the current response map, using 2-D cross-correlation, where $\mathbf{x}_t$ is the vectorized version of S_t: $\mathbf{g}_t = \mathbf{x}_t \otimes \mathbf{w}_t$;
3: Compute the PSR of $\mathbf{g}_t$, and get the object center p_t according to Sect. 2.4;
4: Using PSR to estimate if the object is occluded, according to Sect. 2.4;
5: Sample the positive image patch P_1 in frame t and add it to positive templates. Sample negative patches $P_2, \ldots, P_n$;
6: Using the positive templates, the negative patches, and $\mathbf{w}_t$ to update the model $\mathbf{w}_{t+1}$ according to Sect. 2.3;
7: **return** $\mathbf{w}_{t+1}, \mathbf{p}_t$;

In all of the experiments, the parameters are all fixed. In the training stage, we sample 40 negative image patches within 30 pixels away from the bounding box of the target. And the positive patches are collected from the last 10 frames' target patches. And we choose $\lambda = 0.15$ to balance the correlation results. We will make our experiments in two ways. First, we compare our algorithm with the related algorithms. And then Comparison with other state-of-the-art algorithms are made.

2.1 Pre-processing

The proposed method uses Fourier transform in training process. As Fast Fourier Transform(FFT) is periodic, it is very sensitive to the image boundary. A noisy Fourier representation can be generated if there exists big discontinuity between opposite edges of the images. The effect can be reduced by multiplying a hanning window with the image to gradually reduce the training patches to zero.

2.2 Comparison with MOSSE and SVM

To demonstrate the improvements of our approach in localization and generalization, we first make a experiment comparing our algorithm with the SVM, and also the MOSSE tracker.

For generality, there are many kinds of objects (human body, face, rigid object and toy). And the mean center position error per frame is used as criterion. Table 1 shows the quantitative performance of these algorithms.

It can be seen that the proposed method outperforms other trackers. Figure 3(a–c) shows the results of some typical videos under difficult situations. In the experiment, we also find the Moose filter is sensitive to the initialization bounding box, if the initialization bounding box included much background information, the tracking result can be severally influenced by background, And in SVM tracker, tracking results can be improved by increasing positive and negative samples, but this can increase the computational burden. It can be seen in Fig. 3(a) that under pose changes, MOSSE and our tracker can localize the object correctly, but SVM tracker drifts. Figure 3(b–c) shows with scale changes and out of plane rotation changes, our tracker performs well compared with others.

2.3   Comparison with Other Trackers

In this section, we compare the proposed tracker with other 5 state-of-the-art trackers (the tracking results of them are provided by the benchmark), including the TLD [7], Struck [8], MIL [6], L1APG [15], MTT [16] trackers.

2.4   Quantitative Evaluation

We evaluate the performance of these trackers using the center location error. Table 2 reports the average center location errors in pixels. It can be seen that under different situations, our tracker can locate accurately, it always performs best or second best. Figure 2 shows the tracking results of different trackers.

Table 1. The average center location error of twelve sequences is the distances between the tracking results center and the ground truthes of them. The bold represents for the best tracker.

Sequences	bolt	cardark	Suv	football1	freeman3	sylvster	Trellis	Woman	david	deer	dog1	faceocc2
MOSSE	30	4.3	42	22.6	15	10.8	11.9	16.6	12	7.5	10.3	14.6
SVM	200	6.9	50	49	50	31.7	13.3	71.4	53.5	23	7.2	14.2
Ours	**24**	**2.2**	**4.9**	**14**	**12**	**8**	**6.4**	**10**	**10.1**	**5.8**	**4.8**	**10**

Table 2. Compared average center error(pixels)on twelve sequences. The **bold** represents for the best tracker, and *italic* for the second best.

Sequences	bolt	cardark	SUV	football1	freeman3	Sylvster	Trellis	Woman	david	deer	dog1	faceocc2
TLD	*231*	35	56	9.7	14	77	27	*11*	34	7	22	*18.7*
Struck	250	*3.9*	41	**13**	12	*26*	**6.4**	12.2	*15*	7	*11*	58
MIL	286	48	*12*	32	19	90	135	27	73	9	21	83
L1APG	283	25.2	15	22	30	40	165	71	345	11	14	101.5
MTT	278	20.7	17	18	*12*	58	170	25	350	9	16	119
Ours	**24**	**2.2**	**4.9**	*14*	**12**	**8**	**6.4**	**10**	**10.1**	**5.8**	**4.8**	**10**

2.5   Qualitative Evaluation

Illumination, pose and Scale changes. We evaluate sequences with different kinds of illumination changes. The *david* and *Trellis* contain gradual illumination, pose and scale changes. We can see from Fig. 2 that under illumination changes (e.g. Trellis #51, #228) only our tracker and Struck tracker can

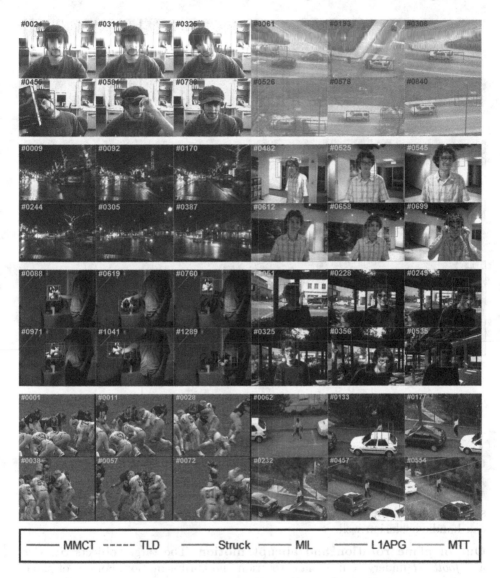

Fig. 2. Tracking results on six of the twelve videos(*faceocc2, SUV, cardark, david, sylvster, football1, Threllis, Woman*).

locate the object accurately, other trackers have drifts to some extent. And when the pose changes a lot (e.g. *Trellis* #356), only our tracker performs well. In the sequence *david*, when the scale changes a lot (e.g. #482, #525), only the proposed algorithm is able to track the object accurately. This can be attributed to that we design a shape peak for the center of the target. So even the object's appearance changes, we not only can classify the object, but also find its accurate center.

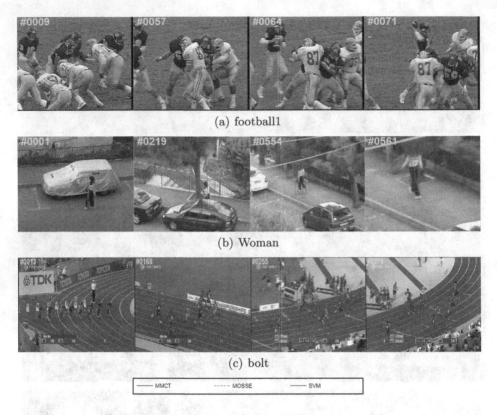

(a) football1

(b) Woman

(c) bolt

———— MMCT   ----- MOSSE   ———— SVM

Fig. 3. The tracking results of different trackers

Occlusion. The target objects are partially occluded in the *Women, Occluded face 2, SUV* sequences. When the target is severely occluded (e.g. *SUV* #526, *Woman* #133), our tracker can still perform well. By using dense sampling strategy around the target, the spatial information is maintained a lot and thus can handle occlusion well.

Out of plane rotation and abrupt motion. The target objects *Sylvster* and *football1* undergo out of plane rotation and abrupt motion. For out of plane rotation (e.g. *Sylvster*#0619, #1041), Most trackers except the proposed tracker and the **Struck** method drift. For abrupt motion and rotation out of plane (e.g. *Football1* #0038), our tracker and TLD perform well.

Background clutters. In the *football1* and *cardark* sequences, the target object undergoes fast movements in cluttered backgrounds. Our tracker performs well where as others fail to locate the target object.

3   Conclusion

In this work, we present a new adaptive tracking-by-detection method based on adaptive correlation filter and the SVM. Unlike existing method using sparse sampling strategy and focusing on classification, thinking from the intension of tracking, localizing the target, we build the model getting good performance in localization and also separate the target from the clutter background. And we transform it to Fourier domain for fast training using FFT. And in training, we do not simply use the weighted sum of the model, which are computed from different frames, representing the new model. But we change the SVM objective criteria to both adapt to new samples and also keep consistence with previous model. Through experiments on public benchmark sequences, we also clearly demonstrated that our algorithm can track objects very well under large pose, scale variation, occlusion and cluttered background. And our MMCT can almost always outperform the state-of-the-art algorithms.

References

1. Comaniciu, D., Ramesh, V., Meer, P.: Real-time tracking of non-rigid objects using mean shift. In: Proceedings of the IEEE Conference on Computer Vision and Pattern Recognition, vol. 2, pp. 142–149. IEEE (2000)
2. Blake, A., Isard, M., Reynard, D.: Learning to track the visual motion of contours. Artif. Intell. **78**, 179–212 (1995)
3. Avidan, S.: Support vector tracking. In: Proceedings of the 2001 IEEE Computer Society Conference on Computer Vision and Pattern Recognition, CVPR 2001, vol. 1, pp. I-184–I-191 (2001)
4. Wu, Y., Lim, J., Yang, M.H.: Online object tracking: a benchmark. In: 2013 IEEE Conference on Computer Vision and Pattern Recognition (CVPR), pp. 2411–2418. IEEE (2013)
5. Avidan, S.: Ensemble tracking. IEEE Trans. Pattern Anal. Mach. Intell. **29**, 261–271 (2007)
6. Babenko, B., Yang, M.H., Belongie, S.: Visual tracking with online multiple instance learning. In: IEEE Conference on Computer Vision and Pattern Recognition, CVPR 2009, PP. 983–990 (2009)
7. Kalal, Z., Matas, J., Mikolajczyk, K.: Pn learning: bootstrapping binary classifiers by structural constraints. In: 2010 IEEE Conference on Computer Vision and Pattern Recognition (CVPR), PP. 49–56. IEEE (2010)
8. Hare, S., Saffari, A., Torr, P.H.: Struck: structured output tracking with kernels. In: 2011 IEEE International Conference on Computer Vision (ICCV), PP. 263–270. IEEE (2011)
9. Bolme, D.S., Beveridge, J.R., Draper, B.A., Lui, Y.M.: Visual object tracking using adaptive correlation filters, pp. 2544–2550 (2010)
10. Rodriguez, A., Boddeti, V.N., Kumar, B.V., Mahalanobis, A.: Maximum margin correlation filter: a new approach for localization and classification. IEEE Trans. Image Process. **22**, 631–643 (2013)
11. Bolme, D., Draper, B., Beveridge, J.: Average of synthetic exact filters. In: IEEE Conference on Computer Vision and Pattern Recognition, CVPR 2009, PP. 2105–2112 (2009)
12. Henriques, J.F., Caseiro, R., Martins, P., Batista, J.: Exploiting the circulant structure of tracking-by-detection with kernels. In: Fitzgibbon, A., Lazebnik, S., Perona, P., Sato, Y., Schmid, C. (eds.) ECCV 2012, Part IV. LNCS, vol. 7575, pp. 702–715. Springer, Heidelberg (2012)

13. Réfrégier, P.: Filter design for optical pattern recognition: multicriteria optimization approach. Opt. Lett. **15**, 854–856 (1990)
14. Platt, J., et al.: Sequential Minimal Optimization: A Fast Algorithm for Training Support Vector Machines. MIT Press, Cambridge (1998)
15. Bao, C., Wu, Y., Ling, H., Ji, H.: Real time robust l1 tracker using accelerated proximal gradient approach. In: 2012 IEEE Conference on Computer Vision and Pattern Recognition (CVPR), pp. 1830–1837. IEEE (2012)
16. Zhang, T., Ghanem, B., Liu, S., Ahuja, N.: Robust visual tracking via multi-task sparse learning. In: 2012 IEEE Conference on Computer Vision and Pattern Recognition (CVPR), pp. 2042–2049. IEEE (2012)

Scene Classification by Feature Co-occurrence Matrix

Haitao Lang[1,2]($\boxtimes$), Yuyang Xi[1], Jianying Hu[1], Liang Du[2], and Haibin Ling[2]

[1] Department of Physics and Electronics, Beijing Key Laboratory
of Environmentally Harmful Chemicals Analysis, Beijing University of Chemical
Technology, Beijing 100029, China
langht@mail.buct.edu.cn
[2] Department of Computer and Information Sciences, Temple University,
Philadelphia 19122, USA

Abstract. Classifying scenes (such as mountains, forests) is not an easy task owing to their variability, ambiguity, and the wide range of illumination and scale conditions that may apply. Bag of features (BoF) model have achieved impressive performances in many famous databases (such as the *15 scene* dataset). A main drawback of the BoF model is it disregards all information about the spatial layout of the features, leads to a limited descriptive ability. In this paper, we use co-occurrence matrix to implant the spatial relations between local features, and demonstrate that feature co-occurrence matrix (FCM) is a potential discriminative character to scenes classification. We propose three FCM based image representations for scenes classification. The experimental results show that, under equal protocol, the proposed method outperforms BoF model and Spatial Pyramid (SP) model and achieves a comparable performance to the state-of-the-art.

1 Introduction

Classifying scenes into semantic categories is a problem of great interest in both research and practice. For example, an online collection of photos needs to be grouped into categories like 'coast', 'highway', and 'office' to support efficient browsing and/or retrieval tasks. At the same time, scene classification is not an easy task owing to their variability, ambiguity, and the wide range of illumination and scale conditions that may apply.

Recently, there is a trend of using low-level image features in classification of imagery data [1–3]. The development and analysis of low-level feature descriptors have been widely considered in the past years. Among the vastly employed methods are the scale-invariant feature transform(SIFT) [4], speeded up robust feature(SURF) [5], histogram of oriented gradients(HOG) [6], gradient location and orientation histogram(GLOH) [7], region covariance matrix(RCM) [8], local binary patterns(LBP) [9] etc. How to organize these local features to construct an robust image descriptor is crucial to the performance of scene classification. The most popular image representation is the bag of features(BoF) [1], which

© Springer International Publishing Switzerland 2015
C.V. Jawahar and S. Shan (Eds.): ACCV 2014 Workshops, Part I, LNCS 9008, pp. 501–510, 2015.
DOI: 10.1007/978-3-319-16628-5_36

describes an image by the overall distribution of low level features. Traditional BoF framework equally encodes all local features and does not emphasize any elements with regard to spatial layout. Hence, spatial pyramid (SP) structure representation is often used to extend the global BoF representation. SP model [2] approximates geometric layout of local features by partitioning the image plane into increasingly fine sub-regions. Due to its better performance and simple implementation, it has become a standard procedure for scene classification.

In this paper, we investigate the relationship between spatial layout of local features and the scene categories. It is shown that when using feature co-occurrence matrix(FCM) to map the original scene image from gray space to features distribution space, from a statistical point, there is an explicit difference between scene categories. Based on this observation, we propose to use co-occurrence matrix to extend the orderless BoF representation and construct three FCM based image representations. We evaluate the proposed method and compared it with original BoF model and SP model on 15 scene database with equal experimental protocol. The experimental results show that the proposed method outperforms BoF model and SP model and achieves a comparable performance to the state-of-the-art. The proposed method is a good alternative to image representation for scene classification tasks.

The remaining of this paper is organized as follows. We briefly review the related works on BoF and its extension models and co-occurrence matrix in Sect. 2. Then we introduce the proposed FCM method and local features used in our work in Sect. 3. In Sect. 4, we propose three FCM based image representation methods. The evaluation to our method and comparisons to others are described in Sect. 5. Finally, we conclude the paper in Sect. 6.

2   Related Works

2.1   Bag of Features Model and Its Extensions

State-of-the-art methods following the bag of features (BoF) framework mainly contain four steps: (1) local feature extraction and description, (2) feature coding/encoding, (3) feature pooling and (4) classifier learning.

The local features are firstly extracted by densely or randomly sampling, or sparse keypoints detector(such as Harris detector [10], scale and affine invariant detector [11] *etc*). SIFT [4], GLOH [7], HOG [6] *etc.* are usually used to build a descriptor to local interesting points. In "coding" step, a clustering method (such as k-means clustering [12]) is conducted over all descriptors to obtain a vocabulary (codebook). "encoding" procedure deals with how to use one or multiple codes from codebook to represent a new descriptor. Hard voting [13], soft voting [14] and reconstruction based methods (LCC [15], sparse coding [16] *etc.*) are three typical methods. In pooling step, the quantization indices of all the local features are summarized to form the global image representation. Histogram is a typical average pooling strategy, which sums up all the occurrences of each index throughout the entire image in an orderless manner. Instead of performing averaging operation, max pooling adopts the element wise maximum values of feature vectors over the whole image as the pooled features. The classifier learning step generally uses the kernel built on matching scores of the global image representations.

To overcome the loss of spatial information in original BoF model, Lazebnik *et al.* [2] propose the spatial pyramid matching (SPM) model. The image is subdivided at three different levels of resolution. For each level of resolution, the features falling in each sub-region (bin) are counted. Finally, each spatial histogram is weighted according to:

$$\kappa^L(X,Y) = \mathcal{I}^L + \sum_{\ell=0}^{L-1} \frac{1}{2^{L-\ell}}(\mathcal{I}^\ell - \mathcal{I}^{\ell+1}) \tag{1}$$

The success of spatial pyramid representation comes from the valid assumption that the images with similar scene and geometry layout possibly belong to the same category. While due to there exists large intra-class variation of same scene categories as well as significant inter-class similarity between different scene categories. In many visual classification tasks, the spatial distribution of discriminative information is non uniform. Thus different parts of image should serve different roles for scene classification. Sharma *et al.* [17] use the saliency maps to weight the corresponding visual features improves the discriminative power of the image representation. Chen *et al.* [18] introduces so-called side information (i.e., prior knowledge such as clues of object layout) for image classification based on BoF representation. Using the side information, the image local feature pool can be clustered into cells and further a coarse to fine hierarchical representation can be generated. Since the partition of the cells is guided with side information more semantically concerned, the encoding within each cell tends to be more semantically matchable and thus is expected to achieve better performance.

2.2   Co-occurrence Matrix

Co-occurrence matrix is essentially a two-dimensional histogram in which the (i,j)th element of the matrix $\mathbf{M}$ is the frequency of *event i* co-occurs with *event j*. Here "event" can be a pixel value and also can be a specific low level feature of image. In texture classification community, gray level co-occurrence matrix (GLCM) is firstly introduced by Haralick [19]. A GLCM is specified by the relative frequencies $\mathbf{M}(i,j,d,\theta)$ in which two pixels, separated by distance d, occur in a direction specified by the angle θ, one with gray level i and the other with gray level j:

$$\mathbf{M}(i,j,d,\theta) = \sum_{p=1}^{I_y}\sum_{q=1}^{I_x} I(p,q) = i \quad and \quad I(p+d_y(\theta), q+d_x(\theta)) = i$$

$$if \quad \theta = 0, \quad d_y = 0 \quad and \quad |d_x| = d$$
$$if \quad \theta = 45, \quad d_x = d_y = d \quad or \quad d_x = d_y = -d$$
$$if \quad \theta = 90, \quad d_x = 0 \quad and \quad |d_y| = d$$
$$if \quad \theta = 135, \quad d_x = -d, \quad d_y = d \quad or \quad d_x = d, \quad d_y = -d \tag{2}$$

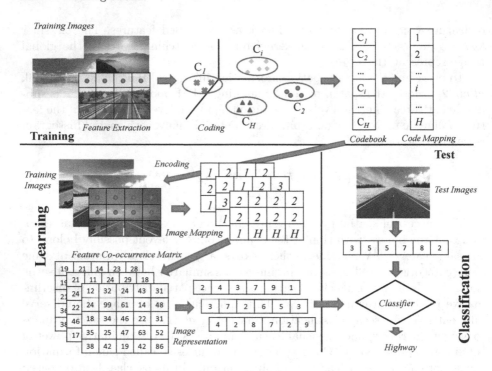

Fig. 1. Toy example of FCM based scene classification framework.

where I_y and I_x represents the row and column number of the image I, $I(p,q)$ is pixel gray value in $p-th$ row and $q-th$ column. i and j is the gray level with the maximum H.

3   Feature Co-occurrence Matrix

3.1   Feature Co-occurrence Matrix

The framework to use FCM method to conduct scene classification is shown in Fig. 1. In training phase, we first use standard BoF method to construct a codebook. Then we build a mapping from codes(visual words) to numerical indexes, *i.e.* we assign a specific number to each code. The maximum number is the size of the vocabulary. By this way, we can calculate the co-occurrence relationship between codes simply. In learning phase, low level features are extracted from each image firstly, then are encoded according to the codebook. Then the image pixel value where the feature is extracted is replaced by the number which represent the corresponding code. By this way, an image is mapped from the gray/color space to code index space. After that, we compute the FCM according to Eq. 2. Once the frequency of each code transition is computed, a normalization is conducted to $\mathbf{M}(i,j,d,\theta)$ based on:

$$M^*(i,j,d,\theta) = \frac{M(i,j,d,\theta)}{\sum_{m=0}^{H}\sum_{n=0}^{H}M(m,n,d,\theta)} \tag{9}$$

where H denotes the size of codebook.

Due to the similarity between two matrix is hard to evaluate, we need transform the FOM to a vector as the image representation. In this paper, we introduce three methods to conduct this operation. The details is described in Sect. 4.

3.2 Discrimination of FCM for Scenes Classification

We investigate the discriminative ability of FCM to different scene categories. As an instance, we plot the mean FCM of four scenes from 15 scene dataset in Fig. 2. From the heat maps, we find when describing scenes with FCMs, there are distinct statistical differences between categories. Based on this observation, we argue that FCM is a potential discriminative features to classify scenes (Table 1).

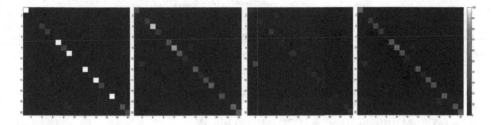

Fig. 2. FCMs of four scene categories, from left to right is coast, bedroom, forest, and kitchen respectively. SIFT is used as low level feature. The codebook size is 200. To obtain better visualization, we segment a subregion, i.e., the up-left corner of whole FCM with the area of 20 × 20.

4   FCM Based Image Representation

We propose three strategies to build an image representation based on FCM.

4.1   Image Representation by Unfolding FCM

A simple method to construct an image representation with FCM is unfolding the matrix to a vector. Considering the dictionary capacity up to hundreds of thousands of levels, we use PCA to reduce the dimension of original unfolded vector to a reasonable level, e.g. 256 etc. Then the compact vectors extracted from four FCMs are concatenated in a single feature vector as the final image representation.

Table 1. Haralick's Statistical Properties of GLCM

Name	Formula
Angular	$f_1 = \sum_i \sum_j \mathbf{M}(i,j)^2$
Contrast	$f_2 = \sum_{n=0}^{H-1} n^2 \sum_{i=1}^{H} \sum_{j=1}^{H} \mathbf{M}(i,j)$
Correlation	$f_3 = \frac{\sum_i \sum_j (i,j)\mathbf{M}(i,j) - \mu_x \mu_y}{\sigma_x \sigma_y}$
Variance	$f_4 = \sum_i \sum_j (i - \mu)^2 \mathbf{M}(i,j)$
Inverse difference moment	$f_5 = \sum_i \sum_j \frac{1}{1+(i-j)^2} \mathbf{M}(i,j)$
Sum average	$f_6 = \sum_{k=2}^{2H} k\mathbf{M}(k)$
Sum variance	$f_7 = \sum_{k=2}^{2H} [(k - f_6)^2 \mathbf{M}_{x+y}(k)]$
Sum entropy	$f_8 = -\sum_{k=2}^{2H} \mathbf{M}_{x+y}(k) \log[\mathbf{M}_{x+y}(k)]$
Entropy	$f_9 = -\sum_i \sum_j \mathbf{M}(i,j) \log[\mathbf{M}(i,j)]$
Difference variance	$f_{10} = \sum_{k=0}^{H-1} [k - \sum_{l=0}^{H-1} l\mathbf{M}_{x-y}(l)]^2 \mathbf{M}_{x-y}(k)$
Difference entropy	$f_{11} = -\sum_{k=0}^{H-1} \mathbf{M}_{x-y}(k) \log[\mathbf{M}_{x-y}(k)]$
Measure of correlation 1	$f_{12} = \frac{f_9 + \sum_{i=1}^{H} \sum_{j=1}^{H} \mathbf{M}(i,j) \log[\mathbf{M}(i)\mathbf{M}(j)]}{\max(HX, HY)}$
Measure of correlation 2	$f_{13} = \sqrt{1 - \exp[2(\sum_{i=1}^{H} \sum_{j=1}^{H} \mathbf{M}(i)\mathbf{M}(j) \log[\mathbf{M}(i)\mathbf{M}(j)] + f_9)]}$
Maximal correlation coefficient	$f_{14} = \sqrt{\langle(\sum_{k=1}^{H} \frac{\mathbf{M}(i,k)\mathbf{M}(j,k)}{\mathbf{M}_x(i)\mathbf{M}_y(j)})\rangle_2}$

Abbreviations:

$\mathbf{M}(i,j)$: (i,h)th entry in $\mathbf{M}$, H: dimension of $\mathbf{M}$, μ is the mean of μ_x and μ_y

$\mathbf{M}(x)_i = \sum_{j=1}^{H} \mathbf{M}(i,j)$, $\mathbf{M}(y)_j = \sum_{i=1}^{H} \mathbf{M}(i,j)$, $\mu_x = \sum_{i=1}^{H} i\mathbf{M}_x(i)$, $\mu_y = \sum_{i=1}^{H} i\mathbf{M}_y(i)$

$\sigma_x = \sqrt{\sum_{i=1}^{H} \mathbf{M}_x(i)(i - \mu_x)^2}$, $\sigma_y = \sqrt{\sum_{i=1}^{H} \mathbf{M}_y(i)(i - \mu_y)^2}$

$\mathbf{M}_{x+y}(k) = \sum_{i=1}^{H} \sum_{j=1, i+j=k}^{H} \mathbf{M}(i,j)$, $\mathbf{M}_{x-y}(k) = \sum_{i=1}^{H} \sum_{j=1, |i-j|=k}^{H} \mathbf{M}(i,j)$

$\langle \cdot \rangle$ denotes 2nd largest eigenvalue

4.2 Image Representation by Properties of FCM

The second method is use the properties contained in the co-occurrence matrices to construct the image representation. In this paper, we use Haralick' [19] 14 statistical properties computed from the co-occurrence matrices i.e., (1) angular second moment, (2) contrast, (3) correlation, (4) sum of squares, (5) inverse difference moment, (6) sum average, (7) sum variance, (8) sum entropy, (9) entropy, (10) difference variance, (11) difference entropy, (12–13) two information measures of correlation, and (14) maximal correlation coefficient. Once the properties extracted from four directional FCMs, we concatenate them in a single image representation vector.

4.3 Image Representation by Singular Value of FCM

In this method, we conduct a singular value decomposition (SVD) to FCM. Formally, the SVD of an $H \times H$ real feature co-occurrence matrix M is a factorization of the form:

$$M = U\Sigma V^* \tag{4}$$

where Σ is an $m \times m$ rectangular diagonal matrix with nonnegative real numbers on the diagonal. The diagonal entries $\Sigma_{i,i}$ of Σ are known as the singular values of M. The non-zero singular values of M are the square roots of the non-zero eigenvalues of both M^*M and MM^*. To construct a image descriptor, we combine all diagonal entries of Σ corresponding to four FCMs.

5   Experiments

5.1   Databset

Scene 15[1] is a dataset containing 15 scene categories, e.g. 'coast', 'beach', 'office', with 4485 images. The task is multi-class classification with the dataset split into 100 random images per class for training and the rest for testing.

5.2   Local Visual Features

Low level feature has significant effect on total performance of algorithm. In this paper, we evaluate and compare three features, raw gray value, SIFT [4] and LBP [9]. For color images, we first transform it to gray image, then use above algorithms to build descriptors to all sampling points in a single gray channel. For SIFT, we use the square root of normalized 128 dimensional descriptor. For LBP, a standard 256-bin histogram is used as a feature descriptor.

5.3   Experimental Protocol

Here we present some details of our experiments. In our experiments, the local features are extracted by dense sampling, the sampling interval in both row and column direction is set to 8 pixel. For gray feature, we use the average of neighboring 4×4 pixels around the sampling point as the descriptor. For SIFT, we use neighboring 16×16 pixels to describe the sampling point. While for LBP, we use neighboring 3×3 pixels to build the descriptor. In "coding" step, k-means clustering is used to construct codebook, while in "encoding" step, minimizing Euclidean distance based hard voting method is used. In this paper, the size of codebook is fixed as 256. For "pooling" step, we compare the proposed method with two baselines, i.e., original BoF model and SP model(three levels). To compute FCM, we use $0, 45, 90, 135$ four transition directions and $d = 1$ as step interval. When image is represented by unfolding FCMs, we use PCA to reduce the dimension of descriptor from $4 \times 200 \times 200$ to 256. When using FCM's properties to represent the image, we build a 4×14 dimensional descriptor. When using SVD to FCM, we build a 4×200 dimensional descriptor. RBF kernel based SVM [20] is used to learn classifier. We conduct 15 binary one-vs-rest classification problems to solve multi-class task. All experiments are repeated 10 times and the mean and standard deviation are reported.

---

[1] http://www-cvr.ai.uiuc.edu/ponce_grp/data/.

Table 2. Evaluation of different methods

Feature	Baseline		FCM		
	BoF	SP	Unfold	Property	SVD
gray	29.8 ± 0.8	34.0 ± 078	51.1 ± 0.6	33.5 ± 0.4	49.8 ± 0.7
SIFT	67.3 ± 0.4	77.9 ± 0.6	73.3 ± 0.9	54.8 ± 0.5	80.2 ± 0.9
LBP	74.3 ± 0.6	79.0 ± 0.5	80.9 ± 0.7	57.3 ± 0.9	82.7 ± 0.8

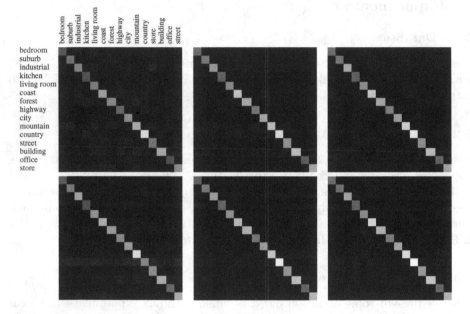

Fig. 3. Confusion matrix for evaluated methods. The top row from left to right are BoF+SIFT, SP+SIFT and SVD+SIFT respectively. The bottom row from left to right are BoF+LBP, SP+LBP and SVD+LBP respectively.

5.4   Results

The experimental results is listed in Table 2. For gray feature, our method Unfold+ FCM achieves 51.1 % improving the better baseline(SP) by 17.1 %. For SIFT and LBP features, our method SVD+FCM achieves 80.2 % and 82.7 % improving the better baseline(SP) by 2.3 % and 3.7 % respectively. We also compare our method with two state-of-the-art methods, [17] (85.5 %) and [21] (88.1 %). The former use the saliency maps to weight the corresponding visual features and the latter combines 14 different low level features to improve the discriminative power of the image representation. While our method use a simple framework and a single feature(LBP) achieves a comparable performance. The confusion matrix is shown in Fig. 3.

6   Conclusion

In this paper, we demonstrate that FCM is a potential discriminative feature to classify scenes. The experimental results show the proposed method outperforms the original BoF model and its popular extension SP model. The proposed method achieves comparable performance to the state-of-the-art on 15 scene dataset. There still is a lot of potential to improve its performance when considering the follows. The first is the size of a codebook which is controlled by the number of keypoint clusters in the clustering process. The second is that we can use the proposed framework to combine multiple complementary features to improve the performance. In future work, we will optimize the proposed method from above two aspects.

Acknowledgement. This work was supported in part by Beijing Higher Education Young Elite Teacher Project under Grants YETP0514, and NSFC under Grants 61471024. Ling was supported in part by the US NSF Grants IIS-1218156 and IIS-1350521.

References

1. Fei-Fei, L., Perona, P.: A bayesian hierarchical model for learning natural scene categories. In: IEEE Conference on Computer Vision and Pattern Recognition, vol. 2, pp. 524–531. IEEE (2005)
2. Lazebnik, S., Schmid, C., Ponce, J.: Beyond bags of features: spatial pyramid matching for recognizing natural scene categories. In: IEEE Conference on Computer Vision and Pattern Recognition, vol. 2, pp. 2169–2178. IEEE (2006)
3. Vogel, J., Schiele, B.: Semantic modeling of natural scenes for content-based image retrieval. Int. J. Comput. Vis. **72**, 133–157 (2007)
4. Lowe, D.: Distinctive image features from scale-invariant keypoints. Int. J. Comput. Vis. **60**, 91–110 (2004)
5. Bay, H., Tuytelaars, T., Van Gool, L.: SURF: speeded up robust features. In: Leonardis, A., Bischof, H., Pinz, A. (eds.) ECCV 2006, Part I. LNCS, vol. 3951, pp. 404–417. Springer, Heidelberg (2006)
6. Dalal, N., Triggs, B.: Histograms of oriented gradients for human detection. IEEE Conference on Computer Vision and Pattern Recognition, vol. 1, pp. 886–893 (2005)
7. Mikolajczyk, K., Schmid, C.: A performance evaluation of local descriptors. IEEE Trans. Pattern Anal. Mach. Intell. **27**, 1615–1630 (2005)
8. Tuzel, O., Porikli, F., Meer, P.: Region covariance: a fast descriptor for detection and classification. In: Leonardis, A., Bischof, H., Pinz, A. (eds.) ECCV 2006. LNCS, vol. 3952, pp. 589–600. Springer, Heidelberg (2006)
9. Ojala, T., Pietikäinen, M., Mäenpää, T.: Multiresolution gray-scale and rotation invariant texture classification with local binary patterns. IEEE Trans. Pattern Anal. Mach. Intell. **24**, 971–987 (2002)
10. Harris, C., Stephens, M.: A combined corner and edge detector. In: Alvey Vision Conference, Manchester, UK, vol. 15, p. 50 (1988)
11. Mikolajczyk, K., Schmid, C.: Scale & affine invariant interest point detectors. Int. J. Comput. Vis. **60**, 63–86 (2004)

12. Hamerly, G., Elkan, C.: Learning the k in k-means. Adv. Neural Inf. Process. Syst. **16**, 281 (2004)
13. Csurka, G., Dance, C., Fan, L., Willamowski, J., Bray, C.: Visual categorization with bags of keypoints. In: Workshop on Statistical Learning in Computer Vision, ECCV, vol. 1, pp. 1–2 (2004)
14. van Gemert, J.C., Geusebroek, J.-M., Veenman, C.J., Smeulders, A.W.M.: Kernel codebooks for scene categorization. In: Forsyth, D., Torr, P., Zisserman, A. (eds.) ECCV 2008, Part III. LNCS, vol. 5304, pp. 696–709. Springer, Heidelberg (2008)
15. Yu, K., Zhang, T., Gong, Y.: Nonlinear learning using local coordinate coding. In: Advances in Neural Information Processing Systems, pp. 2223–2231 (2009)
16. Yang, J., Yu, K., Gong, Y., Huang, T.: Linear spatial pyramid matching using sparse coding for image classification. In: IEEE Conference on Computer Vision and Pattern Recognition, CVPR 2009, pp. 1794–1801. IEEE (2009)
17. Sharma, G., Jurie, F., Schmid, C.: Discriminative spatial saliency for image classification. In: 2012 IEEE Conference on Computer Vision and Pattern Recognition (CVPR), pp. 3506–3513. IEEE (2012)
18. Chen, Q., Song, Z., Hua, Y., Huang, Z., Yan, S.: Hierarchical matching with side information for image classification. In: 2012 IEEE Conference on Computer Vision and Pattern Recognition (CVPR), pp. 3426–3433. IEEE (2012)
19. Haralick, R.M., Shanmugam, K., Dinstein, I.H.: Textural features for image classification. IEEE Trans. Syst. Man and Cybern. **3**, 610–621 (1973)
20. Chang, C., Lin, C.: Libsvm: a library for support vector machines. ACM Trans. Intell. Syst. Technol. (TIST) **2**, 27 (2011)
21. Xiao, J., Hays, J., Ehinger, K.A., Oliva, A., Torralba, A.: Sun database: large scale scene recognition from abbey to zoo. In: IEEE Conference on Computer Vision and Pattern Recognition, CVPR 2010, pp. 1–8. IEEE (2010)

RoLoD: Robust Local Descriptors for Computer Vision

Local Associated Features
for Pedestrian Detection

Song Shao[1,2], Hong Liu[1,2](✉), Xiangdong Wang[1,2], and Yueliang Qian[1,2]

[1] Research Center for Pervasive Computing, Institute of Computing Technology
Chinese Academy of Sciences, Beijing 100190, China
[2] Beijing Key Laboratory of Mobile Computing and Pervasive Device,
Institute of Computing Technology Chinese Academy of Sciences,
Beijing 100190, China
{shaosong,hliu,xdwang,ylqian}@ict.ac.cn

Abstract. Local features are usually used to describe pedestrian appearance. While most of existing pedestrian detection methods don't make full use of context cues, such as associated relationships between local different locations. This paper proposes two novel kinds of local associated features, gradient orientation associated feature (GOAF) and local difference of ACF (ACF-LD), to exploit context information. In our work, pedestrian samples are enlarged to contain some background regions besides human body, and GOAF, ACF and ACF-LD are combined together to describe pedestrian sample. GOAF are constructed by encoding gradient orientation features from two different positions into a single value. These two positions are come from different distance and different direction. For ACF-LD, the sample is divided into several sub regions and the ACF difference matrixes between these areas are computed to exploit the associated information between pedestrian and surrounding background. The proposed local associated features can provide complementary information for detection tasks. Finally, these features are fused with ACF to form candidate feature pool, and AdaBoost is used to select features and train a cascaded classifier of depth-two decision trees. Experimental results on two public datasets show that the proposed framework can achieve promising results compared with the state of the arts.

1 Introduction

Pedestrian detection is one important task in computer vision and pattern recognition, which detects pedestrian's location and size in images or videos. It has wide applications, such as intelligent surveillance, driver assistance and human behavior analysis.

Currently the predominant approaches of pedestrian detection are based on machine learning. The key factors of these methods are feature representation and classifier construction. Most early methods used single features, such as Haar [1], Local Binary Pattern (LBP) [2], Edgelet [3], Shapelet [4] and Histograms of Oriented Gradients (HOG) [5]. Since the description ability of single feature is limited, multiple features appeared in recent years. For example,

© Springer International Publishing Switzerland 2015
C.V. Jawahar and S. Shan (Eds.): ACCV 2014 Workshops, Part I, LNCS 9008, pp. 513–526, 2015.
DOI: 10.1007/978-3-319-16628-5_37

Wang [6] connected HOG with LBP features, Wojek [7] combined Haar, HOG and Shapelet features, Walk [8] added self-similar color features and motion features on the basis of Wojek [7]. Dollar [9] proposed integral channel features (ICF), and gave an optimized channel combination including gradient magnitude, histogram of oriented gradients, and LUV colors. These channel features improved the performance of pedestrian detection and can be computed quickly with integral images technology. Dollar [10] further proposed aggregated channel features (ACF), which uses pixel lookups in aggregated channels to reduce feature extraction time without constructing integral images. Each of the ACF feature represent a local block of the sample, which is simpler and computed more quickly. Effective combination of feature channels and feature selecting strategy by cascaded classifiers make ACF framework perform better than most of other methods both on detection accuracy and speed [10].

However, most above methods usually use features in pedestrian region. Although pedestrian appearance contains abundant information for detection, while in complex scenes, such as occlusion, crowded scenes or poor image resolution, it's hard to detect pedestrian effectively. Context information, such as background region around human body has not been fully used. In many dynamic scenes, such as on-board videos, though the backgrounds are changing as time goes on, the structures of scenes are relatively stable, which contains some useful context information. The associated features between different locations in background and pedestrian region are not fully exploited, which would contain more information than single local features just in human regions. We call these features local associated features in this paper. According to the retrieval reference, up to date, ACF is one of the most successful pedestrian descriptors both in detection accuracy and detection speed [10]. While each ACF feature is one-dimensional channel feature and cannot describe the associated information of different local regions, such as the gradient orientation. Besides, ACF is focus on pedestrian area and the difference between pedestrian and surrounding background is not paid enough attention. Therefore, we try to design some local associated features to exploit the context information between different locations, and fuse with ACF to improve the robustness of feature descriptor.

This paper proposes two novel kinds of local associated features for pedestrian detection: Gradient Orientation Associated Feature (GOAF) and Local Difference of ACF (ACF-LD). In our work, pedestrian samples are enlarged to contain some background regions, and GOAF, ACF and ACF-LD are combined together to describe pedestrian sample. Firstly, GOAF associates gradient orientations of different regions in a certain distance and encodes them into a single value, which can exploit associated information between local regions. Besides, to exploit the associated information between pedestrian and surrounding background, object samples are divided into several sub regions and local difference of ACF are calculated between these regions. The proposed local associated features provide complementary context information for pedestrian detection. In our pedestrian detection framework, ACF and proposed local associated features GOAF and ACF-LD are fused to form candidate feature pool. Then AdaBoost is used to select features and train a cascaded classifier of depth-two decision trees. We evaluate our framework on two different pubic datasets. The experimental results show the effectiveness of our method. For example, the miss rate is reduced from 44.04% to 38.07% at 10^{-1} FPPI on Caltech dataset and from 16.85% to 16.04% at 10^{-1} FPPI on Inria dataset. This means that the proposed local associated features can effectively capture context information of scenes and improve pedestrian detection performance compared with the state of the arts.

2   Related Work

Recent studies show that context information plays an important role in video and image understanding. Researchers have proposed many different types of context information, such as semantic context [11], 3D geometry context [12], local pixels context [5,13] and shape context [14].

Dalal [5] slightly enlarged the detection window to include neighbor pixels around the pedestrian and then extracted HOG features on the enlarged windows. But simply expanding the window just got limited improvement. If the window continues to enlarge, the dimension of features will increase significantly and detection performance will be worse. Neil [15] segmented the image into different kinds of regions such as grass, roads and sky, and then learned the probability of a person appearance in a certain region to adjust the detection results. This approach uses scene context information and the performance relies on image segmentation and probability learn-ing, which is not very effective for complex and dynamic scenes. William [16] pro-posed a feature descriptor called Local Response Context (LRC). This method firstly sampled the detection responses around each detection window to construct a feature vector and then learned a partial least squares regression model as a second classification stage. LRC descriptors reduced the dimensionality of features, but the improvement of performance is limited. Above methods indeed used some context information, while these methods did not exploit the associated relationship between local regions in the same detection window.

Ding [17] computed multi-scale HOG features and put forward a new feature called Local Difference Pattern (LDP), which is similar to LBP. The local region is divided into blocks, and the difference between the average pixel intensity in each block and a reference block forms the LDP feature. In addition, for each detection result, the classifier responses from neighbor locations and scales are incorporated as additional features to join an iterative training process called Context Boost. While the multi-scale features take better advantage of context information, it requires that the pedestrian area have to be half of the detection window. Besides, the whole train-ing and detecting framework is complicated. In our paper, the proposed ACF-LD also adopts region partition method, but the partition schemes are totally different and the segmented sub areas are used to compute the local ACF difference matrixes instead of multi-scale HOGs. Besides, LDP only describe color context cues in each local region, while GOAF connects the gradient orientation values of different local regions.

This paper focuses on constructing novel local associated features, which can provide a new thought for exploiting context information of local features in different regions. The detailed construction of features is introduced in Sect. 3. Section 4 introduces the experiments and discussion. Section 5 summarizes this paper and gives future research work.

3   Our Proposed Method

3.1   The Framework

This paper introduces a pedestrian detection framework based on local associated features and Aggregated Channel Features (ACF) to describing pedestrian object robustly. To exploit and use context information on local features in different regions, we propose two novel kinds of local associated features: Gradient Orientation Associated Feature (GOAF) and Local Difference of ACF (ACF-LD). Then GOAF, ACF and ACF-LD are fused together to form candidate feature pool. Finally, AdaBoost is used to select distinguish features and a cascaded classifier of depth-two decision trees is trained to select and combine these distinguish local features. The whole process of our detection framework is shown in Fig. 1.

Pedestrian samples are enlarged to contain some neighbor background regions. The details of enlarge ratio will discussed in next section. In training stage, GOAP, ACF and ACF-LD are extracted from samples and then a multiple round of bootstrapping is used to train decision trees over these candidate features. Finally, all the weak classifiers are cascaded to form the final detector. In testing process, feature pyramids are constructed for every detection image and then a sliding window is used over multiple scales. The features corresponding to the current detection window are sent to the detector for classification. Then the final results are obtained after non-maximal suppression (NMS).

Our main work is the construction of local associated features. The details about sample enlarging strategy, and constructions of GOAF and ACF-LD will be introduced in next sections.

3.2   Gradient Orientation Association Feature (GOAF)

For machine learning based pedestrian detection method, feature descriptor is the most important key for classification and detection performance. In existing features, up to now, HOG [5] and ACF [10] are useful descriptors and over half of ACF comes from HOG channel features. These channel features describe local gradient magnitude in different orientations and effectively describe the pedestrian characteristics.

We enlarge the sample size to include more nearby background areas as Fig. 2 shows, similar to Dalal's [5] and Dollar's [10] methods. The parameters of pedestrian size $ph \times pw$ and sample size $sh \times sw$ are determined through experiments and discussed in Sect. 4. The background regions can provide useful

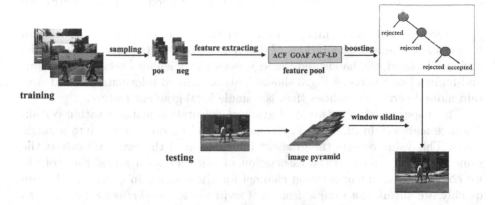

Fig. 1. Our pedestrian detection framework.

Fig. 2. Enlarging samples to contain some neighbor background regions.

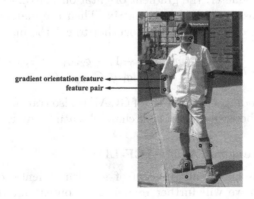

Fig. 3. The sample of different feature pairs for GOAF.

context cues. But just expanding the sample is not enough and some strategy should be used to exploit hidden information.

After observed lots of images, we found that there exists some association between the gradient orientations of different human body parts. For example, such as the head, the hands and the legs, even the neighbor background regions and human body parts as Fig. 3 shows. This associated information would contain more description abilities than the single local gradient feature.

This paper constructs two local gradient orientation features within certain distance and certain direction in the sample, and encoding them into a single value. This value reflects the gradient information of the two positions at the same time and embodies the association of different local areas. First of all, we compute gradient orientation channel for the sample. In order to calculate quickly, we shrink the sample from $h \times w$ pixels to $(h/shrink) \times (w/shrink)$ pixels, and then extract gradient orientation matrix G. Each value of this matrix is between 0 and π To reduce calculation, these values are normalized averagely into integers in $[1, maxVal]$. The value of maxVal decides the total division that gradient orientation will be divided into. Larger maxVal will results in small or fine division of gradient orientation.

Then, using $g_1 = G(x, y)$ denote the gradient orientation value in position (x, y), we can get $g_2 = G(x + xof, y + yof)$ for a giving position offset (xof, yof). Different offsets means different distances and directions.

$$distance = \sqrt{xof^2 + yof2} \tag{1}$$

$$angle = arctan(yof/xof) \tag{2}$$

The two values can make a feature value pair $p(g_1, g_2)$. According to formula (4), this pair can be turned into a single value. So when we choose an offset (xof, yof), we can get a feature matrix F where

$$F(x, y) = f(G(x, y), G(x + xof, y + yof)) \tag{3}$$

The whole construction process of GOAF is shown in the Fig. 4. When we choose an offset parameter, the gradient orientation matrix of each sample will be turned into an associated feature matrix. Then the matrixes from different offsets are normalized and aggregated together to get the final GOAF.

$$f(g_1, g_2) = \begin{cases} g_1(maxVal + 1) + g_2, g_1 \in G, g_2 \in G \\ 0, g_1 \notin G, g_2 \notin G \end{cases} \tag{4}$$

From above encoding, each feature of GOAF is also transferred into a channel feature, which can be easily fused with channel features ACF.

3.3   Local Difference of ACF (ACF-LD)

GOAF contains associated information of gradient orientation from different local regions. Then we will further exploit the context information of difference from various regions, such as pedestrian regions and neighbor background regions.

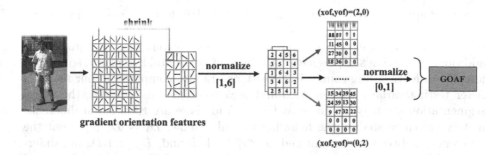

Fig. 4. The computing process of GOAF.

Most of the existing features focus on the pedestrian region, and do not make full use of surrounding background information. In this paper, we enlarge the detection window to contain some background regions. As Fig. 5 shows the detection window or sample is divided into two parts, the target area in center and the surrounding background area. By analyzing abundant samples, we find out that the similarity be-tween human body parts is often higher than the similarity between pedestrian region and background. This contrast can help the detector distinguish the positive and negative samples. So through comparing a human with the neighborhood around it, we try to extract some useful information to describe the difference between them, further enhance the reliability of detection. Details of the construction of ACF-LD are introduced in following.

We divide the background and pedestrian areas into several sub regions as Fig. 5 shows. L1, R1, T1 and B1 are used to represent the nearby background in different directions of pedestrian. L2, R2, T2 and B2 are correspondingly neighboring pedestrian areas. L3 and R3 are pedestrian areas adjacent to L2 and R2. The adjacent pedestrian and background regions, such as L1 and L2, T1 and T2, are very different, but the adjacent pedestrian regions, such as L2 and L3 are more similar to each other.

In order to capture this information effectively, the image should be expressed in reasonable feature space and the similarity between local regions should be

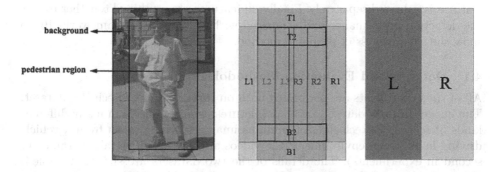

Fig. 5. The segmentation scheme of ACF-LD.

measured appropriately. In our framework, ACF [10] features are used to represent each part. Then difference between these parts is computed, which implies the similarity between the central area and the surrounding background. In addition, pedestrian samples have some level of horizontal symmetry, so the differences of features in symmetrical positions are also computed in our method. After testing various segmentation and area combinations schemes, the above segmentation strategy is chosen as Fig. 5. And there are total seven difference matrix are calculated as the following formulas show. $D_1 \sim D_4$ represent the differences between pedestrian and nearby background, D_5 and D_6 are differences within the pedestrian area, and D_7 reflects the differences of symmetrical positions in image. (Function acf is used to compute ACF matrix of each local region, and function $flip$ means to flip the matrix horizontally.)

$$D_1 = abs\left(acf\left(L_1\right) - acf\left(L_2\right)\right) \tag{5}$$

$$D_2 = abs\left(acf\left(R_1\right) - acf\left(R_2\right)\right) \tag{6}$$

$$D_3 = abs\left(acf\left(T_1\right) - acf\left(T_2\right)\right) \tag{7}$$

$$D_4 = abs\left(acf\left(B_1\right) - acf\left(B_2\right)\right) \tag{8}$$

$$D_5 = abs\left(acf\left(L_2\right) - acf\left(L_3\right)\right) \tag{9}$$

$$D_6 = abs\left(acf\left(R_2\right) - acf\left(R_3\right)\right) \tag{10}$$

$$D_7 = abs\left(acf\left(L\right) - flip\left(acf\left(R\right)\right)\right) \tag{11}$$

These above difference matrixes compose ACF-LD. In the training stage, ACF-LD is computed for every sample and each dimension of ACF-LD is fused with ACF and GOAF to form the final feature pool. To reduce computation cost, only the ACF-LD values used in the detector is needed to compute during detection process.

4   Experiments

This section introduces some experiments to evaluate the proposed local associated features and the whole detection framework. Firstly, a series of experiments are con-ducted to find out appropriate parameters related to GOAF and segmentation strategy of ACF-LD. Then the effectiveness of the two local associated features is valuated separately. Finally all features are combined together to test the detection performance and compared with Dollar's ACF framework [10] to show the effectiveness of proposed method.

4.1   Datasets and Evaluation Methodology

All of the experiments are performed both on Inria [5] and Caltech [18] dataset. The images in Inria dataset are static pictures, which come from many different kinds of scenes. Caltech dataset contains images of video taken from a vehicle driving in an urban environment (the videos are 30 Hz and we take 1 frame per second in experiments). The details of the two datasets are shown in Table 1. P-img, N-img and P-num represent the number of positive images, negative images and labeled pedestrian samples.

Table 1. Details about the datasets.

Dataset	Type	Train			Test			M-Height
		P-img	N-img	P-num	P-img	N-img	P-num	
Inria	Photo	614	1218	1208	288	453	566	279
Caltech	Mobile	67k	61k	192k	65k	56k	155k	48

4.2 Performance of Local Associated Features

GOAF+ACF. To exploit associated information between different local regions, GOAF associates gradient orientation values in a certain distance and direction from different regions and encodes them into a single value. As described in Sect. 3.2, the main parameters for GOAF are the normalization factor max-Val and offset (xof, yof) between local regions. In our experiment, the original gradient orientation $(0 \sim \pi)$ is normalized into integers in $[1, maxVal]$. If $maxVal$ is too small, some gradient orientations may not be possible to distinguish from each other, and if too large, the robust associated relationships can't be described. Finally we set $maxVal$ as 6 in our experiments.

Besides, experimental results show that the value of offset can influence the performance of GOAF. Different offset means different distance and direction between two local gradient orientation values. The selection of different offset in our method is huge. To simplify the parameters selection and feature construction process, we mainly consider the horizontal and vertical offsets, which contain abundant associated information. We use xp to represent horizontal offset combinations and yp to represent vertical offset combinations. For example, $xp \& yp = [1, 2] \& [2, 3]$ means offset $(1, 0)$, $(2, 0)$, $(0, 2)$ and $(0, 3)$ are used to compute GOAF.

To test the selection of different parameters, GOAF and ACF are calculated as candidate feature pool and then a cascaded classifier with 2048 depth-two decision trees is trained by Adaboost. In our experiments, we tested over hundreds of parameter combinations. Finally, we choose offsets $[1, 2, 4, 6] \& [1, 2, 4, 6]$ to compute GOAF on Inria and $[1, 2, 4] \& []$ on Caltech. The offsets are different on the two datasets, which means the different scene, image resolution, pedestrian posture and size would influence the selection of offsets.

As Fig. 6 shows the average miss rates are 40.87 % on Caltech based on our method, obviously better than Dollar's 44.04 %. The result shows that the selected GOAF would capture useful context information hidden in the image and is an effective supplement to ACF. The miss rate 16.25 % on Inria is also slightly better than Dollar's 16.85 %. While the miss rate of Dollar's [10] is already very low, it's harder to further improve much.

ACF-LD+ACF. For another local associated feature, ACF-LD is focus on the difference information between different regions including pedestrian or background regions. As introduced in Sect. 3.3, the samples are divided into several sub regions and local differences of ACF are calculated between these

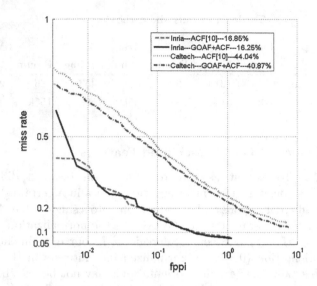

Fig. 6. Experimental results of GOAF with ACF on Caltech and Inria.

regions. The key factors of ACF-LD are segmentation strategy and similarity measurement. We tested different segmentation schemes and found that the difference matrixes computed by sub pedestrian and background regions aligned in horizontal direction can capture more useful information. This could be because the image has some degree of symmetry in horizontal direction, a transition from background to pedestrian and from pedestrian to back-ground again. The experiments show that the partition described in Sect. 3.3 is effective. ACF and ACF-LD features of samples are computed to train a cascaded classifier of 2048 depth-two decision trees. The miss rate of our results on Caltech (39.58 %) is obviously lower than Dollar's (44.04 %). The contrast between pedestrian and nearby background can help identify the targets. The results on Inria also become a litter better from 16.85 % to 16.27 %. ACF-LD exploits context associated information of the difference between different regions of sample, which can complement with ACF to improve the performance of pedestrian detection (Figs. 7 and 8).

GOAF+ACF-LD+ACF. To further improve the robust of feature description, we fuse all the above features, such as GOAF, ACF-LD and ACF, to evaluate the overall detection performance of our framework. We use the parameters discussed in Sects. 4.1 and 4.2. For 12864 samples on Inria, the candidate feature pool size is 15456, including 5120 ACF, 6240 ACF-LD and 4096 GOAF. And for 6432 samples on Caltech, the size of feature pool is 3234, including 1280 ACF, 1570 ACF-LD and 384 GOAF. Experiments show that the miss rate can be further reduced. We get 38.07 % on Caltech, nearly 6 % lower than Dollar's 44.04 %. There is also 0.8 % reduced on Inria from 16.85 % to 16.04 %.

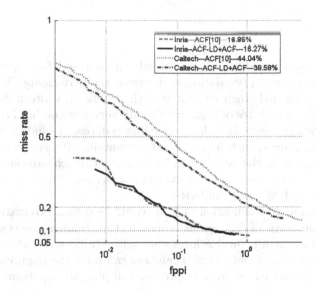

Fig. 7. Experimental results of ACF-LD with ACF on Caltech and Inria.

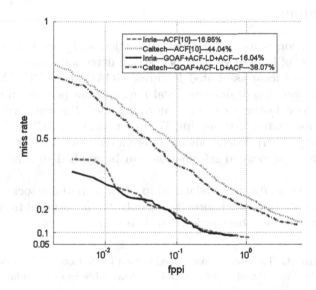

Fig. 8. Experimental results of GOAF, ACF-LF with ACF on Caltech and Inria.

The results mean that local associated features with the channel features can robust describe pedestrian object and improve the performance of pedestrian detection.

4.3   Discussion

The above experimental results show that the proposed local associated features can effectively exploit hidden associated information and further improve the detection performance on two datasets. However, the performance improvement on Inria is very limited when compared with results on Caltech dataset. One reason is that the miss rates of Inria are already very low and hard to be further reduced. On the other side, Inria is a static image dataset including images from various kinds of scenes, such as grassland, mountain, city and country. There are large variations in the scene of Inria dataset. Caltech are on-board video datasets with images from several urban traffic scenes. The scenes of Caltech have certain consistency and similarity.

For detection and classification task, it would be difficult to extract effective local association features for various scenes in the training process. The associated features are more suitable for scenes with stable structure information. Actually most fixed and mobile video cameras can meet the requirement, therefore, the local associated features we present can play an important role in real applications.

5   Conclusions

Context information is useful for object detection and classification, which is not made full use of in most existing pedestrian detection methods. This paper proposes two novel local associated features, GOAF and ACF-LD, to exploit the context in-formation of different local regions from pedestrian or neighbor background. These features combine information of different local regions to capture useful associated relationships. We fuse them with ACF and verify their effectiveness for pedestrian detection on different datasets. The idea and method of extracting local associated information can be applied to other algorithms conveniently.

This paper just analyses the relationship of local spatial associated information, and temporal associated information can be further studied to exploit more context information for robust pedestrian detection.

Acknowledgement. This work is supported in part by National Nature Science Foundation of China: 61202209 and in part by Beijing Natural Science Foundation: 4142051.

References

1. Viola, P., Jones, M.: Rapid object detection using a boosted cascade of simple features. Proc. IEEE Conf. Comput. Vis. Pattern Recognit. **1**, 511–518 (2001)
2. Ojala, M., Pietikainen, M., Maenpaa, T.: Rapid object detection using a boosted cascade of simple features. IEEE Trans. Pattern Anal. Mach. Intell. **24**, 971–987 (2002)

3. Wu, B., Nevatia, R.: Detection of multiple, partially occluded humans in a single image by baycsian combination of edgelet part detectors. Proc. IEEE Int. Conf. Comput. Vis. **1**, 90–97 (2005)

4. Sabzmeydani, P., Mori, G.: Detecting pedestrians by learning shapelet features. In: Proceedings of IEEE Conference on Computer Vision and Pattern Recognition, pp. 1–8 (2007)

5. Dalal, N., Triggs, B.: Histograms of oriented gradients for human detection. Proc. IEEE Conf. Comput. Vis. Pattern Recognit. **1**, 886–893 (2005)

6. Wang, X., Pietikainen, T., Han, T.X., Yan, S.: An hog-lbp human detector with partial occlusion handling. In: Proceedings of IEEE International Conference on Computer Vision, pp. 2–29 (2009)

7. Wojek, C., Schiele, B.: A performance evaluation of single and multi-feature people detection. In: Rigoll, G. (ed.) DAGM 2008. LNCS, vol. 5096, pp. 82–91. Springer, Heidelberg (2008)

8. Walk, S., Majer, N., Schindler, K.: New features and insights for pedestrian detection. Comput. Vis. Pattern Recognit. **24**, 1030–1037 (2010)

9. Dollar, P., Tu, Z., Perona, P., Belongie, S.: Integral channel features. In: Proceedings of British Machine Vision Conference, pp. 1–11 (2009)

10. Dollar, P., Appel, R., Belongie, S.: Fast feature pyramids for object detection. IEEE Trans. Pattern Anal. Mach. Intell. **34**, 743–761 (2014)

11. Carbonetto, P., de Freitas, N., Barnard, K.: A statistical model for general contextual object recognition. In: Pajdla, T., Matas, J.G. (eds.) ECCV 2004. LNCS, vol. 3021, pp. 350–362. Springer, Heidelberg (2004)

12. Hoiem, D., Efros, A.A., Hebert, M.: Putting objects in perspective. Int. J. Comput. Vis. **80**, 3–15 (2008)

13. Wolf, L., Bileschi, S.: A critical view of context. Int. J. Comput. Vis. **69**, 251–261 (2006)

14. Ramanan, D.: Using segmentation to verify object hypotheses. In: Computer Vision and Pattern Recognition (2007)

15. Robertson, N.M., Letham, J.: Contextual person detection in multi-modal outdoor surveillance. In: Signal Processing Conference (EUSIPCO), 2012 Proceedings of the 20th European, pp. 1930–1934 (2012)

16. Schwartz, W.R., Davis, L.S., Pedrini, H.: Local response context applied to pedestrian detection. In: San Martin, C., Kim, S.-W. (eds.) CIARP 2011. LNCS, vol. 7042, pp. 181–188. Springer, Heidelberg (2011)

17. Ding, Y., Xiao, J.: Contextual boost for pedestrian detectiont. In: Computer Vision and Pattern Recognition, pp. 2895–2902 (2012)

18. Dollar, P., Wojek, C., Schiele, B., Perona, P.: Pedestrian detection: an evaluation of the state of the art. IEEE Trans. Pattern Anal. Mach. Intell. **34**, 743–761 (2012)

19. Dollar, P., Belongie, S., Perona, P.: The fastest pedestrian detector in the west. In: Proceedings of British Machine Vision Conference, pp. 1–11 (2010)

20. Tao, X., Hong, L., Yueliang, Q., Zhe, W.: A fast pedestrian detection framework based on static and dynamic information. In: International Conference on Multimedia and Expo, pp. 242–247 (2012)

21. Kim, T.K., Cipolla, R.: Mcboost: multiple classifiers boosting for perceptual co-clustering of images and visual features. In: Proceedings of IEEE Conference on Neural Information Processing Systems, pp. 841–856 (2008)

22. Xu, Y., Cao, X., Qiao, H.: An efficient tree classifier ensemble-based approach for pedestrian detection. IEEE Trans. Syst. Man Cybern. Part B: Cybern. **41**, 107–117 (2011)

23. Hong, L., Tao, X., Xiangdong, W.: Robust human detection based on related hog features and cascaded adaboost and svm classifiers. In: Proceedings of International Conference on Multimedia Modeling, pp. 345–355 (2013)
24. Dollar, P., Appel, R., Kienzle, W.: Crosstalk cascades for frame-rate pedestrian detection. In: Proceedings of European Conference on Computer Vision, pp. 645–659 (2012)

Incorporating Two First Order Moments Into LBP-Based Operator for Texture Categorization

Thanh Phuong Nguyen[✉] and Antoine Manzanera

ENSTA-ParisTech, 828 Boulevard des Maréchaux, 91762 Palaiseau, France
{thanh-phuong.nguyen,antoine.manzanera}@ensta-paristech.fr

Abstract. Within different techniques for texture modelling and recognition, local binary patterns and its variants have received much interest in recent years thanks to their low computational cost and high discrimination power. We propose a new texture description approach, whose principle is to extend the LBP representation from the local gray level to the regional distribution level. The region is represented by pre-defined structuring element, while the distribution is approximated using the two first statistical moments. Experimental results on four large texture databases, including Outex, KTH-TIPS 2b, CUReT and UIUC show that our approach significantly improves the performance of texture representation and classification with respect to comparable methods.

1 Introduction

Texture analysis is an active research topic in computer vision and image processing. It has a significant role in many applications, such as medical image analysis, remote sensing, document vectorisation, or content-based image retrieval. Thus, texture classification has received considerable attention over the two last decades, and many novel methods have been proposed [1–15].

The representation of texture features is a key factor for the performance of texture classification systems. Numerous powerful descriptors were recently proposed: modified SIFT (scale invariant feature transform) and intensity domain SPIN images [4], MR8 [7], the rotation invariant basic image features (BIF) [11], (sorted) random projections over small patches [15]. Most earlier works focused on filter banks and the statistical distributions of their responses. Among the popular descriptors in these approaches are Gabor filters [16], MR8 [7], Leung and Malik's [17] filters, or wavelets [18]. Filter bank approaches are attractive by being expressive and flexible, but they may be hard to work out by being computationally expensive and application dependent. Varma and Zisserman [7] have shown that local intensities or differences in a small patch can produce better performance than filter banks with large spatial support.

Local Binary Patterns (LBP) emerged ten years ago when Ojala et al. [3] showed that simple relations in small pixel neighborhoods can represent texture with high discrimination. They used a binary code to represent the signs of difference between the values of a pixel and its neighbours. Since then, due to its

© Springer International Publishing Switzerland 2015
C.V. Jawahar and S. Shan (Eds.): ACCV 2014 Workshops, Part I, LNCS 9008, pp. 527–540, 2015.
DOI: 10.1007/978-3-319-16628-5_38

great computational efficiency and good texture characterisation performance, the LBPs have been applied in many applications of computer vision and a large number of LBP variants [8,9,14,19] have been introduced. They have been introduced to remedy several limitations of basic LBP including small spatial support region, loss of local textural information, rotation and noise sensitivities. Instead of using central pixel as threshold, several authors used the median [20] or the mean [21,22] value of neighbouring pixels. Similarly, Liao et al. [23] considered LBP on mean values of local blocks. Local variance was used as a complementary contrast measure [3,12]. Gabor filters [24] are widely used for capturing more global information. Guo et al. [9] included both the magnitudes of local differences and the pixel intensity itself in order to improve the discrimination capability. Dominant LBPs (DLBP) [8] have been proposed to deal with the most frequent patterns instead of uniform patterns. Zhao et al. [25] combined with covariance matrix to improve the performance.

We address a new efficient schema to exploit variance information in this paper. Unlike typical methods that considered the joint distribution of LBP codes and local contrast measure (LBP/VAR) [3] or integrated directly variance information into LBP model [12], we capture the local relationships within images corresponding to local mean and variance. We show that this approach is more efficient to exploit the complementary contrast information. Our descriptor enhances the expressiveness of the classic LBP texture representation while providing high discrimination, as we show in a comparative evaluation on several classic texture data sets. It has limited computational complexity and doesn't need neither multiscale processing nor magnitude complementary information (CLBP_M) to obtain state-of-the-art results: CLBP [9], CLBC [14], CRLBP [22], NI/RD/CI [26].

The remaining of the paper is organised as follows. Section 2 presents in more details the other works most directly related to our method. Section 3 elaborates the proposed approach. Experimental results are presented in Sect. 4 and conclusions are finally drawn in Sect. 5.

2   Related Work

2.1   Rotation Invariant Uniform LBP

Ojala et al. [27] supposed that texture has locally two complementary aspects: a spatial structure and its contrast. Therefore, LBP was proposed as a binary version of the texture unit to represent the spatial structure. The original version works in a block of 3×3 pixels. The pixels in the block are coded based on a thresholding by the value of center pixel and its neighbors. A chain code of 8 bit is then obtained to label the center pixel. Hence, there are totally $2^8 = 256$ different labels to describe the spatial relation of the center pixel. These labels define local textural patterns.

The generalised LBP descriptor, proposed by Ojala et al. [3], encodes the spatial relations in images. Let f be a discrete image, modelled as a mapping from $\mathbb{Z}^2$ to $\mathbb{R}$. The original LBP encoding of f is defined as the following mapping from $\mathbb{Z}^2$ to $[0,1]^P$:

$$\text{LBP}_{P,R}(f)(\mathbf{z}) = (s(f(\mathbf{y}_p) - f(\mathbf{z})))_{0 \leq p < P}, \tag{1}$$

$$\text{with } s(x) = \begin{cases} 1, x \geq 0 \\ 0, \text{otherwise.} \end{cases}$$

Here $\mathbf{y}_p$ $(0 \leq p < P)$ are the P neighbours of z, whose values are evenly measured (or interpolated) on the circle of radius R centred on z.

The uniformity measure of an LBP is defined as follows:

$$U(\text{LBP}_{P,R}) = \sum_{p=1}^{P} |\text{LBP}_{P,R}^p - \text{LBP}_{P,R}^{p-1}|, \tag{2}$$

where $\text{LBP}_{P,R}^p$ is the p-th bit of $\text{LBP}_{P,R}$, and $\text{LBP}_{P,R}^P = \text{LBP}_{P,R}^0$. An LBP is called uniform if $U(\text{LBP}_{P,R}) \leq 2$. Ojala et al. observed that, on natural texture images, most patterns are uniform. Finally the rotation invariant uniform LBP is defined as follows:

$$\text{LBP}_{P,R}^{riu2} = \begin{cases} \sum_{p=0}^{P-1} \text{LBP}_{P,R}^p, \text{if } U(\text{LBP}_{P,R}) \leq 2 \\ P + 1, \text{otherwise.} \end{cases} \tag{3}$$

LBP^{riu2} proved [3] a very efficient local texture descriptor and then has been intensively used in texture classification. Uniform patterns are considered as more reliable and more statistically significant. Furthermore, ignoring non-uniform patterns considerably reduces the length of the descriptor, with only $P + 2$ distinct $LBP_{P,R}^{riu2}$ compared to 2^P distinct $LBP_{P,R}$.

2.2 Complementary Information

Local Contrast. Aside from the classic LBP, different authors have addressed the complementary information in order to improve the performance of texture classification. In the first work about LBP [27], a local contrast measure, defined between each block 3×3, has been used in combination with LBP codes. Then, it has been replaced in [3] by local variance (VAR) measured in a circular spatial support just like the LBP:

$$\text{VAR}_{P,R} = \frac{1}{P} \sum_{p=0}^{P-1} (g_p - \mu)^2 \tag{4}$$

where $\mu = \frac{1}{P} \sum_{p=0}^{P-1} g_p$

Because variance measure has a continuous value, in order to construct a joint distribution of LBP and local variance (LBP/VAR), a quantization of variance measure is needed. The efficiency of quantization step depends on the number of bins and also the cut values of the bins of the histogram, hence a training stage is applied to determine such parameters. In [12], Guo et al. addressed a different way that avoids this stage by incorporating variance measure (VAR) into LBP model. It was used as an adaptive weight for adjusting the contribution of LBP code in construction of histogram.

Magnitude Information. Guo et al. [9] presented a state-of-the-art variant by regarding the local differences as two complementary components, signs: $s_p = s(f(\mathbf{y}_p) - f(\mathbf{z}))$ and magnitudes: $m_p = |f(\mathbf{y}_p) - f(\mathbf{z})|$. They proposed two operators, called CLBP-Sign ($CLBP\_S$) and CLBP-Magnitude ($CLBP\_M$) to code these two components. The first operator is identical to the LBP. The second one which measures the local variance of magnitude is defined as follows:

$$\text{CLBP\_}M_{P,R}(f)(\mathbf{z}) = (s(m_p - \tilde{m}))_{0 \leq p < P}, \tag{5}$$

where $\tilde{m}$ is the mean value of m_p for the whole image. In addition, Guo et al. observed that the local value itself carries important information. Therefore, they defined the operator CLBP-Center ($CLBP\_C$) as follows:

$$\text{CLBP\_}C(f)(\mathbf{z}) = s(f(\mathbf{z}) - \tilde{f}), \tag{6}$$

where $\tilde{f}$ is set as the mean gray level of the whole image. Because these operators are complementary, their combination leads to a significant improvement in texture classification, then this variant is also considered as a reference LBP method.

3   Texture Representation Using Statistical Binary Patterns

The Statistical Binary Pattern (SBP) representation aims at enhancing the expressiveness and discrimination power of LBPs for texture modelling and recognition, while reducing their sensitivity to unsignificant variations (e.g. noise). The principle consists in applying rotation invariant uniform LBP to a set of images corresponding to local statistical moments associated to a spatial support. The resulting code forms the Statistical Binary Patterns (SBP). Then a texture is represented by joint distributions of SBPs. The classification can then be performed using nearest neighbour criterion on classical histogram metrics like χ^2. We now detail those different steps.

3.1   Moment Images

A real valued 2d discrete image f is modelled as a mapping from $\mathbb{Z}^2$ to $\mathbb{R}$. The spatial support used to calculate the local moments is modelled as $B \subset \mathbb{Z}^2$, such that $O \in B$, where O is the origin of $\mathbb{Z}^2$.

The r-order moment[1] image associated to f and B is also a mapping from $\mathbb{Z}^2$ to $\mathbb{R}$, defined as:

$$m^r_{(f,B)}(\mathbf{z}) = \frac{1}{|B|} \sum_{\mathbf{b} \in B} (f(\mathbf{z} + \mathbf{b}))^r \qquad (7)$$

where $|B|$ is the cardinality of B. Accordingly, the r-order centred moment image ($r > 1$) is defined as:

$$\mu^r_{(f,B)}(\mathbf{z}) = \frac{1}{|B|} \sum_{\mathbf{b} \in B} \left(f(\mathbf{z} + \mathbf{b}) - m^1_{(f,B)}(\mathbf{z}) \right)^r \qquad (8)$$

From now on, we shall use also the notation m_r, μ_r to indicate r-order moment and r-order centred moment respectively.

3.2   Statistical Binary Patterns

Let R and P denote respectively the radius of the neighbourhood circle and the number of values sampled on the circle. For each moment image M, one statistical binary pattern is formed as follows:

- one $(P + 2)$-valued pattern corresponding to the rotation invariant uniform LBP coding of M:

$$\text{SBP}_{P,R}(M)(\mathbf{z}) = \text{LBP}^{riu2}_{P,R}(M)(\mathbf{z}) \qquad (9)$$

- one binary value corresponding to the comparison of the centre value with the mean value of M:

$$\text{SBP}_C(M)(\mathbf{z}) = s(M(\mathbf{z}) - \tilde{M}) \qquad (10)$$

Where s denote the sign function already defined, and $\tilde{M}$ the mean value of the moment M on the whole image. $\text{SBP}_{P,R}(M)$ then represents the structure of moment M with respect to a local reference (the centre pixel), and $\text{SBP}_C(M)$ complements the information with the relative value of the centre pixel with respect to a global reference ($\tilde{M}$). As a result of this first step, a $2(P+2)$-valued scalar descriptor is then computed for every pixel of each moment image.

3.3   Texture Descriptor

Principles. Let $\{M_i\}_{1 \leq i \leq n_M}$ be the set of n_M computed moment images. $\text{SBP}^{\{M_i\}}$ is defined as a vector valued image, with n_M components such that for every $\mathbf{z} \in \mathbb{Z}^2$, and for every i, $\text{SBP}^{\{M_i\}}(\mathbf{z})_i$ is a value between 0 and $2(P + 2)$.

If the image f contains a texture, the descriptor associated to f is made by the histogram of values of $\text{SBP}^{\{M_i\}}$. The joint histogram H is defined as follows:

$$H : [\![0\,;2(P + 2)[\![^{n_M} \to \mathbb{N}$$

$$H(\mathbf{v}) = |\{\mathbf{z}; \text{SBP}^{\{M_i\}}(\mathbf{z}) = \mathbf{v}\}|$$

The texture descriptor is formed by H and its length is $[2(P + 2)]^{n_M}$.

---

[1] Note that a moment image corresponds to a local filter defined by a statistical moment, and should not be confused with the concept of "image moment".

Implementation. In this paper, we focus on $\text{SBP}_{P,R}^{m_1\mu_2}$, i.e. the SBP patterns obtained with the mean m_1 and the variance μ_2. Using two orders of moments, the size of the joint histogram in the texture descriptor remains reasonable. Figure 1 illustrates the calculation of the texture descriptor using m_1 and μ_2 images. The local spatial structure on each image are captured using $\text{LBP}_{P,R}^{riu2}$. In addition, two binary images are computed by thresholding the moment images with respect to their average values. For each moment, the local pattern may then have $2(P+2)$ distinct values. Finally, the joint histogram of the two local descriptors is used as the texture feature and is denoted $\text{SBP}^{m_1\mu_2}$. Therefore, the feature vector length is $4(P+2)^2$. As we can see in Fig. 1, alongside of the pics corresponding to non-uniform bin, the local structures of the 2D histogram clearly highlight the correlation between the uniform patterns of the two LBP images.

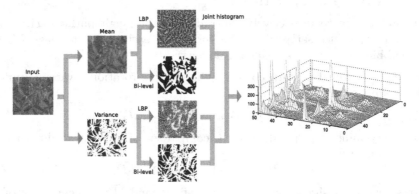

Fig. 1. Texture representation based on a combination of two first order moments. In LBP images, red pixels correspond to non-uniform patterns. The structuring element used here is $\{(1,5),(2,8)\}$ (see Sect. 3.3) while $\text{LBP}_{24,3}^{riu2}$ is applied.

Properties. Several remarks can be made on the properties of $\text{SBP}^{m_1\mu_2}$ and its relation with existing works.

– *Robustness to noise:* m_1 and μ_2 act like a pre-processing step which reduces small local variations and then enhances the significance of the binary pattern with respect to the raw images.
– *Rotation invariance:* Isotropic structuring elements should be used in order to keep the rotation invariance property of the local descriptor.
– *Information richness:* LBPs on moment images capture an information which is less local, and the two orders of moments provide complementary information on the spatial structure.

There are links between $\text{SBP}^{m_1\mu_2}$ and the CLBP descriptors of Guo et al. [9] (see Sect. 2). First, the binary images used in SBP correspond to CLBP_C operator. Second, the respective role of m_1 and μ_2 are somewhat similar to the

CLBP_S and CLBP_M operators. However in CLBP and its variants, CLBC [14] and CRLBP [22], the magnitude component (CLBP_M) is more a contrast information complementing CLBP_S, whereas SBP^{μ_2} represents the local structure of a contrast map, and can be considered independently on SBP^{m_1}.

Parameter Settings. We describe now the different parameters that can be adjusted in the framework, and the different settings we have chosen to evaluate. Two main parameters have to be set in the calculation of the SBP:

- the spatial support B for calculating the local moments, also referred to as structuring element.
- the spatial support $\{P, R\}$ for calculating the LBP.

Although those two parameters are relatively independent, it can be said that B has to be sufficiently large to be statistically relevant, and that its size should be smaller or equal than the typical period of the texture. Regarding $\{P, R\}$, it is supposed to be very local, to represent micro-structures of the (moment) images.

As mentioned earlier, for rotation invariance purposes, we shall use isotropic structuring elements. To be compliant with the LBP representation, we have chosen to define the structuring elements as unions of discrete circles: $B = \{\{P_i, R_i\}\}_{i \in I}$, such that $(P_i)_i$ (resp. $(R_i)_i$) is an increasing series of neighbour numbers (resp. radii). As an example, Fig. 2 shows the filtered images using a structuring element $B = \{(1,4),(2,8)\}$.

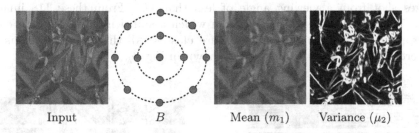

| Input | B | Mean (m_1) | Variance (μ_2) |

Fig. 2. Computation of moment images using structuring element $B = \{(1,4),(2,8)\}$.

3.4 Texture Classification

Two texture images being characterised by their respective histogram (descriptor) F and G, their texture dissimilarity metrics is calculated using the classical χ^2 distance between distributions:

$$\chi^2(F, G) = \sum_{i=1}^{d} \frac{(F_i - G_i)^2}{F_i + G_i}$$

where d is the number of bins (dimension of the descriptors). Our classification is then based on a nearest neighbour criterion. Every texture class with label λ is characterised by a prototype descriptor K_λ, with $\lambda \in \Lambda$. For an unknown texture image f, its descriptor D_f is calculated, and the texture label is attributed as follows:

$$l(f) = \arg\min_{\lambda \in \Lambda} \chi^2(D_f, K_\lambda)$$

4   Experimentations

We present hereafter a comparative evaluation of our proposed descriptor.

4.1   Databases and Experimental Protocols

The effectiveness of the proposed method is assessed by a series of experiments on four large and representative databases: Outex [28], CUReT (Columbia-Utrecht Reflection and Texture) [29], UIUC [4] and KTH-TIPS2b [30].

The Outex database (examples are shown in Fig. 5) contains textural images captured from a wide variety of real materials. We consider the two commonly used test suites, Outex_TC_00010 (TC10) and Outex_TC_00012 (TC12), containing 24 classes of textures which were collected under three different illuminations ("horizon", "inca", and "t184") and nine different rotation angles ($0°$, $5°$, $10°$, $15°$, $30°$, $45°$, $60°$, $75°$ and $90°$). Each class contains 20 non-overlapping 128×128 texture samples.

The CUReT database contains 61 texture classes, each having 205 images acquired at different viewpoints and illumination orientations. There are 118 images shot from a viewing angle of less than $60°$. From these 118 images, as in [7,9], we selected 92 images, from which a sufficiently large region could be cropped (200×200) across all texture classes. All the cropped regions are converted to grey scale (examples are shown in Fig. 3).

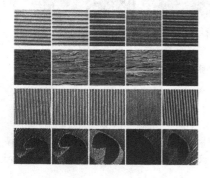

Fig. 3. CUReT dataset.

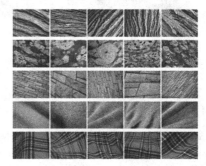

Fig. 4. UIUC dataset.

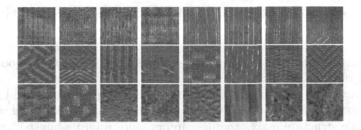

Fig. 5. Example of texture images from Outex dataset.

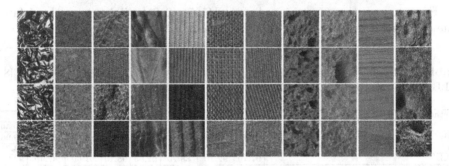

Fig. 6. Example of texture images from KTH-TIPS 2 dataset.

The UIUC texture database includes 25 classes with 40 images in each class. The resolution of each image is 640 × 480. The database contains materials imaged under significant viewpoint variations (examples are shown in Fig. 4).

The KTH-TIPS2b database contains images of 11 materials. Each material contains 4 physical samples taken at 9 different scales, 3 viewing angles and 4 illuminants, producing 432 images per class (see [30] for more detailed information). Figure 6 illustrates an example of the 11 materials. All the images are cropped to 200 × 200 pixels and converted to grey scale. This database is considered more challenging than the previous version KTH-TIPS. In addition, it is more completed than KTH-TIPS2a where several samples have only 72 images.

4.2 Results on the Outex Database

Two common test suites TC10 and TC12 are used for evaluated our method. Following [28], the 24 × 20 samples with illumination condition "inca" and rotation angle 0° were adopted as the training data. Table 1 reports the experimental results of $SBP_{P,R}^{m_1\mu_2}$ in comparison with different methods.

From Table 1, we can get several interesting findings:

- The proposed descriptor largely improves the results on the two test suites TC10 and TC12 in comparison with state-of-the-art methods in all configurations and with all structuring elements. Related with the most significant

LBP-based variant (CLBP_S/M/C), the average improvement on three exprements (TC10, TC12t and TC12h) can reach 3.5% in all configurations of (P, R).

– The structuring elements having single circular neighborhood give good results for test suite TC10.
– The structuring elements having two circular neighborhoods lead to good results for the two test suites TC10 and TC12. It could be explained that their multiscale structure makes the descriptors more robust against illumination changes addressed in TC12 test suite.

From now on, the structuring element $\{(1,5),(2,8)\}$ will be chosen by default due to its good results.

Table 1. Results obtained by different methods: LBP^{riu2} [3], LTP [31], DLBP [8], DLBP_NGF [8], CLBP_S_M/C [9], CLBC [14], CRLBP [22] on Outex dataset.

Method	(P,R)=(8,1)				(P,R)=(16,2)				(P,R)=(24,3)			
	TC10	TC12 t	TC12 h	Mean	TC10	TC12 t	TC12 h	Mean	TC10	TC12 t	TC12 h	Mean
LBP^{riu2}	84.81	65.46	63.68	71.31	89.40	82.26	75.20	82.28	95.07	85.04	80.78	86.96
LTP	94.14	75.88	73.96	81.33	96.95	90.16	86.94	91.35	98.20	93.59	89.42	93.74
DLBP					97.70	92.10	88.70	92.83	98.10	91.60	87.40	92.37
DLBP_NGF					99.10	93.20	90.40	94.20	98.20	91.60	87.40	92.40
CLBP	96.56	90.30	92.29	93.05	98.72	93.54	93.91	95.39	98.93	95.32	94.53	96.26
CLBC	97.16	89.79	92.92	93.29	98.54	93.26	94.07	95.67	98.78	94.00	93.24	95.67
CRLBP	96.54	91.16	92.06	93.25	98.85	96.67	96.97	97.50	99.48	97.57	97.34	98.14
Our proposed descriptor: SBP$_2$=SBP$^{m_1\mu_2}$												
Structuring element	(P,R)=(1,8)				(P,R)=(2,16)				(P,R)=(3,24)			
	TC10	TC12 t	TC12 h	Mean	TC10	TC12 t	TC12 h	Mean	TC10	TC12 t	TC12 h	Mean
{(1,6)}	99.06	93.79	95.81	95.55	**99.77**	96.64	97.80	98.07	99.66	96.78	97.01	96.82
{(1,8)}	98.96	93.66	93.79	95.47	99.61	96.30	97.64	97.85	99.53	96.69	97.27	97.83
{(1,4),(2,8)}	97.60	92.36	95.14	95.03	99.58	97.20	98.91	98.56	99.66	**97.68**	99.02	98.79
{(1,5),(2,6)}	97.97	92.27	95.02	95.09	99.40	96.74	98.17	98.10	99.56	97.41	98.26	98.41
{(1,5),(2,8)}	98.25	93.45	96.83	96.18	99.61	**97.57**	**99.24**	**98.81**	99.71	**97.68**	**99.07**	**98.82**
{(1,5),(2,10)}	99.03	93.70	96.94	96.56	99.66	**97.57**	99.10	98.78	**99.74**	97.66	98.96	98.79
{(1,6),(2,10)}	98.83	**93.80**	96.97	96.53	99.69	97.27	98.77	98.58	**99.74**	97.64	98.93	98.77
{(1,6),(2,12)}	**99.09**	93.61	**97.06**	**96.59**	99.63	97.38	98.96	98.66	99.69	97.50	98.89	98.69

4.3   Results on the CUReT Database

Following [4,9], in order to get statistically significant experimental results, N training images were randomly chosen from each class while the remaining $92 - N$ images per class were used as the test set. Table 2 shows an experiment of our method on CUReT database from which we can make the following remarks.

– SBP works well in all configurations of (P, R).
– SBP is comparable to VZ_MR8 and it outperforms other LBP-based descriptors.

Table 2. Results obtained on CURET dataset.

Method	(Γ,Π) $(0,1)$				$(\Gamma,\Pi)=(10,0)$				$(\Gamma,\Pi)=(01,5)$			
	46	23	12	6	46	23 t	12	6	46	23	12	6
$LBP^{riu2}/VAR_{P,R}$ [3]	93.87	88.76	81.59	71.03	94.20	89.12	81.64	71.81	91.87	85.58	77.13	66.04
CLBP_S/M [9]	93.52	88.67	81.95	72.30	94.45	90.40	84.17	75.39	93.63	89.14	82.47	73.26
CLBP_S/M/C [9]	95.59	91.35	84.92	74.80	95.86	92.13	86.15	77.04	94.74	90.33	83.82	74.46
	N=46				N=23				N=12			N=6
$dis(S+M)^{ri}_{N,R}$ [33] [2]	98.3				96.5				91.9			83.0
VZ_MR8 [34]	97.8				95.0				90.5			82.9
VZ_Joint [34]	97.7				94.6				89.4			81.1
Our proposed descriptor												
Structuring element	$(P,R)=(8,1)$				$(P,R)=(16,3)$				$(P,R)=(24,5)$			
	46	23	12	6	46	23	12	6	46	23	12	6
$\{(1,5),(2,8)\}$	97.32	93.49	88.23	78.37	98.01	94.91	90.60	81.98	96.89	92.95	88.37	78.29

The results are obtained with a multi-scale approach: $(P,R) \in \{(8,1),(8,3),(8,5)\}$

4.4 Results on the UIUC Database

As in [4], to eliminate the dependence of the results on the particular training images used, N training images were randomly chosen from each class while the remaining $40-N$ images per class were used as test set. Table 3 shows the results obtained by our approach on UIUC dataset in comparison with other state-of-the-art methods. As can be seen from this table, our proposed descriptor largely outperforms CLBP.

Table 3. Experimentation on UIUC dataset.

Method	$(P,R)=(8,1)$				$(P,R)=(16,2)$				$(P,R)=(24,3)$			
	20	15	10	5	20	15	10	5	20	15	10	5
CLBP_S/M [9]	81.80	78.55	74.8	64.84	87.87	85.07	80.59	71.64	89.18	87.42	81.95	72.53
CLBP_S/M/C [9]	87.64	85.70	82.65	75.05	91.04	89.42	86.29	78.57	91.19	89.21	85.95	78.05
Our proposed descriptor												
Structuring element	$(P,R)=(8,1)$				$(P,R)=(16,2)$				$(P,R)=(24,3)$			
	20	15	10	5	20	15	10	5	20	15	10	5
$\{(1,5),(2,8)\}$	91.31	89.56	85.97	78.19	95.52	94.34	91.87	85.69	96.55	95.40	93.07	87.11

4.5 Experiment on KTH-TIPS 2b Dataset

At the moment, KTH-TIPS 2b can be seen as the most challenging dataset for texture recognition. For the experiments on this dataset, we follow the training and testing scheme used in [30]. We perform experiments training on one, two, or three samples; testing is always conducted only on un-seen samples. Table 4 details our results on this dataset.

5 Conclusions and Discussions

We have presented a new approach for texture representation called Statistical Binary Patterns (SBP). Its principle is to extend the LBP representation from

Table 4. Classification rates obtained on the KTH-TIPS 2b database.

(P,R)	(8,1)			(16,2)			(24,3)		
N_{train}	1	2	3	1	2	3	1	2	3
LBP [3]	48.1	54.2	52.6	50.5	55.8	59.1	49.9	54.6	57.8
NI/RD/CI [26]	56.6	61.9	64.8	57.7	62.5	65.1	52.4	57.5	61.7
(P,R)	(8,1)+(16,2)			(16,2)+(24,3)			(8,1)+(16,2)+(24,3)		
NI/RD/CI [26]	58.1	62.9	66.0	55.9	61.0	64.2	56.7	61.7	65.0
Our proposed descriptors									
Structuring element	(P,R)=(8,1)			(P,R)=(16,2)			(P,R)=(24,3)		
	1	2	3	1	2	3	1	2	3
{(1,5),(2,8)}	57.68	63.24	66.03	59.03	64.91	68.60	59.04	65.20	68.81

the local gray level to the regional distribution level. In this work, we have used single region, represented by pre-defined structuring element and we have limited the representation of the distribution to the two first statistical moments. While this representation is sufficient for single mode distributions such as Gaussian, it won't be convenient for more complex distributions. Hence, investigating higher order moments in SBP would be highly desirable. The proposed approach has been experimented on different large databases to validate its interest. It must be mentioned that the complementary information of magnitude (CLBP_M), which is the principal factor for boosting classification performance in many other descriptors in the literature, has not been used in our approach. Our method also opens several perspectives that will be addressed in our future works.

- Can we apply efficiently our descriptor in other domains of computer vision such as face recognition, dynamic texture, ...?
- Can we still improve the performance by integrating other complementary information such as magnitude (CLBP_M), multi-scale approach, ...or combining with other LBP variants?

References

1. Cula, O.G., Dana, K.J.: Compact representation of bidirectional texture functions. In: CVPR, vol. 1, pp. 1041–1047 (2001)
2. Zhang, J., Tan, T.: Brief review of invariant texture analysis methods. Pattern Recognition **35**, 735–747 (2002)
3. Ojala, T., Pietikainen, M., Maenpaa, T.: Multiresolution gray-scale and rotation invariant texture classification with local binary patterns. IEEE Trans. PAMI **24**, 971–987 (2002)
4. Lazebnik, S., Schmid, C., Ponce, J.: A sparse texture representation using local affine regions. IEEE Trans. PAMI **27**, 1265–1278 (2005)
5. Varma, M., Zisserman, A.: A statistical approach to texture classification from single images. Int. J. Comput. Vis. **62**, 61–81 (2005)

6. Permuter, H., Francos, J., Jermyn, I.: A study of gaussian mixture models of color and texture features for image classification and segmentation. Pattern Recognit **39**, 695–706 (2006)
7. Varma, M., Zisserman, A.: A statistical approach to material classification using image patch exemplars. IEEE Trans. PAMI **31**, 2032–2047 (2009)
8. Liao, S., Law, M.W.K., Chung, A.C.S.: Dominant local binary patterns for texture classification. IEEE Trans. Image Process. **18**, 1107–1118 (2009)
9. Guo, Z., Zhang, L., Zhang, D.: A completed modeling of local binary pattern operator for texture classification. IEEE Trans. Image Process. **19**, 1657–1663 (2010)
10. Chen, J., Shan, S., He, C., Zhao, G., Pietikäinen, M., Chen, X., Gao, W.: WLD: a robust local image descriptor. IEEE Trans. PAMI **32**, 1705–1720 (2010)
11. Crosier, M., Griffin, L.D.: Using basic image features for texture classification. Int. J. Comput. Vis. **88**, 447–460 (2010)
12. Guo, Z., Zhang, L., Zhang, D.: Rotation invariant texture classification using LBP variance (LBPV) with global matching. Pattern Recognit. **43**(3), 706–719 (2010)
13. Puig, D., Garcia, M.A., Melendez, J.: Application-independent feature selection for texture classification. Pattern Recognit. **43**, 3282–3297 (2010)
14. Zhao, Y., Huang, D.S., Jia, W.: Completed local binary count for rotation invariant texture classification. IEEE Trans. Image Process. **21**, 4492–4497 (2012)
15. Liu, L., Fieguth, P.W., Clausi, D.A., Kuang, G.: Sorted random projections for robust rotation-invariant texture classification. Pattern Recognit. **45**, 2405–2418 (2012)
16. Manjunath, B.S., Ma, W.Y.: Texture features for browsing and retrieval of image data. IEEE Trans. PAMI **18**, 837–842 (1996)
17. Leung, T.K., Malik, J.: Representing and recognizing the visual appearance of materials using three-dimensional textons. Int. J. Comput. Vis. **43**, 29–44 (2001)
18. Sifre, L., Mallat, S.: Rotation, scaling and deformation invariant scattering for texture discrimination. In: CVPR, pp. 1233–1240 (2013)
19. Liu, L., Zhao, L., Long, Y., Kuang, G., Fieguth, P.: Extended local binary patterns for texture classification. Image Vis. Comput. **30**, 86–99 (2012)
20. Hafiane, A., Seetharaman, G., Zavidovique, B.: Median binary pattern for textures classification. In: Kamel, M.S., Campilho, A. (eds.) ICIAR 2007. LNCS, vol. 4633, pp. 387–398. Springer, Heidelberg (2007)
21. Jin, H., Liu, Q., Lu, H., Tong, X.: Face detection using improved lbp under bayesian framework. In: ICIG. (2004) 306–309
22. Zhao, Y., Jia, W., Hu, R.X., Min, H.: Completed robust local binary pattern for texture classification. Neurocomputing **106**, 68–76 (2013)
23. Liao, S.C., Zhu, X.X., Lei, Z., Zhang, L., Li, S.Z.: Learning multi-scale block local binary patterns for face recognition. In: Lee, S.-W., Li, S.Z. (eds.) ICB 2007. LNCS, vol. 4642, pp. 828–837. Springer, Heidelberg (2007)
24. Zhang, W., Shan, S., Gao, W., Chen, X., Zhang, H.: Local gabor binary pattern histogram sequence (LGBPHS): a novel non-statistical model for face representation and recognition. In: ICCV, pp. 786–791 (2005)
25. Zhao, G., Pietikainen, M., Chen, X.: Combining lbp difference and feature correlation for texture description. IEEE Trans. Image Process. **23**, 2557–2568 (2014)
26. Liu, L., Zhao, L., Long, Y., Kuang, G., Fieguth, P.W.: Extended local binary patterns for texture classification. Image Vision Comput. **30**, 86–99 (2012)
27. Ojala, T., Pietikäinen, M., Harwood, D.: A comparative study of texture measures with classification based on featured distributions. PR **29**(1), 51–59 (1996)

28. Ojala, T., Mäenpää, T., Pietikäinen, M., Viertola, J., Kyllönen, J., Huovinen, S.: Outex - new framework for empirical evaluation of texture analysis algorithms. In: ICPR, pp. 701–706 (2002)
29. Dana, K.J., van Ginneken, B., Nayar, S.K., Koenderink, J.J.: Reflectance and texture of real-world surfaces. ACM Trans. Graph. **18**, 1–34 (1999)
30. Caputo, B., Hayman, E., Fritz, M., Eklundh, J.O.: Classifying materials in the real world. Image Vision Comput. **28**, 150–163 (2010)
31. Tan, X., Triggs, B.: Enhanced local texture feature sets for face recognition under difficult lighting conditions. IEEE Trans. Image Process. **19**, 1635–1650 (2010)
32. Guo, Y., Zhao, G., Pietikäinen, M.: Discriminative features for texture description. Pattern Recognit. **45**, 3834–3843 (2012)

Log-Gabor Weber Descriptor
for Face Recognition

Jing Li, Nong Sang[✉], and Changxin Gao

National Key Laboratory of Science and Technology on Multi-spectral
Information Processing, School of Automation, Huazhong University
of Science and Technology, Wuhan, China
nsang@hust.edu.cn

Abstract. It is well recognized that image representation is the most
fundamental task of the face recognition, effective and efficient image
feature extraction not only has small intraclass variations and large
interclass similarity but also robust to the impact of pose, illumination,
expression and occlusion. This paper proposes a new local image descriptor for face recognition, named Log–Gabor Weber descriptor (LGWD).
The idea of LGWD is based on the image Log-Gabor wavelet representation and the Weber local binary pattern(WLBP) features. The main
motivation of the LGWD is to enhance the multiple scales and orientations Log-Gabor magnitude and phase feature by applying the WLBP
coding method. Histograms extracted from the encoded magnitude and
phase images are concatenated into one to form the image description
finally. The experimental results on the ORL, Yale and UMIST face
database verify the representation ability of our proposed descriptor.

1 Introduction

Face recognition, as one of the most focused research topic in image processing,
pattern recognition and computer vision, has been widely applied in many fields,
such as access control, video surveillance and human-computer interaction etc.
Although numerous approaches have been proposed and tremendous progress
has been made, during the past decades, it could still not perform as well as
desired under uncontrolled conditions. Therefore, how to extract robust and
discriminative features is of vital importance to face recognition.

In the literature, the two-dimensional image feature extraction approaches
for face recognition can be mainly divided into two categories: holistic feature
methods and local feature methods. Holistic feature methods take a single feature, which is extracted from the whole face image, as image description for
face recognition. Principal component analysis (PCA) [1], linear discrimination
analysis (LDA) [2], independent component analysis (ICA) [3], locality preserving projection (LPP) [4] and local linear embedding (LLE) [5] are the typical
ones of this kind. They can be unified into a general framework known as graph
embedding [6]. However, the performance of this class of approaches depends
greatly on the training set and is liable to be influenced by the expression, pose,

© Springer International Publishing Switzerland 2015
C.V. Jawahar and S. Shan (Eds.): ACCV 2014 Workshops, Part I, LNCS 9008, pp. 541–553, 2015.
DOI: 10.1007/978-3-319-16628-5_39

illumination, misalignment, occlusions and so on. On the other side, local feature methods are more robust in uncontrolled conditions. They generally divide the images into several sub-images, extract the features of every one separately. Then combine the feature of each sub-image into a single feature vector by adopting the information fusion methods for further recognition or to combine the recognition result of each sub-image. Gabor filters based method is one of the most representative local feature extraction methods. It has been widely investigated owing to its superior performances in uncontrolled environments [7]. However, the properties of Gabor filters principally involve two drawbacks. One is that, in order to prevent a too high DC component, the bandwidth of a Gabor filter is typically limited to one octave. Hence, a larger number of filters are needed to cover the desired spectrum. The other is that their response is symmetrically distributed around the center frequency, which results in redundant information in the lower frequencies that could instead be devoted to capture the tails of images in the higher frequencies.

An alternative to the Gabor filter is the Log-Gabor filter. It has all the merit of Gabor filter and additionally can be constructed with arbitrary bandwidth and the bandwidth can be optimized to produce a filter with minimal spatial extent. Hence, in this work we prefer to use Log-Gabor filter to extract multiple scales and orientations image information. Log-Gabor filter based feature extraction methods have been excellently applied to image enhancement [8], segmentation [9], edge detection [10] and so on. The existing Log-Gabor transform based image representation methods are mainly categorized into three classes. The first class tries to devises a high dimensional Log-Gabor magnitude feature vector and then reduces its dimension using feature dimension reduction methods PCA and ICA [11]. The second class attempts to divide the image into small patch and its Log-Gabor magnitude mean and standard deviation are used to represent image [12]. The third class applied phase quantization to extract the phase information of the resultant Log-Gabor transform image and generate the binary face image template for recognition [13]. The local magnitude of Log-Gabor transform indicates the energetic information of the image, while the local phase is independent of the local magnitude and it can be used to distinguish between different local structures. To the best of our knowledge, the complementary effect taken by combining magnitude and phase feature simultaneously on the image feature extraction problem has not been systematically explored in the current work.

Derived from Weber's law, Weber local binary pattern(WLBP) [14] is a powerful local descriptor and it has exhibited impressively performance than other widely used descriptors. To further fully exploit the potential rich discrimination texture information embedded in the magnitude and phase feature of the image Log-Gabor transform, in this paper, we propose an image representation scheme, namely Log-Gabor Weber Descriptor (LGWD). LGWD encodes the local pattern of Log-Gabor magnitude and phase feature by using the WLBP. Firstly, we use the Log-Gabor transform to extract the magnitude and phase feature of the image. Secondly, the WLBP descriptor is used to encode information of the magnitude feature, while the phase quantization and the local XOR coding method based WLBP descriptor is utilized to encode the phase feature. Lastly, histogram feature extracted from magnitude and phase are concatenated to one to form the final image representation feature vector; chi-square distance is adopted to measure the similarity between two different LGWD histograms. The experimental results on three benchmark face databases achieved competitive performance compared to other methods. This verified the efficiency and effectiveness of the proposed LGWD based face image representation method.

The remaining part of the paper is organized as follows. In Sect. 2, we give a brief review of related Log-Gabor transform and WLBP descriptor. Our Log-Gabor Weber Descriptor based face image representation method is described in detail in Sect. 3. Experimental results are presented and discussed in Sect. 4. Finally, Sect. 5 contains our conclusions and plans for future research work.

2 Related Works

2.1 Log-Gabor Transform

Log-Gabor filters [15,16] have Gaussian transfer functions when viewed on the logarithmic frequency scale. Due to the singularity of log function, the two-dimensional Log-Gabor filter needs to be constructed in the frequency domain and can only be numerically constructed in the spatial domain via the inverse Fourier transform. In polar coordinates system, it comprises two components, namely the radial filter component and the angular filter component. The frequency response of the two compontents are described as following two expressions respectively.

$$G_r(r) = \exp(\frac{\log(r/f_0)}{2 \cdot \sigma_r^2}) \tag{1}$$

$$G_\theta(\theta) = \exp(-\frac{(\theta - \theta_0)^2}{2 \cdot \sigma_\theta^2}) \tag{2}$$

The transfer function of the overall Log-Gabor filter is constructed by multiplying the frequency response of the two components together as

$$G(r,\theta) = G_r(r) \cdot G_\theta(\theta) = \exp(\frac{\log(r/f_0)}{2 \cdot \sigma_r^2}) \cdot \exp(-\frac{(\theta - \theta_0)^2}{2 \cdot \sigma_\theta^2}) \tag{3}$$

where (r, θ) represents the polar coordinates, f_0 is the center frequency of the filter and it is related to our current scale n by $f_0 = minWave \times mult^n$, in which $minWave$ is the wavelength of smallest scale filter, $mult$ is the scaling factor between successive filters. θ_0 is the orientation angle of the filter, σ_r and σ_θ determine the scale bandwidth and the angular bandwidth respectively. In our experiments, the spatial frequency domain is divided into 6 orientations $(m = 0, 1, \ldots, 5)$ for each of 4 scales $(n = 0, 1, \ldots, 3)$ resulting in a filter bank of $6 \times 4 = 24$ filters. $minWave = 3.0$, $mult = 1.7$, $\sigma_r = 0.65$. The parameters were chosen such that the Log-Gabor filter bank spanned roughly two octaves with some degree of overlap between successive filters. The primary effect of adjusting these parameters is to vary the scale of regions which respond strongly to symmetry processing - thus they were chosen to compromise between small and large sub-patterns.

The image Log-Gabor transform is implemented in the frequency domain. First, using the Fast Fourier Transform(FFT), transforms the image from the spatial domain to the frequency domain. Then, multiply fourier transformed image with the Log-Gabor frequency response, Log-Gabor transformed image is obtained by taking the inverse Fourier transform of multiplied resultant as following:

$$I_{o-m,n} = IFFT(I_F(\mu, \nu) \cdot G_{m,n}(\mu, \nu)) \qquad (4)$$

where $I_F(\mu, \nu)$ is the Fourier transform of the input image, $G_{m,n}(\mu, \nu)$ is the frequency response of Log-Gabor filter with orientation m and scale n, $I_{o-m,n}$ is the Log-Gabor transformed image with the filter $G_{m,n}$. $I_{o-m,n}$ is a complex with two parts, i.e., real part $I_{re-m,n}$ and imaginary part $I_{im-m,n}$. Based on these two parts, the magnitude and phase feature of image Log-Gabor transform can be computed by the following two formulas respectively.

$$I_{Mag}(z) = \sqrt{I_{re-m,n}^2(z) + I_{im-m,n}^2(z)} \qquad (5)$$

$$I_{Phas}(z) = atan2(I_{re-m,n}(z), I_{im-m,n}(z)) \qquad (6)$$

2.2 Local Weber Descriptor

The Weber Local Descriptor(WLD) [17] is derived from Weber's Law, which was proposed by the German physiologist Ernst Weber in 1834. The Weber's law [18] states that the smallest change in the intensity of a stimulus capable of being perceived is proportional to the intensity of the original stimulus. This implies that the ratio of the change in the intensity of the stimulus reflects the degree of human perception of the stimulus. The WLD was proposed to characterize texture information of an image by considering the ratio of changes in pixel intensity which can be considered as stimulus information for visual perception [17].

As an improvement on the WLD, WLBP [14] contains differential excitation component and Local Binary Pattern(LBP) component. These two components are complementary to each other. Specifically, differential excitation preserves the local intensity information but omits the orientations of edges. On the contrary, LBP describes the orientations of the edges but ignore the intensity information.

Differential excitation measures the ratio of change in pixel intensity between a center pixel against its neighbors. It captures the local salient visual patterns.

For example, a high differential excitation value indicates that center pixel potentially belongs to an edge or a spot as there is a strong difference in pixel intensity
between center pixel and its neighbors.

In the case of 3×3 neighborhoods, x_i denotes the $i - th$ neighbors of central
point x_c and p is the number of neighbors, here $p = 8$. For simplicity, in this
work, the differential excitation is computed following the defination of original
WLD [17]:

$$\alpha = arctan(\sum_{i=0}^{p-1} \frac{x_i - x_c}{x_c}) \tag{7}$$

where the $arctan$ function is applied to prevent the output from being too large
and thus could partially suppress the side-effect of noise. Then α is linearly
quantized into T dominant differential excitations as following:

$$\xi_i = floor(\frac{\alpha + \pi/2}{\pi/T}) \quad i = 0, 1, 2, \cdots, T - 1 \tag{8}$$

where $floor(x)$ is a function, which returns the largest integer less than or equal
to x. The differential excitations α within $\left[(\frac{-\pi}{2} + \frac{(i-1)\pi}{T}), (\frac{-\pi}{2} + \frac{i\pi}{T})\right]$ are consequently quantized to ξ_i. In this work, we set $T = 8$.

LBP operator, proposed by Ojala et al. [19], is a powerful means of texture
description. Compared to the orientation component of WLD, LBP can extract
more local structure information and it has been proven to be highly discriminative. With the LBP component, local micro-patterns corresponding to spots,
edges and flat areas are all extracted. The formulas for computing LBP is shown
as the following:

$$LBP(x_c) = \sum_i s(x_i - x_c)2^i \tag{9}$$

$$s(x) = \begin{cases} 1 & \text{if } x \geq 0 \\ 0 & \text{if } x < 0 \end{cases} \tag{10}$$

In which s_i and x^c are the value of neighbor and center points respectively. When
8 neighbour points are chosen, it will be produced 2^8 different binary patterns.
In order to reduce the LBP histogram dimension, we use the rotation invariant
uniform mapping method proposed in [19] to reduce the number of bins from
256 to 10.

After coding the image with WLBP, a two-dimensional histogram of differential excitation and LBP of the image can be defined as

$$H_{WLD}(t, c) = \sum_{i=0}^{I-1} \sum_{j=0}^{J-1} I(\xi(x_{ij}) = t)I(LBP(x_{ij}) = c), \quad c \in C, t \in T \tag{11}$$

where $I \times J$ is the dimensionality of the image, $x_{i,j}$ is the pixel at location (i, j) in the image coordinates, T is the number of intervals of differential excitation, C is the number of the LBP code, and

$$I(A) = \begin{cases} 1 & \text{if } A \text{ is true} \\ 0 & \text{otherwise} \end{cases} \qquad (12)$$

Note that in this two-dimensional histogram, each column corresponds to a certain LBP coding, and each cell $H(t, c)$ corresponds to the frequency of a certain differential excitation interval on a LBP code. The two-dimensional histogram is further reshaped into a one-dimensional histogram by concatenating all the elements of $H(c, t)$. Therefore, the size of the final descriptor is $T \times C$.

3   The Proposed Approach

In this section, we describe the proposed LGWD, which contains two parts: Log-Gabor magnitude Weber descriptor (LGMWD) and Log-Gabor phase Weber descriptor (LGPWD). The first part encodes the variation of image Log-Gabor magnitude feature between the central and its surrounding pixels, whereas the second part encodes the variation of Log-Gabor phase feature. Figure 1 illustrates the flowchart of the proposed LGWD.

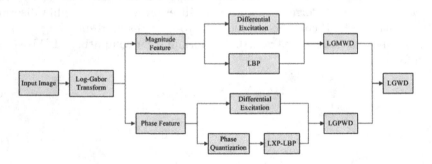

Fig. 1. Flowchart of the proposed LGWD.

3.1   Log-Gabor Magnitude Weber Descriptor (LGMWD)

In image Log-Gabor transform, the magnitude feature is a measurement of local energetic information. For example, high magnitude usually indicates higher energetic local features (e.g., edges, lines, textures). Apply the WLBP operator over each Log-Gabor magnitude feature map to encode the variation of local energy. Suppose that LGMWD histogram extracted from magnitude image $I_{Mag-m,n}$ is $H_{M-m,n}$, the overall LGMWD of the input image is

$$H_M = [H_{M-0,0}, H_{M-0,1}, \dots, H_{M-3,5}] \qquad (13)$$

3.2 Log-Gabor Phase Weber Descriptor (LGPWD)

To encode the Log-Gabor phase feature map use the WLBP operator, one difference with LGMWD is that the binary sequence of LBP component of LGPWD is generated by judging whether the phase of center pixel and its neighbours belong to the same interval(e.g., $[90°, 180°]$).

Briefly speaking, when compute the LBP component of LGPWD, phases are firstly quantized into different range, then local XOR coding method is applied to the quantized phases of the central pixel and each of its neighbors, and finally the resulting binary labels are concatenated together as the local pattern of the central pixel. The LBP component of LGPWD in binary and decimal form is defined as follows:

$$LBP_{m,n}(z_c) = \left[B_{m,n}^P, B_{m,n}^{P-1}, \ldots, B_{m,n}^1 \right]_{binary} = \left[\sum_{i=1}^{P} 2^{i-1} B_{m,n}^i \right]_{decimal} \quad (14)$$

where z_c denotes the central pixel position in the Log-Gabor phase feature map with scale n and orientation m , P is the size of neighborhood, and $B_{m,n}^i$ denotes the pattern calculated between z_c and its neighbor z_i, which is computed as follows:

$$B_{m,n}^i = q(\Phi_{m,n}(z_c)) \otimes q(\Phi_{m,n}(z_i)) \quad i = 1, 2, \ldots, P \quad (15)$$

where $\Phi_{m,n}(z_c)$ denotes the phase value of pixel z_c, $\otimes$ denotes the local XOR coding operation, which is based on XOR operator, as defined in (16); $q(\cdot)$ denotes the quantization operator, which calculates the quantized code of phase according to the number of phase ranges N, as defined in (17)

$$c_1 \otimes c_2 = \begin{cases} 0 & \text{if } c_1 = c_2 \\ 1 & \text{otherwise} \end{cases} \quad (16)$$

$$q(\Phi_{m,n}(\cdot)) = floor(\frac{\Phi_{m,n}(\cdot)}{\pi/N}) \quad (17)$$

Apply the WLBP operator over each Log-Gabor phase map to encode the variation of local phase. Suppose that LGPWD histogram extracted from phase image $I_{Pha-m,n}$ is $H_{P-m,n}$, the overall LGPWD of the input image is

$$H_P = [H_{P-0,0}, H_{P-0,1}, \ldots, H_{P-3,5}] \quad (18)$$

3.3 LGWD Image Representation and Classification

Once the multiple scales and orientations LGMWD and LGPWD histogram features are extracted from each transformed image, it is necessary to combine them in a manner to take advantage of the magnitude and phase feature. In this work, LGMWD and LGPWD histograms are simply concatenated into a single feature vector to represent the image as follows

$$H = [H_M, H_P] \quad (19)$$

Nearest Neighbor with chi-squared distance is used for classification. Suppose H_1 and H_2 are two normalized LGWD histogram, the chi-square distance between two histograms is defined using the following form:

$$\chi^2(H_1, H_2) = \frac{1}{2} \sum_i \frac{(H_{1i} - H_{2i})^2}{H_{1i} + H_{2i}} \qquad (20)$$

4   Experiments

4.1   Face Databases

Three benchmark face database: ORL face database, Yale face database and UMIST face database are used in the experiments to evaluate the performances of the proposed face image representation method.

ORL face database [20] contains 10 different images of each of 40 distinct subjects. The size of each image is 92×112 pixels, with 256 grey levels per pixel. For some subjects, the images were taken at different times, varying the lighting, facial expressions and facial details.

Yale face database [2] contains 165 grayscale images of 15 individuals, 11 images per subject, where there are rich illumination, expression and occlusion variations. The size of each image is 100×100 pixels, with 256 grey levels per pixel.

UMIST face database [21] consists of 564 images of 20 people. Each subject covers a range of poses from profile to frontal views and a range of race, sex and appearance. For simplicity, the pre-cropped version of the UMIST database is used in this experiment. The size of cropped image is 92×112 pixels with 256 gray levels.

In a word, face images used in the experiments have a large variation in terms of pose, illumination, expression, occlusion, race and time lapse. Test on these images can have a comprehensive evaluation of the robustness of the image representation method to these factors. Figure 2 illustrates some example facial images of three subjects from three face databases.

For each database, we randomly partitioned it into K $(K = 10, 8, 5)$ subsets. Among the K subsets, **one is used as training data, and the remaining K-1 ones as validation data for testing**. The final results is the average recognition accuracy of the K iterations. For equal comparison, we collect the histogram from the whole image which will result in lower recognition rate than conventional sub-image based method.

4.2   Investigating the Effectiveness of Log-Gabor Magnitude and Phase Feature

Recall from Sect. 3 that the proposed LGWD feature consists of two parts: LGMWD and LGPWD. In this section, we have performed experiments to investigate the following issues: 1) which part (i.e., LGMWD versus LGPWD parts)

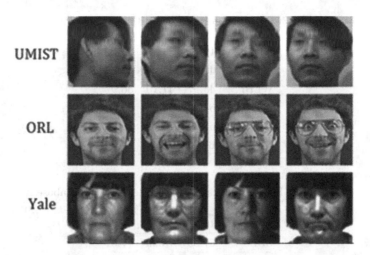

Fig. 2. Face image from the three databases.

contributes more to face recognition performance; and 2) the feasibility for the usage of combined LGMWD and LGPWD features against a framework utilizing separately LGMWD and LGPWD features.

Table 1. Recognition rate (%) on the ORL database

K	10	8	5
LBP	44.69	47.25	55.44
WLBP	53.42	56.64	69.50
LGMWD	73.22	80.21	86.13
LGPWD	72.19	75.82	85.44
LGWD	**78.64**	**81.21**	**89.88**

The results are given in the Tables 1, 2 and 3. Note that for comparison purpose, the recognition rates obtained using both LBP and WLBP features are provided as baseline performances. From the Tables 1, 2 and 3, we can arrive at the following two conclusions. First, LGMWD contribute much more than the LGPWD for face recognition performance; In addition, compared with LBP and WLBP features, the recognition rates obtained for the use of LGPWD are better than both feature extraction algorithms. This demonstrates high discriminating capabilities of LGPWD. Second, the combination of both LGMWD and LGPWD parts achieves better results, compared with the cases of separately using them; this indicates that LGMWD and LGPWD parts are able to provide different information and to be mutually compensational in terms of boosting face recognition performance.

Table 2. Recognition rate (%) on the yale database

K	10	8	5
LBP	40.40	41.21	47.88
WLBP	43.57	47.10	56.97
LGMWD	65.25	69.00	78.33
LGPWD	62.49	67.79	76.67
LGWD	**66.73**	**70.91**	**81.52**

Table 3. Recognition rate (%) on the UMIST database

K	10	8	5
LBP	67.69	70.29	77.61
WLBP	71.65	75.30	86.17
LGMWD	81.43	85.04	91.09
LGPWD	77.16	82.36	89.17
LGWD	**84.04**	**87.79**	**93.48**

Table 4. Recognition rate (%) on the ORL database

K	10	8	5
LBP	44.69	47.25	55.44
WLBP	53.42	56.64	69.50
Gabor–WLBP	71.78	74.18	83.88
Log-Gabor Magnitude PCA	61.34	65.11	74.94
Log-Gabor Statistic	67.28	69.89	81.06
Log-Gabor Phase	62.46	66.73	75.88
MBC	70.61	72.75	83.00
LGWD	**78.64**	**81.21**	**89.88**

4.3 Comparisons with Other Methods

In this section, we compare the face recognition performance of our methods LGWD with those closely related representative methods, including LBP [22], WLBP, Gabor transform based WLBP(Gabor-WLBP), Log-Gabor magnitude PCA method [11], Log-Gabor statistic method [12](all the images are divided into 8×8 sub-image), Log-Gabor phase template method [13] and monogenic binary coding(MBC) [23]. Experimental results of these methods on three databases are illustrated in the Tables 4, 5 and 6 respectively.

From the obtained results, we have the following observations. First, WLBP outperforms LBP, the proposed LGWD outperforms benchmark methods LBP,

Table 5. Recognition rate (%) on the yale database

K	10	8	5
LBP	40.40	41.21	47.88
WLBP	43.57	47.10	56.97
Gabor–WLBP	53.80	59.74	67.42
Log-Gabor Magnitude PCA	41.75	43.25	51.31
Log-Gabor Statistic	43.37	47.71	57.58
Log-Gabor Phase	41.92	43.36	51.44
MBC	53.94	57.14	66.06
LGWD	**66.73**	**70.91**	**81.52**

Table 6. Recognition rate (%) on the UMIST database

K	10	8	5
LBP	67.69	70.29	77.61
WLBP	71.65	75.30	86.17
Gabor–WLBP	77.80	82.01	89.87
Log-Gabor Magnitude PCA	65.06	69.49	79.39
Log-Gabor Statistic	78.47	81.84	88.78
Log-Gabor Phase	62.61	66.86	77.87
MBC	74.11	78.19	86.78
LGWD	**84.04**	**87.79**	**93.48**

WLBP and Gabor-WLBP. It reveals the Log-Gabor transform contains richer image descriminant information than image grayscale and Gabor transform representation. Second, LGWD outperforms three existing Log-Gabor based image representation methods, which shows that WLBP is more effective to extract the information of Log-Gabor transform than mean and standard deviation based statistic method, dimension reduction method and phase coding method. Third, LGWD outperforms monogenic signal based representation MBC. All of these observations definitly support the effectiveness of the techniques proposed in this study. And implies that it is possible to present one effective image descriptor based on Log-Gabor transform information and WLBP encoding method.

5 Conclusion

Image representation is increasingly accepted as a difficult and challenging computer vision problem. In this paper, we have investigated a novel image representation approach for face recognition, namely Log–Gabor Weber Descriptor

(LGWD). The LGWD absorbs the merit of both image Log–Gabor transform information and local Weber descriptor method. Experimental results on ORL, Yale and UMIST database showed LGWD outperformed these closely related image feature extraction methods. Therefore confirmed the proposed approach can extract more discriminative information for face recognition.

Although high performance is achieved by the proposed method, it should be pointed out that our method has a drawback of high dimensionality(The dimension of our LGWD is $4 \times 6 \times 2 \times 8 \times 10 = 3840$). How to reduce the LGWD feature dimension or design a compact local Log-Gabor feature descriptor of better performance will be considered in our future work.

References

1. Turk, M., Pentland, A.: Eigenfaces for recognition. J. Cogn. Neurosci. **3**, 71–86 (1991)
2. Belhumeur, P.N., Hespanha, J.P., Kriegman, D.J.: Eigenfaces vs. fisherfaces: recognition using class specific linear projection. IEEE Trans. Pattern Anal. Mach. Intell. **19**, 711–720 (1997)
3. Bartlett, M.S., Movellan, J.R., Sejnowski, T.J.: Face recognition by independent component analysis. IEEE Trans. Neural Networks **13**, 1450–1464 (2002)
4. Lu, J., Tan, Y.: Regularized locality preserving projections and its extensions for face recognition. IEEE Trans. Syst. Man Cybern. Part B: Cybern. **40**, 958–963 (2010)
5. Li, X., Lin, S., Yan, S., Xu, D.: Discriminant Locally Linear Embedding With High-Order Tensor Data. IEEE Transactions on Systems, Man, and Cybernetics, Part B: Cybernetics **38**, 342–352 (2008)
6. Yan, S., Xu, D., Zhang, B., Zhang, H., Yang, Q., Lin, S.: Graph embedding and extensions: a general framework for dimensionality reduction. IEEE Trans. Pattern Anal. Mach. Intell. **29**, 40–51 (2007)
7. Liu, C., Wechsler, H.: Gabor feature based classification using the enhanced fisher linear discriminant model for face recognition. IEEE Trans. Image Process. **11**, 467–476 (2002)
8. Wang, W., Li, J., Huang, F., Feng, H.: Design and implementation of Log-Gabor filter in fingerprint image enhancement. Pattern Recogn. Lett. **29**, 301–308 (2008)
9. Cline, M.T., Bernard, G.: Character segmentation-by-recognition using log-gabor filters. In: International Conference on Pattern Recognition, pp. 901–904 (2006)
10. Gao, X., Sattar, F., Venkateswarlu, R.: Multiscale corner detection of gray level images based on Log-Gabor wavelet transform. IEEE Trans. Circuits Syst. Video Technol. **17**, 868–875 (2007)
11. Chen, X., Jing, Z.: Infrared face recognition based on Log-Gabor wavelets. Int. J. Pattern Recognit. Artif. Intell. **20**, 351–360 (2006)
12. Arrospide, J., Salgado, L.: Log-Gabor filters for image-based vehicle verification. IEEE Trans. Image Process. **22**, 2286–2295 (2013)
13. Singh, R., Vatsa, M., Noore, A.: Textural feature based face recognition for single training images. Electron. Lett. **41**, 640–641 (2005)
14. Liu, F., Tang, Z., Tang, J.: WLBP: weber local binary pattern for local image description. Neurocomputing **120**, 325–335 (2013)
15. David, J.F.: Relations between the statistics of natural images and the response properties of cortical cells. J. Opt. Soc. Am. A **4**, 2379–2394 (1987)

16. Kovesi, P.: Image features from phase congruency. Videre: A J. Comput. Vis. Res. 1 1–26 (1999)
17. Chen, J., Shan, S., He, C., Zhao, G., Pietikainen, M., Chen, X., Gao, W.: WLD: a robust local image descriptor. IEEE Trans. Pattern Anal. Mach. Intell. **32**, 1705–1720 (2010)
18. Jain, A.K.: Fundamentals of Digital Signal Processing. Prentice-Hall, Englewood Cliffs (1989)
19. Ojala, T., Pietikainen, M., Maenpaa, T.: Multiresolution gray-scale and rotation invariant texture classification with local binary pattern. IEEE Trans. Pattern Anal. Mach. Intell. **24**, 971–987 (2002)
20. Samaria, F.S., Harter, A.C.: Parameterisation of a stochastic model for human face identification. In: IEEE Workshop on Applications of Computer Vision, pp. 138–142 (1994)
21. Graham, D.B., Allinson, N.M.: Characterizing virtual eigensignatures for general purpose face recognition. In: Wechsler, H., Jonathon Phillips, P., Bruce, V., Fogelman Soulié, F., Huang, T.S. (eds.) Face recognition: from theory to applications. NATO ASI Series F, Computer and Systems Sciences, vol. 163, pp. 446–456. Springer, Heidelberg (1998)
22. Ahonen, T., Hadid, A., Pietikanen, M.: Face description with local binary patterns: application to face recognition. IEEE Trans. Pattern Anal. Mach. Intell. **28**, 2037–2041 (2006)
23. Yang, M., Zhang, L., Shiu, S.C.K., Zhang, D.: Monogenic binary coding: an efficient local feature extraction approach to face recognition. IEEE Trans. Inf. Forensics Secur. **7**, 1738–1751 (2012)

Robust Line Matching Based on Ray-Point-Ray Structure Descriptor

Kai Li, Jian Yao[✉], and Xiaohu Lu

School of Remote Sensing and Information Engineering, Wuhan University,
Wuchang District, Wuhan, Hubei, People's Republic of China
jian.yao@whu.edu.cn

Abstract. In this paper, we propose a novel two-view line matching method through converting matching line segments extracted from two uncalibrated images to matching the introduced Ray-Point-Ray (RPR) structures. The method first recovers the partial connectivity of line segments through sufficiently exploiting the gradient map. To efficiently matching line segments, we introduce the Ray-Point-Ray (RPR) structure consisting of a joint point and two rays (line segments) connected to the point. Two sets of RPRs are constructed from the connected line segments extracted from two images. These RPRs are then described with the proposed SIFT-like descriptor for efficient initial matching to recover the fundamental matrix. Based on initial RPR matches and the recovered fundamental matrix, we propose a match propagation scheme consisting of two stages to refine and find more RPR matches. The first stage is to propagate matches among those initially formed RPRs, while the second stage is to propagate matches among newly formed RPRs constructed by intersecting unmatched line segments with those matched ones. In both stages, candidate matches are evaluated by comprehensively considering their descriptors, the epipolar line constraint, and the topological consistency with neighbor point matches. Experimental results demonstrate the good performance of the proposed method as well as its superiority to the state-of-the-art methods.

1 Introduction

Image matching is an indispensable procedure in almost all applications which require recovering 3D scene structure from 2D images such as 3D reconstruction, scene interpretation, robotic navigation, structure from motion, etc. A lot of objects in real scenes can be outlined easily by line segments. So, recovering 3D scene structure from line matches has advantages over that from point matches. In some cases, for example, the scenes are poorly-textured, recovering their 3D structures from line matches seems the only choice because point matches are often insufficient in this kind of scenes. Despite that recovering 3D scene structure from line segment matches seems a better choice than that from point matches, both the instability of endpoints of line segments and the lost of connectivity of line segments complicate the matching of line segments.

© Springer International Publishing Switzerland 2015
C.V. Jawahar and S. Shan (Eds.): ACCV 2014 Workshops, Part I, LNCS 9008, pp. 554–569, 2015.
DOI: 10.1007/978-3-319-16628-5_40

Line matching methods can be roughly divided into two categories: methods that match individual line segments and methods that match a group of line segments. Some methods matching individual line segments take advantages of the photometric information associated with the line segments, such as intensity [1,2], color [3], or/and gradient [4,5] in the regions around the line segments. All of these methods are based on the assumption that there are considerable overlaps between corresponding line segments. But if two line segments in correspondence don't share sufficient corresponding part, it is hardly possible to match them correctly. Moreover, these methods tend to produce false matches in regions with repeated textures because of the lack of variation in the photometric information. Another group of methods matching individual line segments incorporate point matches into line matching [6–9]. These methods first find a large group of point matches using existing point matching methods [10,11], and then exploit invariants between coplanar points and line(s) under certain image transformations. Line segments which meet the invariants are regarded to be in correspondence. All these methods share the same disadvantage that they tend to fail when scenes captured are poorly-textured since there are often not sufficient point matches to be found in these scenes.

Matching group of line segments is more complex, but more constraints are available for disambiguation. Most of these methods [12–14] first use some strategies to intersect line segments to form junctions and then utilize features associated with the junctions for line matching. In [15,16], line segments are not intersected to form junctions, but the stability of the relative positions of the endpoints of a group of line segments in a local region under various image transformations is exploited. Their method is robust in some very challenging cases. But the dependence on approximately corresponding relationship between the endpoints of line matches leads to the tendency of the method to produce false matches when substantial disparity exists in the locations of the endpoints.

Our proposed line matching method exploits features of junctions too, but in a quite different way. At the place of the junctions, we form a Ray-Point-Ray (RPR) structure, consisting of a junction point and two rays (line segments) connected to the point. RPRs are described with a robust SIFT-like descriptor. Through exploiting photometric and geometric constraints associated with RPR matches as well as their topological relationship with neighbor point matches, we propagate RPR matches in an iterative scheme. Experimental results show our method generates more correct line matches with higher accuracy than the state-of-the-art methods in most cases.

2 Partial Line-Connectivity Recovery

Line segments extracted by the existing line segment detectors [17,18] are often separated with each other since the segmentation procedure. We refine them by partially recovering their connectivity through exploiting the gradient map of the original image. Given a line segment l_1, we search its neighbors using two circles centered at its two endpoints with the same radius, d. Other line segments fall inside these two circles are neighbors of l_1. For a searched neighbor line segment l_2, if the orientation difference between l_1 and l_2 is less than a given threshold α (α was set to 5° in this paper), we tentatively merge the two line segments to a long one and adjust the endpoints of the new line segment to maximize its Gradient Magnitudes (GM), which is the mean of gradient magnitudes of all pixels in the line segment. The direction of a line segment is the direction of the vector from one endpoint of the line segment to another one. The way of adjusting the endpoints of a merged line segment to maximize its GM is illustrated in Fig. 1 where l_1 and l_2 are two line segments to be merged, l_3 is the merged line segment, l_3' is an example of adjusting the endpoints of l_3 to find the line segment with maximal GM, and the red dots are pixels in the lines orthogonal with l_3 and passing through its endpoints. 5 pixels for each endpoints is used, generating 25 candidate line segments by linking any pixel in one side to all pixels in the other side. The one with maximal GM is selected as the merged line segment of l_1 and l_2. If the GM of the merged line segment is greater than 80 % of the sum of that of l_1 and l_2, we accept it and use it to replace l_1 and l_2 for further steps. If the direction difference between l_1 and l_2 is above α, we intersect them and generate a junction. If the distance between the junction and the endpoint of l_1 is less than d, we accept the junction and extend l_1 to the junction. After all line segments being processed by above steps, we obtain a new set of line segments in which some of which are connected with each other. The steps above are conducted iteratively until no more line segment could be merged or extended. Figure 2 is a demonstration of our line refinement method on a real scene. Comparison between the second image and the third one shows that some line segments are extended to be longer while some line segments are merged with others.

Fig. 1. An illustration of merging two line segments and adjusting the endpoints of the merged line segment to maximize the gradient magnitude of the line segment.

Fig. 2. An example of the method of recovering the partial connectivity of line segments on a real scene: (Left) the original image; (Middle) the line segments detected by EDLines [17]; (Right) the refined line segments by our method. The small circles in the right image denote the added junctions.

3   Line Matching

After recovering the partial connectivity of line segments, some line segments are connected with others from which we construct two sets of initial RPRs from two images respectively. Through describing and matching these RPRs, we find a set of initial RPR matches, which are the basis for the propagation of RPR match in an iterative scheme.

3.1   RPR Construction

Line segments may connect with each other in four forms as shown in Fig. 3. The numbers of RPRs to be constructed are different in different forms. The principle of forming RPRs from connected line segments is that any two line segments which connect with a common joint point but are not in the same line can be used to form a RPR. Under this principle, 1, 2 and 4 RPRs can be constructed in Figs. 3(a)–(c), respectively. In Fig. 3(d), the number of RPRs to be constructed is dependent on the number of line segments connected to the intersection and their configuration.

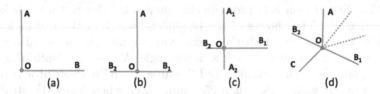

Fig. 3. Four forms of line segments connecting with each other.

3.2   RPR Descriptor

A RPR consists of a point and two rays (line segments) connected with the point. Such relationship between the point and the two rays is stable under image transformations. Inspired by SIFT [10], we use the directions of both rays as dominant directions of the point and generate orientation histograms. Entries in both groups of orientation histograms are concatenated to get the descriptor of the RPR.

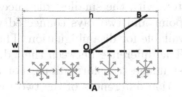

Fig. 4. An illustration of generating orientation histogram using one ray of a RPR.

The way of generating orientation histograms using one of the rays of a RPR is illustrated in Fig. 4, in which the direction of ray **OA** is currently used as dominant direction. The region is a rectangle centered at the point with the width along the ray of 8 pixels and the height of 16 pixels. The region is evenly split into 4 subregions along the height, generating 4 rectangles with the same size of 8×4. In each subregion, an orientation histogram containing 8 bins is constructed. Thus, using one ray as dominant direction results in a vector of $8 \times 4 = 32$ dimensions, and the final descriptor of a RPR is a vector of 64 dimensions. For each pixel in the rectangle, its gradient magnitude is weighted and added to corresponding entry of certain histogram according to its gradient orientation. The weighted gradient magnitude of a pixel is calculated as:

$$G(x, y) = |g(x, y)| \frac{1}{\sqrt{2\pi\sigma^2}} \exp\left(-\frac{(\Delta y)^2}{2\sigma^2}\right), \tag{1}$$

where $g(x, y)$ denotes the gradient magnitude of the pixel located at (x, y), Δy is the distance of the pixel to the ray, σ is a parameter determining the weighting function, and $G(x, y)$ is the weighted gradient magnitude.

We utilize these ways to construct histograms base on the following two reasons. The first reason is to promote efficiency of the descriptor while keeping its distinctiveness. The choices of the height and width of the rectangle derive from SIFT descriptor in which the region forming an orientation histogram is a 4×4 square, while ours is a 8×4 rectangle. We do not split the region into subregions along the ray like SIFT because the RPR descriptor is only 64 dimensions in our way, rather than 256 dimensions when using the strategy of SIFT. It is certainly more efficient. Besides, in most cases, the numbers of RPRs to be matched are far less than the numbers of detected keypoints of SIFT. A descriptor with lower dimension is distinctive enough for matching. The second one is to avoid the possible shift of the point along the ray. The point in each RPR results from intersecting two lines in neighbor. Its position may shift slightly along the ray since the positions of the two lines forming the point may vary slightly under image transformations. If we split the region along the ray and use the point as the basis for assigning weights, the histograms of regions in correspondence would differ with each other mistakenly.

Two sets of RPR descriptors would be obtained after describing all RPRs constructed from two images. The general way of matching these descriptors is to like the SIFT matching procedure by computing their Euclidean distances and selecting pairs of descriptors with the smallest distances as matches. But since each RPR consists of a point and two rays connected with the point, there is additional information available for disambiguation. The two rays in each RPR locate in a local region. The difference between their directions should vary at a small range under most image transformations. For a test pair of RPRs, if θ_1 and θ_2 denote the direction differences of the two rays of the pair of RPRs, then, $|\theta_1 - \theta_2|$ should be a small value if the RPR pair is a correct match. This constraint can be used to discard many false candidates before evaluating their descriptor distances and thus contributes to better matching results.

3.3   Global Image Scale Change Estimation

Our strategy for global image scale change estimation is based on the fact that two images in the (approximately) same scale produce the most putative RPR matches. We first build Gaussian pyramids for original reference and query images. At the same time, line segments extracted from original images are accordingly adjusted to fit the new images. In this paper, the pyramids have 4 octaves with 4 layers in each octave. After that, the original reference image is to match all images in the pyramid built for the query image with their RPRs. The same procedure is applied to the original query image and all images in the pyramid built for the reference image. The pair of images producing the most putative RPR matches is regarded to be in the same scale and will be used in further steps.

3.4   RPR Match Propagation

Point matches obtained along with RPR matches can be used for the estima-tion of the fundamental matrix. To achieve a stable and precise fundamental matrix, we first use RANSAC [19] to estimate an initial fundamental matrix and refine it by the Normalized 8-point Method [20] and Levenberg-Marquardts optimization [21] in order. After that, we obtain the fundamental matrix and the corresponding group of RPR matches, from which we commence propagat-ing RPR matches. The RPR match propagation is achieved by progressively increasing the threshold for the distance of an accepted point match according to the fundamental matrix, which is defined as follows:

$$d\left(\mathbf{x}_i, \mathbf{x}_i'\right) = \left(\mathbf{x}_i'^{\top}\mathbf{F}\mathbf{x}_i\right)^2 \times \left(\frac{1}{(\mathbf{F}\mathbf{x}_i)_1^2 + (\mathbf{F}\mathbf{x}_i)_2^2} + \frac{1}{(\mathbf{F}^{\top}\mathbf{x}_i')_1^2 + (\mathbf{F}^{\top}\mathbf{x}_i')_2^2}\right), \quad (2)$$

where $\mathbf{x}_i$ and $\mathbf{x}_i'$ represent the i-th candidate corresponding points in the refer-ence image and query image respectively, and $(\mathbf{F}\mathbf{x})_k^2$ denotes the square of the k-th entry of the vector $\mathbf{F}\mathbf{x}$. The distance will later be referred in a simplified manner, as Distance according to Fundamental Matrix (DFM). RPR pairs whose DFMs are smaller will be matched first and then serve as the basis for the next iteration to introduce new RPR matches. Unmatched RPRs are first grouped according to matched RPRs and then matched in corresponding groups. Point matches are introduced to filter false matches while guide the process of adding new matches.

Point Match Expanding. There are 3 pairs of points in each RPR match: one pair of junctions and two pairs of endpoints of the two pairs of rays. The pair of junctions are matched along with the RPR match. Point match expanding process aims only at the two pairs of endpoints. For a test pair of endpoints, we check if they can be regarded as a match based on the following criteria. First, the distance of their descriptors should be less than a given threshold η ($\eta = 0.5$

in this paper). We describe the pair of endpoints using the similar way as that describing RPRs. But since only one ray is connected with each endpoint, their descriptors are therefore only 32 dimensions by using the direction of one ray as the dominant direction. Second, the DFM of the test endpoint pair should be less than a pre-defined threshold (18 was used in this paper).

RPR Match Filtering. False RPR matches with the two points locating near the corresponding epipolar lines may have been accepted by the fundamental matrix mistakenly. These false matches can be eliminated by exploiting the topological consistency between RPR matches and their neighbor point matches.

For a RPR match $(\mathbf{M}_1, \mathbf{M}_2)$, we first find out a certain number of the nearest matched points to the points of $\mathbf{M}_1$ and $\mathbf{M}_2$. The number is set as 8 to make a balance between efficiency and in-sensitiveness with noises among the point match group. After that, we obtain two sets of neighbor matched points, $\tilde{\mathcal{M}}_1$ and $\tilde{\mathcal{M}}_2$ for $\mathbf{M}_1$ and $\mathbf{M}_2$ respectively. If $(\mathbf{M}_1, \mathbf{M}_2)$ is a correct match, elements in $\tilde{\mathcal{M}}_1$ and $\tilde{\mathcal{M}}_2$ should meet the following two conditions. First, a proportion of elements in $\tilde{\mathcal{M}}_1$ and $\tilde{\mathcal{M}}_2$ should be correspondences. The proportion was set as 0.5 in this paper. Second, the distribution of elements in $\tilde{\mathcal{M}}_1$ and $\tilde{\mathcal{M}}_2$ and their correspondences according to $\mathbf{M}_1$ and $\mathbf{M}_2$ should also be consistent with each other. Refer to Fig. 5, **AOB** and **A'O'B'** are a pair of RPRs in correspondence from two images. The other RPRs in yellow are matched RPRs nearby. Matched points generated by RPR matches are represented by yellow dots, which can be the junctions or the endpoints of the matched rays. The junction and the two rays in addition to their reverse extensions in each matched RPR form a coordinate-like structure. Matched points distribute in different quadrants of the coordinate. This kind of distribution is invariant under projective transformation if the matched points are coplanar with the matched RPR in 3D space. So, a proportion of elements in $\tilde{\mathcal{M}}_1$ and $\tilde{\mathcal{M}}_2$ and their correspondences should lie in the same quadrants of the two coordinates formed by $\mathbf{M}_1$ and $\mathbf{M}_2$. The proportion was set as 0.8 empirically in this paper. The kind of topological consistency between RPR matches and their neighboring point matches will be referred in a simplified manner, as Point Distribution Consistency (PDC) for subsequent use.

RPR Match Propagation. RPR match propagation consists of two main steps: unmatched RPRs grouping and unmatched RPRs matching. For each unmatched RPR, we find 3 of its nearest matched RPRs whose points are the nearest 3 points to the point of the unmatched RPR among all points in matched RPRs. The unmatched RPR is then put into the 3 groups. The reason that each unmatched RPR is redundantly put into multiple groups is to ensure potential matches will be distributed into at least one pair of groups in correspondence. After that, there are one matched RPR and some unmatched RPRs in each group. As shown in Fig. 5, unmatched RPRs, marked in blue, are distributed in the four quadrants of the coordinates formed by the matched RPR. These unmatched RPRs are divided into 4 groups according to the quadrants their points belong.

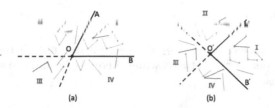

Fig. 5. An illustration of a pair of RPRs in correspondence and the distribution of their neighboring line segments and RPRs. **AOB** and **A′O′B′** are a pair of RPRs in correspondence from two images. The other matched RPRs are marked in yellow. Matched single line segments are marked in green. Unmatched single line segments and RPRs are marked in blue. The yellow and green dots are matched points generated by RPR matches and single line matches respectively (Color figure online).

After grouping unmatched RPRs in two images, the matching process is performed in each pair of groups in correspondence. For each test RPR pair, it will be accepted as a candidate match if it meets all these requirements. First, the distance of their descriptors is less than a given threshold η_1 (η_1 equals 0.5 in the paper). Second, the DFM of the pair of points is less than η_2. The value of η_2 changes in every iteration. Third, the two RPRs meet the PDC.

Iterative Scheme. Some new RPR matches has been generated in previous step. The point match group can be expanded again from the newly obtained RPR matches. Under the new point match group, some false RPR matches may be filtered out by rechecking their PDC since their neighbor matched points may have varied. After that, it is necessary to update the point match group and line match group since some RPR matches may have been deleted.

Up to now, an iteration of RPR match propagation has been fulfilled. Before increasing the value of η_2 and starting a new iteration, we need to conduct the same iteration again without changing η_2 to add those RPR matches neglected in the previous iteration. In the previous iteration, several RPRs in one image may be matched to the same RPR in another image as their best correspondence. We select the pair whose descriptor distance is the smallest and reject others. The left RPRs may find their correct correspondences under the new pair of groups of unmatched RPRs. After that, we increase the value of η_2 and begin a new iteration.

The way of tuning η_2 at different iterations determines the number of RPR matches to be obtained and the times of iterations. So, it is necessary to explain it separately. The same strategy is employed in the propagation of RPR match among single line segments discussed in next section. DFM for a pair of point, as defined in Eq. (2), is actually the quadratic sum of the distances of a pair of points to corresponding epipolar lines. For a pair of points, $\mathbf{P}_1$ and $\mathbf{P}_2$, their DFM can be represented as $\sigma = d_1^2 + d_2^2$ where d_1 denotes the distance from $\mathbf{P}_2$ to the epipolar line determined by $\mathbf{P}_1$ and d_2 denotes the distance from $\mathbf{P}_1$ to the epipolar line determined by $\mathbf{P}_2$. Point matches in our case are not in precise correspondence, but in an approximate manner. Their DFM is a relative great value. At the initial iteration, we accept two points as a match if η_2 is less than α. Then, we progressively increase η_2 at a certain step until it reaches the upper limit β. In this paper, α was set as 8 while β equals 18 and the step is 5 for two adjacent iterations.

3.5   Single Line Segment Matching

RPR matches among initially formed RPRs are mostly found after propagation. Each RPR match brings two line matches. Besides these line matches, there exist a considerable amount of single line segments which do not connect with other line segments but can find their correspondences in the other image. To find these line matches, we first group single line segments based on matched RPRs and then match them in corresponding groups.

Single Line Segment Grouping. For each single line segment, we find 3 nearest RPRs whose points are the nearest 3 points to the midpoint of line segment among all points in matched RPRs. The line segment is then distributed into the corresponding 3 groups. After that, in each group, there are a matched RPR and some unmatched single line segments. Refer to Fig. 5, unmatched single line segments, marked in blue, distribute in different quadrants of the coordinates formed by the matched RPR. These unmatched line segments are then divided into four groups according to the quadrants their endpoints belong to. For each line segment, if any of its two endpoints lies in a certain quadrant, the line segment is put into the corresponding group. Under this grouping strategy, each line segment may be put into several groups. Despite, in most cases, this may lead to multiple evaluations of the same pair of line segments, it is still necessary to do so to ensure potential line matches will are distributed into at least one pair of groups in correspondence.

Single Line Segment Matching. The matching process is conducted in each pair of groups in correspondence. For such a group pair $(\mathcal{G}_1, \mathcal{G}_2)$, there are a RPR match denoted as $(\mathbf{M}_1, \mathbf{M}_2)$ and two sets of single line segments. Each single line segment in $\mathcal{G}_1$ will be evaluated with all single line segments in $\mathcal{G}_2$. For a test pair $(\mathbf{l}_1, \mathbf{l}_2)$, we first check whether the direction difference of $\mathbf{l}_1$ and $\mathbf{l}_2$ is consistent with the direction differences of the two line segment matches generated by $(\mathbf{M}_1, \mathbf{M}_2)$. Correct line segment matches in local region share similar direction differences under image transformations. So we calculate the mean value of the direction differences of the two pairs of matched rays, denoted as σ_1. Suggest the direction difference between $\mathbf{l}_1$ and $\mathbf{l}_2$ is σ_2. If $|\sigma_2 - \sigma_1| < \epsilon$, where ϵ is a user-defined threshold set as $20°$ in this paper, we accept the test pair temporarily and take it for further evaluation.

Then, we use the epipolar constraint to check the approximate correspondence of the endpoints of $\mathbf{l}_1$ and $\mathbf{l}_2$. It is unreasonable to count on the correspondence of the endpoints of corresponding line segments to match them.

However, we notice almost all line segment matches own at least one pair of endpoints which approximately correspond with each other. Thus, if neither the DFMs of the two pairs of endpoints of l_1 and l_2 is below a relatively small value (100 is used in the paper), we regard (l_1, l_2) as a false match.

As mentioned before, the fundamental matrix is incapable to be used to discard false corresponding points that approach the corresponding epipolar lines. So some pairs of line segments may have passed previous constraint mistakenly. We employ topological consistency between line segment matches and their neighbor point matches to avoid this problem. For l_1 and l_2, we find n (set as 8 in this paper) of the nearest matched points to the midpoints of the two line segments, generating two point sets, S_1 and S_2. If (l_1, l_2) is a correct match, S_1 and S_2 should meet the following two conditions. First, a proportion of elements in S_1 and S_2 should be correspondences. The proportion was set as 0.5 in this paper. Second, a certain ratio, set as 0.8 in this paper, of elements in S_1 and S_2 and their correspondences should lie on the same sides of l_1 and l_2. The side of a line segment is defined as clockwise or anticlockwise direction relative to the line segment based on the endpoint whose x-coordinate is smaller.

If (l_1, l_2) has passed all tests above, we use the two line segments to intersect with the two pairs of rays of (M_1, M_2) to form new RPRs. Through evaluating the newly formed RPR pairs, we determine whether (l_1, l_2) can be accepted or not. The way of matching the newly formed RPR pairs is same as that presented in the step of RPR match propagation. If any of the two pairs of newly formed RPRs is finally accepted as a RPR match. We regard (l_1, l_2) as a candidate match.

After matching single line segments in all corresponding groups, we obtain a set of new line segment matches and the corresponding set of RPR matches. Among these RPR matches, there may exist several RPRs in one image match with the same RPR in another image. We select the best one with the smallest descriptor distance.

From the new line segment matches and RPR matches, we can expand the point match group again using the same strategy as that presented in Sect. 3.4. False RPR matches used to be accepted can possibly be discerned by rechecking the PDC under the expanded point match group. Some false RPR matches may have been removed after previous step. The line match group and the point match group should accordingly be updated again.

Up to now, an iteration of propagating RPR matches among single line segments has been finished. Before increasing the threshold for DFM and conducting a new iteration, the iteration without changing the threshold should be conducted again to pick up matches neglected in previous iteration. After that, we increase the value of the threshold and begin a new iteration to get more line segment matches. The iteration will halt if the threshold reaches the upper limit. The way of tuning the threshold at different iterations is the same as that presented in Sect. 3.4.

4 Experimental Results

Experiments on a set of representative image pairs were conducted to substantiate the robustness of the method and to prove its superiority by comparing with the state-of-the-art line matching methods.

(a) Light change: (253, 3, 98.8%)   (b) Low texture: (26, 0, 100%)

(c) Rotation change: (240, 0, 100%)   (d) Scale change: (134, 10, 92.5%)

(e) Viewpoint change: (548, 2, 99.6%)   (f) Image blur: (120, 2, 98.3%)

(g) Viewpoint change: (152, 3, 98.0%)   (h) JPEG compression: (266, 5, 98.1%)

Fig. 6. Results of the proposed line matching method on some representative image pairs. The eight image pairs will be represented as $\mathcal{P}_1$–$\mathcal{P}_8$ in order for subsequent use. See the text for details.

Line Matching Results. Figure 6 shows the line matching results on some representative image pairs, denoted as $\mathcal{P}_1$–$\mathcal{P}_8$ in order. Two line segments in correspondence are drawn in the same color with the same label. The statistical results shown below each sub-figure in Fig. 6 is a triple consisting of the number of total matches, the number of false matches and the accuracy in order.

It is noticed that our algorithm is robust under common image transformations, namely light, scale, rotation, viewpoint changes, image blur, JPEG compression, and in poorly-textured scene. The accuracy is above 98 % on all image pairs except P_4, where there exists great scale change between the two images. The reason that the accuracy of our method on image pairs with great scale change is relative lower is that after being adjusted to the same scale, line segments lying closely in the original image become so adjacent with each other that it is very hard to pick out the correct one.

Evaluation of the RPR Descriptor. The RPR descriptor describes the local regions formed by a point and two rays connected with the point. The relationship between the point and the two rays is fully exploited, which results in the descriptor is more robust and efficient than other famous local region descriptors which directly describe the local regions centered at the points. We use the same way presented in [22] to evaluate our descriptor and SIFT descriptor [10]. We describe initially formed RPRs in image pairs P_1–P_8 by using our proposed RPRdescriptor and the SIFT descriptor. The RPRs are matched by evaluating their descriptor distances under the same threshold. For each image pair, we count the number of correct matches, the number of total matches, and the 1-precision, which is the ratio between the number of false matches and the number of total matches. The comparative results are shown in Table 1. Note that any pair of descriptors whose distance below the given threshold is regarded as a match. So, one RPR in one image may match with several RPRs in another image and all these matches are included into the total matches. This is why the 1-precisions of the matching results on some image pairs are so high. From the table, it can be noticed that on some image pairs, our RPR descriptor generates more correct matches, while the SIFT descriptor generates more correct matches on the others. But on all image pair, the 1-precisions of our descriptor are lower than that of the SIFT descriptor, which indicates our RPR descriptor is better than the SIFT descriptor for the specific local regions.

Table 1. Comparative results of our RPR descriptor and SIFT descriptor on image pairs P_1–P_8 by describing the local regions formed by RPRs. The triple elements shown in the table represent the number of the correct matches, the number of total matches, and the corresponding 1-precision.

	P_1	P_2	P_3	P_4
RPR	(116, 163, 28.8 %)	(12, 16, 25.0 %)	(241, 330, 27.0 %)	(53, 77, 31.2 %)
SIFT	(113, 217, 47.9 %)	(16, 26, 38.5 %)	(170, 271, 37.3 %)	(72, 134, 46.3 %)
	P_5	P_6	P_7	P_8
RPR	(460, 601, 23.5 %)	(17, 132, 87.1 %)	(25, 110, 77.3 %)	(23, 34, 32.4 %)
SIFT	(325, 493, 34.1 %)	(10, 259, 96.1 %)	(41, 424, 90.3 %)	(47, 95, 50.5 %)

Comparison with Single Line Based Methods. Two methods are selected for comparison. They are Lines-Points Invariants (LPI) [8] and Line Signature

(LS) [16]. Their implementations are provided by their own authors. Except experimenting on image pairs $\mathcal{P}_1$–$\mathcal{P}_8$ using the three methods (LPI, LS and RPR), we do another two groups of experiments on the same image pairs to eliminate the influence of different line detection methods on the line matching results. The first is that we take line segments used in LPI, detected by LSD [18], as input for our method. The second is that we use our line segments, detected by EDLines [17], as input for LPI. All comparative results are listed in Table 2. From the table it can be concluded that no matter using line segments detected by EDLines or LSD as input, our proposed RPR method generates much better results than that of LPI. Our method can find quite more line matches with higher accuracy in almost all cases. When comparing with LS, our method produces comparable number of matches but with higher accuracy on the image pairs $\mathcal{P}_1$–$\mathcal{P}_3$. On the image pairs $\mathcal{P}_4$–$\mathcal{P}_8$, the accuracy of the two methods is similar, but our method produces more correct matches.

Table 2. Comparative results of the proposed RPR algorithm, LPI [8] and LS [16]. The columns from the left to right are the results of our RPR method using line segments detected by EDLines [17], our RPR method based on line segments detected by LSD [18], LPI using line segments detected by LSD, LPI based on line segments detected by EDLines, and LS. The triple elements shown in the table represent the number of line matches, the number of false matches, and the accuracy respectively. The last row represents the average accuracy of each method.

	RPR (EDLines)	RPR (LSD)	LPI (LSD)	LPI (EDLines)	LS
$\mathcal{P}_1$	(253, 3, 98.8 %)	(295, 5, 98.3 %)	(219, 2, 99.1 %)	(185, 9, 95.1 %)	(248, 5, 98.0 %)
$\mathcal{P}_2$	(26, 0, 100 %)	(30, 0, 100 %)	(12, 0, 100 %)	(15, 0, 100 %)	(53, 12, 77.4 %)
$\mathcal{P}_3$	(240, 0, 100 %)	(219, 0, 100 %)	(227, 2, 99.1 %)	(235, 0, 100 %)	(242, 1, 99.6 %)
$\mathcal{P}_4$	(134, 10, 92.5 %)	(153, 5, 96.7 %)	(78, 3, 96.2 %)	(92, 8, 91.3 %)	(40, 3, 92.5 %)
$\mathcal{P}_5$	(548, 2, 99.6 %)	(582, 3, 99.5 %)	(390, 3, 99.2 %)	(356, 2, 99.4 %)	(271, 2, 99.3 %)
$\mathcal{P}_6$	(120, 2, 98.3 %)	(287, 3, 99.0 %)	(116, 5, 95.7 %)	(142, 4, 97.2 %)	(81, 2, 97.5 %)
$\mathcal{P}_7$	(152, 3, 98.0 %)	(123, 1, 99.2 %)	(90, 11, 87.8 %)	(107, 8, 92.5 %)	(139, 3, 97.8 %)
$\mathcal{P}_8$	(266, 5, 98.1 %)	(292, 3, 99.0 %)	(148, 5, 96.6 %)	(193, 7, 96.4 %)	(205, 4, 98.0 %)
	98.2 %	99.0 %	96.7 %	96.5 %	95.0 %

Fig. 7. Three pairs of images used to compare our method with the method presented in [14]. From left to right, the three image pairs will be represented as "apt", "tcorner" and "valbonne" in order for later use.

Comparison with Junction Based Methods. Our method exploits features of junctions for matching line segments. So comparisons with junction-based line matching methods are necessary to comprehensively evaluate our method. The method introduced in [14], represented as LICF, is used for the comparison. We cannot obtain the implementation of the method, so the comparison is conducted by evaluating the different performances of the two methods on same image pairs. Figure 7 shows three image pairs used in the published paper. The corresponding comparative results are shown in Table 3. Comparing the results of the two methods shows that our method produces fairly more matches with similar accuracy, which means quite more correct matches are obtained by our method.

Table 3. Comparative results of our RPR method and LICF [14] on some image pairs. Refer to Table 2 for details.

	apt	tcorner	valbonne
RPR	(190, 8, 95.8%)	(122, 3, 97.5%)	(123, 5, 95.9%)
LICF	(53, 3, 94.3%)	(70, 8, 88.6%)	(34, 1, 97.1%)

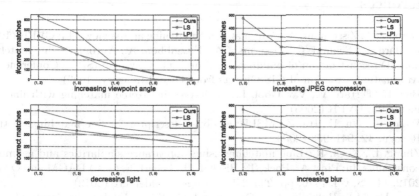

Fig. 8. The results of our proposed RPR method, LS [16], and LPI [8] on image datasets with various image transformations.

Comparative Performances under Image Transformations. Compared with other line matching methods, one remarkable merit of our method is that it produces more correct matches with high accuracy under various image transformations. Figure 8 shows the results of our method, LPI [8], and LS [16] on the famous image datasets, "graffiti", "ubc", "leuven", and "bikes"[1], in which viewpoint change, JPEG compression, light change and image blur between images exist respectively. Each dataset contains 6 images and the image transformations from the second to the last image relative to the first image is stronger and stronger. Image pairs $\mathcal{P}_7$, $\mathcal{P}_8$, $\mathcal{P}_1$, and $\mathcal{P}_6$ are token from the four datasets

---

[1] http://lear.inrialpes.fr/people/mikolajczyk/Database/index.html.

respectively. Figure 8 shows the matching results between the first and the second to sixth image in each dataset. It is obvious that our method produces quite more correct matches under these image transformations in most cases.

5   Conclusions

We propose a novel line matching method through converting matching line segments to the newly introduced Ray-Point-Ray (RPR) structures. A SIFT-like robust descriptor is proposed to match RPRs under an efficient iterative scheme by progressively adding new matches while deleting false matches. Experimental results demonstrate the robustness of the method and its superiority to the state-of-the-art methods for the larger group of correct matches and the higher accuracy in most cases.

Acknowledgement. This work was supported by the National Basic Research Programme of China (Project No. 2012CB719904) and the National Natural Science Foundation of China (Project No. 41271431).

References

1. Schmid, C., Zisserman, A.: Automatic line matching across views. In: CVPR (1997)
2. Baillard, C., Schmid, C., Zisserman, A., Fitzgibbon, A.: Automatic line matching and 3D reconstruction of buildings from multiple views. In: ISPRS Conference on Automatic Extraction of GIS Objects from Digital Imagery (1999)
3. Bay, H., Ferrari, V., Van Gool, L.: Wide-baseline stereo matching with line segments. In: CVPR (2005)
4. Wang, Z., Wu, F., Hu, Z.: MSLD: a robust descriptor for line matching. Pattern Recogn. **42**, 941–953 (2009)
5. Zhang, L., Koch, R.: Line matching using appearance similarities and geometric constraints. In: Pinz, A., Pock, T., Bischof, H., Leberl, F. (eds.) DAGM and OAGM 2012. LNCS, vol. 7476, pp. 236–245. Springer, Heidelberg (2012)
6. Lourakis, M.I., Halkidis, S.T., Orphanoudakis, S.C.: Matching disparate views of planar surfaces using projective invariants. Image Vis. Comput. **18**, 673–683 (2000)
7. Fan, B., Wu, F., Hu, Z.: Line matching leveraged by point correspondences. In: CVPR (2010)
8. Fan, B., Wu, F., Hu, Z.: Robust line matching through line-point invariants. Pattern Recogn. **45**, 794–805 (2012)
9. Chen, M., Shao, Z.: Robust affine-invariant line matching for high resolution remote sensing images. Photogram. Eng. Remote Sens. **79**, 753–760 (2013)
10. Lowe, D.G.: Distinctive image features from scale-invariant keypoints. Int. J. Comput. Vis. **60**, 91–110 (2004)
11. Winder, S., Hua, G., Brown, M.: Picking the best daisy. In: CVPR (2009)
12. Bay, H., Ess, A., Neubeck, A., Van Gool, L.: 3D from line segments in two poorly-textured, uncalibrated images. In: 3DPVT (2006)
13. Micusik, B., Wildenauer, H., Kosecka, J.: Detection and matching of rectilinear structures. In: CVPR (2008)

14. Kim, H., Lee, S.: Simultaneous line matching and epipolar geometry estimation based on the intersection context of coplanar line pairs. Pattern Recogn. Lett. **33**, 1349–1363 (2012)
15. Wang, L., Neumann, U., You, S.: Wide-baseline image matching using line signatures. In: ICCV (2009)
16. Wang, L., Adviser-Neumann, U.: Line segment matching and its applications in 3D urban modeling. Ph.D. thesis, University of Southern California (2010)
17. Akinlar, C., Topal, C.: EDLines: a real-time line segment detector with a false detection control. Pattern Recogn. Lett. **32**, 1633–1642 (2011)
18. Von Gioi, R.G., Jakubowicz, J., Morel, J.-M., Randall, G.: LSD: A fast line segment detector with a false detection control. IEEE Trans. Pattern Anal. Mach. Intell. **32**, 722–732 (2010)
19. Zuliani, M.: RANSAC for dummies with examples using the RANSAC toolbox for Matlab and more (2009)
20. Hartley, R., Zisserman, A.: Multiple View Geometry in Computer Vision. Cambridge University Press, Cambridge (2003)
21. Gavin, H.P.: The Levenberg-Marquardt method for nonlinear least squares curve-fitting problems. Department of Civil and Environmental Engineering, Duke University, Technical report (2013)
22. Mikolajczyk, K., Schmid, C.: A performance evaluation of local descriptors. IEEE Trans. Pattern Anal. Mach. Intell. **27**, 1615–1630 (2005)

Local-to-Global Signature Descriptor for 3D Object Recognition

Isma Hadji and Guilherme N. DeSouza$^{(\boxtimes)}$

Vision-Guided and Intelligent Robotics Lab (ViGIR)
Electrical and Computer Engineering Department,
University of Missouri,
Columbia, MO 65211, USA
ih9p5@mail.missouri.edu, DeSouzaG@missouri.edu

Abstract. In this paper, we present a novel 3D descriptor that bridges the gap between global and local approaches. While local descriptors proved to be a more attractive choice for object recognition within cluttered scenes, they remain less discriminating exactly due to the limited scope of the local neighborhood. On the other hand, global descriptors can better capture relationships between distant points, but are generally affected by occlusions and clutter. So, we propose the Local-to-Global Signature (LGS) descriptor, which relies on surface point classification together with signature-based features to overcome the drawbacks of both local and global approaches. As our tests demonstrate, the proposed LGS can capture more robustly the exact structure of the objects while remaining robust to clutter and occlusion and avoiding sensitive, low-level features, such as point normals. The tests performed on four different datasets demonstrate the robustness of the proposed LGS descriptor when compared to three of the SOTA descriptors today: SHOT, Spin Images and FPFH. In general, LGS outperformed all three descriptors and for some datasets with a 50–70% increase in Recall.

1 Introduction

Object recognition is arguably the most important topic in the field of computer vision because of the different applications that this task can serve. Such applications include, scene understanding, robot navigation, tracking, assistive technology and many others. This importance has also grown in the past decade as research has steered towards the use of 3D instead of 2D information for the task of object recognition. But as in the years of 2D object recognition, designing good feature vectors, or descriptors, is still the most critical step involved in 3D object recognition. Indeed, the descriptors have the greatest effect on the overall recognition result, as argued in [1].

The techniques adopted for object description can be divided into two main categories; global or local. At first, global descriptors seem a more natural choice, since they seek to encode the entire set of keypoints into a single feature vector describing the object. This makes global features far more discriminating given

© Springer International Publishing Switzerland 2015
C.V. Jawahar and S. Shan (Eds.): ACCV 2014 Workshops, Part I, LNCS 9008, pp. 570–584, 2015.
DOI: 10.1007/978-3-319-16628-5_41

that the entire geometry of the object is taken into account. However, being able to observe as much as possible of this same geometry becomes a condition for the success of global descriptors. On the other hand, local descriptors rely only on the local neighborhood of each keypoint, making the recognition more robust in scenes with clutter and occlusion [1], but for the price of increased sensitivity to changes in those neighborhoods (e.g. instrument noise). Another advantage of local descriptors is in the ability to perform point to point correspondences, which makes pose estimation using, for example RANSAC, much easier.

In this research we propose a new descriptor built at the up-to-now unseemly intersection between these two paradigms. The Local-to-Global Signature – or LGS descriptor – is local in the sense that each feature vector describes a single keypoint, rather than the entire object, but at the same time global as it looks beyond the local neighborhood (support regions) to describe the properties of keypoints with respect to the entire object. Unlike traditional global descriptors, the LGS overcomes issues related to occlusion and clutter by constructing signature vectors instead of histogram-based vectors. The advantages of using signatures to reduce the effects of occlusion are further detailed in Sect. 3, but again, that is not the only contribution of the proposed descriptor. In summary, the LGS was built around the following five main ideas: (1) Relying on surface point classification to capture the entire geometry of the object (global property); (2) Describing keypoints (local property), but using both local and global support regions grouped by the same surface class; (3) Using signatures (global property) to avoid loss of information and mitigate the effects of occlusion; (4) Using distributions of L2-distances to encode the relationship between keypoint and support regions to increase robustness to noise and eliminate the use of sensitive features such as surface normals; and (5) Using confidence on the relationship above to improve the matching during object recognition.

Figure 1 illustrates the overall construction of the LGS descriptor. These ideas will be further detailed in the remaining of the paper which is organized

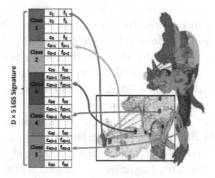

Fig. 1. Steps of the construction of the LGS descriptor: yellow dots represent keypoints to be described; points are color coded according to the surface class to which they belong, from sharp regions in red to smooth surfaces in light blue; the signature consists of the distances f between the keypoint and its local and global support regions, plus the corresponding confidences in their assignments (Color figure online).

as follows: Sect. 2 briefly surveys the most prominent 3D descriptors used in the field. In Sect. 3, we introduce the LGS descriptor and highlight the different steps involved in constructing it. Section 4 describes in details the experiments, the datasets used and presents all the results obtained. Finally, we conclude with a discussion on the advantages of the proposed approach, its limitations as well as possible directions for future work.

2   Background and Related Work

As briefly mentioned in Sect. 1, 3D descriptors belong to two main categories; global and local. Local descriptors focus on the local neighborhood of keypoints, while Global descriptors represent the complete object by either using all or a subsample of the detected keypoints. The techniques used to encode the relationships between keypoints can also be divided in two sub-categories: signature-based and histogram-based. A detailed comparison between these two sub-categories can be found in [2].

Among the first successful local 3D descriptors employing histograms we find the Point Feature Histogram (PFH) [3]. PFH describes the angular variations between each two surface normals in the k-neighborhood of a region of interest. Because this descriptor models variations between normals at all pairs of points, it is very computationally intensive. For that reason, Fast PFH (FPFH) was introduced in [4] to speed up the process, but in detriment to its power to discriminate objects. An alternative was then introduced in [5], where the authors proposed a 3D extension of the 2D Shape Context descriptor [6]. This descriptor relies on the position of keypoints in a local neighborhood (support region) with respect to a virtual spherical grid. Each spherical grid accumulates a weighted sum of the points falling in that grid. In order to achieve repeatability, the north pole of the spherical grid is first aligned with the direction of the surface normal at the point being described. This alignment is sensitive to the choice of the reference frame – a major issue for any descriptor based on surface normals. So, a major breakthrough in terms of 3D descriptors was introduced by the SHOT descriptor [2], where the authors addressed this problem with the reference frame repeatability by attaching a unique reference frame to each point. Once the reference frame is determined, the local support of each point is discretized in a way similar to that in the 3D Shape Context descriptor. Another contribution of SHOT was that the histogram of each grid is concatenated to form the final signature. This technique for defining local repeatable reference frames was later used to improve the 3D Shape Context in [7] by taking the initial representation proposed for 3D Shape Context and disambiguate it using the unique reference frame introduced by the SHOT descriptor. Although quite different in terms of their implementations, the afore mentioned techniques are very similar in concept. Their reliance on local neighborhoods and low-level features, such as surface normals, compromise their ability to discriminate objects and make them quite sensitive to viewing perspectives and noise.

As far as signature-based descriptors are concerned, Point Signatures (PS) [8] is possibly the best known descriptor. It relies on point positions also within a local neighborhood, but it encodes these positions in the local sphere in terms of the angle between the surface normal at the point and the signed distance of the points to the plane separating the sphere into two halves. While this method captures exactly the structure of the local neighborhood, it remains very sensitive to small changes in the normal estimation and the reference frame. Another signature-based descriptor is proposed in [9], where signatures are built from the depth values of the local surface after its normal vector has been aligned with the Z axis. Also, in order to reduce the dimensionality of the signature, the method resorts to a PCA subspace of the feature vector. Although signatures usually lead to a better discriminating power, the reliance on local neighborhoods causes low repeatability in keypoint matching since many keypoints in one object can have similar local structures.

Finally, when it comes to global descriptors, the View-point Feature Histogram (VFH) [10], which is an extension to the FPFH, is one of the most widely used. Instead of describing relationships between points in local neighborhoods, VFH does so for every point in the cloud with respect to their centroid. In addition to this, VFH introduces view point variance – an important addition when it comes to estimating object pose. Another attempt of generalizing FPFH was provided in Global FPFH [11]. In this case, FPFH descriptors are used locally for every point in the object and a conditional random field is trained in order to classify the collected local descriptors into a set of primitives. Next, the relationships between keypoints are encoded by counting the number and type of transitions between primitives while traversing keypoints in an octree. Later, in Global Radius-based Surface Descriptor, GRSD [12], a modification to GFPFH was proposed by eliminating the first classifier and adding a geometric solution to the shape primitives. A similar approach called Global Structure Histogram, or GSH, was also proposed in [13] with the goal of capturing the global structure of the object. In order to do so, the authors followed the same procedure proposed in [11], but using a clustering algorithms to learn the points classes and by encoding the relationships between keypoints in all clusters using geodesic distances.

In this work, we maintain that adopting the strengths of both local and global descriptors can lead to highly discriminating features. These features can be successfully used in the task of object recognition and pose estimation within scenes, even if they suffer from clutter and occlusion.

3 Proposed Method

3.1 Motivation

The proposed LGS descriptor can be regarded as a bridge between local and global paradigms. Recently, global 3D descriptors introduced interesting ideas involving the classification of object surfaces based on their shapes. These ideas should ultimately increase the discriminating power of the descriptor. However, it is clear that a great amount of information is lost in the process of encoding the relationships between all keypoints of an object into a single feature vector. In addition to that, global descriptors are very sensitive to clutter and they require a good segmentation algorithm as a pre-processing step [1]. Also, even if a perfect segmentation could be obtained, occlusions can have a major effect on the performance of global descriptors. On the other hand, a common problem that may affect all local descriptors is the inability to discriminate similar support regions not only within the same object but also across different objects. To illustrate this idea, we extracted both the SHOT and the LGS descriptors from two different keypoints coming from two completely different objects. We set the radius size of the support region used to estimate the SHOT descriptors to 25 times the mesh resolution – i.e. almost twice the recommended size of 15 times the mesh resolution [2]. As Fig. 2 clearly shows, the SHOT descriptors for these two keypoints are almost identical even though they come from very different objects, while the LGS makes clear distinction between the same two keypoints.

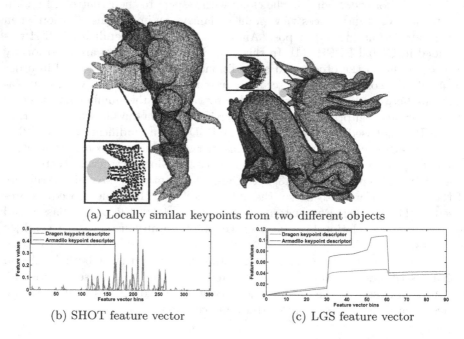

(a) Locally similar keypoints from two different objects

(b) SHOT feature vector   (c) LGS feature vector

Fig. 2. Contrast between the SHOT and the proposed LGS descriptor: (a) two keypoints (shown in yellow) falling on similar local neighborhoods, but coming from two completely different objects. Notice the similarity between the two, red and blue, SHOT descriptors in (b) as opposed to the clear contrast between the LGS descriptors in (c), also in red and blue (Color figure online).

For these reasons, the proposed descriptor starts from a local approach, where each keypoint is represented with a unique feature vector – i.e. descriptor. Then, it turns global by looking at the entire object while capturing the structure of the object with respect to that keypoint.

3.2   LGS Descriptor

As we just mentioned, the main advantage of the LGS descriptor resides in the idea of looking beyond the local support. In fact, it looks at different regions representing different properties of the object. This is accomplished by: (i) assigning classes to all the points in the cloud (global property); (ii) describing keypoints (local property), but using both local and global support regions; (iii) using signatures to describe the relationships between keypoints and selected points from all the assigned classes; (iv) using L2 norm to robustly encode such relationships; and (v) using confidence on the relationship above during matching.

Point Classification. To classify the points we use the radius-based surface classification proposed in the RSD descriptor [12]. Our technique starts form the assumption that every two points fall on a sphere. Therefore, within each neighborhood the radii of all virtual spheres are estimated using points locations. Then, the maximum and minimum sphere radii present in the local neighborhood are derived. For more details we refer the reader to [12].

In the original proposal of RSD, the authors used both the minimum and maximum radii to classify points as belonging to one of the geometric primitive shapes. In our implementation however, we chose to classify points from very sharp to very smooth. Our decision for this classification approach is motivated by one main argument: to be able to find a pre-defined number of classes independently of the object considered. This may not be feasible when using shape primitives where some specific shapes may not appear in all objects. In addition, since this classification scheme is independent of primitive shapes, it allows the algorithm to vary the number of classes by simply varying the ranges of sharpness and smoothness. For example, we cannot expect to find a toroid shape in all objects, but we can reasonably assign different levels of sharpness or smoothness to any surface based on its curvature.

In our implementation, we rely only on the minimum radius from the RSD method described above, where a very small radius is an indication of a very sharp surface, and a large radius represents a smooth one. After deriving the minimal radii associated with each point on the surface of the object, the algorithm can split the radii values into N different ranges, representing the N classes of the object's surfaces. If for example $N = 3$, any point with radius $r < \alpha * mesh\,resolution$ would be assigned to class 1; a point with radius $\alpha * mesh\,resolutionr < r < (\alpha + 5) * mesh\,resolution$ is assigned to class 2; and all other points are assigned to class 3. In our experiments, the value of α was empirically set to 6.

Since LGS uses continuous ranges of radii values for surface point classification, fuzzy regions may emerge. These fuzzy regions contain points with values of radii that are close to the end of one range and the beginning of the next one. Therefore, these points could belong to any of the two consecutive ranges. It is easy to understand that these regions are unstable and points can move from one class to another as noise is added or removed. Also, noise can cause spikes to appear on otherwise smooth surfaces. Therefore, to cope with these potential instability in surface classification, we propose the assignment of both a class and a membership to the class for each point in the cloud. Basically, after the initial crisp classes are assigned to each point, the algorithm searches over each point and calculates a coefficient of confidence that the specific point belongs to the assigned class. These confidences c approximate the probability that a current point p belongs to class n given the number of points from that class in its local support region; that is:

$$c = \frac{\#\,of\,points\,in\,class\,n\,in\,the\,neighborhood\,of\,p}{Total\,number\,of\,points\,in\,the\,neighborhood\,of\,p} \tag{1}$$

The rational behind these confidences is the assumption that in any small neighborhood points are more likely to belong to the same surface class. Also, a low confidence indicates that the point belongs to a fuzzy or noisy region, where multiple classes may be assigned. This determines when the algorithm is to give high or low weights to the points used in the next steps of the algorithm.

Signature Construction. Once all points are assigned to one of the N classes and their confidences are calculated, the LGS descriptor is constructed in a signature-based fashion. Figure 1 should help the reader in understanding the following steps. First, for every keypoint, the algorithm finds its corresponding k-nearest neighbors within each class. Next, the k-neighbors in each class are sorted based on their distance from the keypoint and divided into D clusters. Then, the median L2-distances from the keypoint to the points falling in each cluster form a D dimensional feature vector. These distances are the actual features f_l of the LGS descriptor. Finally, feature vectors representing the neighborhoods of the keypoint in each one of the N classes are concatenated to form the final signature whose length is equal to $D \times N$. Again, Fig. 1 illustrates the construction of a simplified signature in the LGS descriptor. It is important to mention again that the main motivation for using a signature-based descriptor is its potential robustness to occlusions. In fact, if parts of the object happen to be occluded in the scene, only the entries corresponding to those parts of the object will be altered in the LGS descriptor, while the rest is unchanged. On the other hand, a histogram-based descriptor would be completely affected if the support region of a keypoint is partially occluded.

Descriptor Matching. In parallel to constructing the LGS descriptor, the algorithm builds a second feature vector filled with the confidences c_l corresponding to each feature point f_l. Once again, this idea is illustrated in Fig. 1 for a very simplified case. These confidences are used as weights during the matching stage, when the LGS computes the distance d_{ij} between a pair of signatures (i, j). In other words, the distance between each entry of the pair of signatures is multiplied by the corresponding minimum confidence. This allows LGS to reduce the effect of unstable points located on fuzzy regions as discussed in the beginning of this section. Mathematically, the weighted distance used to compare LGS descriptors is given by:

$$d_{ij} = sqrt(\sum_{l=1}^{(D*N)} min(c_{li}, c_{lj}) * (f_{li} - f_{lj})^2) \tag{2}$$

4 Experimental Results and Discussion

In order to evaluate the discriminating power of the proposed LGS signature, we devised testing scenarios that highlight the robustness of LGS for the case of a model-scene matching framework.

In that sense, we present two main experiments to validate our work, using two well-recognized datasets from the literature [2,14]. The first dataset referred in [14] as Retrieval dataset consists of synthetic data with added noise at three levels equal to 10 %, 30 % and 50 % of the mesh resolution. The second dataset, and the most challenging one, contains real data acquired using stereo cameras. It presents the biggest challenge for any descriptor due to: the level of occlusion and clutter; the presence of smoother and more similar objects, and the larger reconstruction errors between scenes and models. Figure 3 presents examples of models and scenes from these two datasets.

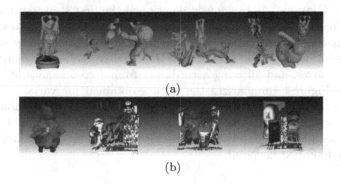

(a)

(b)

Fig. 3. Sample models and scenes. (a) Stanford synthetic dataset. (b) UniBo Vision Lab Stereo dataset [14].

4.1 Experiment 1 - Parameter Selection

As explained in Sect. 3, the proposed LGS descriptor may be affected by the choice of two parameters: the number of classes N used to construct the signature and the number k of neighbors used in each class. In this section, we present an experiment highlighting the effect of each one of these parameters on the performance of the LGS descriptor. This also provides the reader with an intuition regarding how they can be set.

In general, a descriptor T is considered robust if for any given keypoint in the scene, its exact correspondence can be found in the model. The steps followed for establishing correspondences are as follows:

1. For each model-scene pair (m, s), find the set of keypoints k_m and k_s and describe each keypoint using the T descriptor.
2. Compare the descriptor for each keypoints in the scene against all the descriptors in the model and take as a correspondence the descriptor from the model with the shortest distance – for the LGS, that is given by Eq. (2).

3. Find the true correspondences using the provided ground-truth transformation between model and scene.
4. Compare the correspondences from step 2 and step 3 and count total number of true/false matches.

In order to abstract the performance of the descriptor from the effect of non-repeatable keypoints, we followed the same framework proposed in [2] for keypoint detection. More specifically, we down-sampled each model so that only 5 % of the original cloud was kept. These will be the model keypoints. Then, using the ground-truth transformation, we found the corresponding keypoints in the scene. These steps guaranteed 100 % repeatable keypoints.

We opted for down-sampling the keypoints to ensure that these are uniformly distributed over the object and that the algorithm is not biased by the type or location of the keypoints.

Number of Classes. As previously mentioned the first step towards building the LGS signature is to classify points based on the surface type to which they belong. In particular, we classify surfaces from very sharp to very smooth. In this experiment, we highlight the effect of the number of classes by varying them from 1 to 5 classes, and allowing accordingly from 1 to 5 shades of sharpness/smoothness. Figure 4 summarizes the results obtained for variable number of classes when the number of neighbors is fixed at 300 nearest neighbors in each class. It is worth noting that for a number of classes equal to 1, the LGS descriptor approaches a local signature-based descriptor where all nearest neighbors are close to the described keypoint.

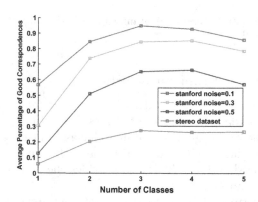

Fig. 4. Average percentage of good correspondences versus number of classes.

As we can see from Fig. 4, the number of classes plays an important role in the performance of the LGS. This becomes particularly clear for the Retrieval dataset, where changing from two to three classes improves the percentage of good correspondences by as much as 15 % (at noise level equal to 0.5). This dataset contains objects with highly different surfaces and shapes, therefore a small number of classes is not discriminating enough. As one would expect, a small number of classes does not capture enough of the variations in the structure of these objects. On the other hand, adding too many classes also leads to reduced discriminating power. We attribute this to the drop in the classification accuracy that affects the regions used in building the signature. In fact, increasing the number of classes involves using smaller ranges to assign points to different surface types. Therefore, forcing the presence of more fuzzy and unstable regions whose points can be assigned to any one of the neighboring ranges.

In order to support this claim, we performed an evaluation of the classification accuracy versus the number of classes. For each point in the model, we let the algorithm find its corresponding point in the scene and we checked whether they had been assigned to the same class (see Fig. 5). As we claimed, by adding more classes we observed a drop in the classification accuracy, which ultimately caused a drop in the number of good correspondences.

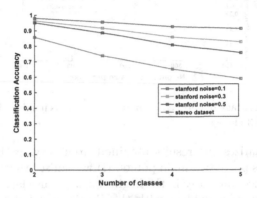

Fig. 5. Classification accuracy versus number of classes

K-Neighborhoods. The LGS descriptor consists of concatenating features from the k-closest neighbors in each class. So, the number of neighbors does play a role in the stability of the signature. In fact, a very small neighborhood implies relying more heavily on points that fall close to fuzzy regions – i.e. junction regions between two classes. As mentioned earlier in Sect. 3, these are non-stable regions since points on those regions can switch classes very easily in the presence of sensor noise. On the other hand, neighborhoods too large can cause the algorithm to look beyond stable regions, likely falling on the next fuzzy region. In addition to that, very large neighborhoods may imply using much bigger parts of the object which can therefore cause the LGS signature to become more affected by occlusions.

Once again we validate these claims by evaluating the effect of different neighborhood sizes on the discrimination of the LGS descriptor. Given the results obtained in the previous experiment, we fixed the number of classes to 3 for this test. We varied the neighborhood sizes from 100- to 1000-nearest neighbors per class. It should be noted here that given the definition of the LGS signatures, varying the neighborhoods sizes implies varying the length of the signatures. In particular, as previously mentioned, in the algorithm for the LGS, each neighborhood is split into smaller clusters and the median distance of each small cluster is used. For example, in our implementation, we used clusters with 10 points each, therefore for a number of classes $N = 3$ and a neighborhood size $k = 100$, we have the number of clusters in each class $D = 100/10$, leading to a signature with dimension $D * N = 30$.

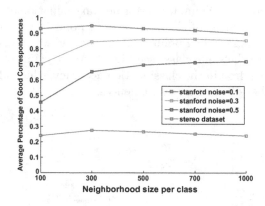

Fig. 6. Average percentage of good correspondences versus neighborhood size per class. In this test we used 3 classes.

Figure 6 summarizes the results obtained from varying the neighborhoods sizes. These results prove two main points: (i) we can see that the percentage of good correspondences increases in the beginning as we use larger neighborhoods. Again here, this is more evident with the Retrieval dataset where increasing the neighborhood size from 100 to 300 per class improves the percentage of correspondences by about 20 % in some cases (at noise level equal to 0.5). This confirms our initial statement regarding the importance of looking beyond the local neighborhood when constructing local descriptors; (ii) interestingly enough, we can see that the LGS descriptor is not too sensitive to the use of larger neighborhoods. We attribute this to the advantages brought by using a signature-based feature vectors, as well as the use of class confidences as weights during the matching stage. This could be seen as making the LGS less sensitive to occlusions.

4.2 Experiment 2 - Matching Capability

In this experiment, we validate our proposal in terms of matching capability, using the *Recall* versus $1 - Precision$ metric that captures both true and false positives as argued in [2,15,16]. In order to find potential correspondences, for every feature in the scene, the algorithm computes the first and second nearest neighbors in the model. Then, a match is established between the scene feature vector and its nearest neighbor in the model if the ratio between the first two nearest neighbors is bellow a certain threshold, as suggested in [2,17]. This threshold is the value that is varied from $0\,to\,1$ in order to produce the *Recall* versus $1 - Precision$ curves in Fig. 7. A correct match is counted as a true positive, and as a false positive otherwise. The total number of correspondences is known from the ground truth and therefore *Recall* and $1 - Precision$ are calculated as follows:

$$Recall = \frac{True\,Positives}{Total\,number\,of\,correspondences} \tag{3}$$

$$1 - Precision = \frac{False\,positives}{False\,positives + True\,Positives} \tag{4}$$

We include in this experiment a quantitative comparison against three SOTA descriptors: (i) FPFH (ii) Spin Images and (iii) SHOT. We chose FPFH and Spin Images as representatives of histogram-based local 3D descriptor and also because they are arguably the most widely used descriptors in the field. Also, we selected SHOT for comparison with a descriptor based on a signature of histograms, and again because it represents one of the state-of-the-art local 3D descriptors, achieving the best results on the datasets used here.

For the LGS descriptor, we picked the best parameters learnt from the previous experiment. In particular, we use number of classes $N = 3$ with the number of neighbors per class k set to 300. Given that we split each neighborhood to clusters of 10 points, this lead to a 90-dimensional signature. For the other descriptors used in this comparison, the main parameter to be set is the radius size of the local support region of the keypoints. In this case, we used the same size recommended in [2] – i.e. 15 times the mesh resolution.

As the results demonstrate, in most cases the LGS proved to be more discriminating than the different local descriptors tested in this paper. This should support the claim regarding the importance of looking beyond the local neighborhood for keypoint description. In addition, the use of signature proved to hold a higher discriminating power since LGS and SHOT always outperformed the histogram-based approaches. Also, given that the LGS descriptor does not directly rely on sensitive features, such as point normals, it was less affected by noise than SHOT as Figs. 7 (a) through (c) demonstrate – we discuss case (d), next.

One limitation of using the proposed approach is highlighted in the last experiment using the Stereo Dataset. In this case, the results for the LGS where only better than the Spin Images. In fact, as previously mentioned in Sect. 4.1, for the Stereo dataset, both models and scenes are reconstructed from different stereo pairs, which causes differences in the shapes of the objects found in the scene and the model. This in turn leads to a less accurate surface classification (Fig. 5), which ultimately affects the LGS signature. In fact, many of the points in the model are assigned different classes from the ones in the scene. As a consequence, the support regions found by the algorithm in each class of the model are also different from the ones found in the scene.

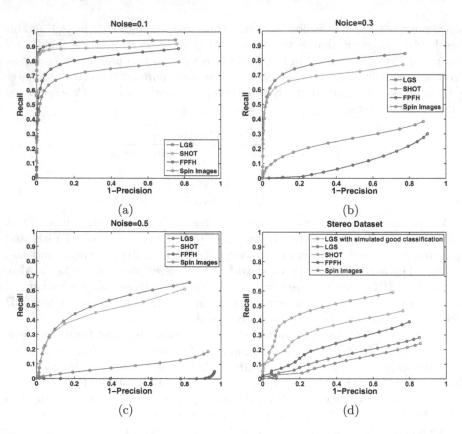

Fig. 7. Quantitative comparison between descriptors using the Recall vs 1-Precision metric for (a) Retrieval Dataset with noise=0.1, (b) noise=0.3, (c) noise=0.5, and (d) The Stereo dataset.

In order to further illustrate the effect of this severe mis-classification observed on this last dataset, we simulated 100 % good classification of the points. Specifically, we first classified points in the models into 3 classes as described in Sect. 3. Then, using the ground-truth transformation, we assigned the same class to the corresponding points in the scenes. Finally, using this simulated classification results, we constructed the LGS descriptor as usual. As can be seen from Fig. 7(d), provided a good classification, the LGS signature can still achieve higher discrimination than any of the tested descriptors. While this can be seen as a limitation of our method and indeed requires future improvements, it also further proves the benefits of looking beyond the local neighborhood and encoding more of the structure of objects in describing each keypoint. In addition, the four experiments together prove that relying on a signature instead of a histogram allows LGS to capture more details, increasing therefore the discriminating capability of the descriptor.

5 Conclusion

In this paper we proposed a novel signature-based 3D keypoint descriptor that bridges the gap between local and global descriptors. That the LGS addresses some of the difficulties inherent to each of these paradigms became clear when we compared the proposed LGS to the SOTA descriptors. In addition, we highlighted the benefits of using signatures and their role in avoiding great loss of information – as opposed to what happens with histogram-based approaches. Finally, we showed that the relative positions of the keypoints with respect to local and global support regions hold enough discriminating power while they replace low-level features such as point normals, which are very sensitive to noise. Our results clearly demonstrate the relevance of the proposed LGS descriptor while highlighting some of its limitations. Future directions for this research include addressing the problems caused by data acquisition and sensor noise when it comes to mis-classification of points in the support regions. In particular, we believe that a better surface classification scheme can lead to much more robust signatures as demonstrated in Fig. 7 (d). An automatic selection of the number of classes N, and the testing of LGS descriptor for the entire object-recognition pipeline against the descriptors used here, and potentially others (e.g. GSH, GFPFH, and GRSD) should be also part of our future work.

References

1. Aldoma, A., Marton, Z.C., Tombari, F., Wohlkinger, W., Potthast, C., Zeisl, B., Rusu, R.B., Gedikli, S., Vincze, M.: Tutorial: Point cloud library: three-dimensional object recognition and 6 dof pose estimation. IEEE Robot. Automat. Mag. **19**, 80–91 (2012)
2. Tombari, F., Salti, S., Di Stefano, L.: Unique signatures of histograms for local surface description. In: Daniilidis, K., Maragos, P., Paragios, N. (eds.) ECCV 2010, Part III. LNCS, vol. 6313, pp. 356–369. Springer, Heidelberg (2010)
3. Rusu, R.B., Marton, Z.C., Blodow, N., Beetz, M.: Persistent point feature histograms for 3d point clouds. In: Proceedings of the 10th International Conference on Intelligent Autonomous Systems (IAS-10), Baden-Baden, Germany (2008)
4. Rusu, R., Blodow, N., Beetz, M.: Fast point feature histograms (fpfh) for 3d registration. In: IEEE International Conference on Robotics and Automation, ICRA 2009, pp. 3212–3217 (2009)
5. Frome, A., Huber, D., Kolluri, R., Bülow, T., Malik, J.: Recognizing objects in range data using regional point descriptors. In: Pajdla, T., Matas, J.G. (eds.) ECCV 2004. LNCS, vol. 3023, pp. 224–237. Springer, Heidelberg (2004)
6. Belongie, S., Malik, J., Puzicha, J.: Shape matching and object recognition using shape contexts. IEEE Trans. Pattern Anal. Mach. Intell. **24**, 509–522 (2002)
7. Tombari, F., Salti, S., Di Stefano, L.: Unique shape context for 3d data description. In: Proceedings of the ACM Workshop on 3D Object Retrieval, pp. 57–62. ACM, New York (2010)
8. Chua, C.S., Jarvis, R.: Point signatures: a new representation for 3d object recognition. Int. J. Comput. Vision **25**, 63–85 (1997)

9. Mian, A., Bennamoun, M., Owens, R.: On the repeatability and quality of keypoints for local feature-based 3d object retrieval from cluttered scenes. Int. J. Comput. Vis. **89**, 348–361 (2010)
10. Rusu, R., Bradski, G., Thibaux, R., Hsu, J.: Fast 3d recognition and pose using the viewpoint feature histogram. In: 2010 IEEE/RSJ International Conference on Intelligent Robots and Systems (IROS), pp. 2155–2162 (2010)
11. Rusu, R.B., Holzbach, A., Bradski, G., Beetz, M.: Detecting and segmenting objects for mobile manipulation. In: Proceedings of IEEE Workshop on Search in 3D and Video (S3DV), held in conjunction with the 12th IEEE International Conference on Computer Vision (ICCV), Kyoto, Japan (2009)
12. Marton, Z.C., Pangercic, D., Blodow, N., Beetz, M.: Combined 2D–3D categorization and classification for multimodal perception systems. Int. J. Robot. Res. **30**, 1378–1402 (2011)
13. Madry, M., Ek, C., Detry, R., Hang, K., Kragic, D.: Improving generalization for 3d object categorization with global structure histograms. In: 2012 IEEE/RSJ International Conference on Intelligent Robots and Systems (IROS), pp. 1379–1386 (2012)
14. Tombari, F., Salti, S., Di Stefano, L.: Performance evaluation of 3d keypoint detectors. Int. J. Comput. Vis. **102**, 198–220 (2013)
15. Mikolajczyk, K., Schmid, C.: A performance evaluation of local descriptors. IEEE Trans. Pattern Anal. Mach. Intell. **27**, 1615–1630 (2005)
16. Ke, Y., Sukthankar, R.: Pca-sift: A more distinctive representation for local image descriptors. In: Proceedings of the 2004 IEEE Computer Society Conference on Computer Vision and Pattern Recognition, CVPR'04, pp. 506–513. IEEE Computer Society (2004)
17. Lowe, D.: Distinctive image features from scale-invariant keypoints. Int. J. Comput. Vis. **60**, 91–110 (2004)

Evaluation of Descriptors and Distance Measures on Benchmarks and First-Person-View Videos for Face Identification

Bappaditya Mandal[1]([✉]), Wang Zhikai[2], Liyuan Li[1], and Ashraf A. Kassim[2]

[1] Visual Computing Department,
Institute for Infocomm Research, Singapore, Singapore
{bmandal,lyli}@i2r.a-star.edu.sg
[2] Electrical and Computer Engineering,
National University of Singapore, Singapore, Singapore
{a0080959,ashraf}@nus.edu.sg

Abstract. Face identification (FI) has made significant amount of progress in the last three decades. Its application is now moving towards wearable devices (like Google Glass and mobile devices) leading to the problem of FI on first-person-views (FPV) or ego-centric videos for scenarios like business networking, memory assistance, etc. In the existing literature, performance analysis of various image descriptors on FPV data are little known. In this paper, we evaluate four popular image descriptors: local binary patterns (LBP), scale invariant feature transform (SIFT), local phase quantization (LPQ) and binarized statistical image features (BSIF) and ten different distance measures: Euclidean, Cosine, Chi square, Spearman, Cityblock, Minkowski, Correlation, Hamming, Jaccard and Chebychev with first nearest neighbor (1-NN) and support vector machines (SVM) as classifiers for FI task on both benchmark databases: FERET, AR, GT and FPV database collected using wearable devices like Google Glass (GG). Comparative analysis on these databases using various descriptors shows the superiority of BSIF with Cosine, Chi square and Cityblock distance measures using 1-NN as classifier over other descriptors and distance measures and even some of the current state-of-art benchmark database results.

1 Introduction

The rise of wearable technology has opened up numerous opportunities to further improve our lifestyles with technological advancements. Bulky medical equipments used to measure our vital statistics can be replaced with watches or handphones and heavy cameras replaced with GoPros [1] and Google Glass [2]. With facial recognition technology emerging in the past decade, wearable cameras such as the GG allow for amazing possibilities. These cameras can recognize daily activities and detect social interactions effortlessly; atomic actions such as turning left and right can be detected from first-person camera movement, while group activities can be recognized based on individual actions and pairwise context [3]. Faces can be used as a significant source of information as

C.V. Jawahar and S. Shan (Eds.): ACCV 2014 Workshops, Part I, LNCS 9008, pp. 585–599, 2015.
DOI: 10.1007/978-3-319-16628-5_42

their attention patterns play a huge role in identification, recognizing each other in social interactions, business networking, visual memory assistant and many other applications that are popular nowadays [4,5].

There are some unique challenges of performing FI on FPVs data generated from wearable devices. One of them is that in FPV, both wearable camera and the subject are moving or jittering, so the images are often blurry in nature and mug shot (studio or controlled condition) image of the person is not readily available/possible. Also, it is difficult to obtain large number of images of the person to be recognized because the person might not stay in the view for a long time. Moreover, in wearable devices the computation resources are limited so the algorithm should be fast enough to be executed under constrained mobile environment. So it is important that we perform the evaluations of local descriptors on FPV (face data obtained from GG) as well as benchmark face image databases under the same framework. This will help us to understand what features along with the distance measures are beneficial for wearable devices and in general FPV data.

In this paper, we don't want to use any training/learning algorithms, but use raw local descriptor features for FI on FPV data. We evaluate various descriptors and distance measures in order to determine the optimal configurations for which we can obtain good FI accuracies in FPV video data. For us to achieve a fair comparison, we test 4 very popular descriptors: LBP, LPQ, BSIF and SIFT in computer vision on numerous benchmark databases including AR, FERET, GT databases, as well as our own collected data from wearable devices like the GG. We then rigorously test these descriptors in conjunction with the chosen distance measures, as well as support vector machines in every possible combination in order to obtain reliable results. Through this process, we hope to pave the way for future work which utilizes FI in wearable devices such as GoPros and the GG, by providing reliable and efficient descriptors and distance measures.

In the following subsection we first study the related work and then in Sect. 2, we describe the local descriptors and distance measures along with 1-NN and SVM as classifiers. In Sect. 3, we present the experimental results and analysis. In Sect. 4 we present the summary of the experimental results and finally conclude in Sect. 5.

1.1   Related Work

Face recognition (FR) is a very challenging problem. Despite extensive research over the last four decades, the problem is still far from solved in unconstrained environments [6,7]. A systematic independent evaluation of recent face recognition algorithms from commercial and academic institutions can be found in the face recognition vendor test (FRVT) 2013 report [8]. Using it on the wearable devices like GG, makes the problem still more challenging. In addition to all the traditional problems of FR (like uneven illumination condition, pose, expression and aging) [6,9], FPVs possess blurry or jittering and out of focus images. This is because both the camera and target (face) are always on moving or non-stationary platforms.

FI on wearable devices is gaining very popularity because of its wide range of applications like an assistant in social interaction, memory aid for people who cannot remember faces or unable to recall names of the person whom he/she meet before, business networking and profession cooperation [10,11]. In such situations, keeping a log about the people you interacted with during the day to augment your memory and help you remember better is useful. Till date, all the FR studies have been focused on benchmark face images third person view (TPV) data but with the availability of wearable devices, many more FPV are generated, stored, used and shared by the users. So it is important that we evaluate various local features and distance measures for FPV videos.

Utsumi et al. proposed a wearable FR system in [12], which uses a course-to-fine recognition method. Their system requires a desktop PC because of the high computational cost of the various algorithms. Krishna et al. [13], developed an iCare Interaction Assistant device for helping visually impaired individuals for social interactions. Their evaluations are limited to only 10 subjects' face images captured under tightly controlled and calibrated face images using classical subspace methodologies like principal component analysis (PCA) [14], linear discriminant analysis (LDA) [15] and Bayesian interpersonal classifier (BIC) [16] and have not evaluated the performance using recent local descriptors.

Face detection and recognition involving various scales and orientation are proposed in [17]. They use a color camera and an infrared camera for capturing face images. They have used hidden markov model on a very small number of subjects to perform FR. Their system is bulky and cumbersome, far away from any practical system. Wang et al. proposed a FR system for improving social lives of Prosopagnosics (people with inability to recognize faces or distinguish facial features that differentiate people) [10]. They have used LBP features for development of their FR system, however, LBP is very sensitive to noises and blurry images. For this system also, the performance evaluation is limited to 20 subjects only. A well-known performance evaluation of local descriptors has been done in [18]. Their evaluations are performed on images of natural scenery, texts and buildings, etc. and not on face images.

Many researchers have used LBP, SIFT, LPQ and BSIF for solving various computer vision problems. For example, LBP has been used for FR [19]. SIFT has been used for detecting and extracting scale invariant features for face classification [20]. LPQ is used for texture classification in blurry images [21]. Recently, Kannala et al. proposed BSIF in [22] to perform the texture recognition better than LBP and LPQ. To the best of our knowledge BSIF has never been used for FR. In this work, we evaluate these 4 popular local features with 10 different distance measures/classifier like SVM to find out which combination works best for improving FR accuracy in FPV videos and also for benchmark face image databases. In the next section, we first discuss each descriptors briefly and then perform their evaluations on various databases to study the effectiveness of these descriptors in FR.

2   Local Descriptors, Distance Measures and Classifiers

In order to compare and differentiate between faces belonging to different individuals, one can use descriptors to describe each face. There are 4 very common local descriptors being used in FR: LBP, SIFT, LPQ and BSIF and 10 different distance measures and SVM in the literature. Below we describe each of these local descriptors, distance measures and SVM classifier briefly.

2.1   Face Image Descriptors

Local Binary Patterns. Since faces can be seen as a composition of micropatterns, we can describe these micropatterns using the local binary patterns (LBP) operator. This operator assigns a label to every pixel of an image by thresholding the 3×3 (*i.e.* for a 3 by 3 sized filter; this can be predefined by the user) neighborhood of each pixel with the center pixel value and considering the result as a binary number. Finally, we obtain the histogram of the labels as a structure which can be used as a texture descriptor [19].

Scale Invariant Feature Transform. With wearable devices, we often have different views of the same subject, which can result in reduction of classification accuracy as a different view can cause a subject's identity to be misinterpreted. With Scale Invariant Feature Transform (SIFT), we have features that are invariant to scale and orientation and thus are highly distinctive of the subject. This allows us to extract features which provides us with reliable matching between varying views of the same subject. This operator first computes the locations of potential interest points in the image by obtaining the maxima and minima of a set of Difference of Gaussian filters applied at varying scaled throughout the image. Next, we discard points of low contrast in order to refine the locations. We then assign an orientation to each key point based on local image features. Finally, we compute a local feature descriptor at every key point, which is based on the local image gradient and transformed according to the orientation of the key point, allowing us to have orientation invariance [20].

Local Phase Quantization. In FPV, images are often blurry because of the camera motion and the target (face) object motion. Also, the images obtained are often out of focus because the wearable camera (such as GG) takes some time to adjust/focus to the object in view. Since image deblurring is difficult and introduces new artifacts, we use a blur insensitive descriptor, the local phase quantization (LPQ) operator [21]. This operator first decorrelates the image, as information is maximally preserved in scalar quantization if the samples to be quantized are statistically independent. Next, short-term Fourier Transform is performed on the image in every 3×3 (*i.e.* for a 3 by 3 sized filter; this can be predefined by the user) neighborhood of each pixel. Again, we obtain the histogram of the result as a structure which can be used as a texture descriptor. The resultant code is insensitive to centrally symmetric blur [21].

Binarized Statistical Image Features. Unlike LBP and LPQ where we are required to manually define the filters, binarized statistical image features (BSIF) has pre-defined texture filters which we can utilize [22]. The BSIF operator first convolves the image with the pre-defined texture filters and binarizes the filter responses using a threshold of zero. The texture filters are learnt from a training set of natural image patches by maximizing the statistical independence of the filter responses. Research also suggests that the results of BSIF can be further improved if the pre-defined texture filters can be tailored to fit images that have unusual characteristics, such as certain medical images of specific sections of the human anatomy [22].

2.2 Distance Measures and Classifiers

In this work, we do not intend to use any training or learning mechanism. Rather, we want to evaluate the effectiveness of various local features with different distance measures for FI task. We evaluate two classifiers: (i) various distance measures with 1-nearest neighbor (1-NN) and (ii) features from local descriptors with SVM.

Distance Measures. We extract the features from face images using various descriptors and then use distance measures to define faces belonging to different and same people. This is done by computing the distance between two distinct faces. In this paper, we study and analyze Euclidean, Cosine (angle-based), Chi square, Spearman, Cityblock, Minkowski, Correlation, Hamming, Jaccard and Chebychev distance measures [23]. Each of these is rigorously tested with the 4 local descriptors previously mentioned and the results are provided in the Experimental Results and Analysis section.

Support Vector Machines. Another method for determining if faces belong to the same or different person using various descriptors is support vector machines (SVM) [24]. SVMs are a useful technique for data classification. A data classification task involves separating data into training and testing sets. Each instance in the training set contains one "target value" (i.e. the identity of the person which the specific face belongs to) and several "attributes" (i.e. the features of the specific face). SVM produces a model based on the training data which predicts the target values of the test data given only the test data attributes [24].

3 Experimental Results and Analysis

We evaluate the performance of 4 local descriptors using 10 distance measures and classifiers on 4 benchmark databases (TPV face data) and 1 our own collected wearable device database (FPV face data) for face identification purpose. The databases used to test these methods are 2 sets of Facial Recognition Technology (FERET) database [25], Aleix Martinez and Robert Benavente (AR) database [26], Georgia Tech (GT) database [27] and our own collected wearable device database [28]. In all the experiments, images are preprocessed following the CSU Face Identification Evaluation System [29]. For all the local descriptors, default parameters are set accordingly to the results reported in their respective references. They are kept same for all experiments across all the databases reported in this paper. Out of 10 distance measures mentioned before, we filter and present only top 6 best performing distance measures on each of the databases.

3.1   Results on FERET Database 1

There are 2,388 images comprising of 1,194 persons (two images FA/FB per person) selected from the FERET database [25]. Images are cropped into the size of 33×38 similar to [30–33]. We evaluate the performance of 4 local descriptors used in conjunction with 10 distance measures as well as SVMs. We use the first image of each subject as the gallery image and the second image as the probe image. So there are 1,194 gallery images and 1,194 probe images. After using the LBP and LPQ descriptors on the face images, we have an array of 1194 by 256 for both the gallery and probe image sets. For BSIF, we obtain an array of 1194 by 4096 for both the sets. We then use each descriptor on the images and match them by calculating the distance between the gallery image and all the probe images. Next, we apply a simple first nearest neighbor (1-NN) classifier to test the efficiency of the descriptors by calculating the recognition rates. We also apply SVM on the features obtained from the local descriptors to calculate the recognition rates. The recognition rates with top 6 best performing distance measures are recorded in Table 1.

Table 1. Recognition rates (%) using various image descriptors vs. different distance measures/classifier for face recognition on FERET database 1.

Descriptors	Distance Measures with 1-NN as classifier						Classifier
	Euclidean	Cosine	Chi square	Cityblock	Correlation	Spearman	SVM
LBP	43.2	43.2	50.8	50.1	42.5	7.7	18.7
LPQ	60.6	63.2	65.7	66.4	64.1	65.0	64.5
BSIF	80.2	84.8	90.6	91.0	86.8	86.9	73.0

We use the SIFT descriptor on the images and compare the features of two face images by finding the closest descriptor between the two and record the distance between the pair. Using this method, we calculate the distances between the gallery images of each individual and all the test images. A simple first nearest neighborhood classifier (1-NN) is applied to test the face descriptors using SIFT. We also filter the matches for uniqueness by adding a threshold to the comparison algorithm ("the uniqueness is measured as the ratio of the distance between the best matching key point and the distance to the second best key point" [34]). This threshold improves the classification accuracy by rejecting matches which are too ambiguous [34]. The recognition rates against the threshold values are shown in Fig. 1.

Without tuning any parameters[1], both the raw SIFT and LBP features on this database with large number of subjects perform very poorly as shown in Fig. 1 and Table 1. Also, BSIF with SVM as classifier does not perform well on this database as shown in Table 1. This is probably because SVM, which was originally meant for binary classification, in general, does not perform well on databases with large number of classes or for large multi-class problem (this is also evident in other experimental results). However, Table 1 shows that BSIF with Cityblock distance measures and 1-NN as classifier can achieve around 91 % accuracy, outperforming all other features with different distance measures.

3.2   Results on FERET Database 2

This database is a subset of the original FERET database [25], created by choosing 256 subjects with at least four images per subject. However, we use the same number of images (four) per subject for all subjects similar to that used in [30,32,35,36]. We use the first image of every individual as the gallery/training set and the remaining three images as the probe/testing set. Since there are 256 people in the dataset, there are 256 images in the training set and 768 images in the testing set. For this database, we apply different feature extractors such as LBP, SIFT, LPQ and BSIF operators with the various distance measures. Using the LBP/LPQ descriptor on the images, we have an array of 256×256 for the training set and an array of 768×256 for the testing set as the LBP descriptor uses 256 features, whereas BSIF resulted in obtaining 4096 features from each of the face images. The results using various descriptors are shown in Table 2.

Similar to the previous experiment with SIFT, we use this operator on this database to obtain the features and then perform recognition of individuals with varying threshold values. The results are shown in Fig. 1.

It can be seen from Table 2 and Fig. 1 that LBP, LPQ and SIFT features performs similar. For SIFT features the thresholding plays an important role for this database. However, the highest recognition rates can be obtained using BSIF features with Chi square and Cityblock distance measures and 1-NN as classifier. It is also notable that using BSIF features with such distance measures

---

[1] A test was performed with LBP and the Chi square distance measure with a different filter size and 1-NN as classifier, producing a classification accuracy of 60 %. The difference of 10 % (in Table 1) showcases the significance of fine-tuning the parameters in LBP. However, in this work, we are not focusing on fine tuning the parameters for LBP, but use same default parameters for all the experiments. This is also same for all other descriptors including SIFT, LPQ and BSIF.

Table 2. Recognition rates (%) using various image descriptors vs. different distance measures/classifier for face recognition on FERET database 2.

| Descriptors | Distance Measures with 1-NN as classifier | | | | | | Classifier |
	Euclidean	Cosine	Chi square	Cityblock	Correlation	Spearman	SVM
LBP	81.8	82.8	87.9	86.7	82.8	4.5	32.6
LPQ	46.5	92.4	93.2	93.8	92.0	89.3	64.1
BSIF	95.9	96.9	99.0	98.8	96.7	98.4	82.3

and classifier, the recognition rates outperform all other state-of-art methods on this database [30, 35, 36].

3.3   Results on AR Database

The AR database has frontal view faces with varying facial expressions and illumination conditions. The color images in AR database [26] are converted to gray scale and cropped into the size of 120×170, same as the image size used in [26, 32, 37]. In this database, we have 75 subjects, with 14 images each. We evaluate the performance of 4 local descriptors used in conjunction with 10 distance measures as well as SVMs. We store 7 images from each subject in the gallery set while the remaining 7 images per subject are used as probe images [26, 32, 37]. We then use each descriptor on the images and match them by calculating the distance between the gallery images and all the probe images. After using the LBP and LPQ descriptors on the face images, we have an array of 525 by 256 for both the gallery and probe image sets, which are then subjected to the distance measures as well as SVM. For BSIF, we obtain an array of 525 by 4096 for both the sets.

Table 3. Recognition rates (%) using various image descriptors vs. different distance measures/classifier for face recognition on AR database.

| Descriptors | Distance Measures with 1-NN as classifier | | | | | | Classifier |
	Euclidean	Cosine	Chi square	Cityblock	Correlation	Spearman	SVM
LBP	63.2	64.4	73.5	72.8	64.4	19.6	70.1
LPQ	84.4	84.4	88.2	88.4	85.5	83.0	94.9
BSIF	92.8	94.7	98.7	98.7	94.7	97.5	99.8

We apply a simple first nearest neighbor (1-NN) classifier to test the efficiency of the descriptors by calculating the recognition rate. The recognition rates for various local features vs. different distance measures and SVM are recorded in Table 3. Similar to the previous experiments with SIFT, we use this operator on this database to obtain the features and then perform recognition of individuals with varying threshold values. The results are shown in Fig. 1.

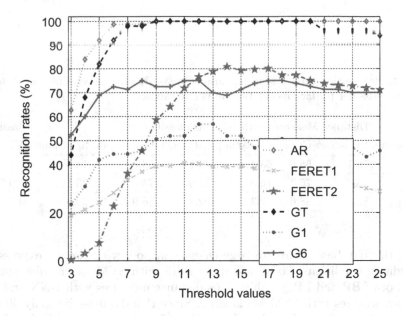

Fig. 1. Recognition rates vs Threshold values used in the matching on all six databases: AR, FERET 1, FERET 2, GT, G1 and G6 using SIFT features (best viewed in color) (Color figure online).

From Fig. 1, it is evident that SIFT with thresholds 6 and above and BSIF features with Chi square and Cityblock distance measures with 1-NN as classifier in Table 3, perform best on this AR dataset. Also, BSIF features with SVM as classifier (in Table 3) perform very good probably because there are no changes in pose for this database. They also outperform the present state-of-art results of FR on this AR database [30,36,38].

It is also notable that the recognition accuracy does not vary much with change in the distance measures. The change in the accuracy occurs across different features obtained from the descriptors. Perhaps this is because the database has variations of illuminations and expressions with same frontal pose face images. Table 3 and Fig. 1 show that BSIF and SIFT are superior local descriptors of face images when there are no changes in pose as compared to LBP and LQP.

3.4 Results on GT Database

The Georgia Tech (GT) Face Database [27] consists 750 color images of 50 subjects (15 images per subject). These images have large variations in both pose and expression and some illumination changes. Images are converted to gray scale and cropped into the size of 92 × 112. The first eight images of all subjects are used in the training and the remaining seven images serve as testing images. This protocol is same as done in [30,31,39]. The testing results are numerically recorded in Table 4. Similar to the previous experiments with SIFT, we use this operator on this database to obtain the features and then perform recognition of individuals with varying threshold values. The results are shown in Fig. 1.

Table 4. Recognition rates (%) using various image descriptors vs. different distance measures/classifier for face recognition on GT database.

Descriptors	Distance Measures with 1-NN as classifier						Classifier
	Euclidean	Cosine	Chi square	Cityblock	Correlation	Spearman	SVM
LBP	59.1	58.6	65.4	65.7	58.6	13.4	61.7
LPQ	75.4	76.3	76.6	76.6	74.0	73.1	84.0
BSIF	88.3	88.6	92.3	92.3	86.9	90.0	94.9

For this database, there are large changes in pose with varying expressions and lighting conditions. Local descriptor BSIF with SVM as classifier outperforms both LBP and LPQ with various distance measures with 1-NN and also their raw features with SVM as classifier. Since this database has only 50 subjects, it probably shows that when the number of subjects are few, SVM performs better than 1-NN classifier for all local descriptors, except LBP. Similar to the AR database, SIFT outperforms all other features with all distance measures and also SVM on BSIF features as recorded in Table 4 and shown in Fig. 1.

3.5 Wearable Device Database

This database is collected to study the problems of FR in wearable devices and it is publicly available at [28]. It contains faces of persons observed from FPV in natural social interactions, where people are involved in group meetings, indoor social interactions, business networking and all other activities in indoor office environment. There are large changes in poses, expressions, illuminations and jitters because of head and/or camera movement. Collected between Sep 2012 to Aug 2014, it comprises of 7075 images of 88 subjects (average 80.4 images per subject). Out of which 46 subjects are collected using head mounted Logitech C190 webcam connected to a tablet and rest 42 subjects by using first version of the Google Glass [2]. The database is composed of 9 females and 79 males across 9 races. Face and eye detections [40, 41] are applied to the color images captured by the wearable devices. Face images are then converted to gray scale and cropped into the size of 67×75. One sample image captured by GG and the extracted and normalized face images are shown in Fig. 2. The red box shown is the face image where both the eye coordinates are successfully detected and blue box shows a face in which either one of the eye coordinates is not detected.

Protocol. We evaluate the performance of 4 local descriptors using various distance measures for two applications of FR. In the first scenario, only 1 frontal face image per subject is available in the gallery, while remaining all face images

Fig. 2. Left, original image captured by Google Glass. Right, extracted normalized face images (red boxes: both eye coordinates are detected, blue box: either of the eyes is not detected). (Best viewed in color) (Color figure online).

are used as probes (termed as G1). This is similar to the commercial database of personal information containing only one mug shot image for each person. As mentioned previously and presented in the recent state-of-the-art wearable FR devices [12,13], that keeping one mug shot image in the gallery may not be suitable for wearable FR for natural social interactions. However, in this work we perform experiments for such challenging scenarios. The FI accuracy with various features and distance measures are recorded in Table 5.

Table 5. Recognition rates (%) using various image descriptors vs. different distance measures/classifier for face recognition on wearable device database with G1 scenario.

| Descriptors | Distance Measures with 1-NN as classifier | | | | | | Classifier |
	Euclidean	Cosine	Chi square	Cityblock	Correlation	Spearman	SVM
LBP	72.8	65.4	72.8	59.3	70.4	60.5	32.3
LPQ	63.0	61.7	71.6	66.7	69.1	58.0	32.0
BSIF	76.5	71.6	71.6	75.3	67.9	69.1	35.2

In the second scenario, 6 images with varying pose, expression and illumination per subject are stored in the gallery, while remaining all images are used as probes (termed as G6). The recognition rate of various features with different distance measures/classifier are presented in Table 6.

Similar to the previous experiments with SIFT, we use this operator on our wearable device database to obtain the features and then perform recognition of individuals with varying threshold values. We tested the threshold values over

Table 6. Recognition rates (%) using various image descriptors vs. different distance measures/classifier for face recognition on wearable device database with G6 scenario.

	Distance Measures with 1-NN as classifier						Classifier
Descriptors	Euclidean	Cosine	Chi square	Cityblock	Correlation	Spearman	SVM
LBP	82.7	81.5	85.2	86.4	82.7	82.7	58.7
LPQ	76.5	80.2	82.7	84.0	81.5	80.2	61.2
BSIF	86.4	87.7	87.7	86.4	84.0	84.0	59.4

a range and obtained the results on both the above scenarios G1 and G6. The recognition rates against the threshold values are shown in Fig. 1.

For this database, all the descriptors in G6 scenario have better performances than G1 scenario because of the availability of more number of samples in the gallery. It is evident from Fig. 1 that the SIFT features do not perform good on this unconstrained FPV face images. This is because the noises and artifacts in FPV face images are far more than face images in the standard benchmark databases (like FERET, AR and GT). This probably shows that SIFT features are very sensitive to blurry and jittery images, like that in the FPV videos.

In general, BSIF with various distance measures and 1-NN as classifier, has superior performances as compared to LBP and SIFT on this wearable device database for both the G1 and G6 scenario, while LPQ's performance is comparable to LBP. This shows that the features from BSIF are less sensitive to various unconstrained face image conditions, such as blurry, jittery and out of camera focus face images (in addition to the traditional FR problems such as pose, illumination and expression). BSIF with Euclidean, Cosine and Cityblock distance measures outperform all other features for this database for both G1 and G6 scenarios as shown in Tables 5 and 6.

4   Summary of the Experimental Results

We have performed comprehensive evaluation of 4 local descriptors in combination with 10 distance measures with 1-NN and SVM as classifiers on 6 databases. Without using any learning/training mechanism, these raw descriptors are evaluated for FI (multi-class classification) task. Each of these databases has its own challenges for performing FI. BSIF features when used with SVM as classifier, is observed to perform well on databases with small number of subjects, such as the AR and GT databases. Also, this is same for SIFT features, which performs well when the number of subjects in the database is small as shown in Fig. 1. One notable fact is that, the recognition performance is largely dependent on features that are selected rather than the distance measures. For example, in GT database (Table 4), LPQ outperforms LBP largely because of nature of the features. The performance does not change much with varying distance measures. Similar observations are noted between BSIF and LPQ, as shown in Table 4 and FERET databases in Tables 1 and 2.

In FERET database 2 (Table 2), BSIF using Chi square and City block distance measures with 1-NN as classifier outperform all other state of art FR results on this database [30,35,36]. Also, on AR database (Table 3 and Fig. 1), SIFT with thresholds 6 and above and BSIF features with Chi square and City-block distance measures with 1-NN as classifier and raw BSIF features with SVM as classifier, outperform the present state-of-art results of FR on this AR database [30,36,38].

From Tables 1–6, it is evident that both BSIF and LPQ, in general, perform better with different distance measures on most of the databases as compared to LBP. This shows that unlike LPQ and BSIF, LBP is very much sensitive to its parameters tuning (which is not done in this work). It is also evident that BSIF with Cosine, Chi square and Cityblock distance measures, in general, outperforms all other features on all databases. For wearable device database, it seems SVM does not provide good results as the face images are captured in unconstrained environment. BSIF is shown to be more robust to blurry, jittery and out of camera focus face images (in addition to the traditional FR problems such as pose, illumination and expression), as it exhibits superior performance to all other local descriptors in all the databases with 1-NN as classifier.

5 Conclusions

In the past few decades, many researchers have evaluated local descriptors like LBP, LPQ and SIFT for different computer vision problems including FI. To the best of our knowledge, BSIF has never been used for FI task. Also, the evaluations of these four local descriptors for FI in FPV or ego-centric views data are largely unknown in the literature. In this paper, we have evaluated local descriptors using various distance measures and classifiers 1-NN and SVM on wearable devices and benchmark face databases. This helps us to understand the performance of various local descriptors for FI task under the common framework. Through this process, we hope to pave the way for future work which utilizes FI in wearable devices such as GoPros and the GG, by providing reliable and efficient descriptors and distance measures. Among these descriptors, BSIF with Cosine, Chi square and Cityblock distance measures using 1-NN as classifier are superior to all other descriptors and distance measures on both benchmark and FPV video data.

References

1. GoPro (2014). http://gopro.com/
2. Google: Google glass (2014). http://www.google.com/glass/start/
3. Mandal, B., Eng., H.L.: 3-parameter based eigenfeature regularization for human activity recognition. In: IEEE International Conference on Acoustics Speech and Signal Processing (ICASSP), pp. 954–957 (2010)
4. TedBlog: The future of facial recognition: 7 fascinating facts. http://blog.ted.com/2013/10/17/the-future-of-facial-recognition-7-fascinating-facts/ 2014

5. Mandal, B., Eng, H.L.: Regularized discriminant analysis for holistic human activity recognition. IEEE Intell. Syst. **27**, 21–31 (2012)
6. Zhao, W., Chellappa, R., Phillips, P.J., Rosenfeld, A.: Face recognition: a literature survey. ACM Comput. Surv. **35**, 399–458 (2003)
7. Phillips, P.J.: Face & ocular challenges (2010). Presentation: http://www.cse.nd.edu/BTAS_10/BTAS_Jonathon_Phillips_Sep_2010_FINAL.pdf
8. Grother, P., Ngan, M.: Face recognition vendor test (frvt) performance of face identification algorithms (2014). Technical Report: http://biometrics.nist.gov/cs_links/face/frvt/frvt2013/NIST_8009.pdf
9. Mandal, B., Jiang, X.D., Kot, A.: Multi-scale feature extraction for face recognition. In: IEEE International Conference on Industrial Electronics and Applications (ICIEA), pp. 1–6 (2006)
10. Wang, X., Zhao, X., Prakash, V., Shi, W., Gnawali, O.: Computerized-eyewear based face recognition system for improving social lives of prosopagnosics. In: Proceedings of the 7th International Conference on Pervasive Computing Technologies for Healthcare, pp. 77–80 (2013)
11. Mandal, B., Ching, S., Li, L., Chandrasekha, V., Tan, C., Lim, J.H.: A wearable face recognition system on google glass for assisting social interactions. In: 3rd International Workshop on Intelligent Mobile and Egocentric Vision, ACCV (2014)
12. Utsumi, Y., Kato, Y., Kunze, K., Iwamura, M., Kise, K.: Who are you?: A wearable face recognition system to support human memory. In: ACM Proceedings of the 4th Augmented Human International Conference, pp. 150–153 (2013)
13. Krishna, S., Little, G., Black, J., Panchanathan, S.: A wearable face recognition system for individuals with visual impairments. In: ACM SIGACCESS Conference on Computer and Accessbility, pp. 106–113 (2005)
14. Turk, M., Pentland, A.: Eigenfaces for recognition. J. Cogn. Neurosci. **3**, 71–86 (1991)
15. Swets, D.L., Weng, J.: Using discriminant eigenfeatures for image retrieval. IEEE PAMI **18**, 831–836 (1996)
16. Moghaddam, B., Jebara, T., Pentland, A.: Bayesian face recognition. Pattern Recogn. **33**, 1771–1782 (2000)
17. Singletary, B.A., Starner, T.E.: Symbiotic interfaces for wearable face recognition. In: HCII2001 Workshop On Wearable Computing (2001)
18. Mikolajczyk, K., Schmid, C.: A performance evaluation of local descriptors. IEEE PAMI **27**, 1615–1630 (2005)
19. Ahonen, T., Hadid, A., Pietikainen, M.: Face description with local binary patterns: application to face recognition. IEEE PAMI **28**, 2037–2041 (2006)
20. Aly, M.: Face recognition using sift features. CNS/Bi/EE report 186 (2006)
21. Ojansivu, V., Heikkilä, J.: Blur insensitive texture classification using local phase quantization. In: Elmoataz, A., Lezoray, O., Nouboud, F., Mammass, D. (eds.) ICISP 2008 2008. LNCS, vol. 5099, pp. 236–243. Springer, Heidelberg (2008)
22. Kannala, J., Rahtu, E.: Bsif: Binarized statistical image features. In: ICPR, pp. 1363–1366 (2012)
23. Perlibakas, V.: Distance measures for pca-based face recognition. Pattern Recogn. Lett. **25**, 711–724 (2004)
24. Hsu, C., Chang, C., Lin, C.: A practical guide to support vector classification (2010)
25. Phillips, P.J., Moon, H., Rizvi, S., Rauss, P.: The feret evaluation methodology for face recognition algorithms. IEEE PAMI **22**, 1090–1104 (2000)
26. Martinez, A.M.: Recognizing imprecisely localized, partially occluded, and expression variant faces from a single sample per class. IEEE PAMI **24**, 748–763 (2002)

27. Nefian, A.V.: Georgia tech face database (2014). http://www.anefian.com/research/face_reco.htm
28. Mandal, B., Ching, S., Li, L.: Werable device database (2014). https://sites.google.com/site/bappadityamandal/human-detection-and-fr
29. Beveridge, R., Bolme, D., Teixeira, M., Draper, B.: The csu face identification evaluation system users guide: Version 5.0 (2013). Technical Report: http://www.cs.colostate.edu/evalfacerec/data/normalization.html
30. Jiang, X.D., Mandal, B., Kot, A.: Eigenfeature regularization and extraction in face recognition. IEEE PAMI **30**, 383–394 (2008)
31. Jiang, X.D., Mandal, B., Kot, A.: Face recognition based on discriminant evaluation in the whole space. In: IEEE 32nd International Conference on Acoustics, Speech and Signal Processing (ICASSP 2007), Honolulu, Hawaii, USA, pp. 245–248 (2007)
32. Mandal, B., Jiang, X., Eng, H.L., Kot, A.: Prediction of eigenvalues and regularization of eigenfeatures for human face verification. Pattern Recogn. Lett. **31**, 717–724 (2010)
33. Mandal, B., Jiang, X.D., Kot, A.: Dimensionality reduction in subspace face recognition. In: IEEE ICICS, pp. 1–5 (2007)
34. VLFEAT: Vlfeat open source (2014). http://www.vlfeat.org/overview/sift.html#tut.sift.param
35. Lu, J., Plataniotis, K.N., Venetsanopoulos, A.N., Li, S.Z.: Ensemble-based discriminant learning with boosting for face recognition. IEEE TNN **17**, 166–178 (2006)
36. Jiang, X.D., Mandal, B., Kot, A.: Complete discriminant evaluation and feature extraction in kernel space for face recognition. Mach. Vis. Appl. **20**, 35–46 (2009). (Springer)
37. Park, B.G., Lee, K.M., Lee, S.U.: Face recognition using face-arg matching. IEEE Trans. Pattern Anal. Mach. Intell. **27**, 1982–1988 (2005)
38. Geng, C., Jiang, X.: Fully automatic face recognition framework based on local and global features. Mach. Vis. Appl. **24**, 537–549 (2013)
39. Mandal, B., Jiang, X.D., Kot, A.: Verification of human faces using predicted eigenvalues. In: 19th International Conference on Pattern Recognition (ICPR), Tempa, Florida, USA (2008)
40. Viola, P., Jones, M.: Robust real-time face detection. IJCV **57**, 137–154 (2004)
41. Yu, X., Han, W., Li, L., Shi, J.Y., Wang, G.: An eye detection and localization system for natural human and robot interaction without face detection. In: Groß, R., Alboul, L., Melhuish, C., Witkowski, M., Prescott, T.J., Penders, J. (eds.) TAROS 2011. LNCS, vol. 6856, pp. 54–65. Springer, Heidelberg (2011)

Local Feature Based Multiple Object Instance Identification Using Scale and Rotation Invariant Implicit Shape Model

Ruihan Bao[✉], Kyota Higa, and Kota Iwamoto

Information and Media Processing Laboratories,
NEC Corporation, Kawasaki, Japan
r-bao@ay.jp.nec.com

Abstract. In this paper, we propose a Scale and Rotation Invariant Implicit Shape Model (SRIISM), and develop a local feature matching based system using the model to accurately locate and identify large numbers of object instances in an image. Due to repeated instances and cluttered background, conventional methods for multiple object instance identification suffer from poor identification results. In the proposed SRIISM, we model the joint distribution of object centers, scale, and orientation computed from local feature matches in Hough voting, which is not only invariant to scale changes and rotation of objects, but also robust to false feature matches. In the multiple object instance identification system using SRIISM, we apply a fast 4D bin search method in Hough space with complexity $O(n)$, where n is the number of feature matches, in order to segment and locate each instance. Furthermore, we apply maximum likelihood estimation (MLE) for accurate object pose detection. In the evaluation, we created datasets simulating various industrial applications such as pick-and-place and inventory management. Experiment results on the datasets show that our method outperforms conventional methods in both accuracy (5 %–30 % gain) and speed (2x speed up).

1 Introduction

Locating and identifying multiple objects in an image is important for robotics [1,2] and automation [3]. Furthermore, it also attracts attentions for industrial applications such as inventory management and planograms [4,5]. Figure 1(a) shows an example of multiple object identification. In such applications, instead of understanding general object classes [6], recognizing specific object instances and there poses (e.g. its location, orientation and relative scale) is of interest. Though the definition of the object instance is varied in researches [4,7,8], in this paper we are interested in the problem like [2,4], in which an instance is a particular object example that has identical texture (i.e. appearance) with the database object.

For object instance detection, local feature based methods using SIFT [9] and SURF [10] are very popular. A classic process includes feature extraction (i.e. keypoint detection and local descriptor generation), feature matching,

© Springer International Publishing Switzerland 2015
C.V. Jawahar and S. Shan (Eds.): ACCV 2014 Workshops, Part I, LNCS 9008, pp. 600–614, 2015.
DOI: 10.1007/978-3-319-16628-5_43

(a) Example of multiple object identification   (b) Results of object center estimation

Database image

Fig. 1. Multiple object identification.

and geometric verification by Hough transform or RANdom SAmple Consensus (RANSAC). When there are few or no repeated instances in the query image, the problem can be simply treated as detecting objects by identifying true feature matches from false feature matches caused by background or irrelevant objects in the foreground. However, in the case when many repeated instances are present in an image (e.g. images of production lines or store shelves), in addition to identifying true matches from false matches, since all instances generate true feature matches, it is also crucial to segment those correct feature matches individually and locate each instance accordingly.

In order to locate and identify each instance in an image containing multiple object instances, [2, 4, 11] propose methods that cluster keypoint coordinates in query images. Specifically, the method in [4] applies windows to locate possible positions of object instances. In contrast, [11] applies graph based method using Markov Random Field (MRF) on the feature matches to segment object instances. Finally, in [2], a scalable and low latency object recognition system called MOPED is introduced. The system locates object instances by roughly clustering keypoints coordinates using mean-shift after feature matching, and then applying coarse-to-fine object detection steps using RANSAC iteratively. In order to improve the speed for the process, the system is carefully implemented by taking advantages of parallel computing technology such as OpenMP and GPU. These methods, however, have a common problem. Since keypoints of an instance are sparsely distributed and do not form dense clusters for each instance, it is therefore very difficult to achieve high detection accuracy by clustering on keypoint coordinates in complex scenes (e.g. many repeated instances or cluttered background).

Alternatively, [1, 3, 5] propose Hough voting based methods. These methods allow each feature match to cast a vote for the common object center position estimated using keypoint scale, orientation and the coordinates, and then locating object instances by clustering object centers using mean-shift [1,3] or grid voting [5]. Methods employing object centers are effective in locating instances since object centers are more densely clustered in Hough space. However, when applied to real world applications, these methods are still problematic due to large number of false matches. Figure 1(b) illustrates the difficulties in detecting multiple objects in a complex scene, in which we plot the object centers computed from local features using methods in [1,3,5]. It shows that though the estimated object centers from true feature matches form clusters, they are overwhelmed by object centers estimated from false matches, thus making it difficult to locate individual instance. In [9], a 4D Hough voting method is introduced combined with an iterative outlier removal scheme. However, as will be discussed later, the object location vote (instead of object center vote) in [9] is very sensitive to both scale changes and rotation of objects. When repeated instances are present, location votes from different instances will be excessively overlapped in the Hough space, making it hard to differentiate and locate objects by Hough voting. Therefore, this method is not suitable for multiple instance detection.

Recently, in a related area of object category detection, Implicit Shape Model (ISM) [6] has been proposed and received a lot of attentions. It successfully combines feature matching, codebook learning and Hough voting into the same framework and produces promising results. ISM adopts scale invariant object centers in Hough voting and extends the voting space to include scale changes as the third dimension. Nevertheless, ISM is not invariant to object rotation and thus can only be used under the assumption that all objects in query images have no rotation.

In this paper, we apply the idea of ISM to instance identification and extend it to accommodate scale and rotation changes by proposing Scale and Rotation Invariant Implicit Shape Model (SRIISM). Specifically, we compute object centers using keypoint scale, orientation and coordinates so that they are invariant to object centers compared to original ISM. In addition, we add object scale and orientation votes to make the Hough voting more robust to false matches compared with conventional methods [1,3,5]. This is equivalent to weighting object centers according to the distribution of object rotation and scale. The main contributions of this paper are:

(1) We propose a method called SRIISM that models the joint distribution of object centers, scale and orientation in Hough voting. The proposed method is not only invariant to object scale changes and rotation, but also very robust to false matches caused by cluttered background and irrelevant objects.
(2) We apply the model of SRIISM and develop a system for multiple object instance identification, which includes 4D Hough voting, fast 4D bin search of complexity $O(n)$, and pose estimation using maximum likelihood estimation (MLE). The system is tested on datasets simulating various applications such as pick-and-place and inventory management, and we show that superior performance in both speed and accuracy can be achieved.

The paper is organized in the following way. In Sect. 2, the proposed model of SRIISM is discussed. In Sect. 3, the details of the multiple object instance identification system using SRIISM is introduced. Finally, the evaluation datasets and experiment results are described and discussed in Sect. 4.

2   Scale and Rotation Invariant Implicit Shape Model (SRIISM)

In this section, we introduce the Scale and Rotation Invariant Implicit Shape Model (SRIISM) for instance identification, inspired by ISM [6,12]. Different from the original ISM using visual words which can be seen as a special type of local feature, our model uses original local features extracted from images. Let f_j be the observed local feature (represented by descriptor) in the query image and l_j be the associated parameters (feature pose) of the feature, which is the 2D coordinates, orientation and scale of the feature. l_j can be easily obtained from scale and rotation invariant local features such as SIFT and SURF. Let $p(O, x)$ be the probability of the presence of object O with pose x. x includes object center, orientation and scale. We denote a local feature entry as D_i, which contains the local descriptor as well as associated coordinates, orientation and scale. The SRIISM then computes the probability of $p(O, x)$ by marginalizing through local features $(p(O, x, f_j, l_j))$ in an query image, i.e.

$$p(O, x) = \sum_j p(O, x, f_j, l_j) \tag{1}$$

$$= \sum_j p(f_j, l_j) p(O, x | f_j, l_j) \tag{2}$$

Assume that the prior term $p(l_j, f_j)$ over features and feature pose are uniformly distributed, we marginalize again for the feature entries (D_i) in the database, and get the following equations,

$$p(O, x) \propto \sum_j p(O, x | f_j, l_j) \tag{3}$$

$$= \sum_{i,j} p(O, x | D_i, f_j, l_j) p(D_i | f_j, l_j) \tag{4}$$

$$= \sum_{i,j} p(O, x | D_i, l_j) p(D_i | f_j) \tag{5}$$

$$= \sum_{i,j} p(x | O, D_i, l_j) p(D_i | f_j) p(O | D_i) \tag{6}$$

From Eqs. (4) to (5), we used the fact that $p(D_i | f_j, l_j) = p(D_i | f_j)$, which means that an observed local feature f_j is matched to the feature entries D_i only by its local feature. Moreover, $p(O, x | D_i, f_j, l_j) = p(O, x | D_i, l_j)$ is based on the fact that object pose x is only inferred by the feature coordinates, scale and orientation from query images (l_j) and database images (D_i). Finally, applying

Bayes rule to $p(O, x|D_i, l_j)$ and assuming that $p(O|D_i, l_j) = p(O|D_i)$ (i.e. coordinates, scale and orientation of local features are independent to the presence of objects), we obtain Eq. (6).

In Eq. (6), $p(x|O, D_i, l_j)$ is the probabilistic Hough vote. The original ISM defined voting elements as object center and object scale, computed from 2D coordinates and scales. Instead, we propose the following voting elements that can elegantly handle scale and rotation changes to the object.

$$\begin{bmatrix} x_{obj} \\ y_{obj} \end{bmatrix} = \begin{bmatrix} x_{img} \\ y_{img} \end{bmatrix} - \frac{s_{img}}{s_{db}} \times \begin{bmatrix} \cos\theta_{obj} & -\sin\theta_{obj} \\ \sin\theta_{obj} & \cos\theta_{obj} \end{bmatrix} (\begin{bmatrix} x_{db} \\ y_{db} \end{bmatrix} - \begin{bmatrix} x_c \\ y_c \end{bmatrix}) \tag{7}$$

$$s_{obj} = s_{img}/s_{db} \tag{8}$$

$$\theta_{obj} = \theta_{img} - \theta_{db} \tag{9}$$

Here, $x = (x_{obj}, y_{obj}, s_{obj}, \theta_{obj})$ is the proposed 4D scale and rotation invariant Hough vote for object center, scale and orientation. $(x_{img}, y_{img}, s_{img}, \theta_{img})$ and $(x_{db}, y_{db}, s_{db}, \theta_{db})$ are the 2D coordinates, scales and orientations for local features from query image and database, respectively. (x_c, y_c) are registered object centers (or any reference points) from the database images. Figure 2(b) shows an example of the proposed Hough vote. In practice, since scale votes computed by taking ratio (Eq. (8)) are sensitive to even small changes in divisor, we thus take the logarithm of the values to convert the computation from division to the substraction.

In Eq. (6), the term $p(D_i|f_j)$ is the matching quality between feature f_j and database entry D_i. One way to define this probability is to assign matching score based on the feature distance. A more general treatment in object identification [5,9] is to perform exhaustive search between features f_j of query image and each database image, then find the closest D_i in the feature space for further processing. This is equivalent to assigning $p(D_i|f_j)$ to 1 if D_i is matched to f_j, and otherwise to 0.

Finally, term $p(O|D_i)$ represents the confidence of making inference of object O when observed D_i. One way is assigning term frequency inverse document frequency (tf-idf) [13,14] to this probability. For simplicity, we assume local features have the same chances to be observed in each database object, thus we assign $1/M$ to this probability, where M is the number of database objects.

Here, we would like to compare our method with that in [9]. In [9], a Hough transform based method is mentioned in which object location, scale and orientation are used in Hough voting. Though the calculation of location vote is not clearly specified, as written in [15], the location is computed as the difference of 2D coordinates of keypoints. This means that the estimated location is easily affected by scale change and rotation of the object, thus the estimated locations from each keypoint will scatter in the voting space. For multiple object identification, especially when repeated instances are close together, this scattering will cause excessive overlap in the Hough space, making it extremely difficult to identify and locate each object instance (shown in Fig. 2(a)). In our method, we instead estimate the object center position, which is calculated by using keypoint scale and orientation in addition to the 2D coordinates (in Eq. (7)), so that it

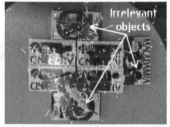

(a) Hough votes computed in [9]

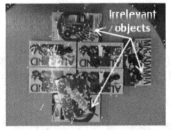

(b) Hough votes computed by this paper

Database image

Fig. 2. Example of 4D Hough votes proposed in [9] (a) and in this paper (b) when instances in query images have different rotation and scales compared to objects in database images. The root position of the arrow represents the object location (object center), direction of the arrow represents object orientation and the length of the arrow represents object scales. Notice that our Hough votes shown in (b) are clustered in object centers, and scale and orientation are consistent.

is invariant to the scale change and rotation of the object. Thus the estimated object center position is much more consistent among keypoints, forming a much tighter cluster in the voting space (shown in Fig. 2(b)). This is very effective for multiple object identification, since it can help to accurately identify and locate each object instance.

It is also important to compare our model with other multiple object identification methods in [1,3,5]. In those methods, only the coordinates of object center computed by Eq. (7) are used for Hough voting, whereas in our method we additionally use scale s_{obj} and orientation θ_{obj} of the object. In order to illustrate the difference, we denote object centers as (x_{obj}, y_{obj}), object scale as s_{obj}, object orientation as θ_{obj}. Since object center, scale and orientation are independent, the joint distribution of object centers, scale and orientation can be written as,

$$p(O, x_{obj}, y_{obj}, s_{obj}, \theta_{obj}) = p(O, x_{obj}, y_{obj})p(O, s_{obj})p(O, \theta_{obj}) \qquad (10)$$

The term $p(O, x_{obj}, y_{obj}, s_{obj}, \theta_{obj})$ on the left-hand side is our proposed joint distribution for Hough voting containing object centers, scale and orientation, and the term $p(O, x_{obj}, y_{obj})$ on the right-hand side is the object center vote used in the conventional methods. Therefore, by modeling the joint distribution of object centers, scale and orientation instead of object centers alone, our method can be seen as assigning weights to object centers by the distribution of object orientation $p(O, \theta_{obj})$ and scale $p(O, s_{obj})$, while conventional methods implicitly assume uniform distribution of scale and orientation. Figure 3 shows an example of distribution of the orientation and scale computed by Eqs. (8) and (9) from a query image. It shows that the distribution of scale and orientation exhibit bell shape property (peaked at the object scales and orientation). Thus, this distribution can help to perform more accurate Hough voting.

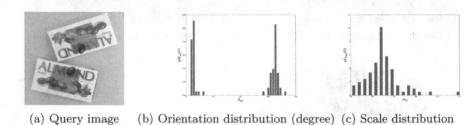

(a) Query image   (b) Orientation distribution (degree) (c) Scale distribution

Fig. 3. Example of scale distribution and orientation distribution for multiple object identification.

3 Multiple Object Instance Detection and Identification

3.1 Overview

In this section, we explain how to apply the model of SRIISM to detect and identify multiple object instances in query images. Figure 4 shows the block chart of the proposed system to detect multiple objects. For object images in database, we extract local features and create feature entries D_i, including local descriptors, scale and orientation extracted at keypoints. For each object, we also save its object center position (x_c, y_c) to the database.

For the query image, local features are first extracted and matched with features of the database. After feature matching, scale and rotation invariant Hough voting are carried out in the 4D space (Algorithm 1). Then a fast 4D bin search is employed and object pose (represented by bounding box) are recovered using maximum likelihood estimation. Finally, post processing is performed by which over detection is removed.

3.2 Scale and Rotation Invariant Hough Voting

Algorithm 1 illustrates how to compute the $p(O, x | f_j, l_j)$ in Eq. (3). After feature matching, each matched feature pair votes for the possible object center, scale and orientation of the objects. Since the probability should be summed to one, we assign the term $p(x | O, D_i, l_j)$ to $1/N_f$, where N_f is the number of local features in the query image.

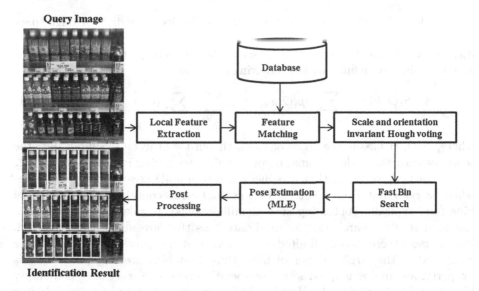

Fig. 4. Overview of the proposed algorithm.

Algorithm 1. 4D Hough voting

---

for all features in query image (f_j, l_j) do
    for all feature entries in database D_i do //compute scale and rotation invariant
    vote according to Eqs. (7),(8),(9)
        $x \leftarrow (x_{obj}, y_{obj}, s_{obj}, \theta_{obj})$
        $p(x|O, D_i, l_j) \leftarrow 1/N_f$ //N_f is the number of features in query images
        $p(O, x|f_j, l_j) \leftarrow p(x|O, D_i, l_j)p(D_i|f_j)p(O|D_i)$.
    end for
end for

3.3 Fast 4D Bin Search

After Hough voting is carried out, we then detect the possible position (represented by object center) and associated pose of the object. Conventionally for this purpose, methods such as in [1] apply density estimation methods such as mean-shift to locate object instances and recover their poses. Nevertheless, these methods using mean-shift are not only time-consuming (takes $O(n^2)$, where n is the number of feature matches), but also can only recover at most four degree-of-freedom approximation of the object pose (position, scale and orientation). This is not enough for applications requiring more accurate object pose such as robotic vision. Therefore, we propose a fast 4D bin search method (takes $O(n)$) to first locate objects in the Hough space and then apply maximum likelihood estimation directly on feature matches to recover affine or higher order poses (6 degree-of-freedom or more) of the object.

In order to find the possible feature matches of objects, we divide Hough space into 4D bins, namely, for object centers (x_{obj}, y_{obj}), scale s_{obj} and orientation θ_{obj}. Then we select bins that have scores larger than a threshold. Here, the scores of bins are defined by the following equation,

$$S(O, k) = \sum_{x_m \in V(k)} p(O, x_m) = \sum_{x_m \in V(k)} \sum_j p(O, x_m | f_j, l_j) \qquad (11)$$

where, $S(O, k)$ means the score of the k-th bin for object O. Simply put, the score for each bin is the summation of Hough votes falling into it ($x_m \in V(k)$).

In practice, we found that it is inefficient to use 4D array to store bin votes when carrying out Hough voting. Because 4D array contains large number of bins (e.g. typically containing several million bins), it takes a lot of time in the final step to search for the candidate bins that are above the threshold. But in fact, there is only limited number of matches producing Hough votes compared to the large number of bins, thus most bins are empty. Using this property, we employ map structure (associative array) to store only non-empty bins. Specifically, when a 4D Hough vote is computed from a feature match, it is quantized and converted to a unique key representing the index of the bin. Then the probabilistic votes associated with the key is incremented if the entry for the key exists in the associative array, otherwise the entry for the key is created with an initial vote. Finally, our method iterates through all the entries and those bins with votes above threshold are selected. This implementation performs much faster than using 4D array since it does not search through empty bins.

In the experiment, bins for object centers (in pixel), orientation (in radians) and scale (ratio in logarithmic scale) are equally partitioned. Furthermore, in order to reduce the influence caused by quantization errors, we allow those bins to be overlapped (e.g. 50 %).

3.4   Object Pose Estimation

Once candidate bins are selected based on the bin scores, we estimate the 6 degree-of-freedom (or more) object pose by maximum likelihood estimation (MLE).

Our method uses the local feature coordinates of feature matches belonging to selected bins. We denote the coordinate of a feature match of query and database by $t = [x_t, y_t, 1]$ and $q = [x_q, y_q, 1]$. Then their relationship, given an affine model $A \in R^{3 \times 3}$ can be expressed as:

$$t = Aq + \xi . \qquad (12)$$

Here ξ is the error term due to the image noise. We assume that ξ has the form of Gaussian distribution with zero mean and variance of σ^2, that is, $\xi \sim N(0, \sigma^2)$, we can then write the likelihood term for t given q and A,

$$p(t|q, A, \sigma) \sim N(Aq, \sigma^2) \qquad (13)$$

which is also a Gaussian distribution with the mean Aq and variance σ^2.

Given all the match pairs $(t_n, q_n)_{n=1,2,0,\dots}$ in a bin, we can apply maximum likelihood estimation to find out the affine pose for the objects.

$$A = \arg\max_{A} \prod_n p(t_n|q_n, A, \sigma^2) \tag{14}$$

In order to solve Eq. (14), we can take the logarithm of the likelihood function and reformulate it to an equivalent form:

$$A = \arg\min_{A} \sum_n \|t_n - Aq_n\|^2 \tag{15}$$

Then Eq. (15) can be easily solved using least square solver.

In practice, we found that when the threshold for the bin score is set low, the bin may contain votes from false matches. Therefore, the estimated results are not the correct object poses. In order to solve this problem, we apply additional verification method for bins containing few Hough votes.

Assume affine pose A are estimated, it can be decomposed into $A = TrR_2SR_1$ using SVD [16], where Tr, R_i and S are translation, rotation and scale matrix. We then compare the product of scale and rotation matrix ($Q = R_2SR_1$) from affine model with the value of $\bar{s}_{obj}R(\bar{\theta}_{obj})$, in which $\bar{s}_{obj}$ and $\bar{\theta}_{obj}$ are average scale and orientation of Hough votes in each grid and $R(\cdot)$ is the rotation matrix. When the elements of two matrix are not agreed to an extent, we re-set the score of $S(O, k)$ to 0. The mathematical explanation of this process is to add a prior term to A (i.e. $p(A)$) in MLE according to the evidence from Hough votes.

3.5   Post Processing

Finally, we annotate detected objects by projecting bounding boxes using estimated affine model. Since nearby bins for the same object produce overlapping bounding box, we keep the one with the maximum bin score if they are overlapped. In order to compute the overlapped area, we employ Sutherland-Hodgman algorithm [17] to find out corresponding vertices of the overlapping polygons and then apply cross product to compute corresponding areas.

4   Experiment

In order to evaluate the proposed method, we reproduced experiment from related papers [1,3,5] and added real world datasets taken from supermarket and convenience stores.

The first dataset (shown in Fig. 5) simulates pick-and-place applications for industrial automation. We evaluated our method for repeated instance detection tasks similar to [3], where repeated instances have different rotation and scale. Furthermore, the datasets also contain partial occlusions. In order to test the performance to the objects with various texture levels, we divided objects into

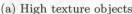

(a) High texture objects   (b) Medium texture objects   (c) Low texture objects

Fig. 5. Examples of objects used in repeated instance dataset (dataset 1).

Fig. 6. Examples of cluttered environment dataset (dataset 2).

three sets according to the texture level they have (high texture, medium texture, and low texture). In this experiment, we collected 87 query images (512 × 384 pixels) with 556 instances in all. In such tasks (e.g. industrial automation), false positive (i.e. over detection or false detection) rate must be kept low. Therefore, we evaluated the detection rate (recall) under the condition that the precision is 100 % (by chosing proper threshold during Hough voting), and compared with conventional methods.

In the second dataset (shown in Fig. 6), we reproduced the task of [1,5], in which multiple objects (include repeated instances) are to be detected in a cluttered environment. The dataset also includes a collection of challenging real world images taken from the super market (see Fig. 6). In order to test the robustness to the perspective changes for our proposed method, we took query images at different perspectives. We also included occlusions in the query images. The database contains 221 objects. For the query images, there are 28 images in total, which the width and height are ranging from 2000 to 3000 pixels, and the task is to identify total of 502 objects (targets) from over a thousand of objects (irrelevent objects).

4.1 Experiment Settings

In all experiments, we compared the proposed method with conventional methods using mean-shift (denoted as 2D mean-shift) [1,3] and grid voting (denoted as 2D gridvoting+RANSAC) [5] on object centers. Moreover, in order to show that our method is independent to local descriptors used, we implemented and tested our method using SIFT and BRIGHT [18], which is a binary local descriptor used in [5]. In order to count the true positive, we applied 50 % overlapping criterion, that is, for correct detection (true positive), its bounding box should be at least 50 % overlapped with that of ground truth. In addition, we also pose constraints when counting true positives, that is, scale and rotation of the bounding box should be consistent with ground truth.

The experiment has been conducted on Windows7 PC with Core i7-2700 K CPU@3.50 GHz.

4.2 Experiment Results

Figure 7 shows detection results for the repeated instance detection tasks (dataset 1). It shows that the proposed method outperforms object center based methods for all three types of objects (high texture, medium texture and low texture). The result also shows that our method in overall achieves better performance both for SIFT and BRIGHT. Especially, our method outperformed object center based methods by 30 % for low texture objects. This is because while objects with low texture generate only few correct feature matches, they are easily contaminated by the false matches and resulted in detection failures when applying conventional methods.

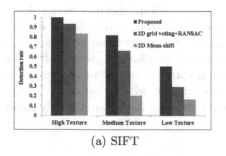
(a) SIFT

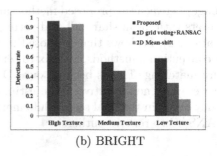
(b) BRIGHT

Fig. 7. Results on the repeated instance detection (dataset 1).

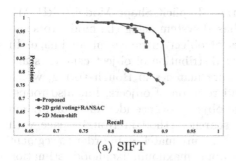
(a) SIFT

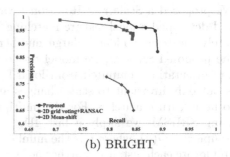
(b) BRIGHT

Fig. 8. Results on the cluttered environment dataset (dataset 2).

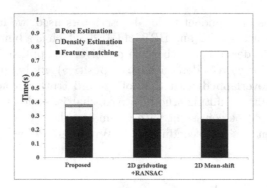

Fig. 9. Matching time between proposed method and conventional methods (dataset 2).

Figure 8 shows results on the cluttered environment dataset (dataset 2). In order to compare our method with conventional methods, we compute the recall and precision rate for all three methods (proposed, 2D mean-shift and 2D grid voting+RANSAC). It shows that our method has the best performance among all three methods. At 95 % precision rate, our methods is 5 % better in recall rates compared to 2D grid voting methods both for SIFT and BRIGHT as local descriptor.

Figure 9 shows the average processing time (matching with one database image) between the proposed method and the conventional methods using object centers with mean-shift and grid voting [5] on the cluttered environment dataset (dataset 2). It shows that proposed method in total works twice as fast as that of conventional methods, and six to senven times faster for the process after feature matching. This is because our 4D Hough voting is robust to false matches so we apply non-iterative affine estimation combined with a geometric consistency check in the final step, while methods such as [5] have to iteratively apply RANSAC to remove outliers.

5   Conclusion

We proposed a Scale and Rotation Invariant Implicit Shape Model (SRIISM), and developed a local feature matching based system using the model to accurately locate and identify large numbers of object instances in an image. In the proposed SRIISM, we model the joint distribution of object centers, scale, and orientation computed from local feature matches in Hough voting, which is not only invariant to scale changes and rotation of objects, but also robust to false feature matches. For the multiple object instance identification system using SRIISM, we apply a fast 4D bin search method in Hough space with complexity $O(n)$, where n is the number of feature matches, in order to segment and locate each instance. Furthermore, we apply maximum likelihood estimation (MLE) for accurate object pose detection. In the evaluation, we created datasets simulating various industrial applications such as pick-and-place and inventory management. Experiment results on the datasets showed that our method outperforms conventional methods in both accuracy (5%–30% gain) and speed (2x speed up). In the future works, we will extend our research to the non-rigid object identification by considering more flexible local transformation models.

References

1. Zickler, S., Veloso, M.: Detection and localization of multiple objects. In: 2006 6th IEEE-RAS International Conference on Humanoid Robots, pp. 20–25 (2006)
2. Collet, A., Martinez, M., Srinivasa, S.S.: The moped framework: Object recognition and pose estimation for manipulation. Int. J. Robot. Res. **30**, 1–23 (2001). 0278364911401765
3. Piccinini, P., Prati, A., Cucchiara, R.: Real-time object detection and localization with sift-based clustering. Image Vis. Comput. **30**, 573–587 (2012)
4. Lin, F.E., Kuo, Y.H., Hsu, W.H.: Multiple object localization by context-aware adaptive window search and search-based object recognition. In: Proceedings of the 19th ACM International Conference on Multimedia, MM 2011, pp. 1021–1024. ACM, New York (2011)
5. Higa, K., Iwamoto, K., Nomura, T.: Multiple object identification using grid voting of object center estimated from keypoint matches. In: 2013 20th IEEE International Conference on Image Processing (ICIP), pp. 2973–2977 (2013)
6. Leibe, B., Leonardis, A., Schiele, B.: Robust object detection with interleaved categorization and segmentation. Int. J. Comput. Vis. **77**, 259–289 (2008)
7. Liu, M.Y., Tuzel, O., Veeraraghavan, A., Chellappa, R.: Fast directional chamfer matching. In: 2013 IEEE Conference on Computer Vision and Pattern Recognition, pp. 1696–1703 (2010)
8. Barinova, O., Lempitsky, V., Kholi, P.: On detection of multiple object instances using hough transforms. IEEE Trans. Pattern Anal. Mach. Intell. **34**, 1773–1784 (2012)
9. Lowe, D.G.: Distinctive image features from scale-invariant keypoints. Int. J. Comput. Vis. **60**, 91–110 (2004)
10. Bay, H., Ess, A., Tuytelaars, T., Gool, L.V.: Speeded-up robust features (surf). Comput. Vis. Image Underst. **110**, 346–359 (2008)
11. Wu, C.C., Kuo, Y.H., Hsu, W.: Large-scale simultaneous multi-object recognition and localization via bottom up search-based approach. In: Proceedings of the 20th ACM International Conference on Multimedia, MM 2012, pp. 969–972. ACM, New York (2012)
12. Maji, S., Malik, J.: Object detection using a max-margin hough transform. In: IEEE Conference on Computer Vision and Pattern Recognition, CVPR 2009, pp. 1038–1045. IEEE (2009)
13. Sivic, J., Zisserman, A.: Video google: A text retrieval approach to object matching in videos. In: Proceedings of the Ninth IEEE International Conference on Computer Vision, pp. 1470–1477. IEEE (2003)
14. Arandjelovic, R., Zisserman, A.: Three things everyone should know to improve object retrieval. In: 2012 IEEE Conference on Computer Vision and Pattern Recognition (CVPR), pp. 2911–2918. IEEE (2012)
15. Perona, P.: David lowe's recognition system (2004)
16. Korman, S., Reichman, D., Tsur, G., Avidan, S.: Fast-match: Fast affine template matching. In: 2013 IEEE Conference on Computer Vision and Pattern Recognition (CVPR), pp. 1940–1947. IEEE (2013)

17. Sutherland, I.E., Hodgman, G.W.: Reentrant polygon clipping. Commun. ACM **17**, 32–42 (1974)
18. Iwamoto, K., Mase, R., Nomura, T.: Bright: A scalable and compact binary descriptor for low-latency and high accuracy object identification. In: 2013 20th IEEE International Conference on Image Processing (ICIP), pp. 2915–2919 (2013)

Efficient Detection for Spatially Local Coding

Sancho McCann[(✉)] and David G. Lowe

University of British Columbia, Vancouver, Canada
{sanchom,lowe}@cs.ubc.ca

Abstract. In this paper, we present an efficient detector for the Spatially Local Coding (SLC) object model. SLC is a recent, high performing object classifier that has yet to be applied in a detection (object localization) setting. SLC uses features that jointly code for both appearance and location, making it difficult to apply the existing approaches to efficient detection. We design an approximate Hough transform for the SLC model that uses a cascade of thresholds followed by gradient descent to achieve efficiency as well as accurate localization. We evaluate the resulting detector on the Daimler Monocular Pedestrian dataset.

1 Introduction

Spatially Local Coding (SLC) [1] has been recently proposed as an alternative way of including spatial information for object recognition. By using features that code for both appearance and location, SLC avoids the need to use fixed grids in the spatial pyramid model and uses a simple, whole-image region during the pooling stage. It outperforms modern variants of the spatial pyramid at equivalent model dimensionalities, and achieved better classification performance than all previous single-feature methods when tested on the Caltech 101 and 256 object recognition datasets [1].

Given SLC's high performance as a classifier, it is natural to adopt it for use as a detector to find the location of objects within images of natural scenes.

Our contribution in this exploratory paper is the design and evaluation of an efficient detector for Spatially Local Coding. Given the novel way that SLC includes location information, designing this detector is not a straight-forward application of an existing detector framework. We demonstrate the promise of this detector by evaluating on the Daimler monocular pedestrian dataset [2].

We start with a short review of previous detector work, then review the SLC model in detail, describe the design and optimization of our detector, and finish with evaluation and discussion.

2 Related Work

There is a large diversity of approaches to object detection, and we review in this section a small sampling to highlight the range of methods that have shown success.

© Springer International Publishing Switzerland 2015
C.V. Jawahar and S. Shan (Eds.): ACCV 2014 Workshops, Part I, LNCS 9008, pp. 615–629, 2015.
DOI: 10.1007/978-3-319-16628-5_44

One family of approaches starts with local feature patches such as SIFT [3] and combines them in bags-of-words [4] or spatial pyramids [5] for classifications of image sub-windows. These have been applied to detection through the use of the spatial pyramid as an exemplar model [6].

Another line of work starts with larger-scale features representing the overall object appearance or parts. Examples here include histograms of oriented gradients (HOG) [7] and the deformable parts model [8].

Any classifier can be used as a detector by treating the detection problem as localized classification: sliding the classifier across the image at different scales and finding maxima of the classification function. However, this can be expensive, especially as one increases the resolution of the search space. More efficient alternatives has been proposed. Efficient sub-window search [9,10] avoids exhaustive evaluation of every sub-window by using a branch-and-bound search.

Most related to our proposed method is the implicit shape model (ISM) [11], which uses a generalized Hough transform to vote for object centers. ISM's object model is generative, and its probabilistic Hough voting *is* part of the model. In our detector, the Hough voting (see Sect. 5) is only an approximation to the SLC model.

Higher order information such as object interrelationships [12] and tracking (in the case of video or image sequences) could also be leveraged to aid in detection performance, but these approaches are orthogonal to basic detector design.

Recent work on deep learning [13,14] has demonstrated excellent classification performance when very large training sets and computational resources are used to learn diverse sets of features for recognition. When such large training sets are unavailable, there continues to be a need for systems that can use existing standard feature sets such as are explored in this paper.

3   Spatially Local Coding

We now review the Spatially Local Coding (SLC) classifier [1] with a focus on the aspects relevant to detection. We present SLC as a variant within the coding/pooling framework of Boureau *et al.* [15].

Given an image $\mathcal{I}$, let feature extraction be represented as:

$$\Phi(\mathcal{I}) : \mathcal{I} \mapsto \{(\phi_1, x_1, y_1), \ldots (\phi_{n_\mathcal{I}}, x_{n_\mathcal{I}}, y_{n_\mathcal{I}})\},$$

with feature i having a local appearance described by ϕ_i, and centered at (x_i, y_i).

Features are coded through a coding function $g((\phi_i, x_i, y_i))$. In bags-of-words [4] and spatial pyramid [5] approaches, coding functions code only the appearance portion of the descriptor ϕ_i, such that $g((\phi_i, x_i, y_i)) = \hat{g}(\phi_i)$. Any location information is included later, at the pooling stage.

SLC differs from previous coding/pooling methods in that it uses a coding function g that directly handles spatial locality, and uses a single, whole-image pooling region during the pooling stage. Instead of choosing $g(\phi_i, x_i, y_i) = \hat{g}(\phi_i)$,

SLC simultaneously codes ϕ_i and the location (x_i, y_i) by using a location-augmented descriptor: $\phi_i^{(\lambda)} = [\phi_{i1}, \phi_{i2}, \ldots, \phi_{id}, \lambda x_i, \lambda y_i]$, where $\lambda \in \mathbb{R}$ is a location weighting factor giving the importance of the location in feature matching.

SLC chooses localized soft-assignment [16] for $\hat{g}$ to map each feature onto codewords. Let $\mathrm{NN}_{(\kappa)}(\phi_i)$ be the set of κ nearest neighbors to ϕ_i in a dictionary $\mathbf{D}^{(\lambda)}$ of location-weighted codewords. Then, the localized soft assignment coding is:

$$\hat{g}(\phi_i^{(\lambda)}) = \mathbf{u}_i = [u_{i1}, u_{i2}, \ldots, u_{ik}] : \tag{1}$$

$$u_{ij} = \frac{\exp(-\beta d(\phi_i^{(\lambda)}, \mathbf{D}_j^{(\lambda)}))}{\sum_{a=1}^{k} \exp(-\beta d(\phi_i^{(\lambda)}, \mathbf{D}_a^{(\lambda)}))}$$

$$d(\phi_i^{(\lambda)}, \mathbf{D}_j^{(\lambda)}) = \begin{cases} \|\phi_i^{(\lambda)} - \mathbf{D}_j^{(\lambda)}\|^2 & \text{if } \mathbf{D}_j^{(\lambda)} \in \mathrm{NN}_{(\kappa)}(\phi_i) \\ \infty & \text{otherwise} \end{cases}$$

The final SLC histogram representation $\mathbf{h}$ of a subregion $\mathcal{S}$ is a max-pooling histogram [15,17] of all $\mathbf{u}_i$ in that region:

$$\mathbf{h}_{\max} = [h_1, h_2, \ldots, h_k] \quad \text{where}$$

$$h_j = \max\{u_{ij} | (x_i, y_i) \in \mathcal{S}\} \tag{2}$$

The resulting histogram is used as the input layer to a linear SVM. Reference [1] showed that a linear SVM obtained higher performance on the SLC model than a histogram intersection kernel SVM (Fig. 1).

In summary, SLC uses location-augmented feature vectors, localized soft-assignment coding, and a single, whole-image max-pooling region. SLC moves the task of maintaining spatial locality into the coding stage, whereas previously, this had been left for the pooling stage.

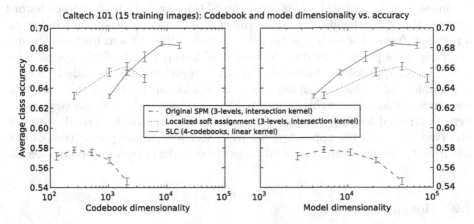

Fig. 1. Reference [1] demonstrated that SLC performed better than the basic spatial pyramid model and better than more recent state-of-the-art refinements in the context of single-image classification (Caltech 101 and 256). Figure reproduced from [1].

We avoid the issue of λ selection by use of the multi-level SLC variant as suggested by [1]. Multi-level SLC codes across several dictionaries at once, each with a different λ. Not only does this avoid having to fine-tune a parameter, McCann et al. [1] showed that a combination of several dictionaries with different λ gave better classification performance than any of the individual dictionaries alone.

4   Detection Methods

This section reviews two methods that seek to avoid having to perform an exhaustive sliding window detection. A 320×240 image "contains more than one billion rectangular sub-images" [9]. As individual evaluations of the SLC classifier are relatively expensive, it is important to avoid unnecessary evaluations.

4.1   Efficient Sub-window Search

Efficient sub-window search (ESS) was presented by Lampert et al. [9, 10] as a way of effectively performing exhaustive search of all image sub-windows in time that is sub-linear relative to the number of possible sub-windows. ESS is a branch-and-bound algorithm that relies on efficiently computing an upper bound for the scores of sets of sub-windows.

The efficiency of the original ESS bound results from a one-time quantization from features into codewords and then accumulation of those codewords' linear SVM weights into integral histograms [9]. This step is not possible with SLC. Without committing to a particular sub-window reference frame, the quantization from a feature into an SLC codeword is not defined, because location relative to the sub-window is a component of the feature.

The alternative, feature-centric efficient sub-window search proposed by Lehmann et al. [18, 19] is also inadequate for SLC because the feature-to-codeword quantization still depends on a commitment to a particular detection window as a reference frame. For each possible sub-window, a different feature-to-codeword mapping takes place, rendering the bound of feature-centric ESS expensive to compute. An approximate bound similar to the approximation we make in Sect. 5 is possible, and we include an implementation with our code, but there are two reasons we decide against this approach. First, as observed by [18], the large number of extracted features results in a computation bottleneck. Second, we have observed that using the approximation from Sect. 5 when computing the bounds results in traversing many unfruitful paths through the branch-and-bound search space.

4.2   Hough Transform

The detection framework we develop builds upon the Generalized Hough transform [11, 19]. While it lacks the theoretical guarantees of the efficient sub-window search, the Hough transform is compatible with approximations to SLC, and can still quickly suggest promising regions of the image over which to focus expensive evaluation of the full SLC model. It handles large numbers of features well (the voting phase scales linearly with the number of features). We are able to greatly speed up detection over the sliding window approach, without sacrificing performance.

We do not use a standard Hough transform. Our approach differs in two ways. First, we perform *max-pooling voting* prior to multiplying those intermediate votes by weights derived from the linear SVM. Second, our initial votes are based on an *approximation* to the SLC model. This is due to the way that SLC coding depends on committment to a particular detection hypothesis. It is not obvious that this approximation will result in useful peaks in the Hough space, and we evaluate the suitability of these peaks in Sect. 5.2.

After obtaining these preliminary hypotheses from the peaks in Hough space, we apply a cascade of thresholds and refinement to focus the SLC classifier on only the most promising regions. The design and optimization of this pipeline is described next.

5   The Detection Pipeline

5.1   Approximate SLC Hough Transform

We follow the approach of Lehmann *et al.* [19] in considering hypothesis footprints. Each hypothesis stamps out a footprint over which evidence for an object detection is accumulated. However, instead of accumulating votes hypothesis-by-hypothesis as in a sliding window approach, the Hough transform has each feature cast a weighted vote for (or against) hypotheses consistent (or inconsistent) with that feature's occurrence.

The vote weights associated with each codeword are derived from the linear SVM weights. As in Lampert *et al.* [10], we re-write the linear SVM decision function as a sum of per-codeword weights. The SVM decision function is $f(\mathbf{h}) = \beta + \sum_i \alpha_i \langle \mathbf{h}, \mathbf{h}^{(i)} \rangle$, where $\mathbf{h}$ is the SLC histogram being classified, $\mathbf{h}^{(i)}$ are the training histograms, and α_i are the learned per-example SVM weights. We can extract per-codeword weights $w_j = \sum_i \alpha_i \mathbf{h}_j^{(i)}$, and re-write the decision function as:

$$f(\mathbf{h}) = \beta + \sum_j \mathbf{h}_j \cdot w_j \qquad (3)$$

and drop the bias term β because only the relative scores matter.

Since SLC is based on max-pooling rather than sum-pooling, we need to perform the Hough transform in two phases. The first phase, where most of the work is done, involves building a max-pooling histogram at each bin in the discretized Hough space.

This phase occurs in a 5D Hough space: (s, a, x, y, c), with s being scale, represented by hypothesis window width, a being aspect ratio, x and y being the hypothesis center, and c being the codeword. See Fig. 2 for a visualization.

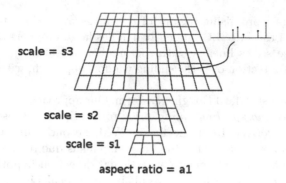

Fig. 2. For each $(a, s) \in \mathcal{A} \times \mathcal{S}$, we build an x, y grid of Hough bins, each of which tracks per-codeword votes. After max-pooling voting is complete, we collapse each histogram into a Hough score for each bin using Eq. 3.

To determine the region over which a feature will cast a vote, we use a location distribution for each SLC codeword c: $(\mu^{(c)}, \sigma^2_{(c)})$ (with $\mu = (\mu_x, \mu_y) \in [-0.5, 0.5]^2$). The location distribution is only explicitly used during this approximate Hough transform step and is not part of the final SLC model. It is a means of approximating the region over which a given codeword is likely to contribute to the hypothesis footprint.

SLC quantizes using feature location relative to a reference frame. However, there is no such reference frame when quantizing features prior to voting in Hough space. Thus, we make the following approximation. We quantize using localized soft assignment [16] based solely on appearance information (we drop the λ in Eq. 1). This maps a feature to the codewords that it *might* be matched to under SLC. We use this tentative matching based on appearance along with the learned location distributions to determine the bins in which to cast Hough votes.

Given feature center (f_x, f_y), an appearance-only soft-coding that results in non-zero weight for codeword c, and learned location distribution for that codeword, $(\mu^{(c)}, \sigma^2_{(c)})$, we compute for each $(a, s) \in \mathcal{A} \times \mathcal{S}$ the Hough bins that this features should vote in as follows:

$$\hat{\mu}_x = f_x - \mu_x^{(c)} \cdot s; \quad \hat{\mu}_y = f_y - \mu_y^{(c)} \cdot \frac{s}{a} \tag{4}$$

$$\hat{\sigma}_x = \sigma_x^{(c)} \cdot s; \quad \hat{\sigma}_y = \sigma_y^{(c)} \cdot \frac{s}{a} \tag{5}$$

Equations 4 and 5 invert the learned location distribution into Hough space at the appropriate scale (width) and aspect ratio. We update the max-pooling histogram for codeword c in all Hough bins within $3\hat{\sigma}$ of $\hat{\mu}$.

After all votes have been cast, we apply Eq. 3 to each bin to turn the histograms into scores.

5.2 Optimizing Hough Predictions

In this section, we coarsely optimize parameters of our approximate Hough transform on a subset of the VOC 2012 *car* category. We train on all non-truncated, non-difficult, non-occluded cars in the *train* portion of the training set, and test on all images in the *validation* set that include a car.

Our goal in using the Hough transform and cascade of refinements is to reduce the number of times we need to evaluate using the full model. First, we check that the approximation we use when computing the Hough votes is appropriate. Do the Hough scores reflect roughly the regions of the image that are more likely to contain the object of interest? We run a simple thresholding algorithm (Algorithm 1) for this check. Figure 3 shows that the Hough scores are meaningful. Almost every Hough bin has a score greater than -1, so the recall when thresholding at -1 is effectively the maximum recall we can achieve with a Hough transform at this bin density. We retain almost 100 % of that recall and eliminate 60 % of the bins from consideration by thresholding at zero. (Recall achieves a maximum of only 86 % in this series of experiments due to our choice of search grid resolution and minimum scale. We make the same choices for the sliding window baseline we compare against later in Fig. 5.)

Algorithm 1. Thresholding bins

Data: 4D Hough map m, threshold t
results ← {};
for *bin* $b \in m$ **do**
    if $b.score >= t$ **then**
        add b to results ;
    end
end

We can do better. Instead of selecting all Hough bins that pass a threshold, what if we select only local *peaks* in Hough space that pass the threshold? Figure 4 shows the result of running Algorithm 2 with a varying threshold. There are on average 2528 local Hough peaks in each image. This results in recall of 81 % compared to exhaustive consideration of all candidate windows (86 %), as shown at the left extreme of Fig. 3. Again, by thresholding at zero, we remove even more windows from consideration without losing further recall.

This is additional evidence that the signal from our Hough transform, even though based on an approximation of our full model, is meaningful. The score well separates candidate regions from background.

Many windows still remain, and many candidates overlap significantly with one-another. We perform non-maximum suppression at this point to produce the final predictions based solely on the Hough scores (Algorithm 3). This results in poor detection performance (13 % average precision) compared to the sliding window detector.

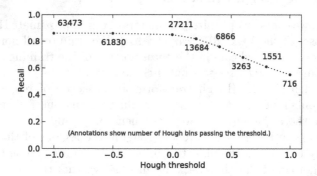

Fig. 3. We are able to maintain high recall while eliminating more than half of the candidate windows by discarding Hough bins receiving negative scores. This experiment runs Algorithm 1 with a varying threshold. (Recall in this experiment only reaches 86 % due to our choice of grid resolution and minimum scale.)

Algorithm 2. Thresholding peaks

Data: 4D Hough map m, threshold t
results ← {};
for *bin* $b \in m$ **do**
    if *b.score* $>= t$ *and IsLocalMax(b)* **then**
        add b to results ;
    end
end

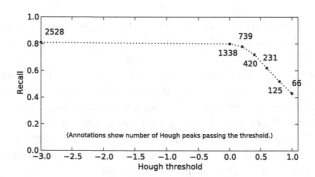

Fig. 4. By focusing only on peaks, we eliminate many candidate locations. Thresholding at zero further reduces the number of candidate windows without reducing recall. This experiment runs Algorithm 2 with a varying threshold.

Algorithm 3 Predict from thresholded peaks

Data: 4D Hough map m, threshold t
results ← {};
for *bin* $b \in m$ **do**
  if *b.score* $>= t$ *and IsLocalMax(b)* **then**
    | add b to results ;
  end
end
NonMaximumSuppression(results)

The discrepancy can be explained by the Hough scores being only an approximation to the full model. The scores of the candidate windows at this stage are not as accurate as what the full model would provide, and they aren't as precisely localized. Even if we did re-evaluate each of these peaks with the true model (Algorithm 4), the fact that they aren't well-localized means that those scores will still not be as informative, boosting performance to only 19 % average precision. Figure 5 (*Hough peaks* and *Re-scored Hough peaks*) shows the performance of these two methods.

Algorithm 4. Predict from re-scored peaks

Data: 4D Hough map m, threshold t
results ← {};
for *bin* $b \in m$ **do**
  if *b.score* $>= t$ *and IsLocalMax(b)* **then**
    b.score = EvaluateSLC(b);
    if *b.score* $>= t$ **then**
      | add b to results ;
    end
  end
end
NonMaximumSuppression(results)

One last step is necessary to nearly recover the performance of the sliding window approach. After re-scoring the Hough peaks, so that we know their score under the true model, we again discard peaks with negative scores, suppress strong overlaps, and finally refine the remaining peaks using a gradient descent procedure [19]. Our gradient descent uses a finite difference approximation of the gradient, followed by line search. We simultaneously refine the (x, y) location, the aspect ratio, and scale until we reach a local maximum.

As a result, we consider a much finer set of potential candidate windows around the peaks than the sliding window approach does, but only evaluate a small number of windows in total using the full SLC model.

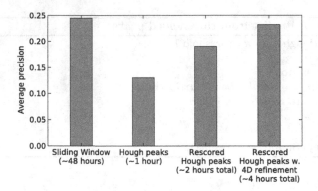

Fig. 5. By re-scoring the Hough peaks, and then refining their locations using gradient descent, we retrieve the performance of the sliding window detector while gaining a large speedup. All methods were subject to non-maximum suppression to eliminate overlapping predictions.

Algorithm 5. Full detection pipeline

Data: 4D Hough map m, threshold t
results $\leftarrow$ {};
for $bin\ b \in m$ **do**
 if $b.score >= t\ and\ IsLocalMax(b)$ **then**
  b.score = EvaluateSLC(b);
  if $b.score >= t$ **then**
   add b to results ;
  end
 end
end
NonMaximumSuppression(results);
for $b \in results$ **do**
 b $\leftarrow$ GradientDescentSLC(b);
end
NonMaximumSuppression(results);

Now that we have largely recovered the performance of the brute force sliding window detector using a much faster approximation, we turn our attention to the design of the training phase.

5.3   Mining Hard Negative Examples

For the above experiments, we used a model that was trained with a single pass through the training set, collecting all non-difficult, non-truncated, non-occluded positive examples and 10x that many negative examples selected randomly from regions of the training data that did not overlap with any positive example.

A technique used by many others in the past [7, 8, 20] is to mine the training data for hard negative examples: negative windows that the classifier erroneously

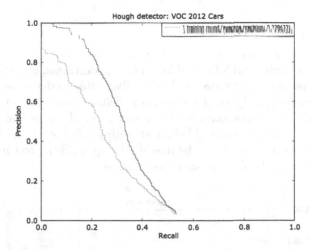

Fig. 6. Three rounds of training with hard negative mining results in a significant improvement in classifier accuracy.

predicts as belonging to the class. Walk *et al.* [20] observe that without hard negative selection, performance is extremely sensitive to the randomness in selecting examples for the negative training set, and that at least 2 re-training rounds are required in order to reach full performance using HOG + linear SVM. We follow this practice by running the detector over the training set, and adding the highest scoring false positives into our negative training set. Figure 6 shows the effect of hard negative mining on the car class. The effect of additional hard negative examples is clear.

6   Evaluation

We evaluate our detector on the Daimler monocular pedestrian dataset [2]. This dataset presents a challenging real-world problem and has been used to test and compare among other methods. Additionally, its training data is consistant in its cropping, alignment, and aspect ratio, which allows us to focus solely on the localization performance of our search strategy, and not introduce the confounding factors of handling of widely varying aspect ratio or viewpoint.

The Daimler pedestrian training set comprises 15660 pedestrian training examples, each presented in a 48×96 cropped image, and 6744 images containing no pedestrians. We extract multi-scale SIFT and build three SLC dictionaries with $\lambda = \{0.0, 1.5, 3.0\}$. We perform three training rounds. During the first training round, we extract random negative training windows such that the number of negatives is $7\times$ the number of positives (chosen to fit within memory constraints). We train a linear SVM using k-fold cross validation for selection of the hyper-parameters. Before re-training in the next round, we run our detector across all 6744 negative training images to mine for the most difficult (most highly scored) negative examples, and replace the 25 % easiest examples from the previous training round with an equivalent number of new difficult examples.

For testing, we followed the evaluation protocol described by Dollár *et al.* [21]. We evaluate against all fully-visible, labeled pedestrians, ignoring bicyclists, motorcyclists, pedestrian groups, and partially visible pedestrians. Detections and failed detections of ignored annotations neither count for or against our detector. As in [21], we standardize the aspect ratio of all ground truth boxes to 0.41 by retaining the annotated height and adjusting the width of the ground truth bounding box to 0.41 times the height. We up-scale the test images by 1.6 to allow detection of the smaller scale pedestrians.

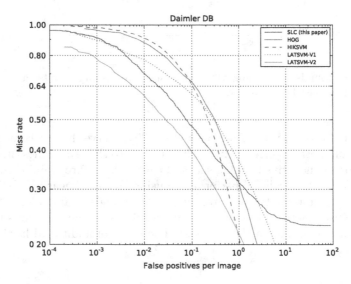

Fig. 7. Results on the Daimler monocular pedestrians dataset. (The data is taken from [21]. We have extracted the performance curves for the single-feature, non-motion methods that they evaluate.)

Enzweiler *et al.* [2] initially proposed reporting recall vs. false positives per frame. However, Dollar *et al.* [21] point out that false positives per image is a more useful measure of false positives in this setting. They observe that the difference is in whether one is evaluating the performance of the classifier underlying the detector, or evaluating the performance of the entire detector pipeline as a complete system. "Choices made in converting a binary classifier to a detector, including choices for spatial and scale stride and non-maximal suppression, influence full image performance." [21]. We follow them in reporting miss-rate vs. false positives per image. This is similar to reporting precision vs. recall as in the Visual Object Classes Challenge, but false positives per images is perhaps more important in an automotive pedestrian detection setting. Figure 7 shows our results compared against the single-feature, non-motion results reported by [21].

(a) Two successful detections

(b) Two successful detections

(c) A success, and a false positive

(d) A missed detection (the child)

Fig. 8. Examples of typical detections made by our detector on Daimler monocular pedestrian dataset. We've displayed detections above the threshold associated with the equal error rate. The numbers attached to each detection report the SLC score. True positives are outlined in green. Ground truth annotations are outlined in blue. False positives are outlined in red (Color figure online).

Our detector outperforms HOG [7], histogram intersection kernel SVM [22], and an early version of the deformable parts model [23] throughout a wide range of false-positive rates. The one method that has an advantage is the latent SVM [8] that learns explicit subparts. The recall of our detector does saturate at higher false positive rates, but Dollár *et al.* identify the region between 10^{-2} and 10^{0} false positives per window as the region most relevant for comparison. This is supported by Hussein *et al.* [24] who also use a score that focuses more on the region of the curve with low false alarms. They say, "this is useful since in many applications we are more interested in the low false alarm rate range". Nevertheless, the saturation of recall in our detector is a point for future investigation. We suspect this is due to some of the true detections not being covered by the initial set of Hough peaks, and our refinement phase is not able to recover from these poor local maxima. There is obvious room for improvement, but we believe these results demonstrate that application of the Spatially Local Coding feature to the detection problem is a fruitful area for future research.

Figure 8 shows typical output of our detector. Common error cases include reporting false positives on regions containing vertical structures and merging two pedestrians into a single detection.

7 Conclusion

We've engineered a method of localization using Spatially Localized Features through use of a Hough transform approximation, followed by non-maximum suppression and gradient descent refinement. The approach was demonstrated on the widely used Daimler monocular pedestrian dataset, achieving a good level of detection accuracy. We believe this demonstrates that further research towards using SLC in the detection context is warranted.

There is considerable scope for further research to extend this approach. One component of the current pipeline needing improvement is in the selection of Hough peaks. Initializing the gradient descent procedure from a broader selection of locations would lead to improved recall. There is also considerable scope for expanding the set of features being used and location weights to achieve higher detection performance.

References

1. McCann, S., Lowe, D.G.: Spatially local coding for object recognition. In: Lee, K.M., Matsushita, Y., Rehg, J.M., Hu, Z. (eds.) ACCV 2012, Part I. LNCS, vol. 7724, pp. 204–217. Springer, Heidelberg (2013)
2. Enzweiler, M., Gavrila, D.M.: Monocular pedestrian detection: survey and experiments. PAMI **31**, 2179–95 (2009)
3. Lowe, D.G.: Distinctive image features from scale-invariant keypoints. IJCV **60**, 91–110 (2004)
4. Csurka, G., Dance, C., Fan, L., Willamowski, J., Bray, C.: Visual categorization with bags of keypoints. In: Workshop on Statistical Learning in Computer Vision, ECCV (2004)
5. Lazebnik, S., Schmid, C., Ponce, J.: Beyond bags of features: spatial pyramid matching for recognizing natural scene categories. In: CVPR (2006)
6. Chum, O., Zisserman, A.: An exemplar model for learning object classes. In: CVPR (2007)
7. Dalal, N., Triggs, B.: Histograms of oriented gradients for human detection. In: CVPR (2005)
8. Felzenszwalb, P.F., Girshick, R.B., McAllester, D., Ramanan, D.: Object detection with discriminatively trained part based models. PAMI **32**, 1627–1645 (2009)
9. Lampert, C.H., Blaschko, M.B., Hofmann, T.: Beyond sliding windows: Object localization by efficient subwindow search. In: CVPR (2008)
10. Lampert, C.H., Blaschko, M.B., Hofmann, T.: Efficient subwindow search: a branch and bound framework for object localization. PAMI **31**, 2129–42 (2009)
11. Leibe, B., Leonardis, A., Schiele, B.: Robust object detection with interleaved categorization and segmentation. IJCV **77**, 259–289 (2008)

12. Wohlhart, P., Donoser, M., Roth, P.M., Bischof, H.: Detecting partially occluded objects with an implicit shape model random field. In: Lee, K.M., Matsushita, Y., Rehg, J.M., Hu, Z. (eds.) ACCV 2012, Part I. LNCS, vol. 7724, pp. 302–315. Springer, Heidelberg (2013)

13. Krizhevsky, A., Sutskever, I., Hinton, G.: Imagenet classification with deep convolutional neural networks. In: NIPS (2012)

14. Szegedy, C., Liu, W., Jia, Y., Sermanet, P., Reed, S., Anguelov, D., Erhan, D., Vanhoucke, V., Rabinovich, A.: Going deeper with convolutions. arXiv:1409.4842 [cs.CV] (2014)

15. Boureau, Y.L., Bach, F., LeCun, Y., Ponce, J.: Learning mid-level features for recognition. In: CVPR (2010)

16. Liu, L., Wang, L., Liu, X.: In defense of soft-assignment coding. In: ICCV (2011)

17. Yang, J., Yu, K., Gon, Y., Huang, T.: Linear spatial pyramid matching using sparse coding for image classification. In: CVPR (2009)

18. Lehmann, A., Van Gool, L., Leibe, B.: Feature-centric efficient subwindow search. In: CVPR (2009)

19. Lehmann, A., Leibe, B., van Gool, L.: PRISM: Principled implicit shape model. In: British Machine Vision Conference (2009)

20. Walk, S., Majer, N., Schindler, K., Schiele, B.: New features and insights for pedestrian detection. In: CVPR (2010)

21. Dollár, P., Wojek, C., Schiele, B., Perona, P.: Pedestrian detection: an evaluation of the state of the art. PAMI **34**, 743–61 (2012)

22. Maji, S., Berg, A., Malik, J.: Classification using intersection kernel support vector machines is efficient. In: CVPR (2008)

23. Felzenszwalb, P., McAllester, D., Ramanan, D.: A discriminatively trained, multi-scale, deformable part model. In: CVPR (2008)

24. Hussein, M., Porikli, F., Davis, L.: A comprehensive evaluation framework and a comparative study for human detectors. IEEE Trans. Intell. Transp. Syst. **10**, 417–427 (2009)

Performance Evaluation of Local Descriptors for Affine Invariant Region Detector

Man Hee Lee and In Kyu Park[✉]

Department of Information and Communication Engineering,
Inha University, 100 Inha-ro, Incheon 402-751, Korea
maninara@hotmail.com, pik@inha.ac.kr

Abstract. Local feature descriptors are widely used in many computer vision applications. Over the past couple of decades, several local feature descriptors have been proposed which are robust to challenging conditions. Since they show different characteristics in different environment, it is necessary to evaluate their performance in an intensive and consistent manner. However, there has been no relevant work that addresses this problem, especially for the affine invariant region detectors which are popularly used in object recognition and classification. In this paper, we present a useful and rigorous performance evaluation of local descriptors for affine invariant region detector, in which MSER (maximally stable extremal regions) detector is employed. We intensively evaluate local patch based descriptors as well as binary descriptors, including SIFT (scale invariant feature transform), SURF (speeded up robust features), BRIEF (binary robust independent elementary features), FREAK (fast retina keypoint), Shape descriptor, and LIOP (local intensity order pattern). Intensive evaluation on standard dataset shows that LIOP outperforms the other descriptors in terms of precision and recall metric.

1 Introduction

Visual feature detection and description are widely used in most computer vision algorithms including visual SLAM (simultaneous localization and mapping) [1], structure from motion [2], object recognition [3], object tracking [4], and scene classification [5]. Various feature detectors and descriptors have been developed such as SIFT (scale invariant feature transform) [6] and SURF (speeded up robust features) [7]. SIFT is the one of the state-of-the-art algorithms with good repeatability and matching accuracy. SIFT detects local features using scale space extrema of DoG (difference of Gaussians) and describes feature point using HOG (histogram of oriented gradients). In addition to them, a considerable number of previous work have done to describe local features effectively. A rigorous survey of performance evaluation of local descriptor can be found in Mikolajczyk and Schmid's work [8].

On the other hand, robust region detectors have been developed such as Harris affine [9], Hessian affine [10], and MSER (maximally stable extremal regions) [11]

© Springer International Publishing Switzerland 2015
C.V. Jawahar and S. Shan (Eds.): ACCV 2014 Workshops, Part I, LNCS 9008, pp. 630–643, 2015.
DOI: 10.1007/978-3-319-16628-5_45

detectors. Many computer vision applications utilize these detectors because affine invariant region is robust to affine transformation and has better reputability than local feature detector under significant viewpoint changes. Conventional affine invariant region detectors do not have their own inherent descriptors. Consequently, traditional feature descriptors have to be utilized to describe and match detected regions. Note that customized descriptors for describing affine invariant regions have been introduced, *e.g.* shape descriptor[12].

In this paper, we evaluate the performance of local descriptors for affine invariant region detectors. To the best knowledge of the authors, there has been no previous work that addresses this problem in recent years. While employing MSER detector for the affine invariant region detection, we compare standard local patch based descriptors (SIFT and SURF) as well as the state-of-the-art binary descriptors including BRIEF (binary robust independent elementary features) [13], FREAK (fast retina keypoint) [14], and LIOP (local intensity order pattern) [15]. The performance of those descriptors is evaluated and compared in various scenes with different zooming, rotation, large viewpoint changes, object deformation, and large depth variation.

This paper is organized as follows. In Sect. 2, the existing performance evaluation of feature descriptors is introduced. Section 3 describes the evaluation framework and criteria with brief summary of the evaluated region detectors and descriptors. The experimental result and discussion are presented in Sect. 4. Finally, we give a conclusive remark in Sect. 5.

2   Related Work

Table 1 shows the summarization of the previous performance evaluation of feature descriptors.

Mikolajczyk and Schmid [8] evaluated the performance of local feature descriptors in various geometric and photometric transformations, which is known to be the most exhaustive work. In addition, they proposed GLOH (gradient location and orientation histogram) descriptor which was the extension of SIFT descriptor using log-polar location grid. They concluded that GLOH and SIFT obtained the best performance to handle image rotation, zoom, viewpoint change, image blur, image compression, and illumination change.

Moreels and Perona [18] compared the feature detectors and descriptors for diverse 3D objects. They generated database which consists of 144 different objects with viewpoint and illumination changes. Several combinations of feature detectors and descriptors were evaluated, which shows that Hessian-affine detector combined with SIFT descriptor demonstrated the best performance.

Dickscheid et al. [21] measured the completeness of local features for image coding. They proposed the qualitative metric for evaluating the completeness of feature detection using feature density and entropy density. In their experiment, MSER detector achieved the best performance.

Dahl et al. [19] compared different pairs of local feature detectors and descriptors on the multi-view dataset. It was observed that MSER and DoG detectors with SIFT descriptor obtained the best performance.

Table 1. Previous works on performance evaluation of feature detectors/descriptors.

Author	Type	Environment	Best result
Mikolajczyk [8]	local descriptor	geometric + photometric transform	GLOH, SIFT
Miksik [16]	local descriptor	accuracy and speed	LIOP, BRIEF
Restrepo [17]	shape descriptor	object classification	FPFH
Moreels [18]	detector + descriptor	3D object	Hessian-affine + SIFT
Dahl [19]	detector + descriptor	multi-view dataset	MSER + SIFT
Mikolajczyk [10]	affine region detector	geometric + photometric transform	MSER
Haja [20]	region detector	texture + structure	MSER
Dickscheid [21]	local detector	image coding	MSER
Canclini [22]	local detector	image retrieval	BRISK

The performance of local shape descriptors for object classification task was evaluated by Restrepo and Mundy [17]. The local shape descriptors were extracted from the probabilistic volumetric model. They compared several shape descriptors to classify object categories using *Bag of Words* model from large scale urban scenes. FPFH (fast point feature histogram) obtained good performance in their experiments.

Miksik and Mikolajczyk [16] evaluated the trade off between accuracy and speed of local feature detectors and descriptors. They evaluated the performance of several binary descriptors and local intensity order descriptors. It was shown that binary descriptors outperformed other descriptors in time-constrained applications with low memory requirement.

Canclini *et al.* [22] evaluated the performance of feature detectors and descriptors for image retrieval application. They compared several low-complexity feature detectors and descriptors, which concluded that binary descriptors achieved better performance than non-binary descriptors in terms of matching accuracy and computational complexity.

Although the previous works have addressed the problem of performance evaluation of different feature detectors and descriptors, they are out of dated and do not consider the recently proposed descriptors. Recently computer vision applications constantly need the performance evaluation of contemporary state-of-the-art algorithms, which is the main motivation of this paper. In this paper, rather than providing too general performance evaluation which can be vague in practical point of view, we narrow the focus down to performance evaluation of descriptors combined with affine invariant region detection. This combination has not been addressed in the previous literatures. Furthermore, recent state-of-the-art descriptors are fully covered in this paper.

3   Performance Evaluation Framework

3.1   Affine Invariant Region Detector

Several techniques have been proposed on affine invariant region detector such as Harris affine, Hessian affine, and MSER detectors. Harris affine region detector [9] detects interest points using multi-scale Harris detector, which is invariant to scale and affine transformation. The extremum in the scale space of Laplacian of Gaussian is selected as the proper scale of an interest point. The elliptical region at the interest point is estimated iteratively using second moment matrix.

Similar to Harris affine region detector, Hessian affine region detector [10] is also known to be invariant to scale and affine transformation. The interest points are detected using Hessian matrices which have strong response on blobs and ridges. The scale is estimated using the Laplacian over scale space. To estimate the elliptical affine shape, the second moment matrix is used too.

MSER [11] are the regions defined as connected components which are obtained by thresholding. The detected extremal regions are either darker or brighter than surrounding region. In addition, MSER can extract important regions regardless of the threshold. MSER has the following desirable properties.

- Invariance to affine transformation and image intensity changes
- Adjacency of neighboring components is preserved in continuous geometric transformation
- Multi-scale detection
- Approximately linear complexity

The performance of various affine region detectors was compared by Mikolajczyk et al. [10], which shows that MSER detector has better repeatability than others in many cases. In addition, Haja et al. [20] compared the performance of different region detectors in terms of shape and position accuracy. They showed that MSER obtained the best accuracy than other detectors.

Based on the conclusion of those literatures, MSER is employed for the affine invariant region detector in this paper. Figure 1 shows the typical result of affine invariant region detection using MSER detector.

3.2   Selected Descriptors to be Evaluated

Mikolajczyk and Schmid [8] evaluated the performance of several local descriptors including SIFT, GLOH (gradient location and orientation histogram), shape context, PCA-SIFT, spin images, steerable filters, differential invariants, complex filters, moment invariants, and cross-correlation of sampled pixel values. Their evaluation showed that SIFT outperformed other descriptors. The proposed evaluation framework is designed to provide valuable performance comparison of recent descriptors while avoiding duplicated evaluation with previous literatures. Therefore, among the descriptors that Mikolajczyk and Schmid evaluated, only SIFT is selected in the proposed evaluation framework. In addition

(a)   (b)

Fig. 1. Example of MSER detection. (a) Affine invariant regions detected by MSER. Each region is visualized as different color. (b) Each region is fit to elliptical shape.

to SIFT, recent state-of-the-art feature descriptors, *i.e.* SURF [7], BRIEF [13], FREAK [14], Shape descriptor [12], and LIOP [15], which were published after Mikolajczyk and Schmid's work, have been included in the proposed evaluation framework. Therefore, a total of six descriptors are evaluated in this paper.

SIFT [6] descriptor is a distinctive local descriptor which is invariant to the scale and illumination changes. In our implementation, gradient magnitude and orientation is sampled in a 16×16 region around the keypoint. Then, orientation histograms (quantized to 8 directions) are generated over 4×4 subregion of the original sampling region. To increase the robustness to small location changes, the magnitude of each sample is weighted by Gaussian weighting function. Since there are 4×4 histograms with 8 orientation bins, the descriptor is represented by 128 dimensional feature vector at each keypoint.

SURF [7] descriptor is the speeded up version of SIFT descriptor. SURF descriptor is used widely in the feature matching as well as SIFT descriptor. The integral image and binary approximated integer Gaussian filter are utilized to approximate SIFT descriptor with significantly low computation. Because the gradient values within a subpatch are integrated to generate a SURF descriptor, it is more robust to image noise than SIFT.

BRIEF [13] is the notable binary descriptor that is computed by pairwise intensity comparison. To reduce the influence of noise and therefore to increase stability and the repeatability, local patches are first smoothed using Gaussian filter. Then, binary test samples are selected from an isotropic Gaussian distribution. Hamming distance is used to compute the distance between BRIEF descriptors.

FREAK [14] is the up-to-date binary descriptor that is biologically motivated by human retinal structure. The sequence of binary string is computed by pairwise comparison of image intensities over a retinal sampling pattern which are obtained by training data. Each sample point is smoothed with different size of Gaussian kernel so that it becomes less sensitive to noise. In the matching procedure, most of the outliers are removed by comparing first 16 bytes which represents coarse information

Shape descriptor [12] is an affine invariant descriptor designed for MSER, which uses the shape of the detected MSER itself as the descriptor. In each local patch, the detected region and the background are converted to the white and black, respectively. The gradient histogram based descriptor is constructed similar with SIFT descriptor.

LIOP [15] descriptor is known as the state-of-the-art feature descriptor. LIOP is also invariant to image rotation and intensity change by encoding local ordinal information of each pixel. A patch is divided to subregions using the intensity order based region division method. In each subregion, intensity relation between each pixel is mapped to the appointed value. The histogram of these values is used as the descriptor of the subregion. The overall descriptor is constructed by accumulating the histogram.

The original implementations of these algorithms are provided by the authors directly or indirectly via their contribution in OpenCV[1] and VLFeat[2].

3.3 Descriptor Matching

In this paper, each region is fit to an elliptical shape which is subsequently warped to a square patch. The orientation of normalized patch is estimated from the histogram of gradient directions. Then, nearest neighbor thresholding method is utilized to match each descriptor, in which matching pair is identified if the distance is smaller than a threshold. Note that, thresholding based matching is commonly used to evaluate the performance of descriptor, because it can explain well how many descriptors are similar to each other. However, the distinctiveness of each descriptor has been already shown in their original papers. Since this paper is intended to investigate the performance itself of each descriptor, we utilize the high performance matching method, i.e. nearest neighbor thresholding method. Correct matching is determined by the overlapping ratio of the reprojected region [9].

3.4 Dataset

In the proposed evaluation framework, test images are selected from three popular dataset, including Mikolajczyk and Schmid's dataset [8], Salzmann's dataset[3] [23], and Moreels's dataset[4] [18].

Mikolajczyk and Schmid's dataset [8] is used for measuring the performance under viewpoint change, image zoom/rotation. The images in the dataset have other imaging conditions such as illumination change, image blur, and

---

[1] http://www.opencv.org.

[2] http://www.vlfeat.org.

[3] http://cvlab.epfl.ch/data/dsr.

[4] http://vision.caltech.edu/pmoreels/Datasets/TurntableObjects/.

JPEG compression. However, they are not included in our experiments since we focus on the affine invariant property of MSER. For in-plane image rotation and scale change, *boat* and *bark* datasets are used. And we utilize *graffiti* and *bricks* datasets for viewpoint change. Salzmann's dataset has useful test images with deformable objects. For deformable objects, we test *bed sheet* and *cushion* datasets. In each dataset, we select several frames to evaluate the matching performance. Moreels's dataset consists of 144 different 3D objects with calibrated viewpoints under 3 different lighting conditions. We test *potato* and *volley ball* dataset for 3D objects. In each dataset, we select one image pair with 45 degree viewpoint change.

For quantitative evaluation of descriptor matching, the standard metric (recall vs. 1-precision plot) is utilized which was proposed in Mikolajczyk and Schmid's work [8].

4   Performance Evaluation Results

In this section, we summarize and compare the performance of SIFT, SURF, BRIEF, FREAK, Shape descriptor, and LIOP descriptors combined with MSER detector. Each warped patch is organized 144 × 144 pixels including 16 border pixels. The size of patch for BRIEF is set to 82 × 82 with same border. The scale of each scale invariant descriptor is fit to the patch size. Matching performance is measured by (number of inlier / number of outlier). In the following figures, green and red ellipses denote the correct and incorrect correspondences, respectively. Also in the following plots, horizontal and vertical axes represent 1-precision and recall, respectively. The experiment is carried out on Intel Core i7 2.7 GHz processor with 16 GB memory.

4.1   Image Rotation and Scale Changes

boat and *bark* datasets [8] are used for in-plane image rotation and scale change. The correct match is determined by reprojecting each region using ground truth homography matrix. Figure 2 shows the visual comparison of the matching results and the recall vs. 1-precision plot for *boat* and *bark* dataset. In each descriptor, we change the distance threshold for the nearest neighbor matching to measure the variation of the performance. As shown in Fig. 2, LIOP descriptor outperforms other descriptors in image zoom and rotation. Also, it is observed that SIFT descriptor achieves better performance than others including SURF descriptor. For binary descriptors only, BRIEF outperforms FREAK descriptor.

4.2   Viewpoint Changes

To evaluate the performance for viewpoint change, we utilize *graffiti* and *bricks* datasets [8] which varies their viewpoint approximately 50 degrees apart. The visual comparison of the matching result is shown in Fig. 3 (a)~(l). In the

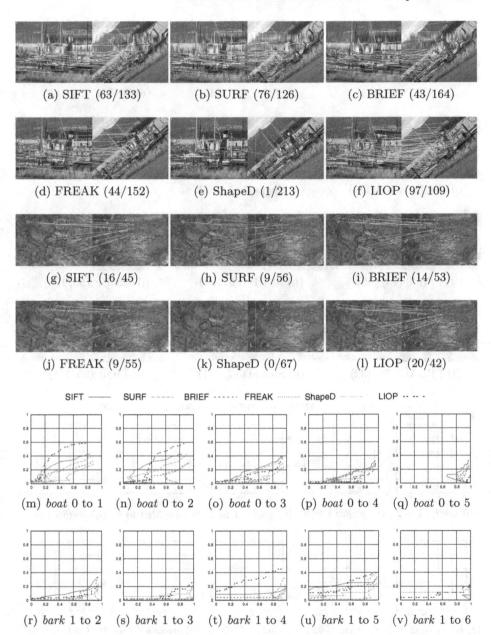

Fig. 2. Matching performance evaluation for image rotation and scale change (*boat* and *bark* dataset). (a)~(l) (number of inlier / number of outlier) of evaluated descriptors. (m)~(v) recall vs. 1-precision plots.

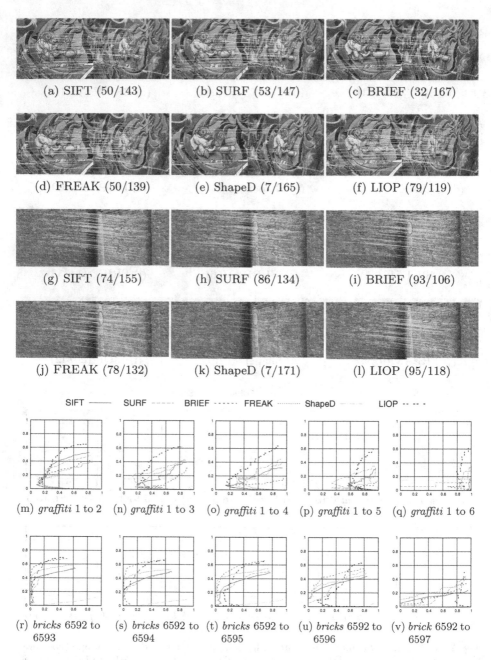

Fig. 3. Matching performance evaluation for viewpoint change (*graffiti* and *bricks* dataset). (a)∼(l) (number of inlier / number of outlier) of evaluated descriptors. (m)∼(v) recall vs. 1-precision plots.

scene with significant viewpoint change, LIOP descriptor also outperforms other descriptors. Figure 3 (m)~(v) shows the recall vs. 1-precision comparison of graffiti and *bricks* datasets. As shown in Fig. 3, the best performance is archived again by LIOP descriptor. Note that, since *bricks* image contains repeated patterns, MSER detector extracts uniform regions. In this case, all descriptors show higher performance than other dataset with nonuniform MSER regions.

4.3 Deformable Objects

To evaluate the performance for deformable objects, we test *bed sheet* and *cushion* dataset [23]. They provide 3D coordinates of ground truth 3D mesh and corresponding 2D coordinates of mesh vertices in images. Therefore, true correspondence can be estimated using 2D coordinate pairs. Figure 4 shows the visual comparison of the matching results. As shown in Fig. 4, all descriptors show poor performance, which can be explained as follows. MSER is an affine invariant region detector which detects maximal or minimal connected regions in the image. If the detected region is changed to different elliptical shape due to the scene deformation, the normalized patch of each region is also changed. In that case, local descriptor matching is not going to be accurate. Nevertheless, SIFT and LIOP achieve relatively better performance than others.

4.4 3D Objects

In Moreels's dataset [18], we test *potato* and *volley ball* models which have heavily textured objects. Figure 5 shows the visual comparison of the matching results, which shows that, a few correspondences are matched with few inliers. Since the test image has significant viewpoint change, the matching performance decreases even though MSER is robust to the affine transform. Therefore, it is difficult to match features using MSER in the scene with significant viewpoint change and the deformable object.

4.5 Processing Time

Table 2 presents the computational time for descriptor generation and nearest neighbor matching. In our MSER implementation, approximately 500 regions are detected from two images. Our observation can be summarized as follows.

- Binary descriptors show the fastest description and matching speed. Binary descriptors are described with 256 and 512 bits and the difference of descriptors is calculated using Hamming distance.
- As is observed in other literatures, SURF is 5.6× faster than SIFT.
- Shape descriptor is slower than SIFT descriptor because SIFT can be constructed from gray scale image directly. On the other hand, in Shape descriptor generation, binary image of MSER regions has to be computed.
- The slowest descriptor is LIOP. If we reduce the patch size to half, the computational time of LIOP descriptor decrease to 3,344 ms but it is still slow.

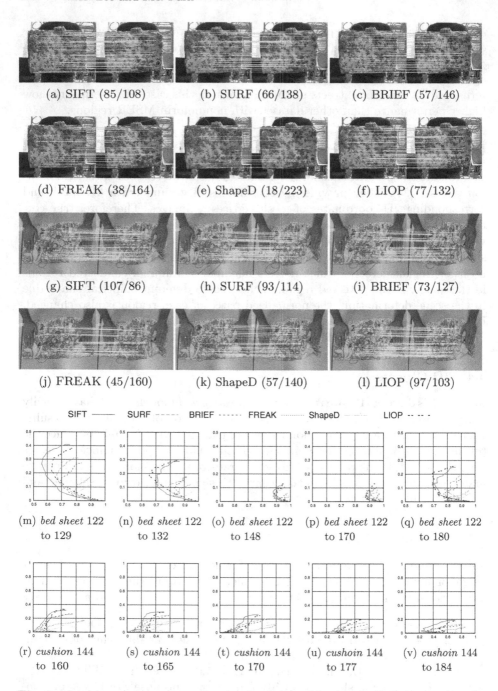

(a) SIFT (85/108)   (b) SURF (66/138)   (c) BRIEF (57/146)

(d) FREAK (38/164)   (e) ShapeD (18/223)   (f) LIOP (77/132)

(g) SIFT (107/86)   (h) SURF (93/114)   (i) BRIEF (73/127)

(j) FREAK (45/160)   (k) ShapeD (57/140)   (l) LIOP (97/103)

SIFT ——— SURF - - - - BRIEF · · · · · · FREAK ········· ShapeD — — LIOP -- -- -

(m) *bed sheet* 122 to 129   (n) *bed sheet* 122 to 132   (o) *bed sheet* 122 to 148   (p) *bed sheet* 122 to 170   (q) *bed sheet* 122 to 180

(r) *cushion* 144 to 160   (s) *cushion* 144 to 165   (t) *cushion* 144 to 170   (u) *cushoin* 144 to 177   (v) *cushoin* 144 to 184

Fig. 4. Matching performance evaluation for deformable objects (*bed sheet* and *cushion* dataset). (a)~(l) (number of inlier / number of outlier) of evaluated descriptors. (m)~(v) recall vs. 1-precision plots.

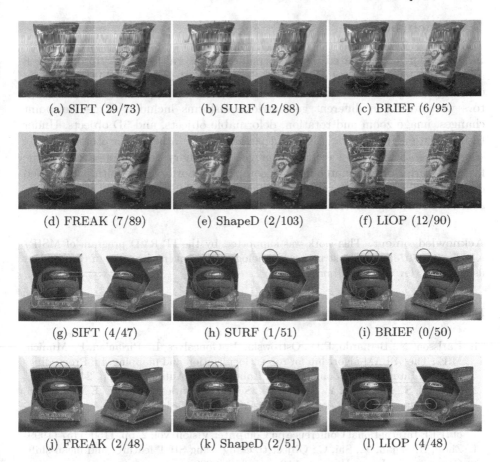

(a) SIFT (29/73)　　　(b) SURF (12/88)　　　(c) BRIEF (6/95)

(d) FREAK (7/89)　　　(e) ShapeD (2/103)　　　(f) LIOP (12/90)

(g) SIFT (4/47)　　　(h) SURF (1/51)　　　(i) BRIEF (0/50)

(j) FREAK (2/48)　　　(k) ShapeD (2/51)　　　(l) LIOP (4/48)

Fig. 5. Matching performance evaluation for 3D objects (*potato* and *volley ball* dataset) measured by (number of inlier / number of outlier).

Table 2. Computation time in milliseconds (500 × 500 matching).

Descriptor (dimension)	Descriptor generation	Nearest neighbor matching
SIFT (128)	2,189	142
SURF (128)	388	157
BRIEF (256)	43	84
FREAK (512)	84	163
Shape descriptor (128)	2,888	112
LIOP (255)	42,827	267

- Because SIFT, SURF, and Shape descriptor have same dimension of descriptor vector, the matching time is almost similar.
- The matching with LIOP descriptor is two times slower than SIFT because LIOP has twice size of descriptor than SIFT.

5   Conclusion

In this paper, we evaluated the performance of several local descriptors as well as binary descriptors for the affine invariant region detector, *i.e.* MESR. A total of six (SIFT, SURF, BRIEF, FREAK, Shape descriptor, and LIOP) descriptors were tested in different geometric transforms including large viewpoint changes, image zoom and rotation, deformable objects, and 3D objects. Under the evaluation framework, LIOP outperformed the other descriptors in image zoom and rotation, and viewpoint change. The binary descriptors archived the fastest description and matching with comparable performance with patch based descriptor. The experimental result indicated that MSER detector was not suitable for describing deformable object and 3D object.

Acknowledgement. This work was supported by the IT R&D program of MSIP/ KEIT. [10047078, 3D reconstruction technology development for scene of car accident using multi view black box image].

References

1. Karlsson, N., Bernardo, E.D., Ostrowski, J., Goncalves, L., Pirianian, P., Munich, M.E.: The vSLAM algorithm for robust localization and mapping. In: Proceedings of IEEE International Conference on Robotics and Automation, pp. 24–29 (2005)
2. Agarwal, S., Furukawa, Y., Snavely, N., Simon, I., Curless, B., Seitz, S.M., Szeliski, R.: Building Rome in a day. Commun. ACM **54**, 105–112 (2011)
3. Lowe, D.G.: Object recognition from local scale-invariant features. In: Proceedings of IEEE International Conference on Computer Vision, vol. 2, pp. 1150–1157 (1999)
4. Zhou, H., Yuan, Y., Shi, C.: Object tracking using SIFT features and mean shift. Comput. Vis. Image Underst. **113**, 345–352 (2009)
5. Serrano, N., Savakis, A.E., Luo, J.: Improved scene classification using efficient low-level features and semantic cues. Pattern Recogn. **37**, 1773–1784 (2004)
6. Lowe, D.G.: Distinctive image features from scale-invariant keypoints. Int. J. Comput. Vis. **60**, 91–110 (2004)
7. Bay, H., Ess, A., Tuytelaars, T., Gool, L.V.: Speeded-up robust features. Comput. Vis. Image Underst. **110**, 346–359 (2008)
8. Mikolajczyk, K., Schmid, C.: A performance evaluation of local descriptors. IEEE Trans. Pattern Anal. Mach. Intell. **27**, 1615–1630 (2005)
9. Mikolajczyk, K., Schmid, C.: Scale and affine invariant interest point detectors. Int. J. Comput. Vis. **60**, 63–86 (2004)
10. Mikolajczyk, K., Tuytelaars, T., Schmid, C., Zisserman, A., Matas, J., Schaffalitzky, F., Kadir, T., Gool, L.V.: A comparison of affine region detectors. Int. J. Comput. Vis. **65**, 43–72 (2005)
11. Matas, J., Chum, O., Urban, M., Pajdla, T.: Robust wide baseline stereo from maximally stable extremal regions. In: Proceedings of the British Machine Vision Conference, vol. 1, pp. 384–393 (2002)
12. Forssen, P.E., Lowe, D.G.: Shape descriptors for maximally stable extremal regions. In: Proceedings of IEEE International Conference on Computer Vision, pp. 1–8 (2007)

13. Calonder, M., Lepetit, V., Strecha, C., Fua, P.: BRIEF: Binary robust independent elementary features. In: Proceedings of European Conference on Computer Vision, pp. 778–792 (2010)
14. Alahi, A., Ortiz, R., Vandergheynst, P.: FREAK: Fast retina keypoint. In: Proceedings of IEEE Conference on Computer Vision and Pattern Recognition, pp. 510–517 (2012)
15. Wang, Z., Fan, B., Wu, F.: Local intensity order pattern for feature description. In: Proceedings of IEEE International Conference on Computer Vision, pp. 603–610 (2011)
16. Miksik, O., Mikolajczyk, K.: Evaluation of local detectors and descriptors for fast feature matching. In: Proceedings of International Conference on Pattern Recognition, pp. 2681–2684 (2012)
17. Restrepo, M.I., Mundy, J.L.: An evaluation of local shape descriptors in probabilistic volumetric scenes. In: Proceedings of the British Machine Vision Conference, pp. 46.1–46.11 (2012)
18. Moreels, P., Perona, P.: Evaluation of features detectors and descriptors based on 3D objects. Int. J. Comput. Vis. **73**, 263–284 (2007)
19. Dahl, A.L., Aanaes, H., Pedersen, K.S.: Finding the best feature detector-descriptor combination. In: Proceedings of International Conference on 3D Imaging, Modeling, Processing, Visualization and Transmission, pp. 318–325 (2011)
20. Haja, A., Jahne, B., Abraham, S.: Localization accuracy of region detectors. In: Proceedings of IEEE Conference on Computer Vision and Pattern Recognition, pp. 1–8 (2008)
21. Dickscheid, T., Schindler, F., Forstner, W.: Coding images with local features. Int. J. Comput. Vis. **94**, 154–174 (2011)
22. Canclini, A., Cesana, M., Redondi, A., Tagliasacchi, M., Ascenso, J., Cilla, R.: Evaluation of low-complexity visual feature detectors and descriptors. In: Proceedings of International Conference on Digital Signal Processing, pp. 1–7 (2013)
23. Salzmann, M., Moreno-Noguer, F., Lepetit, V., Fua, P.: Closed-form solution to non-rigid 3D surface registration. In: Proceedings of European Conference on Computer Vision, pp. 581–594 (2008)

Unsupervised Footwear Impression Analysis and Retrieval from Crime Scene Data

Adam Kortylewski[✉], Thomas Albrecht, and Thomas Vetter

Departement of Mathematics and Computer Science,
University of Basel, Basel, Switzerland
{adam.kortylewski,thomas.albrecht,thomas.vetter}@unibas.ch

Abstract. Footwear impressions are one of the most frequently secured types of evidence at crime scenes. For the investigation of crime series they are among the major investigative notes. In this paper, we introduce an unsupervised footwear retrieval algorithm that is able to cope with unconstrained noise conditions and is invariant to rigid transformations. A main challenge for the automated impression analysis is the separation of the actual shoe sole information from the structured background noise. We approach this issue by the analysis of periodic patterns. Given unconstrained noise conditions, the redundancy within periodic patterns makes them the most reliable information source in the image. In this work, we present four main contributions: First, we robustly measure local periodicity by fitting a periodic pattern model to the image. Second, based on the model, we normalize the orientation of the image and compute the window size for a local Fourier transformation. In this way, we avoid distortions of the frequency spectrum through other structures or boundary artefacts. Third, we segment the pattern through robust point-wise classification, making use of the property that the amplitudes of the frequency spectrum are constant for each position in a periodic pattern. Finally, the similarity between footwear impressions is measured by comparing the Fourier representations of the periodic patterns. We demonstrate robustness against severe noise distortions as well as rigid transformations on a database with real crime scene impressions. Moreover, we make our database available to the public, thus enabling standardized benchmarking for the first time.

1 Introduction

Footwear impressions are one of the most frequently secured types of evidence at crime scenes. For the investigation of crime series they are among the major investigative notes, permitting the discovery of continuative case links and the conviction of suspects. In order to simplify the investigation of cases committed by suspects with the same footwear, the crime scene impressions are assigned to a reference impression (see Fig. 1). Through the assignment process, the noisy and incomplete evidence becomes a standardized information with outsole images, brand name, manufacturing time, etc. Currently, no automated systems exist that can assess the similarity between a crime scene impression

© Springer International Publishing Switzerland 2015
C.V. Jawahar and S. Shan (Eds.): ACCV 2014 Workshops, Part I, LNCS 9008, pp. 644–658, 2015.
DOI: 10.1007/978-3-319-16628-5_46

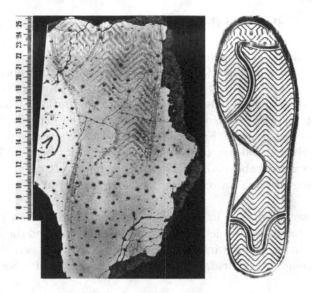

Fig. 1. A crime scene impression and the corresponding reference impression. A main challenge for pattern recognition systems is to isolate the shoe sole pattern from the structured noise. Moreover, the impression is incomplete and the zigzag line is shifted between the impressions. Therefore no point-wise similarity measure can be applied. Furthermore, the relative orientation and translation between the images is unknown.

and reference impressions, due to the severe image degradations induced by the impression formation and lifting process. The main challenges here are the combination of unknown noise conditions with rigid transformations and missing data (see Fig. 1). Moreover, the image modalities could be inverse, meaning that the impression information could be white for one impression and black for the other. Furthermore, training and testing data are scarce, because usually no or few crime scene impressions are available per reference impression. Additionally, in many cases no point-to-point correspondence exists between the impressions because different parts of the shoe sole are produced independently of each other. This results for example in a phase shift of the zigzag line between the impressions in Fig. 1. Therefore, a higher level understanding of the pattern is necessary. In this work, we introduce an unsupervised image retrieval algorithm that overcomes the limitations of existing work by detecting and analyzing periodic patterns in the footwear impressions under unconstraine noise conditions. The only assumption being, that the noise signal is not strictly periodically structured. The basic idea behind our approach is that periodic structures are the most preserved information under unknown noise conditions because of their inherent redundancy. In our reference impression database containing 1175 images, about 60 % of the images show periodic patterns. Given the challenging application scenario, solving the recognition task for this subset of the data is a valuable contribution.

The main contribution of this work is to extend the periodic pattern model from Lin et al. [8], such that the localization and analysis of periodic patterns under unconstrained, structured noise becomes possible. The reliable extraction of the pattern model enables the compensation of rigid transformations, the estimation of the scale of the repeating texture tile, and the exclusion of background information and noise from the pattern representation. After calculating a local Fourier representation, we robustly segment the periodic pattern from other structures in the image. Finally, during the image retrieval process we use the extracted Fourier features to compute the similarity between images. Our footwear impression database is available to the public at http://gravis.cs.unibas.ch/fid/, enabling research on real case data and standardized comparisons for the first time.

In Sect. 2, we will discuss in detail how the feature extraction is performed. Afterwards, we introduce the detection of periodic patterns and the image retrieval algorithm. In Sect. 3, we introduce our footwear impression database and perform a thorough experimental evaluation. We conclude with Sect. 4.

1.1 Previous Work

Early work on automated shoeprint recognition approached the problem by either analyzing the frequency spectra of the whole images using the Fourier transform [5,6], or by describing the image with respect to its axes through Hu-Moment invariants [1]. Such global shoe print processing methods are particularly sensitive to noise distortions and incomplete data. Pavlou et al. [14,15] and Su et al. [17] proposed to classify shoe print images based on local image features. Their approaches are based on combinations of local interest point detectors and SIFT feature descriptors. However, in general such gradient-based methods are not sufficient for the application of crime scene impression retrieval (see the experiments in Sect. 3). The reason is that the image gradient of the noisy data is strongly distorted compared to clean reference data. Therefore, a reliable gradient-based detection of interest points or the correct rotational normalization based on the image gradient are difficult. Patil et al. [13] divide the image into a block structure of constant size and process each block individually with Gabor features. Their approach is too rigid to capture different shoe print orientations because of the constant sized block grid. Nevertheless, they show that local normalization is a crucial step in processing noisy images. Regarding rotational invariance, DeChazal et al. [5] are tolerant to rotation by brute force rotating the images in one-degree steps. However, this is not usefull in practice because of the computational costs. Nibouche et al. [12] use SIFT combined with RANSAC to compensate for the rotation, but since the feature descriptors are heavily distorted by the structured noise, the method is only applicable to noiseless data. These and most other works in the field work on synthetically generated training and testing data, assuming a very simple noise model such as Gaussian or salt and pepper noise. However, one key challenge of real data is that the noise is unconstrained and therefore cannot be simulated by such simple noise distributions. Dardi and Cervelli [2,4] published algorithms applied to real data. However, the approach in [4] is based on a rigid partitioning of images and thus is sensitive to image transformations and the approach in [2] is rotationally invariant, but is not robust against noise and incomplete data. Another approach was introduced by Tang et al. [18]. They extract basic shapes (circles, lines and ellipses) out of the shoe print image with a modified Hough transform and store these shapes in a graph representation. Attributes such as distance, relative position, dimension and orientation are encoded into the graph structure, making their recognition algorithm robust to image transformations. But many shoe soles are comprised of more complex patterns which cannot be described by such basic shapes. Additionally, in their experiments, crime scene impressions are mixed with synthetic data, giving no clear performance statements for the real case scenario. A general review on shoeprint recognition methods has been presented by Luostarinen and Lehmussola [11]. However, the experimental setup is very restricted, because they assume known correspondance between crime scene impression and reference impression. Since that information can only be provided after already knowing the correctly matching reference, that work has only limited relevance for the task of footwear impression retrieval.

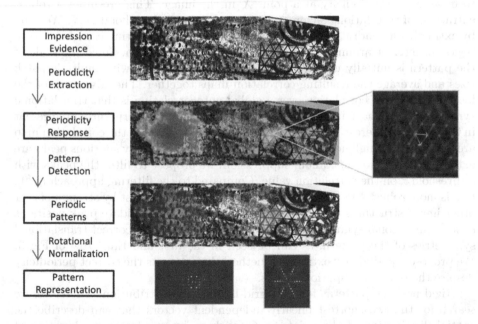

Fig. 2. Graphical abstract of the feature extraction procedure. We first extract the periodicity at each point in the impression evidence. Then, we compute Fourier features and use these to detect the periodic patterns. Finally, the periodic patterns are represented by the rotationally normalized Fourier features.

2   Impression Analysis Based on Periodic Patterns

The goal of this work is to compare images in the presence of unconstrained noise, rigid transformations, without point-to-point correspondence and across different modalities, based on their periodic patterns. A periodic pattern is fully defined by a basic texture tile that is spatially distributed in two dimensions according to two fixed distribution vectors. We begin with periodicity detection by fitting a model of translational symmetries to local autocorrelation responses. Then, we calculate translation-invariant Fourier descriptors and separate the periodic patterns from other structures in the image by point-wise classification. Afterwards, the individual periodic patterns can be rotationally normalized with respect to their inherent translational symmetry distribution. Finally, the comparison between impressions is achieved by first recomputing the Fourier representation of the crime scene pattern based on the extracted periodicity models of the reference impressions, and the subsequent comparison of the feature representations. An overview about our approach is depicted in Fig. 2.

2.1   Periodicity Extraction

The basic building block of the proposed image retrieval pipeline is the ability to measure the periodicity at a point X in the image. This presumes a robust extraction of translational symmetries in the local neighbourhood of X. We start by extracting a quadratic image patch around X and correlating it within a local region of interest around X by normalized cross correlation. Since the scale of the pattern is initially unknown, we repeat this procedure with multiple patch sizes and average the resulting correlation maps together. The advantage of the local, patch-based correlation over a global autocorrelation is that translational symmetries stay sharp in the correlation map, even if other structures are present in the region of interest. Afterwards, the positions of peaks in the correlation map are extracted through non-maximum-suppression. Since many spurious peaks are extracted by the non-maximum-suppression, we propose to filter the peaks with a threshold τ on the correlation value. Compared to the filtering approach in [9], this is more robust against distortions in the region of interest by strong noise or other image structures (Fig. 3). The resulting list C of candidate peak positions c_k for translational symmetries, still does not solely contain correct translational symmetries of the pattern, as can be seen in the first and third row of Fig. 3. We propose a periodicity extraction method that extracts the correct periodicity despite the remaining spurious peaks.

Rigid periodic patterns follow a grid-like spatial distribution. Therefore, we search for the two shortest linearly independent vectors that can describe the spatial distribution of the points in C, with an integer linear combination of themselves. To this end, we build on the Hough transform approach of [8] and constrain the candidate vectors to originate at the center of the region of interest and point to one of the candidate peaks c_k. Thus introducing a vector space with the origin set at the center of the region of interest. Additionally, the number of possible candidate vectors is reduced to the number N of points in C. We modify

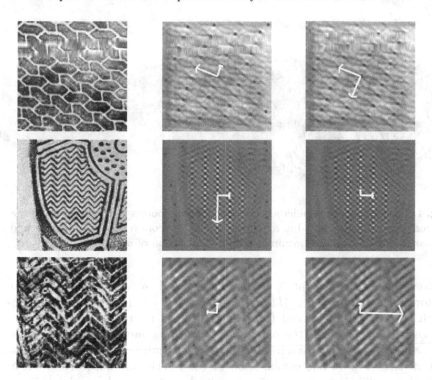

Fig. 3. Detection of translational symmetries. Left column: region of interest; middle column: result of the approach of [9]; right column: result of our approach. The results show the autocorrelation maps in gray, the candidate peaks for translational symmetries in red and the detected periodicity indicated by two vectors in yellow. The proposed approach extracts peaks more reliably and gives better periodicity estimates compared to the approach of [9].

the scoring function of [8] by incorporating the correlation values at the peak positions $NCC(c_k)$ and by penalizing the length of both distribution vectors, instead of just the largest one (Eq. 1). The function $nint(x)$ in Eq. 1 rounds a scalar x to the next integer.

$$h(c_p, c_q, C) = \alpha \cdot (NCC(c_p) + NCC(c_q))$$
$$+ \sum_{c_k \in C,\, k \neq p,q} \frac{(1 - 2\max(|a_k - nint(a_k)|, |b_k - nint(b_k)|))\, NCC(c_k)}{\|c_p\| + \|c_q\|}$$
$$\begin{bmatrix} a_k \\ b_k \end{bmatrix} = [c_p, c_q]^{-1} [c_k]; \quad c_k, c_p, c_q \in C \qquad (1)$$

The variables a_k and b_k are coefficients related with the linear combination of c_p and c_q for the peak c_k. The factor α performs the tradeoff between the importance of the correlation values and the shortness of the distribution

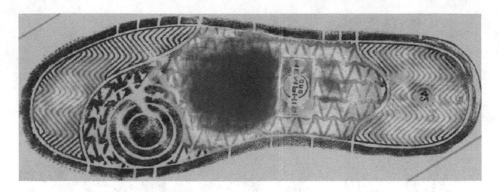

Fig. 4. Illustration of periodicity scores for different points in a reference impression; the maximal value is colored in red. The periodicity score reflects the amount and the similarities of translational symmetries in a local region of interest.

vectors. The periodicity at X is then determined by searching for the pair of vectors $(c_p, c_q) = \text{argmax}_{p,q}\{h(c_p, c_q, C)|p = 1..N, q = 1..N\}$. In Fig. 4 the periodicity score h is illustrated for each point in a reference impression. It can be seen, that maxima of the score are reached predominantly at the centers of the periodic patterns. We have illustrated the results of our approach and the one of [9] in Fig. 3. The combination of the multi-scale autocorrelation maps with the modified scoring function is robust against local correlation maxima as well as distortions through noise and additional structures in the region of interest.

2.2 Periodic Pattern Detection and Representation

On the basis of the periodicity measurement introduced in Sect. 2.1, we compute a low-dimensional representation of the periodic patterns by pooling redundancies into one feature descriptor and subsequently determine the number of periodic patterns in the image. We propose to extract Fourier based features to encode the appearance and the periodicity of a periodic pattern. However, the computation of discriminative local Fourier features presumes the selection of the right window size. Especially for small window sizes this is critical, since image structures that do not belong to the periodic pattern have a great distortive impact on the frequency spectrum, as do the discontinuities at the window boundary. In the following, we describe a method to determine a suitable window for the Fourier transform based on the periodicity information. The Fourier features are then applied to detect the number of periodic patterns contained in the image in an iterative procedure.

We start by computing the periodicity at each point $P = (i, j)$ in the image $I \in \mathbb{R}^{n \times m}$. The maximal periodicity scores h and the corresponding distribution vectors v_1 and v_2 are stored in a Matrix $H = \{H(P) = [h(P), v_1(P), v_2(P)] \mid i = 1..n, j = 1..m\}$. Based on that information, we apply an iterative grouping procedure to determine if and how many periodic patterns are available in the image.

Algorithm 1. Pseudocode of the Periodic Pattern Detection

```
1:  Input: H : Periodicity, I : Image
2:  PPatterns ← Empty
3:  while max(H,h) > 0 do
4:      X ← maxpos(H,h)
5:      f_X ← CalculateFourierDescriptor(X,H,I)
6:      H(X)=0;
7:      for (Y)∈I do
8:          if s(f_X,CalculateFourierDescriptor(Y,H,I))>ρ then
9:              H(Y)=0;
10:     PPatterns ← f_X
11: Output: PPatterns

12: function CALCULATEFOURIERDESCRIPTOR
13:     Input: X, H, I
14:     [v1,v2] ← H[X]
15:     R_{I,X} ← Rect(X,v1,v2,I)
16:     f_X ← |F(G(X,Σ_{R_{I,X}}) · R_{I,X})|
17:     Output ← f_X
```

The procedure is summarized in pseudocode in Algorithm 1. An iteration starts at the point X in the image with maximal periodicity score (see Algorithm 1, l.4). High periodicity indicates that translational symmetries exist in the local neighborhood. This also implies that other image structures are less likely to occur in the surrounding area. Based on the corresponding distribution vectors $v_1(X)$ and $v_2(X)$, a window R is determined as the smallest rectangle containing the points $\{X-v_1(X)-v_2(X); X-v_1(X)+v_2(X); X+v_1(X)-v_2(X); X+v_1(X)+v_2\}$. Then, the image patch $R_{I,X}$ of size R centered at position X is extracted. An important advantage of determining the window R based on the periodicity of the pattern is that we are able to capture approximately an integer number of periods of the pattern. Additionally, no other structures are contained in $R_{I,X}$, thus making it a particularly good basis for further feature extraction processes. In practice it is still beneficial to reduce distortions in the frequency spectrum through small discontinuities at the boundary, by multiplying $R_{I,X}$ with a Gaussian window of size R centered at X before the Fourier transformation. We denote the Gaussian window by $G(X, \Sigma_{R_{I,X}})$. Two periods of the pattern are extracted in each direction of repetition because that increases the sampling rate during the Fourier transformation and thus leads to more discriminative Fourier features. A central property of periodic patterns is that only the phase of the frequency spectrum varies throughout the pattern, but the magnitude stays approximately constant. We exploit this fact by using just the magnitude of the frequency spectrum of $R_{I,X}$ as a feature descriptor for the periodic pattern:

$$f(R_{I,X}) = |F(G(X, \Sigma_{R_{I,X}}) \cdot R_{I,X})|. \tag{2}$$

With this translationally invariant descriptor we are able to classify if another point Y belongs to the same pattern as X by evaluating the normalized

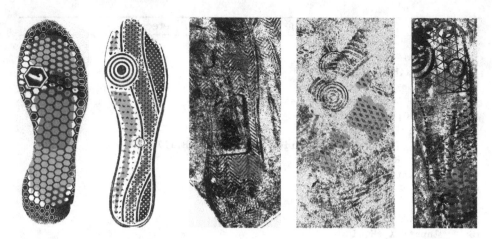

Fig. 5. Results of the grouping procedure on reference images as well as crime scene impressions. Different patterns in the same image are coded in different colors. Periodic patterns on very different scales are well separated from other structures in the images even under strong noise conditions.

cross-correlation of the Fourier descriptors at these points:

$$s(f(R_{I,X}), f(R_{I,Y})) = \frac{(f(R_{I,X}) - \mu_{f(R_{I,X})})(f(R_{I,Y}) - \mu_{f(R_{I,Y})})}{\sigma_{f(R_{I,X})}\sigma_{f(R_{I,Y})}}. \quad (3)$$

An important detail to notice is that both Fourier descriptors are calculated with the same window size R, ensuring that the feature descriptors have the same dimensionality. By ignoring the phase of the frequency spectrum, the feature descriptor gets invariant to inverse image modalities.

Based on the extracted Fourier feature $f(R_{I,X})$ we classify each point in the image by computing the similarity measure of Eq. 2, followed by a thresholding with a constant value ρ (see Algorithm 1, l.8). For the points with a similarity greater than ρ, we set the corresponding periodicity score in H to zero. This procedure is repeated until the maximum remaining periodicity score is zero. After this procedure, the image is decomposed into a set of periodic patterns, each represented by its corresponding Fourier feature. In Fig. 5, example results illustrate the extracted patterns for different impressions containing periodic patterns from a wide variety of scales and appearances. It can be seen that different periodic patterns are well separated from the background even under structured noise, indicating that the extracted Fourier features are a good representation of the corresponding patterns.

2.3   Similarity and Footwear Retrieval

We propose to compute the similarity between a crime scene impression and a reference impression by means of the similarity between their periodic patterns. Before two patterns are compared, we first actively normalize their rotation based on their translational symmetry structures computed above. We do so instead of e.g. making the feature descriptor rotationally invariant, in order to retain the discriminative power of the descriptors. The normalization is achieved by rotating the image so that the smallest distribution vector points upright. In cases where both vectors have the same length we align to the vector along which the image gradient is greatest. After this, we recalculate the Fourier features on the rotationally normalized images. In practice, the estimated window sizes R for the Fourier transformation do not always have exactly the same size for the same patterns in crime scene impressions and reference impressions. This is due to differences in the noise conditions and the pixel discretization of the image. By using the window size of the reference pattern for both impressions, we ensure that the features have the same dimensionality. Although this can lead to distortions of the frequency spectrum when the patterns differ, the effect is negligible for similar patterns. Afterwards, we compare the features as described in Eq. 2. The feature extraction for the crime scene impressions can also be interpreted as an interest point detection with rotational normalization, as the Fourier features are recomputed during the matching with the window sizes of the respective reference impressions. The similarity between a reference impression $A = \{f(R^1_{I_A, X_1}), ..., f(R^n_{I_A, X_n})\}$ and a crime scene impression $B = \{I_B; Y_1, ..., Y_m\}$ can now be computed by the following similarity measure:

$$S(A, B) = \frac{1}{m} \sum_{j=1}^{m} \max_i \; s(f(R^i_{I_A, X_i}), \; f(R^i_{I_B, Y_j})). \tag{4}$$

As the crime scene impression may not contain all periodic patterns from the shoe outsole, only the subset of periodic patterns from the reference with maximal similarity is evaluated.

3 Experiments

We test our approach on a database of real crime scene impressions. The database consists of 170 crime scene impressions, among which 102 show periodic patterns. Out of the 170 impressions the pattern extraction algorithm detects for 133 impressions periodic patterns, including all true periodic patterns and 31 false positives. During the evaluation, we show the performance on all 133 impressions with detected periodic pattern in order to evaluate a fully automated retrieval setting. Those 37 impressions that are not included in the evaluation are also to be published with the database and are marked to enable a correct performance comparison of future works. The reference database consists of 1175 reference impressions. In a random subset of 230 impressions, 142 showed a periodic pattern, thus we assume that roughly 60 % of the reference impressions show periodic patterns.

3.1 Setup

In our approach, we do not account for scale changes, because this information is provided by a ruler on each impression image and can thus be given manually.

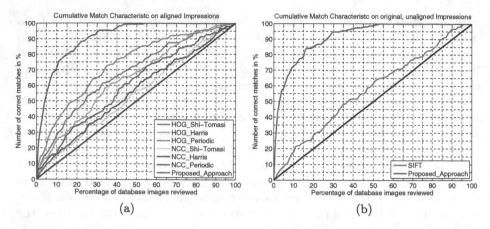

Fig. 6. Cumulative match characteristic for (a) crime scene impressions aligned to the corresponding reference impression and (b) the original unaligned data.

For the experiments, the images are scaled to 10 pixels per centimeter. Also, the crime scene images are cropped so that the impression is roughly centered in the image frame (see Fig. 7). During the experiments we choose the parameter $\alpha = 2$. The patch sizes for the localized multi-scale autocorrelation are chosen to vary from 11 to 37 pixels in a four pixel step size. With this configuration, we capture all scales of periodic patterns that show at least two repetitions in both directions. We constrain the angle between the distribution vectors to be between 60 and 90 degrees as in [9]. This reduces the number of possible distribution vector combinations and thus acts as a regularization and saves computational time. Despite this constraint, we are still able to describe the translational symmetries of all periodic patterns in the database. During the experiments, we first compute the similarity between the query impression and all reference impressions in the database. We sort the reference images by similarity, thus producing a ranking list of the most similar reference impressions. As a performance measure, we apply the cumulative match characteristic. This score reflects the question: "What is the probability of a match if I look at the first n percent of the ranking list?" (adapted from [5]). We split the experiments into two parts. For the first part, we manually align all crime scene impressions to the corresponding reference impressions. Thus, the remaining difference between the impressions is of a structural nature. The normalization makes it possible to compare our approach with robust methods that are not invariant to rigid transformations, such as the normalized cross correlation (NCC) or the histogramm of oriented gradients (HOG) [3]. In the second part, we compare the SIFT algorithm [10] with the proposed approach on the original unaligned data.

3.2   Footwear Impression Retrieval Performance

The comparison methods in the first experiment are histogramms of oriented gradients and normalized cross correlation. We combine HOG and NCC with

Table 1. Summary of the experiments in terms of footwear impression retrieval performance measured by the cumulative match characteristic.

	Method	Recall@				
		1 %	5 %	10 %	15 %	20 %
Aligned	HOG + Shi-Tomasi	3.8	15.8	25.6	37.6	44.4
	HOG + Harris	4.5	13.5	24.8	30.1	37.6
	HOG + Periodic	9.8	22.6	36.8	45.9	50.4
	NCC + Shi-Tomasi	3.8	9.8	15.0	19.6	26.3
	NCC + Harris	7.5	12.8	17.3	22.6	28.6
	NCC + Periodic	2.3	10.5	17.3	30.1	36.8
	This work	**27.1**	**56.4**	**70.0**	**76.7**	**85.0**
Unaligned	SIFT	1.5	8.3	18.1	24.1	28.6
	This Work	**27.1**	**59.4**	**74.4**	**79.7**	**85.7**

interest point detections by the Harris [7] and Shi-Tomasi [16] corner detectors and with the feature locations detected by our approach. For the corner detection we use the algorithms implemented in Matlab. The patch size is fixed to 21×21 pixels since that leads to the best performance in the experiments. Our approach estimates the window size by itself during the feature extraction. Although the alignment to the correct reference impression is not possible in forensic practice, this experiment is of interest because it measures the performance of HOG and NCC and interest point detectors under structured noise. During the experiments, we detect interest points on the crime scene impressions and subsequently compute the descriptors at the interest point locations. Since the crime scene impression and the reference impression are in correspondence, we repeat this procedure at the same locations in the reference impressions. The number of extracted features on a query impression is fixed to the number of features detected by our approach, in order to allow for a comparison of the results. The results are illustrated in Fig. 6a. NCC performs nearly equally for all interest points, as it does not account for any structural information in the image patches. HOG features perform better than NCC on average as they utilize local structural information to a certain degree and gain robustness to small transformational deviations compared to NCC by pooling information locally into histrogramms. Combined with the periodic interest points detected by our approach, it also performs better than on the Harris and Shi-Tomasi interest points. That is mainly because the periodic interest points always lie inside the shoe impression, whereas the others also detect less discriminative points on the impression boundary. Through the criterion of high periodicity during the interest point extraction, these points are also less likely to be distorted by noise. Despite that the alignment clearly introduces a positive bias on the matching performance, the results in general show that neither HOG features nor NCC perform well under the noisy environment. Our approach performs significantly better compared to the other feature detector and descriptor combinations. Especially under the first 10 % in the ranking list, the performance gain is about 33 %. Important is that our procedure does not profit from the alignment of the data since the rotaion is normalized during feature extraction. In Fig. 6b, we compare our approach on the original unaligned data with the SIFT feature extraction and description approach. For the experiments we use the SIFT implementation from Andrea Vedaldi and Brian Fulkerson [19]. Although, SIFT performs above chance rate, especially for the more important first fifth of the ranking list, the performance is still not satisfactory. Our proposed approach performs significantly better, so that e.g. the Recall@10 % is improved from 18.1 % to 74.4 % compared to SIFT. The performance results are summarized in Table 1. Another observation is that our approach reaches faster to 100 % matching score than the other approaches. This is because the algorithm does not detect a periodic pattern for 558 reference impressions and can thus exclude these from the candidate list early. Note that the performance results of the proposed approach are nearly the same for aligned and unaligned data, underlining the rotational invariance of the feature extraction. The low result for the Recall@1 % originates from the fact the proposed approach focuses on the periodic patterns in the impressions and no other structures. Since many reference impression have e.g. grid like patterns, these are all grouped together at the front of the ranking list (see Fig. 7). But since no other structures are included in the similarity measure, their order is not clearly defined and depends on nuances in the scale or noise of the impressions. We have illustrated two image retrieval results in Fig. 7. Despite strong structured noise and even different modalities in the first example, the grid-like and circular structures of the periodic patterns are clearly reflected in the retrieval results. The missing 25.6 % at Recall@10 % can be ordered in two categories. One part are impressions that are smeared through liquid on the ground, such that the rigid transformation assumption in the feature extraction does not hold anymore. The other part are double-prints, meaning that two impressions are overlayed on top of each other such that the algorithm is not able to extract the correct periodicity.

4   Conclusion

In this work, we have proposed an image retrieval algorithm based on periodic patterns. The algorithm is robust under unconstrained noise conditions by separating the meaningful pattern information from the structured background. Additionally, it is robust against incomplete data and it overcomes the problem of absence of point-to-point correspondence between impressions by extracting a translation-invariant pattern representation. Furthermore, it is able to match rotated data by actively normalizing the pattern representations with respect to the intrinsic tranlational symmetry structure of the periodic patterns. Our experiments demonstrate a significant performance gain over standard image retrieval techniques for the task of footwear impression retrieval. By making the database with real crime scene impressions and reference impressions publicly available, we open a new application to the field of computer vision, concerning the issue of how to separate patterns from structured noise despite incompleteness and spatial transformations. Thus, our publication enables standardized benchmarking in the field for the first time. In the future, we plan to make the approach scale invariant and, since regular patterns are only available on about 60 % of the data, we will also focus on how to robustly incorporate other structures from footwear impressions into the retrieval process.

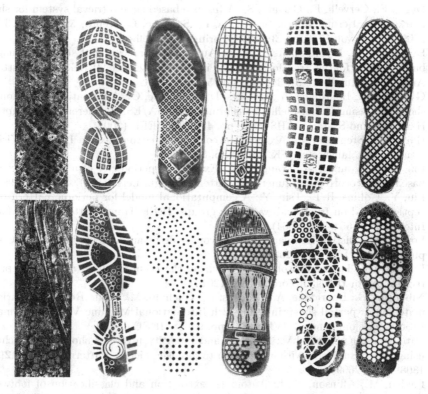

Fig. 7. Results of the image retrieval algorithm for two crime scene images. The column shows the query image. The second to fifth columns show the top results in the ranking list. And the last column shows the correct references. The correct references are found at position five and 13 in the ranking lists.

Acknowledgement. This Project was supported by the Swiss Comission for Technology and Innovation (CTI) project 13932.1 PFES-ES. The authors thank the German State Criminal Police Offices of Niedersachsen and Bayern and the company forensity ag for their valuable support.

References

1. AlGarni, G., Hamiane, M.: A novel technique for automatic shoeprint image retrieval. Forensic Sci. Int. **181**(1), 10–14 (2008)
2. Cervelli, F., Dardi, F., Carrato, S.: A translational and rotational invariant descriptor for automatic footwear retrieval of real cases shoe marks. Eusipco (2010)
3. Dalal, N., Triggs, B.: Histograms of oriented gradients for human detection. In: IEEE Computer Society Conference on Computer Vision and Pattern Recognition, CVPR 2005, vol. 1, pp. 886–893. IEEE (2005)
4. Dardi, F., Cervelli, F., Carrato, S.: A texture based shoe retrieval system for shoe marks of real crime scenes. In: Foggia, P., Sansone, C., Vento, M. (eds.) ICIAP 2009. LNCS, vol. 5716, pp. 384–393. Springer, Heidelberg (2009)
5. De Chazal, P., Flynn, J., Reilly, R.B.: Automated processing of shoeprint images based on the fourier transform for use in forensic science. IEEE Trans. Pattern Anal. Mach. Intell. **27**(3), 341–350 (2005)
6. Gueham, M., Bouridane, A., Crookes, D., Nibouche, O.: Automatic recognition of shoeprints using fourier-mellin transform. In: NASA/ESA Conference on Adaptive Hardware and Systems, AHS 2008, pp. 487–491. IEEE (2008)
7. Harris, C., Stephens, M.: A combined corner and edge detector. In: Alvey Vision Conference, Manchester, UK, vol. 15, p. 50 (1988)
8. Lin, H.-C., Wang, L.-L., Yang, S.-N.: Extracting periodicity of a regular texture based on autocorrelation functions. Pattern Recogn. Lett. **18**(5), 433–443 (1997)
9. Liu, Y., Collins, R.T., Tsin, Y.: A computational model for periodic pattern perception based on frieze and wallpaper groups. IEEE Trans. Pattern Anal. Mach. Intell. **26**(3), 354–371 (2004)
10. Lowe, D.G.: Distinctive image features from scale-invariant keypoints. Int. J. Comput. Vis. **60**(2), 91–110 (2004)
11. Luostarinen, T., Lehmussola, A.: Measuring the accuracy of automatic shoeprint recognition methods. J. Forensic Sci. (2014)
12. Nibouche, O., Bouridane, A., Crookes, D., Gueham, M., et al.: Rotation invariant matching of partial shoeprints. In: 13th International Machine Vision and Image Processing Conference, IMVIP 2009, pp. 94–98. IEEE (2009)
13. Patil, P.M., Kulkarni, J.V.: Rotation and intensity invariant shoeprint matching using gabor transform with application to forensic science. Pattern Recogn. **42**(7), 1308–1317 (2009)
14. Pavlou, M., Allinson, N.M.: Automatic extraction and classification of footwear patterns. In: Corchado, E., Yin, H., Botti, V., Fyfe, C. (eds.) IDEAL 2006. LNCS, vol. 4224, pp. 721–728. Springer, Heidelberg (2006)
15. Pavlou, M., Allinson, N.M.: Automated encoding of footwear patterns for fast indexing. Image Vis. Comput. **27**(4), 402–409 (2009)
16. Shi, J., Tomasi, C.: Good features to track. In: 1994 IEEE Computer Society Conference on Computer Vision and Pattern Recognition Proceedings CVPR 1994, pp. 593–600. IEEE (1994)
17. Su, H., Crookes, D., Bouridane, A., Gueham, M.: Local image features for shoeprint image retrieval. In: British Machine Vision Conference, vol. 2007 (2007)
18. Tang, Y., Srihari, S.N., Kasiviswanathan, H., Corso, J.J.: Footwear print retrieval system for real crime scene marks. In: Sako, H., Franke, K.Y., Saitoh, S. (eds.) IWCF 2010. LNCS, vol. 6540, pp. 88–100. Springer, Heidelberg (2011)
19. Vedaldi, A., Fulkerson, B.: Vlfeat: An open and portable library of computer vision algorithms. In: Proceedings of the International Conference on Multimedia, pp. 1469–1472. ACM (2010)

Reliable Point Correspondences in Scenes Dominated by Highly Reflective and Largely Homogeneous Surfaces

Srimal Jayawardena[1]($\boxtimes$), Stephen Gould[2], Hongdong Li[2], Marcus Hutter[2], and Richard Hartley[2]

[1] Autonomous Systems Laboratory, CSIRO, Brisbane, Australia
srimal.jayawardena@csiro.au
[2] Research School of Computer Science, The ANU, Canberra, Australia

Abstract. Common Structure from Motion (SfM) tasks require reliable point correspondences in images taken from different views to subsequently estimate model parameters which describe the 3D scene geometry. For example when estimating the fundamental matrix from point correspondences using RANSAC. The amount of noise in the point correspondences drastically affect the estimation algorithm and the number of iterations needed for convergence grows exponentially with the level of noise. In scenes dominated by highly reflective and largely homogeneous surfaces such as vehicle panels and buildings with a lot of glass, existing approaches give a very high proportion of spurious point correspondences. As a result the number of iterations required for subsequent model estimation algorithms become intractable. We propose a novel method that uses descriptors evaluated along points in image edges to obtain a sufficiently high proportion of correct point correspondences. We show experimentally that our method gives better results in recovering the epipolar geometry in scenes dominated by highly reflective and homogeneous surfaces compared to common baseline methods on stereo images taken from considerably wide baselines.

1 Introduction

Structure from Motion (SfM) tasks that recover geometric scene information from a set of images obtained from different views typically require reliable point correspondences across the images (or tracks in the case of videos) as a prerequisite. Such SfM tasks range from complete 3D scene reconstruction to stereo matching performed on uncalibrated images. Typically keypoints in images are detected and matched in order to obtain point correspondences. Much research has been done in this area and popular applications which use feature correspondences include aligning tourist photos from the Internet [1].

Motivation. Scenes dominated by highly reflective and largely homogeneous surfaces such as the body of a car [3–5], buildings with a lot of glass panes (*e.g.* failure case of [6] in Fig. 7(g) and (i)), medical images [7–9] *etc.* tend to

© Springer International Publishing Switzerland 2015
C.V. Jawahar and S. Shan (Eds.): ACCV 2014 Workshops, Part I, LNCS 9008, pp. 659–674, 2015.
DOI: 10.1007/978-3-319-16628-5_47

Fig. 1. Best point correspondences obtained from naive SIFT [2] matching do not give a sufficient spatial spread to recover the epipolar geometry. The reliable matches are concentrated around relatively non-reflective areas. Best viewed in color. Images may be cropped for clarity (Color figure online).

generate unreliable point correspondences. The number of iterations required for convergence of subsequent model fitting algorithms such as estimating the fundamental matrix using RANSAC increase exponentially and the task becomes intractable as the level of noise in the point correspondence grows. Typically noise ratios of more than 50 % tend to be impractical [10]. The high amount of noise in point correspondences obtained from scenes dominated by highly reflective and largely homogeneous surfaces can be due to the following reasons.

1. Reflections in common reflective surfaces do not represent a physical artifact in 3D space. Therefore in general they do not conform to the EPG (Epipolar Geometry) of the 3D scene. A special case is for rectilinear camera motion [11] where the epipolar deviations of specularities on surfaces that are convex and not highly undulating are usually quite small. Additionally ideal planar reflective surfaces are a limiting case where there is no epipolar deviation.
2. Parts of the same reflection may not appear the same in all images taken from different views. They are often distorted, broken up or missing in the other images. Therefore keypoints on reflections in the image tend to introduce spurious matches.
3. Large homogeneous surfaces such as the panel of a car are in general texture impoverished. Therefore descriptors evaluated on such surfaces are not sufficiently discriminate. On the other hand, textured non-reflective areas of the scene which need not necessarily be spatially well distributed across the images may generate more descriptive keypoints and therefore stronger point matches. An example is shown in Fig. 1 where SIFT keypoints were matched using SIFT descriptors with SIFT matching (nearest neighbor/ratio test [2]). The strongest matches are localized to a corner of the image containing the wheel of the car which is comparatively less reflective and better textured. Since the matches are not spatially well distributed such matches can produce degenerate configurations in subsequent SfM tasks such as estimating the fundamental matrix.

Contributions. We propose a method to obtain reliable point correspondences in scenes dominated by highly reflective and largely homogeneous surfaces.

The noise level in correspondences obtained from our method are sufficiently low to perform subsequent SfM tasks such as recovering the epipolar geometry. Our method is able to obtain a sufficient amount of representative matches (inliers) which can be used to recover the epipolar geometry of the scene from images where baseline methods fail (Sect. 4). Unlike existing methods [12,13] our method does not place any restrictions on the camera (*e.g.* affine camera, small motions). Moreover, it works on scenes with highly specular and reflective surfaces of vehicles, glass paneled buildings *etc.*, which create a lot of inter-object reflections. Instead of detecting keypoints, we propose to consider all points along image edges. Most of such edge points are usually disregarded in conventional keypoint detection and matching methods which are known to give good results in non-reflective and well textured scenes. We match all edge points employing a dense descriptor, DAISY [14], which can be computed quickly at all pixels in the image and therefore at all edge points. Additionally, a spatial constraint is enforced by dividing the image into a grid of buckets and selecting only the best k putative matches from each bucket, to ensure better spatial distribution of the point matches. Although it is possible to use SIFT to densely compute descriptors for all pixels, dense SIFT (DSIFT [15]) needs to be computed at a predetermined scale since there is no keypoint detection step to determine the scale. Using a hand tuned fixed scale will calculate feature descriptors that do not properly describe point features that are of a different scale, resulting in low matching scores for potential inlier point matches. Alternatively, computing dense SIFT descriptors at a range of scales for all edge points and performing subsequent matching would make the computation complexity prohibitively high and we have not attempted this in our investigation. On the other hand the DAISY descriptor naturally incorporates gradient histograms computed at a range of scales for each pixel at locations radially distributed around the interest point. DAISY has been shown to be more computationally efficient [14] than SIFT. Hence we used DAISY descriptors for our method.

2   Related Work

Most SfM and multi-view stereo algorithms which work on images of several views of a 3D scene require finding some form of correspondences between the views. Much work has been done on detecting and identifying correspondences between multiple views of a 3D scene. However, much of this work is targeted towards images of non-reflective and well textured objects and scenes.

For example, work which has received much attention include *Photo Tourism* [1], which employs the SIFT [2] key point detection and matching algorithm to find point correspondences. This work was originally intended for tourist images commonly found on the Internet which include outdoor landscapes and historic buildings. As such it does not work well with images of highly reflective objects containing largely homogeneous regions.

It is worth noting at this point that feature detection and description are two separate tasks although some algorithms such as SIFT [2], SURF [16] and

BRISK [17] tend to do both. Other methods such as the key point and edge detector by Harris *et al.* [18] focus only on the detection aspect. Some common feature descriptors include the use of a histogram of oriented gradients (HoG) and Phog/Phow descriptors [19] which are commonly used in image classification and recognition.

Detecting regions covariant with a certain class of transformations can be useful in finding correspondences between views and [20] compares some common affine region detectors including MSER, IBR, EBR, Hessian-Affine and Harris-Affine. Wide-baseline correspondences have been found by [21] using MSER, [22] using edge descriptors and more recently by [14] using DAISY descriptors. However, these methods by themselves are not well suited for images of highly reflective objects with largely homogeneous regions such as cars, reflective buildings *etc.*.

Recent work by Lin *et al.* [23] finds correspondences and camera pose using motion coherence on scenes which were previously regarded as feature impoverished SfM scenes; containing largely edge cues but few corners. However, their method seems to be intended primarily for scenes consisting of long edges and few corners such as images of buildings and cupboards. Edge based features have also been used by [24] for shape recognition. However, their work seems to be focused on simple shapes such as bicycles and tennis rackets, where edges tend to give strong cues in otherwise poorly textured scenes. Our car images on the other hand, do not guarantee reliable edges that can be matched across images as edges since the edges are often fragmented and noisy.

Although we do not directly match edges, our proposed methodology matches points along image edges. Shape contexts [25] use points along object edges for matching shapes and object recognition but not for obtaining point correspondences, which is the focus of our work. Since we match all image edge points, we need a dense descriptor which can be quickly evaluated over all edge points. Although dense implementations of SIFT [2] and SURF [16] exist, we prefer to use the DAISY [14] descriptor which is faster and also better suited for wide-baseline images. Faster rotation invariant GPU implementations of the DAISY also exist [26], although we have not used it in our work.

Reflections are not necessarily harmful for the recovery of the epipolar geometry (EPG) between two images. Work done by Saminathan *et al.* [11] shows that the epipolar deviations of specularities on convex surfaces which are not highly undulating are usually quite small.

Prior work by [12] estimates the EPG using apparent contours for the limited case of affine and circular motions. Reference [13] use straight line edges for EPG and point matching. Our method does not place such restrictions on the camera motion or type, nor on the type of edges.

3   Problem Formulation and Proposed Solution

Our goal is to obtain point correspondences from two images of an object with highly reflective and largely homogeneous regions. The obtained correspondences should be good enough for SfM tasks such as recovering the epipolar geometry of the scene or estimating a homography transform for near planar objects in the scene. Our proposed method for obtaining reliable point correspondences is as follows.

3.1 Putative Point Matches

Given two images I and I' with point sets P and P', we wish to find the correct mapping $m(p) = p'$ for points $p = (u, v) \in P = \{p_1, p_2, ..., p_{n_1}\}$ and $p' = (u', v') \in P' = \{p_1', p_2', ..., p_{n_2}'\}$. Suppose we have a feature descriptor $\phi(p)$ evaluated on point p and a suitable distance measure $d(.)$ to compare two descriptors. An optimal assignment for $p \in P$ would be

$$m(p) = \underset{p' \in P'}{\operatorname{argmin}} d\left(\phi(p), \phi(p')\right) \tag{1}$$

Selecting candidate points. It is common practice [10] to use salient feature points (key-points) in the image as candidate points p, p' to perform matching. Commonly used methods to obtain key-points are as follows.

Harris corner points [18] are obtained in a gray-scale image I by considering the sum of squared differences (SSD) of a 2D patch at location (u, v) and shifting it by (x, y). Let I_x and I_y be the partial derivatives of I such that

$$I(u + x, v + y) \approx I(u, v) + I_x(u, v)x + I_y(u, v)y \tag{2}$$

The weighted SSD between these two patches is given by

$$S(x, y) = \sum_u \sum_v w(u, v) \left(I(u + x, v + y) - I(u, v)\right)^2 \tag{3}$$

A corner (or an interest point) is characterized by a large variation of S in all directions of the vector (x, y). The Harris matrix is defined as

$$A = \sum_u \sum_v w(u, v) \begin{bmatrix} I_x^2 & I_x I_y \\ I_x I_y & I_y^2 \end{bmatrix} = \begin{bmatrix} \langle I_x^2 \rangle & \langle I_x I_y \rangle \\ \langle I_x I_y \rangle & \langle I_y^2 \rangle \end{bmatrix} \tag{4}$$

where angle brackets denote averaging (i.e. summation over (u, v)). The Harris matrix A should have two "large" eigenvalues to be an interest point. Since computing eigenvalues is computationally expensive, interest points are obtained using

$$M_c = \lambda_1 \lambda_2 - \kappa (\lambda_1 + \lambda_2)^2 = \det(A) - \kappa \operatorname{trace}^2(A) \tag{5}$$

where κ is a tunable sensitivity parameter.

The SIFT [2] keypoint detector efficiently searches over different scales and image locations using a difference-of-Gaussian function. At each candidate location, key-points are detected based on measures of their stability. Thereby, after the detection step, the scale is known for each key-point. The SIFT descriptor $\phi(p)$ is obtained at key-point p using local image gradients measured at the scale obtained from the key-point detection step.

We show experimentally in Sect. 4 that key-points from the above methods do not in general result in reliable point correspondences across photographs dominated by large reflective and homogeneous regions. Figure 1 shows an example where SIFT [2] key-points and SIFT nearest neighbor matching [2] result in point matches which are concentrated towards a corner of the image which has relatively non-reflective regions. Such point matches are unsuitable to recover the epipolar geometry of the scene as it is not spatially well distributed to describe the 3D scene. Often strong key-points in reflective homogeneous surfaces are caused by reflections which are may not be present in the other view and hence cannot be matched. On the other hand, the homogeneous surface itself does not have strong features that can be detected as key-points, apart from points along edges of the surface. Hence it makes sense to simply focus on points along image edges.

Image edges have been known to be helpful when working with feature impoverished imagery. For example, [27] have used edge features to improve the performance of visual tracking in the presence of motion blur, in a simultaneous localization and mapping (SLAM) application using video sequence of mostly non-reflective scenes. Also, [22] have used an edge based descriptor to obtain wide baseline correspondences to perform structure from motion (SfM) on imagery of scenes mostly dominated by straight line edges. In a similar spirit, we found that the quality of the obtained point correspondences and the structural information obtained subsequently can be greatly improved by restricting the candidate points to points lying along image edges. Therefore, we select image point sets P and P' such that the points lie on edges in the image which are defined as follows.

Edges in the image. We define image edges as sharp changes in contrast occurring in an image which could be caused due to a genuine artifacts on the surface of an object or due to reflections caused from surrounding objects (Fig. 2).

Fig. 2. Various edges in the car door image include edges caused by reflections (red), edges caused by the surface of the car body (blue) and other edges (green). Best viewed in color (Color figure online).

Let the set $e_i = \{p_j, p_k, \ldots\}$ be an image edge segment in image I containing a set of edge points. We obtain the set of all edge points $E = \bigcup_i e_i$ in image I and similarly E' in I'. Our goal then, is to find point matches as per Eq. 1 considering only points which lie on image edges such that $p \in E$ and $p' \in E'$. We used the popular Canny [28] edge detector which has been shown to perform well experimentally [29] with parameters adopted to the data. We used the MATLAB Canny implementation which uses a standard deviation $\sigma = \sqrt{2}$ and computes the two hysteresis thresholds relative to the highest gradient magnitude in the image.

Matching edge points require feature descriptors to be evaluated at each edge point rather than on sparse key-points. It is convenient to use a dense feature descriptor to this end. Owing to its speed and use with wide baseline stereo images, we chose the DAISY [14] feature descriptor for our work.

DAISY Descriptor. Inspired by SIFT and GLOH, the DAISY [14] descriptor uses histograms of gradients. However, rather than a Difference of Gaussian (DOG), DAISY uses a Gaussian weighting and a circularly symmetric kernel, making it much faster to compute densely. Gradients are calculated at locations radially distributed around each pixel with larger regions and increasing levels of smoothing as the radial distance increases as shown in Fig. 3. For a given image I, an H number of *orientation maps* $G_o(u, v)$ for $1 \leq o \leq H$ are computed where G_0 is the image gradient norm at location (u, v) in direction o such that $G_o = max\left(\frac{\partial I}{\partial o}, 0\right)$. Each orientation map is convoluted several times with Gaussian kernels G_Σ of different standard deviations Σ to obtain convoluted orientation maps for different sized regions $G_o^\Sigma = G_\Sigma * G_o$. The size of the region is controlled by Σ. As convolutions with a large Gaussian kernel can be obtained by consecutive convolutions with smaller Gaussian kernels, orientation maps at different scales can be obtained very efficiently as $G_o^{\Sigma_2} = G_{\Sigma_2} * G_0 = G_\Sigma * G_{\Sigma_1} * G_o = G_\Sigma * G_o^{\Sigma_1}$ where $\Sigma = \sqrt{\Sigma_2^2 = \Sigma_1^2}$.

Fig. 3. DAISY [14,30] descriptor orientation map regions (circles) about pixel 'X' for $H = 8$

Sift on Edge Points. As SIFT [2] is limited to key-points, we considered Dense SIFT (DSIFT [15]) on edge points. However, the scale which is computed automatically during key-point detection in SIFT [2], needs to given explicitly with DSIFT. On the contrary, the DAISY descriptor, which is also a dense descriptor, incorporates a range of scales by definition. Additionally, as per the computation complexity evaluation in [14,30], DAISY is also a lot faster than SIFT. Hence we used DAISY.

Edge Feature Ambiguities. Indeed edge points on their own are not as discriminative as corners and blobs. However, images of highly reflective objects with large homogeneous regions lack discriminative corners/blobs with sufficient spatial distribution to recover the EPG. For such images, DAISY descriptors evaluated over edge points give better results (Sect. 4). The spatial constraint Sect. 3.2 also reduces the ambiguity of edge point matches. Sophisticated techniques such as graphical models [31] could also be utilized to employ smoothness constraints enforcing points on the same edge in I to match to points on a single edge in I'. In practice however, detected edges are often noisy and tend to fragment in an unpredictable manner. Hence, we found the simple greedy matching in Eq. 1 to be more effective. Next we describe estimating the EPG using putative point correspondences.

3.2 Recovery of the Epipolar Geometry (EPG)

Given two images that describe a 3D scene, its epipolar geometry (EPG) gives information about the camera setup in a projective sense. The EPG can be used to infer knowledge about the 3D scene via triangulation or stereo matching. In the case of an uncalibrated and unknown camera setup, image rectification may be performed prior to stereo matching.

A given point in one image will lie on its epipolar line in the second image, which is actually the projection of the back projected ray from the first image on to the second image. The epipolar geometry is described algebraically by the *Fundamental Matrix* F [10], which is based on this relationship. To be more specific, suppose two corresponding points $p, p' \in \mathbb{R}^2$ on I and I' have homogeneous coordinates $x, x' \in \mathbb{P}^2$ and F is the 3×3 fundamental matrix of rank 2, then $x'^T F x = 0$ for all correct point correspondences $x \leftrightarrow x'$.

RANSAC. Given a set of noisy point correspondences, the EPG may be robustly found using RANdom SAmple Consensus (RANSAC) [32] based methods. The essence of these methods is to find a fundamental matrix F such that $x'^T F x = 0$ for a random subset of the given points such that it agrees with the largest number of the remaining points. This is repeated for a given number of iterations and the best solution is selected. The RANSAC approach is robust in the presence of noisy outliers with considerable errors.

PROSAC has been shown to perform better than RANSAC by assuming that putative matches with higher quality (i.e. with a lower matching distance $d(.)$) are more likely to be inliers [33]. In our case however, inter object reflections on the reflective surfaces (e.g. reflections of trees on vehicle panels and glass) may generate high quality matches which are outliers to the EPG of the main scene. Therefore we do not consider the matching distance in the RANSAC step.

We use the normalized 8pt algorithm for model fitting in each RANSAC iteration and a distance threshold of 0.01 to filter outliers [10]. We use M-estimator

SAmple Consensus (MSAC) [34] as it is known to converge faster than standard RANSAC. However, the number of samples required to ensure with a given probability, that at least one sample has no outliers for a given sample size, increase exponentially as shown by [10]. This makes images with highly reflective and homogeneous regions which give very noisy point correspondences, very challenging to work with. Selecting points along edges in the image gives more reliable matches for images with largely homogeneous regions and reflections.

Spatial Constraint. To obtain an EPG which is representative of the actual 3D scene, it is important to have matching inlier points which are spatially well distributed across the images. However, naive feature matching of reflective images tend to concentrate correct point matches over areas which are relatively less reflective as shown in Fig. 1. To avoid this problem, we enforce a *spatial constraint* inspired by [35–37]. The complete matching algorithm with the spatial constraint for obtaining putative point correspondences is given in Algorithm 1.

Input : Images I and I'
Output: Putative point correspondences
1) Find the set of edge points E in image I and E' in image I'
2) **Match edge points (asymmetric):** For each edge point $p_i \in E$ find the matched edge point $p_j' \in E'$ as $p' = m(p) = \mathrm{argmin}_{p' \in E'} \, d\left(\phi(p), \phi(p')\right)$
3) **Enforce spatial constraint:** Consider a rectangular grid of $b_W \times b_H$ spatial buckets over I. Pick the best k matches with the lowest $d(.)$ from each bucket

Algorithm 1. Matching algorithm with the spatial constraint

We are not limited to small camera motions or scale changes as we match edge points asymmetrically from I to I' and only consider buckets over I. As an extreme case, consider the bucket at the top-left corner of the bucketed image I. We may well pick a matching point from the bottom-right corner in the other image I' (which is not bucketed) as long as the matching distance is within the lowest k distance values for the bucket.

We present next, an experimental evaluation of our method along with baseline comparisons.

4   Experiments

We compare our method quantitatively and qualitatively against baseline methods. Experiments were done on the standard DAISY dataset [14,30] and our own car dataset of over 70 images of highly reflective car images. We experimentally found that for the spatial constraint parameters in Algorithm 1, $b_W = I_W/16$, $b_H = I_H/16$ and $k = 2$ gave good results for image pairs of size $I_W \times I_H$ pixels each.

Table 1. Description of methods

Method	Description
deg	DAISY descriptors on edge points - with spatial constraint (ours)
de	DAISY descriptors on edge points - no spatial constraint (ours)
dhg	DAISY descriptors on Harris corner points - with spatial constraint
dh	DAISY descriptors on Harris corner points - no spatial constraint
dsg	DAISY descriptors on SIFT key-points - with spatial constraint
ds	DAISY descriptors on SIFT key-points - no spatial constraint
sg	SIFT descriptors on SIFT key-points and SIFT matching - with spatial constraint
s	SIFT descriptors on SIFT key-points and SIFT matching - no spatial constraint

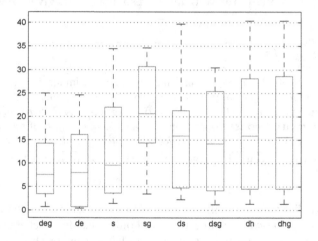

Fig. 4. The box plots show the *comparison measure* of Zhang [36] for our method *de* and baseline methods *s, ds and dh* with *g* at the end indicating tests where the spatial constraint was enforced. We used the DAISY dataset [14,30]. A lower *comparison measure* indicates better performance. Our method *deg* has the lowest median and inter quartile range (IQR) for the *comparison measure*.

4.1 Quantitative Results and Comparison with Baseline Methods

We quantitatively evaluate the quality of the EPG recovered from our method and baseline methods as follows. We use the method adopted by [36] to measure the similarity between the recovered EPG and the ground truth EPG. The method gives a *measure of comparison* between the recovered fundamental matrix F and the ground truth fundamental matrix F_{gt}. As described in [36], the measure is obtained by considering the perpendicular distances between points and corresponding epipolar lines obtained using both fundamental matrices in a symmetric manner. Simillar fundamental matrices give a lower value for the

measure. In our experiments, the better method should recover an EPG closer
to the ground truth EPG and therefore give a lower comparison measure. We
use the wide baseline *fountain* and *herzjesu* images with ground truth camera
calibration information from the DAISY dataset [14,30]. We use the provided
ground truth projection matrices to compute a ground truth fundamental matrix
F_{gt} for a given image pair.

Figure 4 shows box plots of the *comparison measure* using the method by
[36] explained above. A summary of the methods evaluated in this paper along
with acronyms used are given in Table 1. All images in the datasets have the
same dimentions. Although the *comparison measure* evaluated using only the
inlier point correspondences is in the order of sub pixels, we evaluate the *comparison measure* [36] over all point correspondences when comparing methods in Fig. 4. This is because even an incorrectly estimated epipolar geometry
will still give a very low *comparison measure* for degenerate cases since the
inlier point matches satisfy the incorrectly estimated fundamental matrix (*e.g.*
coplanar inlier points). We see in Fig. 4 that our method with the spatial constraint *deg* gives the lowest median *comparison measure* (indicated by the horizontal line in the middle of each box) and also has the lowest dispersion or
spread as seen by the interquartile range indicated by the ends of each box. The
baselines *ds* and *dh* improve marginally with the spatial constraint (*dsg* and
dhg). However, the spatial constraint causes a significant performance drop with
SIFT key-points and SIFT matching (*s* vs *sg*). The SIFT distance ratio (nearest neighbor test) already filters out matches with features that are not very
discriminative but could have supported the correct EPG. Enforcing the spatial
constraint in (*sg*) further reduces the number of matches which may support
the correct EPG. Therefore, enforcing the spatial constraint in *sg* yields poor
quality matches which significantly affects the EPG computation. The results in
Fig. 4 show that our methods *deg* and *de* continue to perform better than the
baselines, even with the spatial constraint. Qualitative results shown in Sect. 4.2
indicate the same.

Matching Distance. Putative matches were found using the SIFT distance
ratio (nearest neighbor test) [2] for baseline methods *s* and *sg*. Results on relatively non-reflective images were comparable with our method *deg* (Fig. 4).
However, such matches are not very reliable with very reflective images (methods *s* and *sg* in Figs. 5 and 8). For matching DAISY descriptors in *de*, *deg*, *ds*,
dsg, *dh* and *dhg*, we used the L2 norm for $d(.)$ in Eq. 1 as per [14,30].

Descriptor Scale. The scale was computed automatically from SIFT key-point
detection in baseline *s*. For the other methods, we used the DAISY descriptor [14,30] which has image differences obtained at radially distributed positions
about the initial point/pixel, computed by applying increasingly larger Gaussian
kernels when moving away from the point. We used $R = 15, Q = 3, T = 8, H = 8$
as per [14,30] and Sect. 3.1.

4.2   Qualitative Results

To get a sense of the recovered EPG we present some qualitative results. We evaluated our method and baselines qualitatively on our car dataset containing over 70 image pairs. Results on the entire dataset are provided as supplementary material. A typical result is shown in Figs. 5 and 6 with methods denoted as per Table 1. The recovered EPG from the baseline methods in Fig. 5 are clearly wrong as the epipolar lines seem to indicate that the photographer has walked towards the car where as in reality the photographer has moved side ways. As the recovered epipoles are incorrectly located inside the images, uncalibrated rectification [10] cannot be performed. On the other hand, our method *deg* in Fig. 6(a) recovers a significantly better EPG in Fig. 6(b) for the same image pair. The near horizontal direction of the epipolar lines correctly reflect the movement of the camera. In fact, it is possible to perform uncalibrated stereo rectification (Fig. 6(e)). Not enforcing the spatial constraint (*de*) gives poorer results (Fig. 6(c) and (d)) in this instance, which are not suitable for stereo rectification.

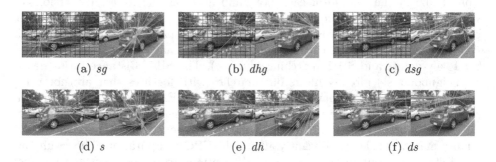

(a) *sg*   (b) *dhg*   (c) *dsg*

(d) *s*   (e) *dh*   (f) *ds*

Fig. 5. EPG and inliers for the baseline methods discussed in Sect. 4.2. Notation *c.f.* Table 1 and Sect. 4.1. Color code: cyan lines - epipolar lines, yellow dots - matched points, magenta lines - point correspondences. Best viewed in color. Images may be cropped for clarity (Color figure online).

Note that the uncalibrated rectification process introduces a projective distortion in the transformed images which is as expected [10]. Hence the apparent disparities between the blue points in the rectified left image and the superimposed red points from the rectified right image (Fig. 6(e)) may not correctly indicate inverse depth as with calibrated rectification. EPG and uncalibrated rectification results on the image pair in Fig. 1 are shown in Fig. 7.

We further verify our method on a highly reflective image pair of a building in Fig. 8. The camera motion between the two photographs is clearly horizontal. Hence the recovered epipolar (EP) lines (shown in cyan) should be horizontal. This is reflected correctly in our methods *deg* and *de*. However, the EPG recovered using the baseline methods do not indicate this and is clearly wrong. Among the baselines, *dhg* performs better but EP lines (particularly at the top) are not horizontal.

(a) Putative matches with *deg* (with spatial constraint)

(b) Plausible EPG and inliers from (a)

(c) Putative matches with *de* (no spatial constraint)

(d) Incorrect EPG and inliers from (c)

(e) Uncalibrated stereo rectification using EPG in (b). Left image has points from right image superimpose in red.

Fig. 6. Results from our method discussed in Sect. 4.2 showing the effect of the spatial constraint. Notation *c.f.* Table 1 and Sect. 4.1. Color code: cyan lines - epipolar lines, yellow dots - matched points, magenta lines - point correspondences. Best viewed in color. Images may be cropped for clarity (Color figure online).

(a) EPG and inliers

(b) Uncalibrated stereo rectification

Fig. 7. Results using our method *deg* on photographs of a very reflective car door. Notation *c.f.* Table 1 and Sect. 4.1. Color code: cyan lines - epipolar lines, yellow dots - matched points, magenta lines - point correspondences. Best viewed in color. Images may be cropped for clarity (Color figure online).

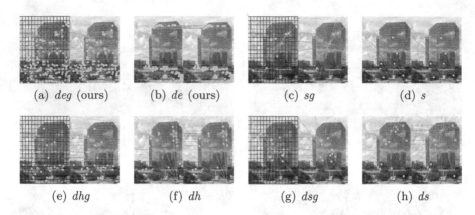

(a) *deg* (ours)	(b) *de* (ours)	(c) *sg*	(d) *s*
(e) *dhg*	(f) *dh*	(g) *dsg*	(h) *ds*

Fig. 8. Recovered EPG and inlier point correspondences from a photograph of a highly reflective building. Since the camera motion between the two images is horizontal, the recovered epipolar lines should be horizontal. This is reflected correctly in our methods *deg* and *de*, unlike with the baseline methods. The upper EP lines with *dhg* are not horizontal. The spatial grid is overlaid on the left image where the spatial constraint was enforced. Notation *c.f.* Table 1 and Sect. 4.1. Color code: cyan lines - epipolar lines, yellow dots - matched points, magenta lines - point correspondences. Best viewed in color. Images may be cropped for clarity (Color figure online).

5   Discussion

We present a method for finding reliable point correspondences in images of scenes dominated by highly reflective and largely homogeneous surfaces. Conventional methods for finding point correspondences are mainly designed for textured and non-reflective surfaces. As such they generate a lot of spurious matches from images with highly reflective and homogeneous surfaces and give poor results when recovering the epipolar geometry of the scene. We have proposed a novel method of combining established computer vision techniques by matching points along image edges and enforcing a spatial constraint to obtain reliable point correspondences from such images, resulting in sufficiently low noise levels. In addition to providing theoretical intuition, we have experimentally showed that our approach gives good results on images with highly reflective and homogeneous surfaces where baseline methods fail. An interesting future direction would be to explore QDEGSAC [38] to avoid potentially degenerate configurations with unknown camera calibration. An interesting application would be to detect the reflections in the images based on depth cues after performing a 3D reconstruction of the scene based on the obtained point correspondences and recovered EPG.

References

1. Snavely, N., Seitz, S.M., Szeliski, R.: Modeling the world from internet photo collections. Int. J. Comput. Vis. (IJCV) **80**, 189–210 (2008)

2. Lowe, D.: Distinctive image features from scale-invariant keypoints. Int. J. Comput. Vis. (IJCV) 60, 91–110 (2004)
3. Jayawardena, S., Yang, D., Hutter, M.: 3D model assisted image segmentation. In: Proceedings of the International Conference on Digital Image Computing Techniques and Applications (DICTA). IEEE (2011)
4. Jayawardena, S., Hutter, M., Brewer, N.: A novel illumination-invariant loss for monocular 3D pose estimation. In: Proceedings of the International Conference on Digital Image Computing Techniques and Applications (DICTA). IEEE (2011)
5. Jayawardena, S.: Image based automatic vehicle damage detection. Ph.D. thesis, The Australian National University (2013)
6. Pylvanainen, T., Berclaz, J., Korah, T., Hedau, V., Aanjaneya, M., Grzeszczuk, R.: 3D city modeling from street-level data for augmented reality applications. In: 3DIMPVT. IEEE (2012)
7. Greenspan, H., Gordon, S., Zimmerman, G., Lotenberg, S., Jeronimo, J., Antani, S., Long, R.: Automatic detection of anatomical landmarks in uterine cervix images. IEEE Trans. Med. Imaging 28, 454–468 (2009)
8. Zimmerman-Moreno, G., Greenspan, H.: Automatic detection of specular reflections in uterine cervix images. In: Medical Imaging, International Society for Optics and Photonics, p. 61446E (2006)
9. Wu, T.T., Qu, J.Y.: Optical imaging for medical diagnosis based on active stereo vision and motion tracking. Opt. Express 15, 10421–10426 (2007)
10. Hartley, R.I., Zisserman, A.: Multiple View Geometry in Computer Vision. Cambridge University Press, Cambridge (2000)
11. Swaminathan, R., Kang, S.B., Szeliski, R., Criminisi, A., Nayar, S.K.: On the motion and appearance of specularities in image sequences. In: Heyden, A., Sparr, G., Nielsen, M., Johansen, P. (eds.) ECCV 2002, Part I. LNCS, vol. 2350, pp. 508–523. Springer, Heidelberg (2002)
12. Mendonça, P.R., Cipolla, R.: Estimation of epipolar geometry from apparent contours: Affine and circular motion cases. In: Proceedings of Computer Vision and Pattern Recognition (CVPR) (1999)
13. Schmid, C., Zisserman, A.: The geometry and matching of lines and curves over multiple views. Int. J. Comput. Vis. (IJCV) 40, 199–233 (2000)
14. Tola, E., Lepetit, V., Fua, P.: DAISY: an efficient dense descriptor applied to wide baseline stereo. IEEE Trans. Pattern Anal. Mach. Intell. (PAMI) 32, 815–830 (2010)
15. Vedaldi, A., Fulkerson, B.: VLFeat: An open and portable library of computer vision algorithms (2008). http://www.vlfeat.org/
16. Bay, H., Tuytelaars, T., Van Gool, L.: SURF: speeded up robust features. In: Leonardis, A., Bischof, H., Pinz, A. (eds.) ECCV 2006, Part I. LNCS, vol. 3951, pp. 404–417. Springer, Heidelberg (2006)
17. Leutenegger, S., Chli, M., Siegwart, R.: BRISK: Binary robust invariant scalable keypoints. In: Proceedings of the International Conference on Computer Vision (ICCV) (2011)
18. Harris, C., Stephens, M.: A combined corner and edge detector. In: Alvey Vision Conference (AVC) (1988)
19. Bosch, A., Zisserman, A., Muoz, X.: Image classification using random forests and ferns. In: Proceedings of the International Conference on Computer Vision (ICCV) (2007)
20. Mikolajczyk, K., Tuytelaars, T., Schmid, C., Zisserman, A., Matas, J., Schaffalitzky, F., Kadir, T., Gool, L.V.: A comparison of affine region detectors. Int. J. Comput. Vis. (IJCV) 65, 43–72 (2005)

21. Matas, J., Chum, O., Urban, M., Pajdla, T.: Robust wide-baseline stereo from maximally stable extremal regions. Image Vis. Comput. (IVC) **22**, 761–767 (2004)
22. Meltzer, J., Soatto, S.: Edge descriptors for robust wide-baseline correspondence. In: Proceedings of Computer Vision and Pattern Recognition (CVPR) (2008)
23. Lin, W., Cheong, I., Tan, P., Dong, G., Liu, S.: Simultaneous camera pose and correspondence estimation with motion coherence. Int. J. Comput. Vis. (IJCV) **96**, 145–161 (2012)
24. Mikolajczyk, K., Zisserman, A., Schmid, C., et al.: Shape recognition with edge-based features. In: Proceedings of the British Machine Vision Conference (BMVC) (2003)
25. Belongie, S., Malik, J., Puzicha, J.: Shape matching and object recognition using shape contexts. IEEE Trans. Pattern Anal. Mach. Intell. (PAMI) **24**, 509–522 (2002)
26. Fischer, J., Ruppel, A., Weißhardt, F., Verl, A.: A rotation invariant feature descriptor O-DAISY and its FPGA implementation. In: Proceedings of the International Conference on Intelligent Robots and Systems (IROS) (2011)
27. Klein, G., Murray, D.: Improving the agility of keyframe-based SLAM. In: Forsyth, D., Torr, P., Zisserman, A. (eds.) ECCV 2008, Part II. LNCS, vol. 5303, pp. 802–815. Springer, Heidelberg (2008)
28. Canny, J.: A computational approach to edge detection. IEEE Trans. Pattern Anal. Mach. Intell. (PAMI) **8**, 679–698 (1986)
29. Heath, M.D., Sarkar, S., Sanocki, T., Bowyer, K.W.: A robust visual method for assessing the relative performance of edge-detection algorithms. IEEE Trans. Pattern Anal. Mach. Intell. (PAMI) **19**, 1338–1359 (1997)
30. Tola, E., Lepetit, V., Fua, P.: A fast local descriptor for dense matching. In: Proceedings of Computer Vision and Pattern Recognition (CVPR) (2008)
31. Boykov, Y., Veksler, O., Zabih, R.: Fast approximate energy minimization via graph cuts. IEEE Trans. Pattern Anal. Mach. Intell. (PAMI) **23**, 1222–1239 (2001)
32. Fischler, M.A., Bolles, R.C.: Random sample consensus: a paradigm for model fitting with applications to image analysis and automated cartography. Assoc. Comput. Mach. (ACM) **24**, 381–395 (1981)
33. Aghazadeh, O., Sullivan, J., Carlsson, S.: Novelty detection from an ego-centric perspective. In: Proceedings of Computer Vision and Pattern Recognition (CVPR) (2011)
34. Rogers, M., Graham, J.: Robust active shape model search. Proc. of the European Conference on Computer Vision (ECCV) (2006)
35. Zhang, Z., Deriche, R., Faugeras, O., Luong, Q.T.: A robust technique for matching two uncalibrated images through the recovery of the unknown epipolar geometry. Artif. Intell. (AI) **78**, 87–119 (1995)
36. Zhang, Z.: Determining the epipolar geometry and its uncertainty: A review. Int. J. Comput. Vis. (IJCV) **27**, 161–198 (1998)
37. Kitt, B., Geiger, A., Lategahn, H.: Visual odometry based on stereo image sequences with ransac-based outlier rejection scheme. In: Intelligent Vehicles Symposium (IV) (2010)
38. Frahm, J.M., Pollefeys, M.: RANSAC for (Quasi-)Degenerate data (QDEGSAC). In: Proceedings of Computer Vision and Pattern Recognition (CVPR) (2006)

ORB in 5 ms: An Efficient SIMD Friendly Implementation

Prashanth Viswanath$^{(\boxtimes)}$, Pramod Swami, Kumar Desappan,
Anshu Jain, and Anoop Pathayapurakkal

Texas Instruments India Private Ltd, Bangalore, Karnataka, India
p-viswanath@ti.com

Abstract. One of the key challenges today in computer vision applications is to be able to reliably detect features in real-time. The most prominent feature extraction methods are Speeded up Robust Features (SURF), Scale Invariant Feature Transform(SIFT) and Oriented FAST and Rotated BRIEF(ORB), which have proved to yield reliable features for applications such as object recognition and tracking. In this paper, we propose an efficient single instruction multiple data(SIMD) friendly implementation of ORB. This solution shows that ORB feature extraction can be effectively implemented in about 5.5 ms on a Vector SIMD engine such as Embedded Vision Engine(EVE) of Texas Instruments(TI). We also show that our implementation is reliable with the help of repeatability test.

1 Introduction

Keypoint detectors and descriptors play an important role in computer vision applications such as object recognition, image stitching, structure from motion etc. The most commonly used methods are Scale Invariant Feature Transform(SIFT) [1], Speeded up Robust Features(SURF) [2] and Oriented FAST and Rotated BRIEF(ORB) [3]. These methods have been proven to be reliable in detecting features in real world images. However, the biggest challenge is to meet the real-time performance requirements. In most applications, keypoint extraction is followed by computationally intensive processing such as tracking the features or recovering 3D points and so on. Ethan Rublee et al. [3] shows that it takes about 15.3 ms to compute ORB descriptors for roughly 1000 keypoints in a 640 × 480 image on an Intel i7 2.8 GHz processor. Kwang-yeob Lee and Kyung-jin Byun [4] proposes a hardware accelerator for ORB whose run time is 18 ms for an input image of 640 × 480. Our solution takes about 5.5 ms to process 3 levels of the input image whose base resolution is 400 × 400 and compute 500 descriptors on Embedded Vision Engine(EVE) of Texas Instruments(TI) running at 650MHz [5].

In this paper, we propose a computationally efficient implementation of ORB which is suited for single instruction multiple data(SIMD) architecture. Our contributions are as follows:

© Springer International Publishing Switzerland 2015
C.V. Jawahar and S. Shan (Eds.): ACCV 2014 Workshops, Part I, LNCS 9008, pp. 675–686, 2015.
DOI: 10.1007/978-3-319-16628-5_48

- An alternate SIMD friendly implementation of FAST9 keypoint detection
- An alternate approach to computing FAST9 score instead of the iterative approach keeping the definition of FAST9 score the same, which is the highest threshold at which a keypoint still remains a keypoint [6]
- Sparse point non-maximal suppression based on the FAST9 score
- SIMD friendly implementation of the Harris score and rBRIEF descriptor computation

We also share the performance details of the implementation on EVE which is part of the TDA2x device of TI. EVE is a fully programmable accelerator created specifically to enable the processing, latency and reliability needs found in computer vision applications. The EVE includes one 32-bit Application-Specific RISC Processor(ARP32) and one 512-bit Vector Coprocessor(VCOP) with built-in mechanisms and unique vision-specialized instructions for concurrent, low-overhead processing. The VCOP is a SIMD engine with built-in loop control and address generation. It is a dual 8-way SIMD engine. It has certain special properties such as transpose store, de-interleave load, interleaving store and so on. The VCOP also has specialized pipelines for accelerating lookup tables and histograms [5]. To validate the ORB implementation, we perform the repeatability test for images with varying view points and blurring.

The rest of the paper is organized as follows: Sect. 2 provides the details of our implementation, Sect. 3 shows the results and performance of the implementation on EVE and Sect. 4 offers conclusions as to effectiveness of our implementation.

2   Proposed Solution

2.1   Overview

The ORB algorithm flow is as shown in Fig. 1. The input to our algorithm is an 8-bit image of size WxH. The output is a list of 256-bit ORB descriptor, the corresponding XY co-ordinates and the level of image at which the feature was detected. Since FAST is not a multi-scale algorithm, we obtain different levels of the image using image pyramid and employ FAST9 feature detection on each level. The FAST9 detector outputs the list of XY co-ordinates in 32-bit packed format indicating the location of the keypoints. For these keypoints, we compute the FAST9 score as explained in Sect. 2.3. We compute FAST9 score only for the keypoints and not every pixel of the image. This results in significant reduction of memory bandwidth and computational requirements. After the FAST9 score computation, we apply non-maximal suppression as detailed in Sect. 2.4. Since FAST9 keypoints are clustered together, we apply 4-way non-maximal suppression which suppresses non-maximas considering neighbors in four directions: top, bottom, left and right. The traditional approach of non-maximal suppression involves applying a sliding 3×3 window across the entire image and determine the non-maximas. However, our implementation of non-maximal suppression operates on the sparse keypoints directly. This further results in significant reduction of memory bandwidth and computational requirements. After suppression, we sort the non-suppressed keypoints based on the FAST9 score and output the best $2N$ keypoints. For these best keypoints, we compute the Harris score. At this stage, we have multiple lists of co-ordinates and their Harris score, each corresponding to a particular image level. We then sort the lists across multiple levels based on the Harris score and output the best N keypoints for which the rBRIEF descriptors are computed.

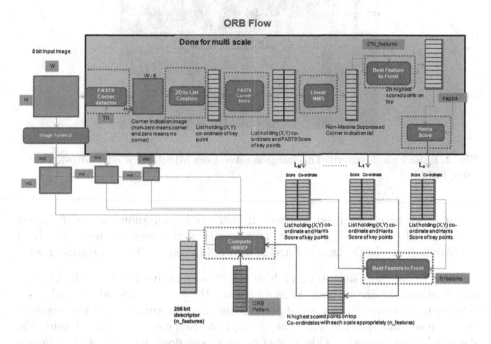

Fig. 1. ORB algorithm flow.

2.2 FAST9 Keypoint Detection

FAST is a popular feature detector in real time systems. FAST algorithm picks a 7×7 window around each pixel p as shown in Fig. 2. It takes intensity threshold between the center pixel p and those in a circular ring around p as an input parameter. The algorithm checks if there is a contiguous arc of K or more pixels in the circular ring which satisfy either the bright or dark condition. We use FAST9 (K = 9), which has been shown to have good performance [6,7].

Although the algorithm is simple, it poses following challenges for a Vector SIMD engine:

- Though the pixel access pattern around each pixel in a 7×7 window is fixed, these are non-sequential locations and hence not friendly towards a simple vector load instruction.

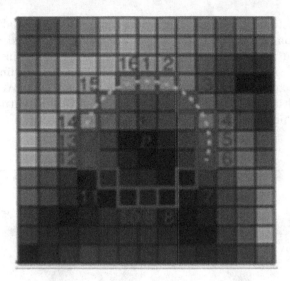

Fig. 2. FAST9 pixel pattern. This image is taken from http://www.edwardrosten.com/work/fast.html.

- For every pixel, we need to check if there is a contiguous arc of K pixels that satisfy the FAST property. This results in the need to check $16+(K\text{-}1)$ combinations. In the case of FAST9, we would need to check 24 combinations.

Traditional approach to check how many consecutive pixels in the circular ring that are similar involves running a loop $16+(K\text{-}1)$ times which updates a counter to indicate the number of pixels satisfying the condition. It also needs to reset the counter selectively for the appropriate elements while maintaining the status of other elements. This level of control logic does not work well in Vector SIMD engines. Hence, we propose a simple 'SHIFT' and 'AND' based technique to find K consecutive pixels which are similar.

We store the comparison of 16 offset pixels with the center pixel in a bit packed format such that we form a 16-bit mask for each center pixel. If bit number 5 is 0, it means that pixel 5 in the circular ring is not similar to center pixel and so on. This kind of mask is simple to generate on a Vector SIMD engine. Since the contiguous arc of K similar pixels can be in any location (i.e. last bit followed by first $K\text{-}1$ bits), we need to do a wrap around search. This can be easily accommodated by duplicating the mask and generate a 32-bit mask. For FAST9, it is sufficient if we duplicate only lower 8 bits of the mask and place it from bit 17–24 to obtain a 24-bit mask X. Now, we can perform the 'SHIFT' and 'AND' logic as follows:

Pseudo code to find 9 consecutive similar pixels

```
begin
// Will tell if there are 2 consecutive similar pixels
```

```
    X1 = X >> 1
    X2 = X1 & X
// Will tell if there are 4 consecutive similar pixels
    X3 = X2 >> 2
    X4 = X3 & X2
// Will tell if there are 8 consecutive similar pixels
    X5 = X4 >> 4
    X6 = X5 & X4
// Will tell if there are 9 consecutive similar pixels
    X7 = X6 >> 1
    X8 = X7 & X6
    If X8 != 0, mark pixel as a corner
end.
```

As is apparent, this technique has logarithmic convergence. This approach requires just 4 steps against the $16 + (9-1) = 24$ steps required in the traditional approach for FAST9 keypoint detection. For FAST12, we would also need just 4 steps with the change in the shift factor from 1 to 4 in the last step.

In order to address the first challenge, we use 17 vector load instructions of SIMD width, one vector load for the center pixels, and 16 vector loads for the offset pixels in the circular ring. Although this approach has certain short comings such as minimal data reuse, it is efficient since we are operating on SIMD width elements at a time. In EVE, the SIMD width is 8. Hence, we can work on 8 pixels at a time. Hence, there is an 8x performance benefit.

The ORB requires orientation of FAST keypoints. This is computed at a later stage while computing the descriptor rather than at FAST keypoint detection stage, since the detecting stage operates at every pixel in the image as opposed to the descriptor stage which is operating only on the best N key points.

As mentioned earlier, the input to the FAST9 keypoint detector is an input image or a level of the image. The output is the XY co-ordinate list in 32-bit packed format (16-bit X followed by 16-bit Y) indicating the location of the keypoint. It is important that the XY co-ordinate list is generated in the raster scan order of the image for applying the sparse point non-maximal suppression. This is explained in Sect. 2.4.

This method of FAST9 detection has been implemented on EVE and the core compute performance is 5.3 cycles/pixel. This implementation has the same cycle count for every pixel as opposed to other approaches such as machine learning approach whose cycle count is highly data dependent [6,7].

2.3 FAST9 Score Computation

FAST9 score is defined as the threshold for which, a FAST9 key point still remains a key point [6]. For a given FAST9 key point, its score is given by the highest threshold for which the key point still has 9 contiguous similar pixels around it. The traditional approach of computing the score would involve iteratively incrementing the threshold and check if a key point still remains a key point [6]. The challenges posed by this approach are:

- The keypoints given by FAST9 detector are sparse in nature and hence poses challenges in vectorizing the operations
- The iterative approach of computing the score and conditionally exiting is not suitable for Vector SIMD engines

FAST9 detector gives the XY location of the key points. We pick 7×7 window of pixels around each of the key point. In order to be able to vectorize our computations, we re-order the data by picking only the 16 pixels of the circular ring from the 7×7 window and place them consecutively. Again, in order to take into account the wrap around nature of FAST9, we pick the lower 8 pixels again and place them as well. Hence, for every key point we have 24 offset pixels placed consecutively.

Next, we compute maximum and minimum intensity values of the offset pixels taking them 9 at time, i.e. max(0–8), max(1–9)...max(15–23) and min(0–8), min(1–9)..min(15–23). Hence, for every center pixel key point, we have 16 maxima and minima values. Maxima are used if the center pixel satisfies the dark condition(center pixel is darker compared to the offset pixels) of the FAST9 algorithm and Minima are used if the center pixel satisfies the bright condition(center pixel is brighter compared to the offset pixels) of the FAST9 algorithm. Since we do not indicate whether the keypoint is bright or dark during the FAST9 keypoint detection stage, we compute both maxima and minima in this stage. Out of the 16 maxima and minima, only one of them is the score. In order to compute that, we find minimum value $Vmin$ from the maxima values and maximum value $Vmax$ from the minima values. We compare the $Vmin$ and $Vmax$ with the center pixel. There are only two possibilities: either $Vmin$ and $Vmax$ are greater than center pixel intensity or $Vmin$ and $Vmax$ are lesser than center pixel intensity. The other two possibilities where $Vmin$ is greater than center pixel intensity while $Vmax$ is lesser than center pixel intensity and vice versa are not possible by the definition of FAST9. Hence, the final score is computed as follows:

- If $Vmax$ and $Vmin$ are greater than center pixel, the score is the minimum of difference of $Vmax$ and $Vmin$ with the center pixel intensity minus one. This represents the dark condition.
- If $Vmax$ and $Vmin$ are lesser than center pixel, the score is the maximum of difference of the center pixel with $Vmax$ and $Vmin$ minus one. This represents the bright condition.

Pseudo code to compute FAST9 score

```
begin
        cb = center pixel;
        for(startpos=0; startpos<16; startpos++)
        {
            pMin = co[startpos]; // co[] = offset pixel array
            pMax = pMin;
```

```
        for(i-1; i<9; i++)
        [
            if(pMax < co[startpos+i])
                pMax = co[startpos+i];
            if(pMin > co[startpos+i])
                pMin = co[startpos+i];
        }
        Bscore[startpos] = pMax; // Bscore[] = array of maxima
        Dscore[startpos] = pMin; // Dscore[] = array of minima
    }
    score_b = Bscore[0];
    score_d = Dscore[0];
    for(i=1; i<16; i++)
    {
        if(score_b > Bscore[i])
            score_b = Bscore[i];
        if(score_d < Dscore[i])
            score_d = Dscore[i];
    }
    if((score_b > cb) && (score_d > cb))
    {
        if(score_b > score_d)
            score = score_d - cb - 1;
        else
            score = score_b - cb - 1;
    }
    else if ((score_b < cb) && (score_d < cb))
    {
        if(score_b > score_d)
            score = cb - score_b - 1;
        else
            score = cb - score_d - 1;
    }
end.
```

The above approach is not iterative and involves simple operation such as max and min which are supported by most of Vector SIMD engines. The core compute performance of this approach is 31.5 cycles/keypoint on the EVE engine of TI. For the data rearrangement, we use the table look up hardware and transpose store property which are supported by EVE [5]. The speed up of computing FAST9 score is due to:

- Vectorizing the operations to compute FAST9 score and non-iterative approach
- Operating on keypoints only and not every pixel of the image.

2.4   Sparse Point Non-maximal Suppression

FAST9 keypoints obtained for an image are generally clustered. Hence, non-maximal suppression is an important step to obtain the most reliable feature points. Traditional approach of non-maximal suppression involves applying a 2D 3×3 running window across the input and retain only the maxima in the window. This approach has the following disadvantages:

- FAST9 score has to be computed for every pixel of the input image in order apply the 2D suppression which results in significant wastage of compute cycles and memory bandwidth
- Assuming that FAST9 score is computed only for the keypoints, it has to be mapped back into the 2D image structure which is again a challenging task

Hence we propose an approach which operates on sparse keypoints directly without the 2D notion and can be vectorized easily. This approach is split into two stages: horizontal non-maximal suppression and vertical non-maximal suppression.

Horizontal non-maximal suppression: In this stage, we find the maxima along the X direction. For every keypoint, we check if it has a neighbor in either the left or right direction and then compare their FAST9 scores to suppress the non-maximas. As it was mentioned earlier, it is important that the XY co-ordinate list of keypoints are in raster order. This would mean that the keypoints would be listed in buckets of Y, i.e., same Y but different X (example XY list - 0×0303, 0×0403, 0×0803, 0×0404.. and so on). Once we have the data in this format, we can easily determine if a neighbor exits in the right or left by checking against $(X-1, Y)$ and $(X+1, Y)$. We can also compare the FAST9 scores accordingly and suppress the non-maximum keypoints along the horizontal direction. While storing the co-ordinates of the non-suppressed keypoints, we pack them with ID such that the 32-bit output is 10-bit X, followed by 10-bit Y and 12-bit ID. This ID is used in the next stage of vertical non-maximal suppression. As you can see, this implementation is simple and vector friendly.

Pseudo code to for Horizontal non-maximal suppression

```
begin
// i = 1 to num_corners-1 since we need 1 pixel border in each side
    for(i=1; i<(num_corners-1); i++)
    {
        left_xy = corners[i-1];
        center_xy = corners[i];
        right_xy = corners[i+1];
        left_scr = scores[i-1];
        center_scr = scores[i];
        right_scr = scores[i+1];
        left_xy += 0x10000;
        right_xy -= 0x10000;
```

```
//Generate right and left neighbor mask
        Vnf1 = (left_xy == center_xy);
        Vnf2 = (right_xy == center_xy);
//Generate score mask
        Vsf1 = (center_scr <= left_scr);
        Vsf2 = (center_scr <= right_scr);
        Vf1 = Vnf1 & Vsf1;
        Vf2 = Vnf2 & Vsf2;
//Final mask indicating the neighbor and the maximum score
        Vf1 |= Vf2;
        x = center_xy & 0xFFFF0000;
        y = center_xy & 0x0000FFFF;
        // pack X, Y and ID: 10 bit X, 10 bit Y and 12 bit ID
        nms_x_corners[i] = (x << 6) | (y << 12) | (i);
        if(Vf1)
            nms_x_score[i] = 0;
        else
            nms_x_score[i] = scores[i];
    }
end.
```

Vertical non-maximal suppression: In order to suppress along the Y direction (top and bottom), it would be easy if the co-ordinate list is arranged in raster order of Y, i.e. buckets of X, same X and different Y (example XY list - 0×0303, 0×0304, 0×0306, 0×0403.. and so on). In order to obtain the data in this format, we sort the XY-ID output from the horizontal non-maximal suppression stage in ascending order. Since the XY-ID list is now in different order, we need to obtain the corresponding score values. This is done using the ID to look-up the score values of the corresponding XY. Once the score values are re-ordered, we can follow the same approach as in horizontal suppression to suppress the non-maximas in top and bottom direction.

In order to obtain the best $2N$ keypoints, we then sort the non-suppressed keypoints based on the FAST9 score and output it. The horizontal non-maximal suppression and vertical non-maximal suppression takes 1.8 cycles/keypoint and 1.3 cycles/keypoint respectively on EVE. 32-bit sort takes 4.78 cycles/point for a 2048-point sort on EVE. The speed up of is due to:

– Vectorizing the suppression operations
– Operating on sparse keypoints only and not every pixel of the image.

2.5 Harris Score Computation and rBRIEF Descriptor Computation

Since literature suggests that FAST does not produce a measure of cornerness and has large responses along edges, Harris score is used to order the FAST keypoints [3]. Harris score is computed by picking a 7×7 window around each keypoint and computing the gradient and the tensor matrix in that region.

Again, there is significant reduction in computational requirements and memory bandwidth by computing gradients only in the region around the keypoint instead of operating on the entire image. Once the score is computed, it is sorted to output the best N keypoints and the image level in which they were found. On these N keypoints, the rBRIEF descriptors are computed.

The rBRIEF descriptor is based on the rotated BRIEF algorithm [8,9]. The descriptor is computed by picking a 48×48 window around each keypoint. We first compute the orientation of the keypoint by computing the moments as in [3]. We use the table lookup to generate the moment mask. Before computing the descriptor, the image is smoothened to reduce the effect of noise. We apply the 5×5 smoothing function only within the 48×48 window around each keypoint. This further reduces computational requirements and memory bandwidth compared to applying 5×5 smoothing function on the entire image. We use the table lookup to generate the 256 pairs of source-destination pattern needed to compute the descriptor.

The performance of the sparse point Harris score computation is 39 cycles/keypoint on the EVE. The rBRIEF descriptor computation takes 2192 cycles/keypoint on the EVE. The rBRIEF descriptor cycle count is higher as it also includes the smoothing filter and the moment computation.

3   Experiments and Results

In this section, we provide performance details of the algorithm implementation on the EVE engine of TI. We also show that it is reliable by providing the results of repeatability test.

3.1   Processing Time

For our experiments, we consider an image of size 400×400. We compute the ORB descriptors on 3 levels of the input image i.e. 400×400, 200×200 and 100×100. The maximum number of keypoints that can be detected in each level of the image is 2048. We then output the best 500 keypoints across the 3 levels and compute the descriptors for those. The algorithm has been implemented on EVE which is running at 650 MHz.

In Table 1, column 1 indicates the stage of the algorithm, column 2 indicates the processing time taken in *cycles/pixel* and column 3 indicates total time taken in *ms* to process the 3 levels of 400×400 input image. Column 2 not only includes the computation cycles, but also the direct memory access(DMA) cycles required to move data between memories and other system level overheads. The total time taken is around 5.5 ms. This is atleast 3 times faster compared to previous implementations mentioned in [3,4].

3.2   Repeatability

The detector is evaluated with the repeatability metric defined as the percentage of points simultaneously present in two images [10,11]. The higher the repeatability rate between two images, better the correspondence between the keypoints

Table 1. Average processing time of various stages of ORB.

Stage	Processing time cycles/pix	Time taken in ms
Pyramid	0.43	0.138
FAST9	8.28	2.67
FAST9 score + NMS + Sort	181.8	0.572
Harris score + Sort	157.68	0.496
rBRIEF descriptor	2192	1.68
Total	2540.19	5.56

in the two images. We evaluate our detector for viewpoint changes and blurring. We use the images, Matlab code to carry out performance tests, and binaries of other detectors from http://www.robots.ox.ac.uk/vgg/research/afne for our evaluation.

Figure 3a shows the repeatability rate of different detectors for viewpoint changes. Our implementation(tiorbf) outperforms other methods such as Edge based detector(ebraff), Intensity extrema based detector(ibraff), Maximally stable extremal regions(mseraf), Harris-Affine(haraff) and Hessian-Affine(hesaff) upto viewpoint changes of less than 30 degrees. Figure 3b shows the repeatability rate of different detectors for blurred images. Blurred images were obtained by changing the focus of the camera. It can be seen that our implementation outperforms other methods for blur factors less than 5.

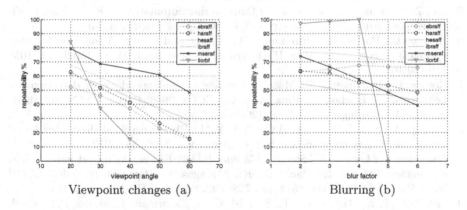

Viewpoint changes (a)   Blurring (b)

Fig. 3. Repeatability test viewpoint change and blurring.

4   Conclusion

In this paper, we have presented an efficient SIMD friendly implementation of ORB. We have provided key optimization techniques used in the implementation such as detecting 9 consecutive pixels, alternate non-iterative solution to compute FAST9 score and applying non-maximal suppression on sparse points. The main factor that contributes to the performance gain in our solution is that most of the processing is done on the sparse keypoints directly and not on every pixel of the image. We have shown that our solution is atleast 3 times faster compared to other approaches. Further, we have run the repeatability test to show that our solution is reliable despite the optimizations and modifications done.

One of the challenges that we have not addressed adequately is the approach taken during the non-maximal suppression. Our implementation can only perform 4-way non-maximal suppression, while the traditional approach uses 8-way non-maximal suppression. However, this can be addressed by performing an additional step post the 4-way non-maximal suppression to suppress based on neighbors along the diagonal direction.

References

1. Lowe, D.: Distinctive image features from scale-invariant keypoints. Int. J. Comput. Vis. **60**, 91–110 (2004)
2. Bay, H., Tuytelaars, T., Van Gool, L.: SURF: speeded up robust features. In: Leonardis, A., Bischof, H., Pinz, A. (eds.) ECCV 2006, Part I. LNCS, vol. 3951, pp. 404–417. Springer, Heidelberg (2006)
3. Rublee, E., Rabaud, V., Konolige, K., Bradski, G.: Orb: An efficient alternative to sift or surf. In: Internation Conference on Computer Vision, pp. 2564–2571 (2011)
4. Lee, K., Byun, K.: A hardware design of optimized orb algorithm with reduced hardware cost. Adv. Sci. Technol. Lett. **43**, 58–62 (2013)
5. Lin, Z., Sankaran, J., Flanagan, T.: Empowering automotive with ti's vision accelerationpac (2013). http://www.ti.com/lit/wp/spry251/spry251.pdf
6. Rosten, E., Porter, R., Drummond, T.: Faster and better: A machine learning approach to corner detection. IEEE Trans. Pattern Anal. Mach. Intell. **32**, 105–119 (2010)
7. Rosten, E., Drummond, T.: Machine learning for high-speed corner detection. In: Leonardis, A., Bischof, H., Pinz, A. (eds.) ECCV 2006, Part I. LNCS, vol. 3951, pp. 430–443. Springer, Heidelberg (2006)
8. Huang, W., Wu, L.D., Song, H.C., Wei, Y.M.: Rbrief: a robust descriptor based on random binary comparisons. IET Comput. Vis. **7**, 29–35 (2013)
9. Calonder, M., Lepetit, V., Strecha, C., Fua, P.: BRIEF: binary robust independent elementary features. In: Daniilidis, K., Maragos, P., Paragios, N. (eds.) ECCV 2010, Part IV. LNCS, vol. 6314, pp. 778–792. Springer, Heidelberg (2010)
10. Mikolajczyk, K., Tuytelaars, T., Schmid, C., A Zisserman, J., Matas, F.T., Gool, L.: A comparison of affine region detectors. Int. J. Comput. Vis. **65**, 43–72 (2005)
11. Mikolajczyk, K., Schmid, C.: A performance evaluation of local descriptors. IEEE Trans. Pattern Anal. Mach. Intell. **27**, 1615–1630 (2005)

Hierarchical Local Binary Pattern for Branch Retinal Vein Occlusion Recognition

Zenghai Chen[1](✉), Hui Zhang[2], Zheru Chi[1], and Hong Fu[1,3]

[1] Department of Electronic and Information Engineering,
The Hong Kong Polytechnic University, kowloon, Hong Kong
zenghai.chen@connect.polyu.hk, chi.zheru@polyu.edu.hk
[2] Yancheng Institute of Health Sciences, Yancheng, Jiangsu, China
zhanghui@ycmc.edu.cn
[3] Department of Computer Science,
Chu Hai College of Higher Education, kowloon, Hong Kong
hongfu@chuhai.edu.hk

Abstract. Branch retinal vein occlusion (BRVO) is one of the most common retinal vascular diseases of the elderly that would dramatically impair one's vision if it is not diagnosed and treated timely. Automatic recognition of BRVO could significantly reduce an ophthalmologist's workload, make the diagnosis more efficient, and save the patients' time and costs. In this paper, we propose for the first time, to the best of our knowledge, automatic recognition of BRVO using fundus images. In particular, we propose Hierarchical Local Binary Pattern (HLBP) to represent the visual content of an fundus image for classification. HLBP is comprised of Local Binary Pattern (LBP) in a hierarchical fashion with max-pooling. In order to evaluate the performance of HLBP, we establish a BRVO dataset for experiments. HLBP is compared with several state-of-the-art feature presentation methods on the BRVO dataset. Experimental results demonstrate the superior performance of our proposed method for BRVO recognition.

1 Introduction

Branch retinal vein occlusion (BRVO) is the second most common retinal vascular disease after diabetic retinopathy, with a prevalence range from 0.6 % to 1.1 % in the population [1]. BRVO is a blockage of the small veins in the retina. It usually occurs in the elderly, and can be caused by hypertension, cardiovascular disease, obesity, etc. Without timely treatment, BRVO would lead to macular edema, intraretinal hemorrhage, surface wrinkling retinopathy, and vitreous hemorrhage, which can then cause vision impairment or even blindness to the patients. These severe complications can be prevented or alleviated if BRVO is diagnosed early and treated timely. The diagnosis of BRVO is mainly made by analyzing a patient's fundus images or fluorescein angiography images. Although fluorescein angiography images provide more details about one's retinal conditions, the acquisition process is invasive and costly. By contrast, the acquisition of fundus images is non-invasive and inexpensive [2]. Moreover, an ophthalmologist

C.V. Jawahar and S. Shan (Eds.): ACCV 2014 Workshops, Part I, LNCS 9008, pp. 687–697, 2015.
DOI: 10.1007/978-3-319-16628-5_49

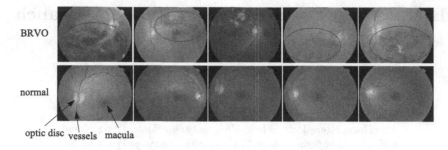

BRVO

normal

optic disc vessels macula

Fig. 1. BRVO fundus images and normal fundus images.

can diagnose BRVO effectively by analyzing fundus images only. The recognition of BRVO is thus based on fundus images in this paper.

Analyzing every fundus image by an ophthalmologist can get high diagnosis accuracy. But this would increase the ophthalmologist's workload, in particular when the ophthalmologist has a large number of images to examine every day. Furthermore, the patients need to pay more money and spend more time to waiting for the results if all the images are analyzed by the ophthalmologist. If the fundus images can be automatically recognized and diagnosed with an acceptable accuracy, the workload of an ophthalmologist and the cost and time of the patients can be significantly reduced. Automatic recognition of fundus images can be a powerful auxiliary diagnosis tool to an ophthalmologist. For instance, the automatic recognition system could make preliminary results for all the patients' cases first. The ophthalmologists could then concentrate on cases whose results look suspicious or cases specially required by the patients. It has been shown that automatic retinopathy recognition could significantly reduce the workload of manual image graders (usually ophthalmologists) by 50 % [3]. In addition to that, automatic recognition systems would help improve the diagnosis conditions of rural areas that are short of ophthalmologists. Other advantages of an automatic recognition system for fundus images include release from fatigue and improved repeatability [4]. It is meaningful, therefore, for us to propose an automatic BRVO recognition method based on fundus images in this paper.

Figure 1 shows some color fundus images of BRVO in the first row, and some normal eyes' images in the second row for a comparison. The main components of a normal retina include blood vessels, optic disc and macula. For BRVO, the blockage of small veins in the retina causes retinal hemorrhages, retinal edema and intraretinal microvascular abnormalities etc. There is why we can see from Fig. 1 that BRVO fundus images have abnormal regions, as pointed out by red circles. The purpose of our work is to automatically and precisely distinguish the BRVO images from the normal images. The fundus images are captured in different positions and illuminations, which would increase the recognition difficulty. There are some research papers working on the detection of microaneurysms and diabetic retinopathy [2, 4–6]. To the best of our knowledge, however, no paper working on the automatic recognition of BRVO has been published. One main method to process fundus images, such as the detection of diabetic retinopathy [4–6], performs the detection or segmentation of vessels and/or optic disc first, and then conducts further recognition based on the detection or segmentation result. One drawback of this method is that the performance of the detection or segmentation of vessels and/or optic disc remarkably affects the final recognition result. In light of that, we would like to process a fundus image by extracting its global visual features, without any detection or segmentation of vessels and/or optic disc.

Computer vision achieved rapid developments in the past two decades. Numerous successful feature representation methods have been proposed, e.g., Local Binary Pattern (LBP) [7], Histograms of Oriented Gradients (HOG) [8], SIFT [9], Spatial Pyramid Matching (SPM) [10], Gist [11], and CENTRIST [12] etc., among which LBP is a simple and efficient texture descriptor. LBP has been successfully applied to various computer vision tasks, e.g., pedestrian detection [13], segmentation [14], face analysis and recognition [15–17], biomedical image analysis [18], etc. Different variations of LBP have been proposed as well [19–22]. As shown in Fig. 1, the BRVO images have strong texture features. It is well known that LBP is a powerful texture descriptor. Therefore, we choose LBP for solving the BRVO recognition problem. In this paper, we propose Hierarchical Local Binary Pattern (HLBP) for BRVO recognition. The architecture of HLBP is motivated by deep learning, a type of promising machine learning algorithm boomed in recent years.

Deep learning started to attract increasingly significant attention in both academic and industrial communities since 2006 when Hinton and Salakhutdinov proposed a novel algorithm to train deep neural networks effectively [23]. It has been demonstrated that deep architectures can extract high-level and more abstract features than shallow architectures [24], making deep architectures a good solver for image recognition problems. Convolutional neural networks (CNNs) are one of the most successful deep architectures for image recognition. They have shown state-of-the-art performance for different problems, e.g., image classification [25], scene labeling [26], facial point detection [27], video classification [28], etc. A CNN consists of multiple (usually two or three) stages. Each stage has a convolution layer and a subsampling layer. In general, a CNN is a hierarchical combination of coding (convolution) and pooling (subsampling) operators. By following this idea, we constructs HLBP, where LBP acts as the coding operator, for BRVO recognition. The architecture of HLBP will be elaborated in Sect. 2.2. In order to evaluate HLBP, we establish a BRVO dataset for experiments, and compare the performance of HLBP with several state-of-the-art feature representation methods. The rest of this paper is organized as follows. Section 2 describes LBP and the proposed HLBP. Section 3 introduces a BRVO dataset, and reports experimental results. Section 4 concludes this paper, and points out future research work. Before ending this introductory section, it is worth mentioning the contributions of this paper as follows.

1. We propose for the first time, to the best of our knowledge, automatic recognition of BRVO using fundus images. The automatic recognition of BRVO could significantly reduce an ophthalmologist's workload, improve the diagnosis conditions of rural areas, and save the patients' time and costs.

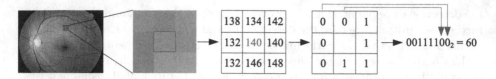

Fig. 2. Calculation of local binary pattern.

2. We propose a new feature representation method, termed hierarchical local binary pattern, to effectively characterize the visual content of a fundus image for recognition.
3. We establish a BRVO dataset to evaluate the performance of the proposed method and other state-of-the-art methods. The dataset is made public for free use, such that people can conduct experiments and comparisons conveniently.

2 Methodology

This section presents the methodology used in our work. In particular, LBP is introduced first, and then the proposed HLBP is elaborated.

2.1 Local Binary Pattern (LBP)

LBP is a local texture descriptor. LBP is calculated by comparing a central pixel's intensity with its neighboring pixels. Figure 2 illustrates the calculation procedure of LBP. For each 3×3 gray patch, compare the intensity value of the central pixel and its eight neighboring pixels. If a neighbor is bigger than or equal to the central pixel, a bit 1 is set in the corresponding neighbor's position. Otherwise, a bit 0 is set. After that, all the eight neighbors are set to binary bits. The eight binary bits are then concatenated in a clockwise or counterclockwise order, to form a 8-bit binary number. The binary number can be finally converted to a decimal number in [0 255]. The decimal number is the LBP value of the central pixel. Mathematically, the LBP of a central pixel c can be formulated as follows [7]:

$$LBP_c = \sum_{p=0}^{7} 2^p s(g_p - g_c), \quad s(x) = \begin{cases} 1, \, if \; x \geq 0; \\ 0, \, otherwise. \end{cases} \tag{1}$$

where g_c is the intensity of the central pixel c, and g_p is the intensity of the neighboring pixel p. Using Eq. (1), we can have LBP values for all the pixels (the boundary pixels could be removed), and calculate a histogram for the LBP values as the representation of an image.

Figure 2 and Eq. (1) show the calculation of LBP with eight neighboring pixels. There are also schemes that consider circles with larger radiuses as neighborhood and use more than eight neighbors for the calculation of LBP [7]. In this paper, however, we only take into account eight-pixel neighborhood, as this is the most simple and popular scheme for the calculation of LBP.

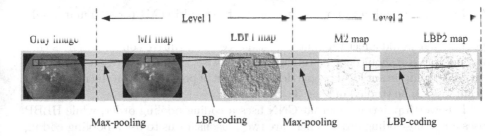

Fig. 3. The proposal hierarchical local binary pattern method.

LBP is a local texture coding method. It can be roughly regarded as a convolution or filtering operator as well. The convolution of CNN for the central pixel of a 3×3 patch can be defined as follows:

$$y_c = f(\sum_{p=1}^{9} k_p x_p + b),$$ (2)

where b is a bias, and $f(\cdot)$ is a transfer function. By observing Eqs. (1) and (2), we can find that when the bias $b = 0$ and $f(\cdot)$ is a linear function, LBP is similar to a convolution operator. The main differences include: (1) LBP performs "convolution" on the differences between the central pixel and its neighbors, rather than on the original pixels; (2) LBP uses fixed weights 2^p, while the weights k_p in CNN are trainable. Intuitively, we can construct hierarchical local binary pattern by following the architecture of CNN: hierarchical combination of coding and pooling.

2.2   Hierarchical Local Binary Pattern (HLBP)

The proposed HLBP is shown in Fig. 3. HLBP consists of two levels, each of which has a max-pooling layer and a LBP-coding layer. HLBP process an image as follows. Firstly, Level 1 receives an gray image as input. Max-pooling is first performed on the input image to generate a feature map M1. LBP is then conducted on M1 map in the LBP-coding layer to generate a LBP1 feature map. Secondly, LBP1 map is fed to Level 2 to perform max-pooling and LBP-coding, as a result of which M2 map and LBP2 map are produced. Thirdly, the histogram of LBP1 map and the histogram of LBP2 map are calculated, respectively. The two histograms are concatenated to form a feature vector as the representation of the input image.

The formulation of LBP-coding has specified in Eq. (1); while the max-pooling with a $m \times m$ window is defined as follows:

$$y = \max(x_i), \quad i \in \{1, 2, \cdots, m \times m\},$$ (3)

where x_i is the gray intensity of pixel i. The reason for us to choose max-pooling rather than other pooling schemes is that max-pooling is invariant to small translations [26]. The max-pooling window slips over the whole image or LBP1 map with a step size of one pixel to generate M1 or M2 map. Thus, given an $n \times n$ image (or LBP1 map) and $m \times m$ pooling window, the resulting M1 (or M2) map would be in size $(n - m + 1) \times (n - m + 1)$. The pooling window size is an essential component that would affect the final performance. An intuitive consideration is that the pooling size should not be too large. Otherwise, the pooling result would lose an image's important local properties. The effect of pooling size on the recognition accuracy will be studied in Sect. 3.

It is worth pointing out that CNN uses a coding-pooling order, while HLBP uses a pooling-coding order. There are two reasons for us to use a pooling-coding order. Firstly, it was demonstrated that patch-based LBP achieves better performance than LBP [29]. Performing LBP on max-pooling results is like performing LBP on image patches, as each max-pooling value is calculated in a small window (patch). Secondly, according to our observation, HLBP using pooling-coding (max-LBP) order performs much better than that using coding-pooling (LBP-max) order. One possible explanation could be that max-pooling-based LBP or patch-based LBP is able to incorporate spatial information of neighboring small regions/patches, since LBP is run on small regions/patches. The extracted features are hence more discriminative. The operation of max-pooling increases the input values. So the feature maps in Level 2 are much whiter (higher gray values) than that of Level 1, as shown in Fig. 3. After two levels, all of the feature maps' values tend to be 255 (the highest value of an 8-bit gray image). In other words, it is hardly to extract useful features after Level 2. There is why we only consider two levels in HLBP. The working mechanism of HLBP could be as follow. LBP2 can be regarded as a LBP of LBP, or a feature of feature. It is more discriminative but sensitive. In the recognition phase, LBP1 is to classify most of the relatively easy samples; while for samples that seem difficult to LBP1, LBP2 is expected to provide better results.

3   Experiments

This section presents the experiments. A BRVO datasest is first described, and then the experimental results and corresponding analysis are presented.

3.1   Experimental Dataset

With the help of a hospital, we establish a BRVO dataset for our experiments.[1] The dataset has in total 200 fundus images acquired from 200 persons, 100 BRVO fundus images and 100 normal fundus images. All the images are of size 768×576. Some of the images are shown in Fig. 1. All the original fundus images are in color. But only gray information is used in this paper. We use Matlab built-in function rgb2gray to convert the original RGB images into gray images.

We adopt ten-fold evaluation scheme. That is, the dataset is randomly partitioned into ten folds. Each fold has ten BRVO images and ten normal images. Each time one fold is used for testing, and the remaining nine folds for training.

---

[1] The dataset can be downloaded on: http://pan.baidu.com/s/1ntohK5V.

Table 1. Accuracy rates (%) of different methods using SVM with linear kernel.

Fold no.	Gist [11]	HOG [8]	SPM [10]	CENTRIST [12]	LBP [7]	HLBP
1	85	90	95	95	95	95
2	90	90	95	85	85	100
3	95	100	100	95	100	100
4	95	100	95	90	90	95
5	100	100	95	90	95	95
6	100	85	100	85	100	100
7	95	95	85	90	80	95
8	95	95	95	85	85	95
9	100	95	100	90	85	100
10	100	90	90	85	85	95
Std	4.97	5.16	4.71	3.94	7.07	**2.58**
Mean	95.5	94.0	95.0	89.0	90.0	**97.0**

As a result, ten results of ten testing folds are obtained. The average accuracy of the ten results is used as the final performance for a comparison. SVM classifier [30] is employed for classification. The parameters of SVM is determined by two-fold cross-validation on the training data. Both linear kernel and RBF kernel will be investigated.

3.2 Experimental Results

HLBP will be compared with LBP, CENTRIST, Gist, HOG, and SPM. The two histograms of LBP1 and LBP2 of HLBP are normalized to sum to one, respectively. Thus, each HLBP feature vector sums to two. Both LBP and CENTRIST are normalized to sum to one as well. Note that CENTRIST highly relates to LBP, with a difference in bit ordering. The features of Gist, HOG, and SPM are normalized to $[-1\ 1]$, respectively.

The recognition accuracy rates of different methods for ten testing folds and the overall performances are tabulated in Tables 1 and 2. Table 1 uses linear kernel, while Table 2 uses RBF kernel. The HLBP compared in Tables 1 and 2 utilizes a 4×4 max-pooling size. As can be seen, HLBP outperforms the other methods in terms of mean accuracy rate and standard deviation, for both linear kernel and RBF kernel. In particular, the improvement of HLBP over LBP is significant, indicating that the hierarchical architecture could enhance the discriminative power of LBP. As CENTRIST is similar to LBP, they achieve similar overall accuracy rates. The standard deviation of LBP with linear kernel is relatively high. The main reason is that it performs poor on Fold 7. Gist, HOG, and SPM achieve better performance than LBP and CENTRIST. One reason could be that their feature coding algorithms encode spatial information between regions. For example, SPM exploits spatial pyramid to encode spatial relationships of regions in different scales. HOG organizes oriented gradient using cells and blocks. Gist is

Table 2. Accuracy rates (%) of different methods using SVM with RBF kernel.

Fold no.	Gist [11]	HOG [8]	SPM [10]	CENTRIST [12]	LBP [7]	HLBP
1	95	90	95	95	95	95
2	90	90	95	85	85	100
3	90	100	95	90	100	100
4	95	100	100	95	95	95
5	100	100	95	90	95	95
6	100	90	100	90	95	100
7	95	95	85	95	80	90
8	95	95	95	85	90	95
9	95	95	100	95	90	100
10	100	95	90	80	90	100
Std	3.69	4.08	4.71	5.27	5.80	**3.50**
Mean	95.5	95.0	95.0	90.0	91.5	**97.0**

originally designed to represent the dominant spatial structure of an image. Thanks to the max-pooling and the hierarchical architecture, HLBP is able to incorporate the spatial information of an image, as a result of which high performance can be achieved. By comparing Tables 1 and 2, we can see that RBF kernel cannot always guarantee better results than linear kernel.

In order to investigate the effect of max-pooling size on the recognition accuracy, we test several pooling sizes, and the results of using linear kernel are depicted in Fig. 4. Figure 4 shows the curves of recognition accuracy rates for three cases: (1) using LBP1 only; (2) using LBP2 only; and (3) using both LBP1 and LBP2 (i.e., HLBP). As shown in Fig. 4, LBP1 always outperforms LBP2. Their combination LBP1+LBP2 can obtain better results for a moderate pooling size. This indicates that LBP2 is complementary to LBP1. LBP2 could recognize samples that seem difficult to LBP1. By comparing Fig. 4 and Table 1, we can see that for pooling sizes from 3×3 to 6×6, LBP1 achieves higher accuracy than LBP, meaning that performing LBP on the max-pooling result is a better choice than performing LBP on the original image pixels. HLBP (LBP1 + LBP2) achieves the best performance on the BRVO dataset when 4×4 max-pooling size is adopted. The performance gradually reduces when the pooling size decrease or increase from 4×4 size. This is reasonable, because if the size is too small (e.g., 2×2), the spatial information incorporated is not significant enough; while if the size is too large, local spatial properties would lose. When the pooling size is too large (e.g., 7×7 or 8×8 in Fig. 4), the accuracy rate of LBP2 drops remarkably, and the combination of LBP1+LBP2 does not make a better result than LBP1. Figure 4 demonstrates that it is crucial to adopt a moderate (e.g., from 3×3 to 5×5) max-pooling size, if HLBP is expected to achieve a good performance.

Figure 5 shows the six images that are misclassified by HLBP using a 4×4 pooling size and linear kernel. As can be seen, the two misclassified BRVO images

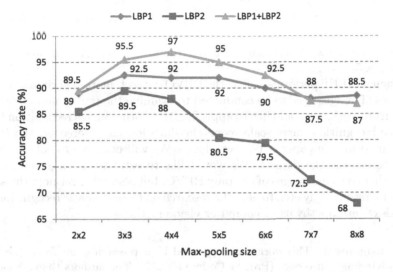

Fig. 4. Accuracy rates (%) of HLBP using different max-pooling sizes and linear kernel.

have only a few irregular dark ares. They look similar to normal images. For the four normal images, different illuminations make them difficult to recognize.

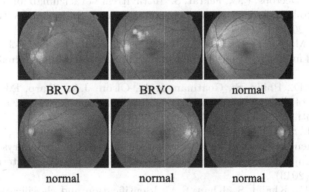

Fig. 5. Images misclassified by HLBP using a 4 × 4 pooling size and linear kernel. The caption underneath each image is the ground truth.

4   Conclusion and Future Work

We propose a new feature representation method HLBP, based on LBP and max-pooling, for automatic recognition of BRVO in this paper. The proposed method is simple but efficient. We establish a BRVO dataset to evaluate the proposed HLBP. The performance of HLBP is compared with widely-used feature presentation methods. Experimental results demonstrate that HLBP with a 4 × 4 max-pooling size outperforms other methods. The improvement of HLBP over LBP is significant. The performance of HLBP reduces as the max-pooling size decreases or increases from 4×4. When the max-pooling size is too large, combining LBP1 and LBP2 would not produce a better result. It is, therefore, important to choose a proper pooling size. We recommend that the max-pooling size should be from 3×3 to 5×5.

Although HLBP performs well on the BRVO dataset, more experiments on various types of images need to be conducted to examine the performance of HLBP. In the future work, we would like to apply HLBP to other image recognition tasks, e.g., face recognition, large-scale scene classification etc. Moreover, the BRVO dataset used in this paper is not very large. We will continue to collect fundus images to enlarge the dataset. We would like to establish a large retinopathy dataset that contains images of not only BRVO, but also other retinopathies such as diabetic retinopathy etc., to help the research on the automatic recognition and diagnosis of retinopathy using computer vision methods.

Acknowledgement. This work was supported by a research grant from The Hong Kong Polytechnic University (Project Code: G-YL77). The authors thank Yancheng Third People's Hospital for providing the BRVO and normal color fundus images.

References

1. Ehlers, J.P., Decroos, F.C., Fekrat, S.: Intravitreal bevacizumab for macular edema secondary to branch retinal vein occlusion. Retina J. Retinal and Vitreous Dis. **31**, 1856–1862 (2011)
2. Walter, T., Massin, P., Erginay, A., Ordonez, R., Jeulin, C., Klein, J.C.: Automatic detection of microaneurysms in color fundus images. Med. Image Anal. **11**, 555–566 (2007)
3. Fleming, A.D., Philip, S., Goatman, K.A., Olson, J.A., Sharp, P.F.: Automated assessment of diabetic retinal image quality based on clarity and field definition. Invest. Ophthalmol. Vis. Sci. **47**, 1120–1125 (2006)
4. Tavakoli, M., Shahri, R.P., Pourreza, H., Mehdizadeh, A., Banaee, T., Toosi, M.H.B.: A complementary method for automated detection of microaneurysms in fluorescein angiography fundus images to assess diabetic retinopathy. Pattern Recogn. **46**, 2740–2753 (2013)
5. Akram, M.U., Khalid, S., Khan, S.A.: Identification and classification of microaneurysms for early detection of diabetic retinopathy. Pattern Recogn. **46**, 107–116 (2013)
6. Zhang, B., Karray, F., Li, Q., Zhang, L.: Sparse representation classifier for microaneurysm detection and retinal blood vessel extraction. Inf. Sci. **200**, 78–90 (2012)
7. Ojala, T., Pietikainen, M., Maenpaa, T.: Multiresolution gray-scale and rotation invariant texture classification with local binary patterns. TPAMI **24**, 971–87 (2002)
8. Dalal, N., Triggs, B.: Histograms of oriented gradients for human detection. In: CVPR, pp. 886–893 (2005)
9. Lowe, D.G.: Distinctive image features from scale-invariant keypoints. IJCV **60**, 91–110 (2004)
10. Lazebnik, S., Schmid, C., Ponce, J.: Beyond bags of features: spatial pyramid matching for recognizing natural scene categories. In: CVPR, pp. 2169–2178 (2006)
11. Oliva, A., Torralba, A.: Modeling the shape of the scene: a holistic representation of the spatial envelope. IJCV **42**, 145–175 (2001)

12. Wu, J., Rehg, J.M.: CENTRIST: a visual descriptor for scene categorization. TPAMI
    33, 1480 1501 (2011)
13. Wang, X., Han, T.X., Yan, S.: An HOG-LBP human detector with partial occlusion
    handling. In: ICCV, pp. 32–39 (2009)
14. Li, M., Staunton, R.C.: Optimum Gabor filter design and local binary patterns for
    texture segmentation. Pattern Recogn. Lett. **29**, 664–672 (2008)
15. Heusch, G., Rodriguez, Y., Marcel, S.: Local binary patterns as an image prepro-
    cessing for face authentication. In: FG, pp. 9–14 (2006)
16. Moore, S., Bowden, R.: Local binary patterns for multi-view facial expression recog-
    nition. Comput. Vis. Image Underst. **115**, 541–558 (2011)
17. Zhang, B.C., Gao, Y.S., Zhao, S.Q., Liu, J.Z.: Local derivative pattern versus local
    binary pattern: face recognition with high-order local pattern descriptor. TIP **19**,
    533–544 (2010)
18. Sorensen, L., Shaker, S., de Bruijne, M.: Quantitative analysis of pulmonary emphy-
    sema using local binary patterns. IEEE Trans. Med. Imaging **29**, 559–569 (2010)
19. He, Y., Sang, N., Gao, C.: Pyramid-based multi-structure local binary pattern for
    texture classification. In: Kimmel, R., Klette, R., Sugimoto, A. (eds.) ACCV 2010,
    Part III. LNCS, vol. 6494, pp. 133–144. Springer, Heidelberg (2011)
20. Zhu, C., Bichot, C.E., Chen, L.: Multi-scale color local binary patterns for visual
    object classes recognition. In: ICPR, pp. 3065–3068 (2010)
21. Chen, J., Kellokumpu, V., Zhao, G., Pietikainen, M.: RLBP: robust local binary
    pattern. In: BMVC (2013)
22. Guo, Y., Zhao, G., Zhou, Z., Pietikainen, M.: Video texture synthesis with multi-
    frame LBP-TOP and diffeomorphic growth model. TIP **22**, 3879–3891 (2013)
23. Hinton, G.E., Salakhutdinov, R.R.: Reducing the dimensionality of data with neural
    networks. Science **313**, 504–507 (2006)
24. Bengio, Y.: Learning deep architectures for AI. Found. Trends Mach. Learn. **2**, 1–127
    (2009)
25. Krizhevsky, A., Sutskever, I., Hinton, G.E.: ImageNet classification with deep con-
    volutional neural networks. In: NIPS (2012)
26. Farabet, C., Couprie, C., Najman, L., LeCun, Y.: Learning hierarchical features for
    scene labeling. TPAMI **35**, 1915–1929 (2013)
27. Sun, Y., Wang, X., Tang, X.: Deep convolutional network cascade for facial point
    detection. In: CVPR, pp. 3476–3483 (2013)
28. Karpathy, A., Toderici, G., Shetty, S., Leung, T., Sukthankar, R., Fei-Fei, L.: Large-
    scale video classification with convolutional neural networks. In: CVPR, pp. 1725–
    1732 (2014)
29. Wolf, L., Hassner, T., Taigman, Y.: Descriptor based methods in the wild. In: ECCV
    Workshop on Faces in Real-Life Images (2008)
30. Chang, C.C., Lin, C.J.: LIBSVM: a library for support vector machines. ACM Trans.
    Intell. Syst. Technol. **2**, 27:1–27:27 (2011)

Extended Keypoint Description
and the Corresponding Improvements
in Image Retrieval

Andrzej Śluzek[(✉)]

Khalifa University, Abu Dhabi, United Arab Emirates
andrzej.sluzek@kustar.ac.ae

Abstract. The paper evaluates an alternative approach to BoW-based image retrieval in large databases. The major improvements are in the re-ranking step (verification of candidates returned by BoW). We propose a novel keypoint description which allows the verification based only on individual keypoint matching (no spatial consistency over groups of matched keypoints is tested). Standard Harris-Affine and Hessian-Affine keypoint detectors with SIFT descriptor are used. The proposed description assigns to each keypoint several words representing photometry and geometry of the keypoint in the context of neighbouring image fragments. The words are Cartesian products of typical SIFT-based words so that huge vocabularies can be built. The preliminary experiments on several popular datasets show significant improvements in the pre-retrieval phase combined with a dramatically lower complexity of the re-ranking process. Because of that, the proposed methodology is particularly recommended for the retrieval in very large datasets.

1 Introduction

Keypoint-based image matching is one of the fundamental tools in CBVIR. Even though the reported solutions differ in keypoint detectors, keypoint descriptors and the vocabulary sizes (matching using the original descriptor vectors is computationally inefficient) typical approaches to the retrieval of similar images or sub-images generally follow the same two-step scheme. First, the candidate images are pre-retrieved. One of the standard techniques is BoW (e.g. [1]) where the sparse histograms of visual words are matched to find similar images. This model ignores the spatial distributions of keypoints so that the second step of geometric/configurational verification is needed to re-rank the pre-retrieved candidates, i.e. to identify the most similar images (or similar fragments within them) from the pool of candidates. Computational complexity of the second step is high, and many attempts have been reported (e.g. [2–4], etc.) to simplify it. Nevertheless, the retrieval algorithms cannot be considered sufficiently scalable as long as spatial distributions of matching keypoints have to be analyzed.

In this paper, we evaluate an alternative approach where the first level of BoW-based pre-retrieval is retained, but the complexity of the second level is dramatically reduced. This approach is based on the concept of *contextual keypoint*

© Springer International Publishing Switzerland 2015
C.V. Jawahar and S. Shan (Eds.): ACCV 2014 Workshops, Part I, LNCS 9008, pp. 698–709, 2015.
DOI: 10.1007/978-3-319-16628-5_50

descriptors (preliminarily introduced in [5]) which are built using dependencies between keypoints extracted by two complementary detectors, e.g. Harris-Affine and Hessian-Affine. Each keypoint is represented by several descriptors, i.e. by several words, so that keypoint matching is more flexible (the level of similarity is defined by the number of words shared by two descriptions). Another advantage of such extended descriptions is that similarities between larger image fragments can be established using only matches between individual keypoints (no spatial analysis needed!). It is believed the presented approach may contribute to development of fully scalable CBVIR algorithms.

In Sect. 2, the background works are briefly overviewed, in particular the works related to the proposed description. Section 3 explains (and illustrates on selected examples) advantages of the method, which is experimentally verified using several popular datasets. Concluding remarks and observations are included in Sect. 4.

2   Background Works

2.1   Keypoint Matching

Performances of keypoint matching depend on the quality of keypoint detectors (this aspect is not discussed in the paper) and on credibility of the matching scheme. The *mutual nearest neighbour* O2O scheme is generally considered (see [6]) the most credible one so that we use its results as the benchmark to evaluate matching based on visual words. Two standard detectors, i.e. Harris-Affine (*haraff*) and Hessian-Affine (*hesaff*) [7] are selected (the reasons for this choice are later explained in detail) and SIFT [8] is the selected keypoint descriptor because of its popularity and high repeatability. Actually, we use its RootSIFT variant whis was reported superior in [9]. Performances of keypoint matching are evaluated on a popular benchmark dataset of diversified images[1]. The dataset provides homographies between *the-same-category* images, so that the ground truth keypoint correspondences can be identified similarly to [10].

Table 1 summarizes the results of keypoint matching using both O2O and visual vocabularies of diversified sizes. The ranges of values have been obtained by using several alternative i.e. generated from different populations of images) vocabularies of each size and/or using sets of keypoints extracted by two detectors. It should be also highlighted that (unlike in most work using the same benchmark dataset) *all* pairs of images are compared to better reflect scenarios of larger-scale image retrieval.

Although satisfactory *recall* can be achieved by using small vocabularies, low *precision* values (which further deteriorate if more images are added to the dataset) confirm a well-known fact that credible image retrieval based only on individual keypoint matches is not reliable. It can be also noticed that the overall performance of keypoint matching (represented by *F-measure*) improves with the size of vocabulary, but this size cannot be indiscriminately increased. If a

---

[1] http://www.robots.ox.ac.uk/~vgg/research/affine/.

Table 1. Performances of keypoint matching using *haraff* and *hesaff* keypoints (see also [5]).

Measure	O2O	2^{10} words	2^{16} words	2^{20} words	2^{25} words	2^{30} words
Recall	0.571–0.587	0.584–0.608	0.266–0.281	0.209–0.219	0.173–0.176	0.141–0.142
Precision	0.104–0.114	0.002–0.003	0.003–0.004	0.035–0.076	0.071–0.124	0.132–0.178
F-measure	0.177–0.190	0.003–0.005	0.005–0.008	0.060–0.112	0.080–0.144	0.137–0.158

vocabulary grows too large, the quantization intervals become smaller that the natural fluctuations of descriptor values, and very few (if any) matches would be found even in pairs of highly similar images. Some sources (e.g. [11,12]) indirectly indicate that several millions is the maximum practical size of visual vocabularies.

2.2   Extended Descriptors

Descriptions of keypoints would be obviously enriched, if some data about the keypoint context can be incorporated. Intuitively, the context of a region-based feature can be defined as a collection of neighbouring contour-based features (and another way around). We propose, therefore, to use two complementary keypoint detectors (e.g. *hesaff* detecting blob-like features and *haraff* detecting corner-like features) and to combine their SIFT descriptors in a way explained in the definitions below and illustrated in Figs. 1 and 2.

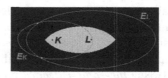

Fig. 1. Examples of a *haraff* keypoint (K) and a *hesaff* keypoint (L) extracted from a simple image.

Def.1 Given a *hesaff(haraff)* keypoint K and its neighbouring *haraff (hesaff)* keypoint L (with the corresponding ellipses E_K and E_L, see Fig. 1), the **CONSIFT** descriptor of K in the context of L is defined by a $384D$ vector which is a concatenation of three $128D$ SIFT descriptors: (a) the original SIFT computed over E_K ellipse, (b) SIFT computed over E_K ellipse with $\overrightarrow{K,L}$ vector as the reference orientation, and (c) SIFT computed over E_L ellipse with $\overrightarrow{L,K}$ vector as the reference orientation (see Fig. 2).

Thus, the first part of CONSIFT descriptors characterizes local properties of keypoints, while the remaining parts provide some data about photometric and geometric properties of keypoint neighbourhood.

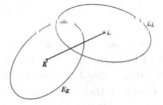

Fig. 2. A configuration of two keypoint ellipses for computing CONSIFT descriptor.

Def.2 Given a *hesaff(haraff)* keypoint K, its *extended description* consists of CONSIFT descriptors computed in the context of *haraff(hesaff)* L_i keypoints belonging to the neighbourhood of K. The neighbourhood is defined to follow a common-sense idea that a blob feature should be surrounded by a number of similar-scale corner features distributed approximately around the perimeter of the blob feature (or another way around).

Thus, L_i keypoints are considered neighbourhood keypoints if:

1. The Mahalanobis distances D_M between K and L_i satisfy:

$$1/\sqrt{2} \leq D_M \leq 2, \tag{1}$$

where the unit distance is defined by the shape of E_K ellipse.
2. The areas of E_K and E_i ellipses are similar (i.e. the ratio is between 0.5 and 2).

Using a large set of test images, we have verified that the average size of such neighbourhoods is 8–10 (both for *haraff* and *hesaff* keypoints). If necessary, the maximum size can be constrained (e.g. not more than 20).

The extended descriptions are not particularly suitable for a direct keypoint matching (because of a high dimensionality of CONSIFT vectors). However, they can be conveniently used for matching by visual words. Each CONSIFT descriptor is actually a union of three SIFTs. Therefore, CONSIFT vocabularies can be built as Cartesian products of the original SIFT vocabularies. Even if those SIFT vocabularies are relatively small (i.e. the quantization is coarse) the resulting CONSIFT vocabularies are huge. For example, 1000-word SIFT vocabularies generate a billion-word CONSIFT counterpart ($10^3 \times 10^3 \times 10^3 = 10^9$). Such a vocabulary is expected to combine high *precision* (a large number of words) with high *recall* (coarse quantization of the contributing words).

In the scheme based on extended descriptions two keypoints match, if their descriptions share at least one CONSIFT word. However, more flexible conditions can be easily defined, e.g. only keypoints sharing at least N (where $N > 1$) CONSIFT words in their extended descriptions are considered a match. This is a significant advantage over traditional word-based matching, where keypoints can share at most a single word.

Superficially, the proposed method may look similar to image matching by *visual phrases* (e.g. [13]) since both approaches consider keypoints in a wider context. However, visual phrases (as defined in [13]) need spatial analysis, first in the histogram building and later in the consistency verification phase. Thus, for the practicality over large databases the class of transformations consider in visual phrases (similarly to other methods incorporating geometric verification) is constrained, e.g. allowing only shifts and/or scale changes between matching images.

Higher performances of CONSIFT-based matching (using ether 1 or 2 two shared CONSIFT words) are shown in Table 2 on the same dataset. The performances are better even than the O2O scheme based on full SIFT vectors (compare to Table 1). Additionally, the table indicates that vocabularies with approx. 1 billion CONSIFT words (i.e. the underlying SIFT vocabularies have only 1000 words) have superior *F-measures* than 64-billion word vocabularies. It suggests that also the size of CONSIFT vocabularies cannot grow indiscriminately. However, for huge databases of images with significant numbers of keypoints, larger CONSIFT vocabularies are still recommended because they provide very high *precision* while the level of *recall* is less critical in such cases.

Table 2. Keypoint matching using extended descriptions of *haraff* and *hesaff* keypoints (N indicates the minimum number of shared CONSIFTs).

Measure	$2^{10+10+10}$ words $(N = 1)$	$2^{10+10+10}$ words $(N = 2)$	$2^{16+10+10}$ words $(N = 1)$	$2^{16+10+10}$ words $(N = 2)$
Recall	0.355–0.371	0.285–0.314	0.166–0.175	0.138–0.158
Precision	0.201–0.245	0.332–0.402	0.398–0.440	0.477–0.612
F-measure	0.265–0.290	0.322–0.334	0.241–0.243	0.225–0.237

It can be, therefore, claimed that the use of extended keypoint descriptions is justified even though the memory resources are significantly increased; a single keypoint has, in average, 8–10 CONSIFT words instead of a single SIFT word (some keypoints have multiple SIFT descriptors - see [8]).

An illustrative example comparing keypoint matching by O2O scheme, SIFT words and CONSIFT words is given in Fig. 3.

3   Image Retrieval

3.1   Bag-of-Words Pre-retrieval

The ultimate objective of extended keypoint descriptions is to simplify image retrieval in very large databases. However, we do not intend to change the principles of BoW-based pre-retrieval returning images ranked by the BoW similarity.

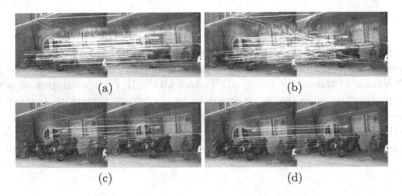

Fig. 3. Keypoint matching results using: (a) O2O, (b) 2^{16} SIFT words, (c) 2^{25} SIFT words and (d) 2^{30} CONSIFT words.

In BoW representation (i.e. sparse histograms of word distributions in images) image similarities are approximated by the similarities between those histograms. Because our approach is proposed for databases of unknown and unpredictable sizes, the popular techniques of BoW normalization which require database statistics (e.g. *td-idf*, [14]) cannot be applied, and we use histograms of *absolute* word frequencies in images.

Numerous measures of histogram similarities exist (e.g. [15]) but not all of them are applicable to BoW matching. Under the assumptions regarding BoW building in this work, we eventually selected a simple *histogram intersection* measure (proposed in [16]), where the distance between two histograms H_A and H_B over *Voc* vocabulary is defined by

$$d(H_A, H_B) = \sum_{w \in Voc} min(H_A(w), H_B(w)). \qquad (2)$$

Such a measure nicely corresponds to the intuitive notion of similarity between both full images and sub-images (including textured images).

Additionally, to normalize the results over images with diversified numbers of keypoints, the Eq. 2 similarity between a query image A and a database image B is weighted by the factor S_F

$$S_F = 2^{(1-n_B/n_A)}, \qquad (3)$$

where n_A and n_B are the numbers of keypoints in the corresponding images. Such a normalization allows for more realistic pre-retrieval results in case of databases containing images with dramatically diversified numbers of keypoints (note that $S_F = 1$ for a pair of images with the same numbers of keypoints, $S_F < 1$ when the query has fewer keypoints, and $2 > S_F > 1$ for database images with fewer keypoints).

Performances of BoW pre-retrieval using SIFT and CONSIFT words have been tested on four popular datasets of very diversified characteristics, i.e. Oxford5k[2], UKB[3], Visible[4] and Caltech-Faces1999[5]. They represent diversified aspects of image retrieval, i.e. full images retrieval (UKB) or sub-image retrieval in easier (Oxford5k) and more complicated (Visible) scenarios, etc.

The results are shown in Table 3, which contains *mean average precision* (mAP) values obtained by using SIFT and CONSIFT vocabularies of several sizes. Because the objective is to evaluate performance variations (rather than the absolute values of mAP), the mAP values for the 64k-word SIFT vocabulary are used as the reference (unit score). The presented scores are the average results for *haraff* and *hesaff* keypoints, and for several alternative vocabularies of each size.

Table 3. Relative *mean average precisions* (mAP) of image retrieval using SIFT and CONSIFT vocabularies of various sizes.

Dataset	64k words SIFT	1M words SIFT	32M words SIFT	1G words CONSIFT
Oxford5k	1.0	1.03	1.19	1.53
UKB	1.0	1.22	1.32	1.59
Visible	1.0	1.06	1.17	1.32
Faces1999	1.0	1.54	2.54	2.88

The content of Table 3 indicates that performances of BoW-based image pre-retrieval can be significantly improved if SIFT vocabularies are replaced by their CONSIFT counterparts. As an illustration, Figs. 4, 5, 6, and 7 show the top rank returns by SIFT and CONSIFT for an exemplary queries from each tested dataset. We deliberately select not too successful examples to discuss improvements that can be subsequently introduced to in the second step (re-ranking pre-retrieved candidates).

3.2   Verification of Pre-retrieved Results

The content of Table 3 indicates that BoW pre-retrieval with CONSIFT words provides better performances (in terms of mAP values) than with SIFT words, but such results are not considered final. In other words, the verification step is still applied.

For the selected datasets, many works on consistency verification exist (e.g. [17] for Oxford5k, [4] for UKB and Oxford5k, [18] for Visible or [19] for

---

[2] http://www.robots.ox.ac.uk/~vgg/data/oxbuildings/.

[3] http://www.vis.uky.edu/~stewe/ukbench/.

[4] http://156.17.10.3/~visible/data/upload/FragmentMatchingDB.zip.

[5] http://www.vision.caltech.edu/html-files/archive.html.

Fig. 4. Top rank BoW-returned images for an exemplary query in Visible dataset, using a CONSIFT vocabulary of $1G$ words (top row) and a SIFT vocabulary of $64k$ words (bottom row).

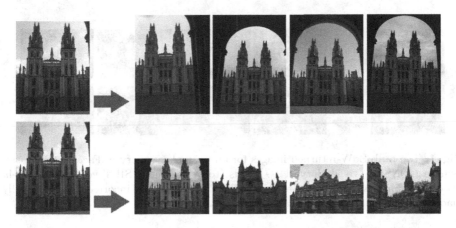

Fig. 5. Top rank BoW-returned images for an exemplary query in Oxford5k dataset, using a CONSIFT vocabulary of $1G$ words (top row) and a SIFT vocabulary of $64k$ words (bottom row).

Faces1999). They generally apply solutions which are computationally intensive (in spite of the reported simplification efforts). The only known example of a solution working without geometric verification, in [4], requires query expansion which should be considered computationally intensive as well.

We propose to reduce the verification step to a straightforward matching of extended keypoint descriptors, where a match between two keypoints is accepted if their descriptions share at least N (the recommended values of N are 3 or more) CONSIFT words (see Subsect. 2.2). Such a mechanism effectively identifies pairs of keypoints which have sufficiently similar neighbourhoods. In other words, pre-retrieved images are accepted only if they contain larger fragments similar to some fragments of the query. In this mechanism, there is no difference between full and partial similarity of images (image retrieval *versus* sub-image retrieval) which in our opinion is another advantage of the method.

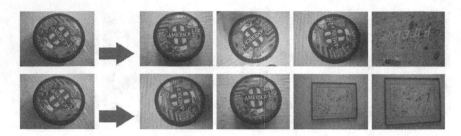

Fig. 6. Top rank BoW-returned images for an exemplary query in UKB dataset, using a CONSIFT vocabulary of $1G$ words (top row) and a SIFT vocabulary of $64k$ words (bottom row). Note that in UKB dataset each query has only three relevant returns, so that at least one return must be always incorrect.

Fig. 7. Top rank BoW-returned images for an exemplary query in Faces1999 dataset, using a CONSIFT vocabulary of $1G$ words (top row) and a SIFT vocabulary of $64k$ words (bottom row). Note that for the CONSIFT retrieval the incorrect face is actually placed on the same background as the query face.

Examples are provided in Fig. 8 which illustrates how selected images from Figs. 4, 5, 6, and 7 are matched to the queries. Oviously, only images pre-retrieved by CONSIFT vocabularies are taken into account because for SIFT-based ranking it is generally impossible to distinguish between correct and incorrect pre-retrievals based on the spatial distributions of matching keypoints (an illustrative example is given in Fig. 9).

As shown in Fig. 8, the proposed method returns rather small numbers of keypoint correspondences pointing to the most similar fragments in both images. Images which are incorrectly pre-retrieved usually have no keypoint matches. If, however, some matches are found between the query and an (allegedly) incorrect image, those keypoint correspondences identify fragments which are, nevertheless, visually similar (although sometimes a careful inspection is needed to notice the actual existence of such a similarity).

The pre-retrieved images are subsequently re-ranked (similarly to the most popular solutions, e.g. [3,4,12,17], etc.) based on the number of keypoint correspondences found. The experiments on the performances of re-ranked retrieval are still under way. Nevertheless, the preliminary results indicate that the

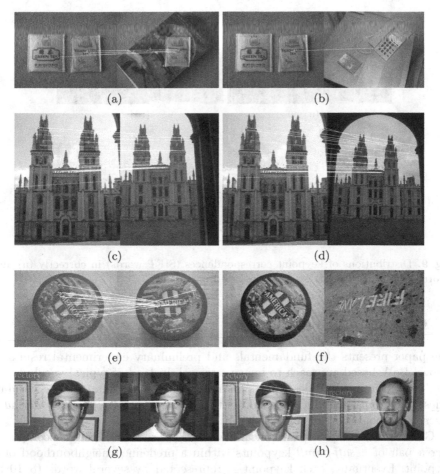

Fig. 8. Verification of BoW-pre-retrieved images (CONSIFT vocabulary) by matching extended descriptions of individual keypoints. Examples are from: (a,b) Visible dataset (Fig. 4), (c,d) Oxford5k dataset (Fig. 5), (e,f) UKB dataset (Fig. 6) and (g,h) Faces1999 dataset (Fig. 7).

performances are comparable to those reported in works using geometric verification. For example, the mAP improvements in the re-ranked lists over Oxford5k dataset are very similar to the improvements presented in [17].

Similarly to most works, we re-rank only a fixed number of top pre-retrievals (e.g. 300 threshold for Oxford5k dataset) even though more candidates are usually returned. However, with a huge size of CONSIFT vocabularies there are often cases when BoW pre-retrieval returns fewer images than the threshold number. This can be considered another advantage of the proposed approach (especially for large-scale applications).

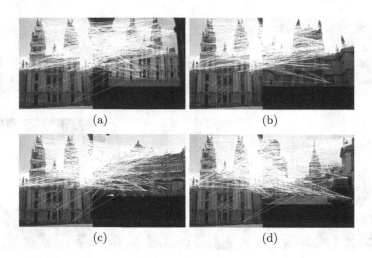

Fig. 9. Distributions of keypoint correspondences (SIFT words) in correctly (a) and incorrectly (b,c,d) pre-retrieved images (see Fig. 5).

4   Concluding Remarks

The paper presents the fundamentals and preliminary experimental results of a novel BoW-based approach to image retrieval. Instead of typical vocabularies (e.g. derived from SIFT descriptors on *haraff/hesaff* keypoints) we use vocabularies built from CONSIFT descriptors to represent each keypoint by *extended descriptions* consisting of several CONSIFT-based words.

CONSIFT descriptors are concatenations of three SIFT vectors computed over a pair of *hesaff-haraff* keypoints within a predefined neighbourhood of a keypoint. Eventually, each keypoint is represented by several words (8–10 in average) so that keypoint matching is more flexible (the level of similarity can be estimated by the number of CONSIFT words shared by two descriptions).

A standard two-level model of retrieval is assumed, i.e. the BoW-based pre-retrieval of candidate images is followed by the verification of configuration constraints in the candidate images. No changes are introduced to the first level. Nevertheless, it has been preliminarily verified on popular datasets that performances of BoW pre-retrieval are improved if CONSIFT words are used. The major improvement, however, is at the second level. By using CONSIFT words, we can replace the spatial consistency verification (which is the bottleneck of existing methods) by a simple matching of individual keypoint without any significant deterioration of performances (as shown in the preliminary experiments). Therefore, the approach seems particularly suitable for scalable applications of CBVIR (e.g. image retrieval in huge databases).

Additionally, the analysis of implementation details suggests that both levels of image retrieval can be prospectively merged into a single process based on inverted indexing (similarly to [12]). Nevertheless, this issue is not discussed in this paper.

References

1. Csurka, G., Bray, C., Dance, C., Fan, L., Wilamowski, J.: Visual categorization with bags of keypoints. In: Proceedings of the ECCV 2004, Workshop on Statistical Learning in Computer Vision, Prague, pp. 1–22 (2004)
2. Chum, O., Perdoch, M., Matas, J.: Geometric min-hashing: Finding a (thick) needle in a haystack. In: Proceedings of the IEEE Conference on CVPR 2009, pp. 17–24 (2009)
3. Jegou, H., Douze, M., Schmid, C.: Improving bag-of-features for large scale image search. Int. J. Comput. Vis. **87**, 316–336 (2010)
4. Tolias, M., Jegou, H.: Visual query expansion with or without geometry: refining local descriptors by feature aggregation. Pattern Recogn. **47**, 3466–3476 (2014)
5. Śluzek, A.: Visual categorization with bags of contextual descriptors improving credibility of keypoint matching. In: Proceedings of ICARCV 2014, Singapore (2014) (in print)
6. Zhao, W.L., Ngo, C.W., Tan, H.K., Wu, X.: Near-duplicate keyframe identification with interest point matching and pattern learning. IEEE Trans. Multimedia **9**, 1037–1048 (2007)
7. Mikolajczyk, K., Schmid, C.: Scale and affine invariant interest point detectors. Int. J. Comput. Vis. **60**, 63–86 (2004)
8. Lowe, D.G.: Distinctive image features from scale-invariant keypoints. Int. J. Comput. Vis. **60**, 91–110 (2004)
9. Arandjelovic, R., Zisserman, A.: Three things everyone should know to improve object retrieval. In: Proceedings of the IEEE Conference on CVPR 2012, pp. 2911–2918 (2012)
10. Mikolajczyk, K., Schmid, C.: A performance evaluation of local descriptors. IEEE Trans. PAMI **27**, 1615–1630 (2005)
11. Nistér, D., Stewénius, H.: Scalable recognition with a vocabulary tree. In: Proceedings of the IEEE Conference on CVPR 2006, vol. 2, pp. 2161–2168 (2006)
12. Stewénius, H., Gunderson, S.H., Pilet, J.: Size matters: exhaustive geometric verification for image retrieval Accepted for ECCV 2012. In: Fitzgibbon, A., Lazebnik, S., Perona, P., Sato, Y., Schmid, C. (eds.) ECCV 2012, Part II. LNCS, vol. 7573, pp. 674–687. Springer, Heidelberg (2012)
13. Zhang, Y., Jia, Z., Chen, T.: Image retrieval with geometry-preserving visual phrases. In: Proceedings of the IEEE Conference CVPR 2011, pp. 809–816 (2011)
14. Sivic, J., Zisserman, A.: Video google: a text retrieval approach to object matching in videos. In: Proceedings of 9th IEEE Conference on ICCV 2003, Nice, vol. 2, pp. 1470–1477 (2003)
15. Cha, S.H., Srihari, S.: On measuring the distance between histograms. Pattern Recogn. **35**, 1355–1370 (2002)
16. Swain, M., Ballard, D.: Color indexing. Int. J. Comput. Vis. **7**, 11–32 (1991)
17. Philbin, J., Chum, O., Isard, M., Sivic, J., Zisserman, A.: Object retrieval with large vocabularies and fast spatial matching. In: Proceedings of the IEEE Conference on CVPR 2007, pp. 1–8 (2007)
18. Paradowski, M., Śluzek, A.: Local keypoints and global affine geometry: triangles and ellipses for image fragment matching. In: Kwaśnicka, H., Jain, L.C. (eds.) Innovations in Intelligent Image Analysis. SCI, vol. 339, pp. 195–224. Springer, Heidelberg (2011)
19. Śluzek, A., Paradowski, M.: Visual similarity issues in face recognition. Int. J. Biometrics **4**, 22–37 (2012)

An Efficient Face Recognition Scheme Using Local Zernike Moments (LZM) Patterns

Emrah Basaran$^{(\boxtimes)}$ and Muhittin Gokmen

Department of Computer Engineering, Istanbul Technical University,
Istanbul, Turkey
basaranemrah@itu.edu.tr

Abstract. In this paper, we introduce a novel face recognition scheme using Local Zernike Moments (LZM). In this scheme, we follow two different approaches to construct a feature vector. In our first approach, we use Phase Magnitude Histograms (PMHs) on the complex components of LZM. In the second approach, we generate Local Zernike Xor Patterns (LZXP) by encoding the phase components, and we create gray level histograms on LZXP maps. For both of these methods, firstly, we divide images into sub-regions, then we construct the feature vectors by concatenating the histograms calculated in each of these sub-regions. The dimensionality of the feature vectors constructed in this way may be very high. So, we use a block based dimensionality reduction method, and with this method, we obtain higher performance. We evaluate our method on FERET database and achieve significant results.

1 Introduction

Due to the increasing use of face recognition applications in many fields such as security, access control and content based image retrieval, face recognition has an important place among the topics studied in computer vision. However, this problem has not been resolved fully yet because of some factors, e.g., illumination, pose and expression changes. So, many different feature extraction methods have been proposed in order to perform face recognition under different conditions.

In literature, there are two main approaches to reveal the features of face images: holistic feature extraction and local feature extraction. The most widely used holistic methods are Eigenfaces [1] and Fisherfaces [2]. In these methods, features are extracted after the projection of images into a subspace, and this projection is performed using Principal Component Analysis (PCA) in Eigenfaces and Linear Discriminant Analysis (LDA) in Fisherfaces. Although these holistic methods are useful under controlled environments, recognition performance is easily affected from misalignment of face images [3].

Muhittin Gokmen—This work is supported by The Scientific and Technological Research Council of Turkey with the grant number 112E201.

C.V. Jawahar and S. Shan (Eds.): ACCV 2014 Workshops, Part I, LNCS 9008, pp. 710–724, 2015.
DOI: 10.1007/978-3-319-16628-5_51

In recent years, local feature extraction methods have frequently been used. With these methods, features of the local structures of facial regions are obtained. Because these local features have a discriminative power between different human faces, promising results have been achieved with local feature extraction methods. Also, these methods are robust to illumination, partial occlusions and expression changes [4]. The best known local methods are Gabor [5] and Local Binary Patterns (LBP) [6]. In Gabor transformation, phase and magnitude components are obtained by convolving the images with a set of filters, and the local features are extracted using these components [3, 4, 7]. LBP is a simple method but with this method powerful features can be extracted by encoding the local neighborhood variations. There are also some studies that use an operator similar to LBP: Local Ternary Patterns (LTP) [8], Local Salient Patterns (LSP) [9] and Multi-scale Block LBP (MB-LBP) [10]. To benefit from the advantages of Gabor and LBP, these methods are used together in some studies. In [11], the magnitude components of Gabor wavelets are encoded using LBP, and in [12], the feature sets that are constructed using these methods are combined. Also, in [13], a new operator called Local Xor Patterns (LXP) is introduced and this operator is used to encode Gabor phase components.

Another local feature extraction method is proposed by Sariyanidi *et al.* [14] called Local Zernike Moments (LZM). For fingerprint recognition [15] and character recognition [16] problems, Zernike Moments (ZMs) have been used successfully. LZM transformation is developed to benefit from the power of ZMs also in local scale. As a result of LZM, similar to Gabor transformation, a different number of complex components are generated, and the features are extracted from these components. Simple and powerful descriptors can be constructed using LZM, but, a dimensional reduction method is needed due to the dimension of the feature vectors being very high. For this purpose, a common method LDA can be used. However, with LDA, successful results cannot be obtained when there are few samples for each class, as encountered in face recognition problem frequently. To overcome this problem, in [17], after using PCA, the feature vectors are randomly divided into sub-segments, and LDA is applied to each segment. Then, classification is performed separately for each one of these segments. The final result is determined using a fusion method. In [13], a different method is described. In this method, spatial blocks are placed on the images generated by Gabor transformation. Then, for each one of these blocks, a feature vector is obtained, and Fisher Linear Discriminant (FLD) is applied on these vectors. In addition, in [18], to reduce the dimensionality, Whitened PCA (WPCA) is applied after taking the square root of the feature vectors.

In this paper, we propose a new face recognition scheme using Local Zernike Moments. This scheme consists of two different feature extraction methods and a block based dimensionality reduction method. In the first feature extraction method, Phase Magnitude Histograms (PMHs) are constructed on the complex images generated by LZM transformation. In the second method, Local Zernike Xor Pattern (LZXP) maps are generated by applying LXP operator on the phase components of LZM, then the feature vectors are created using gray level histograms. The dimensionality of these vectors may be very high, so we use WPCA to reduce the dimensionality. Rather than using WPCA on these vectors directly, we create sub-segments using spatial blocks, then we apply WPCA for each of these segments separately. In the proposed scheme, we also use a score-level fusion technique to increase the recognition performance.

The rest of this paper is organized as follows. Section 2 briefly describes LZM transformation. Section 3 presents the block based dimensionality reduction. Section 4 introduces LZXP maps. Experimental results are given in Sect. 5, and in Sect. 6, the paper is concluded.

2   LZM Transformation

In pattern recognition problems such as fingerprint and character recognition, images are successfully represented using ZMs since the images contain explicit shape information. However, in face recognition, micro structures of the face images are more distinctive than explicit shape information. For this reason, LZM transformation has been proposed in [14] to take the advantage of ZMs in local scale, and promising results are obtained using this transformation for the face recognition problem.

2.1   Zernike Moments

In image analysis applications, ZMs of an image are stated as the projection of the image over Zernike polynomials. These Zernike polynomials are defined in polar coordinates within a unit circle and written as:

$$V_{nm}(\rho, \theta) = R_{nm}(\rho)e^{-jm\theta}. \tag{1}$$

Here, $R(\rho)$ is a radial polynomial and it is defined as:

$$R_{nm}(\rho) = \sum_{k=0}^{\frac{n-|m|}{2}} (-1)^k \frac{(n-k)!}{k!(\frac{n+|m|}{2} - k)!(\frac{(n-|m|)}{2} - k)!} \rho^{n-2k}. \tag{2}$$

In (1) and (2), n and m parameters indicate the order and repetition of ZMs respectively, and these values must be selected as $0 \leq n$, $0 \leq |m| \leq n$ and $n - |m|$ is even. Using (1) and (2), ZMs of an image are calculated as:

$$Z_{nm} = \lambda(n, N) \sum_{i=0}^{N-1} \sum_{j=0}^{N-1} V_{nm}(\rho_{ij}, \theta_{ij}) f(i, j), \tag{3}$$

where $\lambda(n, N) = \frac{2(n+1)}{\pi N^2}$, $\rho_{ij} = \sqrt{x_i^2 + y_j^2}$, $\theta_{ij} = tan^{-1}(y_i/x_i)$ and $f(i, j)$ is the image function with x_i and y_i values scaled to range of $[-1, 1]$.

2.2 Local Zernike Moments

The main feature of LZM transformation is that the calculations are performed around each pixel to expose the local variations in an image. For these calculations, Zernike polynomials are converted to $k \times k$ filtering kernels:

$$V^k_{nm}(i,j) = V_{nm}(\rho_{ij}, \theta_{ij}), \tag{4}$$

and the main formula of LZM is defined as follows

$$Z^k_{nm}(i,j) = \sum_{p,q=-\frac{k-1}{2}}^{\frac{k-1}{2}} f(i-p, j-q)V^k_{nm}(p,q). \tag{5}$$

In LZM transformation, ZMs are calculated within the $k \times k$ kernels and around each pixel. Using these ZMs, as a result of the transformation, a different number of complex images are created. The number of these images is related to n, and it is calculated using (the moments are not calculated for $m = 0$):

$$K(n) = \begin{cases} \frac{n(n+2)}{4} & \text{if n is even,} \\ \frac{(n+1)^2}{4} & \text{if n is odd.} \end{cases} \tag{6}$$

2.3 LZM Feature Vectors

In [14], PMHs are used to construct a feature vector, and these histograms are calculated on the complex components created by LZM transformation. As a first step, LZM components are divided into sub-regions. Then, a PMH is calculated in each sub-region, and a feature vector is obtained by concatenating these PMHs. To divide the components into sub-regions, two different grids are used; the first one is the same size as the face image and has $N \times N$ sub-regions. The second one is a half sub-region shifted version of the first one and has $(N-1) \times (N-1)$ sub-regions. Also in [14], it is indicated that better results are obtained if the components are separated into real and imaginary parts then LZM transformation is applied again for these components and the PMHs are constructed using the components created by second transformation. In Figs. 1 and 2, an illustration of cascaded LZM transformation and steps of the construction of a feature vector are shown respectively. In Fig. 3, some sample complex images produced using LZM transformation are represented.

3 Local Zernike XOR Patterns

Due to the success of LBP in the face recognition problem, a few methods based on the local patterns of Gabor features have been introduced, and successful results are obtained with these methods. In [11], the phase and magnitude components are encoded by using LBP operator, and in [7], a new operator is introduced called Local Xor Patterns (LXP). This operator is used to encode the real and imaginary components.

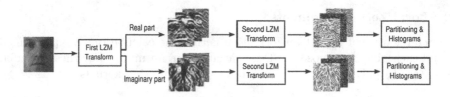

Fig. 1. A general illustration of cascaded LZM transformation [14].

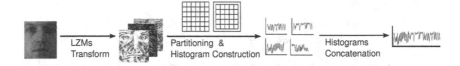

Fig. 2. Construction steps of a LZM feature vector [14].

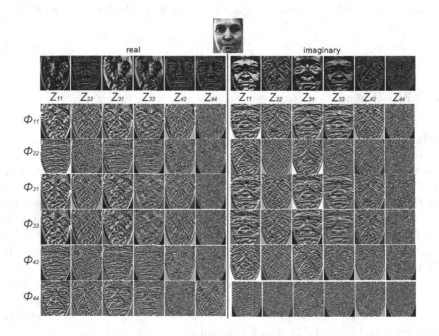

Fig. 3. Complex images by LZM transformation. First row shows real and imaginary components produced by first transformation. Second and subsequent rows show the phase components produced by second transformation. These images are produced with $k_1 = 5$ and $k_2 = 7$

While encoding the local patterns with LXP operator, a pixel is compared with its neighbor pixels using *exclusive or* operator, and a *bit string* is created. Then, this bit string is converted to a decimal value. In [13], a different approach, called Local Gabor XOR Patterns (LGXP), based on LXP operator is proposed.

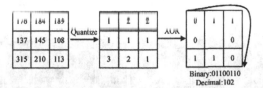

Fig. 4. Encoding of a phase image with LXP operator after the quantization of the phase values into 4 ranges [13].

With LGXP, as shown in Fig. 4, the phase components are encoded after the quantization of the phase values into different ranges.

For LBP, if the compared phase values are close to each other, any change on these phases easily effects the pattern. But, for LGXP, the pattern is more robust to the phase changes because of the quantization of the phase values, and in [13], it is shown that the results obtained with LGXP are successful than with LBP. So, in this study, the approach of LGXP is used with the phase components generated using LZM transformation, and this method is called Local Zernike Xor Patterns (LZXP).

In LZXP, as a first step, the phase components are quantized into different ranges. This quantization operation can be defined as

$$q(\Phi_{nm}(\bullet)) = i;$$

$$\text{if} \quad \frac{360*i}{r} \le \Phi_{nm}(\bullet) < \frac{360*(i+1)}{r}, i = 0, 1, \cdots, r-1 \qquad (7)$$

where $\Phi_{nm}(\bullet)$ indicates the phase components, $q(\bullet)$ indicates the quantization operator and r indicates the number of phase ranges. Then, LXP operator is used with $q(\Phi_{nm}(\bullet))$ as follows

$$LZXP_{nm}^i = q(\Phi_{nm}(z_c)) \otimes q(\Phi_{nm}(z_i)), i = 0, 1, \cdots, P. \qquad (8)$$

Here, $\otimes$ indicates LXP operator, z_c indicates a central pixel in $q(\Phi_{nm}(\bullet))$, z_i indicates a neighbor pixel of z_c, and P denotes the size of the neighborhood (e.g., $P = 8$ for 3×3 neighborhood, and $P = 16$ for 4×4 neighborhood). Finally, LZXP maps are constructed by calculating decimal pixel values using binary labels. This calculation can be defined as

$$LZXP_{nm}(z_c) = \sum_{i=1}^{P} 2^{i-1} * LZXP_{nm}^i. \qquad (9)$$

where $LZXP_{nm}(z_c)$ is decimal values and $LZXP_{nm}^i$ is binary labels. In Fig. 5, $LZXP$ maps of the images, which are given in Fig. 3, are shown.

4 Block Based Dimensionality Reduction

As a result of the operations described in Sect. 2, high dimensional feature vectors may be obtained according to the parameters (the dimension of the feature

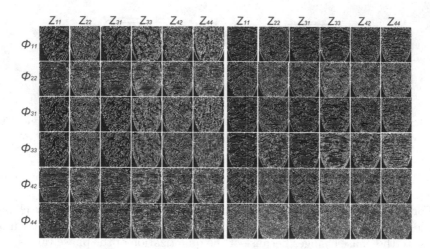

Fig. 5. LZXP maps of the images given in Fig. 3. These maps are produced with $r = 4$ and $P = 8$.

vector is 312768, in [14]). The dimension of these feature vectors can be reduced with the commonly used dimension reduction methods such as PCA and LDA. Nevertheless, if there are only a few samples in each class used for training, successful results can not be obtained with these methods [17]. To overcome this problem, in [13], a dimension reduction method called Block Based Fisher's Linear Discriminant (BFLD) is introduced. Firstly, in BFLD, images are divided into sub-regions using spatial blocks, and a feature vector is constructed for each block. Then, low dimensional feature vectors are obtained by using FLD on these vectors.

Using Gabor transformation, a different number of complex images are produced, as in LZM transformation. In [13], these complex images are divided into non-overlapping sub-regions, and a histogram is calculated within each sub-region. Then, by concatenating these histograms, high dimensional feature vectors are obtained. As mentioned above, in [13], a block based dimensionality reduction method is proposed to get low dimensional feature vectors from high dimensional vectors. In this method, as a first step, M spatial blocks are placed on the complex images as each block has an equal number of sub-regions. For each of these blocks, a feature vector is constructed by concatenating the histograms calculated in the sub-regions. Then, by concatenating these feature vectors corresponding to the same blocks over all complex images, M feature vectors are obtained. These operations are shown in Fig. 6. After the construction of M feature vectors, in BFLD, the dimensionality of these feature vectors is reduced using FLD. In this study, we use the complex images generated by LZM transformation, and we achieve better results when we use WPCA, which enables the construction of more compact and distinctive representations [18], in dimensionality reduction step. So, we use WPCA instead of FLD, and we named this method Block Based WPCA (BWPCA). In BWPCA, as illustrated

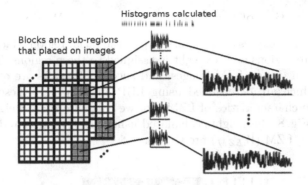

Fig. 6. Block based construction of feature vectors.

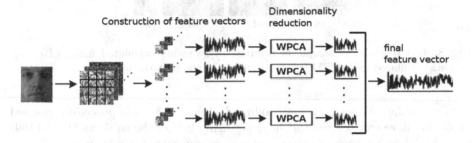

Fig. 7. Construction of final feature vector by concatinating low dimensional vectors.

in Fig. 7, the low dimensional feature vectors are obtained using WPCA, and the final feature vector for an image is constructed by concatenating these low dimensional vectors.

The methods described above are also repeated for test face images. Firstly, the feature vectors of images are constructed by using the block-based method, then the dimensionality of these feature vectors is reduced using the WPCA transformation matrices which are created in the training step.

5   Experimental Results

We have used FERET [19] face database to test the proposed methods. In this database, there is a "gallery" set consists of 1196 images of 1196 subjects, and there are four test sets: "Fb" (1195 images of 1195 subjects), "Fc" (194 images of 194 subjects), "Dup1" (722 images of 243 subjects) and "Dup2" (234 images of 75 subjects). In FERET, there is also a training set consists of 1002 images of 429 subjects, and we have used this set to learn the parameters of the methods. Before using these images, as a preprocessing operation, they are cropped and resized to 130×150. Then, they are normalized to have unit variance and zero mean.

5.1   Determination of the Weights of the Facial Regions

In many studies, to express the importance of the facial regions, face images are divided into sub-regions and a weight is assigned to these regions. This method is also applied in [14] on LZM images, and successful results are obtained with the weights which are determined using LBP [6]. However, these weights do not involve the characteristics of LZM. So, we calculate these weights again in this study. In Fig. 8, the weights calculated using LBP (W_{LBP}) and the weights calculated using LZM (W_{LZM}) are shown.

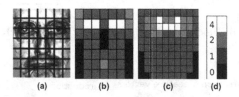

(a)   (b)   (c)   (d)

Fig. 8. Left to right: A sample face image (a), weights calculated using LBP (b), weights calculated using LZM (c), colors used to represent the weights (d).

To calculate W_{LZM}, we have followed almost the same procedure applied in [6]. We have constructed the *subfc* set using half of the subjects (1013-1109) from *Fc*. Then, we have calculated the recognition rates for *subfc* using only one of the sub-regions at a time. As illustrated in Fig. 8, the weights are determined for 7×7 sub-regions in [6]. However, because the best results are obtained in [14] for 10×10 sub-regions, we have determined the weights for 10×10 sub-regions. And, these weights are calculated using

$$w_k = \begin{cases} 0, & r_k \leq \%10, \\ 1, & \%10 < r_k \leq \%40, \\ 2, & \%40 < r_k \leq \%60, \\ 4, & \%60 \leq r_k \end{cases} \tag{10}$$

where w_k is the weight and r_k is the recognition rate of kth sub-region. In Table 1, the best results obtained in [14] using W_{LBP} and the results obtained using W_{LZM} are presented. We calculate these results with the same distance measure, L_1, and with the same parameter values: the number of sub-regions $N \times N = 10 \times 10$, the number of bin count used in histograms $b = 24$, the kernel

Table 1. Results obtained using W_{LBP} and W_{LZM}.

Weights	Fb	Fc	Dup1	Dup2
W_{LBP}	98.7	99.5	84.8	82.5
W_{LZM}	98.9	99.5	85.9	84.6

since and the degree of moments used in first and second LZM transformations $k_1 = 5, k_2 = 7, n_1 = 4$ and $n_2 = 4$. As can be seen from the results given in Table 1, W_{LZM} is more efficient than W_{LBP}. So, we have used W_{LZM} in the remainder of this study.

5.2 Experiments on LZXP

To see the effect of LXP operator on the phase components, we have made some tests using LZXP maps without any dimensionality reduction operation. We create the LZXP feature vectors concatenating the histograms calculated in the sub-regions, and we determine these sub-regions using two different grids which consist of $N \times N$ and $(N-1) \times (N-1)$ sub-regions as described in Sect. 2.3.

While generating the LZXP maps, as a first step, the parameters of LZM transformation k_1, k_2, n_1 and n_2 must be determined. Then, the number of phase ranges (r) to quantize the phase components and the size of the neighborhood (P) to apply the LXP operator must be specified. Also, the number of sub-regions $(N \times N)$ and the number of bin count (b) of histograms must be determined. As a result of our experiments, we have achieved the best results using L_1 distance measure, and when setting $n_1 = 4$, $n_2 = 4$, $k_1 = 7$, $k_2 = 7$, $r = 2$, $P = 16$, $N = 10$ and $b = 16$. These results are shown in Table 2.

Table 2. Recognition rates with LZXP maps.

Dataset	Fb	Fc	Dup1	Dup2
Recognition rate	100	99.5	83.8	83.8

5.3 Construction of Feature Vectors Using BWPCA

As mentioned in Sect. 4, the LZM feature vectors used in [14] are high dimensional and in this study, we use a block based dimensional reduction method, BWPCA, to obtain low dimensional and also more distinctive feature vectors. In Sect. 2, it is expressed that the images generated by LZM transformation are divided into sub-regions using two different grids. So, in this study, we place the blocks on both of these grids. But, the number of the blocks on these grids and the number of cells in each block are not equal, since the first and second grids are not of equal size. In this section, we have also given the results obtained with BFLD in order to show that the results obtained with BWPCA are better.

Firstly, we have tested BFLD and BWPCA with L_1 distance measure and with the parameters used to obtain the results shown in Tables 1 and 2 to directly observe the effect of these methods. But, for the number of the blocks placed on the first and the second grids ($B_1 \times B_1$ and $B_2 \times B_2$), we use $B_1^{lzm} = (2, 2)$, $B_2^{lzm} = (3, 3)$, $B_1^{lzxp} = (2, 2)$ and $B_2^{lzxp} = (3, 3)$ (the first number in parenthesis is the value used for BFLD, the other one is for BWPCA). These values are determined according to the experiments we have made, and the results are shown

Table 3. Recognition rates of LZM and LZXP maps using BWPCA with the parameters given in Sect. 5.1 and 5.2.

Method	Fb	Fc	Dup1	Dup2
LZM+BFLD	99.4	99.5	89.9	89.7
LZXP+BFLD	99.7	100	89.1	88.5
LZM+BWPCA	99.7	99.5	92.9	91.5
LZXP+BWPCA	99.7	100	92.1	90.5

in Table 3. As a result of the dimensionality reduction, the dimension of feature vectors is reduced to $(B_1^{lzm} \times B_1^{lzm} + B_2^{lzm} \times B_2^{lzm}) * (NumberOfClasses - 1) =$ 15535 from 312768 for LZM, and $(B_1^{lzxp} \times B_1^{lzxp} + B_2^{lzxp} \times B_2^{lzxp}) * (NumberOf$ $Classes - 1) = 15535$ from 208512 for LZXP. Also, better results are obtained as can be seen in Table 3.

In BFLD and BWPCA, the feature vectors are constructed using a block based method, then dimensionality reduction operations are performed. For this reason, rather than using the parameters given in Sects. 5.1 and 5.2, we have made many experiments using only the training set to determine the optimal parameter values. We have also tested some different distance and similarity measures. As a result of the experiments, we have achieved best results, which are shown in Table 4, using *L1 distance* for BFLD and *cosine similarity* for BWPCA and when setting $n_1^{lzm} = (4, 4)$, $n_2^{lzm} = (4, 4)$, $k_1^{lzm} = (7, 5)$, $k_2^{lzm} = (7, 7)$, $N^{lzm} = (9, 9)$ $b^{lzm} = (24, 16)$, $B_1^{lzm} = (3, 3)$, $B_2^{lzm} = (2, 2)$, $n_1^{lzxp} = (4, 4)$, $n_2^{lzxp} = (4, 4)$, $k_1^{lzxp} = (5, 5)$, $k_2^{lzxp} = (7, 7)$, $N^{lzxp} = (10, 9)$, $b^{lzxp} = (16, 16)$, $r^{lzxp} = (3, 2)$, $P^{lzxp} = (16, 16)$, $B_1^{lzxp} = (2, 3)$ and $B_2^{lzxp} = (3, 2)$.

Different Block Types. To obtain the LZM+BWPCA and LZXP+BWPCA results given in Table 4, the first and the second grids are divided into 3×3 and 2×2 blocks respectively, as shown in Fig. 9. We have also tried different block types and different grids. According to our experiments, we obtain the best results for LZM and LZXP when the block types are determined as illustrated in Fig. 10. The results obtained using these block types $(B_2 WPCA)$ are given in Table 5.

Table 4. Recognition rates of LZM and LZXP maps using BWPCA with the optimal parameters.

Method	Fb	Fc	Dup1	Dup2
LZM+BFLD	99.3	100	90.4	89.3
LZXP+BFLD	99.6	100	89.8	91.0
LZM+BWPCA	99.8	99.5	93.4	93.2
LZXP+BWPCA	99.8	100	93.2	91.0

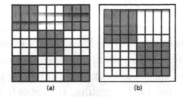

Fig. 9. First (a) and second (b) grids placed on the images. These grids consist of 9×9 and 8×8 sub-regions respectively, and on these grids 3×3 and 2×2 blocks are determined

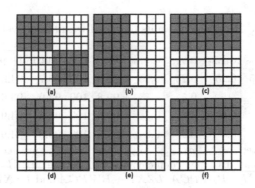

Fig. 10. First row: first (a), second (b) and third (c) grids used to divide LZM maps into sub-regions. These grids consist of 10×10, 8×8 and 8×8 sub-regions respectively. On these grids 2×2, 1×2 and 2×1 blocks are determined. Second row: first (d), second (e) and third (f) grids used to divide LZXP maps into sub-regions. These grids consist of 8×8, 8×8 and 8×8 sub-regions respectively. On these grids 2×2, 1×2 and 2×1 blocks are determined

Table 5. Recognition rates of LZM and LZXP maps using B_2WPCA.

Method	Fb	Fc	Dup1	Dup2
$LZM + B_2WPCA$	99.8	100	94.6	94.0
$LZXP + B_2WPCA$	99.7	100	94.5	94.4

5.4 Fusion of LZM and LZXP

In literature, descriptors which are created in different ways are frequently combined using different fusing methods (e.g., [11, 13, 20]) to increase the recognition performance. Generally, these fusion methods are performed in feature level or score level, and in this study, a score-level fusion approach is used. As described in the previous section, we create four different feature vectors ($LZM + BWPCA$, $LZM + B_2WPCA$, $LZXP + BWPCA$ and $LZXP + B_2WPCA$) using LZM and $LZXP$ maps, and using different block types. To combine the scores of these vectors, as a first step, we calculate the distances between the test and the training images, then multiply each distance value with a coefficient (α). In this way, the effect of the distance values to the recognition result is weighted. Finally, these distance values are summed, and so, combined scores are obtained.

We have made experiments to effectively fuse the different combinations of features. We conclude from these experiments that if we give nearly equal importance to all block types we get better results. In Table 6, different coefficient sets and the results obtained using these coefficients are given. We also compare our best results with state-of-the-art approaches in Table 7. As can be seen from this table, our results show an improvement on Dup1 and Dup2 relative to others.

Table 6. Recognition results obtained with score-level fusion.

α_{lzm}	α_{lzm_2}	α_{lzxp}	α_{lzxp_2}	Fb	Fc	Dup1	Dup2
0.25	0.25	0.25	0.25	99.8	100	95.7	94.9
0.5	-	0.5	-	99.8	100	94.9	93.6
-	0.5	-	0.5	99.8	100	95.6	94.9
0.2	0.3	0.2	0.3	99.8	100	96.0	94.9

Note: α_{lzm}, α_{lzm_2}, α_{lzxp} and α_{lzxp_2} are correspond to the coefficients of $LZM + BWPCA$, $LZM + B_2WPCA$, $LZXP + BWPCA$ and $LZXP + B_2WPCA$ respectively.

Table 7. Recognition rates of different face recognition methods on FERET database.

Method	Fb	Fc	Dup1	Dup2	Overall
HGPP [7]	97.5	99.5	79.5	77.8	90.2
S[LGBP+LGXP] [13]	99.0	99.0	94.0	93.0	96.9
G-LQP [21]	99.9	100	93.2	91.0	97.0
MBC-F [22]	99.7	99.5	93.6	91.5	97.0
LMEGW//LN+LGXP [3]	99.9	100	94.7	91.9	97.5
s-POEM+POD+WPCA [18]	99.7	100	94.9	94.0	97.7
GOM [4]	99.9	100	95.7	93.1	97.9
LMEGW//LN+LGBP [3]	99.9	100	95.6	93.6	98.0
Our method	**99.8**	**100**	**96**	**94.9**	**98.2**

6   Conclusion

In this work, we propose an efficient face recognition scheme. We have developed this scheme mainly based on LZM transformation. LZM is a local feature extraction method, and the efficiency of this method for face recognition has been shown in previous studies. In the proposed scheme, we extract local features from face images using two different methods. In the first method, we construct PMHs on the complex images which are generated by cascaded LZM transformation. In the second method, we construct gray level histograms on LZXP maps which are produced by encoding the phase components of LZM transformations. While encoding these phase components, we use LXP operator after quantization of the phase values into different ranges. To construct LZM and LZXP feature vectors, face images are first divided into sub-regions, then the histograms are calculated in each sub-region. Due to the fact that dimensionality of these feature vectors may be very high, we apply a block based dimensionality reduction method called BWPCA. Using this method, we have obtained low dimensional and also more discriminative feature vectors. When we use these low dimensional feature vectors, for both LZM and LZXP, we have observed that the recognition performance is increased for all prob sets, especially Dup1 and Dup2. In addition, we use a simple score level fusion method to further increase the performance. We multiply each distance value, calculated using LZM and LZXP features, with a coefficient, then we fuse these values as the final distance. With this fusion approach, we have also achieved a small performance increment for Dup1.

Discriminative and powerful features can be obtained using LZM transformation. In this study, we improve these properties of the features by using dimensionality reduction methods and generating these features in different ways. In future work, we will examine different operators to encode LZM components and apply different fusing methods. In addition, in order to better understand the ability of LZM transformation, we will test the proposed methods on different face databases.

References

1. Turk, M., Pentland, A.: Face recognition using eigenfaces. In: Proceedings of IEEE Computer Society Conference on Computer Vision and Pattern Recognition, CVPR 1991, pp. 586–591 (1991)
2. Belhumeur, P., Hespanha, J., Kriegman, D.: Eigenfaces vs. fisherfaces: recognition using class specific linear projection. IEEE Trans. Pattern Anal. Mach. Intell. **19**, 711–720 (1997)
3. Cament, L.A., Castillo, L.E., Perez, J.P., Galdames, F.J., Perez, C.A.: Fusion of local normalization and gabor entropy weighted features for face identification. Pattern Recogn. **47**, 568–577 (2014)
4. Chai, Z., Sun, Z., Mendez-Vazquez, H., He, R., Tan, T.: Gabor ordinal measures for face recognition. IEEE Trans. Inf. Forensics Secur. **9**, 14–26 (2014)
5. Serrano, A., de Diego, I.M., Conde, C., Cabello, E.: Recent advances in face biometrics with gabor wavelets: a review. Pattern Recogn. Lett. **31**, 372–381 (2010)
6. Ahonen, T., Hadid, A., Pietikäinen, M.: Face recognition with local binary patterns. In: Pajdla, T., Matas, J.G. (eds.) ECCV 2004. LNCS, vol. 3021, pp. 469–481. Springer, Heidelberg (2004)
7. Zhang, B., Shan, S., Chen, X., Gao, W.: Histogram of gabor phase patterns (hgpp): a novel object representation approach for face recognition. IEEE Trans. Image Process. **16**, 57–68 (2007)

8. Tan, X., Triggs, B.: Enhanced local texture feature sets for face recognition under difficult lighting conditions. IEEE Trans. Image Process. **19**, 1635–1650 (2010)
9. Chai, Z., Sun, Z., Tan, T., Vazquez, H.M.: Local salient patterns - a novel local descriptor for face recognition. In: ICB 2013, pp. 1–6 (2013)
10. Liao, S., Zhu, X., Lei, Z., Zhang, L., Li, S.Z.: Learning multi-scale block local binary patterns for face recognition. In: Lee, S.-W., Li, S.Z. (eds.) ICB 2007. LNCS, vol. 4642, pp. 828–837. Springer, Heidelberg (2007)
11. Zhang, W., Shan, S., Gao, W., Chen, X., Zhang, H.: Local gabor binary pattern histogram sequence (lgbphs): a novel non-statistical model for face representation and recognition. In: Tenth IEEE International Conference on Computer Vision, ICCV 2005, vol. 1, pp. 786–791 (2005)
12. Tan, X., Triggs, B.: Fusing Gabor and LBP feature sets for kernel-based face recognition. In: Zhou, S.K., Zhao, W., Tang, X., Gong, S. (eds.) AMFG 2007. LNCS, vol. 4778, pp. 235–249. Springer, Heidelberg (2007)
13. Xie, S., Shan, S., Chen, X., Chen, J.: Fusing local patterns of gabor magnitude and phase for face recognition. IEEE Trans. Image Process. **19**, 1349–1361 (2010)
14. Sariyanidi, E., Dagli, V., Tek, S.C., Tunc, B., Gokmen, M.: Local zernike moments: a new representation for face recognition. In: 2012 19th IEEE International Conference on Image Processing (ICIP), pp. 585–588 (2012)
15. Zhai, H.L., Hu, F.D., Huang, X.Y., Chen, J.H.: The application of digital image recognition to the analysis of two-dimensional fingerprints. Anal. Chim. Acta **657**, 131–135 (2010)
16. Kan, C., Srinath, M.D.: Invariant character recognition with zernike and orthogonal fouriermellin moments. Pattern Recognit. **35**, 143–154 (2002). Shape representation and similarity for image databases
17. Wang, X., Tang, X.: Random sampling lda for face recognition. In: Proceedings of the 2004 IEEE Computer Society Conference on Computer Vision and Pattern Recognition, CVPR 2004, vol. 2, pp. II-259–II-265 (2004)
18. Vu, N.S.: Exploring patterns of gradient orientations and magnitudes for face recognition. IEEE Trans. Inf. Forensics Secur. **8**, 295–304 (2013)
19. Phillips, P., Moon, H., Rizvi, S., Rauss, P.: The feret evaluation methodology for face-recognition algorithms. IEEE Trans. Pattern Anal. Mach. Intell. **22**, 1090–1104 (2000)
20. Xie, X., Lam, K.M.: Gabor-based kernel pca with doubly nonlinear mapping for face recognition with a single face image. IEEE Trans. Image Process. **15**, 2481–2492 (2006)
21. Hussain, S.U., Napolon, T., Jurie, F.: Face recognition using local quantized patterns. In: Proceedings of the British Machine Vision Conference, pp. 99:1–99:11. BMVA Press (2012)
22. Yang, M., Zhang, L., Shiu, S.K., Zhang, D.: Monogenic binary coding: an efficient local feature extraction approach to face recognition. IEEE Trans. Inf. Forensics Secur. **7**, 1738–1751 (2012)

Face Detection Based on Multi-block Quad Binary Pattern

Zhubei Ge[1], Canhui Cai[1($\boxtimes$)], Huanqiang Zeng[1], Jianqing Zhu[2],
and Kai-Kuang Ma[3]

[1] School of Information Science and Engineering, Huaqiao University,
Quanzhou, China
chcai@hqu.edu.cn

[2] Institute of Automation, Chinese Academy of Sciences, Beijing, China

[3] School of EEE, Nanyang Technological University, Singapore, Singapore

Abstract. A novel local texture descriptor, called *multi-block quad binary pattern* (MB-QBP), is proposed in this paper. To demonstrate its effectiveness on local feature representation and potential usage in computer vision applications, the proposed MB-QBP is applied to face detection. Compared with the *multi-block local binary pattern* (MB-LBP), MB-QBP has more features to conduct a better training process to refine the classifier. Consequently, the over-fitting problem becomes much smaller in the MB-QBP-based classifier. Extensive simulation results conducted by using the test images from the BioID and CMU+MIT databases have clearly shown that the proposed MB-QBP-based face detector outperforms the MB-LBP-based approach by about 6 % on the correct detection rate under the same training conditions.

1 Introduction

Face detection, face recognition, and facial expression analysis have become active research topics for a wide range of image-based and computer-vision related applications, such as video surveillance, human-computer interaction, to name a few. A common processing step to accomplish these tasks is how to extract salient features from the input image data effectively, with low algorithmic and computational complexities. For that, several local feature extraction methods have been proposed and shown their advantages in the application of face detection and recognition, such as *local gradient pattern* (LGP) [1], *local texture feature* (LTP) [2], Haar-like feature [3], and *local binary pattern* [4], and so on.

Besides feature extraction, *learning* is another essential process in face detector, and the developed learning algorithm is used in the training stage to select salient features from a large feature set generated at the feature extraction stage. That is, the feature descriptor and the learning algorithm are intimately related to each other—i.e., the simplicity and effectiveness of the feature descriptor directly affects the speed and efficiency of detector training and face detection. For example, the well-known Viola and Jones face detector [3] exploits the Ada-Boost learning algorithm to construct a cascade detector based on a Haar-like feature set for quickly identifying and eliminating those non-face sub-images.

© Springer International Publishing Switzerland 2015
C.V. Jawahar and S. Shan (Eds.): ACCV 2014 Workshops, Part I, LNCS 9008, pp. 725–734, 2015.
DOI: 10.1007/978-3-319-16628-5_52

However, the Haar-like features might be too simple and ineffective for representing faces. Consequently, training a detector based on a large number of Haar-like features will inevitably yield a very high computational complexity. Besides, since each node classifier is trained independently, a substantial amount of feature redundancy exists in different nodes of the cascade classifier. To reduce this feature redundancy, Wu *et al.* [5] proposed a nested cascaded detector by using the last trained node classifier as the first weak classifier of the current training node so as to reduce the number of weak classifiers and thus speed up the training process. To further reduce the required feature numbers, Zhang *et al.* [6] proposed the *multi-block local binary pattern* (MB-LBP) by extending the *local binary pattern* (LBP) [4] to multi-block LBP and further proposed a MB-LBP-based face detection method. The number of exhaustive set of MB-LBP features is much smaller than that of Haar-like features; consequently, the training process becomes much quicker. Subsequently, many variants of LBP face detectors have been proposed, such as *local gradient pattern* (LGP) [1] and *multi-scale enhanced local texture feature* (MS-LTP) [7]. However, the feature sets of these methods are too small for training a face detector to reach satisfactory face detection rates and the corresponding weak classifiers are too complex that may result in over-fitting problem. Besides, the computational complexity for classifier training in these methods is still larger than expected.

Since the training process is highly dependent on the description ability and computational complexity of local texture descriptors, it is essential to develop an efficient texture pattern that is able to improve both the efficiency and convergence speed of the learning process. For that, a new local feature descriptor, called *multi-block quad binary pattern* (MB-QBP), is introduced and applied on face detection application in this paper. Experimental results have shown that the proposed MB-QBP is able to achieve a superior performance to the MB-LBP.

The rest of this paper is organized as follows. Section 2 describes our proposed local texture descriptor, MB-QBP, in detail. Section 3 presents the face detection based on our developed feature descriptor, MB-QBP. Section 4 provides the experimental results and discussions. Section 5 concludes the paper.

2 Proposed Multi-Block Quad Binary Pattern (MB-QBP)

2.1 Quad Binary Pattern (QBP)

The proposed *quad binary pattern* (QBP) is a local texture feature descriptor. It basically performs a local transformation for each pixel location to capture the local feature. The computation conducted for the pixel location (i, j) is based on a 2×2 block, which consists of the current pixel (i, j), its right pixel $(i, j + 1)$, its lower pixel $(i + 1, j)$, and its lower-right $(i + 1, j + 1)$. The QBP at the pixel location (i, j) is defined as a 4 bit binary descriptor:

$$\alpha(i, j) = \sum_{n=0}^{3} 2^n \cdot s(g_n, \mu, t) \tag{1}$$

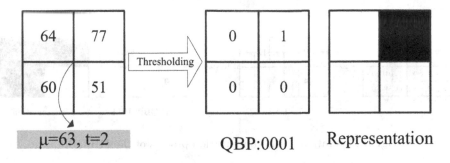

Fig. 1. Illustration of the computation process of the QBP.

where g_n represents the n-th pixel in the 2×2 block, μ is the arithmetic mean of the above-defined four pixels in the current block and can be computed by

$$\mu = \frac{1}{4} \sum_{n=0}^{3} g_n \tag{2}$$

and

$$s(g_n, \mu, t) = \begin{cases} 1, & g_n - \mu \geq t \\ 0, & g_n - \mu < t \end{cases} \tag{3}$$

where $t \geq 0$ is a threshold that is empirically determined and used to combat the effect due to the pixel-intensity fluctuation that might occur by noise interferences.

Figure 1 shows an example of a QBP computation for a pixel location (i, j) that has a gray-scale pixel intensity value 64 and a threshold $t = 2$. Based on Eqs. (1)–(3), the QBP descriptor "0001" or its decimal value $\alpha(i, j) = 1$ can be obtained.

2.2 Multi-Block Quad Binary Pattern (MB-QBP)

To expand the local feature representation ability of the QBP, the computation framework of the QBP as described in Sect. 2.1 is extended from the *pixel* domain to the *block* domain and called *multi-block quad binary pattern* (MB-QBP). The details are provided as follows.

To illustrate, Fig. 2 shows an example of the MB-QBP computation process, which is quite similar to that of QBP, except that the four pixels in the QBP computation are now replaced by four "block-pixels", respectively, where each block-pixel is the average of the pixel-intensity values computed over a variable-sized block (e.g., 3×2 in Fig. 2) of the original input image. Based on the computation as illustrated in Fig. 2, some other possible MB-QBP feature descriptor representation patterns can be appreciated from Fig. 3. The advantage of the proposed MB-QBP is that it offers more pattern descriptors than MB-LBP to select during the training stage. For example, considering a 24×24 small image, there are totally 20736 MB-QBP feature patterns, which is about 2.5 times of

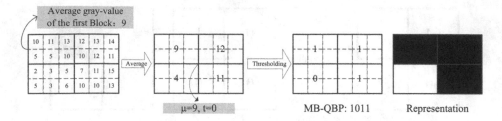

Fig. 2. Illustration of the computation process of the MB-QBP.

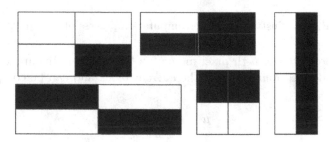

Fig. 3. Some examples of the MB-QBP.

MB-LBP feature patterns (8464) [6]. Therefore, by using MB-QBP, the learning algorithm is able to select the better feature combination and thus improve the classification accuracy of the resulted classifier. Consequently, the effect of over-fitting problem often encountered in the classifier training becomes much smaller in the proposed MB-QBP, compared with that of MB-LBP.

3   Face Detection Based on MB-QBP

To show the superiority of the proposed MB-QBP, we apply it on face detection. Based on the framework of local texture model-based face detection [6], the proposed MB-QBP-based face detection approach uses the MB-QBP as the feature to construct the weak classifier and adopts the Gentle AdaBoost algorithm [8] to obtain the strong classifier and further construct the cascade classifier.

3.1   Weak Classifiers

For each MB-QBP feature, we adopt multi-branch tree structure to generate weak classifiers. The resulted multi-branch tree has $2^4 = 16$ branches and the weak classifier is defined as:

$$f_m(x) = \begin{cases} a_0, x^k = 0 \\ \quad \cdots \\ a_j, x^k = j \\ \quad \cdots \\ a_{15}, x^k = 15 \end{cases} \tag{4}$$

where x^h denotes the kth element of the feature vector x and a_j can be obtained by using the criterion of minimum weighted square error. $J - \sum_{i=1}^{N} w_i$ $(y_i - f_m(x_i))^2$, in which w_i is the weight of current training sample and y_i is the sample's label. Hence, a_j can be defined as

$$a_j = \frac{\sum_i w_i y_i \delta\left(x_i^k = j\right)}{\sum_i w_i \delta\left(x_i^k = j\right)} \tag{5}$$

3.2   Strong Classifiers

In this work, the Gentle AdaBoost [8] is used to select strong classifiers due to its generalization ability and robustness. More specifically, let F_{max} and D_{min} denote the maximum false positive rate and minimum detection rate of each node, respectively. Given a set of training samples $\{(x_1, y_1), (x_2, y_2), \cdots, (x_N, y_N)\}$, where x_i is the inputting sample and $y_i \in \{+1, -1\}$ is the sample's label, the training algorithm of a strong classifier $F(x)$ with M numbers of weak classifiers can be described as follows:

(1) Initialize the weight distribution of the samples: $w_i = \frac{1}{N}, i = 1, 2, \cdots, N,$
    $F(x) = 0$
(2) For $(m = 1, 2, \cdots, M)$
    (2.1) Normalize the training set's weight distribution: $w_m(x_i) = w_m(x_i)/$
        $\sum_i w_m(x_i)$
    (2.2) Pick out the weak classifier $f_m(x)$ and update the current strong classifier $F(x) = F(x) + f_m(x)$ to ensure the detection rate of $F(x)$ can meet the condition $D_m > D_{min}$ under the positive samples. If the current strong classifier's detection rate meet the condition $F_m \leq F_{max}$, go to Step 3.
    (2.3) Update the weight distribution: $w_{m+1}(x_i) = w_m(x_i) exp\left(-y_i f_m(x_i)\right)$
(3) Output the strong classifier: $F(x) = \sum_{m=1}^{M} f_m(x)$

3.3   Cascade Detector

To construct the cascade classifier, we input N face training samples and M non-face training samples to the Gentle AdaBoost to train a strong classifier. This strong classifier is then used to update the positive and negative training samples, that is, choosing N face training samples that can be correctly classified and M non-face training samples that fail to be correctly classified from face training samples set. Consequently, the next training stage can be more concerned on those samples that fail to be correctly classified by previous strong classifier so as to eliminate the negative training samples accurately. By training a cascade detector with nodes, we can obtain a detector with false positive rate $F = \prod_{i=1}^{K} F_{max}$ and detection rate $D = \prod_{i=1}^{K} D_{min}$.

4 Experimental Results and Discussions

In this work, we compare the proposed MB-QBP-based face detection with MB-LBP-based approach [6]. Considering different training sets, parameters and criterion will lead to different results, we use the same training parameters and samples to individually train MB-QBP based face detector and MB-LBP based face detector for having a fair comparison. Moreover, we also use the same search strategy and evaluation criterion in the detection test as follows [9]:

- The Euclidian distance between the center of a detected window and actual face window shall be less than 30 % of the width of the actual face window.
- The width of the detected face window shall be in the range of ±50 % of the width of the actual face window.

Every non-face sample, which was mistakenly detected as a face, was counted as a false alarm. Hit rates are reported in percentage, while the false alarms are specified by their values. In our experiments, the training samples are 10000 frontal face images with the resolution of 24 × 24, which mainly collected from FERET database [10] and LFW database [11]. The non-face background library comes from Wu [12]. The test databases are BioID and CMU+MIT. The BioID database contains 1521 gray level images with a resolution of 384 × 286, and each image shows the frontal view of a face from one of 23 different person. Since they are captured under the "real world" conditions, these test images have a large variety of illuminations, backgrounds, face sizes and face expressions. The CMU+MIT database consists of 130 gray scale images containing 507 faces. It was originally created for evaluating different algorithms on the detection of frontal views of human faces. The experimental platform and software are listed as follows: OS–Windows 7, CPU–Intel(R) Core(TM) 2 Quad CPU Q9400 2.66 GHz, Visual Studio 2010, OpenCV 2.4.3 and Matlab 8.1. The performance is shown by using *Receiver Operating Characteristic* (ROC) curves. In these ROCs, each operation point is obtained by evaluating cascade instances with different numbers of layers. In other words, the parameter to be changed for obtaining the ROCs is the number of the layers of the cascades.

The experiments consist of two parts: (1) We randomly divide the 10000 frontal face images into 5 equal groups (i.e., each group has 2000 frontal face images), and use 5-fold cross-validation to compare the detector performance based on MB-QBP features and MB-LBP features. Then the resulted 5 MB-LBP-based and MB-QBP-based face detector pairs are re-tested on the BioID database. (2) We use all the 10000 frontal face images to individually train the face detectors based on MB-QBP and MB-LBP features and further evaluate them on CMU+MIT database.

4.1 5-Fold Cross-Validation Comparison

To have a more fair comparison, we use cross-validation. More specifically, the 10,000 frontal face images are randomly divided into 5 equal groups, in which 4

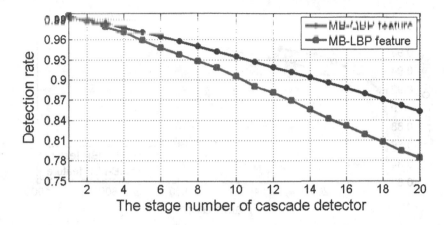

Fig. 4. The curve of detection rate vs. the number of the cascade stages.

groups are used for training and the remaining one is utilized for testing. The training parameters are set as follows: $D_{min} = 0.995, F_{max} = 0.5$, cascade stage number is 20, and each stage classifier's training set contains 8000 face samples and 8000 non-face samples.

Figure 4 shows the curves of the average detection rate of the five face detectors individually resulted from MB-QBP and MB-LBP versus the number of the cascade stages. It can be observed that the detection rate of MB-LBP-based face detector reduces faster than that of MB-QBP-based face detector with the increment of the number of classifiers. For example, in the cascade stage number 2, two types of face detectors have similar detection rate. However, in the cascade stage number 14, MB-QBP-based face detector can achieve the detection rate 90 % while MB-LBP-based face detector only has the detection rate 85 %. This is because the population of MB-QBP features is larger than that of MB-LBP and the MB-QBP feature is simpler than the MB-LBP feature, the MB-QBP needs more weak classifiers than MB-LBP in each node classifier, and the difference in number of weak classifiers used is increased with the stage number, which makes the MB-QBP-based face detector less possible to be over-fitted than MB-LBP-based one.

In order to obtain the detection rate under different false alarms, we add about 4000 non-face images with the resolution of 24 × 24 into each testing set and plot the ROC curves by averaging the detection rate of the above five face detectors individually resulted from MB-QBP-based face detector and MB-LBP-based one in Fig. 5. These curves clearly show that when the number of the false alarms is equal to 5, MB-QBP-based face detector has a higher detection rate (89 %) than MB-LBP-based face detector (83 %). In other words, the proposed MB-QBP-based face detector yields 6 % higher hit rate than the MB-LBP-based face detector. In addition, we further evaluate the performance of the above five face detectors resulted from MB-QBP and MB-LBP on the BioID database. The corresponding results are shown in Fig. 6. One can further see that our proposed

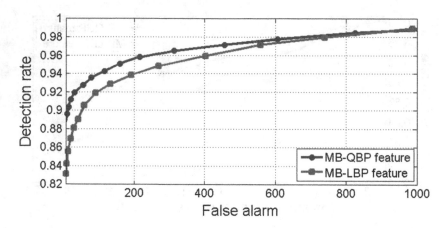

Fig. 5. ROC curves of MB-QBP-based face detector and MB-LBP-based face detector on self-made training set.

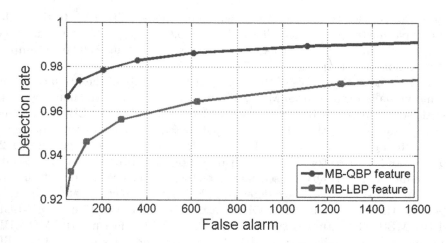

Fig. 6. ROC curves of MB-QBP-based face detector and MB-LBP-based face detector on the BioID database.

MB-QBP-based face detector has a better performance than the MB-LBP-based face detector.

4.2   Performance Evaluation on CMU+MIT Database

In this experiment, we individually train MB-QBP-based face detector and MB-LBP-based face detector, and then test them on the CMU+MIT database, in which each node classifiers training set contains 10000 non-face samples and 9900 face samples. Since the performance of the face detector is highly dependent on the training process, both MB-QBP-based and MB-LBP-based face detectors are trained with the same training parameters. Figure 7 shows the ROC curves of

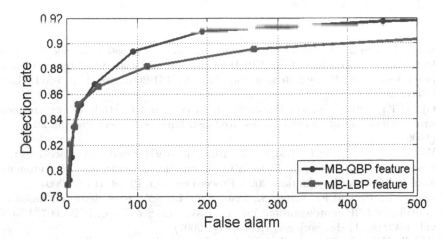

Fig. 7. The ROC curves of MB-QBP-based face detector and MB-LBP-based face detector on the CMU+MIT database.

MB-QBP-based and MB-LBP-based face detectors on the CMU+MIT database. From the results, we can see that MB-QBP-based face detector is superior to the MB-LBP-based face detector. Moreover, since CMU+MIT database contains some generalized "faces", which are quite different from the training set used in our training phase. Figure 7 also shows the generation ability of MB-QBP-based face detector is better than that of MB-LBP-based one.

5   Conclusion

In this paper, a novel local texture descriptor, called *multi-block quad binary pattern* (MB-QBP), is presented and applied to face detection. Since MB-QBP has more features to conduct a better training process to refine the classifier, the over-fitting problem of MB-QBP-based classifier becomes much smaller, compared with the MB-LBP. Experimental results conducted by using the test images from the BioID and CMU+MIT databases have clearly demonstrated that the proposed MB-QBP-based face detector is able to achieve a high correct detection rate.

Acknowledgement. This work was supported in part by the National Natural Science Foundation of China under the Grants 61250009 and 61372107, in part by the Xiamen Key Science and Technology Project Foundation under the Grant 3502Z20133024, and in part by the High-Level Talent Project Foundation of Huaqiao University under the Grants 14BS201 and 14BS204.

References

1. Jun, B., Kim, D.: Robust face detection using local gradient patterns and evidence accumulation. Pattern Recogn. **45**, 3304–3316 (2012)

2. Tan, X., Triggs, B.: Enhanced local texture feature sets for face recognition under difficult lighting conditions. IEEE Trans. Image Process. **19**, 1635–1650 (2010)
3. Viola, P., Jones, M.: Rapid object detection using a boosted cascade of simple features. In: Proceedings of the 2001 IEEE Computer Society Conference on Computer Vision and Pattern Recognition, 2001, CVPR 2001, vol. 1, p. I-511. IEEE (2001)
4. Ojala, T., Pietikäinen, M., Harwood, D.: A comparative study of texture measures with classification based on featured distributions. Pattern Recogn. **29**, 51–59 (1996)
5. Wu, B., Ai, H., Huang, C., Lao, S.: Fast rotation invariant multi-view face detection based on real adaboost. In: Sixth IEEE International Conference on Automatic Face and Gesture Recognition, 2004, Proceedings, pp. 79–84. IEEE (2004)
6. Zhang, L., Chu, R.F., Xiang, S., Liao, S.C., Li, S.Z.: Face detection based on multi-block LBP representation. In: Lee, S.-W., Li, S.Z. (eds.) ICB 2007. LNCS, vol. 4642, pp. 11–18. Springer, Heidelberg (2007)
7. Wei, Z., Dong, Y., Zhao, F., Bai, H.: Face detection based on multi-scale enhanced local texture feature sets. In: 2012 IEEE International Conference on Acoustics, Speech and Signal Processing (ICASSP), pp. 953–956. IEEE (2012)
8. Friedman, J., Hastie, T., Tibshirani, R., et al.: Additive logistic regression: a statistical view of boosting (with discussion and a rejoinder by the authors). Ann. Statist. **28**, 337–407 (2000)
9. Lienhart, R., Kuranov, A., Pisarevsky, V.: Empirical analysis of detection cascades of boosted classifiers for rapid object detection. In: Michaelis, B., Krell, G. (eds.) DAGM 2003. LNCS, vol. 2781, pp. 297–304. Springer, Heidelberg (2003)
10. Phillips, P.J., Moon, H., Rizvi, S.A., Rauss, P.J.: The feret evaluation methodology for face-recognition algorithms. IEEE Trans. Pattern Anal. Mach. Intell. **22**, 1090–1104 (2000)
11. Huang, G.B., Mattar, M., Berg, T., Learned-Miller, E.: Labeled faces in the wild: A database for studying face recognition in unconstrained environments, pp. 07–49. University of Massachusetts, Amherst, Technical Report (2007)
12. Wu, J., Brubaker, S.C., Mullin, M.D., Rehg, J.M.: Fast asymmetric learning for cascade face detection. IEEE Trans. Pattern Anal. Mach. Intell. **30**, 369–382 (2008)

Author Index